Food Processing

Food Processing

Principles and Applications

Second Edition

Edited by

Stephanie Clark, Stephanie Jung, and Buddhi Lamsal

Department of Food Science and Human Nutrition, Iowa State University, Iowa, USA

Library of Congress Cataloging-in-Publication Data

Food processing : principles and applications / [compiled by] Stephanie Clark, Stephanie Jung, and
Buddhi Lamsal. – Second Edition.
    pages   cm
  Includes bibliographical references and index.
  ISBN 978-0-470-67114-6 (cloth)
   1. Food industry and trade.   I. Clark, Stephanie, 1968– editor of compilation.   II. Jung, Stephanie,
editor of compilation.   III. Lamsal, Buddhi, editor of compilation.
  TP370.F6265 2014
  338.4′7664–dc23

                              2013048364

Cover images: Food factory ©iStock/leezsnow, Dumping of wheat grains ©iStock/jovanjaric, Fruits and
vegetables ©iStock/aluxum, Worker processing fresh cheese ©iStock/hemeroskopion, Milk production
line ©iStock/jevtic
Cover design by Andy Meaden

Set in 9.25/11.5pt MinionPro by SPi Publisher Services, Pondicherry, India

1   2014

Contents

List of Contributors

George Amponsah Annor
Department of Nutrition and Food Science
University of Ghana
Legon-Accra, Ghana

V.M. (Bala) Balasubramaniam
Department of Food Science and Technology
The Ohio State University
Columbus, OH, USA
and
Department of Food Agricultural and Biological
Engineering
The Ohio State University
Columbus, OH, USA

Sheryl A. Barringer
Department of Food Science and Technology
The Ohio State University
Columbus, OH, USA

Gleyn Bledsoe
College of Agricultural and Life Sciences
University of Idaho
Moscow, ID, USA

Joyce Irene Boye
Food Research and Development Centre
Agriculture and Agri-Food Canada
Québec, Canada

R.C. Chandan
Global Technologies, Inc.
Minneapolis, MN, USA

Haiqiang Chen
Department of Animal and Food Sciences
University of Delaware
Newark, DE, USA

Stephanie Clark
Department of Food Science and Human Nutrition
Iowa State University
Ames, IA, USA

Valérie Conway
Department of Food Science and Nutrition
Université Laval
Québec, Canada

Ali Demirci
Department of Agricultural and Biological Engineering
Pennsylvania State University
University Park, PA, USA

Robert H. Driscoll
School of Chemical Engineering
The University of New South Wales
Sydney, Australia

Susan E. Duncan
Department of Food Science and Technology
Virginia Tech
Blacksburg, VA, USA

Duygu Ercan
Department of Agricultural and Biological Engineering
Pennsylvania State University
University Park, PA, USA

Colette C. Fagan
Department of Food and Nutritional Sciences
University of Reading
Reading, Berkshire, UK

Renée M. Goodrich-Schneider
Food Science and Human Nutrition Department
University of Florida
Gainesville, FL, USA

Sundaram Gunasekaran
Food and Bioprocess Engineering Laboratory
University of Wisconsin-Madison
Madison, WI, USA

Federico Harte
Department of Food Science and Technology
University of Tennessee
Knoxville, TN, USA

Jonathan Hopkinson
Danisco USA
New Century, KS, USA

Gulten Izmirlioglu
Department of Agricultural and Biological Engineering
Pennsylvania State University
University Park, PA, USA

Christian James
Food Refrigeration and Process Engineering
Research Centre
The Grimsby Institute
Grimsby, UK

Stephen J. James
Food Refrigeration and Process Engineering
Research Centre
The Grimsby Institute
Grimsby, UK

Stephanie Jung
Department of Food Science and Human Nutrition
Iowa State University
Ames, IA, USA

Prabhat Kumar
Research and Development
Frito Lay
Plano, TX, USA

Buddhi P. Lamsal
Department of Food Science and Human Nutrition
Iowa State University
Ames, IA, USA

Miguel A. Landeros-Urbina
Coca-Cola FEMSA
Mexico City, Mexico

Pierre-Louis Leclerc
Department of Food Science and Nutrition
Université Laval
Québec, Canada

Janet H. Lillemo
Lillemo & Associates, LLC
Plymouth, MN, USA

Zhen Ma
Food Research and Development Centre
Agriculture and Agri-Food Canada
Québec, Canada

Robert Maddock
Department of Animal Sciences
North Dakota State University
Fargo, ND, USA

Kevin McDonnell
Bioresources Research Centre
School of Biosystems Engineering
University College Dublin
Dublin, Ireland

Maneesha S. Mohan
Department of Food Science and Technology
University of Tennessee
Knoxville, TN, USA

Fionnuala Murphy
Bioresources Research Centre
School of Biosystems Engineering
University College Dublin
Dublin, Ireland

Hudaa Neetoo
Faculty of Agriculture
University of Mauritius
Réduit, Mauritius

Mahmoudreza Ovissipour
School of Food Science
Washington State University
Pullman, WA, USA

Sung Hee Park
Department of Food Science and Technology
The Ohio State University
Columbus, OH, USA

Yves Pouliot
Department of Food Science and Nutrition
Université Laval
Québec, Canada

Barbara Rasco
School of Food Science
Washington State University
Pullman, WA, USA

Amy S. Rasor
Department of Food Science and Technology
Virginia Tech
Blacksburg, VA, USA

Kent D. Rausch
Department of Agricultural and Biological Engineering
University of Illinois
Urbana, IL, USA

José I. Reyes-De-Corcuera
Department of Food Science and Technology
University of Georgia
Athens, GA, USA

K.P. Sandeep
Department of Food, Bioprocessing and Nutrition
Sciences
North Carolina State University
Raleigh, NC, USA

Tonya C. Schoenfuss
Department of Food Science
University of Minnesota
St Paul, MN, USA

Susan E.M. Selke
School of Packaging
Michigan State University
East Lansing, MI, USA

Joongmin Shin
Packaging, Engineering and Technology
University of Wisconsin-Stout
Menomonie, WI, USA

Vijay Singh
Department of Agricultural and Biological Engineering
University of Illinois
Urbana, IL, USA

Douglas P. Smith
Prestage Department of Poultry Science
North Carolina State University
Raleigh, NC, USA

Nutsuda Sumonsiri
Department of Food Science and Technology
The Ohio State University
Columbus, OH, USA

Johan T. van der Veen
Ten Kate Holding
Musselkanaal, The Netherlands

Stephen L. Woodgate
Beacon Research
Clipston, Leicestershire, UK

Jianping Wu
Department of Agricultural, Food and Nutritional
Science
University of Alberta
Edmonton, AB, Canada

1 Principles of Food Processing

Sung Hee Park,[1] Buddhi P. Lamsal,[2] and V.M. Balasubramaniam[1,3]
[1]*Department of Food Science and Technology, The Ohio State University, Columbus, Ohio, USA*
[2]*Department of Food Science and Human Nutrition, Iowa State University, Ames, Iowa, USA*
[3]*Department of Food Agricultural and Biological Engineering, The Ohio State University, Columbus, Ohio, USA*

1.1 Processing of foods: an introduction

Processing of foods is a segment of manufacturing industry that transforms animal, plant, and marine materials into intermediate or finished value-added food products that are safer to eat. This requires the application of labor, energy, machinery, and scientific knowledge to a step (unit operation) or a series of steps (process) in achieving the desired transformation (Heldman & Hartel, 1998). Value-added ingredients or finished products that satisfy consumer needs and convenience are obtained from the raw materials.

The aims of food processing could be considered four-fold (Fellows, 2009): (1) extending the period during which food remains wholesome (microbial and biochemical), (2) providing (supplementing) nutrients required for health, (3) providing variety and convenience in diet, and (4) adding value.

Food materials' shelf life extension is achieved by preserving the product against biological, chemical, and physical hazards. Bacteria, viruses, and parasites are the three major groups of biological hazards that may pose a risk in processed foods. Biological hazards that may be present in the raw food material include both pathogenic microorganisms with public health implications and spoilage microorganisms with quality and esthetic implications. Mycotoxin, pesticide, fungicide, and allergens are some examples of chemical hazards that may be present in food. Physical hazards may involve the presence of extraneous material (such as stones, dirt, metal, glass, insect fragments, hair). These hazards may accidentally or deliberately (in cases of adulteration) become part of the processed product. Food processing operations ensure targeted removal of these hazards so that consumers enjoy safe, nutritious, wholesome foods. With the possibility of extending shelf life of foods and advances in packaging technology, food processing has been catering to consumer convenience by creating products, for example, ready-to-eat breakfast foods and TV dinners, on-the-go beverages and snacks, pet foods, etc. Food processing, as an industry, has also responded to changes in demographics by bringing out ethnic and specialty foods and foods for elderly people and babies. Nutrition fortification, for example, folic acid supplementation in wheat flour, is another function of processing food.

The scope of food processing is broad; unit operations occurring after harvest of raw materials until they are processed into food products, packaged, and shipped for retailing could be considered part of food processing. Typical processing operations may include raw material handling, ingredient formulation, heating and cooling, cooking, freezing, shaping, and packaging (Heldman & Hartel, 1998). These could broadly be categorized into primary and secondary processing. Primary processing is the processing of food that occurs after harvesting or slaughter to make food ready for consumption or use in other food products. Primary processing ensures that foods are easily transported and are ready to be sold, eaten or processed into other products (e.g. after the primary processing of peeling and slicing, an apple can be eaten fresh or baked into a pie). Secondary processing turns the primary-processed food or ingredient into other food products. It ensures that foods can be used for a number of purposes, do not spoil quickly, are healthy and wholesome to eat, and are available all year (e.g. seasonal foods).

Food Processing: Principles and Applications, Second Edition. Edited by Stephanie Clark, Stephanie Jung, and Buddhi Lamsal.
© 2014 John Wiley & Sons, Ltd. Published 2014 by John Wiley & Sons, Ltd.

In the previous example, baking of the pie is a secondary processing step, which utilizes ingredient from primary processing (sliced apple).

The food and beverage manufacturing industry is one of the largest manufacturing sectors in the US. In 2011, these plants accounted for 14.7% of the value of shipments from all US manufacturing plants. Meat processing is the largest single component of food and beverage manufacturing, with 24% of shipments in 2011. Other important components include dairy (13%), beverages (12%), grains and oilseeds (12%), fruits and vegetables (8%), and other food products (11%). Meat processing is also the largest component (17%) of the food sector's total value added, followed by beverage manufacturing (16%) (Anonymous, 2012; USDA Economic Research Service, 2013). California has the largest number of food manufacturing plants (www.ers.usda.gov/topics/food-markets-prices/processing-marketing.aspx), followed by New York and Texas. Demand for processed foods tend to be less susceptible to fluctuating economic conditions than other industries.

Some basic principles associated with processing and preservation of food are summarized in this chapter. In-depth discussion can be found elsewehwere (Earle & Earle, 2012; Fellows, 2009; Gould, 1997; Heldman & Hartel, 1998; Saravacos & Kostaropoulos, 2002; Smith, 2003; Toledo, 2007; Zhang et al., 2011), including various chapters in this book.

1.2 Unit operations in food processing

Most food processes utilize six different unit operations: heat transfer, fluid flow, mass transfer, mixing, size adjustment (reduction or enlargement), and separation. A brief introduction to these principles is given in this chapter; more detailed information about the theory behind the principles and applications can be found in standard food or chemical engineering textbooks, including Singh and Heldman (2009), Welti-Chanes et al. (2005), and McCabe et al. (2001).

During food processing, food material may be combined with a variety of ingredients (sugar, preservatives, acidity) to formulate the product and then subjected to different unit operations either sequentially or simultaneously. Food processors often use process flow charts to visualize the sequence of operations needed to transform raw materials into final processed product. The process flow diagrams often include quality control limits and/or adjustment and description of any hazards. Figure 1.1 shows a sample process flow diagram for making Frankfurter comminuted sausage.

1.2.1 Heat transfer

Heat transfer is one of the fundamental processing principles applied in the food industry and has applications in various unit operations, thermal processing, evaporation (concentration) and drying, freezing and thawing, baking, and cooking. Heating is used to destroy microorganisms to provide a healthy food, prolong shelf life through the destruction of certain enzymes, and promote a product with acceptable taste, odor, and appearance. Heat transfer is governed by heat exchange between a product and its surrounding medium. The extent of heat transfer generally increases with increasing temperature difference between the product and its surrounding.

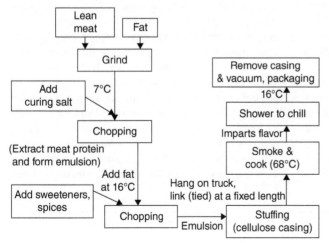

Figure 1.1 Process flow diagram of Frankfurter comminuted sausage manufacturing.

Conduction, convection, and radiation are the three basic modes of heat transfer. Conduction heat transfer occurs within solid foods, wherein a transfer of energy occurs from one molecule to another. Generally, heat energy is exchanged from molecules with greater thermal energy to molecules located in cooler regions. Heat transfer within a potato slice is an example of conduction heat transfer.

Heat is transferred in fluid foods by bulk movement of fluids as a result of a temperature gradient, and this process is referred to as convective heat transfer. Convective heat transfer can be further classified as natural convection and forced convection. Natural convection is a physical phenomenon wherein a thermal gradient due to density difference in a heated product causes bulk fluid movement and heat transfer. Movement of liquids inside canned foods during thermal sterilization is an example of natural convection. If the movement and heat transfer are facilitated by mechanical agitation (such as use of mixers), this is called forced convection.

Radiation heat transfer occurs between two surfaces as a result of the transfer of heat energy by electromagnetic waves. This mode of heat transfer does not require a physical medium and can occur in a vacuum. Baking is one example of heat transfer via radiation from the heat source in the oven to the surface of bread. However, heat propagates via conduction within the body of the bread.

1.2.2 Mass transfer

Mass transfer involves migration of a constituent of fluid or a component of a mixture (Singh & Heldman, 2009) in or out of a food product. Mass transfer is controlled by the diffusion of the component within the mixture. The mass migration occurs due to changes in physical equilibrium of the system caused by concentration or vapor pressure differences. The mass transfer may occur within one phase or may involve transfer from one phase to another. Food process unit operations that utilize mass transfer include distillation, gas absorption, crystallization, membrane processes, evaporation, and drying.

1.2.3 Fluid flow

Fluid flow involves transporting liquid food through pipes during processing. Powders and small-particulate foods are handled by pneumatic conveying, whereas fluids are transported by gravity flow or through the use of pumps. The centrifugal pump and the positive displacement pump are two pumps commonly used for fluid flow (Figure 1.2).

Centrifugal pumps utilize a rotating impeller to create a centrifugal force within the pump cavity, so that the fluid is accelerated until it attains its tangential velocity

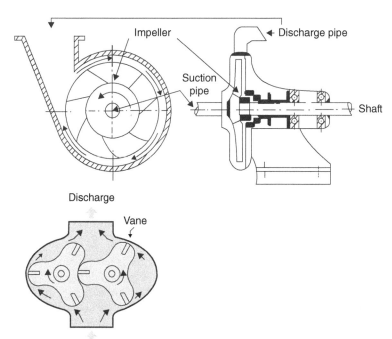

Figure 1.2 Schematic of centrifugal (*top*; from Food and Agriculture Organization of the United Nations 1985 *Irrigation Water Management: Training Manual No. 1 – Introduction to Irrigation*, www.fao.org/docrep/R4082E/R4082E00. htm), and positive displacement pumps (http://en.citizendium.org/wiki/Pump).

close to the impeller tip. The flow is controlled by the choice of impeller diameter and rotary speed of the pump drive. The product viscosity is an important factor affecting centrifugal pump performance; if the product is sufficiently viscous, the pump cavity will not fill with every revolution and the efficiency of the pump will be greatly reduced. Centrifugal pumps are used for transportation of fluids from point A to point B as in transporting fluid for cleaning operations. Centrifugal pumps do not have a constant flow rate.

A positive displacement pump generally consists of a reciprocating or rotating cavity between two lobes or gears and a rotor. Fluid enters by gravity or a difference in pressure and the fluid forms the seals between the rotating parts. The rotating movement of the rotor produces the pressure to cause the fluid to flow. Because there is no frictional loss, positive pumps are used where a constant rate of flow is required (timing pump), for high-viscosity fluids or for transporting fragile solids suspended in a fluid (such as moving cottage cheese curd from a vat to a filler).

1.2.4 Mixing

Mixing is a common unit operation used to evenly distribute each ingredient during manufacturing of a food product. Mixing is generally required to achieve uniformity in the raw material or intermediate product before it is taken for final production. Mixing of cookie or bread dough is an example, wherein required ingredients need to be mixed well into a uniform dough before they are portioned into individual cookies or loaves. Application of mechanical force to move ingredients (agitation) generally accomplishes this goal. Efficient heat transfer and/ or uniform ingredient incorporation are two goals of mixing. Different mixer configurations can be used to achieve different purposes (for detailed information, please refer to Fellows, 2009). The efficiency of mixing depends upon the design of impeller, including its diameter, and the speed baffle configurations.

1.2.5 Size adjustment

In size adjustment, the food is reduced mostly into smaller pieces during processing, as the raw material may not be at a desired size. This may involve slicing, dicing, cutting, grinding, etc. However, increasing a product size is also possible. For example, aggregation, agglomeration (instant coffee), and gelation are examples of size adjustment that result in increase in size. In the case of liquid foods, size

reduction is often achieved by homogenization. During milk processing, fats are broken into emulsions via homogenization for further separation.

1.2.6 Separation

This aspect of food processing involves separation and recovery of targeted food components from a complex mixture of compounds. This may involve separating a solid from a solid (e.g. peeling of potatoes or shelling of nuts), separating a solid from a liquid (e.g. filtration, extraction) or separating liquid from liquid (e.g. evaporation, distillation) (Fellows, 2009). Industrial examples of separation include crystallization and distillation, sieving, and osmotic concentration. Separation is often used as an intermediate processing step, and is not intended to preserve the food.

1.3 Thermophysical properties, microbial aspects, and other considerations in food processing

1.3.1 Raw material handling

Raw material handling is the very first step in the food processing. Raw material handling includes postharvest transportation (farm to plant), sorting, cleaning or sanitizing before loading into equipment in the plant. These could also be considered as part of primary processing of the food materials. Microorganisms could attach to inert non-porous surfaces in raw foods and it has been demonstrated that these cells transfer from one surface to another to another when contact occurs (Zottola & Sasahara, 1994). Appropriate raw material selection and handling affect microbial safety and final product quality. Future food preservation studies need to consider the impact of raw material (including postharvest handling prior to preservation) on the final processed product.

1.3.2 Cleaning and sanitation

Cleaning and sanitation of raw food material could be considered the first step in controlling any contamination of foreign materials or microorganisms during food processing. Cleaning removes foreign materials (i.e. soil, dirt, animal contaminants) and prevents the accumulation of biological residues that may support the growth of harmful microbes, leading to disease and/or the production of toxins. Sanitization is the use of any chemical or other effective method to reduce the initial bacterial load on the surface of raw materials or food processing equipment. Efficient sanitization includes both the outside and the inside of the plant such as specific floor plan, approved materials used in construction, adequate light, air ventilation, direction of air flow, separation of processing areas for raw and finished products, sufficient space for operation and movement, approved plumbing, water supply, sewage disposal system, waste treatment facilities, drainage, soil conditions, and the surrounding environment (Ray, 2004).

1.3.3 Engineering properties of food, biological, and packaging material

Knowledge of various engineering (physical, thermal, and thermodynamic) properties of food, biological, and packaging material is critical for successful product development, quality control, and optimization of food processing operations. For example, data on density of food material are important for separation, size reduction or mixing processes (Fellows, 2009). Knowledge of thermal properties of food (thermal conductivity, specific heat, thermal diffusivity) is useful in identifying the extent of process uniformity during thermal processes such as pasteurization and sterilization. For liquid foods, knowledge of rheological characteristics, including viscosity, helps in the design of pumping systems for different continuous flow operations. Different food process operations (heating, cooling, concentration) can alter product viscosity during processing, and this needs to be considered during design. Phase and glass transition characteristics of food materials govern many food processing operations such as freezing, dehydration, evaporation, and distillation. For example, the density of water decreases when the food material is frozen and as a result increases product volume. This should be considered when designing freezing operations. Thus, food scientists and process engineers need to adequately characterize or gather information about relevant thermophysical properties of food materials being processed. In-depth discussion of different engineering properties of food materials is available elsewhere (Rao et al., 2010).

1.3.4 Microbiological considerations

Most raw food materials naturally contain microorganisms, which bring both desirable and undesirable effects to processed food. For example, many fermented foods (e.g. ripened cheeses, pickles, sauerkraut, and fermented sausages) have considerably extended shelf life, developed aroma, and flavor characteristics over those of the raw materials arising from microorganisms such as *Lactobacillus*, *Lactococcus*, and *Staphylococcus* bacteria (Jay et al., 2005). On the other hand, raw food material also contains pathogens and spoilage organisms. Different foods harbor different pathogens and spoilage organisms. For example, raw apple juice or cider may be contaminated with *Escherichia coli* O157:H7. *Listeria monocytogenes* are pathogens of concern in milk and ready-to-eat meat. The target pathogen of concern in shelf-stable low-acid foods (such as soups) is *Clostridium botulinum* spores. Different pathogenic and spoilage microorganisms offer varied degrees of resistance to thermal treatment (Table 1.1). Accordingly, the design of an adequate process to produce safer products depends in part on the resistance of such microorganisms to lethal agents, food material, and desired shelf life (see section 4 for details).

1.3.5 Role of acidity and water activity in food safety and quality

Intrinsic food properties (e.g. water activity, acidity, redox potential) can play a role in determining the extent of food processing operations needed to ensure food safety and minimize quality abuse.

Higher acidity levels (pH <4.6) are often detrimental to the survival of microorganisms, so milder treatments are sufficient to preserve an acidic food. Low-acid foods (pH ≥4.6) support the growth and toxin production of various pathogenic microorganisms, including *Clostridium botulinum*. Products such as milk, meat, vegetables, and soups are examples of low-acid foods and require more severe heat treatment than acid foods such as orange juice or tomato products. pH of the food material also impacts many food quality attributes such as color, texture, and flavor. For example, pH of the milk used for cheese manufacturing can help determine cheese texture (hard/soft). Similarly, pH of fruit jelly can determine gel consistency.

Knowledge of availability of water for microbial, enzymatic or chemical activity helps predict the stability and shelf life of processed foods. This is often reported as water activity (a_w), and is defined as the ratio between partial pressure of water vapor (p_w) of the food and the vapor pressure of saturated water (p_w') at the same temperature. The water activity concept is often used in food processing to predict growth of bacteria, yeast, and molds. Bacteria grow mostly between a_w values of 0.9 and 1, most enzymes and fungi have a lower a_w limit of 0.8, and for

Table 1.1 Decimal reduction time (D value, min) of selected pathogenic and spoilage microorganisms found in food material

Pathogenic or spoilage microorganisms	D-value, min[*]	Temperature, °C	Food
Bacterial spores			
Bacillus stearothermilophilus	2.1–3.4	121	Phosphate buffer (pH 7.0)
Clostridium botulinum (types A and B)	0.1–0.2	121	
Clostridium butyricum	1.1	90	
Clostridium nigrificans	2–3	121	
Bacillus cereus	1.8–19.1	95	Milk
Bacillus coagulans spores	3.5	70–95	
Clostridium sporogenes	0.1–0.15	121	
Clostridium thermosaccharolyticum	3–4	121	
Clostridium perfringens	6.6	100.4	Beef gravy (pH 7)
Molds			
Byssochlamys nivea ascopores	193	80–90	
Neosartorya fischeri ascopores	15.1	85–93	
Talaromyces flavus ascopores	54	70–95	
Vegetative bacteria			
Campylobacter jejuni	0.74–1	55	Skim milk
Escherichia coli O157:H7	4.5–6.4	57.2	Ground beef
Listeria monocytogenes	6.27–8.32	60	Beef homogenate
Salmonella typhimurium	396	71	Milk chocolate
Vibrio parahaemolyticus	0.02–0.29	55	Clam homogenate

[*]D-value is the time taken to reduce the microbial population by one log-cycle (by 90%) at a given temperature. Adapted from Fellows (2009), Heldman (2003) and Heldman and Hartel (1999).

most yeasts 0.6 is the lower limit. Thus, food can be made safe by lowering the water activity to a point that will not allow the growth of dangerous pathogens (Table 1.2). Foods are generally considered safer against microbial growth at a_w below 0.6. Salt and sugar are commonly used to lower water activity by binding product moisture. In the recent years, there has been increased emphasis on reducing salt in processed foods. However, such changes should be systematically evaluated as salt reduction could potentially compromise microbiological safety and quality of the processed product (Doyle & Glass, 2010). Water activity of food material can also influence various chemical reactions. For example, non-enzymatic browning reactions increase with water activity level of $0.6–0.7 a_w$. Similarly, lipid oxidation can be minimized at about water activity level 0.2–0.3.

1.3.6 Reaction kinetics

During processing, the constituents of food undergo a variety of chemical, biological, physical, and sensory changes. Food scientists and engineers need to understand the rate of these changes caused by applying a given

Table 1.2 Water activity values in different food products

Food product	Water activity, a_w
Fresh meat and fish	0.99
Bread	0.95
Cured ham, medium aged cheese	0.9
Jams and jellies	0.80
Plum pudding, fruit cakes, sweetened condensed milk, fruit syrups	0.80
Rolled oats, fudge, molasses, nuts, fondants	0.65
Dried fruit	0.60
Dried foods, spices, noodles	0.50
Marshmallow	0.6–0.65
Biscuits	0.30
Whole milk powder, dried vegetables, corn flakes	0.20
Instant coffee	0.20
English toffee	0.2–0.3
Hard candy	0.1–0.2

Compiled from various sources, including Fellows (2009), Heldman and Hartel (1998), and Potter and Hotchkiss (1998).

processing agent and the resulting modifications, so that they can control process operations to produce a product with the desired quality. Enzyme hydrolysis, browning, and color degradation are examples of chemical changes while inactivation of microorganisms after heat treatment is an example of a biological change. Food engineers rely on microbial and chemical kinetic equations to predict and control various changes happening in the processed food. Detailed discussion on kinetic changes in food processing systems is available elsewhere (Earle & Earle, 2012: Institute of Food Technologists (IFT), 2000). There is only a limited database on kinetics of destruction of variety of microorganisms, nutrients, allergens, and food quality attributes as a function of different thermal and non-thermal processing variables and more effort should be made to gather such information.

1.4 Common food preservation/processing technologies

1.4.1 Goals of food processing

The food industry utilizes a variety of technologies such as thermal processing, dehydration, refrigeration, and freezing to preserve food materials. The goals of these food preservation methods include eliminating harmful pathogens present in the food and minimizing or eliminating spoilage microorganisms and enzymes for shelf life extension.

The general concepts associated with processing of foods to achieve shelf life extension and preserve quality include (1) addition of heat, (2) removal of heat, (3) removal of moisture, and (4) packaging of foods to maintain the desirable aspects established through processing (Heldman & Hartel, 1998). Many food processing operations add heat energy to achieve elevated temperatures detrimental to the growth of pathogenic microorganisms. Exposure of food to elevated temperatures for a predetermined length of time (based on the objectives of the process at hand) is a key concept in food processing. Pasteurization of milk, fruit and vegetable juices, canning of plant and animal food products are some examples of processing with heat addition. The microbial inactivation achieved is based on exposure of foods to specific time-temperature combinations. Blanching is another example of heat addition, which helps with enzyme inactivation.

Processing of foods by heat removal is aimed more towards achieving shelf life extension by slowing down the biochemical and enzymatic reactions that degrade

foods. Removal of moisture is another major processing concept, in which preservation is achieved by reducing free moisture in food to limit or eliminate the growth of spoilage microorganisms. Drying of solid foods and concentration of liquid foods fall under this category. Finally, packaging maintains the product characteristics established by processing of the food, including preventing postprocessing contamination. Packaging operations are also considered part of food processing.

In recent decades, the food industry has also investigated alternative lethal agents, such as electric fields, high pressure, irradiation, etc., to control microorganisms. Even though it is desirable that the preservation method by itself does not cause any damage to the food, depending upon the intensity of such agents, the quality of the food may also be affected.

Below are some key processing operations commonly used in the food industry. These and other food processing techniques are elaborated upon in various chapters of this book, for example, Chapter 3 (Separation and Concentration), Chapter 4 (Dehydration), Chapter 5 (Chilling and Freezing), Chapter 6 (Fermentation and Enzyme Technologies), Chapter 7 (Alternative and Emerging Food Processing Technologies), Chapter 8 (Nanotechnology), and Chapter 11 (Food Packaging).

1.4.2 Processes using addition or removal of heat

1.4.2.1 Pasteurization and blanching

Thermal pasteurization (named after inventor Louis Pasteur) is a relatively mild heat treatment, in which liquidds, semi-liquids or liquids with particulates are heated at a specific temperature (usually below $100°C$) for a stated duration to destroy the most heat-resistant vegetative pathogenic organisms present in the food. This also results in shelf life extension of the treated product. Different temperature-time combinations can be used to achieve pasteurization. For example, in milk pasteurization, heating temperatures vary widely, ranging from low-temperature, long-time heating (LTLT, $63°C$ for a minimum of 30 min), to high-temperature, short-time heating (HTST, $72°C$ for a minimum of 15 sec), to ultra-pasteurization ($135°C$ or higher for 2 sec to 2 min) (Singh & Heldman, 2009). In addition to destruction of pathogenic and spoilage microorganisms, pasteurization also achieves almost complete destruction of undesirable enzymes, such as lipase in milk. In recent years, the term "pasteurization" is also extended to destroying pathogenic microorganisms in solid foods (such as pasteurization of almonds through oil roasting, dry roasting, and steam processing).

The intensity of thermal treatment needed for a given product is also influenced by product pH; for example, fruit juices (pH <4.5) are generally pasteurized at 65 °C for 30 min, compared to other low-acid vegetables that need to be treated at 121 °C for 20–30 min. As a moderate heat treatment, pasteurization generally causes minimal changes in the sensory properties of foods with limited shelf life extension. Further, pasteurized products require refrigeration as a secondary barrier for microbiological protection.

Blanching, a mild thermal treatment similar in temperature-time intensity to pasteurization, is applied to fruit and vegetables to primarily inactivate enzymes that catalyze degradation reactions. This treatment also destroys some microorganisms. It is achieved by using boiling water or steam for a short period of time, 5–15 min or so, depending on the product. Other beneficial effects are color improvement and reducing discoloration. Blanching is often used as a pretreatment to thermal sterilization, dehydration, and freezing to control enzymes present in the food. Other benefits of blanching include removal of air from food tissue and softening plant tissue to facilitate packaging into food containers.

1.4.2.2 Thermal sterilization

Thermal sterilization involves heating the food to a sufficiently high temperature (>100 °C) and holding the product at this temperature for a specified duration, with the goal of inactivating bacterial spores of public health significance (Pflug, 1998). This is also known as canning or retorting. Prolonged thermal exposure during heating and cooling can substantially degrade product sensory and nutritional quality. Commercial sterility of thermally processed food, as defined by the US Food and Drug Administration (FDA), is the condition achieved by the application of heat that renders the food free of (i) microorganisms capable of reproducing in the food under normal non-refrigerated storage and distribution conditions, and (ii) viable microbial cells or spores of public health significance. Consequently, commercially sterile food may contain a small number of viable, but dormant, non-pathogenic bacterial spores. Traditionally, food processors use severe heat treatment to eliminate 12-log of *C. botulinum* spores (i.e. 12-D processes) to sterilize low-acid (pH ≥4.6) canned foods. Many canned foods have shelf lives of 2 years or longer at ambient storage conditions.

1.4.2.3 Aseptic processing

Aseptic processing, a continuous thermal process, involves pumping of pumpable food material through a set of heat exchangers where the product is rapidly heated under pressure to ≥130 °C to produce shelf-stable foods. The heated product is then passed through a holding tube, wherein the temperature of the product mixture is equilibrated and held constant for a short period as determined by the type of food and microbes present, and passes through set of cooling heat exchangers to cool the product. The sterilized cooled product is then aseptically packaged in a presterilized package (Sastry & Cornelius, 2002). Conventional aseptic processing technologies utilize heat exchangers such as scraped surface heat exchangers. Advanced food preservation techniques may utilize ohmic heating or microwave heating instead (Yousef & Balasubramaniam, 2013).

1.4.2.4 Sous-vide cooking

Sous-vide cooking involves vacuum packaging food before application of low-temperature (65–95 °C) heating and storing under refrigerated conditions (0–3 °C). Meat, ready meals, fish stews, fillet of salmon, etc. are some examples of sous-vide cooked products. This technology is particularly appealing to the food service industry, and has been adopted mainly in Europe. Due to use of modest temperatures, sous-vide cooking is not lethal enough to inactivate harmful bacterial spores. In addition, vacuum packaging conditions could also support potential survival of *Clostridium botulinum* spores.

1.4.2.5 Microwave heating

Microwave energy (300–300,000 MHz) generates heat in dielectric materials such as foods through dipole rotation and/or ionic polarization (Ramaswamy & Tang, 2008). In microwave heating, rapid volumetric heating could reduce the time required to achieve the desired temperature, thus reducing the cumulative thermal treatment time and better preserving the thermolabile food constituents. A household microwave oven uses the 2450 MHz frequency for microwave. For industrial application, a lower frequency of 915 MHz is selected for greater penetration depth. Microwave heating can be operated in both batch and continuous (aseptic) operations. Care must be taken to avoid non-uniform heating and overheating around the edges. In 2010, the FDA accepted an industrial petition for microwave processing of sweet potato puree that is aseptically packaged in sterile flexible pouches.

1.4.2.6 Ohmic heating

Ohmic heating involves electrical resistance heating of the pumpable food to rapidly heat the food material. The heat is generated in the form of an internal energy transformation (from electric to thermal) within the material as a function of an applied electric field (<100 V/cm) and the electrical conductivity of the food. Ohmic heating has been shown to be remarkably rapid and relatively spatially uniform in comparison with other electrical methods. Therefore, the principal interest has traditionally been in sterilization of those foods (such a high-viscosity or particulate foods) that would be difficult to process using conventional heat exchange methods (Sastry, 2008). Another application of ohmic heating includes improvement of extraction, expression, drying, fermentation, blanching, and peeling.

1.4.2.7 Drying

Drying is one of the oldest methods of preserving food. The spoilage microorganisms are unable to grow and multiply in drier environments for lack of free water. Drying is a process of mobilizing the water present in the internal food matrix to its surface and then removing the surface water by evaporation (Heldman & Hartel, 1998). Drying often involves simultaneous heat and mass transfer. Most drying operations involve changing free water present within the food to vapor form and removing it by passing hot air over the product.

During drying, the heat is transferred from an external heating medium into the food. The moisture within the food moves towards the surface of the material due to the vapor pressure gradient between the surface and interior of the product. The moisture is then evaporated into the heat transfer medium (usually air). The heat transfer can be accomplished through conduction, convection or radiation. While convective heat transfer is the dominant mechanism at the surface, heat is transferred through conduction within the food material. The moisture movement within the food material utilizes a diffusion process. There are several drying and dehydration methods frequently used in food processing such as hot air drying, spray drying, vacuum drying, freeze drying, osmotic dehydration, etc. During hot air drying, heat is transferred through the food either from heated air or from heated surfaces. Vacuum drying involves evaporation of water under

vacuum or reduced pressures. Freeze drying involves removing the water vapor through a process called sublimation. Freeze drying helps to maintain food structure.

1.4.2.8 Refrigeration and freezing

Refrigeration and freezing have become an essential part of the food chain; depending on the type of product, they are used in all stages of the chain, from food processing, to distribution, retail, and final consumption at home. These two unit operations take away heat energy from food systems and maintain the lower temperatures throughout the storage period to slow down biochemical reactions that lead to deterioration. The food industry employs both refrigeration and freezing processes where the food is cooled from ambient to temperatures above 0 °C in the former and between −18 °C and −35 °C in the latter to slow the physical, microbiological, and chemical activities that cause deterioration in foods (Tassou et al., 2010).

Chilled or refrigerated storage refers to holding food below ambient temperature and above freezing, generally in the range of −2 to ~16 °C. Removing sensible heat energy from the product using mechanical refrigeration or cryogenic systems lowers the product temperature. Many raw products (such as milk and poultry) are rapidly chilled prior to further processing to minimize any microbial growth in the raw product. After cooking, foods are often kept under refrigerated conditions during storage and retailing. Many of the minimally processed foods (e.g. pasteurization) are promptly refrigerated to prevent growth of the microorganisms that survive processing.

For frozen storage, food products are frozen to temperatures ranging from −12 °C to −18 °C. Appropriate temperature control is important in freezing to minimize quality changes, ice recrystallization, and microbial growth. Food products can be frozen using either indirect contact or direct contact systems (Heldman & Hartel, 1998). In indirect contact systems, there is no direct contact between the product and the freezing medium. Cold air and liquid refrigerants are examples of freezing media used. Cabinet freezing, plate freezing, scraped surface heat exchanger, and indirect contact air-blast systems are different examples of indirect freezing equipment used in the industry. Direct contact freezing systems do not have a barrier between the product and the freezing medium. Direct contact air-blast, fluidized bed, immersion freezing, and spiral conveyor systems are examples of direct contact freezing.

1.4.3 Non-thermal food processing and preservation

During the past two decades, due to increased consumer interest in minimally processed foods with reduced preservatives, several non-thermal preservation methods have been investigated. These technologies often utilize lethal agents (such as pressure, irradiation, pulsed electric field, ultraviolet irradiation, and ultrasound, among others) with or without combination of heat to inactivate microorganisms (Zhang et al., 2011). This helps to reduce the severity of thermal exposure and preserve product quality and nutrients. Since the mechanism of microorganism inactivation by non-thermal lethal agents may be different from that of heat, it is important to understand the synergy, additive, or antagonistic effects of sequential or simultaneous combinations of different lethal agents. Irradiation, high-pressure processing, and pulsed electric field processing are examples of non-thermal processing methods that may be of commercial interest.

1.4.3.1 Irradiation

Irradiation is one of the most extensively investigated non-thermal technologies. Ionizing radiation includes γ-ray and electron beam. During irradiation of foods, ionizing radiation penetrates a food and energy is absorbed. Absorbed dose of radiation is expressed in grays (Gy), where 1 Gy is equal to an absorbed energy of 1 J/kg. Milder doses (0.1–3 kGy), called "radurization," are used for shelf life extension, control of ripening, and inhibition of sprouting. Radicidation is carried out to reduce viable non-spore forming pathogenic bacteria, using a dose between 3 and 10 kGy. Radappertization from 10 kGy to 50 kGy enables the sterilization of bacterial spores. From its beginning in the 1960s, the symbol Radura has been used to indicate ionizing radiation treatment (Figure 1.3).

Radiation is quite effective in penetrating through various packaging materials. However, the radiation dose may cause changes in packaging polymers. Thus, careful choice of packaging material is critical to avoid any radiolytic products from packaging contaminating the food products. Consumer acceptance is one of the barriers to widespread adoption of irradiation for food processing applications (Molins, 2001).

1.4.3.2 High-pressure processing

High-pressure processing, also referred to as "high hydrostatic pressure processing" or "ultra-high pressure

Figure 1.3 International Radura symbol for irradiation on the packaging of irradiated foods (from www.fsis.usda.gov/ Fact_Sheets/Irradiation_and_Food_Safety/index.asp).

processing," uses elevated pressures (up to 600 MPa), with or without the addition of external heat (up to 120 °C), to achieve microbial inactivation or to alter food attributes (Cheftel, 1995; Farkas & Hoover, 2000). Pressure pasteurization treatment (400–600 MPa at chilled or ambient conditions), in general, has limited effects on nutrition, color, and similar quality attributes. Uniform compression heating and expansion cooling on decompression help to reduce the severity of thermal effects such as quality degradation and nutritional loss encountered with conventional processing techniques. Figure 1.4 summarizes typical pressure and temperature levels for various food process operations. Examples of high-pressure pasteurized products commercially available in the US include smoothies, guacamole, deli meat slices, juices, ready meal components, poultry products, oysters, ham, fruit juices, and salsa.

Heat, in combination with pressure, is required for spore inactivation. This process is called pressure-assisted thermal processing (PATP) or pressure-assisted thermal sterilization (PATS). During PATP, preheated (70–85 °C) food material is subjected to a combination of elevated pressure (500–700 MPa) and temperature (90–120 °C) for a specified holding time (Nguyen & Balasubramaniam, 2011). PATP has shown better preservation of textural qualities in low-acid vegetable products. Minimal thermal exposure with a shorter pressure holding time helps to retain product textural quality attributes in comparison with conventional retort processing where the product experiences prolonged thermal exposure. In 2010, the FDA issued no objection to an industrial petition for

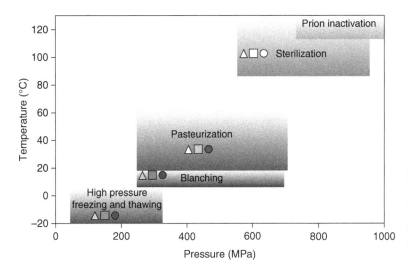

Figure 1.4 Different pressure-temperature regions yield different processing effects. Inactivation of vegetative bacteria, yeast, and mold (□), bacterial spores (○), and enzymes (△) is also shown. A filled symbol represents no effect, an open symbol represents inactivation. Adapted from Nguyen and Balasubramaniam (2011).

sterilizing low-acid mashed potato product through pressure-thermal sterilization.

1.4.3.3 Pulsed electric field processing

During pulsed electric field (PEF) processing, a high-voltage electrical field (20–70 kV/cm) is applied across the food for a few microseconds. A number of process parameters including electric field strength, treatment temperature, flow rate or treatment time, pulse shape, pulse width, frequency, and pulse polarity govern the microbiological safety of the processed foods. Food composition, pH, and electrical conductivity are parameters of importance to PEF processing. During PEF treatment, the temperature of the treated foods increases due to a electrical resistance heating effect. This temperature increase can also contribute to the inactivation of microorganisms and other food quality attributes. The technology effectively kills a variety of vegetative bacteria, but spores are not inactivated at ambient temperatures (Yousef & Zhang, 2006; Zhang et al., 1995). Typical PEF equipment components include pulse generators, treatment chambers, and fluid-handling systems, as well as monitoring and control devices. PEF technology has the potential to pasteurize a variety of liquid foods including fruit juices, soups, milk, and other beverages.

1.4.3.4 Ultrasound

High-power ultrasound processing or sonication is another alternative technology that has shown promise in the food industry (Piyasena et al., 2003), especially for liquid foods, in inactivating spoilage microorganisms. Ultrasound is a form of energy generated by sound waves of frequencies above 16 kHz; when these waves propagate through a medium, compressions and depressions of the medium particles create microbubbles, which collapse (cavitation) and result in extreme shear forces that disintegrate biological materials. Sonication alone is not very efficient in killing bacteria in food, as this would need an enormous amount of ultrasound energy; however, the use of ultrasound coupled with pressure and/or heat is promising (Dolatowski et al., 2007; Piyasena et al., 2003).

1.4.4 Redefining pasteurization

Successful commercial introduction of a number of non-thermal pasteurized products prompted the National Advisory Committee on Microbiological Criteria for Foods (NACMCF) to suggest a new definition for pasteurization (National Advisory Committee on Microbiological Criteria for Foods, 2006). According to NACMCF recommendations, pasteurization is defined as "any process, treatment, or combination thereof, that is applied to food to reduce the most resistant microorganism(s) of public health significance to a level that is not likely to present a public health risk under normal conditions of distribution and storage." High-pressure and PEF processing, ultraviolet processing, γ-irradiation, and other non-thermal processes are examples of processes that potentially satisfy the new definition of pasteurization.

1.5 Other food processing/preservation technologies

1.5.1 Fermentation

Fermentation causes desirable biochemical changes in foods in terms of nutrition or digestion, or makes them safer or tastier through microbial or enzyme manipulations. Examples of fermented foods are cheese, yogurt, most alcoholic beverages, salami, beer, and pickles. Representative vegetative bacteria in the fermentations are *Lactobacillus*, *Lactococcus*, *Bacillus*, *Streptococcus*, and *Pseudomonas* spp. Yeast and fungi (e.g. *Saccharomyces*, *Endomycopsis*, and *Monascus*) are also used for fermentation. Food supports controlled growth of these microorganisms, which modify food properties (texture, flavor, taste, color, etc.) via enzyme secretion.

1.5.2 Extrusion

Extrusion is a process that converts raw material into a product with a desired shape and form, such as pasta, snacks, textured vegetable protein, and ready-to-eat cereals, by forcing the material through a small opening using pressure (Singh & Heldman, 2009). Some of the unique advantages of extrusion include high productivity, adaptability, process scale-up, energy efficiency, low cost, and zero effluents (Riaz, 2000). An extruder consists of a tightly fitting screw rotating within a stationary barrel. Within the extruder, thermal and shear energies are applied to a raw food material to transform it to the final extruded product. Preground and conditioned ingredients enter the barrel where they are conveyed, mixed, and heated by a variety of screw and barrel configurations. Inside the extruder, the food may be subjected to several unit operations, including fluid flow, heat transfer, mixing, shearing, size reduction, and melting. The product exits the extruder through a die, where it usually puffs (if extruded at >100 °C and higher than atmospheric pressure) and changes texture from the release of steam and normal forces (Harper, 1979). Extruded products may undergo a number of structural, chemical, and nutritional changes including starch gelatinization, protein denaturation, lipid oxidation, degradation of vitamins, and formation of flavors (Riaz, 2000). Very limited studies are available to describe kinetics changes in foods during extrusion (Zhao et al., 2011).

1.5.3 Baking

Baking uses dry heat to cook fully developed flour dough into a variety of baked products including bread, cake, pastries, pies, cookies, scones, crackers, and pretzels. The dough needs to undergo various stages (mixing, fermentation, punching/sheeting, panning, proofing among others) before it is ready for baking.

Carbon dioxide gas is produced from yeast fermentation of available sugars, which could either be added or obtained via amylase breakdown of starch. During baking, heat from the source in an oven is transferred to the dough surface by convection; from the surface, it then transfers via conduction. As the heat is conducted through the food, it transforms the batter or dough into a baked food with a firm crust and softer center. During baking, heat causes the water to vaporize into steam. Gelatinization of flour starch in the presence of water occurs. The protein network (gluten) holds the structure, while carbon dioxide gas, that gives the dough its rise, collapses during baking (at ~450 °C). The product increases in size and volume, called leavening. Baking temperatures also cause a number of biochemical changes in the batter and dough, including dissolving sugar crystals, denaturing egg and gluten proteins, and gelatinizing starch. Baking also causes the surface to lose water, and breaks down sugars and proteins on the surface of the baked goods. This leads to formation of a brown color and desired baked flavor (Figoni, 2010).

1.5.4 Hurdle technology

Hurdle technology involves a suitable combination of different lethal agents to ensure microbial safety, quality, and stability of the processed product. Heat, pressure, acidity, water activity, chemical/natural preservatives, and packaging are examples of hurdles that can be combined to improve the quality of the final processed product. The hurdle approach requires the intensity of individual lethal agents (for example, heat or pressure) to be relatively modest, yet is quite effective in controlling microbial risk. Efforts must be made to understand the potential synergistic, additive or antagonistic effects of combining different lethal agents during hurdle technology.

1.5.5 Packaging

Packaging plays a vital role in many food preservation operations. Packaging has many functions, including containment, preservation, communication/education, handling/transportation, and marketing. Packaging helps maintain during storage the quality and properties of foods attained via processing. The packaging protects the food material from microbiological contaminants and other environmental factors. The package also helps prevents light-induced changes in stored food products and minimizes loss of moisture.

Depending upon the intensity of lethal treatments (heat, pressure, radiation dose), processing not only affects the food material but also alters the (moisture and oxygen) barrier properties of packaging materials and possibly induces migration of polymer material into the food. Thus, careful choice of food packaging material is essential for successful food process operation.

1.6 Emerging issues and sustainability in food processing

Modern food processing was developed during the 19th and 20th centuries with the rise of thermal pasteurization and sterilization techniques with the view of the extending shelf life of processed foods. Developments in industrial food processing technologies ensured the availability of a safe, abundant, convenient food supply at reasonable prices. However, the industry is currently undergoing a transformation in response to a variety of societal challenges. In a recent IFT scientific review, Floros et al. (2010) identified three emerging societal issues that will likely shape future developments in food processing.

- **Feeding the world**. The world population today is about 6.8 billion and it is expected to reach about 9 billion by 2050. Sustainable and efficient industrial food processing technologies at reasonable cost are needed to feed this ever-expanding world population.
- **Overcoming negative perceptions about "processed foods."** There has been an increased negative perception towards processed foods in the US. A number of societal factors have contributed to this trend, including negative perceptions towards technology use, diminishing appreciation of scientific literacy, as well as lack of familiarity or appreciation about farming among increasingly urban-based consumers.
- **Obesity**. Overweight and obesity are major health problems in the US and developed countries. Overconsumption of calorie-dense processed food and sedentary consumer lifestyles are some reasons put forward for the increase in obesity.

Apart from these issues, sustainability in the food processing industry is another emerging key societal issue. Sustainability is the capacity to endure; it is utilizing natural resources so that they are not depleted or permanently damaged. Water, land, energy, air, etc. are resources utilized in agriculture and food processing. Environmental concerns related to food production and processing which require consideration include land use change and reduction in biodiversity, aquatic eutrophication by nitrogenous factors and phosphorus, climate change, water shortages, ecotoxicity, and human effects of pesticides, among others (Boye & Arcand, 2013). Sustainable food processing technologies emphasize the efficient use of energy, innovative or alternative sources of energy, less environmental pollution, minimal use of water, and recycling of these resources as much as possible. Sustainable food processing requires processors to maximize the conversion of raw materials into consumer products by minimizing postharvest losses and efficient use of energy and water. Modern food processing plants can contribute to sustainability by utilizing green building materials and practices in their construction, utilizing innovative building designs, using energy-efficient equipment and components, and following efficient practices in routing, storing, and processing of ingredients and distribution and handling of finished products. Most of the food processing operations described in subsequent chapters of this textbook include a short discussion on sustainability.

1.7 Conclusion

Modern food processors can choose from several preservation approaches (heat addition or removal, acidity, water activity, pressure, electric field, among others) to transform raw food materials to produce microbiologically safe, extended shelf life, consumer-desired, convenient, value-added foods. Successful food processing requires integration of knowledge from several disciplines including engineering, chemistry, physics, biology, nutrition, and sensory sciences. The type of food processing operation chosen can influence the extent of changes in product quality (color, texture, flavor) attributes. The extent of nutrient and quality retention in processed food depends upon intensity of treatment applied, type of nutrient or food quality attribute, food composition, and storage conditions.

While industrial food processing provides safe, plentiful, relatively inexpensive food, processed foods are often perceived as unhealthy and not sustainable. Developments in novel "non-thermal" technologies that rely on lethal agents other than heat (pressure, electric field, among others) may partly address this issue by increasing nutrient bioavailability and preserving heat-sensitive phytochemicals. Efforts must be made to introduce lean manufacturing concepts developed in other industries to food processing to reduce energy and water use and allow the production of healthy processed foods at affordable cost.

References

Anonymous (2012) US Department of Commerce Industry Report. Food Manufacturing. http://trade.gov/td/ocg/report08_-processedfoods.pdf, accessed 5 November 2013.

Boye JI, Arcand Y (2013) Current trends in green technologies in food production and processing. *Food Engineering Reviews* 5: 1–17.

Cheftel JC (1995) Review: high pressure, microbial inactivation and food preservation. *Food Science and Technology International* 1: 75–90.

Dolatowski DZ, Stadnik J, Stasiak D (2007) Applications of ultrasound in food technology. *Acta Scientiarum Polonorum Technologia Alimentaria* 6: 89–99.

Doyle ME, Glass KA (2010) Sodium reduction and its effect on food safety, food quality, human health. *Comprehensive Reviews in Food Science and Food Safety* 9: 44–56.

Earle RL, Earle MD (2012) Unit Operations in Food Processing. www.nzifst.org.nz/unitoperations/index.htm, accessed 5 November 2013.

Farkas DF, Hoover DG (2000) High pressure processing. *Journal of Food Science* 65(suppl.): 47S–64S.

Fellows P (2009) *Food Processing Technology Principles and Practice*, 3rd edn. Boca Raton, FL: Woodhead Publishing.

Figoni P (2010) *How Baking Works: Exploring the Fundamentals of Baking Science*, 2nd edn. New Jersey: Wiley.

Floros JD, Newsome R, Fisher W et al. (2010) Importance of Food Science and Technology. An IFT Scientific Review. *Comprehensive Reviews in Food Science and Food Safety* 9: 572–599.

Gould WA (1997) *Fundamentals of Food Processing and Technology*. Cambridge: Woodhead Publishing.

Harper JM (1979) Food extrusion. Critical Reviews in Food Science and Nutrition 11: 155–215.

Heldman DR (2003) *Encyclopedia of Agricultural, Food and Biological Engineering*. New York: Marcel Dekker.

Heldman DR, Hartel RW (1998) *Principles of Food Processing*. New York: Springer.

Institute of Food Technologists (2000) *Kinetics of Microbial Inactivation for Alternative Food Processing Technologies*. A report of the Institute of Food Technologists for the Food and Drug Administration of the US Department of Health and Human Services, IFT/FDA Contract No 223–98–2333. Silver Spring, MD: Food and Drug Administration.

Jay JM, Loessner MJ, Golden DA (2005) *Modern Food Microbiology*, 7th edn. New York: Springer.

McCabe WL, Smith JC, Harriott P (2001) *Unit Operations of Chemical Engineering*, 6th edn. New York: McGraw-Hill.

Molins RA (ed) (2001) *Food Irradiation: Principles and Applications*. New York: John Wiley.

National Advisory Committee on Microbiological Criteria for Foods (2006) Requisite scientific parameters for establishing the equivalence of alternative methods of pasteurization. Journal of Food Protection 69: 1190–1216.

Nguyen LT, Balasubramaniam VM (2011) Fundamentals of food processing using high pressure. In: Zhang HQ, Barbosa-Cánovas GV, Balasubramaniam VM et al. (eds) *Nonthermal Processing Technologies for Food*. Chichester: IFT Press, Wiley-Blackwell, pp. 3–19.

Pflug IJ (1998) *Microbiology and Engineering of Sterilization Processes*, 9th edn. Minneapolis, MN: Environmental Sterilization Laboratory.

Piyasena P, Mohareb E, McKellar RC (2003) Inactivation of microbes using ultrasound: a review. International Journal of Food Microbiology 87: 207–216.

Potter NN, Hotchkiss JH (1998) *Food Science*, 5th edn. New York: Chapman and Hall, p. 242.

Ramaswamy HS, Tang J (2008) Microwave and radio frequency heating. Food Science and Technology International 14: 423–428.

Rao MA, Syed SH, Rizvi A, Datta K (2010) *Engineering Properties of Foods*, 3rd edn. Boca Raton, FL: CRC Press.

Ray B (2004) *Fundamental Food Microbiology*, 3rd edn. Boca Raton, FL: CRC Press, p. 443.

Riaz MN (2000) *Extruders in Food Applications*. Boca Raton, FL: CRC Press.

Saravacos GD, Kostaropoulos AE (2002) *Handbook of Food Processing Equipment*. New York: Kluwer Academic/Plenum Publishers.

Sastry SK (2008) Ohmic heating and moderate electric field processing. Food Science and Technology International 14: 419–422.

Sastry SK, Cornelius BD (2002) *Aseptic Processing of Foods Containing Solid Particulates*. New York: John Wiley.

Singh RP, Heldman DR (2009) *Introduction to Food Engineering*, 4th edn. San Diego, CA: Academic Press.

Smith PG (2003) *Introduction to Food Process Engineering*. New York: Kluwer Academic/Plenum Publishers, p. 318.

Tassou SA, Lewis JS, Ge YT, Hadawey A, Chaer I (2010) A review of emerging technologies for food refrigeration applications. Applied Thermal Engineering 30: 263–276.

Toledo RT (2007) *Fundamentals of Food Process Engineering*, 3rd edn. New York: Springer, pp. 302, 323.

USDA Economic Research Service (2013) Food and Beverage Manufaturing. www.ers.usda.gov/topics/food-markets-prices/

processing-marketing/manufacturing.aspx#.UoZ5mF0o66o, accessed 20 November 2013.

Welti-Chanes J, Vergara-Balderas F, Bermúdez-Aguirre D (2005) Transport phenomena in food engineering: basic concepts and advances. Journal of Food Engineering 67: 113–128.

Yousef AE, Balasubramaniam VM (2013) Physical methods of food preservation. In: Doyle MP, Buchanan RL (eds) *Food Microbiology: Fundamentals and Frontiers*, 4th edn. Washington, DC: ASM Press, pp. 737–763.

Yousef, A.E, Zhang HQ (2006) Microbiological and safety aspects of pulsed electric field technology. In: Juneja VK, Cherry JP, Tunick MH (eds) *Advances in Microbiological Food Safety*. Washington, DC: American Chemical Society, pp. 152–166.

Zhang HQ, Barbosa-Cánovas GV, Swanson BG (1995) Engineering aspects of pulsed electric fields pasteurization. Journal of Food Engineering 25: 261–281.

Zhang HQ, Barbosa-Cánovas GV, Balasubramaniam VM, Dunne CP, Farkas DF, Yuan JTC (eds) (2011) *Nonthermal Processing Technologies for Food*. Chichester: IFT Press, Wiley-Blackwell.

Zhao X, Yimin W, Zhangcun W, Fengliang C, Okhonlaye Ojokoh A (2011) Reaction kinetics in food extrusion: methods and results. Critical Reviews in Food Science and Nutrition 51: 835–854.

Zottola EA, Sasahara KC (1994) Microbial biofilms in the food processing industry – should they be a concern? International Journal of Food Microbiology 23: 125–148.

2 Thermal Principles and Kinetics

Prabhat Kumar[1] and K.P. Sandeep[2]
[1]*Research and Development, Frito Lay, Plano, Texas, USA*
[2]*Department of Food, Bioprocessing and Nutrition Sciences, North Carolina State University, Raleigh, North Carolina, USA*

2.1 Introduction

Thermal processing of food materials is one of the most widely used methods of food preservation. Foods may be thermally processed using numerous heating systems such as retorts (batch or continuous), direct heating systems (steam injection or steam infusion), indirect heating systems (tubular heat exchangers, shell and tube heat exchangers, plate heat exchangers, scraped surface heat exchangers), volumetric heating systems (microwave or ohmic heating), and combinations of these. The choice of the heating system is based on several factors, including the characteristics of the product (pH, water activity, composition, etc.), properties of the product (density, viscosity, specific heat, thermal conductivity, thermal diffusivity, electrical conductivity for ohmic heating, and dielectric properties for microwave heating), quality of the product, need for refrigeration, need or acceptability of moisture addition/removal, and cost.

The extent of thermal treatment required for a food product depends on whether it is an acid product, an acidified product, or a low-acid product. An acid food product is one with a natural pH of less than 4.6. Acid food products include apple juice, orange juice, ketchup, etc. An acidified food product is one with an equilibrium pH of less than 4.6 and a water activity (a_w) greater than 0.85. Examples of acidified foods include peppers treated in an acid brine, pickled foods (excluding foods pickled by fermentation), etc. Acidified food products are typically treated at 90–95 °C for a period of 30–90 sec to inactivate yeasts, molds, and bacteria (usually *Lactobacillus* species). A low-acid food product is any food other than alcoholic beverages, with a natural equilibrium pH greater than 4.6 and a water activity greater than 0.85. These food products include butter, cheese, fresh eggs, pears, papaya, and raisins (Skudder 1993). Low-acid food products are capable of sustaining the growth of *Clostridium botulinum* spores. *Clostridium botulinum* is an anaerobic, gram-positive, heat-resistant spore-forming bacterium that produces a potent neurotoxin. Food-borne botulism is a severe type of food poisoning caused by the ingestion of foods containing the potent neurotoxin formed during the growth of *Clostridium botulinum*. The spores of *Clostridium botulinum* must be destroyed or effectively inhibited to prevent germination and subsequent production of the deadly toxin, which causes botulism.

Low-acid food products come under the regulatory authority of either the Food and Drug Administration (FDA) or the United States Department of Agriculture (USDA), depending on the proportion of meat or poultry in the food product. The FDA regulates most food products, except those containing more than 3% raw or 2% cooked meat or poultry ingredients, which fall under the jurisdiction of the USDA. The general thermal process requirements of both regulatory agencies are similar and they are compiled in Code of Federal Regulations 21 CFR Parts 108 (emergency permit control), 113 (low-acid canned foods), and 114 (acidified foods) for the FDA, and 9 CFR Parts 308, 318 (meat products), 320, 327, and 381 (poultry products) for the USDA. The FDA requires registration of the processing facility (form 2541) and a detailed process filing (form 2541a for acidified and low-acid canned foods and form 2541c for low-acid aseptically processed foods) (Chandarana 1992).

2.2 Methods of thermal processing

There are several methods of thermal processing of foods, with pasteurization and sterilization being the two most widely used. The most common methods of thermal

Food Processing: Principles and Applications, Second Edition. Edited by Stephanie Clark, Stephanie Jung, and Buddhi Lamsal.
© 2014 John Wiley & Sons, Ltd. Published 2014 by John Wiley & Sons, Ltd.

processing include blanching, pasteurization, hot filling, and sterilization. These methods are now described.

2.2.1 Blanching

Blanching is a mild heat treatment commonly applied to fruits and vegetables prior to freezing, drying, or canning. Blanching is performed to inactivate enzymes, enhance drying and rehydration, remove tissue gases, enhance color of green vegetables, and reduce microbial load. The effectiveness of blanching is usually evaluated by assaying for peroxidase and catalase activity. Blanching is usually accomplished by bringing the product into contact with hot water, hot air, or steam for a specified period of time, depending upon the product and/or enzyme of interest. Water blanching can be conducted as a batch operation by dipping a batch of product in hot water for the required time. Continuous hot water blanching can be accomplished using a screw-type, drum-type, or pipe-type blancher. The screw-type blancher consists of a trough fitted with a helical screw. The drum-type blancher consists of a perforated drum fitted internally with a helical screw. The pipe-type blancher can be used for solid products, which can be pumped with water. Similar to water blanching, steam blanching can also be accomplished as a batch or continuous process. The heating time necessary to accomplish blanching depends on the type and size of fruit/vegetable, the method of heating, and the temperature of the heating medium. Typical blanching times at 100 °C for commercial blanching range from 1 to 5 min (Lund, 1975).

2.2.2 Pasteurization

Pasteurization refers to the heat treatment of food products, mostly liquid or liquid with particulates, to inactivate vegetative pathogenic microorganisms. The time-temperature combination for the pasteurization of milk, for instance, is 63 °C for 30 min, referred to as a low-temperature, long-time (LTLT) process, and 72 °C for 15 sec, referred to as a high-temperature, short-time (HTST) process. The heat treatment in pasteurization is not sufficient to inactivate all spoilage-causing vegetative cells or heat-resistant spores. Therefore, the shelf life of pasteurized low-acid products such as milk and dairy products is approximately 2–3 weeks under refrigerated conditions. Ultrapasteurization refers to pasteurization at temperatures of 138 °C or above for at least 2 sec, either before or after packaging. This process further extends the shelf life of the product. Ultrapasteurization

results in destruction of a greater proportion of spoilage microorganisms, leading to an extended shelf life of about 6–8 weeks at refrigeration temperature. This process has been used for flavored milks and non-dairy creamers in portion pack cups (David et al., 1996).

The choice of heating system for pasteurization depends on the characteristics (rheological and thermal properties) of the product, potential for fouling, ease of cleaning, and cost of the heating equipment. A direct type heating system (steam injection and steam infusion) is used for homogeneous and high-viscosity products and is particularly suited for shear-sensitive products such as creams, desserts, and sauces. In a steam injection heating system, liquid product is heated by injection of culinary steam into the product. Rapid heating by steam, combined with rapid methods of cooling, can yield a high-quality product. A steam infusion heating system, similar to steam injection, involves infusing a thin film of liquid product into an atmosphere of steam, which provides rapid heating. A direct heating system (steam injection or steam infusion) adds water to the product due to the condensing steam. The amount of added water should be either accounted for in the product formulation or removed by pumping the heated liquid into a vacuum cooling chamber.

There are four main types of indirect heating systems: tubular, shell and tube, plate, and scraped surface heat exchangers. Tubular heat exchangers are used for homogeneous and high-viscosity products (soups and fruit purees) containing particles of sizes up to approximately 10 mm. The simplest tubular heat exchanger is a double pipe heat exchanger consisting of two concentric pipes. Shell and tube heat exchangers consist of a shell (typically cylindrical in shape) with one or more sets of tubes inside it. The tubes may be coiled in a helical manner or arranged in a trombone fashion. This type of heat exchanger is used when a greater degree of mixing than that achieved in a tubular heat exchanger is desired. Plate heat exchangers are used for homogeneous and low-viscosity (<5 Pa.s) products (e.g. milk, juices, and thin sauces) containing particle sizes up to approximately 3 mm. These heat exchangers consist of closely spaced parallel plates pressed together in a frame. They provide a rapid rate of heat transfer due to the large surface area for heat transfer and turbulent flow characteristics. Scraped surface heat exchangers are used for viscous products (e.g. diced fruit preserves and soups) containing particles of sizes up to approximately 15 mm. These heat exchangers consist of a jacketed cylinder housing with scraping blades on a rotating shaft. The rotating action of the scraping blades prevents fouling on the heat exchanger surface and improves the rate of heat transfer. Fouling is the phenomenon of product build-up on the heat transfer surface caused when a liquid product comes into contact with a heated surface. Fouling increases thermal resistance and thus results in reduced rates of heat transfer. This type of heat exchanger is the best choice for viscous products containing particulates (Skudder 1993).

Apart from tubular, shell and tube, plate, and scraped surface heat exchangers, pasteurization can also be accomplished in a vat or tank-type heat exchanger. In a tank-type heat exchanger, product is pumped into a jacketed vat or tank, heated to pasteurization temperature, held for the required time, and pumped from the vat to the cooling section (Mitten, 1963).

Volumetric heating systems such as microwave and ohmic heating can provide very rapid heating throughout the product, which is desirable for aseptic processing. However, it is challenging to maintain a uniform temperature distribution within the product. Microwave heating systems apply a rapidly changing electromagnetic field to the product. Movement of charged ions and agitation of the small polar molecules within the product (mostly water molecules) due to the changing electromagnetic field generate heat. An ohmic heating system operates by directly passing electric current through a product. The electrical resistance of the product to the passing electric current generates heat (Coronel et al., 2008).

2.2.3 Hot filling

Acid/acidified products such as juices and beverages packed in hermetically sealed containers using an appropriate hot filling process yield commercially sterile shelf-stable products. Hot filling, also known as "hot fill and hold," refers to filling unsterilized containers with a sterilized acid/acidified food product that is hot enough to render the container commercially sterile. A hermetically sealed container is a container that is designed and intended to be secure against contamination by microorganisms and thus to maintain the commercial sterility of its contents after processing.

2.2.4 Sterilization

Sterilization refers to killing of all living microorganisms, including spores, in the food product. Food products are never completely sterilized; instead, they are rendered commercially sterile. Commercial sterility means the condition achieved either by (1) the application of heat, which renders the food free of microorganisms capable of reproducing in the food under normal non-refrigerated conditions of storage and distribution, and viable microorganisms (including spores) of public health significance, or by (2) the control of water activity and the application of heat, which renders the food free of microorganisms capable of reproducing in the food under normal non-refrigerated conditions of storage and distribution. Commercially sterile food products are shelf-stable with a long shelf life (1–2 years) (Anderson et al. 2011; David et al. 1996).

Low-acid food products are rendered commercially sterile to prevent the growth of *Clostridium botulinum* spores (David et al., 1996; Lund, 1975). Commercial sterility can be achieved by in-container sterilization or in-flow sterilization. In-container sterilization generally refers to the retorting process whereas in-flow sterilization refers to aseptic processing.

2.2.4.1 Retorting

Traditionally, retorting has been used to process low-acid food products to ensure destruction of *C. botulinum* spores. Conventional retorting involves filling of the product in metal cans, glass jars, retortable semi-rigid plastic containers or retortable pouches, double seamed or heat sealed, followed by heating, holding, and cooling in a pressurized batch or continuous retort. Retorting of foods in cans, invented by Nicholas Appert in the early 1800s, still remains the gold standard for preservation of foods. Retorts can be operated in either batch or continuous mode. Batch retort is the most versatile sterilization system, with the ability to handle different products (conduction heating and convection heating) and package types. Batch retort can further be classified into still/static (horizontal, vertical, or crateless) retort, and agitating/rotary (end-over-end or axial rotation) retort. When steam is used as the heating medium, it should be introduced into the retort with care such that all the air in the retort is displaced. Inadequate elimination of air may result in understerilization or non-uniform cooking of products in the retort. Removal of air by steam is also known as venting. Cooling is accomplished by shutting off steam and introducing cold water into the retort.

Overpressure is often used to prevent internal pressure inside the container from bursting containers. Thus, overpressure allows thermal processing of a wide variety of containers including glass, rigid plastics, and flexible pouches. The rotary retort agitates the product inside the container by the movement of the air in the headspace, resulting in enhanced heat transfer in the container. A larger headspace results in faster heating/cooling of a product due to efficient mixing. Different heating media and heating methods used in various batch retorts include steam, water, steam-air, water cascading, water spray, or water immersion (Lund, 1975; Weng, 2005).

Continuous retort can also be classified into static (hydrostatic) retort and rotary (hydrolock, sterilmatic, reel and spiral, etc.) retort. Continuous retorts increase throughput and lower manpower costs. Hydrostatic retorts are vertical systems, which use water legs for preheating and cooling, with a central steam heating chamber. Hydrostatic retort is well suited for products that require long cook and cool times along with higher throughput. A continuous rotary retort is a fully automated system designed for high throughput, lower energy consumption, and uniform product quality. These systems require a cylindrical container with limited variation in can diameter and height (Weng, 2005).

2.2.4.2 Aseptic processing

Aseptic processing offers an alternative to conventional retorting to meet the demand for safe, convenient, and high-quality foods. In aseptic processing of foods, the product and the package are sterilized separately and brought together in a sterile environment. This involves sterilization of a food product, followed by holding it for a specified period of time in a holding tube, cooling it, and packaging it in a sterile container. Aseptic processing uses high temperatures for a short period of time, yielding a high-quality (nutrients, flavor, color, or texture) product compared to that obtained by conventional canning. Some of the other advantages associated with aseptic processing include longer shelf life (1–2 years at ambient temperature), flexible package size and shape, less energy consumption, less space requirement, eliminating the need for refrigeration, easy adaptability to automation, and need for fewer operators. However, some of the disadvantages of aseptic processing include slower filler speeds, higher overall initial cost, need for better quality control of raw ingredients, better trained personnel, better control of process variables and equipment, and stringent validation procedures.

Products that are aseptically processed include fruit juices, milk, coffee creamers, purees, puddings, soups, baby foods, and cheese sauces (David et al., 1996). Sterilization of products via aseptic processing can be accomplished using tubular, shell and tube, scraped surface or volumetric heating (microwave and ohmic) systems.

2.3 Microorganisms

The microorganisms of importance in thermal processing are bacteria and fungi because they can grow in foods and cause spoilage or public health issues. Bacteria are a large group of unicellular prokaryotic microorganisms that are found in a wide range of shapes, such as spheres (cocci), rods, and spirals. Bacteria reproduce asexually through binary fission. Under favorable growth conditions, bacteria can grow and divide rapidly. A typical growth cycle of bacteria can be divided into four phases: lag, log, stationary, and death. During the lag phase, bacteria adapt to their new surroundings and multiple slowly. During the log phase, bacteria multiply at an exponential rate. During the stationary phase, growth rate slows down, and eventually they stop multiplying, resulting in their death in the death phase (Tucker & Featherstone, 2011).

Gram-positive bacteria are more resistant to changes in environment because of the thick peptidoglycan layer in the cell wall. The cell wall of gram-positive bacteria consists of peptidoglycan and teichoic acids. Teichoic acids are negatively charged acidic polysaccharides, which may be involved in ion transport. The cell wall of gram-negative organisms consists of a thin peptidoglycan layer, periplasm, and a lipopolysaccharide (LPS) layer. LPS is composed of a lipid A component and a polysaccharide component. Pathogenicity of gram-negative bacteria is usually associated with the lipid A component of LPS, also known as endotoxin. Gram-negative bacteria are less fastidious (grow faster) than gram-positive bacteria.

Some species of rod-shaped bacteria can form highly resistant structures known as spores, which can survive extreme stress conditions such as high heat and pressure. Bacteria in this dormant state may remain viable for thousands of years (Tucker & Featherstone, 2011).

Fungi are a group of eukaryotes that include yeasts and molds. Fungi are neither plants nor animals, as they possess some properties (cell wall) similar to plants and some properties (absence of chlorophyll) similar to animals. Yeasts are unicellular fungi that derive their energy from organic compounds and do not require sunlight to grow. Yeasts (4–8 μm) are larger than bacteria (0.5–5 μm), but smaller than molds (10–40 μm). Yeasts reproduce asexually by budding, when a small bud forms on the parent yeast cell and gradually enlarges into another yeast cell.

Yeasts are either obligate aerobes or facultative anaerobes. They grow best in a neutral or a slightly acidic medium and are generally destroyed above a temperature of 50 °C. Yeasts are used in the food industry for leavening of bread and production of alcohol. However, their ability to grow at low pH and water activity (a_w) makes them organisms of concern for spoilage in fruit products such as juices and jams (Tucker & Featherstone, 2011).

Molds are multicellular fungi that grow in the form of hyphae (multicellular tubular filaments). Molds reproduce sexually and asexually by means of spores produced on specialized structures or in fruiting bodies. All molds are aerobic, but some can grow in low oxygen conditions. They can also grow at extreme conditions (high acid, high salt, low temperature). Molds are used in the food industry for production of soy sauce and certain cheeses. They are also capable of consuming acids, which can remove the acidic conditions that inhibit the growth of *Clostridium botulinum*. Spoilage of food products by mold is primarily due to mycotoxins (aflatoxin, ocratoxin, Patulin, etc.), which are secondary metabolites produced by molds (Tucker & Featherstone, 2011).

2.3.1 Factors affecting microbial growth

Growth of microorganisms is dependent on several factors such as oxygen content, temperature, relative humidity, pH, water activity (a_w), redox potential, and antimicrobial resistance. These factors can be grouped into intrinsic and extrinsic factors. Characteristics of the food itself are known as intrinsic (pH, a_w, redox potential) factors whereas factors external to food are known as extrinsic (oxygen content, temperature, relative humidity) factors.

Most microorganisms grow best at neutral pH and only a few are able to grow at a pH value of less than 4.0. Bacteria are more selective about pH requirements than yeasts and molds, which can grow over a wide range of pH. Microorganisms that can withstand low pH are known as aciduric. Bacteria require higher a_w for growth compared to that required by yeasts and molds. Gram-negative bacteria cannot grow at a_w less than 0.95, whereas most gram-positive bacteria cannot grow at a_w less than 0.90. However, *Staphylococcus aureus* can grow at a_w value as low as 0.85 and halophilic bacteria can grow at a minimum a_w value of 0.75. Halophilic microorganisms are those that require a high salt concentration (3.4–5.1 M NaCl) for growth. Most yeasts and molds can grow at a minimum a_w value of 0.88

and 0.80, respectively. Xerophilic (microorganisms which can grow in low a_w conditions) molds and osmophilic (microorganisms which can grow in high solute concentration) yeasts can grow at a_w as low as 0.61.

Redox potential is the tendency of a substance to convert to its reduced state by acquiring electrons. It is measured in millivolts (mV) relative to a standard hydrogen electrode (0 mV). In general, aerobic microorganisms prefer positive redox potential for growth whereas anaerobic microorganisms prefer negative redox potential (Tucker & Featherstone, 2011). Based on oxygen requirements, microorganisms can be classified into aerobes, anaerobes, facultative anaerobes, and microaerophiles. Aerobes grow in the presence of atmospheric oxygen whereas anaerobes grow in the absence of atmospheric oxygen. Facultative anaerobes are in between these two extremes and can grow in either the presence or absence of atmospheric oxygen. Microaerophiles require a small amount of oxygen to grow (Montville & Matthews, 2008).

Based on the response to temperature, microorganisms can be classified into psychrophilic, psychrotrophic, mesophilic, and thermophilic. Psychrophilic microorganisms have an optimum growth temperature between 12 °C and 15 °C but can grow up to 20 °C. Psychrotrophic microorganisms have an optimum growth temperature between 20 °C and 30 °C but can grow up to 0 °C. Mesophilic microorganisms have an optimum growth temperature between 30 °C and 42 °C but can grow between 15 °C and 47 °C. Thermophilic microorganisms have an optimum growth temperature between 55 °C and 65 °C, but can grow between 40 °C and 90 °C.

Relative humidity is the amount of water vapor present in a mixture of air and water. Relative humidity of the storage environment can affect growth of microorganisms by changing the water activity of the food (Montville & Matthews, 2008; Tucker & Featherstone, 2011).

2.4 Thermal kinetics

2.4.1 Destruction of a microbial population

When a homogeneous microbial population is subjected to a constant temperature, T, the rate of destruction of microbes follows a first-order reaction kinetics as is given by David et al. (1996):

Table 2.1 D and z values of important microorganisms

Microorganisms	Temperature (°C)	D value (min)	z value (°C)	References
Bacillus cereus	121.1	0.0065	9.7	Lund, 1975
Bacillus coagulans	121.1	3		Holdsworth, 2004
Bacillus coagulans	110	6.6–9		Cousin, 1993
Bacillus coagulans var. *thermoacidurans*	96	8		Holdsworth, 2004
Bacillus licheniformis	100	13		Holdsworth, 2004
Bacillus stearothermophilus	121.1	4	7	Lund, 1975
Bacillus stearothermophilus	150	0.008		Cousin, 1993
Bacillus subtilis	121.1	0.48–0.76	7.4–13	Lund, 1975
Bacillus subtilis	140	0.001		Cousin, 1993
Clostridium botulinum	121.1	0.21	9.9	Lund, 1975
Clostridium butyricum	85	8		Holdsworth, 2004
Clostridium sporogenes	121.1	0.15	13	Lund, 1975
Clostridium sporogenes (PA 3679)	121.1	0.3–2.6	10.6	Cousin, 1993
Clostridium thermosaccharolyticum	121.1	16–22	1.7–2.2	Lund, 1975
Desulfotomaculum nigrificans	121.1	3–5		Holdsworth, 2004
Desulfotomaculum nigrificans	121.1	13–54.4		Cousin, 1993

$$\frac{-dN}{dt} = K_T N, \qquad (2.1)$$

where N is the number of microbes surviving after processing time t (s) and K_T is the reaction rate (s^{-1}). Integration of Equation 2.1 from time 0 to time t yields:

$$\frac{N}{N_0} = e^{-K_T t}, \qquad (2.2)$$

where N_0 is the number of viable microorganisms at time $t = 0$.

Equation 2.2 can be rewritten as:

$$\log_{10}\left(\frac{N}{N_0}\right) = \frac{-t}{D}, \qquad (2.3)$$

where $D = 2.303/K_T$. The parameter D is the decimal reduction time, the time required to reduce the size of the surviving microbial population by 90%. The D value is a measure of heat resistance of microorganisms. Microorganisms with a higher D value have a higher heat resistance. The D value determined at a reference temperature (T_{ref}) is denoted by D_{ref}. The effect of temperature on D value is generally described by the following expression (David et al., 1996):

$$\log_{10}\left(\frac{D_T}{D_{ref}}\right) = \frac{T_{ref} - T}{z}, \qquad (2.4)$$

where D_T is the D value at temperature T and z is the change in temperature (°C) required to reduce the D value by 90%. D and z values are the basis of thermal process calculations and are commonly used to design a thermal process. D and z values for most of the common microorganisms are given in Table 2.1. The choice of target microorganisms for designing a thermal process should take into account the characteristics of the product (a_w, pH, etc.) and the storage and transportation conditions.

The ratio of D_{ref} to D is the lethal rate (L_r). Thermal death time (TDT) or F value of a process is defined as the process time at a given temperature required for stipulated destruction of a microbial population, or the time required for destruction of microorganisms to an acceptable level. The F value can be expressed as a multiple of the D value for first-order microbial kinetics. The F value required for a process depends on the nature of food (pH and water activity), storage conditions after processing (refrigerated versus room temperature), target organism, and initial population of microorganisms (Singh, 2007). The F value is usually expressed with a superscript denoting z value and a subscript denoting temperature. It can be computed in terms of lethal rate as:

$$F_{T_{ref}}^z = \int_0^t L_r \, dt = \int_0^t \left(10^{\frac{T-T_{ref}}{z}}\right) dt = -D_{ref} \log \frac{N}{N_0}, \qquad (2.5)$$

where temperature T is a function of time. For a constant temperature process (process temperature remains constant), the above equation for the F value reduces to:

$$F^z_{T_{ref}} = 10^{\frac{T-T_{ref}}{z}}t \qquad (2.6)$$

The F value at a reference temperature of 121.1 °C (250 °F) and a z value of 10 °C (18 °F) is referred to as the F_0 value. The main microorganism of concern for low-acid foods is *C. botulinum*, which has a $D_{121.1}$ value of 0.21 min (see Table 2.1). For processes where *C. botulinum* is the target organism, The F_0 value represents the process time necessary to achieve a 12 log reduction (12D) in microbial population of *C. botulinum* at 121.1 °C. A thermal process designed to reduce the probability of *C. botulinum* spore survival to 10^{-12}, a 12D process, is referred to as the botulinum cook. The F_0 value for a botulinum cook is 2.52 (12 × 0.21) min. An F_0 value of 2.52 min indicates that the process is equivalent to a full exposure of food to 121.1 °C (250 °F) for 2.52 min. Many combinations of time and temperature can yield an equivalent F_0 value of 2.52 min. The F value can be written in terms of the F_0 value as:

$$F^z_T = F_0/L_r = F_0 \times 10^{\frac{T_{ref}-T}{z}} \qquad (2.7)$$

The ratio of F_0 value of the process at a given process time to the F_0 required for commercial sterility is known as lethality. Thus, lethality must be at least 1 for commercial sterility of the product (David et al., 1996). All low-acid food products are processed beyond the minimum botulinum cook in order to eliminate spoilage from mesophilic spore formers. The organism most frequently used to characterize this food spoilage is a strain of *C. sporogenes*, a putrefactive anaerobe (PA), known as PA 3679. The F_0 value required to prevent mesophilic spoilage represents a 5-log reduction in microbial population of *C. sporogenes*. A more severe process may be necessary for situations where thermophilic spoilage could be a concern because of very high heat resistance of thermophilic spores. A 5-log reduction of *Bacillus stearothermophilus* has been used to establish a thermal process to prevent thermophilic spoilage (Teixeira & Balaban, 2011). For foods with a pH value between 4.0 and 4.6, the thermal process is less severe. The microorganisms of concern include *Bacillus coagulans*, *Clostridium butyricum*, and *Clostridium pasteurianum*. The thermal process for foods with a pH value less than 4.0 is designed to inactivate the most resistant yeast, mold, or acid-tolerant bacteria.

Table 2.2 D_c and z_c values for enzymes and quality attributes (Lund, 1975)

Enzyme or quality attribute	Temperature (°C)	D value (min)	z value (°C)
Anthocyanin (in grape juice)	121.1	17.8	23.2
Thiamin (in whole peas)	121.1	164	26.1
Thiamin (in pea puree)	121.1	247	26.7
Thiamin (in peas in brine)	121.1	226.7	27.2
Thiamin (in lamb puree)	121.1	120	25
Lysine (in soya bean meal)	121.1	786	21.1
Chlorophyll a (in spinach: pH = 6.5)	121.1	13	51.1
Chlorophyll b (in spinach: pH = 5.5)	121.1	14.7	79.4
Peroxidase (in peas)	121.1	3	37.2
Chlorophyll (in blanched pea puree)	121.1	14	36.7
Chlorophyll (in unblanched pea puree)	121.1	13.9	45
Color (in peas)	121.1	25	39.4
Organoleptic quality (in peas)	121.1	2.3	28.3
Texture (in peas)	121.1	1.4	32.2
Overall quality (in peas)	121.1	2.5	32.2
Color (in green beans)	121.1	21	38.9
Color (in asparagus)	121.1	17	41.7

Molds and yeasts are easily inactivated by heat but ascospores of yeasts and molds may be more heat resistant (Cousin, 1993).

2.4.2 Destruction of quality attributes

The destruction of nutrients and inactivation of enzymes follow similar kinetics to that of the destruction of microorganisms, which is first-order kinetics. D_c and z_c values for enzymes and quality attributes are given in Table 2.2. Destruction of nutrients in food products is quantified by the term cook value (C), which has been defined (Mansfield, 1962) as:

$$C = \int_0^t 10^{\frac{T-T_{ref}}{z_c}} dt \qquad (2.8)$$

The C value at a reference temperature of $100\,°C$ ($212\,°F$) and a z_c value of $33.1\,°C$ ($59.6\,°F$) is referred to as the C_0 value.

2.4.3 Process optimization

The objective of a food processor is to produce a safe product that retains nutritional and quality attributes at an acceptable level. Therefore, the appropriate combination of time and temperature used for processing is based on factors such as nutrient retention and enzyme inactivation in addition to safety. D_c and z_c values for destruction of nutritional and quality attributes are generally larger than those for microorganisms. This implies that the rate of destruction of microorganisms at higher temperature will be much higher than the rate of destruction of nutritional and quality attributes. Thus, thermal processing of food products at higher temperature can achieve commercial sterility with better retention of nutritional and quality attributes (David et al., 1996).

2.5 Thermal process establishment

The goal in thermal processing is to ensure that the slowest heating point (cold spot) within a product container receives adequate thermal treatment. This involves measurement of product temperature at the slowest heating point. For in-container sterilization processes, there are two main stages in thermal process establishment: the temperature distribution (TD) test to identify the slowest heating zone in the retort and the heat penetration (HP) test to determine the temperature history at the cold spot in prepackaged foods. For in-flow sterilization processes, the TD test is not required.

2.5.1 Temperature distribution test

The temperature distribution inside a retort is not uniform. The location of the slowest heating zone in the retort is determined by performing a TD test. The first step in conducting the test is the selection of the test retort. A survey of the processing room should be done to select the test retort. The survey should include examination of the following factors: steam, air, and water supply to the retort, type and size of each retort in the retort room, purging, drainage, and retort loading considerations (container information, type of product heating, maximum number of containers, etc.). To conduct the TD test, the situation resulting in worst-case conditions for commercial operation should be selected. Containers may be filled with water for convection heating products. For conduction heating products, containers should be filled either with the product or other material that simulates the product (starch solution). Temperature measuring devices (TMD) in sufficient quantity should be used to monitor the temperature of the heating medium within the retort. The most commons TMDs used in thermal processing are duplex type T (copper-constantan) thermocouples with Teflon insulation. Pressure-indicating devices should be used to monitor pressure in the retort shell during the test. Flow meters should be used to measure flow rate of process water during come-up and heating.

The test should be conducted at the maximum retort temperature used during processing. The critical parameters that should be recorded during a TD test include the temperature controller set point, initial temperature (IT), time when steam is turned on, temperature of heating medium, flow rate of heating medium, time when the reference TMD achieves the process set point, and come-up time. Come-up time (CUT) is the time required by a retort to attain a minimum required process temperature with uniform temperature distribution in the retort (IFTPS, 2005; Tucker, 2001).

2.5.2 Heat penetration test

The goal of a heat penetration test is to determine the heating and cooling behavior of a specific product-package combination in a specific retort system for establishment of a safe thermal process. The HP study is conducted before starting production of a new product using a new process. The test involves locating the cold spot in food within the package and establishing the scheduled process time and temperature. For a conduction heating product in a cylindrical can, the cold spot is at the geometric center of the can. For a convection heating product in a cylindrical can, the cold spot is between the geometric center and the base of the container. A study should be conducted to determine the location of the cold spot for a specific product-package-process combination. The cold spot is usually determined by conducting a series of HP tests employing several containers with thermocouples inserted at different locations. The design of an HP test should consider all critical factors to deliver adequate thermal treatment to the slowest heating point within the product. Box 2.1 lists the critical factors that should be considered during an HP study.

An HP test should include at least 10 working thermocouples from each test run. HP tests should also be replicated to account for product, package, and process variability. Once the cold spot and all critical factors are determined, two full replications of each test should be conducted. A third replication should be conducted if results from the first two replications show variation (IFTPS, 1995; Tucker, 2001).

Heat penetration data are evaluated by plotting a heat penetration curve (Figure 2.1). The data are plotted such that there is a linear relationship between the product temperature (T_p) and heating time. A plot of log ($T_r - T_p$) versus time, known as the temperature deficit plot, is linear (see Figure 2.1). T_r is the retort temperature. Heat penetration curves, which are linear throughout the heating time, are referred to as simple heating curves, whereas heat penetration curves showing an abrupt change in heat transfer are referred to as broken heating curves. A heat penetration curve for a broken heating profile has two linear portions due to change in the heating mode from conduction to convection or vice versa. Heat penetration data can also be plotted as an inverted scale plot with T_p on the right-hand axis (see Figure 2.1).

There are two important parameters that define a simple heat penetration curve. The negative inverse of the slope, known as the heating rate factor (f_h), is defined as the time required for the heat penetration curve to traverse one log cycle. A lower f_h value indicates a faster heating rate. The subscript h indicates that the f value is for a heating process. A similar cooling rate factor (f_c) is defined for

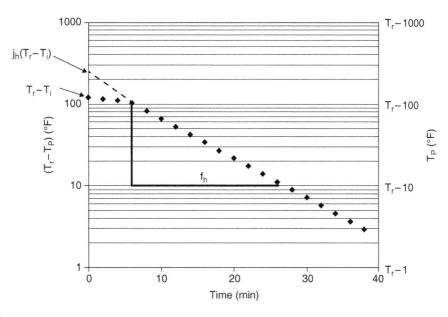

Figure 2.1 Heat penetration curve.

the cooling curve. The other parameter is the intercept of the heat penetration curve. The intercept is obtained by linear extrapolation of the curve back to zero time. The intercept is $T_r - T_{ih}$ where T_{ih} is pseudo initial product temperature determined by linearizing the entire heat penetration curve. The equation of the heat penetration curve can be written as (Lund, 1975):

$$\log(T_r - T_p) = \left(\frac{-t}{f_h}\right) + \log(T_r - T_{ih}), \qquad (2.9)$$

where t is the heating time. A dimensionless heating lag factor, j_h, is defined as (Lund, 1975):

$$j_h = \frac{T_r - T_{ih}}{T_r - T_i}, \qquad (2.10)$$

where T_i is the initial temperature of the product. Substituting $(T_r - T_{ih})$ from Equation 2.10 in Equation 2.9 leads to the following equation:

$$\log(T_r - T_p) = \left(\frac{-t}{f_h}\right) + \log\left[j_h(T_r - T_i)\right] \qquad (2.11)$$

At the end of the process time (B), if the temperature difference $(T_r - T_p)$ is defined as g, then Equation 2.11 can be written as (Lund, 1975):

$$\log(g) = \left(\frac{-B}{f_h}\right) + \log\left[j_h(T_r - T_i)\right] \qquad (2.12)$$

or

$$B = f_h \log\left(\frac{\left[j_h(T_r - T_i)\right]}{g}\right) \qquad (2.13)$$

Equation 2.13, known as the ball formula method, can be used to calculate process time if the value of g is known. The temperature at the cold spot is also determined for the cooling phase. The cooling curve can be characterized in a similar manner by using cooling rate factor (f_c) and cooling lag factor (j_c). Temperature of the cooling water (T_c) is used in place of T_r for plotting the cooling curve (Lund, 1975).

2.6 Thermal process calculation

Many different methods, based on Equations 2.5 and 2.13, have been proposed for thermal process calculations.

The difference between them is the way the temperature at the cold spot is obtained. For some methods, temperature is physically measured whereas some methods use mathematical models for predicting the temperature. The process calculation methods can be divided into three categories: general method, formula method, and numerical method (Weng, 2005).

2.6.1 The general method

The general method, developed by Bigelow in 1920, is the simplest of all methods for the calculation of thermal process. This method involves graphical or numerical integration of Equation 2.5 when the temperature distribution is known either from the heat penetration data or heat transfer equations. The time-temperature graph is converted into a lethal rate (L_r) versus time graph using Equation 2.5. The area under the L_r-t curve is the F value of the process. The area under the L_r-t curve can also be determined using numerical methods such as the trapezoidal rule or Simpson's rule. Simpson's rule is generally more accurate and requires an even number of areas. Thermal process developed by the general method is dependent on product-package-process parameters used during the test. The thermal process is established only for the actual conditions tested (retort temperature and initial temperature). Therefore, this method is not useful to determine F values for different retort temperatures and initial temperatures of product. Thus, the general method does not allow assessment of process deviations (Park, 1996; Weng, 2005).

The general method relies on time-temperature history at the cold spot. Thus, it can be used for both in-container sterilization and in-flow sterilization. For a fluid product in an in-flow sterilization process, process time (t) for a given F value and process temperature (temperature T of product at the exit of the hold tube) can be calculated from Equations 2.5 and 2.6. The required process time must be achieved for the fastest moving portion of the fluid product. The length of the hold tube is calculated as:

$$L = u_{max}t, \qquad (2.14)$$

where u_{max} is the maximum velocity of the product in the holding tube. Maximum velocity occurs at the center of the holding tube. For Newtonian fluids, the magnitude of u_{max} is given as:

$$\begin{aligned} &\text{Laminar flow conditions}: u_{max} = 2\bar{u} \\ &\text{Turbulent flow conditions}: u_{max} = 1.2\bar{u}, \end{aligned} \qquad (2.15)$$

where ū is the average fluid velocity, which is volumetric flow rate divided by cross-sectional area for flow. For non-Newtonian fluids under laminar flow conditions, the magnitude of u_{max} is given as:

$$u_{max} = \frac{3n+1}{n+1}\bar{u}, \qquad (2.16)$$

where n is the flow behavior index. For most fluid products, the limiting case for design of holding tubes is based on $u_{max} = 2\bar{u}$ (Lund & Singh, 1993).

For multiphase food products, calculation of process time is more difficult than for fluid products. The challenges associated include determination of residence time distribution (RTD) of particles and the heat transfer coefficient between particles and fluid. The main problem in process establishment in a multiphase product has been the inability to measure the temperature of particles suspended in a carrier fluid and flowing in a continuous system (Sandeep & Puri, 2001).

2.6.2 The formula method

The formula method uses heat penetration data and mathematical models to establish the heating rate of the product in the form of an equation. Equation 2.13 is the basis for the formula method. The process time can be determined if f_h, j_h, $(T_r - T_i)$, and g are known. The first three parameters can be obtained from a heat penetration curve. However, it is difficult to obtain the value of g (Lund, 1975).

2.6.2.1 The Ball formula method

To overcome the limitation of the general method, C.O. Ball proposed the Ball formula method in 1923. This method allows for extrapolation of process time for different retort temperatures and product initial temperatures. It is the most widely used and accepted method of thermal process calculation in the US. Ball related the value of g to the dimensionless ratio f_h/F_{T_r} (T_r, retort temperature) for various z values and cooling lag factor j_c. If the F value at T_r is U, then the equation for TDT curve can be written as (Lund, 1975):

$$\log(U) = \log\left(F_{250}^z\right) - \frac{T_r - 250}{z} \qquad (2.17)$$

Ball defined a phantom or reference TDT curve as the one with the same slope as TDT curve, but that represents heat resistance of microorganisms if $F_{250} = 1$. By substituting $F_{250} = 1$ min in Equation 2.17, the equation of the phantom TDT curve can be written as (Lund, 1975):

$$\log(F_i) = -\frac{T_r - 250}{z} \qquad (2.18)$$

where F_i is F_{T_r} on the phantom TDT curve. Combining Equations 2.17 and 2.18 yields (Lund, 1975):

$$U = F_i F_{250}^z \qquad (2.19)$$

F_i, which is a function of z value and T_r, can be found in literature (Stumbo, 1973). Thus, U can be calculated if the value of F_{250} is known. Once U is determined, the g value can be determined from f_h/U: g correlations (for various j_c values at a z value of 18 °F) in literature (Stumbo, 1973). Once g is known, the process time (B) can be determined from Equation 2.13. The calculated process time is applicable to processes where the product comes in contact with steam at retort temperature without any lag. This situation holds true only for continuous retorts. In batch retorts, there is a time lag in getting the retort to process temperature. This time lag is known as the come-up time (CUT). The Ball method accommodates for 42% contribution of CUT to the process time. Thus, the process time (P_t) in a batch retort can be given as (Lund, 1975):

$$P_t = B - 0.42 \times CUT \qquad (2.20)$$

The following assumptions are made while calculating the process time using the Ball method: $j_c = 1.41$, $f_c = f_h$ for simple heating curve, no product heating occurs after cooling starts, constant retort temperature, and constant cooling water temperature (100 °C below retort temperature for cans and 72.2 °C below retort temperature for glass containers). These assumptions make the Ball method flexible, but decrease its accuracy. The Ball method often underestimates the F value for conduction heating products. For products packed in thin pouches, the Ball method can overestimate F value, as the j_c value is substantially less than the assumed value of 1.41. The major limitation of the Ball formula method is its inability to handle variable process temperature (Park, 1996; Weng, 2005).

The Ball formula method has been modified by Stumbo and Hayakawa. Stumbo used the same heating factors. However, Stumbo's modified model allowed for j_c values other than 1.41 and considered variation in z values and $(T_r - T_c)$. Hayakawa's modified method provided for a range of j_c values and unlimited range of z value and $(T_r - T_c)$ combinations (Holdsworth & Simpson 2008; Stumbo, 1973). Although the Ball formula method was developed for in-container sterilization processes, it is applicable to in-flow sterilization as well. For in-flow sterilization systems using indirect heat exchangers, time-temperature profile can be described by Equation 2.11 after changing retort temperature to the temperature of the heating medium and product temperature to the mass average temperature of the product (Lund & Singh, 1993).

2.6.3 The numerical methods

The numerical methods have been widely used to establish thermal processes and evaluate process deviations since the 1970s. These methods are sophisticated and can handle variable process temperatures, which makes them very useful for thermal process deviation analysis. NumeriCAL™ and CTemp are two examples of software packages used in the food industry (Weng, 2005).

2.6.3.1 NumeriCAL™

NumeriCAL™ is a software package offered by JBT Corporation (Chicago, IL), which performs thermal process calculations by using a finite difference method to solve partial differential equations of unsteady-state heat transfer. The main advantages of NumeriCAL™ are its accuracy and flexibility. The results obtained by using this software are accepted by both the FDA and USDA. NumeriCAL™ requires the use of heating factors and can be used for products with conduction, convection, or broken heating behavior. NumeriCAL™ consists of two modules: analyze and calculate. The "analyze" module analyzes heat penetration data and develops heating and cooling factors for the worst-case container. The "calculate" module is used for thermal process calculation and thermal process deviation analysis (Weng, 2005).

2.6.3.2 CTemp

CTemp is another software package for calculating the internal temperature of products during thermal processing. It can be used to determine the F value for any in-container sterilization process, optimize the process, analyze process deviations, and efficiently establish process temperature and time (Weng, 2005).

2.6.3.3 AseptiCAL™

NumeriCAL™ and CTemp have been developed for thermal process calculation of an in-container sterilization process. AseptiCAL™ is a software package offered by JBT Corporation (Chicago, IL), which performs thermal process calculations by using advanced finite difference-based mathematical modeling for in-flow (aseptic process) sterilization process of low-acid and high-acid foods with or without particles. This software can be used to determine the center temperature of the fastest moving particle and calculates lethality values and the required length of the hold tube for safety of the process.

2.7 Thermal process validation

Once the thermal process is established, the designed process should be validated during actual process conditions. The methods most commonly used for thermal process validation are temperature measurement, microbiological validation, and the use of time-temperature integrators.

2.7.1 Temperature measurement

The F value delivered during a process can be calculated using Equation 2.5 if the time-temperature history at the cold spot is known. Modern data loggers for temperature measurement are typically multichannel systems with digital output for display that can also record the data. Thermocouples based on type T (copper-constantan) with Teflon insulation are the temperature measurement devices most commonly used. Another type of temperature measurement device is the resistance temperature detector, based on change in electrical resistance with temperature (Tucker, 2001).

2.7.2 Microbiological validation

Microbiological validation is a direct method of thermal process validation. Log reductions achieved during a process are measured using a non-pathogenic microorganism (surrogate microorganism). The log reductions are converted to the F value using Equation 2.5.

2.7.2.1 Count reduction method

In the count reduction method, a known number of microorganisms are exposed to a thermal process.

After processing, the number of surviving micro-organisms is determined. This method requires direct measurement of surviving microorganisms after the treatment in order to determine the F value of the sterilization process. Microbiological validation using count reduction can be conducted using either an inoculated pack or encapsulated spore method. In the inoculated pack method, the entire food is inoculated with a certain number of microorganisms of known heat resistance (D value). The containers are incubated after processing and the surviving microorganisms are counted. The F value received during the process can be calculated from Equation 2.5. For this method to be successful, some microorganisms should survive the process. Thus, a non-pathogenic microorganism with a high $D_{121.1}$ value (*Bacillus stearothermophilus* or *Clostridium sporogenes*) is usually used for an inoculated pack study. Typical levels of inoculum are between 10^3 and 10^5 spores per container (Tucker, 2001).

2.7.2.2 End point method

In the end point method, a known number of microorganisms are exposed to a thermal process. After processing, the presence or absence of surviving microorganisms is determined by cultivation in an appropriate medium. A binary response of growth or no growth is obtained. A no-growth result implies sterility of the sample.

2.7.3 Time-temperature integrators

Time-temperature integrators (TTIs) have received considerable attention as an alternative means of process evaluation to either temperature measurement or microbiological validation. A TTI can be an enzyme that is denatured during heating. If the reaction kinetics of temperature-induced enzyme denaturation match the death kinetics of the target microorganism, that TTI can be used as a non-biological marker of the safety of a process (Tucker, 2001).

2.8 Process monitoring and control

2.8.1 Critical factors in thermal processing

A critical factor is one that may affect the scheduled thermal process and attainment of commercial sterility. Once thermal process validation is done, critical factors should be monitored and controlled to ensure safety of the food product. Some of the critical factors related to the design of a thermal process include factors related to the product, process, and package. Some of the product-related critical factors include microbial load of the raw materials, additives and ingredients used in product formulation, product characteristics (pH, a_w, thermophysical properties, rheological properties, particle size, solid/liquid ratio), fill weight, and pretreatment (blanching, rehydration). Package-related critical factors include type (metal can, glass jars, pouches, semi-rigid containers) and size of the package, headspace, and vacuum. Process-related critical factors include rotation/agitation of containers, initial temperature, CUT, heating method (direct versus indirect), heating medium, process temperature, process time, temperature of cooling medium, residence time distribution (RTD), and flow rate (Zuber et al., 2011).

2.9 Emerging processing technologies

2.9.1 Ohmic heating

Ohmic heating, also known as joule heating, electric resistance heating, and electroconductive heating, is a process in which an alternating current is passed directly through a conductive food product. The heat is generated internally due to the resistance of the food product to the applied electric current. The heating is rapid due to volumetric generation of heat. The heat generated (P) is given as:

$$P = E^2\sigma, \qquad (2.21)$$

where E is the electric field strength (V/m) and σ is the electrical conductivity. Electric field strength can be varied by adjusting the electrode gap or the applied voltage. Electrical conductivity is the measure of how well a substance conducts electricity and is expressed in the units of Siemens per meter ($S.m^{-1}$). The efficiency of ohmic heating is dependent on the electrical conductivity (σ) of the product (Coronel et al., 2008).

2.9.2 Microwave heating

Microwaves are part of the electromagnetic spectrum and have a frequency between 300 MHz and 300 GHz. They lie between the radio (3 kHz–300 MHz) and infrared (300 GHz–400 THz) frequencies of the electromagnetic spectrum. Microwave radiation has the ability to heat materials by penetrating and dissipating heat in them. The important advantages of microwave heating compared with conventional heating include instant start-up, faster heating, and energy efficiency. The main disadvantage associated with microwave heating is non-uniform heating of food products. In the US, only four microwave frequencies (915 ± 13, 2450 ± 50, 5800 ± 75, and $24{,}150 \pm 125$ MHz) are permitted by the Federal Communications Commission (FCC) for industrial, scientific, and medical applications.

Interaction of microwaves with materials depends on their dielectric properties. Dielectric properties determine the extent of heating of a material when subjected to an electromagnetic field. Therefore, knowledge of dielectric properties is important for the design of a microwave heating system. Dielectric properties consist of the dielectric constant (ε') and the dielectric loss factor (ε''). The dielectric constant is a measure of the ability of a material to store electromagnetic energy, whereas the dielectric loss factor is a measure of the ability of a material to convert electromagnetic energy to heat (Metaxas & Meredith, 1983). Dielectric properties can be defined in terms of a complex permittivity (ε) as:

$$\varepsilon = \varepsilon_0(\varepsilon' - j\varepsilon''), \qquad (2.22)$$

where $j^2 = -1$ and ε_0 is the permittivity of free space (8.86×10^{-12} F/m). Loss tangent ($\tan \delta$), a parameter used to describe how well a product absorbs microwave energy, is the ratio of the dielectric loss factor (ε'') to the dielectric constant (ε'). A product with a higher loss tangent will heat faster in a microwave field compared to a product with a lower loss tangent. Dielectric properties of food products depend on the frequency of the microwaves, temperature, composition, and density of the materials (Nelson & Datta, 2001).

The power absorption for volumetric heating using microwave is given as (Coronel et al., 2008):

$$P = 2\pi f \varepsilon_0 \varepsilon'' E^2 \qquad (2.23)$$

where P is power absorbed (W/m^3) per unit volume, f is the microwave frequency (Hz), and E is the electric field strength (V/m). Power penetration depth (δ_p), often used in microwave heating applications, is the distance within the product at which power drops to e^{-1} of its value at the surface of the material and is given by the following equation (Nelson & Datta, 2001):

$$\delta_p = \frac{\lambda}{2\pi \sqrt{2\varepsilon' \left[\sqrt{1 + \left(\frac{\varepsilon''}{\varepsilon'}\right)^2} - 1 \right]}}, \qquad (2.24)$$

where λ is the wavelength of the microwave in free space. Power penetration depth is used to calculate the tube diameter for a continuous flow microwave heating system. For volumetric heating systems such as ohmic heating and microwave heating, the heat transfer model should include the appropriate volumetric heating term.

2.10 Future trends

The principles associated with thermal processing have not changed over the years. However, there have been constant improvements in the technologies used for sustainable food preservation, equipment used for processing (heating and cooling), sensors used for process monitoring, mathematical models developed to characterize the heating within the product, techniques used to improve product quality, protocols developed for handling process deviations, and methodologies used for process validation. We can thus anticipate continued changes in these areas. Some of the expected changes are elaborated in this section.

Microwave, radiofrequency, and ohmic heating technologies are rapid volumetric heating technologies that are gaining popularity in the food industry for thermal processing of hard-to-heat foods (viscous and particulate foods). It is expected that their use will become more prevalent in the next decade. With rapid heating becoming a reality, the lack of a rapid cooling technology is currently an impediment in improving product quality. Thus, it is expected that techniques to rapidly cool products will be developed in the near future. That will also lead to sustainable use of existing energy sources. Another sustainable practice gaining traction in recent years is heat and water recovery of heat transfer equipment. An electronic sensor to monitor the time-temperature history at the cold spot of a particulate product during a continuous retorting operation and an aseptic process will be very valuable in developing a safe process and producing high-quality products. It is thus expected that sensor technologies will be developed to meet these needs. Current mathematical models that can be used to determine the time-temperature history at the cold spot of particulates in a food product during aseptic processing involve several assumptions and are as such complicated to use. Thus, it is expected that better mathematical models that are user-friendlier will be developed in the years to come.

The development of new products and processes relies on appropriate process validation techniques. Further developments in the area of process validation tools (such as enzymatic TTIs, biological indicator units, micro-electronic sensors, and nanosensors) can be expected in the decades to follow. All of these will aid in the development of new processes with improved safety and efficiency and new products with improved quality and functionality.

References

Anderson NM, Larkin JW, Cole MB et al. (2011) Food safety objective approach for controlling *Clostridium botulinum growth* toxin production in commercially sterile foods. *Journal of Food Protection* **74**(11): 1956–1989.

Chandarana DI (1992) Acceptance procedures for aseptically processed particulate foods. In: Singh RK, Nelson PE (eds) *Advances in Aseptic Processing Technologies.* London: Elsevier Applied Sciences, pp. 261–278.

Coronel PM, Sastry S, Jun S, Salengke S, Simunovic J (2008) Ohmic and microwave heating. In: Simpson R (ed) *Engineering Aspects of Thermal Food Processing.* Boca Raton, FL: CRC Press, pp. 73–92.

Cousin MA (1993) Microbiology of aseptic processing and packaging. In: Chambers JV, Nelson PE (eds) *Principles of Aseptic Processing and Packaging.* Washington, DC: Food Processors Institute, pp. 47–86.

David JRD, Graves RH, Carlson VR (eds) (1996) *Aseptic Processing and Packaging of Food: A Food Industry Perspective.* Boca Raton, FL: CRC Press.

Holdsworth SD (2004) Optimising the safety and quality of thermally processed packaged foods. In: Richardson P (ed) *Improving the Thermal Processing of Food.* Boca Raton, FL: CRC Press, p. 8.

Holdsworth D, Simpson R (eds) (2008) *Thermal Processing of Packaged Foods.* New York: Springer.

IFTPS (1995) Protocol for Carrying Out Heat Penetration Studies. www.iftps.org/pdf/heat_pen_6_04.pdf, accessed 5 November 2013.

IFTPS (2005) Protocol for Conducting Temperature Distribution Studies in Water-Cascade and Water-Spray Retorts Operated in a Still Mode, Including Agitating Systems Operated in the Still Mode. www.iftps.org/pdf/water_spray_-final_05.pdf , accessed 5 November 2013.

Lund DB (1975) Heat processing. In: Karel M, Fennema OR, Lund DB (eds) *Principles of Food Science Part II: Physical Principles of Food Preservation.* New York: Marcel Dekker, pp. 31–92.

Lund DB, Singh RK (1993) The system and its elements. In: Chambers JV, Nelson PE (eds) *Principles of Aseptic Processing and Packaging.* Washington, DC: Food Processors Institute, pp. 3–30.

Mansfield T (1962) High temperature/short time sterilization. Proceedings of the First International Congress on Food Science and Technology, vol. 4, pp. 311–316.

Metaxas AC, Meredith RJ (eds) (1983) *Industrial Microwave Heating.* London: Peter Peregrinus.

Mitten HL (1963) Heater-coolers, heat-exchanger equipment, and milk storage tanks. In: Farrall AW (ed) *Engineering for Dairy and Food Products.* New Jersey: John Wiley, pp. 268–296.

Montville TJ, Matthews KR (2008) *Food Microbiology: An Introduction.* Washington, DC: American Society for Microbiology.

Nelson SO, Datta AK (2001) Dielectric properties of food materials and electric field interactions. In: Datta AK, Anantheswaran RC (eds) *Handbook of Microwave Technology for Food Applications.* New York: Marcel Dekker, pp. 69–114.

Park DK (1996) Establishing safe thermal processes by calculation method selection. www.tcal.com/library/Process-CalculationMethod.pdf , accessed 5 November 2013.

Sandeep KP, Puri VM (2001) Aseptic processing of liquid and particulate foods. In: Irudayarai J (ed) *Food Processing Operations Modeling: Design and Analysis.* New York: Marcel Dekker, pp 37–81.

Singh RK (2007) Aseptic processing. In: Tewari G, Juneja VK (eds) *Advances in Thermal and Non-thermal Food Preservation.* New York: Wiley Blackwell, pp. 43–61.

Skudder PJ (1993) Ohmic heating. In: Willhoft EMA (ed) *Aseptic Processing and Packaging of Particulate Foods.* London: Blackie Academic and Professional, pp. 74–89.

Stumbo CR (1973) *Thermobacteriology in Food Processing.* New York: Academy Press.

Teixeira AA, Balaban MO (2011) Computer software for on-line correction of process deviations in batch retorts. In: Sandeep KP (ed) *Thermal Processing of Foods: Control and Automation.* New Jersey: Wiley-Blackwell, pp. 95–130.

Tucker GS (2001) Validation of heat processes. In: Richardson P (ed) *Thermal Technologies in Food Processing.* Cambridge: Woodhead Publishing, pp. 75–90.

Tucker G, Featherstone S (eds) (2011) *Essentials of Thermal Processing.* Oxford: Blackwell Publishing.

Weng ZJ (2005) Thermal processing of canned foods. In: Sun DW (ed) *Thermal Food Processing New Technologies and Quality Issues.* Boca Raton, FL: CRC Press, pp. 335–362.

Zuber F, Cazier A, Larousse J (2011) Optimization, control, and validation of thermal processes for shelf–stable products. In: Sandeep KP (ed) *Thermal Processing of Foods: Control and Automation.* New Jersey: Wiley-Blackwell, pp. 95–130.

3

Separation and Concentration Technologies in Food Processing

Yves Pouliot, Valérie Conway, and Pierre-Louis Leclerc

Department of Food Science and Nutrition, Université Laval, Québec, Canada

3.1 Introduction

Separation and concentration technologies are among the most important unit operations in food processing. From disk-bowl centrifugation for industrial-scale production of skim milk to crystallization for sucrose or ultrafiltration to recover soluble proteins from cheese whey, separation and concentration processes have improved food processing. These technologies have allowed the development of new food products and are being increasingly used for water recycling in food processing. Indeed, food processing consumes large volumes of water. For this reason, waste water treatments use membrane technologies as part of the solution to numerous environmental problems posed by the food industries.

The first processes developed to separate food components selected physical or mechanical means that allowed simple separations involving solid–solid or solid–liquid systems. Screening processes separated solid-solid materials into different sizes, while filtration strictly involved the separation of solid–liquid systems. Among screening methods, the pneumatic separator used compressed gases to separate particles according to their masses; the vibration separator relied on vibration to move particles through different screens separating solids by their sizes or masses; and the magnetic separator applied magnetic fields to remove metallic particles. As for the filtration method, it included conventional filtration, mechanical expression (i.e. for juice extraction from fruits or vegetables), centrifugation, and membrane technologies.

Another group of separation and concentration technologies relied on heat-induced phase changes as the driving force for the separation. From simple evaporation to distillation and solvent extraction, such approaches allowed for the concentration of many liquid foods (i.e. milk, fruit and vegetable juices, etc.) and for as the industrial production of ethanol, liquor, and vegetable oils. The most recent development involving phase change is the use of supercritical carbon dioxide (CO_2), which has found many value-added applications for the food industry over the past decade.

Conventional filtration relies on gravity, pressure or vacuum to create the driving force necessary for the liquid phase to pass throughout different kinds of filters (e.g. perforated plates, cellulose filter papers, glass fiber filters) or granularmaterial (e.g. sand or anthracite). Membrane technologies, including reverse osmosis (RO), nanofiltration (NF), ultrafiltration (UF), and microfiltration (MF), use pressure differences as the driving force of separation. Like conventional filtration, membrane filtration relies on filters but with much smaller pores. It thus covers a wide range of separations from the removal of particles >1 μm (MF) to water purification by the removal of solutes $<10^{-1}$ nm (RO) (Cui et al., 2010). In addition, recent developments in electrically driven separations (electrodialysis), low-pressure separation (pervaporation), and separations using functionalized membranes (ion exchange materials) are opening new horizons in this fascinating field. Future trends in separation and concentration technologies will be characterized by an increased number of integrated combinations, or hybrid processes, combining single units of the before mentioned array of technologies

Food Processing: Principles and Applications, Second Edition. Edited by Stephanie Clark, Stephanie Jung, and Buddhi Lamsal.
© 2014 John Wiley & Sons, Ltd. Published 2014 by John Wiley & Sons, Ltd.

in order to improve energy efficiency and provide environmentally sustainable processes (Muralihara, 2010).

Several challenges must be met in the design of efficient and economical separation and concentration technologies for the food industry.

- **The diversity and complexity of food systems**: each food system has its own physical characteristics whether it is a liquid (viscosity), a suspension (size, shape, concentration of particles) or a solid (mechanical and/or textural properties).

- **The lability of food components during processing conditions**: many food components are prone to chemical changes when exposed to high temperatures, intense shear stress or the presence of oxygen.

- **The need for safe technologies**: the goal of any food processing method is to provide the consumer with nutritious and safe foods, that is, microbiologically clean and free from external contaminants.

- **The need for low-energy, sustainable processes**: the uncertain future of fossil energy, the rising concerns about greenhouse gases and their effects on the population call for new processes that will minimize energy consumption and carbon footprint.

The objectives of this chapter are to provide an overview of separation and concentration technologies used in the food industry and a better understanding of their underlying principles, associated advantages, and limitations. This chapter also introduces new emerging technologies having potential use for the food industry.

3.2 Physical separation of food components

Physical separation methods are generally suitable for removing suspended solids from slurries or for separating solid particles of mixtures.

3.2.1 Filtration

Solid–liquid separations can be performed using two approaches: (1) sedimentation, most often by means of gravity, and (2) filtration. In the food industry, sedimentation is used mainly for waste water treatment and will not be covered in this chapter.

Filtration involves the removal of insoluble particles from a suspension by passing it across a porous material, retaining particles according to their sizes – and shape to some extent. As the filter media retain the larger particles and form a "filter cake", the permeate passes through the

Table 3.1 Summary of the types of filtration equipment used in food processing

Equipment type	Driving force	Advantages // Limitations
Plate-and-frame filter press Horizontal plate filters Shell-and-leaf filters Edge filters	Pressure	Large filtration area, operating at high pressures (up to 25 bar), filter cake easy to remove, high-quality filtration, low capital cost // Batch mode only, automation difficult
Rotary drum filter Rotary vacuum disk filters	Pressure or vacuum	Continuous filtration, low manpower requirement // Lower filtration area and higher capital cost (especially for drum filters)
Centrifugal filters	Centrifugal	Suitable for batch or continuous mode // Complex handling, filter cake (sludge) difficult to remove

Reproduced from Sutherland (2008), with permission from Elsevier.

filter barrier. The mean size particles and their distribution will both have a great influence on the type of filter used. Furthermore, the technique chosen will depend on whether the solid or the liquid is of interest to the manufacturer. Both efficiency and cost-effectiveness should be considered in the type of filter chosen. As shown in Table 3.1, pressure, vacuum or centrifugal forces (discussed in section 3.2.2) can all be used as the driving forces for filtration, which can be achieved using membrane filters, disk filters, cartridges, woven wire screens or packed beds made from organic and inorganic materials (i.e. minerals, carbon, glass, metal, and ceramics) (Sutherland, 2008). Filtration aids, such as diatomaceous earth, cellulose or charcoal, are often used as absorbents to control the formation and the properties of the filter cake, in order to prevent resistance to fluid flow.

Filtration equipments are mainly used in edible oil refining, sugar refining, beer production, wine making, and fruit juice processing. The uses of MF and UF (see section 3.4) for those latter applications are increasing, as more efficient industrial equipment and membranes become available. However, compared to conventional filtration, MF and UF membranes are less inclined to clog and offer greater mechanical strength (e.g. against pressure) and chemical resistance. Furthermore, membrane processes are cleaner, since they avoid the use of filtering aids such as diatomaceous earth (Daufin et al., 2001).

3.2.2 Centrifugation (separators, clarifiers)

A centrifuge is a device that separates particles from suspensions, or even macromolecules from solutions, according to their size, shape, and density. This method subjects dispersed systems to artificially induced gravitational fields (i.e. centrifugal force) as shown in Figure 3.1. Centrifugation can also be used for the separations of immiscible liquids. Industrial centrifugation can be divided into two different classes: (1) sedimenting and (2) filtering.

The sedimenting type of centrifuge relies on density (i.e. specific gravity) differences as a driving force to separate the components of a mixture. Indeed, the rotation of

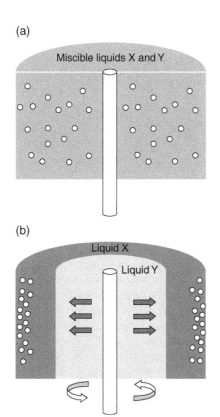

(a)

Miscible liquids X and Y

(b)

Liquid X

Liquid Y

Figure 3.1 Schematic representation of a mixture of two miscible liquids containing particles: (a) in the absence or (b) in the presence of centrifugal force.

materials around a fixed axis generates a centrifugal force as much as 10,000 times greater than gravity, which acts differently on components depending on their specific gravity. Because denser particles require a greater force to stay close to the rotation axis than lighter particles, the denser constituents are forced to the periphery of the centrifuge bowl. This type of centrifuge is suitable for the separation of solid–liquid and liquid–liquid mixtures with very small gravity differences, as low as $100\,kg/m^3$ (Sinnott, 2005).

Filtration centrifuge, in contrast to sedimentation centrifuge, uses perforated bowls and is restricted to the separation of solid–liquid mixtures. This type of centrifugal separation allows the liquid phase to permeate through the porous wall on which solids are retained. Unlike the sedimenting centrifuge, the filtering type of centrifuge has the advantage of allowing the separation of soluble solids without relying on density differences. Although particles must be at least greater than 10 microns for filtration to be practicable (Leung, 2007).

Centrifugation can thus provide solid–liquid and liquid–liquid separations, such as in milk processing, in which milk fat (i.e. cream) removal and clarification can occur simultaneously. Centrifugal clarification is also used to treat oils, juices, and beers. Industrial centrifugation equipment used in food processing plants is mainly of the disk-bowl type, as illustrated in Figure 3.2. In recent years, high-speed centrifugation has been used to reduce bacterial counts in raw milk (i.e. so-called bactofugation) and thus, allow for the production of extended-shelf life (ESL) milk.

3.2.3 Pneumatic separation

In the food industry, air is often used to remove foreign particles (e.g. straw, hull, and chaff) or for classification purposes. Pneumatic separation is one of the oldest separation methods known to humankind. This technique allows the standardization of heterogeneous particle mixtures into uniform fractions based on their density and mass. For example, in the grain industry, by using airflows, grain or flours mixtures can be separated according to their size or chemical composition (e.g. protein content) (Vose, 1978). In oil extraction processes, hulls, because of their low density compared to seeds, can be removed using adjusted-force airflows (Kazmi, 2012). Indeed, small particles fall more slowly, have less inertia and can change direction more easily than larger particles. In principle, the smaller particles (e.g. foreign particles) are dragged along with the air stream while larger elements fall through it. The quality of the separation depends on the behaviour

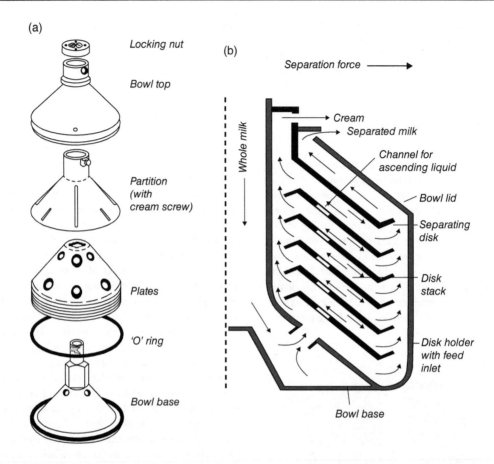

Figure 3.2 Disk-bowl centrifuge: (a) separator components, (b) separation of cream from milk.

of particles in the air stream. Indeed, each material must possess different aerodynamic properties to be accurately separated. For example, the efficiency of pneumatic separation in flour classification, which refers to the recovery of the desired product from the raw material at a given cut-off point, varies from 50% to 80% (Vose, 1978).

3.2.4 Mechanical expression

The extraction of oil or juice from plant materials presents an additional challenge since it requires the disruption of cell structures and hull-protected seeds of varying thickness and resistance. The methods generally used for oil recovery are expression and solvent extraction or the combination of both. Expression is the oldest method applied for oilseeds extraction. The principle of expression is simple. First, the preconditioned oilseeds are pressed in a permeable barrel-like cavity where shear forces, combined or not with heat, squeeze the oil out from the seeds.

Second, the "cake" products, or de-oil material, exit from one side of the press as oil flows throughout the barrel's openings and is collected (Kazmi, 2012).

Mechanical and thermal pretreatments of seed are both essential to enhance oil extraction process performances. Those treatments generally involve dehulling, flaking or grinding of seed, cooking and sometimes enzymatic pretreatment (Kazmi, 2012; Savoire et al., 2013). As dehulling and reduction size steps ensure homogeneous material and increase the surface area, cooking helps break down oil cells, lowers oil viscosity, adjusts moisture, coagulates proteins, and inactivated enzymes (Kazmi, 2012; Khan & Hanna, 1983). The cooking step is usually done using hot steam. However, it can also use dielectric heat generation (i.e. microwave and radiofrequency) or non-thermal processes such as electric pulsed fields (Kazmi, 2012).

Two kinds of presses can be employed for pressing purposes: the screw press and the hydraulic press. Screw presses have the advantage of giving a higher yield

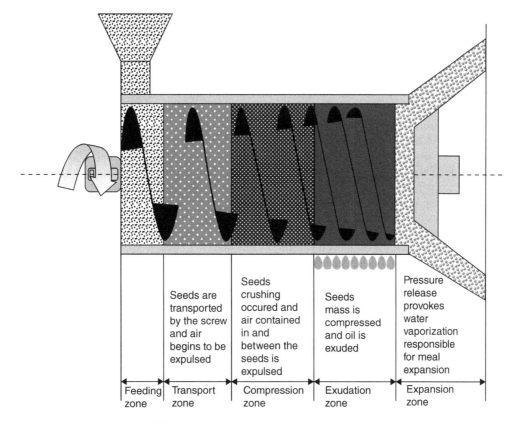

Figure 3.3 Schematic representations of the operating zones of a typical screw press (Savoire et al., 2013).

and are suitable for a continuous mode. For this reason, hydraulic presses are only used for specialty oil, olive oil, and for the pressing of cocoa (Savoire et al., 2013). The different operating zones of a common screw press are illustrated in Figure 3.3.

Efficiency of mechanical oil extraction is low compared to solvent extraction. Indeed, depending on the type of seed and the operation parameters (i.e. temperature, pressure, time, and seed moisture content), maximum attainable yield is limited to around 80% (Hasenhuettl, 2005). Moreover, if a yield of 70–80% is generally obtainable for high-fat oilseeds (e.g. sunflowers, sesame, linseed, rapeseed, palm kernel, etc.), the yield drops to roughly 50–70% for seeds containing less than 20% of oil (e.g. soybeans) (Bargale, 1997). However, this process has the advantage of giving high-quality oils, free from dissolved chemical, and is a safer process than solvent extraction (Fellows, 2009). For this reason, expression is generally limited to a small-capacity plant, to high-fat oilseeds (>35%) or to specialty products (e.g. natural or organic oils, essential oils, etc.) (Lin & Koseoglu, 2005; Rosenthal et al., 1996).

In fruit processing, presses can remove most of the juices from the pulp with minimal undesired components (i.e. phenolic compounds which cause bitterness and browning). Using continuous mode, yield of around 84% of juice is obtainable. The efficiency of juice extraction by mechanical expression depends on the maturity of the raw material, the physical resistance of the structure to mechanical deformation, the time, the pressure, the juice viscosity, as well as the temperature (Fellows, 2009).

3.3 Processes involving phase separation

3.3.1 Liquid-liquid

3.3.1.1 Crystallization

Crystallization is the process by which solid crystals, of a solute, are formed from a solution (Berk, 2009a). In the food industry, products such as sugar, lactose, glucose, and salt are obtained by this process. It may also be used to remove undesirable products. For example, edible oils may be cooled to crystallize high melting-point components by a process called winterizing.

Crystallization generally involves four distinctive steps: (1) the generation of a supersaturated state, (2) nucleation, (3) crystal growth, and (4) recrystallization (Hartel, 2001). A supersaturated solution is generated when a solution has reached the solute's maximum concentration and is then further concentrated, usually by evaporation, or cooled down slowly (Hartel, 2001). This first step is the driving force for crystallization by which the system lowers its energy state. Then nucleation takes place as solutes aggregate to form orderly "clusters" or nuclei. There are two kinds of nucleation: (1) homogeneous and (2) heterogeneous. The first occurs without the presence of foreign particles and the second, more common, is aided by the presence of foreign particles in the solution. The subsequent binding of solute molecules to existing nuclei, thus increasing crystal size, is called crystal growth. This reaction stops as an equilibrium state is reached, if the solution is not kept in a supersaturated state by the constant removal of solvent, usually through constant solvent evaporation (Berk, 2009a). Finally, the recrystallization step naturally takes place by the reorganization of the crystalline structure to a low-energy state. The supersaturation (β), the rate of homogeneous nucleation (J), and the rate of crystal growth (G), are given by the following equations (Berk, 2009a):

$$\beta = \frac{c}{c_s} \tag{3.1}$$

$$J = a\,exp\left[-\frac{16\pi}{3} \times \frac{V^2\sigma^3}{(kT)^3(In\beta)^2}\right] \tag{3.2}$$

$$G = k'(C - Cs)^g \tag{3.3}$$

a = a constant
C = concentration (kg solute per kg solvent)
C_s = saturation concentration (kg solute per kg solvent)
g = a numerical called growth order
k = the Boltzmann constant
k' = empirical coefficient
T = temperature, K
V = molar volume of the solute
s = nucleus-solution surface tension
b = supersaturation ratio

The equipment used for crystallisation is called a crystallizer or pan. Pans can crystallize in batch or continuous mode. They can be equipped with a vacuum or be vented, and generally include a heat exchanger as well as an agitation device.

In the food industry, crystallization is mainly used for sugar refining, most commonly to obtain sucrose. To obtain sucrose, sugar must be removed from sugar canes or sugar beets to produce a diluted liquid which is then clarified to remove impurities (i.e. minerals and organic matters) before further concentration, commonly by evaporation. The sucrose syrup obtained is then subjected to a controlled crystallization process using a vacuum evaporator pan, allowing for the sucrose to separate. Since sucrose solubility in water is high, multiple steps are necessary to remove as much of the sucrose in the solution as possible. Last, sucrose crystals are removed from the mother liquor using centrifugation. This liquor contains a low amount of sucrose and a high concentration of impurities and is further evaporated to obtain molasses (Hartel, 2001).

3.3.1.2 Distillation

Distillation is a physical process whereby volatile components of a mixture are separated based on differences in their volatility – the compounds with the lowest boiling point and the highest vapor pressure are separated first. The equilibrium relationships for two-component vapor–liquid mixtures are governed by the relative vapor pressure of its constituents and can conveniently be presented in a boiling temperature-concentration diagram (Figure 3.4). Indeed, the partial valor of a component (P_A) is proportional to its mole fraction (X_A) at a specific temperature and the total vapor pressure of a mixture is the sum of the partial pressure of its components ($P_A + P_B$) (Fellows, 2009). This equilibrium relationship can also be expressed mathematically as follows:

$$P_A = X_A \times P_A^{\circ} \tag{3.4}$$

$$P_B = X_B \times P_B^{\circ} \tag{3.5}$$

$$\text{Total vapor pressure} = P_A + P_B \tag{3.6}$$

A = component A
B = component B
P = partial pressure, kPa
P° = partial pressure of the pure component, kPa
X = component fraction, mole of (A or B)/mole total (A + B)

The horizontal line in the boiling temperature-concentration diagram (see Figure 3.4) gives the composition of a boiling liquid (x) and of the vapor (y) at a

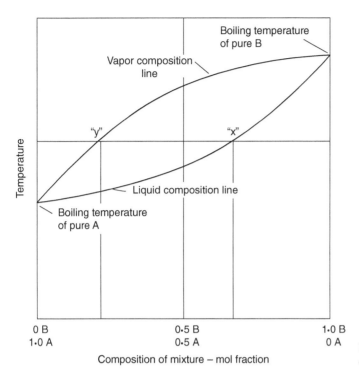

Figure 3.4 Boiling temperature/concentration diagram.

specific temperature. The diagram shows that, at a given temperature, the vapor (y) is richer in the more volatile component than the boiling liquid (x) and this is the basis for separation by distillation.

For example, in an ethanol–water mixture, as the linkages between similar molecules are greater than between dissimilar ones, following heating, the weaker ethanol–water linkages break down more easily as the more volatile compound (i.e. ethanol) is vaporized. Thereby, after condensation of the vapor, the distillate will be concentrated in stronger ethanol–ethanol linkages and further concentration by vaporization will be harder. In fact, some mixtures, like ethanol–water, form azeotropes in which equilibrium prevents further distillation. Indeed, at 95.6% of ethanol (w/w) and 4.4% of water, the mixture boils at 78.15 °C, being more volatile than pure ethanol (i.e. boiling point of 78.5 °C). For this reason, ethanol cannot be completely purified by direct fractional distillation (Fellows, 2009).

The typical distillation equipment used for continuous fractionation of liquids consists of three main items: (1) a boiler, which generates the necessary heat to vaporize the initial mixture; (2) a column, in which the stages for the distillation separation are provided; and (3) a condenser, for condensation of the final product (i.e. distillate) into the upper column. The distillation column contains multiple contact stages, through which liquid moves down and vapor moves up (Figure 3.5b). This method allows the vapor traveling up the column to be cooled by the descending liquid and the liquid to be heated by the ascending vapor. The liquid can thus lose its more volatile components as it travels down the column and the vapor can be enriched as it moves up. A part of liquid obtained by condensation of the vapor is fed back (i.e. the reflux) into the upper part of column (i.e. the rectification zone) in order to provide sufficient liquid for contact with the gas (Berk, 2009b).

Even if most industrial distillation operations use continuous distillation columns, distillation in batch mode is still used in the production of spirits such as whiskey and cognac (see Figure 3.5a). In the food industry, despite its simplicity and low capital cost, distillation is mainly confined to the concentration of alcohol beverages, essential oils, volatile flavors, aroma compounds, and to the deodorization of fats and oils. Along with evaporation, distillation is one of the most energy-consuming processes used in the food industry (Fellows, 2009).

3.3.1.3 Solvent extraction

Solvent extraction is the separation of a soluble compound, the solute, by diffusion from a solid (e.g. plant material) or liquid (e.g. oil) matrix using a volatile solvent. For example,

(a)

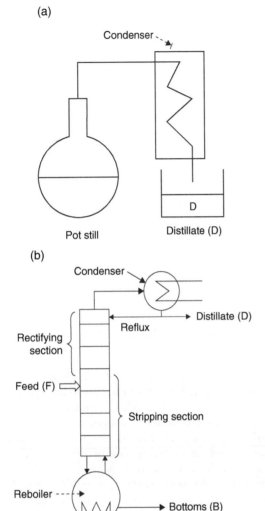

Figure 3.5 Schematic diagram of a typical distillation apparatus: (a) pot still and (b) distillation columns. Adapted from Berk (2009b).

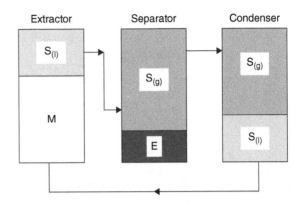

Figure 3.6 Schematic representation of a solvent extraction unit (S = solvent; M = material; E = extract). Adapted from Grandison and Lewis (1996).

in the case of compounds extracted from solid materials such as plants, the solid fragments are mixed with solvent and are held for a predetermined lap of time before the removal of the solvent. This holding process involves two stages: (1) an initiation stage; and (2) a diffusion stage. In the initiation stage, solid fragments swell as they absorb solvent and the soluble components are dissolved. Then, diffusion occurs within the fragments outward. The holding time must be sufficient for solutes to dissolve in the solvent (Fellows, 2009). The fraction extracted is a function of both the distribution ratio (D) and the phase ratio (Θ) and can be expressed mathematically as follows (Rydberg et al, 2004):

$$D_x = \frac{[X_e]}{[X_a]} \tag{3.7}$$

$$\Theta = \frac{C_e}{V_a} \tag{3.8}$$

$$E_x = D_x\Theta \div (D_x\Theta + 1) \tag{3.9}$$

Where;
a = aqueous phase, e = solvent, x = desired component
D_x = distribution ratio
E_x = fraction extracted
V = volume
X = fraction of the desired component
Θ = phase ratio

Solvent extraction can be performed in batch, semi-batch or continuous mode. A continuous solvent extraction process is presented in Figure 3.6. In this process, the material to be extracted is placed in an extraction vessel (i.e. extractor) into which the solvent is introduced at a certain temperature and flow rate. The solvent is then passed into a vessel (i.e. separator) in which the solvent and the extracted compounds are separated, generally by evaporation and finally vacuum distillation. The solvent vapor is then sent to a condenser to be recycled and crude oil is submitted to a refining process (i.e. degumming, alkali refining, bleaching, and deodorization) (Rosenthal et al.,

Table 3.2 Requirements for solvent extraction of oils

Requirements (per tonne of oilseeds)	Batch processing	Continuous processing
Steam, kg	700	280
Power, kW/h	45	55
Water, m^3	14	12
Solvent, kg	5	4

Reproduced from Fellows (2009), with permission from Elsevier.

Table 3.3 Dielectric constant of solvents used for extraction

Solvent	Dielectric constant	Type of molecules dissolved
Hexane	1.9	Oils
Toluene	2.4	Oils
Ethanol	24.3	Polyphenols
Methanol	32.6	Polyphenols
Water	78.5	Salts, sugar

Based on Voet and Voet (2005).

1996). The requirements for batch and continuous extraction mode are presented in Table 3.2.

The efficiency of solvent extraction strongly depends on the solid material condition, the diffusion rate in the solid, the liquid-to-solid ratio, the temperature, the solvent selection (i.e. type, viscosity, and flow rate), the solid's water content, and the presence of competing extractable components (Fellows, 2009; Rydberg et al., 2004: Takeuchi et al., 2008). A pretreatment step like grinding or flaking, prior to extraction, enhances surface contact between the solvent and the solid matrix and thus extraction efficacy. Also, a higher liquid-to-solid ratio provides an increased gradient which facilitates the solute's diffusion. High temperatures increase the solute's solubility and diffusion rate and result in a higher mass transfer rate. Residual water in solid material can negatively affect the solvent's capacity to dissolve solutes and thus, affect the mass transfer. However, the major factor influencing the efficiency of solvent extraction is the nature of the solvent used. For this reason, most extraction techniques manipulate the physical properties of solvents in order to reduce the surface tension, increase the solute's solubility, and promote a higher diffusion rate (Takeuchi et al., 2008).

Solvents range from polar, meaning miscible with water (e.g. ethanol, methanol), to non-polar, which means completely immiscible with water (e.g. hexane). Thus, polar compounds are more soluble in polar solvents whereas non-polar compounds are more readily dissolved in non-polar ones. Selection of the solvent is therefore based on the chemistry of the compound of interest as well as on cost and toxicity. Table 3.3 shows the dielectric constant (DC) of various solvents – a parameter related to polarity or their ability to separate charged molecules (Voet & Voet, 2005). Non-polar solvents (e.g. hexane) have very low DC, while polar solvents (e.g. ethanol, water) have very high DC. For example, water is used to extract sugar, coffee

and tea solutes, but oil and fat extractions require a less polar solvent, generally hexane.

Another important factor influencing the extraction quality is the number of extraction steps or stages. Indeed, efficiency of the extraction increases along with their numbers. Although single stages have low operating costs, this type of extractor produces diluted solutions involving the use of expensive solvent recovery systems (Fellows, 2009). They are rarely used commercially and are restrained to the extraction of specialty oils or to the production of coffee and tea extracts. Multistage apparatus can be viewed as single-stage extractors linked together, allowing the solvent emerging from an extractor's bases to be pumped cross-currently or counter-currently to the next one (Fellows, 2009). Multistage apparatus offers significant advantages such as higher recovery and purity (Rydberg et al., 2004). In the cross-current mode (Figure 3.7a), the feed, and thereafter the raffinate or residue, are treated in successive stages with fresh solvent. Though operation in cross-current mode offers more flexibility, it is not very desirable due to the high solvent requirements and low extraction yields (Kumar, 2009). For larger volume operations and more efficient uses of solvent, a counter-current mode is employed (Figure 3.7b). In counter-current operation, the feed enters the first stage as the final extract leaves. The last stage receive the fresh solvent as the final raffinate leaves. The counter-current operation provides a higher driving force for solute's mass transfer and thus, gives an optimal performance and is the preferred set-up (Berk, 2009c; Kumar, 2009).

Despite its widespread use in food processing, solvent extraction has considerable drawbacks related to solvent costs, toxicity and reactivity. Also, potential environmental problems are associated with their use, storage, and disposal. For these latter reasons, modern environmentally safe and cost-effective extraction techniques are emerging.

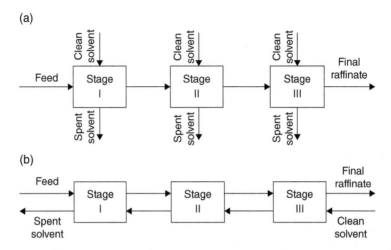

Figure 3.7 Schematic representation of (a) cross-current and (b) counter-current multiple-stage solvent extraction. Adapted from Li et al. (2012).

Amongst these, supercritical fluid extraction (SFE), pressurized liquid extraction, pressurized hot water extraction (i.e. subcritical water extraction (SWE)), microwave-assisted extraction, and membrane-assisted solvent have been proposed (Lin & Koseoglu, 2005). Even if water is the most environmentally friendly solvent possible, extraction methods using water (e.g. SWE) still give lower yields, require high energy for water removal, and generate large amounts of effluents (Rosenthal et al., 1996).

3.3.2 Liquid-gas

3.3.2.1 Evaporation

Evaporation is used in food system to obtain partial separation of a volatile solvent, commonly water, from a non-volatile components. The concentrated liquid obtained possesses an enhanced microbiological and quality stability. It also advantageously reduced storage and transportation costs (Berk, 2009a). This method generally involves the boiling of raw materials and the removal of water vapor. Mass balance (see Equations 3.10–3.11) and heat balance (see Equation 3.12) can be used to calculate the degree of concentration, the operation time, as well as the energy necessary for the process. Indeed, according to the mass balance, "the amount of heat given up by the condensing steam (Q) equals the amount of heat used to raise the feed temperature to boiling point and then to boil off the vapor" as expressed by the following equations and as illustrated in Figure 3.8 (Fellows, 2009).

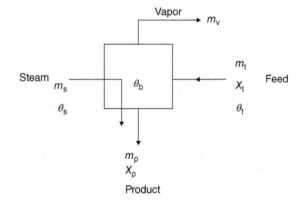

Figure 3.8 Schematic representation of the mass and heat balance in a evaporator unit (Fellows, 2009).

$$m_f X_f = m_p X_p \qquad (3.10)$$

$$m_f = m_p + m_v \qquad (3.11)$$

$$Q = m_s \lambda_s$$
$$Q = m_f C_p \left(\theta_b - \theta_f \right) + m_v \lambda_v \qquad (3.12)$$

Where;
b = food, f = feed, p = product, s = *steam* and v = vapor
C = specific heat capacity, J/kg°C
m = mass transfer rate, kg/s
X = solid fraction
λ = latent heat, J/kg
Θ = boiling temperature, °C

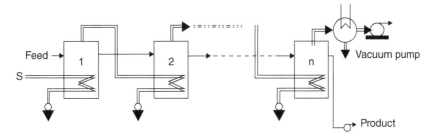

Figure 3.9 Schematic representation of a multiple effect co-current evaporator (Berk, 2009b).

An evaporator is essentially composed of a heat exchanger, or *calandria*, equipped with a device which allows for the separation of vapors from the processed liquid. Heat, commonly from saturated steam, is transferred through a contact surface, generally stainless steel. The driving force for heat transfer is the temperature difference between the steam (s) and the feed (f). The process is completed when water vapor (v) yielded by the product (p) is removed as a condensed liquid. The steam generated during evaporation can be reused to heat several other evaporators or stages, in what is called multieffect evaporation (Figure 3.9). This process is especially interesting since it lowers energy costs related to steam production by using the vapor coming from one effect to heat the product in the next effect. However, to be able to perform this in practice, the boiling point of the substance (θ) must be decreased, stage by stage, in order to maintain the temperature difference between the steam and the feed (Barta et al., 2012; Fellows, 2009). The progressive reduction of pressure by the application of a vacuum state increases the heat transfer and allows the product to boil at lower temperatures. Indeed, while evaporation with single-stage devices requires 1.1 kg of steam to evaporate 1 kg of water, only 0.5–0.6 kg of steam is necessary using a two-stage unit (Barta et al., 2012). The application of vacuum evaporation not only reduces energy consumption, it also preserves quality in heat-sensitive products such as milk. In fact, excessive heating time as well as high temperatures can lead to undesired chemical reactions (e.g. Maillard reactions in milk) or to the degradation of compounds (e.g. vitamins).

The different types of evaporators used in the food industry include batch pans and boiling film evaporators. Batch pans are the simplest and oldest types of evaporators, consisting essentially of a hemispherical, steam-jacketed vessel. Their heat transfer per unit volume is slower, and they thus require long residence times. Because their heat transfer characteristics are poor–using only natural convection– batch pans have been largely replaced by modern film

evaporation systems. In boiling film evaporators, the product flows as a thin film over a heated surface. Several kinds of film evaporators are available, such as: climbing film, falling film (Figure 3.10), and plate evaporators (Figure 3.11). In climbing film evaporators, the liquid moves rapidly upwards by percolation along vertical tubes and upon reaching the top, the concentrates and vapors produced are sent to a separator. In the case of falling film evaporators, a thin liquid film of uniform thickness moves downwards by gravity inside tubes. Compared to the climbing film evaporator, the residence time of the product is shorter, and this type of apparatus allows a great number of effects (Singh & Heldman, 2008). The same rising and falling principles of evaporation can be used on heat exchange surfaces as a series of plates (i.e. plate evaporators), as shown in Figure 3.11.

A good example of the use of evaporation in the food industry is the manufacture of tomato paste. Tomato paste originates from tomato juice (i.e. 5–6% of solids) in which water is removed by evaporation to reach 35–37% of solids (Singh & Heldman, 2008). Condensed milk and concentrated fruit juices are other examples of food products involving evaporation processes. Compared to membrane processes, a higher degree of concentration is attainable using evaporation (i.e. around 85% compared to 30%), but it involves high energy costs and product quality losses (Kazmi, 2012). Compared to freeze-drying processes, evaporation requires 10–15 times more energy for water removal. However, certain technological problems hamper the widespread application of freeze concentration in the food industry. Indeed, freeze concentration involves expensive systems, considerable loss of solids, and low degree of concentration (i.e. maximum of 50–55% of solids).

3.3.2.2 Supercritical fluid extraction (SFE)

Supercritical fluid (SCF) is the state in which the liquid and the gas phases are indistinguishable and in which the compound exhibits properties of both phases.

(a)

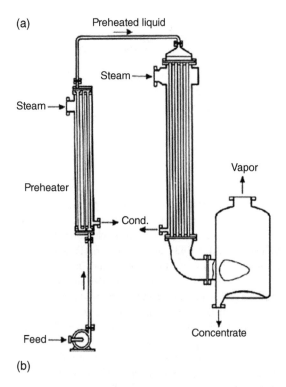

(b)

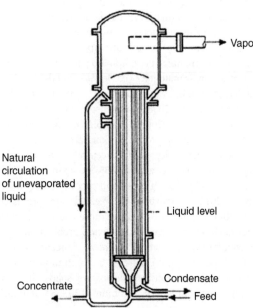

Figure 3.10 Schematic representation of two types of evaporator: (a) falling film and (b) climbing film. Adapted from Berk (2009b).

This phenomenon occurs above a critical pressure and temperature, as seen in a phase diagram (Figure 3.12). As expressed mathematically in Equation 3.13, since at low pressure values the densities (ρ) and gas are low, the solubility parameter (δ) is also low. As pressure increases, the density of gas grows, reaching the critical point (See Figure 3.12) where the densities of the gas and the liquid are the same (Berk, 2009c). The phase diagram shows that there is a continuous transition from the liquid to the SCF state by increasing the temperature at a constant pressure, or from a gas to the SCF state by increasing pressure at a constant temperature (Turner, 2006). The dependent relation between pressure and the SCF state is the phenomenon on which supercritical extraction is based.

$$\delta = 1.25 p_c^{0.5} \left(\frac{\rho_g}{\rho_l} \right) \tag{3.13}$$

Where;

g = gas and l = liquid
p_c = critical pressure
ρ = solvent density
δ = solubility parameter

Compared to a liquid solvent, a SCF has a lower viscosity and diffuses easily like a gas. Its lower surface tension allows for rapid penetration of the food material and thus, increases extraction efficiency. Furthermore, SCF has a higher volatility which allows its complete separation, avoiding the presence of residual solvent in the food material. Favorably, like a liquid, the SCF can dissolve large quantities of desired molecules, and thus extract them effectively (Berk, 2009c).

Supercritical fluid extraction (SFE) is an extraction process carried out using a supercritical fluid as a solvent (Berk, 2009c). In the food industry, carbon dioxide (CO_2) is the most common SCF used for the extraction of compounds. SFE represents an eco-friendly and cost-effective alternative to a chemical solvent, since it is non-toxic, can be used without declaration, is non-flammable, inexpensive, readily available, environmentally acceptable, chemically inert and liquefiable under reasonable pressure (Brunner, 2005). Indeed, CO_2 reaches its SCF state beyond a critical point at a temperature of 31.1°C and a pressure of 73.8 bar, as shown in Figure 3.12 (Fellows, 2009). Below this critical point, liquid CO_2 (e.g. subcritical state) behaves like any other liquid, while above this critical point, it only exists in the SCF state and acts as a gas.

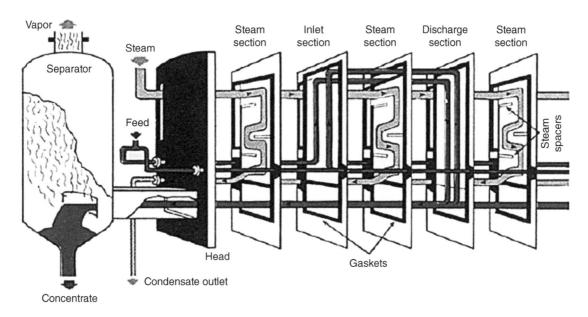

Figure 3.11 Schematic representation of a plate evaporator (Berk, 2009b).

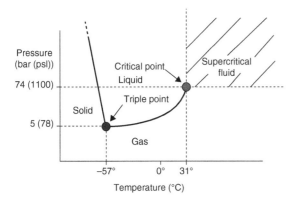

Figure 3.12 Pressure-temperature phase diagram for CO_2. Adapted from Fellows (2009).

The SFE apparatus is similar to the one used for solvent extraction (see section 3.3.1.3). However, the extraction and separation vessels of SFE apparatus are pressurized chambers equipped with heat exchangers since the state of CO_2 is determined by pressure and temperature. The CO_2 is stored in a subcritical state (i.e. liquid state) inside the condenser and pumped into the extraction vessel throughout a heat exchanger by a high-pressure pump. In the extraction vessel, as a result of higher pressure

and temperature, CO_2 reaches the SCF state before being mixed in the food material (Fellows, 2009). Afterwards, SCF-CO_2 passes into the separation vessel where the pressure is lowered, allowing for the return of CO_2 to its gaseous state and thus for the precipitation of the dissolved solutes in the separation vessel (see Figure 3.12). Finally, the extracted compounds are removed from the separation vessel as the gaseous CO_2 is sent to a condenser to be recycled and stored in its subcritical state (e.g. liquid state) by lowering the temperature.

Supercritical fluid extraction can nowadays be performed at an analytical, pilot scale as well as in large-scale industrial plants. More than 100 industrial plants and around 500 pilot plants are using this technology worldwide (Turner, 2006). Supercritical processes can be used to extract a wide variety of molecules such as lipids (e.g. seed oil, fish oil, specific fractions of butter fat or specific essential fatty acids), caffeine from coffee beans and tea leaves (decaffeinated), alcohol from beer and wines, lecithin, bioactive compounds (i.e. antioxidants, phytosterol, and vitamins) and various flavors, colorants, and fragrances (Sahena et al., 2009). This extraction technology offers extraction yields comparable to those of conventional solvent extractions, and can be carried out in different modes of operation (e.g. batch, single stage, multistage, usually in counter-current mode).

In the food industry, large-scale uses are mainly for the decaffeination of coffee beans and black tea as well as for the removal of bitter flavors from hops (Rizvi, 2010). For example, Kaffee HAG AG, originally from Bremen, Germany, is a worldwide brand of SFE-CO_2 decaffeinated coffee owned by US multinational Kraft Foods (50,000 ton/year of coffee) (Otles, 2008). However, extractions of added-value components such as specific fatty acids and essential oils are still limited to smaller scale plants (Brunner, 2005). Overall, even if SFE using CO_2 is advantageous in terms of safety (i.e. food quality and environment) and operating costs, the industrial use of this technology is limited due to high investment costs.

3.4 Membrane separations

3.4.1 Pressure-driven processes

3.4.1.1 Basic principles and separation ranges (RO, NF, UF, MF)

Pressure-driven membrane separation processes have been integrated as unit operations into a large number of food processes and it is one of the fastest growing technologies in the field of separation methods. Membrane technology requires low capital as well as low utility costs, and for this reason membrane separation has replaced the conventional separation technique in many food processes. In fact, conventional separation and concentration techniques usually imply energy-consuming phase changes which can affect both the physical and chemical characteristics of the final product. Furthermore, membrane systems, when compared with the conventional technology of separation and concentration, only require

limited space and thus do not involve expensive installations (Philipina & Syed, 2008).

Membrane processes are classified based on four main ranges of separation, namely: microfiltration (MF) (0.1 –5 μm, 1–10 bar), ultrafiltration (UF) (1–100 nm, 1–10 bar), nanofiltration (NF) (0.5–10 nm, 10–30 bar) and reverse osmosis (RO) (<0.5 nm, 35–100 bar) (Cui et al., 2010) as illustrated in Figure 3.13. Pressure-driven membrane separation operates at a pressure that varies inversely with pore size.

Pressure-driven membrane separation can be performed in dead-end mode or in tangential flow or cross-flow (Figure 3.14). Since in dead-end mode all the fluids to be filtered pass through the membrane's surface, trapped particles can build a filter cake and thus reduce filtration efficiency. For this reason, industrial-scale installations operate in the tangential flow mode in which the solution circulates tangential to the membrane's surface. In tangential mode, the particles rejected by the membrane (i.e. retentate) continue to flow, preventing the formation of thick filter cakes and thus, helping to maintain a more constant flux (J) of fluid passing through the membrane (i.e. permeate) and a more steady retention factor (R).

3.4.1.2 Membrane configurations, operation modes, process design

The availability and characteristics of membrane materials and filtration modules are constantly evolving, making it challenging to select a suitable membrane for a given application. Table 3.4 lists the main characteristics of commercially available membranes. A membrane material can be characterized by either its hydrophobicity, which minimizes fouling, or by its strength and durability

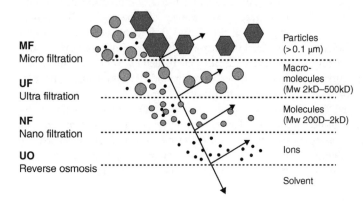

Figure 3.13 Separation range of tangential flow membrane processes.

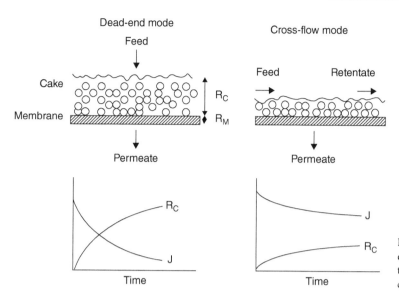

Figure 3.14 Schematic diagram of dead-end and cross-flow filtration modes and their impact on permeation flux (J) and cake formation (Cui et al., 2010).

Table 3.4 Characteristics of commercially available membranes

Material	Range				Resistance			Configuration*			
	MF	UF	NF	RO	Temp	pH	Chlorine**	Sp	Hf	T	Pl
Polysulfone	✓	✓	✓		80	0–14	M	✓	✓		✓
Polyamide	✓	✓	✓	✓	80	0–14	M	✓			✓
Cellulose acetate	✓	✓	✓	✓	80	2–8	L	✓	✓		✓
Ceramic	✓	✓	✓		1000	0–14	H			✓	
Carbon	✓	✓	✓		1000	0–14	H			✓	

*Sp, spiral; Hf, hollow fiber; T, tubular; Pl, planar module.
**L = low; M = medium; H = high.

regarding mechanical breakdowns and intensive cleaning. No material satisfies both characteristics, but polyvinylidene fluoride (PVDF) membranes have gained in popularity for their long life span (i.e. 3–5 years for conventional applications and 5–10 years for water clarification uses) (Kubota et al., 2008). Most membrane materials are available for all four ranges of separation.

Membrane modules are commercially available in four configurations: planar or flat sheet, tubular, hollow fiber, and spiral-wound modules (Sinnott, 2005). The choice of configuration is based on economic considerations associated with performance (i.e. pressure drop, resistance to fouling), surface/volume ratio, cost of cartridge replacement, and ease of cleaning and replacement (Fellows, 2009). Spiral membranes are preferred

for their low cost and high surface/volume ratio compared to tubular elements. However, they can easily be blocked by suspended particles, and thus require relatively clean feeds (Sinnott, 2005). Tubular modules have the lowest surface/volume ratio but their large internal diameter allows for the treatment of feeds containing large particles . Flat sheet modules permit the stacking of several membrane units, although the resulting pressure drop can compromise the process performance. This type of module lies between tubular and spiral-wound modules in term of cost and energy consumption. Despite the fact that hollow-fiber membranes are easy to clean, have the highest surface/volume ratio, and the lowest energy cost among all membrane configurations, this type of module has the disadvantage of confining liquid flow

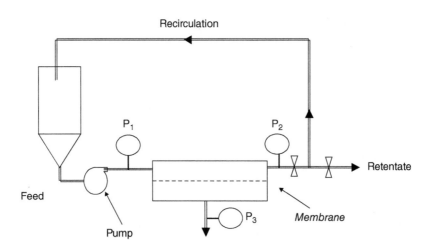

Figure 3.15 Schematic diagram of a simplified filtration system.

Table 3.5 Typical performance parameters of pressure-driven membrane processes

	Membrane type			
	MF	UF	NF	RO
Pore size	0.1–5 μm	1–100 nm	0.5–10 nm	<0.5 nm
Smallest particles removed	Colloids, bacteria	Large organic molecules, viruses	Small organic molecules, divalent ions	All dissolved species
Operating pressure (bar)	1–10	1–10	10–30	35–100
Permeation flux (L/m²·h)	100–1000	50–200	20–50	10–50

Based on Cui et al. (2010) and Jirjis & Luque (2010).
MF, microfiltration; NF, nanofiltration; RO, reverse osmosis; UF, ultrafiltration.

through narrow veins (i.e. inner diameter <1–2 mm) – limiting their use to liquids free from visible suspended particles. For this reason, hollow-fiber membranes are mainly used for RO applications such as desalination (Fellows, 2009) or require a pretreatment to reduce particle sizes to 100 μm (Sinnott, 2005).

As depicted in Figure 3.15, most installations for membrane-based separation processes include the following: (1) a feed tank, (2) the membrane, (3) at least one pump, and (4) two manometers located at the inlet (P_1) and outlet (P_2) of the membrane compartment. The transmembrane pressure (*TMP*) constitutes the driving force of the filtration and indicates the pressure drop associated with permeation. The permeation flux (*J*) provides an estimation of the overall performance of the filtration system by indicating the rate of mass transport across the membrane (Cui et al., 2010).

As expressed mathematically in Equation 3.14, the permeation flux allows for comparison of data from different membrane systems (see Table 3.5). The recirculation speed (*v*) also constitutes a critical parameter of operation, since it can be adjusted to maintain a turbulent flow regime and thereby maximize membrane surface sweeping and slow down membrane fouling.

The separation capacity of UF and NF membranes is determined mainly by their molecular weight cut-off (MWCO). MWCO indicates the molecular weight (Da) of the species rejected in a proportion of 90–95% (Takeuchi et al., 2008). Even though the MWCO and pore diameter constitute reference values for membrane selection, it remains essential to determine the rejection coefficient or retention factor (*R*) for the molecular species or solutes being concentrated

(Equation 3.15) (Cui et al., 2010). For exemple, a retention factor of 1.00 indicates that the membrane rejects 100% of the solute molecules, while a value of 0 indicates that the membrane is totally permeable to them. RO membranes are often graded using salt (e.g. NaCl or $CaCl_2$) passage data. Finally, the volume concentration factor (VCF) represents the ratio of the initial solution volume (V_o) and the final volume of the concentrate obtained (V_r) as expressed in Equation 3.17 (Cui et al., 2010). For example, a VCF value of 3 indicates that 90 L of a solution was concentrated to a final volume of 30 L. In addition, it is useful to estimate the concentration factor (CF) of a product (i.e. the final product concentration/initial product concentration) as a function of VCF for a given species of which the rejection coefficient (R) is known (see Equation 3.18).

$$J = \frac{\Delta V}{\Delta t \times A} \tag{3.14}$$

$$R = 1 - \left(\frac{C_p}{C_r}\right) \tag{3.15}$$

$$TMP = \frac{(p_r - p_p)_{in} - (p_r - p_p)_{out}}{2} \tag{3.16}$$

$$VCF = \frac{V_0}{V_r} \tag{3.17}$$

$$CF = (VCF)^R \tag{3.18}$$

Where;

p	= permeate and r = retentate
A	= membrane area, m^2
C	= concentration
CF	= concentration factor
J	= flux, L/m^2h
R	= retention factor
t	= time, h
TMP	= transmembrane pressure
V	= volume, L
VCF	= volume concentration factor

Membrane separations can be operated in batch or continuous mode. Comparison between the two modes must be made according to several criteria associated with cost and productivity constraints, but also with processing time. In fact, the latter must be minimized to prevent excessive bacterial growth, oxidation of fat, and denaturation of protein constituents due to mechanical shear stress resulting from recirculation.

Figure 3.16 shows a comparison of the batch (a) and continuous modes (b). Batch concentration is schematized by a closed loop in which the retentate is returned to the feed tank and the permeate is removed continuously until the desired VCF is reached. A retentate recirculation loop can be inserted to increase the tangential speed of the fluid and thus maintain a higher mean flux. This also decreases the power required from the feeding pump and decreases general operating costs. However, for high concentration retentate using a recirculation loop, long residence time can lead to microbiological problems since temperature used is often around 50 °C. For this reason, batch mode used is generally well adapted for small-scale applications (Jirjis & Luque, 2010).

The continuous process is characterized by feeding the solution to be treated at the same rate as the concentrate removed in what is termed as: *feed and bleed*. This operation mode offers the main advantages of faster processing, retentate of more uniform quality, continuous production of the final product, and lower feeding tank capacity. In addition, it offers the possibility of juxtaposing multistage filtration loops in which retentate from a stage feeds the next one. Using membranes of different molecular weight cut-off at each stage, the multistage approach allows continuous concentration and purification of molecules of different molecular weights.

3.4.1.3 Polarization and fouling phenomena

With processing time, performance of filtration operations will inevitably decrease as an extra resistance adds up to the membrane resistance. Two phenomena can cause a decrease in the permeation flux (J) during filtration processes: (1) polarization and (2) fouling (Li & Chen, 2010), as represented in Figure 3.17.

Although there is no universal definition, polarization can be described as the reversible accumulation of dissolved or suspended species near the membrane's surface (Li & Chen, 2010). This phenomenon can usually be controlled by adjusting hydraulic parameters (i.e. pressure, speed of recirculation) to re-establish the permeate flux (J). Fouling refers as the irreversible formation of a deposit of retained particles in the membrane pores (i.e. pore blocking) or surface (i.e. absorption). It results in unstable filtration behaviors. In this case, the permeate flux (J) can only be re-established by interrupting the process for membrane cleaning. Since membrane cleaning involves operation as well as energy costs and affects the lifetime of membranes fouling is often the main limitation of membrane applications in the food industries (Li & Chen, 2010). Fouling can be caused by

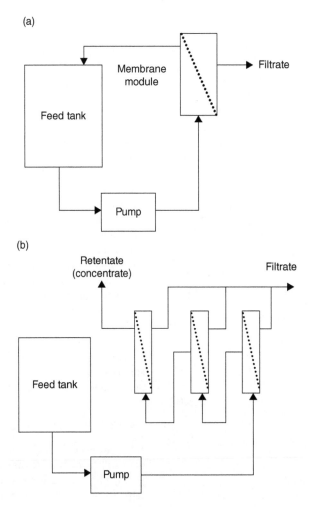

Figure 3.16 Schematic diagrams of (a) batch and (b) continuous filtration operation modes (Raja, 2008).

inorganic compounds (e.g. minerals), microorganisms (e.g. biofilms) or macromolecules (e.g. proteins, carbohydrates, and fats). This undesired phenomenon affecting productivity, as well as membrane selectivity, can however be controlled by several parameters such as: the nature and concentration of the feed, the type of membrane used, the pore size distribution, membrane material, and operational conditions (e.g. filtration mode, transmembrane pressure (TMP), temperature, turbulence, etc.) (Cui et al., 2010; Li & Chen, 2010). Figure 3.18 illustrates the effect of TMP, recirculation speed (v), protein concentration, and temperature on permeation flux (J) in the ultrafiltration of sweet whey.

Amongst all factors influencing fouling phenomena, three major groups are recognized: (1) nature and

concentration of the feed; (2) type of membrane and membrane material; and (3) operational conditions.

3.4.1.3.1 Nature and concentration of the feed

Fouling phenomena are generally increased in high concentration feed, as the amount of proteins decrease permeation flux (J) via an increase in solution viscosity, but mainly by the accumulation of protein in the polarization concentration zone (see Figure 3.17). Indeed, macromolecules (e.g. proteins) can form a gel layer at high concentration or cause cake formation and pore blocking, as hydrophobic molecules (e.g. fatty acids) can be absorbed by the membrane surface. However,

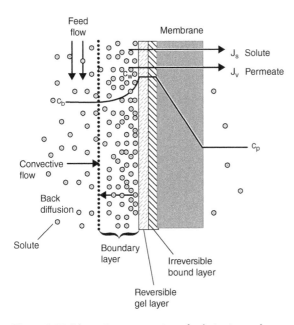

Figure 3.17 Schematic representation of polarization and fouling phenomena (Goosen et al., 2005).

depending of the nature of the feed, high concentration can, in some cases, reduce fouling tendency by promoting molecule aggregation.

It is possible to minimize fouling by pretreating the feed by using operations such as prefiltration, heat treatment, chemical clarification, chlorination, pH adjustment or addition of complexing agents (Beltsios et al., 2008). Those pretreatments affect the structure, ionic conditions, charges and concentration of molecules present in the feed and thus affect the tendency towards fouling. The following examples apply to milk and whey processing, but similar principles of pretreatment can be applied to the filtration of other foodstuffs.

- **Prefiltration/centrifugation**: although the majority of filtration systems include prefilters, fluids such as whey may contain suspended particles (i.e. casein fines) and residual lipids that slip through and foul membranes. Treatment of whey using a conventional clarifier helps to prevent this phenomenon.
- **pH adjustment**: decreasing the pH of dairy fluids between 5.5 and 6.5 can increase permeation flux significantly. Indeed, this keeps calcium phosphate in solution and thus, prevents the collection of precipitate in the membrane structure. It should be noted that decreasing

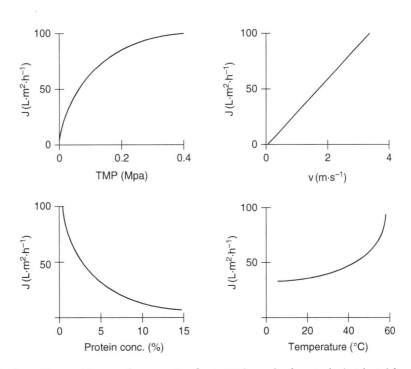

Figure 3.18 Effect of operating conditions on the permeation flux in UF (example of sweet whey). Adapted from Walstra (1999).

pH usually increases the concentration of free calcium ions (Ca^{2+}), which can promote fouling, especially in the case of ceramic membranes.

- **Preheating**: the positive effects of preheating (50–60°C for 30–120 min) on permeation flux are widely exploited in the manufacture of whey protein concentrates by UF. Heating causes excess calcium phosphate to precipitate, the free calcium ion concentration to decrease, and lipoproteins to aggregate. Clarification is often needed to remove precipitated material from preheated whey.
- **Defatting**: defatting consists of decreasing the concentration of residual lipids, for example in a dairy fluid. It is typically a combination of pH adjustments, heating and clarification or centrifugation. It is possible to obtain defatted dairy fluids (including milk and buttermilk) by tangential microfiltration, using a membrane of 1.4 μm pore size.
- **Demineralization/calcium ion sequestration**: demineralization is also an effective means of preventing the precipitation of calcium phosphate. Chemicals that sequester calcium (e.g. EDTA, citric acid) can be used for this purpose, but it must be established first that the sequestering agent has no affinity with the membrane material.

3.4.1.3.2 Type of membrane and membrane material

Membrane properties such as materials, surface morphological structures (e.g. heterogeneity of the pores and pore sizes) and surface properties (e.g. smoothness of the surfaces, hydrophobicity or surface charges) can all modulate the tendency toward fouling (Brunner, 2005). For example, hydrophobic membranes (e.g. polysulfone) will tend to adsorb proteins, as hydrophilic ones (e.g. cellulose acetate) will have a higher affinity for minerals (e.g. calcium) and be less inclined to fouling. For the reason, a hydrophilic coat is frequently applied on hydrophobic membranes in order to minimize fouling phenomena (Li & Chen, 2010).

3.4.1.3.3 Operational conditions

The type of filtration mode and the optimization of operational conditions can also minimize fouling. For example, compared to cross-flow filtration – in which the feed runs tangential to the membrane – dead-end filtration will promote cake formation, as large molecules (i.e. larger than the pores) are stopped at the membrane's surface. Indeed, cross-flow technique prevents filter cake formation (see Figure 3.14).

Adjustments of TMP, temperature of operation as well as the use of turbulence promoters can all decrease the tendency toward fouling. An increased pressure can compact the existing filter cake and thus, negatively affect the fouling phenomena. An higher recirculation speed amplifies the shear rate near the membrane's surface and reduces the risk of protein gelling. Similarly, an increased in TMP can sometimes optimize the permeate flux in cases where recirculation speed is sufficient to maintain a turbulent regime. However, in some case, but in some cases, the resulting increases of the driving force favorably affect polarization by increasing the foulant's compaction (Li & Chen, 2010). An higher temperature can also lead to an improvement in permeate flux by reducing viscosity and increasing the permeability of the membrane material. However, it is important to remember that processing temperatures above 60°C denature many proteins, which not only affects their functional properties but will also make them more liable to participate in irreversible fouling of the membrane. Moreover, the temperature is limited by the thermal resistance of the membrane material (Li & Chen, 2010). Finally, it is possible to reduce membrane fouling with: (1) the use of turbulence promoters, which decreases concentration near the surface, and (2) backflushing, which removes cake layers (Cui et al., 2010). An electric field across the membrane (i.e. electrofiltration), pulsed flow (i.e. fluctuating pressure), and rotating or vibrating dynamic membrane systems are all used to reduce fouling and the polarization phenomena (Fane & Chang, 2008).

In conclusion, control of fouling and polarization phenomena can be achieved by controlling the composition and physicochemical properties of the fluid to be processed and by optimizing the filtration system's performance parameters.

3.4.1.4 Applications of membranes in food processing

Membranes used by the food processing industry represent 20–30% of all worldwide membrane sales. This market is growing at a fast annual rate of around 7.5% (Mohammad et al., 2012). Mainly used in the dairy industry (close to 40% of all use), membrane technology is also used for beverages, sugar refining, and oil processing. Membrane processes are advantageous for environmental, competitive and economic reasons, and they also allow for the production of high-quality food products, from a nutritional and food safety point of view. They are mainly used for concentration purposes in the food industry. Around 58% of the membrane market is represented by MF membrane, followed by UF and RO membranes (around 17% each).

The remaining 8% represents the other membrane technologies, such as NF, pervaporation (2%), and electrodialysis. Moreover, membrane technologies simplify conventional processing methods by removing processing steps, are easy to operate, and can improve process performances (e.g. clarification) (Daufin et al., 2001). RO represents the most economical preconcentration method in food processing and thus can generate enormous energy savings for industries (Munir, 2006).

3.4.1.4.1 Dairy industry (UF or MF of milk, cheese whey, dairy effluents)

Milk is a complex mixture of different types of molecules including proteins, lipids, lactose and minerals, but also contains undesirable components such as bacteria. In the dairy industry, membrane technologies are used to concentrate and separate milk and milk by-product components, thus adding to their commercial value. Actually, whey processing represents the main application of membrane technology in the dairy industry, with more than 75% of all membrane usages (Mohammad et al., 2012).

Tangential flow or cross-flow MF, the uniform transmembrane pressure (UTP) concept, and ceramic-based membrane improvements have all contributed to the increased utilization of MF in dairy processing (Saboya & Maubois, 2000). There are three major applications of MF in the dairy industry: (1) the removal of bacteria from milk; (2) the pretreatment of cheese whey (i.e. defatting and removal of bacteria); and (3) the micellar casein enrichment of cheese milk. Indeed, in the dairy industries, MF is used to produce longer shelf life products (i.e. 16–21 days compared to 6–8 days for conventional pasteurization processes) without cooked off-tastes and with a greater reduction of spore-forming bacteria (Philipina & Syed, 2008). In fact, 99.99% of bacteria can be removed from skim milk using a commercial MF process called Bactocatch® (Tetra Laval, Lund, Sweeden). This process, employing an 1.4 μm MF membrane (50° C, 0.5 bar), has been used commercially mainly in Canada and western Europe (Elwell & Barbano, 2006). It should be noted that, for legal as well as safety considerations, all MF milk commercialized in U.S. and Canada needs to be further pasteurized.

Microfiltration has also been proposed as an effective method to remove fat from whey before further processing by UF. Indeed, compared to the conventional pretreatments used in the industry (i.e. clarification followed by pasteurization), MF removes phospholipoprotein complexes as well as all residual lipids, even the smallest milk fat globules (Li & Chen, 2010). It is possible to increase the UF permeation flux by 30% by pretreating cheese whey using MF before concentration. Moreover, MF can retain valuable whey proteins such as bovine serum albumin and immunoglobulin-G, which can be useful for functional food applications (Morr & Ha, 1993). The pretreatment of cheese whey by MF allows the production of high-quality, low-fat "undenatured" whey protein concentrate (WPC), and whey protein isolate (WPI) (Smithers, 2008). Finally, MF can be utilized in cheese making to change the casein-to-whey protein ratio or the casein-to-fat ratio. In fact, the application of the MF process to whole milk allows for the production of an enriched native and micellar calcium phosphocaseinate retentate which improves rennet coagulation and thus, productivity of cheese manufactures (Li & Chen, 2010). Moreover, by using a 3% phosphocaseinate solution, it is possible to reduce the coagulation time by 53% and increase the firmness (after 30 min) of more than 50% in comparison to raw milk (Daufin et al., 2001).

Ultrafiltration is the most commonly used membrane process in the world's dairy industry (Daufin et al., 2001). The fractionation and concentration of whey protein represent the most important industrial application of UF in the dairy industry. The whey protein UF concentrates can be further purified using DF to obtain high-purity WPC and WPI. Since all proteins in skim are retained and concentrated by UF, concentration and purification of milk protein to produce milk protein concentrates (MPC) and isolates (MPI) also represent a main application of UF in dairy processing. UF-MPC are extensively used by soft cheeses manufacturers as pre-cheese to be coagulated and fermented (Philipina & Syed, 2008). This process allows the full retention of whey proteins in the cheese matrix and thus eliminates whey drainage and reduces rennet requirement by around 80% compared to conventional cheese-making methods (Elwell & Barbano, 2006). However, bitterness problems have been reported in soft cheeses as well as texture defaults in semi-hard and hard cheeses (Elwell & Barbano, 2006). UF of milk generates, as a co-product, protein-free permeate which may subsequently be NF-processed to recover lactose. Finally, UF is also used to standardize milk by the adjustment of the mass ratio to the different milk constituents.

Nanofiltration is somewhat similar to UF and is commonly used in the industry for the processing of UF and MF permeates (Philipina & Syed, 2008). Indeed, the main application of NF in the dairy industry is the desalting of whey permeates (Munir, 2006). As with RO, it can be used for the preconcentration of milk or whey (by the removal of water and minerals) mainly to reduce transportation costs or energy requirements before the evaporation process (Munir, 2006).

3.4.1.4.2 *Fruit and vegetable juices (clarification and concentration of fruit juice)*

Clarification, concentration, and deacidification are the main uses for membrane technology in fruit juices processing (i.e. around 20% of all membrane usages). Concentration of fruit and vegetable juices has economic advantages for packaging, storage, and distribution. Moreover, membrane processes avoid color degradation and cooked off-tastes, as well as the loss of delicate aromas important to fresh juice flavors (Munir, 2006).

Ultrafiltration is used in the processing of multiple fruit and vegetable juices such as orange, lemon, grapefruit, tangerine, tomato, cucumber, carrot, and mushroom (Mohammad et al., 2012). Typically, juice is extracted using a press and passed through a UF module prior to concentration by evaporation or further membrane processes. UF membranes retain the concentrated pulp fraction (i.e. retentate) as well as unwanted enzymes. The UF-clarified permeate obtained can be pasteurized and, if needed, further concentrated. Most of the bioactive compounds are recovered in the permeate using this technology. Thereafter, pasteurized pulp can be reincorporated to the clarified permeate fraction to obtain whole juice. In the industry, pretreatment of fruit by enzymes (e.g. pectinases and cellulases) is commonly applied to improve fruit juice extraction, to reduce juice viscosity, and improve juice yield and color. The use of immobilized pectinases on UF membrane has been proposed as a method to allow the reuse of enzymes while controlling membrane fouling during clarification processes (Giorno & Drioli, 2000).

Reverse osmosis favorably concentrates fruit juices, giving high-quality products in which both nutritional and sensorial qualities are maintained by the low temperature used in the process. However, RO gives a low concentration level (25°Brix) compared to evaporation (42 –65°Brix) (Jesus et al., 2007). Recently, a new method called "high-concentration RO" has been described for the concentration of orange and apple juices, using the combination of two membranes (Echavarría et al., 2011; Munir, 2006). A first membrane retains sugars and aroma components (i.e. retentate), which can then be processed using a second membrane that allows some of the sugars

to pass through. The result is a lower transmembrane osmotic pressure differential and a concentrated retentate (42–60°Brix) with organoleptic properties close to those of fresh juice with no loss of acids, vitamin C, limonene, or pectin (Munir, 2006). This process has some disadvantages including generation of a diluted, low-value by-product, and the need for complex and high-cost systems (Merry, 2010).

3.4.1.4.3 *Sugar refining (concentration, clarification, and purification)*

Traditionally, sugar syrup concentration is performed using evaporation. This is energy consuming and can lower the quality of the sugar and negatively influence its color. Membrane processes represent an alternative to evaporation in sugar refining for concentration purposes. It can also be used for clarification and purification applications. Indeed, raw juices from sugar cane or sugar beet contain not only sucrose, but various polysaccharides, proteins, gums, and other unwanted components. Those impurities are traditionally removed using anionic resins which generate polluting elutes (Daufin et al., 2001). Industrial clarification of raw sugar juices by UF or MF is rapidly growing, due to their greater capacity to remove unwanted macromolecules and microorganisms compared to traditional methods. Membrane processes remove unwanted materials (i.e. retentate) and give a decolorized raw sugar juice (i.e. permeate) ready for concentration and crystallization. Concentration of sugar juices by RO or NF is cost-effective compared to evaporation. The use of NF as a preconcentration step can reduce the loss of sugars in the molasses by 10% (Madaeni et al., 2004; Munir, 2006). RO has been used for the concentration of maple syrup, resulting in more than 30% reduction in processing costs (Munir, 2006).

3.4.1.4.4 *Vegetable oils processing*

The production of edible oils follows a series of steps necessary for proper product quality by the removal of impurities such as water, dust, phospholipids, free fatty acids, gums, waxes, oxidation products, pigments, and trace elements (e.g. iron, copper, and sulfur). These processes include degumming, deacidification/neutralization, bleaching, dewaxing and deodorization. Those latter processing steps mostly involve high temperatures, the use of harsh chemicals and of considerable amounts of energy in the form of steam or electricity. Membrane-based processes (i.e. MF, UF, NF) can practically replace all the steps necessary for edible oils processing in a simple, competitive and eco-friendly way (Ladhe & Krishna Kumar, 2010). However, development of membranes allowing higher flux and selectivity combined to less fouling is still necessary to the replacement of critical energy-costing steps (i.e. degumming, refining, and bleaching) in commercial applications (Lin & Koseoglu, 2005).

Furthermore, membrane-based technologies can advantageously remove oxidation products as well as heavy metal traces. Indeed, with regard to their ability to remove oxidation products, membrane processes have been proposed as a method to extend the life-span of frying oils by the removal of proteins, carbohydrates and their decomposition products, and prevent color changes (Snape & Nakajima, 1996).

Membrane technologies can also be used to recover vegetable protein from oil extracting residues. For example, UF has been successfully applied to obtain soy protein isolates (i.e. 60–65% of proteins) from the defatted soybean meal or "cake" (Ladhe & Krishna Kumar, 2010). The isolate obtained presents good nutritional value and can be used in breakfast cereals, animal nutrition, confectionery, and dairy imitation products, as well as in nutritional and dietary beverages (Koseoglu & Engelgau, 1990).

Although membrane technologies can advantageously replace some of the steps in edible oils processing, only a few commercial applications have been reported. Nitrogen production for packaging uses, waste water treatment using UF/RO and phospholipids removal using MF have been described in litterature (Lin & Koseoglu, 2005).

3.4.1.4.5 Brewing and wine industry

Beer is the second most consumed beverage in the world. This industry is constantly being challenged to create products with consistent quality and unique taste while maintaining low production costs and environmentally friendly process (Carmen & Ernst Ulrich, 2008). Membrane technology applications are emerging as they enhance product quality, are energy saving, and reduce water waste.

The production of beer involves mashing, boiling, fermentation, and maturation steps followed by a pasteurized process to ensure microbiological stability and conservation. During brewing, several filtration operations are essential to remove solids particles (e.g. yeasts, malt, and hops) from the product. This separation of solid from

liquid, traditionally done by dead-end filtration on diatomaceous earth (DE), represents a challenge for economic, environmental, and technical purposes. Indeed, the DE used for traditional filtration is difficult to handle and thus represents a potential health hazard. Also, it needs to be properly disposed after usage and involves additional filtration steps.

Microfiltration is mainly used in the brewing industry for the recovery of beer from fermentation and maturation tank bottoms, and is considered as an industrial standard. Indeed, at the end of the fermentation and maturation of lager type beers, yeast cells sediment at the bottom of tanks to form yeast slurry which represents around 2.5% of the final fermented product volume (Carmen & Ernst Ulrich, 2008). For obvious economic reasons, it is important for manufacturers to recover as much beer as possible from this slurry. Moreover, MF is emerging as a technology used for the separation of wort following the mashing step, for rough beer clarification (currently used by Heineken breweries), and for cold sterilization (i.e. to replace flash pasteurization) (Daufin et al., 2001). Elimination of heating process by the use of MF avoids the occurrence of organoleptic defaults, and it is more and more used in the cool-sterilization of beer and wine (Baker, 2004). MF can be used not only to improve microbiological stability of wine, it can also ensure clarity of the product (Daufin et al., 2001). Moreover, RO can be used for concentration purposes as well as for dealcoholization. In contrast, it can be used to increase alcohol content (Girard et al., 2000). However, RO of wine can produce a concentrate with excessive malic acid, leading to taste faults. Recently, membrane processes have been employed for beer decolorization in the production of clear alcohol to be used in beer-based drink premixes (Carmen & Ernst Ulrich, 2008).

3.4.2 Other membrane-based processes

3.4.2.1 Electrodialysis (electrically assisted processes)

Electrodialysis (ED) is a membrane-type electrochemical process whereby membranes are used to separate ionic molecules from non-ionic ones using electric field as the driving force for separation. ED membranes are made of porous (i.e. nanometer range) ion exchange resins which selectively separate anion and cation species for dilution/concentration applications. In ED processes, anion-permeable or "anionic" membranes are made of resin bearing fixed cationic groups, while cation-permeable or "cationic" membranes are made of resin bearing fixed anionic groups. In practice, anionic and cationic membranes are arranged alternately and are separated by spacers to form thin compartments (the stack unit). Each compartment is either for dilution or for concentration purposes, and electrodes are place at both ends of the system (Figure 3.19). In the industry, 100–200 membranes may be disposed to form a stack unit, and industrial ED systems are composed of one or more stack units. By the application of an electric field, anion species permeate throughout the anionic membranes in the opposite direction to the electric current while cations migrate, in the same direction as the electric current, throughout the cationic membranes (Brennan & Grandison, 2012).

In a simplified version of an ED system illustrated in Figure 3.19a, anions move towards the anode and exit from the left compartment by passing through an anionic membrane, as a cationic membrane keeps them in the center compartment. In the same way, cations leave the right compartment by passing through a cationic membrane but are trapped in the center compartment because of the repletion of the anionic membrane. The result is that, simultaneously, the pumped feed is ion depleted and the final product recovered from the adjacent compartments, called the concentration compartments, is ion-enriched (Munir, 2006). This process has the singularity of leaving the concentration in dissolved non-ionic compounds unchanged. Indeed, even if proteins or other charged macromolecules are attracted by the electric field, their high molecular weights do not allow for permeation throughout the membranes.

Bipolar-membrane ED uses special multilayer membranes to dissociate water in H^+ and OH^- ions as an electric field is applied. Bipolar membranes are composed of three layers: (1) an anion exchange layer; (2) a hydrophilic interface; and (3) a cation exchange layer (Bazinet et al., 1998). As illustrated in Figure 3.19b, when those membranes are added to a conventional ED unit, it allows for the conversion of a neutral salt stream to an acid and base one. This technology is of particular interest to food processing if the product to be treated is an acid or a base.

The main application of ED in the food industry is the desalination of cheese whey, but it can also be used to concentrate, purify, or modify foods (Bazinet et al., 1998). ED applications include the removal of calcium from milk and lactic acid from whey, the control of sugar/acid ratio in wine, the pH control of fruit juices and fermentation reactors, and the purification of bioactive peptides.

(a)

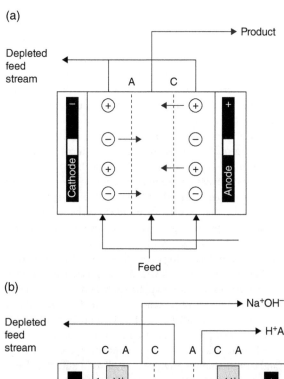

(b)

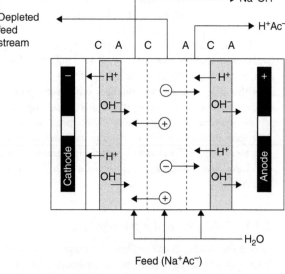

Figure 3.19 Schematic diagrams of (a) conventional electrodialysis unit and (b) bipolar electrodialysis unit (Munir, 2006).

As for all the other membrane processes, ED membranes are susceptible to polarization and fouling and thus, the maximum salt removal attainable by this method is around 90% (Brennan & Grandison, 2012).

3.4.2.2 Pervaporation

Pervaporation is a relatively new membrane separation process and represents one of the most active research

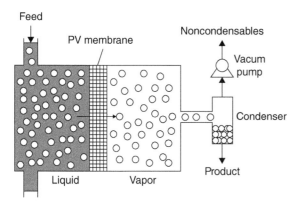

Figure 3.20 Schematic diagram of pervaporation of a two-component system (Munir, 2006).

areas in membrane technology. Unlike MF, UF, NF, RO, and ED, pervaporation is a pressure-driven membrane where the permeate generated is a low-pressure vapor, not a liquid, that can be condensed and collected or further concentrated (Feng & Huang, 1997). Using pervaporation, feed containing volatile components is fractionated when it partially vaporizes and passes through a dense or nonporous selective membrane (Karlsson & Tragardh, 1996). The driving force for separation is the chemical gradient across the membrane, but a vacuum is commonly applied on the permeate side to create an artificial pressure gradient, as shown in Figure 3.20.

Pervaporation is considered as a three-step process: (1) absorption, (2) diffusion, and (3) evaporation (or desorption). In the solution stage, the vaporizing component is drawn into the membrane at the upstream interface by chemical affinity. It then diffuses through the membrane and desorbs from the downstream interface as the membrane selectively allows for the desired components to permeate through, as vapor. The partial vaporization of the feed through the membrane is responsible for the separation ability of the pervaporation (Karlsson & Tragardh, 1996). Since the chemical composition of the membrane determines which components will permeate, a hydrophilic membrane can thus be used to dehydrate organic solutions, while a hydrophobic membrane is suitable for extracting organic components from aqueous solutions.

Pervaporation is a useful alternative for liquid mixtures that are difficult to separate by distillation and is commercially used for alcohol dehydration. In the food industry, pervaporation can be used to concentrate fruit juices, alcohol in fermentation broth, dealcoholization of alcoholic beverages (i.e. final product of 0.5% v/v of ethanol), and recovery and concentration of aroma compounds (Karlsson & Tragardh, 1996). The latter is probably the most promising application of pervaporation in food processing. Conventional concentration processes such as evaporation, inevitably result in aroma compound losses, due to their highly volatile nature. Pervaporation can be used on raw materials, prior to evaporation to concentrate aroma (i.e. at least 100-fold greater than in the raw material), or on the resulting stream to recover lost aroma. Aroma compounds can then be reintroduced in the final product to ensure taste acceptability (Karlsson & Tragardh, 1996). Compared to other aroma recovery processes, pervaporation operates at low temperature (i.e. 20°C), which eliminates damage due to high heat and allows for energy savings.

In the dealcoholization application of pervaporation, an important concern is the loss of aroma compounds in the ethanol phase – due to their very hydrophobic nature. However, new techniques using pervaporation allow for the production of 0.5% (v/v) alcohol wine with more than 80% aroma retention (Karlsson & Tragardh, 1996). Pervaporation has also been proposed as a deodorization process for food industry effluents. It simultaneously allows the recovery, the concentration, and the valorization of flavor compounds from industrial wastes (Galanakis, 2012). However, pervaporation is in a less advanced state than all other membrane processes presented in this section, and understanding of the separation mechanisms is still incomplete. Further research is needed before large-scale pervaporation can be widely promoted (Feng & Huang, 1997).

3.5 Sustainability of separation technologies in food processing

In food industry separation technologies, sustainability problems are mainly related to two factors: (1) energy, and (2) water consumption. As noted by Muralidhara (2010), of the 100 quads (1 quad = 10^{15} BTU/year) of energy consumed in the United States, food processing uses approximately 2 quads. At a cost of $8 per 10^6 BTU, this amounts reaches $16 billion per year. Half of this energy is used for concentration and drying purposes. For example, corn (wet) milling uses 93.7 trillion BTU, grain milling 153.3 trillion BTU, and vegetable oil processing 2.0 trillion BTU each year. Thus, there are opportunities for the integration of membrane technology to the food industry's separation and concentration processes in many kinds of applications. As a purely physical separation, membrane concentration does not require a heat-producing unit, or a phase change (from liquid to vapor), and it can partially or totally replace energy-consuming processes (e.g. distillation and evaporation). It thus results in not only in major energy savings, but in higher product quality. Modern filtration units are compact, efficient, simple, highly selective (i.e. to specific compounds), reduce chemical production, and offer operation flexibility to integrate hybrid processes. They represent part of the solution to quality, energy, space, and environmental concerns (Merry, 2010). Moreover, membrane technologies allow for the re-use and recovery of all process streams, without damage to the environment.

Food processing also consumes large volumes of water which are a major concern for the environment. Advanced membrane technologies represent a solution to polluting, high-energy, space-consuming, conventional waste water complex processes (Madaeni et al., 2004). Indeed, process effluents are today being treated using membrane processes such as RO, UF, and ED and treated effluents are being reused in plants for different purposes such as washing.

In conclusion, membrane technologies have a promising future and may be considered as important tools for the efficient use of processing streams without polluting discharges with regard to energy savings, environmental, and quality regulations.

References

Baker RW (2004) Microfiltration. In: *Membrane Technology and Applications*. New York: John Wiley, pp. 275–300.

Bargale P (1997) *Mechanical Oil Expression from Selected Oilseeds under Uniaxial Compression*. Saskatoon: Department of Agricultural and Bioresource Engineering, University of Saskatchewan, p. 337.

Barta J, Balla C, Vatai G (2012) Dehydration preservation of fruits. In: *Handbook of Fruits and Fruit Processing*. New York: Wiley-Blackwell, pp. 133–151.

Bazinet L, Lamarche F, Ippersiel D (1998) Bipolar-membrane electrodialysis: applications of electrodialysis in the food industry. *Trends in Food Science and Technology* 9(3): 107–113.

Beltsios KG et al (2008) Membrane science and applications. In: *Handbook of Porous Solids*. Weinheim, Germany: Wiley-VCH, pp. 2281–2433.

Berk Z (2009a) Crystallization and dissolution. In: *Food Process Engineering and Technology*. New York: Academic Press, pp. 317–331.

Berk Z (2009b) Evaporation. In: *Food Process Engineering and Technology*. New York: Academic Press, pp. 429–458.

Berk Z (2009c) Extraction. In: *Food Process Engineering and Technology*. New York: Academic Press, pp. 259–277.

Brennan G, Grandison AS (2012) *Food Processing Handbook*. New York: John Wiley, pp. 275–300.

Brunner G (2005) Supercritical fluids: technology and application to food processing. *Journal of Food Engineering* 67(1–2): 21–33.

Carmen M, Ernst Ulrich S (2008) Applications of membrane separation in the brewing industry. In: *Handbook of Membrane Separations*. Boca Raton, FL: CRC Press, pp. 553–579.

Cui ZF, Jiang Y, Field RW (2010) Fundamentals of pressure-driven membrane separation processes. In: Cui ZF, Muralidhara HS (eds) *Membrane Technology*. New York: Elsevier Butterworth-Heinemann, pp. 1–18.

Daufin G, Escudier J, Carrere H et al. (2001) Recent and emerging applications of membrane processes in the food and dairy industry. *Food and Bioproducts Processing* 79(2): 89–102.

Echavarría A, Torras C, Pagan J, Ibarz A (2011) Fruit juice processing and membrane technology application. *Food Engineering Reviews* 3(3): 136–158.

Elwell MW, Barbano DM (2006) Use of microfiltration to improve fluid milk quality. *Journal of Dairy Science* 89(suppl): E20–E30.

Fane A, Chang S (2008) Techniques to enhance performance of membrane processes. In: *Handbook of Membrane Separations*. Boca Raton, FL: CRC Press, pp. 193–232.

Fellows PJ (2009) *Food Processing Technology – Principles and Practice*, 3rd edn. Cambridge: Woodhead Publishing.

Feng X, Huang RYM (1997) Liquid separation by membrane pervaporation: a review. *Industrial and Engineering Chemistry Research* 36(4): 1048–1066.

Galanakis CM (2012) Recovery of high added-value components from food wastes: conventional, emerging technologies and commercialized applications. *Trends in Food Science and Technology* 26(2): 68–87.

Giorno L, Drioli E (2000) Biocatalytic membrane reactors: applications and perspectives. *Trends in Biotechnology* 18(8): 339–349.

Girard B, Fukumoto LR, Koseoglu S (2000) Membrane processing of fruit juices and beverages: a review. *Critical Reviews in Biotechnology* 20(2): 109–175.

Goosen MFA, Sablani S, Al-Hinai H et al. (2005) Fouling of reverse osmosis and ultrafiltration membranes: a critical review. *Separation Science and Technology* 39(10): 2261–2297.

Grandison AS, Lewis MJ (1996) *Separation Processes in the Food and Biotechnology Industries: Principles and Applications*. Cambridge: Woodhead Publishing.

Hartel RW (2001) *Crystallization in Foods*. New York: Aspen Publishers.

Hasenhuettl GL (2005) Fats and fatty oils. In: *Kirk-Othmer Encyclopedia of Chemical Technology*. New York: John Wiley, pp. 801–836.

Jesus DF, Leite M, Silva L et al. (2007) Orange (Citrus sinensis) juice concentration by reverse osmosis. *Journal of Food Engineering* **81**(2): 287–291.

Jirjis BF, Luque S (2010) Practical aspects of membrane system design in food and bioprocessing applications. In: Cui ZF, Muralidhara HS (eds) *Membrane Technology*. New York: Elsevier Butterworth-Heinemann, pp. 179–212.

Karlsson HOE, Tragardh G (1996) Applications of pervaporation in food processing. *Trends in Food Science and Technology* **7**(3): 78–83.

Kazmi A (2012) *Advanced Oil Crop Biorefineries*. London: Royal Society of Chemistry, pp. 102–165.

Khan LM, Hanna MA Expression of oil from oilseeds – a review. *Journal of Agricultural Engineering Research* **28**(6): 495–503.

Koseoglu S, Engelgau D (1990) Membrane applications and research in the edible oil industry: an assessment. *Journal of the American Oil Chemists' Society* **67**(4): 239–249.

Kubota N, Hashimoto T, Mori Y (2008) Microfiltration and ultrafiltration. In: *Advanced Membrane Technology and Applications*. New York: John Wiley, pp. 101–129.

Kumar A (2009) *Bioseparation Engineering: A Comprehensive DSP Volumen*. New Delhi: I.K. International Publishing.

Ladhe AR, Krishna Kumar NS (2010) Application of membrane technology in vegetable oil processing. In: Cui ZF, Muralidhara HS (eds) *Membrane Technology*. New York: Elsevier Butterworth-Heinemann, pp. 63–78.

Leung WWF (2007) *Centrifugal Separations in Biotechnology*. New York: Academic Press.

Li H, Chen V (2010) Membrane fouling and cleaning in food and bioprocessing. In: Cui ZF, Muralidhara HS (eds) *Membrane Technology*. New York: Elsevier Butterworth-Heinemann, pp. pp. 213–254.

Li X, Du Y, Wu G et al. (2012) Solvent extraction for heavy crude oil removal from contaminated soils. *Chemosphere* **88**(2): 245–249.

Lin L, Koseoglu S (2005) Membrane processing of fats and oils. In: *Bailey's Industrial Oil and Fat Products*. New York: John Wiley, pp. 57–98.

Madaeni SS, Tahmasebi K, Kerendi SH (2004) Sugar syrup concentration using reverse osmosis membranes. *Engineering in Life Sciences* **4**(2): 187–190.

Merry A (2010) Membrane processes in fruit juice processing. In: Cui ZF, Muralidhara HS (eds) *Membrane Technology*. New York: Elsevier Butterworth-Heinemann, pp. 33–43.

Mohammad A, Ng CY, Lim YP, Ng GH (2012) Ultrafiltration in food processing industry: review on application, membrane fouling, and fouling control. *Food and Bioprocess Technology* **5**(4): 1143–1156.

Morr CV, Ha EYW (1993) Whey protein concentrates and isolates: processing and functional properties. *Critical Reviews in Food Science and Nutrition* **33**(6): 431–476.

Munir C (2006) Membrane concentration of liquid foods. In: *Handbook of Food Engineering*, 2nd edn. Boca Raton, FL: CRC Press, pp. 553–599.

Muralidhara HS (2010) Challenges of membrane technology in the XXI century. In: Cui ZF, Muralidhara HS (eds) *Membrane Technology*. New York: Elsevier Butterworth-Heinemann, pp. pp. 19–32.

Otles S (2008) *Handbook of Food Analysis Instruments*. New York: Taylor and Francis.

Philipina M, Syed R (2008) Applications of membrane technology in the dairy industry. In: *Handbook of Membrane Separations*. Boca Raton, FL: CRC Press, pp. 635–669.

Raja G (2008) Ultrafiltration-based protein bioseparation. In: *Handbook of Membrane Separations*. Boca Raton, FL: CRC Press, pp. 497–511.

Rizvi SSH (2010) *Separation, Extraction and Concentration Processes in the Food, Beverage and Nutraceutical Industries*. Cambridge: Woodhead Publishing.

Rosenthal A, Pyle DL, Niranjan K (1996) Aqueous and enzymatic processes for edible oil extraction. *Enzyme and Microbial Technology* **19**(6): 402–420.

Rydberg J, Cox M, Musikas C (2004) *Solvent Extraction Principles and Practice*. New York: Marcel Dekker.

Saboya LV, Maubois JL (2000) Current developments of microfiltration technology in the dairy industry. *Lait* **80**(6): 541–553.

Sahena F, Zaidul I, Jinap S et al. (2009) Application of supercritical CO_2 in lipid extraction – a review. *Journal of Food Engineering* **95**(2): 240–253.

Savoire R, Lanoisellé JL, Vorobiev E (2013) Mechanical continuous oil expression from oilseeds: a review. *Food and Bioprocess Technology* **6**(1): 1–16.

Singh RP, Heldman DR (2008) *Introduction to Food Engineering*. New York: Academic Press.

Sinnott RK (2005) *Coulson and Richardson's Chemical Engineering, Volume 6 – Chemical Engineering Design*, 4th edn. New York: Elsevier Butterworth-Heinemann.

Smithers GW (2008) Whey and whey proteins – from 'gutter-to-gold'. *International Dairy Journal* **18**(7): 695–704.

Snape JB, Nakajima M (1996) Processing of agricultural fats and oils using membrane technology. *Journal of Food Engineering* **30**(1–2): 1–41.

Sutherland K (2008) *Filters and Filtration Handbook*. New York: Elsevier Butterworth-Heinemann.

Takeuchi TM, Pereira CG, Braga M et al. (2008) Low-pressure solvent extraction (solid–liquid extraction, microwave assisted, and ultrasound assisted) from condimentary plants. In: Meireles MAA (ed) *Extracting Bioactive Compounds for Food Products*. Boca Raton, FL: CRC Press, pp. 137–218.

Turner C (2006) Overview of modern extraction techniques for food and agricultural samples. In: *Modern Extraction Techniques*. Washington, DC: American Chemical Society, pp. 3–19.

Voet D, Voet JG (2005) *Biochemistry*. New York: John Wiley.

Vose JR (1978) Separating grain components by air classification. *Separation and Purification Reviews* 7(1): 1–29.

Walstra P (1999) *Dairy Technology: Principles of Milk Properties and Processes*. New York: Marcel Dekker.

4 Dehydration

Robert H. Driscoll

School of Chemical Engineering, The University of New South Wales, Sydney, Australia

4.1 Introduction

Dehydration is the removal of water from a product, and is an important unit operation in the food industry. A wide range of dryer types exists, including freeze dryers and fluidized bed dryers, allowing the method of drying to be tailored to a specific product. The reasons for drying are to preserve the product, modify product texture, and reduce transport weight.

The most important of these is preservation, or more precisely extension of shelf life. This is achieved by reducing moisture to a level at which its availability for reactions is reduced, measured by a parameter called product water activity. This fundamental property of a product is related to, but not the same as, moisture content and has a precise chemical definition. As moisture in the product is reduced, the water activity of the product is also reduced. In this way, deterioration reactions are slowed down and the product will last longer.

However, removal of moisture comes at a cost. Drying is expensive, since the energy required to remove water is high. Heat recovery systems (for example, heat pumps) may be used to reduce this cost, but these have higher capital costs and added complexity.

In this chapter, we will consider first the science of moisture movement within the product and its relationship to quality. Then by studying the interaction of the product with air, methods for predicting rate of drying will be developed. Finally, technologies available for drying will be discussed.

4.2 Drying and food quality

Dehydration may change a food product in several ways. The taste, texture, and aroma (called the organoleptic qualities) may be affected, and the heat required for drying may cause chemical reactions such as denaturation of proteins. Drying also affects the physical parameters of the product, as removal of water causes shrinkage (plums become prunes, grapes become raisins). Due to these changes, rehydration after drying may not restore the original product.

4.2.1 Deterioration reactions in foods

Water activity (a_w) is a measure of the availability of water to partake in chemical reactions. Formally, it is defined as the ratio of the water vapor pressure (p_v) in a food to that of pure water vapor (p_s) at the same temperature:

$$a_w = \frac{p_v}{p_s} \qquad (4.1)$$

Water activity can be measured by determining the relative humidity in the headspace above a product in a sealed container. During drying, the fraction of chemically bound water within the product increases, reducing the vapor pressure. Consequently, there is less free water available for chemical and microbial reactions.

Note that changing moisture content is a means to *control* water activity, and therefore it is water activity, not moisture content, which is of first importance in drying. Every product has a unique characteristic relationship between moisture content and water activity, and so the safe moisture content will be different for each product. Grains may be dried to 12–14% moisture, but peanuts and oilseeds must be dried to under 10% for the similar shelf life.

4.2.1.1 Microbial stability

The limits of microbial growth are determined by water activity. For example, most bacteria need $a_w > 0.91$, and

Food Processing: Principles and Applications, Second Edition. Edited by Stephanie Clark, Stephanie Jung, and Buddhi Lamsal.
© 2014 John Wiley & Sons, Ltd. Published 2014 by John Wiley & Sons, Ltd.

molds need $a_w > 0.70$. The exact water activity limit for a specific organism also depends on factors such as pH, oxygen availability, the nature of solutes present, nutrient availability, and temperature. Generally, the less favorable the factors, the higher the value of a_w required for growth.

The effect of microbial action on quality may simply be *economic loss*, for example discoloration, physical damage, off-flavors and off-odors (spoilage microbes), or may be a *food safety issue*, for example with pathogens which cause food-borne diseases. Reduction in a_w will increase the microbiological stability of the product, thus increasing shelf life.

4.2.1.2 Chemical stability

Water may take part in chemical reactions as
• a **solvent**, providing a transport mechanism for reactants to come in contact with each other
• a **reactant**, as a component consumed in the reaction
• a **product**, for example, in non-enzymatic browning reactions, or
• a **modifier**, for example by catalyzing or inhibiting reactions.
During drying, reactions continue until a critical a_w is reached. Drying temperature has an important effect on the rates of chemical reactions, so directly affects the quality of the dried product. A small change in temperature can cause a large change in reaction rate. Higher temperatures also provide energy for a greater range of reactions.

Reactions that depend on moisture to bring reactants together will become increasingly limited during dehydration, due to the reduced molecular mobility of the reactants.

At low moistures, a further preservation mechanism becomes significant. As product moisture drops, solutes become more concentrated and therefore solution viscosity rises. This makes it increasingly difficult for reactants to come together.

Some examples of important food chemical reactions include the following.
• **Enzymatic reactions**, which are not completely understood, but which are very slow at low a_w values, due to lack of mobility of the substrate to diffuse to the active site of the enzyme.
• **Non-enzymatic browning (NEB)**, a water-dependent reaction with maximum reaction rates around $a_w = 0.6-0.7$. Water is also a reaction product. Too much water inhibits the reaction by dilution, too little gives inadequate mobility.

• **Lipid oxidation**, a reaction that is fast at both low and high values of a_w, slow at intermediate values.
• **Loss of nutrients**, for example vitamin B or C losses due to breakdown at high temperatures.
• **Loss of volatiles**, for example flavors and aromas, from the product.
• **Release of structural water**, which changes food texture.

4.2.1.3 Physical stability

Microbiological and chemical stability both correlate with water activity, but physical deterioration correlates with moisture content. Some examples of physical effects include the following.
• **Softening** of texture at high moisture, hardening at low moisture (water acts as a plasticizer of the food material).
• **Differential shrinkage**: outer layers shrink relative to inner layers, leading to either surface cracks or radial cracks.
• **Surface wetting effects**: moisture works on the product surface to expand pores and capillaries.
• **Case hardening**: a hydrophobic layer may be formed in oil-rich or proteinaceous products during rapid drying of outer layers, which traps moisture inside the product.
• **Cell collapse**: cells may collapse if internal moisture is removed, leading to the product shrinking and the surface becoming wrinkled, for example in prunes or sultanas.

4.3 Hot air drying

Most food dryers use heated air passed across a product to remove moisture. Commercially, moisture is normally used as an indication of the progress of drying. However, from the preservation point of view, it is not moisture but water activity which must be controlled.

Air holds a relatively small amount of moisture, normally less than 1% by mass. We can define a relative humidity for the air, which is the ratio of the actual vapor pressure (p_v) to the maximum possible vapor pressure (p_s) at the same temperature. Saturated air has a relative humidity of 1, often written as 100%.

Relative humidity (r) is a measure of the water activity of the air. If a product sample is allowed to equilibrate with a small amount of air in a sealed container, then the air and the product will equilibrate with respect to both temperature and moisture. At equilibrium, the air relative humidity is numerically equal to the product water activity:

$$r = \frac{p_v}{p_s} = a_w \qquad (4.2)$$

If a food product comes in contact with air with $r > a_w$ then it will absorb moisture from the air. Here are some important consequences.

• A product should be stored in a package or a storage environment where the relative humidity is controlled.

• Drying cannot reduce the product water activity below the relative humidity of the air used for drying the product.

Water activity can be controlled by methods other than dehydration. A common method in the food industry is to add humectants, which are chemicals such as sugars, salts and glycerol which bind available water. Freezing a product also reduces the available water for reaction.

Since water activity determines product shelf life, we should dry to a specific water activity. Commercially, however, product drying is based on moisture content, and this can lead to unsafe storage practice. A dryer operator might incorrectly assume that a moisture content that is safe for one product is safe for a second similar product. This may not be valid. In the next section, the relationship between moisture and water activity will be explored.

4.3.1 Product equilibrium

If we place some product in a jar (Figure 4.1) and then seal the jar, the product and air will come to equilibrium over time. There are two forms of equilibrium occurring here: a *thermal* equilibrium, where the temperatures equalize, and a *mass* (or moisture) equilibrium. Mass equilibrium occurs when the rate of evaporation from the surface of the product becomes equal to the rate of condensation. At this point, the equilibrium relative humidity (ERH)

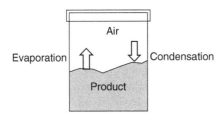

Figure 4.1 Product in a sealed glass jar.

is controlled by the product equilibrium moisture (EMC) and temperature only. By measuring the ERH at different moisture contents, but keeping the temperature constant, a product *isotherm* can be constructed.

Product isotherms represent an important basic property of food products, and help in determining safe storage moistures.

4.3.2 Moisture content definitions

The concentration of water in a product is called the product moisture content. This may be measured in two ways.

• The mass of water divided by the total product mass (*wet basis*, wb).

• The mass of water divided by the dry solids only (*dry basis*, db).

Both definitions are commonly used.

For example, for a product containing 40 kg of water for every 100 kg, the moisture content is expressed as 40% on a wet basis. Since the mass of dry solids is 60 kg, this is equivalent to 40/60 or 67% dry basis.

Using the symbol W for wet basis moisture and M for dry basis moisture:

$$M = \frac{m_w}{m_w + m_s} \quad W = \frac{m_w}{m_s} \qquad (4.3)$$

where m_w is the mass of water and m_s is the dry solids mass in a sample. For example:

• 200 g of water per kg of product is 20% wet basis moisture content

• 200 g of water in 800 kg dry solids is 25% dry basis moisture content.

To convert from wet basis to dry basis:

$$M = \frac{W}{1 - W} \qquad (4.4)$$

• 20% wb is equal to 25% db moisture content.

• 75% wb is 300% db moisture content (which means three parts water to one part dry solids).

To help you understand this concept, exercises 1 and 2 are included.

4.3.2.1 Exercise 1

Using the conversion formulae above, complete the following table.

Total mass product (kg)	Dry product mass (kg)	Dry basis moisture content (%)	Wet basis moisture content (%)
100	50	–	–
100	90	–	–
100	–	20	–
100	–	–	20
–	50	40	–
–	20	–	85

What is the maximum possible wet basis moisture content? Can dry basis moisture content be over 100%?

Let m_p be the initial mass of a product. If we dry the product from an initial moisture content of W_o to W_f (both wet basis), then a useful formula for calculating the mass of water removed is:

$$m_w = m_p \times \left(W_f - W_o\right)/(1 - W_o) \qquad (4.5)$$

4.3.2.2 Exercise 2

How much moisture would I need to remove to dry 100 kg of wet product from 50% to 20% wb?

4.3.3 Evaporation of water

Moisture is removed by evaporation from the surface of a product. Water vapor moves through a boundary layer into a moving air stream that carries the moisture away. The driving force for diffusion through the boundary layer is the difference in vapor pressure between surface moisture and the air. The energy required to evaporate product water must be replaced by heat transferred to the product surface.

4.3.4 Important psychrometric equations

Psychrometry is a branch of physical chemistry that is concerned with the properties of a gas–vapor mixture. The terms "gas" and "vapor" refer to the natural state of a substance at room temperature. For drying, we are concerned with a particular mixture, which is air and its contained water vapor. The gas component is nitrogen, oxygen, and other trace elements, and has an average molecular weight of 29.0 kg/kmol. The term "vapor" refers to the water component, since water is normally a liquid at room temperature.

Both the gas and vapor components are assumed to obey the ideal gas equation which is:

$$PV = nRT_K \qquad (4.6)$$

where P is absolute pressure (Pa), n is number of moles, R is the universal gas constant (8.314 kJ/kmol.K) and T_K is absolute temperature. The total pressure is the sum of the partial pressures exerted by each component in the gas mixture:

$$p_t = \sum_{i=1}^{n} p_i \qquad (4.7)$$

where subscript t indicates total pressure and subscript i refers to the ith component of a mixture of n components.

Two basic properties of an air-vapor mixture are its temperature (as measured by a dry bulb thermometer) and its moisture content, which is defined in exactly the same way as the proportion of water in a product, and is called the absolute humidity, H:

$$H = \frac{m_w}{m_a} \qquad (4.8)$$

where m_w is the weight of water, m_a is the weight of the dry (non-water) air components. The units for absolute humidity may be written as kg/kg d.a. (kilograms per kilogram of dry air), since by convention, air humidity is measured on a dry basis.

Dry bulb temperature T and absolute humidity H are often chosen as the two principal axes for a chart of air-vapor properties, called a psychrometric chart (see Figure 4.7).

Curved lines represent air relative humidity, varying from 0 (for the H = 0 axis) up to 1.0 (the vapor saturation line). Air that contains more moisture than the saturation line is unstable, and will separate into condensate and saturated air.

The total heat content of the air is called enthalpy, and includes both sensible and latent heat. When measured relative to the natural state of its components at 0 °C, the air enthalpy is:

$$h = m_a[c_a T + H(c_v T + \lambda_0)] \qquad (4.9)$$

where c_a is the dry air specific heat, c_v is the specific heat of water vapor (kJ/kg·K), T is air temperature (Celsius), and λ_o is the latent heat of evaporation of pure water at 0 °C. The three components for a mass of air m_a are:
- sensible heat of the dry air, $m_a c_a T$
- sensible heat of the water held in the air, $m_a H c_v T$
- latent heat of the water held in the air, $m_a H \lambda_o$.

Exercise 3 has been included to allow you to practice these concepts.

4.3.4.1 Exercise 3

1. Locate the point $20\,^{\circ}\text{C}$, $10\,\text{g/kg}$ on a psychrometric chart. Then use the chart to find the enthalpy per unit mass of this point.

2. The specific heat of dry air is $1.01\,\text{kJ/kg·K}$, of water vapor is $1.83\,\text{kJ/kg·K}$, of liquid water is $4.18\,\text{kJ/kg·K}$, and the latent heat of water at $0\,^{\circ}\text{C}$ is $2501\,\text{kJ/kg}$ (values at $20\,^{\circ}\text{C}$). From this information, calculate the heat per kg of each term in Equation 4.9 at a temperature of $20\,^{\circ}\text{C}$ and a humidity of $10\,\text{g/kg}$ dry air.

3. Compare your answers to parts 1 and 2. They should be in agreement!

Air can carry very little water vapor, and so the value of air humidity, H, is of the order of 10–$20\,\text{g}$ of water per kg of dry air. Thus, the second term in Equation 4.9 makes little contribution to the total enthalpy. However, since water has a very high latent heat of evaporation, the last term is substantial.

Now we can add lines of constant enthalpy to the psychrometric chart. These lines have a negative slope, and are parallel to the wet bulb temperature lines. The enthalpy lines are useful for representing drying.

4.3.5 Wet bulb temperature

If we place a glass thermometer in an air stream, the temperature measured is called the dry bulb temperature. If the bulb of the thermometer is kept wet, for example with a cotton wick, evaporation of water cools the bulb and the indicated temperature is called the wet bulb temperature (Figure 4.2). Wet bulb temperature is a convenient practical method for determining the amount of moisture in the air.

The rate of heat flow into the cotton wick is given by the convective heat transfer equation:

$$\dot{q} = h_c A (T - T_{wb}) \qquad (4.10)$$

where h_c is the heat transfer coefficient ($\text{W/m}^2\text{K}$). By analogy to convective heat transfer, the rate of evaporation $\dot{m}_e$ is written as:

$$\dot{m}_e = k_y A (H_s - H) \qquad (4.11)$$

where A is the area (in m^2) for evaporation of moisture from the wick, and k_y is the mass transfer coefficient

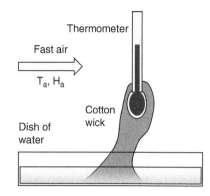

Figure 4.2 Wet bulb temperature.

in $\text{kg/m}^2\text{s}$. H_s is the saturation humidity of the air at the temperature of the air, and λ_T is the latent heat of free water at the evaporation temperature T (in $^{\circ}\text{C}$).

At equilibrium:

$$\dot{q} = \dot{m}_e \lambda_T \qquad (4.12)$$

Thus:

$$T - T_{wb} = (k_y \lambda_T / h)(H_s - H) \qquad (4.13)$$

Equation 4.13 is very close to a linear relationship between H and T, called the wet bulb line, and this line can be plotted on a psychrometric chart. The lines are not perfectly straight, as changes in the value of λ_T with temperature cause small variations from linearity.

Drying is an isoenthalpic process, which means that the heat energy lost from the air to the product equals the heat energy gained by the air by evaporation of moisture from the product. It is common to represent drying processes using constant enthalpy lines on a psychrometric chart. For mixtures of air and water, these lines are very close to the wet bulb temperature lines given by the above equation.

A complete psychrometric chart is given in Appendix A. The following simple drying problem illustrates the use of a psychrometric chart.

4.3.5.1 Problem

Air at $25\,^{\circ}\text{C}$, 40% RH is being used to dry a 5 ton bed of rice at 20% wb moisture. Estimate how much air is required to dry the rice to 14% db, if the air leaves at 84% RH, in equilibrium with the grain.

4.3.5.2 Solution

From the psychrometric chart, the inlet air wet bulb temperature is

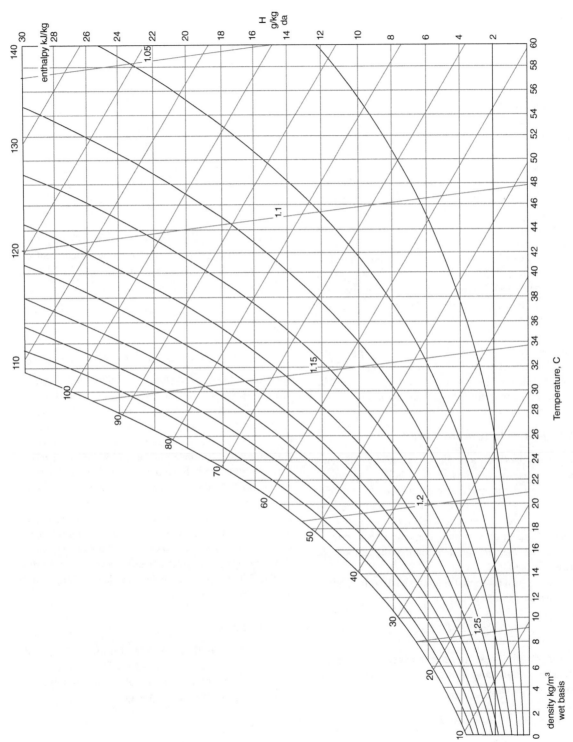

Appendix A: Psychrometric chart, SI units, at one atmosphere pressure.

$$T_{wb} = 16.2\,°C$$

As the air leaves the dryer at 84%, we can trace along the wet bulb line for 16.2 °C, until the air relative humidity is 84%. This point is at 18 °C. Now read off the inlet air moisture content:

$$H_{in} = 0.008 \, kg/kg \, d.a.$$

For the outlet air moisture content:
    $H_{out} = 0.011 \, kg/kg \, d.a.$
    Dry mass = 5000 kg × 80% = 4000.0 kg
    Initial moisture = 5000 kg × 10% = 1000.0 kg
    Final moisture = 14% × 4000 kg = 560.0 kg
    ∴ weight of moisture to remove = 440.0 kg
    air required = 440/(0.011−0.008) = 1470 kg

Note use of *wet basis* when total mass is used, and *dry basis* when solids mass is used as the reference quantity.

4.4 Drying theory

In this section we will develop a basic theory for food drying.

4.4.1 Important moisture definitions

It is convenient to define the moisture in a product in terms of how easy it is to remove this water, as shown in Figure 4.3.
• **Equilibrium moisture content** (EMC) is the moisture level at which a product is in equilibrium with the moisture of its surrounding air.
• **Free moisture** is moisture in excess of the equilibrium moisture content, so can be removed by drying. Further

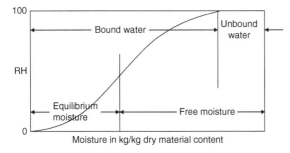

Figure 4.3 Schematic description of the state of water in a food product.

moisture cannot be removed without reducing the air relative humidity, and hence changing the product equilibrium moisture content.
• **Unbound moisture** is moisture in excess of a minimum level (called the critical moisture content) required for the product to exhibit a water activity of 1. Thus, a product with unbound moisture will dry at the same rate as a free water surface, and at a constant rate.
• **Bound moisture** is moisture less than the critical moisture content, so exerts a lower vapor pressure than the saturation vapor pressure. As a result, the product water activity is less than 1 if the product moisture is less than the critical moisture content.

4.4.2 Vapor adsorption theories

Moisture moves through a product to its outer surface, and then evaporates to an air boundary layer. Each water molecule must then diffuse through the air boundary before it is carried away by the air stream. This complete process is called *desorption*. The opposite process, in which water condenses onto the surface, is called *adsorption*, and the movement of water into the product from the surface is called *absorption*.

Models have been proposed to explain adsorption and desorption, leading to equations predicting the water content/water activity equilibrium of the product. At constant temperature, this product-dependent relationship is called an isotherm.
• **Langmuir** (1918). Langmuir studied the bonding of the first water molecules condensing onto a dry product surface (monolayer adsorption), by equating the rate of evaporation at the surface to the rate of condensation. This model describes the shape of the isotherm at low moistures only. Subsequent theoretical models used Langmuir's model as a starting point.
• **Brunnauer, Emmett, and Teller** (BET model) (1938). The BET model was the next major theoretical development, as it extended Langmuir adsorption to multilayer adsorption, summing the contribution of additional layers of moisture. This model results in a typical sigmoidal curve, as commonly found in food products, but only works well up to about 40% RH.
• **Guggenheim-Anderson-deBoer** (GAB model). In recent times, a third major step forward was made with the GAB model, which was developed independently by three researchers, and gives good agreement over the full isotherm curve for most products.
Over the years, many other models have been developed, mostly empirical. Since the GAB model describes product

behavior well, and has a theoretical basis, it has been increasingly applied within the food industry. The constants required for the equation are determined experimentally by placing samples in jars containing saturated salts, and measuring the equilibrium state between the air in the jar and the product sample. The full equation requires six constants.

• **The monomolecular moisture content**, which is the minimum amount of moisture required to occupy all of the surface sites of the product with a single layer of water molecules.

• Two constants related to **equilibration rates** between adsorption and desorption.

• Three **activation energies** which describe the effect of temperature.

It is possible that the large number of constants have helped contribute to the success of the model in describing food product equilibrium characteristics. Nevertheless, the model has achieved general acceptance in the drying community.

4.4.3 Hysteresis

We might assume that drying a product surface is simply the inverse of wetting a surface, so that a plot of the desorption isotherm should look the same as an absorption isotherm, but this is not the case. As moisture adsorbs to the product surface, the energy released may modify the surface structure, affecting the way water molecules bond to the surface. This causes an effect called *hysteresis* in which desorption isotherms differ from absorption isotherms. Hysteresis decreases after successive cycling of the product between low and high moisture (see Figure 4.4).

4.4.4 Theory of moisture movement within a product

Before moisture can evaporate from the surface, it must diffuse to the product surface. There are several possible diffusion mechanisms.

• **Liquid diffusion**, where moisture moves through the product in proportion to the liquid water gradient at any point. This is expressed well by Fick's equation for diffusion.

• **Vapor diffusion**, which assumes that the product is porous, so that vapor can move through connected pores within the material.

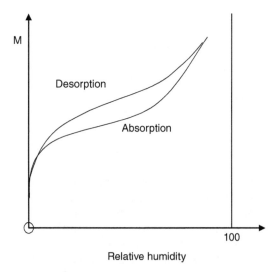

Figure 4.4 Desorption and absorption isotherms.

• **Capillary movement**, wherein liquid water moves by capillary action through pores. This is a good model for high water activity products.

Most theoretical models of diffusion-controlled drying of food products assume liquid diffusion.

4.4.5 The four drying rate periods

The rate at which a thin layer of product dries is an important characteristic. The thin layer drying rate is defined as the rate of water removal from a layer of product that is sufficiently thin that the air leaving the layer is not detectably different from the inlet air. In more practical terms, the entire product encounters identical drying conditions.

The opposite to thin layer drying is deep bed drying, where successive layers of the product encounter air that has interacted with previous layers. A deep bed can be modeled as a succession of thin layers, by equating the exit air from each layer with the inlet to the next layer. Thus, a deep bed does not dry evenly. All dryers can be categorized as either thin layer or deep bed dryers.

Plotting moisture against time for a thin layer gives the drying curve for the product (Figure 4.5). For a high moisture product, we can identify four distinct drying regions.

1. **Initial transient region**: the product equilibrates with the wet bulb temperature of the air, by either evaporation of surface moisture (if the product is initially hotter than the drying air wet bulb temperature) or condensation (if the product starts cooler).

2. **Constant rate region**: during constant rate drying, the process can be modeled using the wet bulb equation:

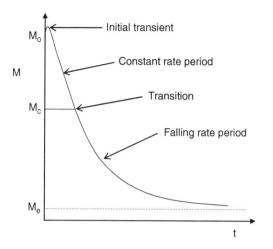

Figure 4.5 Thin layer drying curve: moisture versus time.

$$m_s \frac{dM}{dt} = k_y A (H_s - H_a) \qquad (4.14)$$

where t is drying time, M is the sample moisture content (db), m_s is the dry solids weight of the sample (kg), and other symbols are as defined for the wet bulb temperature equation. During this period, the product dries at the same rate as a free water surface.

3. Transitional region: dry spots start to appear on the surface at the critical moisture content M_c, so that the product no longer behaves like a free water surface, and diffusion of moisture from within the product starts to limit moisture loss at the surface.

4. Falling rate region: once the drying surface no longer has unbound water, moisture removal is limited by diffusion from inside the product:

$$\frac{\partial M}{\partial t} = D(T) \frac{\partial^2 M}{\partial x^2} \qquad (4.15)$$

Equation 4.15 is called Fick's law of diffusion, and the moisture content M is a function of position (x) and time (t). Variation in moisture content within the product causes moisture gradients resulting in moisture movement. The constant D is called the product mass diffusivity and is a measure of how quickly moisture will diffuse through different materials. The diffusivity is affected by temperature, T.

Figure 4.5 shows a thin layer drying curve. M_o is the initial moisture, M_c is called the critical moisture and M_e is the asymptotic equilibrium moisture between the air and product.

The initial transient region is only significant for a short time near the start of drying. During this time, the enthalpy of the product equilibrates with the enthalpy of the air, by changing the product temperature. This does not mean that the air and product come to the same temperature, since surface evaporation is occurring, but that the rate of heat flow into the product (by convection) equalizes with the heat loss by evaporation, exactly as in the wet bulb situation described earlier. This effect can be observed by placing very cold tomatoes into a preheated oven for a very short time (less than a minute), and observing the condensation on the surface. It is also the mechanism for removing heat that we observe when we cool by sweating during exercise!

The constant rate period (from M_o to M_c) is a straight line on a plot of time against moisture (measured on a dry basis). This period is only observed for high-moisture products such as slurries, where the water activity equals 1. Only products with a water activity of 1 will exhibit a constant rate region, and they will dry at the air wet bulb temperature until the critical moisture content is reached. Few food products exhibit a constant rate region.

The transition region occurs around M_c. This region (where the surface starts to dry) is difficult to model and so is not normally included. As a result, many studies of drying use the falling rate period to model the complete drying curve.

The remainder of the curve represents the falling rate period. Many authors divide this into two regions, noting that there seems to be an initial fast falling rate period, and subsequently a slower period leading to equilibration between the air and the product.

4.4.6 Representation of drying on a psychrometric chart

Thin layer drying can be represented on a psychrometric chart (Figure 4.6). The product entering the dryer is plotted according to its temperature and equilibrium relative humidity (point P in Figure 4.6). For example, if the product is at 20 °C and has a water activity of 0.95, then P is the intersection of the 20 °C temperature line and the 95% relative humidity line.

In a dryer, the inlet air is conditioned to be hotter and drier than the product, normally by heating the air (point I in Figure 4.6). The drying process can be represented by a line drawn through the inlet air point at constant enthalpy towards the air saturation line, intersecting with the product relative humidity line (point B).

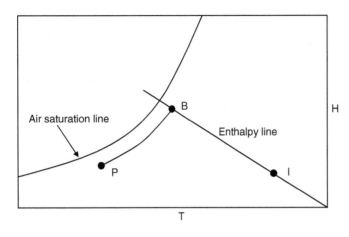

Figure 4.6 Simplified representation of drying on a psychrometric chart.

The curve PB (at constant relative humidity) represents the initial transient region. The line BI (at constant enthalpy) is the falling rate region. If the product enters the dryer with a water activity of 1, then point B represents the constant rate period, and the initial transient will be located on the air saturation line.

This method of presenting drying using a psychrometric chart allows for a deeper appreciation of the two regions most relevant to foods: the initial transient and the falling rate period. By plotting the thin layer drying curve on the chart, useful information on the exit air properties can be calculated directly. Note also that the initial transient can be seen as essentially an enthalpy equilibration between the product and the drying air, and the falling rate period as a water activity equilibration.

4.4.7 Models of the falling rate period

Fick's diffusion equation can be solved for simple regular shapes, homogeneous product, and constant drying conditions (Crank, 1970), resulting in equations of the form:

$$M(x,t) = a_o + \sum_{n=1}^{\infty} d_n \exp a_n x . \exp b_n t \qquad (4.16)$$

where a_n, b_n and d_n are constants. Generally we need only the average moisture content of the product:

$$M(t) = c_0 + c_1 \exp - k_1 t + c_2 \exp - k_2 t + \cdots \qquad (4.17)$$

where c_n and k_n are constants dependent on product shape and inlet air properties, and k_n is related to the mass diffusivity of the product. This is called a multicompartment model.

The simplest useful form of this equation is to keep the first two terms only:

$$M = c_0 + c_1 \exp - k_1 t$$

Substituting M = Mo at t = 0 and M = Me at t = ∞ gives:

$$\frac{(M - M_e)}{(M_o - M_e)} = MR = e^{-k_1 t} \qquad (4.18)$$

where MR is called the moisture ratio, and varies from 1 to 0 during drying. If a constant rate period exists, then time t refers to time *after* the critical moisture is reached, and M_o is replaced with M_c.

Differentiating Equation 4.18 gives:

$$\frac{\partial M}{\partial t} = -k_1 (M - M_e) \qquad (4.19)$$

where k_1 is a drying rate constant in reciprocal time units, showing that the drying rate is proportional to the difference between the present product moisture and its final equilibrium moisture. In many cases, this single term model may be sufficient for predicting product drying times.

The following problems will illustrate solution techniques for the constant rate period and the falling rate period.

4.4.7.1 Example problem for the constant rate period

A 4 m^2 tray is filled with a liquid product with a density of 1000 kg/m^3, to a depth of 7 mm. The product has an initial solids content of 8%, and dries at a rate of 420 g/min. The critical moisture content, marking the end of the constant rate period, is 133% db, after which the drying rate of product starts to fall. Find the drying rate in the constant rate period in terms of moisture content.

4.4.7.2 Solution

Initial weight of product (density × volume) is

$$4\,m^2 \times \left(7 \times 10^{-3}\,m\right) \times 1000\,kg/m^3 = 28\,kg.$$

Since 8% of this is solids, the product dry weight is:

$$28\,kg \times 0.08 = 2.24\,kg.$$

The drying rate is given as 420 g/min or 0.42 kg/min. From the definition of dry basis moisture content, the drying rate in terms of moisture content is:

$$0.42\,kg/min/2.24\,kg = 18.8\%\,db/min.$$

4.4.7.3 Example problem for the falling rate period

A product has a drying rate constant k_1 of $1/120\,min^{-1}$. Calculate the time to dry from 60% to 10% db if the equilibrium moisture content is 4% db.

4.4.7.4 Solution

Solve Equation 4.18 for time:

$$t = (-1/k_1)\ln MR$$
$$\therefore t = -120\ln\left[(10-4)/(60-4)\right] = 268\,min = 4\,h\,28\,min.$$

4.4.8 A complete drying model

For products where the constant rate region is significant (such as slurries and pastes), the drying rate for the constant rate period can be represented by a single constant, k_o. Combining the constant rate and falling rate models gives a simple overall model:

$$\text{For } M \geq M_c: \quad \frac{dM}{dt} = -k_o$$

$$\text{For } M < M_c: \quad \frac{dM}{dt} = -k_1(M - M_e) \qquad (4.20)$$

4.4.9 Effect of airflow

The drying models presented above (Equations 4.14 and 4.18) assume that there is adequate airflow to remove moisture from the product surface. At low air speeds,

the capacity of the air to hold moisture limits the drying rate. Thus for limited airflow, the rate of moisture removal is affected by air speed. Above a critical air speed, heat transfer becomes the rate-limiting step, and air speed has little effect on drying rate. Thus increasing fan power does not necessarily improve the drying rate. This is true for both the constant rate and falling rate periods. As a result, there is little point in increasing drying air speed above this critical level for a thin layer.

However, experimentally a small increase in the rate of drying is observed above the critical air speed, where the rate of drying increases roughly with the cube root of air speed. This small effect is due to an increase in the convective heat transfer coefficient from the air to the product surface as air speed increases, caused by reducing the air boundary layer thickness adjacent to the product.

For deep bed or cross-flow drying, air speed has a major effect on the total drying rate, as the air spends a greater time in contact with the product. Under these conditions, drying capacity is affected strongly by the rate of air supply to the dryer.

4.5 Drying equipment

There is a large range of types of dryers, including drum, rotary, tray, cabinet, vacuum, osmotic, spray, column, and freeze dryers. Dryer designs are selected based on the particular needs of the product requiring drying. Liquids are often partially predried in evaporators to reduce the load on the dryer.

Dryers may be categorized by the following characteristics.

- **Mode of operation**: *batch* means that the dryer is loaded, operated and the dried product unloaded. *Continuous* means that the dryer is loaded and unloaded continuously during operation.
- **Method of heating**: *direct* heating means that the flue gases from combustion come in contact with the product; *indirect* heating means that a heat exchanger is used to transfer heat from the flue gases to the drying air, thus protecting the product from possible contaminants. Heating may also be achieved by electricity, for example air resistance heaters, or by steam. The product may also be directly heated by electricity without air, for example by conduction, infrared, microwave or radiofrequency, but these methods are beyond the scope of this chapter.
- **Nature of product**: product might be loaded into the dryer as *solid*, *liquid*, *slurry* or *granules*, each requiring a different form of dryer. Liquids can be dried in spray

dryers, solids on meshes, granular materials in deep beds or fluidized beds, and slurries in trays.
- **Direction of airflow**: the drying air may be *co-current*, *counter-current* or *cross-flow*. A dryer may have zones with different airflow directions (combination dryers).

In the following section, dryers have been classified as batch or continuous, and the continuous dryers further classified by the direction of airflow.

4.5.1 Batch dryers

Batch dryers are often simpler in design than continuous dryers, but can be difficult to connect with continuous processing lines. In addition, the time required for loading/unloading will reduce the effective time of utilization of the dryer. So batch dryers tend to be used for small-scale production such as rapidly changing product lines, pilot plant processing, rural production, and high-value products.

Kiln dryers are a simple, universal form of dryer used for drying thin layers of product. The term "kiln" refers to a temperature-controlled chamber. A kiln dryer normally consists of a drying tray over some form of heat source, for example a biomass combustor or a furnace, and so are direct-fired. They are inefficient, as the hot air, after passing through a single layer of product, is vented to atmosphere. They are commonly used for drying sliced fruits and vegetables, including hops, apple slices, and pineapple rings.

In-store dryers (bin dryers) can both dry and store product (in a similar way as a cool room both cools and stores). The granular product is placed in bulk on a mesh supporting screen, and air is pumped into a plenum chamber below the product, passing through the screen and then through the product. These dryers have high thermal efficiency, as they operate at near ambient temperatures with the drying front submerged in the product mass for a large proportion of the drying time, so that the air leaves close to saturated. They are suitable for granular products such as grains and nuts.

There are many choices in the operation of an in-store dryer.
- Layer drying, where 1–2 m of wet grains are loaded and partially dried, and then additional harvested grain is placed on top. This increases the efficiency of both the aeration system (increased airflow due to reduced flow resistance) and thermal efficiency (as the exit air now passes through new wet product and so is utilized to its maximum capacity).

- Mixing, for example by unloading a grain bin into a second bin via conveyors, an inexpensive operation which helps mix wet and dry product together. Not all products are suited to being mixed, as migration of moisture from wet grains to adjacent dry grains may cause fissuring (for example, with rice).
- Reverse flow, where the direction of aeration is periodically reversed in order to improve the uniformity of the final product, at the expense of slightly greater energy consumption, caused by moisture having a longer average path before it exits the dryer.

In-store dryers are especially suited to drying seed, as they operate most efficiently at low temperatures. Most seeds should not be heated over about 43 °C in order to retain viability.

Tray dryers (cabinet dryers) require that the product is placed on trays in a closed cabinet. Air enters the dryer, is mixed with recirculated air, reheated, and passed across the trays. A proportion of the heated air is vented from the dryer in order to remove moisture. These dryers are suited to small-scale operations and allow for rapid changes in product line. By recirculating most of the air, heat losses are reduced and the energy efficiency greatly improved. They are very common in food plants, as they are able to handle most products including fruits and vegetables, liquids, fish, and pasta.

Freeze dryers use sublimation rather than evaporation of moisture. Sublimation is the process of direct conversion of ice to vapor, and requires about 2.8 MJ/kg of water. They are expensive, but are suitable for high-value, heat-labile products. The product is placed on heated shelves and the drying chamber evacuated. The effect of reducing the air pressure is to cause the product to cool by sublimation of moisture until the temperature is about −20 °C to −40 °C. Evaporative drying is slow at low temperatures, but sublimation drying is relatively fast. The low temperatures protect the product from changes due to heating and reduce the loss of volatiles such as aroma and flavor. In addition, the structure of the product is generally better preserved by freeze drying. Due to their high capital cost, freeze dryers are normally used for high-value products, such as starter cultures and coffee.

4.5.2 Continuous dryers

Continuous dryers are suited to running for long periods of time with the same product, and may be fitted with feedback controls to maintain drying conditions and/or product exit conditions.

4.5.2.1 Rotary dryers

A rotary dryer consists of a long cylinder, supported on girth rings that are used to rotate the dryer. Product is fed in at one end, and heated air comes in contact with the material as it passes down the cylinder, which is usually slightly inclined to allow the material to flow down. As the cylinder rotates, product may be picked up by flights mounted along the inside surface of the barrel, carried and then dropped, allowing good air-product contact and product mixing. The flights may be straight (aligned along the rotating cylinder) or helical.

The dryer is suited to granular products such as grains and pet food pellets. It may be direct- or indirect-fired, and may be run using either co- or counter-current airflow.

4.5.2.2 Drum dryers

A large range of drum dryer designs exists, but the essential feature is a steam-heated drum being coated with liquid product. As the drum rotates, a thin film of liquid is picked up on the surface of the drum. The thickness of this layer can be controlled by blades close to the drum surface or by a second rotating drum. In the time that the drum rotates, the thin product layer is dried to a solid and scraped off the surface as flakes. The drum dryer is suited to liquids, pastes and purees, such as baby foods, cooked starch slurries, breakfast cereals, vegetable and fruit pulps.

4.5.2.3 Spray dryers

Spray dryers create a fine spray of liquid product in a hot air environment. The liquid is pumped into a nozzle (preferably using a positive displacement pump to ensure a uniform flow of product), which forces the liquid through an atomizer, a device that imposes high shear stresses in the liquid. Examples of atomizers are:
• parallel concentric disks with one rotating, the other stationary. The liquid is forced between the two plates from an axial feed
• perforated plate: the liquid is forced under high pressure through small holes in the plate.
Both methods break the liquid into fine droplets, which assume a spherical shape owing to surface tension. As the liquid leaves the high shear region, it enters a hot air region, which may be designed co- or counter-currently to the product flow. At high temperatures the outside of the droplet dries quickly, forming a hard shell.

Water inside the droplet boils, rupturing the hard shell to create distinctive partial shells or honeycombed patterns, depending on the product properties. This open structure leads to a powdered product that can be easily rehydrated.

The resulting mixture of dried product powder and air exits the dryer and is separated in a cyclone. In some cases, the product may be passed to a fluidized bed dryer to complete the drying process. Examples of products commonly dried in a spray dryer include milk and instant coffee.

Some spray-dried powders do not rehydrate very effectively, due to a process where a clump of powder is wetted on the surface, forming a layer that resists further wetting. To avoid this problem, spray-dried powders may be carefully rewetted (by about 15%) by a steam spray, which allows the powder to coalesce (combine) at a few junction points between the particles. The powder is then redried and sieved. The junction points survive, giving a larger particle structure yet without loss of surface area for rewetting (since the junction points are small). This process is called instantizing, because it allows the powder to be mixed and dissolve rapidly in water, and is commonly used with milk powders.

Since the feed must be atomized and each droplet dried, spray dryers tend to have a low capacity for drying for a given volume, compared with other forms of drying. Their success in the food industry is due to the adoption of predrying by means of evaporators. The evaporator is able to cope with large amounts of product, but requires that the product remains a liquid, preferably of low viscosity, whereas a spray dryer is not able to handle a large amount of liquid, but is perfect for the transition from liquid to solid. The two units, working in series, offer a perfect solution to the overall problem of drying large amounts of liquids to powder.

4.5.2.4 Fluidized bed dryers

Fluidized bed dryers are designed for granular solids such as small grains, powders, and pellets. The material to be dried is placed on a perforated screen, and air is blown through the screen at sufficient speed (typically over 2 m/sec) until the resulting pressure drop across the bed matches the total product weight, so that the product is suspended in the flow of air and behaves like a fluid. Further increases in air speed have little effect on the pressure drop across the bed. The product is successfully fluidizing when aeration cells form in the bed, wherein product and air mix uniformly.

The benefit of fluidization is that the product dries from all sides, creating a more uniform moisture distribution compared with conventional drying in a fixed bed, and reducing the thickness of the vapor boundary layer around the product, thus reducing drying time. The mixing process also encourages greater product moisture uniformity.

Care must be taken to ensure that the dryer design is suited to the specific product. The size, shape, and cohesiveness of the product significantly affect drying efficiency. Also, fluidized bed dryers are susceptible to dead pockets, areas where insufficient air is supplied to achieve particulate fluidization. The stagnant product collects, heats and becomes a contamination or fire hazard.

Large particles can often be successfully fluidized by entrainment with finer particles, reducing the average particle size to a range where fluidization works effectively.

Fluidized bed dryers were developed for the chemical engineering industry but, due to their small footprint, rapid surface drying, high throughput and uniformity of drying, are increasingly finding applications in the food industry. They can be operated as batch or continuous dryers.

4.5.2.5 Spouted bed dryers

These dryers operate on the same principle as a fluidized bed dryer, except that only a central core of product is fluidized; the remaining product forms an annular region around the central air spout. Product entrained in the spout is heated rapidly, before cascading out of the top of the spout and falling (the fountain region) back into the annular region. Since the outside annular region is also aerated, the heated product continues to dry as it falls through the annulus, before being guided back to the air spout.

A simple modification of the unit allows liquids to be dried. The drying chamber is first filled with inert spheres (for example, nylon) and the liquid product is sprayed into the chamber. The liquid coats the spheres, dries rapidly, and then breaks off due to impact between the spheres. The resulting powder is carried out by the air spout and can be collected using a cyclone separator.

Spouted bed dryers are potentially more energy efficient than fluidized bed dryers but in practice, problems with scale-up from design prototypes to full-scale units have limited their application to the food industry. There appear to be multiple problems with scale-up, including

area to volume ratios and the height and stability of the spout.

Spouted bed dryers have been used for drying cereal grains such as wheat, and have successfully dried blood products and other food liquids.

4.5.2.6 Flash dryers

Flash dryers use very high air temperatures for very short residence times while the particles are pneumatically conveyed through the dryer. Thus, the main components of a flash dryer are a large fan and air heater, a cylindrical column through which the product rises with the heated air, a product feeder (such as an Archimedes screw or rotary star valve) to feed the product into the conveying air stream, and a separation unit (such as a cyclone separator) to collect the dried product. They differ from spray dryers in that they handle wet powders and slurries but not liquids. They are used for making dried powders, including drying dietary fiber, starch, casein, and dried gravy. Although the air temperature may be high (typically 150 °C), the product residence time is only a few seconds and its temperature stays low.

4.5.2.7 Multistage (belt) dryers

A belt dryer consists of a drying tunnel and a continuous conveyer belt, which holds the product. For large particles, drying is a slow process, limited by the rate of moisture diffusion from the center of the product to the outside. For this reason, belt dryers are arranged in multiple stages, with the air conditions at each stage being chosen to give the best drying effect and least quality degradation. Often high temperatures are used in the early stages and low temperatures to finish the drying process. The product may take 3–9 hours to pass through the dryer, depending on the type of product and the degree of drying required.

The stages may be arranged sequentially, forming one long continuous dryer, or to save floor space, the belts may be positioned on top of each other, allowing product to tumble from one belt to the next. As it falls to the next belt, fresh product surfaces are exposed, accelerating the drying process.

4.5.2.8 Column dryers

Column dryers are suited to granular products such as corn, rice, and wheat. The central drying chamber is fed continuously with wet product from elevators or buffer bins. The product falls slowly through the dryer as dried product is removed from the base. Hot air is introduced into the product mass through vents or a central air column. Often, a cooling section is used to arrest product thermal damage. The reason for choosing a column design is to save floor space (at the expense of height), in contrast to most dryer designs which take up large amounts of floor space.

Column dryers may be co-current, countercurrent or cross-flow, and modern column grain dryers are mixed flow, meaning that they utilize combinations of flow. Many products need to be dried slowly, for example products which exhibit case-hardening, become excessively brittle if dried quickly, or are so large that internal moisture migration dominates surface evaporation. In these cases tempering bins are often used with the column dryers. These bins collect the exit product and hold it for a period of several hours, allowing the moisture within the product to temper (become more uniform). Each pass through the dryer only removes a fraction of the total free moisture, yet the utilization of the dryer is improved.

4.6 Analysis of dryers

4.6.1 Moisture and heat balances

Analysis of a kiln dryer provides a suitable starting point for dryer analysis (Figure 4.7). This is the simplest dryer configuration, where ambient air is heated in a furnace, then enters the drying chamber, interacts with the product, then exits the chamber.

For the following analysis, assume that the inlet air conditions remain constant. Equating the moisture change of the product to the difference between the inlet and outlet air moisture content gives:

$$\dot{m}_a(H_I - H_E) = m_p \frac{\partial M}{\partial t} \qquad (4.21)$$

where $\dot{m}_a$ is the airflow rate, H_I and H_E are the inlet and outlet absolute humidities, m_p is the mass of product in the dryer and dM/dt is the drying rate (dry basis), and is negative for drying.

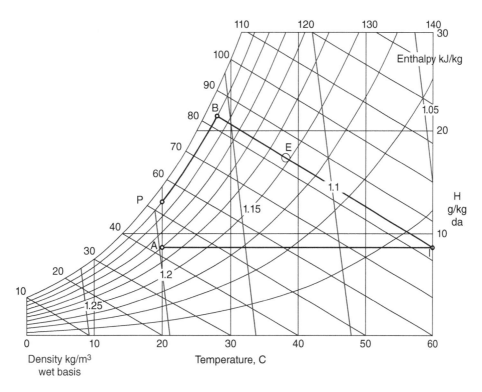

Figure 4.7 Representation of a kiln dryer on a psychrometric chart.

If we know the temperature and relative humidity of both the inlet and outlet air (for example, by measuring them directly using thermometers and relative humidity meters), then we can use the psychrometric chart (see Figure 4.7, where the lines represent a kiln-drying process) to determine the inlet and exit absolute humidities.

4.6.1.1 Exercise 4

Using the psychrometric chart in Figure 4.7:
1. describe what happens to the product
2. describe what happens to the air.
Note that the exit air humidity H_E will be somewhere between H_I and H_B, the exact position depending on factors such as the airspeed and drying time.

Important points on the diagram are denoted by capital letters as follows: ambient air (point A), inlet air to the dryer (point I), exit air from the dryer (point E), initial product state (point P), and conditions after the initial transient (point B). In this diagram, note the following important lines.
- **Heating the air (line AI):** ambient air is heated in the furnace to the required drying temperature.
- **Air and product interaction (line IE):** the air picks up moisture from the product and exits the dryer.
- **Initial transient (line PB):** the product and air come to thermal equilibration but not moisture equilibration.
- **Drying the product (line BI):** the product dries on the tray. As the product dries, its water activity will drop and temperature rise, so the product state changes from B and moves towards I. The product may never reach equilibrium with the drying air, as the kiln dryer will be shut down when the product reaches the required moisture content. For a dryer that incorporates air recirculation, we need to include an additional step – the mixing of the air exiting the chamber with ambient air. This requires both a mass and energy balance to calculate the combined mass flow and new air temperature. Graphically, mixtures can be represented on a psychrometric chart by straight lines joining the two components being mixed. The actual mixture point can then be determined by the mass proportion of components being mixed.

The following problem illustrates analysis of a dryer that recirculates air internally.

4.6.1.2 Problem using recirculation

A cabinet dryer circulates 30 kg/min dry air across a stack of 10 drying trays. Each tray holds 1 kg of product at 95% wb. If the ambient air is at 20 °C and 8 g/kg d.a., and the exit air is at 50 °C and 20 g/kg d.a., estimate:

i. the rate of moisture removal from the trays if the rate of internal air recirculation is 90%
ii. the heater size for the unit and
iii. time to dry in the constant rate period to 45% wb.

4.6.1.3 Solution

i. The amount of air exiting the dryer is:

$$10\% \text{ of } 30\,\text{kg/min} = 3\,\text{kg/min}.$$

Thus the rate of moisture removal is:

$$3\,(\text{kg/min}) \times (20 - 8)/1000 \times 60\,(\text{min/h}) = 2.16\,\text{kg/h}.$$

ii. To find the size of heater required, solve the air mixture equations to find the thermal difference between the air before the heater and the air leaving the heater. If I mix 90% of the exit air with 10% of the inlet air, then:

$$T_m = 90\% \times 50 + 10\% \times 20 = 47\,°C$$
$$H_m = 90\% \times 20 + 10\% \times 8 = 18.8\,\text{g/kg}$$

where T_m and H_m are the mixture point temperature and absolute humidity.

From a psychrometric chart, the enthalpy of air at 47 °C and 18.8 g/kg is 96 kJ/kg. Drawing an enthalpy line through the exit air conditions, locate the inlet conditions with the same enthalpy as the exit air (which is 102.3 kJ/kg) and the same absolute humidity as the mixture air (18.8 g/kg). The difference in enthalpy between these two points is (102.3 - 96) = 6.3 kJ/kg, and this heat must come from the air heater. Thus the required heater size is:

$$6.3\,(\text{kJ/kg}) \times 30\,(\text{kg/min}) \times (1/60)\,(\text{min/sec}) = 3.15\,\text{kW}.$$

iii. The rate of moisture removal is:

$$30\,\text{kg/min} \times (20 - 18.8)/1000 = 0.036\,\text{kg/min}.$$

The amount of moisture to remove from each tray is:

$$1\,\text{kg} \times (1 - 0.95) \times (0.95/0.05 - 0.45/0.55) = 0.91\,\text{kg}.$$

The time required to remove 0.91 kg moisture from 10 trays is:

$$t = 10 \times 0.91/(0.036) = 252.5\,\text{min} = 4\,\text{h}\,12\,\text{min}.$$

So after about 4 hours, the product will enter the falling rate period.

We have analyzed two examples of dryers, the kiln dryer (no recirculation) and the tray dryer (which includes recirculation). The principles that were used to do this can be extended to cover many types of air dryers.

4.7 Sustainability

Dryers may consume up to a quarter of a nation's energy budget (Mujumbdar, 2000) but continue to evolve rapidly in response to industry needs. Some of the pressures that generate changes in drying technology are environmental considerations, quality requirements, and processing innovations due to unit cost, space or time requirements.

Environmental considerations have become an increasingly important issue. Dryers require a high energy input, as the drying air must supply the heat required to evaporate water from the product. If the air passes through the product to the surrounding air, this heat energy is dispersed to the atmosphere. By recirculating the air, we can retain a proportion of the heat put into the air and so substantially reduce energy use. This may be done in several ways.

• Replacement of a small proportion of the air circulating within the dryer with ambient air (as was done with the tray dryer).
• Using a heat recovery system to dehumidify the air, based on refrigeration principles, a system which works best at low temperatures (20–40 °C).
• Chemical desorption of moisture from the exit with replenishment of the absorption material by some other form of drying (such as solar).

This results in less release of heat to the environment and a lower heating cost.

Energy reduction is not the only consideration, however, since there is also a demand for improved quality. The challenge then is to try to meet both objectives. This may be difficult; for example, vacuum and freeze drying allow dehydration at lower temperatures, retaining a greater fraction of the aroma and flavor compounds, but at a much higher equipment cost.

The demand for more economical, higher quality drying motivates technical innovation. New technologies being researched include the following.

• Transmission of energy by forms of electromagnetic radiation, for example radiofrequency drying which transmits the energy uniformly to moisture within the product, so decreasing drying time.
• Heat pump dryers, mentioned in section XXX, where the application of heat energy recycling reduces the energy cost of drying.

• Superheated steam drying.
• Pulse drying, where drying air speed and temperature are switched between low and high levels.
• Replacing heated air with heated granular solids as the heat transfer medium, for example using hot sand to dry a product.
• Combinations of drying technologies, such as fluidized drying combined with superheated steam or heated sand.

Several drying methods have clear advantages in terms of reducing drying cost, especially in-store (or more generally near-ambient) drying and heat pumps. Other technologies offer marginal energy reductions, of the order of 10%, for example pulsed airflow. A number of technologies for improving thermal efficiency have become standard, including impingement drying and combined evaporator/spray/fluidized bed drying. A typical dryer from the 1970s might have a thermal efficiency of around 60%, whereas many modern technologies achieve efficiencies far greater than 100% by using heat recovery systems (such as the heat pump) or natural air drying capacity (for example, the deep bed dryer).

For a company trying to manufacture product in a competitive environment, reductions in energy may be less of an incentive than increasing throughput, since the processing cost is generally a small fraction of the total production cost, and also that margins on food products are generally low and so high production volumes are required. An improvement must satisfy both objectives to be successful – reduce energy but also allow high-volume drying. As the cost of energy increases, sustainable innovations become more appealing to the equipment purchaser.

4.8 Conclusion

In this chapter, we have considered some types of dryers and introduced the study of dryers. Dryer design is dependent on the nature of the product being dried, and a system that works for one product may not be satisfactory for others. Dryers also affect the quality of a product in many different ways, and so choices about temperature and handling inside a dryer are important.

By studying psychrometrics and drying theory, we can better understand how to choose or design a dryer that is suited to the product's needs. Modeling the drying process allows us to estimate the time required for product drying under different conditions, and so choose a dryer capable of handling the required volume of product. We have looked at some simple models of product drying, which allow prediction of drying times and provide a better understanding of the basic processes involved.

References

Crank J (1975) *The Mathematics of Diffusion*, 2nd edn. Oxford: Clarendon Press.

Mujumbdar AS (2000) *Drying Technology in Agriculture and Food Sciences*. Boca Raton, FL: Science Publishers.

Further reading

Baker CGJ (1997) *Industrial Drying of Foods*. London: Blackie Academic and Professional.

Bala BK (1997) *Drying and Storage of Cereal Grains*. New Delhi: IBH Publishing.

Bala BK (1998) *Solar Drying Systems: Simulations and Optimization*. Udaipur, India: Agrotech.

Barr-Rosin (2012) Fluidized Bed Drying. www.barr-rosin.com/products/fluid-bed-dryer.asp, accessed 7 November 2013.

Brooker DB, Bakker-Arkema FW, Hall CW (1992) *Drying and Storage of Grains and Oilseeds*. Westport, CT: Avi Publishing.

Brunnauer S, Emmet PH, Telur E (1938) Adsorption in multimolecular layers. *Journal of the American Chemical Society* **60**: 309–319.

Burmester K, Eggers R (2010) Heat and mass transfer during the coffee drying process. *Journal of Food Engineering* **99**: 430–436.

Chen XD, Mujumdar AS (2008) *Drying Technologies in Food Processing*. Oxford: Blackwell Publishing.

Chung D, Pfost H (1967) Adsorption and desorption of water vapor by ceral grains and their products. *Transactions of the American Society of Agricultural Engineers* **10**: 549–557.

Devahastin D (2000) *Mujumbdar's Practical Guide to Industrial Drying*. Thailand: Exergex Corporation.

Duffie JA, Beckman WA (1991) *Solar Engineering of Thermal Processes*. New York: John Wiley.

Ertekin C, Yaldiz O (2004) Drying of eggplant and selection of a suitable thin layer drying model. *Journal of Food Engineering* **63**: 349–359.

Fellows PJ (2000) *Food Processing Technology: Principles and Practice*, 2nd edn. Boca Raton, FL: CRC Press.

Fennema OR (1997) *Food Chemistry*. Food Science and Technology Series. New York: Marcel Dekker.

Gough MC (1975) A simple technique for the determination of humidity equilibria in particulate foods. *Journal of Stored Products Research* **11**: 161–166.

Halsey G (1985) Physical adsorption on uniform surface. *Journal of Chemical Engineering* **16**: 931–937.

Haynes B (1961) Vapor pressure determination of seed hygroscopicity. Technical Bulletin 1229. Washington, DC: USDA Agricultural Research Station.

Heldman DR (1975) *Food Process Engineering*. Westport, CT: Avi Publishing.

Helrich K (1990) *Official Methods of Analysis*. Gaithersburg, MD: Association of Official Analytical Chemists.

Henderson SM (1952) A basic concept of equilibrium moisture. *Journal of Agricultural Engineering* **33**: 29–32.

Holman JP (1992) *Heat Transfer*, 7th edn. New York: McGraw-Hill.

Howe ED (1980) Principles of drying and evaporating. *Sunworld* **4**: 182–186.

Hukill WV (1954) Grain drying. In: Anderson JA, Alcock AW (eds) *The Storage of Cereal Grains and Their Products*. St Paul, MN: American Society of Cereal Chemists.

Kowlaski SJ (2007) *Drying of Porous Materials*. The Netherlands: Springer.

Langmuir I (1918) The adsorption of gases on plane surfaces of glass and mica and platinum. *Journal of the American Chemical Society* **40**: 1361–1365.

Mohsenin NN (1980) *Thermal Properties of Foods and Agricultural Materials*. New York: Gordon and Breach Science.

Mohsenin NN (1986) *Physical Properties of Plant and Animal Materials*. New York: Gordon and Breach Science.

Mwithiga G, Jindal VK (2004) Cofee drying in a rotary conduction-type heating unit. *Journal of Food Process Engineering* **27**: 143–157.

Oswin CR (1946) The kinetics of package life III. *Isotherm. Journal of the Society of Chemical Industry* **65**: 419–421.

Perry RH, Perry GD (2008) *Chemical Engineers' Handbook*. New York: McGraw-Hill, Chapter 12.

Rahman S (1995) *Food Properties Handbook*. Boca Raton, FL: CRC Press.

Samaniego-Esguerra CM, Boag IF, Robertson GL (1991) Comparison of regression methods for fitting the GAB model to the moisture isotherms of some dried fruit and vegetables. *Journal of Food Engineering* **13**: 115–133.

Singh RP, Heldman RH (2009) Introduction to Food Engineering. *New York: Academic Press, Chapters* **8**, 9.

Smith JE (1947) The sorption of water vapor by high polymers. *Journal of the American Chemical Society* **69**: 646–651.

Toledo RT (1991) *Fundamentals of Food Process Engineering*, 2nd edn. Westport, CT: Van Nostrand Reinhold/AVI, Chapters 3, 4, 5 and 12.

5 Chilling and Freezing of Foods

Stephen J. James and Christian James
*Food Refrigeration and Process Engineering Research Centre,
The Grimsby Institute, Grimsby, UK*

5.1 Introduction to the food cold chain

The principle of the refrigerated preservation of foods is to reduce, and maintain, the temperature of the food such that it stops, or significantly reduces, the rate at which detrimental changes occur in the food. These changes can be microbiological (i.e. growth of microorganisms), physiological (e.g. ripening, senescence, and respiration), biochemical (e.g. browning reactions, lipid oxidation, and pigment degradation), and/or physical (such as moisture loss). An efficient and effective cold chain is designed to provide the best conditions for slowing, or preventing, these changes for as long as is practical. Effective refrigeration produces safe food with a long, high-quality shelf life. In the "chill" chain, the overall objective is to achieve the desired shelf life of a "fresh" product without the commonly perceived quality deterioration resulting from ice crystal formation. Freezing substantially increases the safe storage and distribution life of a food, but ice crystal formation might result in changes that are perceived to be detrimental to the quality of the food.

To provide safe food products of high organoleptic quality (i.e. taste, texture, smell, appearance), attention must be paid to every aspect of the cold chain from initial chilling or freezing of the raw ingredients, through storage and transport, to retail display and domestic handling. The cold chain (which includes both chilled and frozen foods) consists of two distinct types of operation. In processes such as primary and secondary chilling or freezing, the aim is to lower the average temperature of the food. In others, such as chilled or frozen storage, transport, and retail display, the prime aim is to maintain the temperature of the food at a constant optimal value. Removing the required amount of heat from a food is a difficult, time- and energy-consuming operation, but is critical to the operation of the cold chain. As a food moves along the cold chain it becomes increasingly difficult to control and maintain its temperature. This is because the temperatures of bulk packs of refrigerated products in large storerooms are far less sensitive to small heat inputs than single consumer packs in open display cases or in a domestic refrigerator/freezer. Failure to understand the needs of each process results in excessive weight loss, higher energy use, reduced shelf life, and/or deterioration in product quality.

There are, however, examples where maintaining a particular food at temperatures that severely limit, if not completely stop, chemical changes does not achieve the desired final product quality. Examples of this are in the maturing of meat, ripening of fruits, and flavor development (aging) in cheese. In all these cases, the time-temperature history of the food must be carefully controlled so that periods are provided at temperatures where the desired changes can occur. However, the combination of time and temperature needs to be controlled such that undesirable, and especially unsafe, changes, such as the growth of pathogenic bacteria, do not occur.

5.2 Effect of refrigeration on food safety and quality

Refrigeration (cooling) is the total process of reducing the temperature of a food and maintaining that temperature during storage, transport, and retailing. If the temperature of the food does not fall below one where ice is formed in the food, the food is considered chilled and the temperature reduction process is called chilling. If ice is formed then the food is considered frozen, and the temperature reduction process is called freezing.

Food Processing: Principles and Applications, Second Edition. Edited by Stephanie Clark, Stephanie Jung, and Buddhi Lamsal.
© 2014 John Wiley & Sons, Ltd. Published 2014 by John Wiley & Sons, Ltd.

5.2.1 Microbiology and food safety

Temperature is a major factor affecting microbial growth. Microbial growth is described in terms of the lag phase and the generation time. When a microorganism is introduced to a particular environment, there is a time (the lag phase) in which no increase in microbial numbers is apparent, followed by a period when microbial growth occurs exponentially. The generation time is a measure of the time it takes for the population. Microorganisms have an optimum growth temperature at which a particular strain grows most rapidly, i.e. the lag phase and generation times are both at their shortest time. They also have a maximum growth temperature above which growth no longer occurs. Above this temperature, one or more of the enzymes essential for growth are inactivated and the cell is considered to be heat-injured. However, in general, unless the temperature is raised to a point substantially above the maximum growth temperature, the injury is not lethal and growth will recommence as the temperature is reduced. Attaining temperatures substantially above the maximum growth temperature is therefore critical during cooking and reheating operations.

Of most concern during storage, distribution, and retail display of food is a third temperature threshold, the minimum growth temperature for a microorganism. As the temperature of an organism is reduced below that for optimum growth, the lag phase and generation time both increase. The minimum growth temperature can be considered to be the highest temperature at which either of the growth criteria, i.e. lag phase and generation time, becomes infinitely long. The minimum growth temperature is not only a function of the particular organism but also the type of food or growth media that is used for the incubation. Although some pathogens can grow at $0\,^\circ$C or even slightly lower (Table 5.1), from a practical point of view the risks to food safety are considerably reduced if food is maintained below $5\,^\circ$C. However, food spoilage may still be an issue (see later). There are few data available on the impact of the initial freezing process on the safety of foods. However, it is difficult to envisage any sensible freezing process that would result in most foods being held for substantial periods at temperatures that would support a dangerous growth of pathogens. Providing that the food does not rise above $-12\,^\circ$C during frozen storage and display, there should be no issues of food safety with frozen storage and display.

Campylobacter is the most common reported bacterial cause of infectious intestinal disease in many countries.

Table 5.1 Minimum and optimum growth temperatures for pathogens associated with foods

	Minimum temperature (°C)	Optimum temperature (°C)
Campylobacter spp.	30	42–43
Clostridium perfringens	12	43–47
Clostridium botulinum proteolytic	10	–
Staphylococcus aureus	7	–
Pathogenic *Escherichia coli* strains	7	35–40
Escherichia coli O157:H7	6–7	42
Salmonella spp.	5	35–43
Bacillus cereus	5	–
Clostridium botulinum non-proteolytic	3	–
Aeromonas hydrophila	−0.1 to 1.2	–
Listeria monocytogenes	−1 to 0	30–37
Yersinia enterocolitica	−2	28–29

Two species account for the majority of infections: *C. jejuni* and *C. coli*. Illness is characterized by severe diarrhoea and abdominal pain. Freezing and crust-freezing (freezing only the surface of the food) have been suggested as ways to reduce numbers of *Campylobacter* organisms on poultry carcasses, and have been recommended as control measures for reducing *Campylobacter* by the European Food Safety Authority (EFSA). Freezing to below $-20\,^\circ$C has been reported by a number of studies to result in an initial fall in numbers of *Campylobacter* organisms, followed by a slower decline during storage. The mechanism of damage during freezing has been attributed to mechanical damage caused by ice crystals, desiccation due to the reduced water activity, and oxidative damage.

In addition to bacterial pathogens, histamine, viruses, and nematodes can also compromise food safety. Histamine (scombroid or scombrotoxin food) fish poisoning is a food-borne chemical intoxication associated with the consumption of spoiled fish flesh that is high in histidine (from species such as mackerel, sardines, and certain tuna species). In these fish, bacterial histidine decarboxylase converts muscle histidine into histamine. In recent years, viruses have been increasingly recognized as important causes of outbreaks of food-borne disease. While noroviruses and hepatitis A are currently recognized as the most important food-borne viruses, a range of other enteric viruses have also been linked to food-borne illness. Freezing in general has little effect on viruses or on histamine. Inadequate freezing and thawing procedures have been identified as factors in histamine formation.

Nematode parasites are very susceptible to freezing. Nematodes are slender worms, typically less than 2.5 mm (0.10 in) long. The smallest nematodes are microscopic, while free-living species can reach as much as 5 cm (2 in), and some parasitic species are larger still, reaching over a meter in length. Nematodes are responsible for a range of food-borne parasitic diseases in humans. Freezing is a control measure for inactivating trichinae in pork and nematode parasites in seafood (particularly for lightly processed seafood that will receive no cooking before consumption). Freezing is also used as a control measure for inactivating tapeworms (*Taenia saginata*) in beef carcasses.

Even if completely safe, food can become microbiologically unacceptable as a result of the growth of spoilage microorganisms. Their growth, which is highly temperature dependent, can produce unacceptable changes in the sensory quality of many foods. The development of off-odors is usually the first sign of putrefaction, and in meat, it occurs when bacterial levels reach approximately 10^7 cm^{-2} of surface area (Ingram, 1972). When bacterial levels have increased a further 10-fold, slime begins to appear on the surface and meat received in this condition is usually condemned out of hand. At $0\,^{\circ}$C, beef with average initial contamination levels can be kept for at least 15 days before any off-odors can be detected. Every $5\,^{\circ}$C rise in the storage temperature above $0\,^{\circ}$C will approximately halve the storage time that can be achieved. For example, a study by Hong and Flick (1994) reported that the shelf life of blue crab meat, determined by sensory analysis, was 15.5 days at $0\,^{\circ}$C, 12.5 days at $2.2\,^{\circ}$C, and 8.5 days at $5.6\,^{\circ}$C.

5.2.2 Nutritional quality

Microbial safety and spoilage are not the only aspects of food quality that are temperature dependent. The rate of vitamin loss from fruit and vegetables during storage also depends upon the storage temperature. In general, storage just above the freezing point of the fruit or vegetable will have the greatest effect on retarding respiration and transpiration (McCarthy & Matthews, 1994) and maintaining vitamin content (Paull, 1999). Humidity control is also important to prevent wilting and maintain crispness (McCarthy & Matthews, 1994), which in turn prevents the loss of water-soluble vitamins such as vitamin C (Paull, 1999). It is of interest to note that it is not always

a case of the lower the temperature, the better the quality, especially with citrus and tropical fruits. The optimum temperature for orange storage is approximately $12\,^{\circ}$C, with the rate of vitamin loss increasing at temperatures warmer or colder than this value.

There is a wealth of data available that shows that freezing is excellent at retaining nutrient content in comparison to "fresh" (chilled) equivalents (Berry et al., 2008). Although blanching will reduce some of the total initial nutrient content, there is very little reduction in nutrient content during subsequent frozen storage, whereas the nutrient content of many foods drops significantly during chilled storage. In general, more data are available on fruits and vegetables than meat, poultry, and fish (Berry et al., 2008).

5.2.3 Weight loss

Some foods exhibit particular quality advantages as a result of rapid cooling. In meat, the pH starts to fall immediately after slaughter and protein denaturation begins. The result of this denaturation is a pink proteinaceous fluid, commonly called "drip," often seen in prepackaged cuts of meat. The rate of denaturation is directly related to temperature and it therefore follows that the faster the chilling rate, the less the drip. Investigations using pork and beef muscles have shown that faster rates of chilling (e.g. reducing the deep leg temperature of a pig in 10 hours to $5\,^{\circ}$C compared with $14\,^{\circ}$C) can halve the amount of drip loss (Taylor, 1972).

Another quality and economic advantage of temperature control is a reduction in weight loss, which results in a higher yield of saleable material. Meats, fruits, and vegetables, for example, have high water contents and the rate of evaporation depends directly on the vapor pressure at the surface. Vapor pressure increases with temperature and thus any reduction in the surface temperature will reduce the rate of evaporation. The use of very rapid chilling systems (3–4 h compared with overnight) for pork carcasses has been shown to reduce weight loss by at least 1% when compared with conventional systems (James et al., 1983).

Traditionally, the frozen food industry was interested in two particular problems to do with weight loss that were detrimental to the appearance of the frozen food: freezer burn and in-package frosting. Freezer burn is caused by water loss from the surface of the frozen food due to sublimation. The resulting desiccation produces a dry fibrous layer at the surface that has the appearance of a burn. It only occurs in unwrapped or poorly wrapped foods and its development is fastest at high storage temperatures and high air movements. It is not caused by fast freezing. In-package frosting results from a combination of water loss from the surface, loose packaging, and temperature fluctuations during storage. The water lost from the surface is deposited and frozen on the inner surface of the packaging. The use of suitable packaging and good temperature control should eliminate both problems. Temperature fluctuations during storage will enhance both phenomena.

5.2.4 Flavor

The flavor and aroma of fruits and vegetables can be significantly affected by temperature, during both initial cooling and storage. Flavor is determined largely in these foods by the sugar to acid ratio (Paull, 1999). The rate of sugar loss (sweetness) in freshly harvested sweetcorn is very temperature dependent. After 20 h at 30 °C, almost 60% of the sweetness is lost compared with 16% at 10 °C and less than 4% at 0 °C. Prompt cooling is clearly required if this vegetable is to retain its desirable sweetness. Similarly, the ripening of fruit can be controlled by rapid cooling, the rate of ripening declining as temperature is reduced and ceasing below about 4 °C (Honikel, 1986).

Although freezing inhibits the rate of chemical reactions that promote off-flavor formation during frozen storage, it does not prevent these reactions, such as moisture migration, lipid oxidation and protein denaturation, taking place (Ponce-Alquicira, 2005). In many cases it is the production of off-flavors, such as by lipid oxidation in frozen meats, that limits the high-quality storage life of frozen foods.

5.2.5 Texture

Fish passing through rigor mortis above 17 °C are to a great extent unusable because the fillets shrink and become tough (Morrison, 1993). A relatively short delay of an hour or two before chilling can demonstrably reduce shelf life. However, chilling has serious effects on the texture of meat if it is carried out rapidly when the meat is still in the pre-rigor condition; that is, before the pH of the meat has fallen below about 6.2 (Bendall, 1972). A phenomenon known as cold shortening occurs when chilled pre-rigor, resulting in the production of very tough meat after cooking. As a "rule of thumb," cooling to temperatures not below 10 °C in 10 h for beef and lamb, and in 5 h for pork can prevent cold shortening (James & James, 2002). Poultry meat is generally less prone to cold

Table 5.2 Fruits and vegetables susceptible to chilling damage (sources: Hardenburg et al., 1986; IIR, 2000; McGlasson et al., 1979; McGregor, 1989)

Commodity	Lowest safe temperature (°C)	Damage
Apples		
certain varieties	1–2	Internal browning, brown core
Avocados		
West Indian	11	Pitting, internal browning
Other varieties	5–7	Pitting, internal browning
Bananas	12–13	Dull color, blackening of skin
Beans	7–10	Pitting and russeting
Cucumbers	7–10	Pitting, water-soaked spots, decay
Grapefruit	7	Scald, pitting, watery breakdown, internal browning
Lemons	13–14	Internal discoloration, pitting
Mangoes	5–10	Internal discoloration, abnormal ripening
Melons		
Cantaloupe	7	Pitting, surface decay
Honeydew	4–10	Pitting, surface decay
Watermelons	2–4	Pitting, objectionable flavor
Oranges	3	Pitting, brown stains
Papaya	6	Pitting, water soaking of flesh, abnormal ripening
Potatoes	3–4	Mahogany browning, sweetening
Tomatoes	7–10	Water soaking and softening

shortening. However, electrical stimulation, i.e. passing an electrical current through the carcass within a short time of death, can be utilized to enable more rapid cooling to be carried out without the occurrence of cold shortening. Electrical stimulation shortens the time to the onset of rigor mortis by the acceleration of glycolysis with the pH falling by the order of 0.7 units during the first 2 min.

For several fruits and vegetables (Table 5.2), exposure to temperatures below a critical limit, but above the initial freezing temperature, may result in chilling injury. Typical symptoms of chilling injury are internal or external browning, superficial spots, failure to ripen, development of off-flavors, etc. It is primarily fruits and vegetables from the tropical and subtropical zones that are susceptible to chilling injury, but several Mediterranean products are also susceptible (International Institute of Refrigeration, 2000). The extent of damage depends on the temperature, duration of exposure, and sensitivity of the fruit and vegetable. Some commodities have high sensitivity, while others have moderate or low sensitivity. For each commodity, the critical temperature depends on species and/or variety. In some cases, unripe fruits are more sensitive than ripe fruits.

One critical quality factor affected by freezing is food texture (Kerr, 2005). Loss of water-holding capacity due to tissue damage during freezing is a problem for many frozen foods. Such foods exhibit excessive drip loss on thawing and may lack proper juiciness when chewed (Kerr, 2005). Due to the chemical and structural differences in different food groups, each has unique issues associated with changes in textural quality due to freezing.

There is a general view that fast freezing offers some quality advantage, with "quick frozen" appearing on many frozen foods, with the expectation that consumers will pay more for a "quick frozen" product. This is because the rate of freezing has an effect on how and where ice crystals are formed within a food and the size of the ice crystals. Fast freezing rates will promote more nucleation and a greater number of crystals of smaller size which should result in less physical damage to the structure of the food.

However, the rate of freezing is not important for all foods. Foods may be classified into four groups (Poulsen, 1977; Spiess, 1979) according to their sensitivity regarding freezing rate.

1. Products that remain practically unaffected by variations in freezing rate, i.e. products with a high content of dry matter. For example, peas, meat products with a high fat content, and certain ready meals.

2. Products which require a minimum freezing rate (0.5–1 °C/min), but are relatively unaffected by higher freezing rates. For example, fish, lean meat, and many starch- and flour-thickened ready meals.

3. Products whose quality improves when freezing rates are increased (3–6 °C/min). For example, many fruits, egg products, and flour-thickened sauces.

4. Poulsen (1977) defined Group 4 as "products that are sensitive to too high freezing rates and tend to crack." For example, fish and many fruits. Spiess (1979) refined this definition as covering "products in which high freezing rates are advantageous for the product quality, but where temperature tensions in the product result in a destruction of the tissue." For example, fruits and vegetables such as raspberries, tomatoes, and cucumbers.

Terms such as slow, rapid, ultra-rapid, etc. often used in industry have no strict definitions, particularly since in many foods conduction within the food itself significantly slows the rate of freezing. Though rapid freezing may not offer particular advantages in terms of quality or extended shelf life for all foods, it may in terms of throughput. Though not always their sole consideration, throughput is of considerable importance to most producers.

There is evidence that since many plant tissues (fruit and vegetable) have a semi-rigid cellular structure that has less resistance to the expansion of ice crystals in volume, they are more prone to irreversible freezing damage than muscle tissues (Reid, 1994). In respect of the mechanisms of freezing damage in plant tissues, four contributory processes have been proposed: chill damage, solute-concentration damage, dehydration damage, and mechanical damage from ice crystals (Reid, 1994).

1. Chilling damage is a result of exposing the plant tissue to low temperature.

2. Solute-concentration damage is due to the increase in the concentration of solutes in the unfrozen liquid with the formation of ice crystals.

3. Dehydration damage results from increased solute concentration in the unfrozen liquid and the osmotic transfer of water from a cell interior.

4. Mechanical damage occurs as a result of the formation of hard ice crystals.

These types of damage in plant tissues would result in loss of function in cell membrane, disruption of metabolic systems, protein denaturation, permanent transfer of intracellular water to the extracellular environment, enzyme inactivation, and extensive cell rupture (Reid, 1994). Properties that reflect freshness and turgidity would also be lost in frozen food, because they depend largely upon the structural arrangement and chemical composition of the cell wall and the intercellular spaces where pectic substances are the primary constituents (Cano, 1996).

5.3 Blanching

Four groups of enzymes (Table 5.3) are primarily responsible for quality deterioration in vegetables. For vegetables, and to a lesser degree fruits, where enzymic deterioration during frozen storage is a problem, blanching is required to deactivate the undesirable enzymes. This is usually done with a mild heat treatment, typically by treating the product with steam or hot water for 1–10 min at 75–95 °C (International Institute of Refrigeration,

Table 5.3 Enzymes responsible for quality deterioration in unblanched vegetables (Williams et al., 1986)

	Enzyme
Off-flavors	Lipoxydenases, lipases, proteases
Textural changes	Pectic enzymes, cellulases
Color changes	Polyphenol oxidase, chlorophyllase, peroxidase
Nutritional changes	Ascorbic acid oxidase, thiaminase

2006), the time-temperature combination depending on the specific product. Blanching provides a number of other advantages, including improving color and flavor of some vegetables. However, it can also cause undesirable and irreversible textural changes and degrade heat-labile nutrients.

5.4 Principles of refrigeration systems

A detailed analysis of refrigeration cycles and systems may be found in numerous refrigeration textbooks (ASHRAE, 2006; Gosney, 1982; Trott, 1989). Nevertheless, a basic understanding of the principles of how a refrigeration system works is necessary for all users of refrigeration equipment. There are two main types of refrigeration system: total loss refrigeration systems, and mechanical refrigeration systems.

The rate of heat removed from the food ($Q_{product}$, J/s) accounts for the majority of the refrigeration load and can be determined from:

$$Q_{product} = \frac{mc_p\left(T_{initial} - T_{final}\right)}{\Delta t} \quad (5.1)$$

where:

m = total mass, kg
c_p = specific heat capacity, J/kg°C
$T_{initial}$ = initial temperature of the food, °C
T_{final} = final temperature of food, °C and Δt is the time taken to remove the heat

5.4.1 Total loss refrigeration systems

Total loss refrigeration systems utilize the direct contact of a phase-changing refrigerant with a food. The refrigerant changes state during use, which requires latent heat and provides part of the heat extraction. The other part is provided by the heat required to warm the resulting cold gas. The refrigerant is released to the atmosphere/environment and not recovered (hence the term "total loss").

One of the most common total loss refrigerants is ice, with the majority of fish and shellfish being cooled and transported using it. During melting, ice requires 333 kJ·kg^{-1} of latent heat energy to convert from the solid to liquid phase.

Liquid nitrogen and solid carbon dioxide are also common total loss refrigerants. These refrigerants are usually referred to as "cryogens" and refrigeration using these as "cryogenic." The term "cryogenic" simply means very low temperatures. Liquid air was first used commercially in the 1930s (Willhoft, 1991) as a cryogen. However, liquid air contains a high proportion of liquid oxygen, which is a powerful oxidizing agent and its use has been superseded by less harmful liquid nitrogen (LN) and liquid or solid carbon dioxide (CO_2). What distinguishes cryogenic refrigeration from standard liquid immersion/spray systems is that a proportion of the heat removal is accomplished by a change in state of the heat transfer medium (Fennema, 1975). As well as using the latent heat absorbed by the boiling liquid, sensible heat is absorbed by the resulting cold gas.

Nitrogen at atmospheric pressure liquefies at a temperature of −196 °C, giving a refrigerating capacity of 378 kJ per kg of liquid nitrogen. It is usually supplied and stored at a pressure of 3–6 bar, with corresponding boiling points of −185 °C to −177 °C (Heap & Mansfield, 1983; Hoffmanns, 1994). A useful rule of thumb is that 1 ton·h^{-1} of liquid nitrogen is approximately equivalent to 100 kW of mechanical refrigeration.

Carbon dioxide, if stored as a pressurized liquid and released into the atmosphere, changes partly to gas and partly to a frozen solid at −78.5 °C, which sublimates directly into gas without going through a liquid phase. Liquid carbon dioxide is generally supplied either at ambient temperature (e.g. 25 °C and 65 bar), giving a refrigerating capacity of 199 kJ per kg of liquid carbon dioxide, or at −16 °C and 22 bar, giving a refrigerating capacity of 311 kJ per kg of liquid carbon dioxide. Solid carbon dioxide has a refrigerating capacity of 620 kJ per kg of solid carbon dioxide.

Due to very low operating temperatures and high surface heat transfer coefficients between product and medium, cooling rates of cryogenic systems are often substantially higher than other refrigeration systems. Cryogenic systems are best suited to cooling thin products with a high surface area to weight ratio in which heat conduction within the product is not rate limiting, such as pizza, seafood, sliced/diced meats, and vegetables. Cryogens can be used to chill foods, but their very low temperature can lead to undesirable surface freezing of the product. This can be avoided by indirect use (using the cryogen to cool down a secondary refrigeration medium, such as air) and/or by careful control of processing temperatures, such as ramp control in which the refrigeration medium temperature rises as the product temperature falls.

5.4.2 Mechanical refrigeration systems

Mechanical refrigeration systems operate using the same basic refrigeration cycle (Figure 5.1). At its simplest, it utilizes four interlinked components: the evaporator, compressor, condenser, and expansion valve. A low-pressure cold liquid refrigerant is allowed to evaporate to a gas within the "evaporator" coil. This process requires heat, which is extracted (e.g. ultimately from the product), thus cooling any medium surrounding the evaporator (e.g. air). The low-pressure hot gas from the evaporator is compressed in the "compressor" to a high-pressure hot gas. This high-pressure hot gas is then passed through another coil, where it condenses back to a high-pressure cold liquid. This process releases heat into any medium surrounding

the "condenser" coil (e.g. outside air). This high-pressure cold liquid refrigerant then passes through a valve, the "expansion valve," to a lower pressure section. The now low-pressure liquid then passes back to the evaporator.

In a direct expansion system, the evaporator coil is in direct contact with either the food to be refrigerated or the cooling medium (i.e. air, brine, etc.) surrounding the food. In a secondary refrigeration system, liquid is cooled by passing over the evaporator coil, which in turn is used to cool the cooling medium (i.e. air, water, brine, etc.) surrounding the food. The energy used by a refrigeration system depends on its design, but generally the larger the temperature difference between the evaporator and condenser, the greater the energy used in the compressor for a given amount of cooling duty.

Traditionally, fluorocarbons, especially hydrochlorofluorocarbons (HCFCs) such as R22 (chlorodifluoromethane), were used as refrigerants but they are being phased out because of their ozone-depleting potential (ODP), as discussed later in this chapter. Thus, R22 is being replaced with refrigerants such as R404A (a mixture of refrigerants R125 pentafluoroethane, R143a 1,1,1-trifluoroethane and R134a 1,1,1,2-tetrafluoroethane) and R407c (a mixture of refrigerants R32 difluoromethane, R125 pentafluoroethane and R134a 1,1,1,2-tetrafluoroethane), ammonia, and propane.

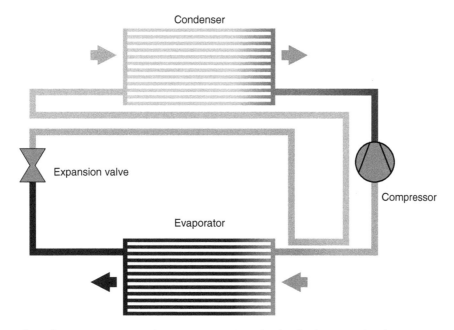

Figure 5.1 A mechanical vapor compression refrigeration system. For color details, please see color plate section.

5.5 Heat transfer during chilling and freezing

Heat transfer is a dynamic process that concerns the exchange of thermal energy from one physical system with a higher temperature to another with a lower temperature. It should be remembered that heat only flows from hot to cold. There are four modes of heat transfer: conduction, convection, radiation, and evaporation/condensation. In practice, they can occur individually or combined. In many applications, all four occur simultaneously but with different levels of importance. During chilling or freezing, heat is removed from the food usually by the combined mechanisms of convection, radiation, and evaporation. The rate of heat transfer ($Q_{convection/radiation/evaporation}$, J/s) between the food and the surrounding medium at any time can be expressed as:

$$Q_{convection/radiation/evaporation} = hA\Delta T = hA\left(T_{surface} - T_{ambient}\right) \quad (5.2)$$

where:

h = surface heat transfer coefficient for combined convection, radiation and evaporation, W/m^2°C
A = exposed surface area of the food, m^2
$T_{surface}$ = surface temperature of the food, °C
$T_{ambient}$ = temperature of the heat transfer medium in contact with the food surface, °C.

The surface heat transfer coefficient (h) is not a property of the food. Its value will depend on the shape of the food and its surface roughness, the type of heat transfer medium, the velocity of the heat transfer medium, and the flow regime. Typical surface heat transfer values for different chilling/freezing methods are shown in Figure 5.1.

5.5.1 Conduction

Conduction, or diffusion, is the transfer of energy between objects that are in physical contact. In conduction, the heat is transferred by means of molecular agitation within a material without any motion of the material as a whole. The heat transfer is made from points with greater energy to points with less energy in an attempt to establish a thermal equilibrium (Zheleva & Kamburova, 2009). Conduction is the main mode of heat transfer within a solid food, such as meats and vegetables, or between solid objects in thermal contact, such as plate conduction chillers. The rate of heat transfer

($Q_{conduction}$, J/s) within the food through conduction at any time can be expressed as:

$$Q_{conduction} = kA\frac{\Delta T}{\Delta x} \quad (5.3)$$

where:

k = thermal conductivity of the food, W/m°C
A = surface area of the food, m^2
ΔT = temperature difference within the food, °C
Δx = thickness of the food, m.

5.5.2 Convection

Convection is the transfer of heat by circulation or movement of hot particles to cooler areas. It is restricted to liquids and gases, as mass molecular movement does not occur at an appreciable speed in solids. Molecular motion is induced by density changes associated with temperature differences at different points in the liquid or gas (natural convection), or when a liquid or gas is forced to pass over a surface by mechanical means (forced convection) (Toledo, 2007). When liquid is cooled, some particles become denser and consequently drop. The surrounding warmer fluid particles move to replace them and are also cooled. The process continues repeatedly, forming convection currents. Convection is the main mode of heat transfer within a liquid food, such as a milk or a sauce, or within liquid or gaseous heat transfer media, such as air-blast, spray, and immersion chillers.

5.5.3 Radiation

Radiation is the transfer of heat by electromagnetic waves. Unlike conduction and convection, radiation does not require a medium because electromagnetic waves can travel through a vacuum. When a suitable surface intercepts these waves, it will absorb them, raising its energy level. Radiation depends upon the relative temperatures, geometric arrangements, and surface structures of the materials that are emitting or absorbing heat (Earle, 1983). The sun and a simple flame are examples of radiating objects producing heat. To achieve substantial rates of heat loss by radiation, large temperature differences are required between the surface of the product and that of the enclosure. Such differences are not normally present during food chilling operations except in the initial chilling of bakery and cooked products. When used, it is usually in combined systems, for example with contact chilling on metal surfaces under a cold surface absorbing radiated heat, or with natural convection (Ciobanu, 1976b).

5.5.4 Evaporation

Evaporation is the transfer of energy required to change a liquid to a vapor. Latent heat is the heat energy required to change a substance from one state to another. For the majority of foods, the heat lost through evaporation of water from the surface is a minor component of the total heat loss, though it is the major component in vacuum cooling. Evaporation from a food's surface reduces yield and is not desirable in many refrigeration operations, but can be useful in the initial cooling of unpackaged cooked food products.

5.6 Chilling and freezing systems

There are a large number of different chilling and freezing systems for food, based on moving air, wet air, direct contact, immersion, ice, cryogenics, vacuum and pressure shift, some of which are described below.

For the majority of chilled and frozen foods, air systems are used primarily because of their flexibility and ease of use. However, other systems such as immersion, contact and cryogenics can offer much faster and more controlled chilling or freezing.

From a hygiene/safety-based approach, prepacking the food prior to chilling or freezing will lower the risk of contamination/cross-contamination during the chilling process. However, in most cases it will significantly reduce the rate of cooling, due to the insulating effect of the packaging, and this may allow the growth of any microorganisms, if present. Provided the cooling medium (air, water, etc.) and refrigeration equipment used are kept sufficiently clean, no one cooling method can be said to be intrinsically more hygienic than any other. For unwrapped food, a rapid cooling system allows less time for any contamination/cross-contamination to occur than in slower cooling systems.

It is not unusual for food products (or ingredients found in food products) to be chilled or frozen a number of times before they reach the consumer. For example, during industrial processing, frozen raw material is often thawed or tempered before being turned into meat-based products, (e.g. pies, convenience meals, burgers, etc. or consumer portions, fillets, steaks, etc.). These consumer-sized portions are often refrozen before storage, distribution, and sale. As long as the maximum

food temperatures reached are below those that support pathogen growth, and the exposure times are short, food safety should not be compromised. However, there will be some reduction in the quality of the food. Whether this is commercially important and reduces the practical shelf life will depend on the product and the conditions of exposure.

5.6.1 Air systems

Air is by far the most widely used method of chilling and freezing food, as it is economical, hygienic, and relatively non-corrosive to equipment. Air systems range from the most basic, in which a fan draws air through a refrigerated coil and blows the cooled air around an insulated room, to purpose-built conveyorized air-blast tunnels or spirals. They range in size from 2 to 3 m³ batch cabinets (Figure 5.2), used in catering operations, to large continuous chilling systems capable of holding 15,000 whole poultry carcasses (Figure 5.3). Relatively low rates of heat transfer (Figure 5.4) are attained from product surfaces in

Figure 5.2 Simple batch air-cooling system for cooling trays of product.

Figure 5.3 Continuous air-chilling system for whole poultry carcasses.

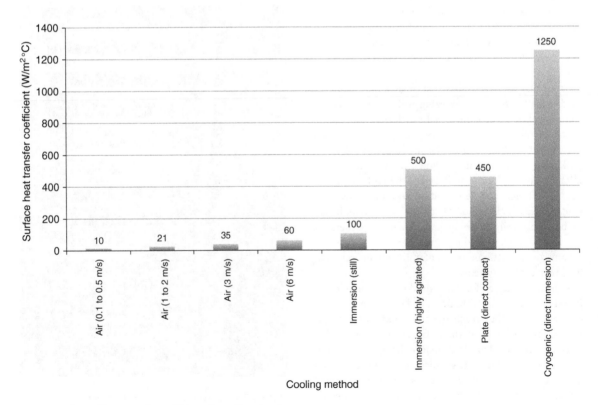

Figure 5.4 Typical heat transfer coefficients in cooling situations.

air-cooled systems. The big advantages of air systems are their lower cost and versatility compared to immersion, contact and cryogens, especially when there is a requirement to cool a variety of irregularly shaped products.

In general, relatively low rates of heat transfer (see Figure 5.1) are attained from product surfaces in air-cooled systems. In standard systems, air speeds are seldom faster than 6 ms^{-1}, but far higher air speeds (up to 30 ms^{-1}) are achievable in impingement systems, and thus surface heat transfer rates are far higher in impingement chillers. Impingement systems are best suited for products with high surface area to weight ratios (for example, products with one small dimension such as hamburger patties, pizzas, etc.). Testing has shown that products with a thickness less than 20 mm chill/freeze most effectively in an impingement heat transfer environment.

Fluidized bed freezing is a modification of air-blast freezing. The principle behind fluidization is that fairly uniform particles are subjected to an upward air stream. At a certain velocity, the particles will float in the air stream, each separated from the other, surrounded by air and free to move. In this state, the particles act in a similar fashion to a fluid, thus the term "fluidized." Products are fed into the higher end of a sloping tunnel, where they are simultaneously conveyed and frozen by the same air. Fluidized bed freezers can also be combined with a conveyorized system. Additional agitation, in the form of a movable base, may be required for some irregular products such as French fries (International Institute of Refrigeration, 2006).

Fluidized bed freezers are used to produce free-flowing products, most notably vegetables (such as peas, sliced carrots, green beans, etc.) and fruit, but can also be used for peeled cooked shrimps, diced meats, etc. (Hodgins, 1990; Persson & Löndahl, 1993). Products frozen by this method, as well as in other in-line blast freezers, are commonly referred to as individually quick frozen (IQF).

Higher air speeds are not suited for thick products where heat transfer within the product is the rate-limiting factor rather than that between the heat transfer medium and the product. For example, while increasing the air velocity during chilling of beef sides substantially reduces chilling times at low air velocity, the effect is smaller at higher velocities. Also, the power required by the fans to move the air within a chill room increases with the cube of the velocity. Thus, while a four-fold increase in air velocity from 0.5 to 2 ms^{-1} will result in a 4.4 h (18%) reduction in chilling time for a 140 kg side, it requires a 64-fold increase in fan power. In most practical situations, where large items are being cooled it is doubtful whether an air velocity greater than 1 ms^{-1} can be justified.

One of the principal disadvantages of air-cooling systems is their tendency to dehydrate unwrapped products. One solution to this problem is to saturate the air with water. Wet-air cooling systems recirculate air over ice-cold water so that the air leaving the cooler is cold (0–1 °C) and virtually saturated with water vapor (100% relative humidity, RH). An ice-bank chiller uses a refrigeration plant with an evaporator (plate or coil) immersed in a tank of water that chills the water to 0 °C. During times of low load, and overnight use of off-peak electricity, a store of ice is built up on the evaporator, which subsequently melts to maintain temperatures during times of high load.

5.6.2 Contact systems

Contact refrigeration methods are based on heat transfer by contact between products and metal surfaces, which in turn are cooled by either primary or secondary refrigerants. Contact cooling offers several advantages over air cooling, such as much better heat transfer (Figure 5.4) and significant energy savings. Contact cooling systems include plate coolers, jacketed heat exchangers, belt coolers, and falling film systems. Vertical plate freezers (Figure 5.5) are commonly used to freeze fish at sea, while horizontal systems are commonly used for meat blocks and ready meals.

Good heat transfer is dependent on product thickness, good contact, and the conductivity of the product. Plate freezers are often limited to a maximum thickness of 50–70 mm (Ciobanu, 1976c; International Institute of Refrigeration, 2006; Persson & Löndahl, 1993). Good contact is a prime requirement. Air spaces in packaging and fouling of the plates can have a significant effect on cooling time; for example, a water droplet frozen on the plate can lengthen the freezing time in the concerned tray by as much as 30–60% (Ciobanu, 1976c).

5.6.3 Immersion/spray systems

Immersion/spray systems involve dipping products into a cold liquid or spraying a cold liquid onto the food. Cooling using ice or direct contact with a cryogenic substance is essentially an immersion/spray process. These systems range in size from 2–3 m^3 tanks used to cool small batches of cooked products to large, continuous chilling systems capable of cooling 10,000 poultry carcasses per hour. This produces high rates of heat transfer due to the intimate

Figure 5.5 Vertical plate freezer.

Figure 5.6 Ice/water immersion cooling of whole fish.

contact between product and cooling medium. Both immersion and spray methods offer several inherent advantages over air cooling in terms of reduced dehydration and coil frosting problems (Robertson et al., 1976). Clearly, if the food is unwrapped, the liquid has to be a substance that is safe to ingest. The freezing point of the cooling medium used dictates its use for chilling or freezing. Obviously, any immersion/spray-freezing process must employ a medium at a temperature substantially

below 0 °C. This necessitates the use of non-toxic salt, sugar or alcohol solutions in water, or the use of cryogens.

5.6.3.1 Ice systems

Chilling with crushed ice, or an ice/water mixture, is simple, effective and commonly used for the cooling of fish (Figure 5.6), turkeys (Figure 5.7), and some fruits and vegetables. Cooling is more attributable to the contact

Figure 5.7 Immersion cooling of turkey carcasses.

between the fish and the cold melt water percolating through it (i.e. hydrocooling) than with the ice itself. The individual fish are packed in boxes between layers of crushed ice, which extract heat from the fish and consequently melt. Ice has the advantage of being able to deliver a large amount of refrigeration in a short time as well as maintaining a very constant temperature, 0 °C to −0.5 °C (where sea water is present).

5.6.3.2 Cryogenic systems

Direct spraying of liquid nitrogen onto a food product whilst it is conveyed through an insulated tunnel is one of the most commonly used methods of applying cryogens. The method of cooling is essentially similar to water-based evaporative cooling, the essential difference being the temperature required for boiling. As well as using the latent heat absorbed by the boiling liquid, sensible heat is absorbed by the resulting cold gas. Due to very low operating temperatures and high surface heat transfer coefficients between product and medium, cooling rates of cryogenic systems are often substantially

higher than other refrigeration systems. For example, an individual 1 cm thick beefburger takes less than 3 min to freeze in a cryogen system compared with up to 100 min in a cold store.

5.6.4 Vacuum systems

Food products having a large surface area to volume ratio (such as leafy vegetables) and an ability to readily release internal water are amenable to vacuum cooling. The products are placed in a vacuum chamber (typically operating at 530–670 N·m^{-2}), and the resultant evaporative cooling removes heat from the food. Evaporative cooling is quite significant; the amount of heat released through the evaporation of 1 g of water is equivalent to that released in cooling 548 g of water by 1 °C. Suitable products, such as lettuce, can be vacuum cooled in less than 1 h. In general terms, a 5 °C reduction in product temperature is achieved for every 1% of water evaporated. Since vacuum cooling requires the removal of water from the product, prewetting is commonly applied to prevent the removal of water from the tissue of the product.

Traditionally, this method of cooling has been relatively common for removing "field heat" of leafy vegetables immediately after harvest, but it is also suitable for many other foods, such as baked products, sauces, soups, particulate foods, and meat joints (James & James, 2002; Zheng & Sun, 2005). It is particularly useful for cooked fillings, stews, sauces and casseroles since pressure cooking and vacuum cooling can be combined in the same vessel, reducing both cooking and cooling times and saving space.

5.6.5 Scraped surface freezers

Scraped surface, or cylindrical, freezers are designed for freezing liquid products on either the inner or the outer surface of a cooled cylinder. The layer of frozen product formed on the surface of the cylinder is continuously scraped from the cylinder surface, thus achieving high heat transfer and a rapid freezing rate (Ciobanu, 1976c; International Institute of Refrigeration, 2006). Scraped surface freezers are used for manufacturing ice creams and similar products.

5.6.6 High-pressure freezing systems

High-pressure freezing, and in particular "pressure shift" freezing, is attracting considerable scientific interest (Otero & Sanz, 2012). The food is cooled under high pressure to sub-zero temperatures but does not undergo a phase change and freeze until the pressure is released. Rapid nucleation yields small homogeneous ice crystals. However, studies on pork and beef have failed to show any real commercial quality advantages (Fernandez-Martin et al., 2006).

5.7 Chilled and frozen storage systems

A typical United Kingdom cold store usually has 75,000 m^3 of storage space and is fitted with 10–14 m long mobile racks. A typical European system is almost three times as large (200,000 m^3) and has 32–38 m long automated racks. The size of a cold store has an effect on the overall heat load through the insulation. A 2830 m^3 cold store uses 124 kWh per m^3 per year, whereas a 85,000 m^3 store uses 99 kWh/ m^3. Large refrigerated (chilled and frozen) distribution centers (Figure 5.8) are increasingly used by large food retailers and serve the dual purpose of a food store and a marshaling yard. At the other end of the cold chain, many millions (over 500,000 in the UK alone) of refrigerated commercial service cabinets are used to store food and/or drink in commercial catering facilities. The majority of the market is for chilled or frozen upright cabinets with one or two doors (between 400 and 600 L for single-door cabinets and 1300 L for double-door cabinets) or undercounter units with up to four doors (150–800 L).

There are clear differences between the environmental conditions required for cooling (chilling or freezing), which is a heat removal/temperature reduction process, and those required for storage, where the aim is to maintain a set product temperature. Three factors during storage, the storage temperature, the degree of fluctuation in the storage temperature and the type of wrapping/packaging in which the food is stored, are commonly believed to have the main influence on storage life.

Publications such as the International Institute of Refrigeration (IIR) *Recommendations for the Chilled Storage of Perishable Produce* (2000) provide data on the storage life of many foods at different temperatures. The storage life of most chilled foods is limited by the growth of spoilage microorganisms. However, with unwrapped food, dehydration of the surface layers may lead to unacceptable color changes. Poor temperature control can also lead to color problems. Rosenvoid and Wiklund (2011) stated that "an increase in the storage temperature from the ideal temperature of −1.5 °C to 2 °C significantly decreased the color stability of lamb loins. Even one week at 2 °C at the end of the storage period had a substantial negative impact on the retail color display life."

Although freezing inhibits some enzymes, a considerable number, such as invertases, lipases, lipoxidases, catalases, peroxidases, etc., remain active (Ciobanu, 1976a). Because frozen foods are stored for relatively long periods, these enzymic reactions, although slow (reactions that take 45 min at 37 °C will take over a week at −29 °C),

Figure 5.8 Refrigerated distribution store serving retail shops.

can cause significant problems. For example, lipases, responsible for breaking down fats, remain active at very low temperatures. The storage life of many frozen foods is limited by these changes, for example rancidity development in the fat of meat. Publications such as the IIR *Recommendations for the Processing and Handling of Frozen Foods* (2006) provide data on the storage life of many foods at different temperatures. Storage lives for frozen food can be as short as 3–4 months for individually quick-frozen, polybag-packed shrimps at −18 °C. On the other hand, lamb stored at −25 °C can be kept for over 2.5 years.

Frozen foods are often stored at or near −18 °C; the reason for the use of this temperature is mainly historic since it is 0 °F. There is an assumption that a low storage temperature is always beneficial to storage life. This is seen to be an oversimplification by some (Jul, 1982), and is not always so. There is a growing realization that storage lives of several foods can be less dependent on temperature than previously thought. Some products such as cured meats often have "reverse stability," i.e. sliced bacon may keep longer at −12 °C than at −20 or −30 °C. While other products may benefit from temperatures as low as −30 °C, storage below such temperatures may not have a substantial effect on storage life. Since research has shown that many food products, such as red meats, often produce non-linear time-temperature curves, there is probably an optimum storage temperature for a particular food product. Improved packing and preservation of products can also increase storage life and may allow higher storage temperatures to be used.

5.8 Chilled and frozen transport systems

Over a million refrigerated road vehicles, 400,000 refrigerated containers and many thousands of other forms of refrigerated transport systems are used to distribute refrigerated foods throughout the world (Gac, 2002). All these transportation systems are expected to maintain the temperature of the food within close limits to ensure its optimum safety and high-quality shelf life. Developments in temperature-controlled transportation systems for chilled products have led to the rapid expansion of the chilled food market.

It is particularly important that the food is at the correct temperature before loading since the refrigeration systems used in most transport containers are not designed to extract heat from the load but to maintain its temperature. Irrespective of the type of refrigeration equipment used,

the product will not be maintained at its desired temperature during transportation unless it is surrounded by air or surfaces at or below the maximum transportation temperature. This is usually achieved by a system that circulates air, either forced or by gravity, around the load. Inadequate air distribution is a major cause of product deterioration and loss of shelf life during transport. If products have been cooled to the correct temperature before loading and do not generate heat then they only have to be isolated from external heat ingress. Surrounding them with a blanket of cooled air achieves this purpose. Care has to be taken during loading to prevent any product from touching the inner surfaces of the vehicle because this would allow heat ingress by conduction during transport.

In the large containers used for long-distance transportation, food temperatures can be kept within ±0.5 °C of the set point. With this degree of temperature control, transportation times of 8–14 weeks (for vacuum-packed meats stored at −1.5 °C) are possible while still retaining a sufficient chilled storage life for retail display. Products such as fruits and vegetables that produce heat by respiration, or products that have to be cooled during transit, also require circulation of air through the product. This can be achieved by directing the supply air through ducts to channels at floor level or in the floor itself. In general, it is not advisable to rely on product cooling during transportation.

5.8.1 Sea transport

Recent developments in temperature control, packaging, and controlled atmospheres have substantially increased the range of foods that can be transported around the world in a chilled condition. Control of the oxygen and carbon dioxide levels in shipboard containers has allowed fruits and vegetables, such as apples, pears, avocados, melons, mangos, nectarines, blueberries and asparagus, to be shipped (typically 40 days in the container) from Australia and New Zealand to markets in the USA, Europe, Middle East, and Japan (Adams, 1988). If the correct varieties are selected and rapidly cooled immediately after harvest, the product arrives in good condition and has a long subsequent shelf life. With conventional vacuum packaging, it is difficult to achieve a shelf life in excess of 12 weeks with beef and 8 weeks for lamb (Gill, 1984). However, a shelf life of up to 23 weeks at −2 °C can be achieved in cuts of lamb individually packed in evacuated bags of linear polyethylene, and then placed in gas-flushed foil laminate bags filled with a volume of

Figure 5.9 Refrigerated container, or reefer, showing refrigeration unit (courtesy of Sarah Klinge, Klinge Corporation).

CO_2 approximately equal to that of the meat (Gill & Penney, 1986). Similar storage lives are currently being achieved with beef primals transported from Australia and South Africa to the European Union (EU).

Most International Standard Organization (ISO) containers are either "refrigerated" or "insulated." The refrigerated containers (reefers) have refrigeration units built into their structure (Figure 5.9). The units operate electrically, either from an external power supply on board the ship or dock or from a generator on a road vehicle. Insulated containers either utilize the plug-type refrigeration units already described or may be connected directly to an air-handling system in a ship's hold or at the docks. Close temperature control is most easily achieved in insulated containers that are placed in insulated holds and connected to the ship's refrigeration system. However, suitable refrigeration facilities must be available for any overland sections of the journey. When the containers are fully loaded and the cooled air is forced uniformly through the spaces between cartons, the maximum difference between delivery and return air can be less than $0.8\,^{\circ}C$ (Heap, 1986). The entire product in a container can be maintained to within $\pm 1.0\,^{\circ}C$ of the set point.

Refrigerated containers are easier to transport overland than the insulated types, but have to be carried on deck when shipped because of problems in operating refrigeration units within closed holds. Therefore, on board ship they are subjected to much higher ambient temperatures, and consequently larger heat gains, than containers held below deck, which makes it far more difficult to control product temperatures.

5.8.2 Air transport

Airfreighting is increasingly being used for high-value perishable products, such as strawberries, asparagus, and live lobsters (Sharp, 1988; Stera, 1999). However, foods do not necessarily have to fall into this category to make air transportation viable since it has been shown that "the intrinsic value of an item has little to do with whether or not it can benefit from air shipment, the deciding factor is not price but mark-up and profit" (ASHRAE, 2006). Perishables account for approximately 14–18% of the total worldwide air cargo traffic. Although airfreighting of foods offers a rapid method of serving distant markets, there are many problems because the product is unprotected by refrigeration for much of its journey. Up to 80% of the total journey time is made up of waiting on the tarmac and transport to and from the airport. During flight, the hold is normally between $15\,^{\circ}C$ and $20\,^{\circ}C$. Perishable cargo is usually carried in standard containers, sometimes with an insulating lining and/or ice or dry ice, but is often unprotected on aircraft pallets (Sharp, 1988). Thus it is important that:
- the product be transported in insulated containers to reduce heat gain
- the product be precooled and held at the required temperature until loading
- containers be filled to capacity
- a thermograph should accompany each consignment.

5.8.3 Land transport

Overland transportation systems range from 12 m refrigerated containers for long-distance road or rail movement of bulk chilled or frozen products, to small uninsulated vans supplying food to local retail outlets or even directly to the consumer. Most current road transport vehicles for refrigerated foods are refrigerated using mechanical, eutectic plates or solid carbon dioxide or liquid nitrogen cooling systems.

Many advantages are claimed for liquid nitrogen transport systems, including minimal maintenance requirements, uniform cargo temperatures, silent operation, low capital costs, environmental acceptability, rapid temperature reduction, and increased shelf life due to the modified atmosphere (Smith, 1986). Some studies have claimed costs to be comparable with mechanical systems (Smith, 1986), whilst others have reported costs to be up to 2.2 times higher (Nieboer, 1988).

In eutectic refrigeration systems, a freezing mixture (such as a salt solution) with a known freezing point (its eutectic point) is contained in plates that are typically fitted to the ceiling of the container or vehicle. Power is used to freeze the eutectic whilst the vehicle is in a static situation, such as overnight. During the vehicle's working day, no power is required, nor noise created, as the frozen solid gradually thaws until reaching its eutectic point. At this stage, large amounts of heat are absorbed as the mixture begins to change its state whilst remaining at a constant temperature.

Large retailers supply their stores in a single load, using large vehicles with variable temperature (ambient, chilled, and frozen) compartments, from central distribution depots. However, many small retail stores, garage outlets, etc. are supplied by sales vans. These are small to medium size refrigerated vehicles that are loaded with products in the morning and travel around to a series of retail outlets selling to each in turn. They therefore have a large number of stops when the doors are opened and food is removed from the van. Sometimes, food that has passed its sell-by date and empty trays are returned from the shops to the vans. The insulation, door protection, and refrigeration plant fitted to the vans have sometimes proved inadequate to maintain food temperature as cold as required. It is a problem for operators of the vans to know in advance whether a particular van, on a particular round, under given ambient conditions, will be able to deliver food at the correct temperature. The rise in supermarket home delivery services, where there are requirements for mixed loads of products that may each require different storage temperatures, is introducing a new complexity to local land delivery (Cairns, 1996).

5.9 Refrigerated retail display systems

Refrigeration systems that include display fixtures in the sales area and systems serving the cold rooms are the major energy-consuming equipment in supermarkets. Refrigerated display equipment in supermarkets and other smaller food retail outlets (Figure 5.10) can be classified as "integral," where all the refrigeration components are housed within the stand-alone fixture, or "remote," where the evaporator or cooling coils within the display fixtures are served by refrigeration equipment located remotely in a plant room. The main advantages of integral units are the flexibility they offer in merchandizing, their relatively low cost and their relatively low refrigerant inventory and much lower potential leak rate compared to centralized systems. Their main disadvantage is the low efficiency of the compressors compared to large centralized compressors, noise and heat ejection in the store, which increases cooling requirements in the summer. Although small food retail outlets invariably use "integral" refrigeration equipment, larger food retail stores predominantly use centralized equipment of much more

Figure 5.10 Typical serve-over refrigerated retail cabinet display.

sophisticated technology plus a small number of integrals for spot merchandizing.

Centralized systems provide the flexibility of installing the compressors and condensers in a centralized plant area, usually at the back of the store or on a mezzanine floor or roof. The evaporators in the refrigerated display fixtures and cold rooms are fed with refrigerant from the central plant through distribution pipework installed under the floor or along the ceiling of the sales area. In the plant room, multiple refrigeration compressors, using common suction and discharge manifolds, are mounted on bases or racks normally known as compressor "packs" or "racks," which also contain all the necessary piping, valves, and electrical components needed for the operation and control of the compressors. Air-cooled or evaporative-cooling condensers used in conjunction with the multiple compressor systems are installed remotely from the compressors, usually on the roof of the plant room. A schematic diagram of the direct expansion (DX) centralized system is shown in Figure 5.11. Separate compressor packs are used for chilled and frozen food applications. Most large supermarkets will have at least two packs to serve the chilled food cabinets and one or two packs to serve the frozen food cabinets.

A major disadvantage of the centralized DX system is the large quantity of refrigerant required, 4–5 kg/kW refrigeration capacity and the large annual leakage rates of between 10% and 30% of total refrigerant charge. One way of significantly reducing the refrigerant charge in supermarket refrigeration systems is to use a secondary or indirect system arrangement. With this arrangement, shown schematically in Figure 5.12, a primary system can be located in a plant room or the roof and can use natural refrigerants such as hydrocarbons or ammonia to cool a secondary fluid, which is circulated to the coils in the display cabinets and cold rooms. Separate refrigeration systems and brine loops are used for the medium- and low-temperature display cabinets and other refrigerated fixtures.

The temperature of individual consumer packs, small individual items and especially thin sliced products responds very quickly to small amounts of added heat. All these products are commonly found in retail display cabinets and marketing constraints require that they have maximum visibility. Maintaining the temperature

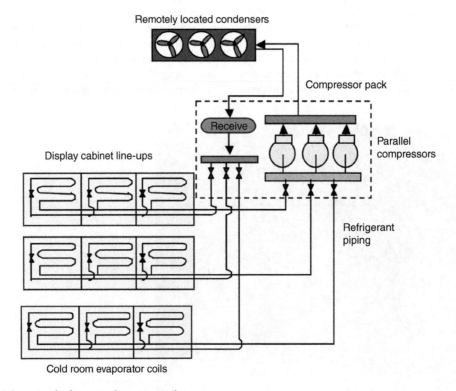

Figure 5.11 Schematic of refrigeration layout in retail store.

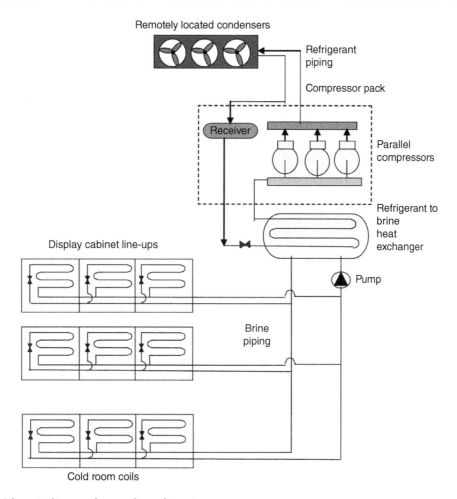

Figure 5.12 Schematic diagram of a secondary refrigeration system.

of products below set limits while they are on open display in a heated store will always be a difficult task. Attempts to improve the maintenance of food temperature include the use of phase change materials and heat pipes in the display shelves (Lu et al., 2010). These improve temperature distribution throughout the cabinet. The use of a heat pipe shelf is claimed to reduce food temperature by between 3.0 °C and 5.5 °C (Lu et al., 2010).

Average air temperatures in chilled displays can vary considerably from cabinet to cabinet, with inlet and outlet values ranging from −6.7 °C to +6.0 °C and −0.3 °C to +7.8 °C, respectively, in one survey (Lyons & Drew, 1985), while other studies reported temperatures ranging from −1.2 °C to 19.2 °C in refrigerated displays for fruit and vegetables (Nunes et al., 2009). A recent survey (Morelli et al., 2012) of 20 Parisian food retail shops

reported that 70% of the time-temperature profiles obtained exceeded 7 °C. Evans et al. (2007) carried out an analysis of the performance of well freezers, chest freezers, frozen and chilled door cabinets (solid or glass door), and open-fronted chilled cabinets under EN441 standard test conditions. This revealed that maximum air and product temperatures in cabinets were generally in the most exposed (to ambient) areas and that minimum temperatures were located in the least exposed areas.

The temperature performance of an individual display cabinet does not only depend on its design. Its position within a store and the way the products are positioned within the display area significantly influence product temperatures. In non-integral (remote) cabinets (i.e. those without built-in refrigeration systems), the design and performance of the central refrigeration system are also critical to effective temperature control.

The desired chilled display life for wrapped meat, fish, vegetables, and processed foods ranges from a few days to weeks and is primarily limited by microbiological considerations. Retailers of unwrapped fish, meat, and delicatessen products normally require a display life of one working day, which is often restricted by appearance changes. Frozen food can potentially be displayed for many weeks.

Reducing energy consumption in a chilled multideck cabinet is substantially different from reducing it in a frozen well cabinet (James et al., 2009). Improvements have been made in insulation, fans, and energy-efficient lighting but only 10% of the heat load on a chilled multideck comes from these sources, compared with 30% on the frozen well. Research efforts are concentrating on minimizing infiltration through the open front of multideck chill cabinets, by the optimization of air curtains and airflows (Gaspar et al., 2011), since this is the source of 80% of the heat load. Another effective method of reducing the refrigeration load is to fit doors to the cabinets. In frozen well cabinets, reducing heat radiation onto the surface of the food, accounting for over 40% of the heat load, is a major challenge.

5.9.1 Unwrapped products

Display cabinets for delicatessen products (see Figure 5.10) are available with gravity or forced convection coils and the glass fronts may be nearly vertical or angled up to 20°. In the gravity cabinet, cooled air from the raised rear-mounted evaporator coil descends into the display well by natural convection and the warm air rises back to the evaporator. In forced circulation cabinets, air is drawn through an evaporator coil by a fan and then ducted into the rear of the display, returning to the coil after passing directly over the products, or forming an air curtain, via a slot in the front of the cabinet and a duct under the display shelf.

Changes in product appearance are normally the criteria that limit the display life of unwrapped foods, with consumers selecting newly loaded product in preference to that displayed for some time. Deterioration in appearance has been related to degree of dehydration in red meat (Table 5.4) and is likely to occur similarly in other foods. Apart from any relationship to appearance, weight loss is of considerable importance in its own right. Relative humidity in delicatessen retail cabinets has been shown to have more effect on weight loss from displayed

Table 5.4 Relationship between weight loss and change in appearance of beef topside

Loss (g · cm^{-2})	Change in appearance
<0.01	Red, attractive and still wet, some brightness loss
0.015–0.025	Surface becoming drier, still attractive but darker
0.025–0.035	Distinct obvious darkening, becoming dry and leathery
0.05	Dry blackening
0.05–0.10	Black

products than air speed or temperature. Reducing the RH from 95% to 40% increases weight loss over a 6 h display period by a factor of between 14 and 18 (James & Swain, 1986).

There is a conflict between the need to make the display attractive and convenient to increase sales appeal and the optimum display conditions for the product. High lighting levels increase the heat load and the consequent temperature rise dehumidifies the refrigerated air. The introduction of humidification systems can significantly improve display life of unwrapped products (Brown et al., 2005) and studies have been carried out to optimize the humidification process (Moureh et al., 2009).

5.9.2 Wrapped products

To achieve the display life of days to weeks required for wrapped chilled foods, the product should be maintained at a temperature as close to its initial freezing point as possible to prevent microbial spoilage. In some cases, e.g. particular cheeses, dairy products and tropical fruits, quality problems may limit the minimum temperature that can be used, but for the majority of meat, fish and processed foods, the range −1 °C to 0 °C is desirable.

Air movement and relative humidity have little effect on the display life of a wrapped product, but the degree of temperature control can be important, especially with transparent, controlled atmosphere packs. Large temperature cycles will cause water loss from the product and this water vapor will condense on the inner surface of the pack, consequently reducing consumer appeal. Although cabinets of the type described for delicatessen products can be used for wrapped foods, most are sold from multideck cabinets with single or twin air curtain systems. Twin air curtains tend to provide more constant product temperatures but are more expensive. It is important that the front edges of the cabinet shelves do not project through the air curtain since the refrigerated air will then be diverted out of the cabinet. On the other hand, if narrow shelves are used, the curtain may collapse and ambient air can be drawn into the display well.

To maintain product temperatures close to 0 °C, the air off the evaporator coil must typically be −4 °C and any ingress of humid air from within the store will quickly cause the coil to ice up. Frequent defrosts are often required and even in a well-maintained unit, the cabinet temperature may rise to 10–12 °C and the product by at least 3 °C (Brolls, 1986). External factors such as the store ambient temperature, the position of the cabinet and poor pretreatment and placement of products substantially affect cabinet performance. Warm and humid ambient air, and loading with insufficiently cooled products, can also overload the refrigeration system. Even if the food is at its correct temperature, uneven loading or too much product can disturb the airflow patterns and destroy the insulating layer of cooled air surrounding the product. An in-store survey of 299 prepackaged meat products in chilled retail displays found product temperatures in the range −8.0 °C to 14.0 °C, with a mean of 5.3 °C and 18% above 9 °C (Rose, 1986). Other surveys (Bøgh-Sørensen, 1980; Malton, 1971) have shown that temperatures of packs from the top of stacks were appreciably higher than those from below due to radiant heat pick-up from store and cabinet lighting. It has also been stated that products in transparent film overwrapped packs can achieve temperatures above that of the surrounding refrigerated air due to radiant heat trapped in the package by the "greenhouse" effect. However, specific investigations have failed to demonstrate this effect (Gill, 1988).

5.9.3 Frozen foods

No frozen food, with the possible exception of ice cream, should be unwrapped when in a retail display cabinet. Traditionally, frozen foods were displayed in a "well-type" cabinet with only the top faces of food packs being exposed. In many cases the cabinets were fitted with a see-through insulated lid to further reduce heat infiltration. Increasingly, there is marketing pressure to display more frozen food in open multideck display cabinets. The rate of heat gain in a multideck cabinet, and consequently the energy consumption, is much higher than in a well cabinet.

Maintaining the temperature of frozen products below set limits while they are on open display in a heated store will always be a difficult task. Radiant heat gain on the surfaces of exposed packs can result in the food thawing in extreme cases. During display, temperature, temperature fluctuations and packaging are the main parameters that control quality.

Temperature fluctuations can increase the rate of weight loss from wrapped meat. Higher rates of dehydration have been measured in a retail cabinet operating at −15 °C than another cabinet operating at −8 °C. Fluctuations in air temperature in the −15 °C cabinet ranged from −5 °C to −21 °C compared with ±1.5 °C in the −8 °C cabinet. Successive evaporation and condensation (as frost) caused by wide temperature differentials resulted in exaggerated in-package dehydration (Cutting & Malton, 1972).

The extent of temperature fluctuations will be dependent upon the air temperature over the product, the product packaging, and the level of radiant heat. Retail display packs have a relatively small thermal mass and respond relatively quickly to external temperature changes. These can be from store and display lighting, defrost cycles, and heat infiltration from the store environment. In products where air gaps exist between the packaging and the meat, sublimation of ice within the product leads to condensation on the inside of the packaging, resulting in a build-up of frost. This dehydration causes small fissures in the surface of the food, allowing the ingress of any packaging gases into the food. This can aid the acceleration of oxidative rancidity within the product. Minor product temperature fluctuations are generally considered to be unimportant, especially if the product is stored below −18 °C and fluctuations do not exceed 2 °C.

5.10 Recommended temperatures

Publications such as the IIR *Recommendations for the Chilled Storage of Perishable Produce* (2000) and *Recommendations for the Processing and Handling of Frozen Foods* (2006) provide data on the storage life of many foods at different temperatures. The Agreement on the International Carriage of Perishable Foodstuffs (ATP Agreement; United Nations, 2012) specifies maxima temperatures for the transportation of chilled and frozen foods (Table 5.5). These temperatures are also a good guideline to follow during storage and retail display of such foods.

Table 5.5 Maxima temperatures for the transportation of chilled and frozen foods specified in the Agreement on the International Carriage of Perishable Foodstuffs (ATP Agreement)

	Maximum temperature (°C)	Description
Chilled foods	7	Red meat and large game (other than red offal)
	6	Raw milk
	6*	Meat products, pasteurized milk, fresh dairy products (yoghurt, kefir, cream, fresh cheese), ready-cooked foodstuffs (meat, fish, vegetables), ready-to-eat (RTE) prepared raw vegetables and vegetable products, concentrated fruit juice and fish products not listed
	4	Poultry, game (other than large game), rabbits
	3	Red offal
	2*	Minced meat
	0**	Untreated fish, molluscs, and crustaceans
Frozen foods	−20	Ice cream
	−18	Frozen or quick (deep)-frozen fish, fish products, molluscs and crustaceans and all other quick (deep)-frozen foodstuffs
	−12	All other frozen foods (except butter)
	−10	Butter

* Or at temperature indicated on the label and/or on the transport documents; ** at temperature of melting ice.

5.11 Refrigeration and the environment

The dominant types of refrigerant used in the food industry in the last 60 years have belonged to a group of chemicals known as halogenated hydrocarbons. Members of this group, which includes the chlorofluorocarbons (CFCs) and the hydrochlorofluorocarbons (HCFCs), have excellent properties, such as low toxicity, compatibility with lubricants, high stability, good thermodynamic performance, and relatively low cost, which make them excellent refrigerants for industrial, commercial, and domestic use. However, their high chemical stability leads to environmental problems when they are released and rise into the stratosphere. Scientific evidence clearly shows that emissions of CFCs have been damaging the ozone layer and contributing significantly to global warming. With the removal of CFCs from aerosols, foam blowing and solvents, the largest single application sector in the world is refrigeration, which accounts for almost 30% of total consumption.

Until recently, R12 (dichlorodifluoromethane), R22 (chlorodifluoromethane) and R502 (a mixture of R22 and R115 chloropentafluoroethane) were the three most common refrigerants used in the food industry. R12 and R502 have significant ozone depletion potential (ODP) and global warning potential (GWP); although R22 has less ODP and GWP than the other refrigerants,

it is still considered to have a significant impact in the long term. Two international agreements, the Montreal and the Kyoto Protocols, have had substantial effects. The Montreal Protocol was originally signed in 1987, and substantially amended in 1990 and 1992; it banned CFCs, halons, carbon tetrachloride, and methyl chloroform by 2000 (2005 for methyl chloroform). The Kyoto Protocol was negotiated by 160 nations in December 1997 and reduced net emissions of certain greenhouse gases, primarily CO_2 (plus methane, nitrous oxide, HFCs, perfluorocarbons, sulfur hexafluoride). Consequently, as a result of these international agreements, pure CFCs (e.g. R12, R502) have been completely banned. Pure HCFCs (mainly R22) are banned in new industrial plant and are being phased out completely in existing plant. HCFC blends and HFC blends originally introduced as CFC replacements are covered by the F-Gas regulations that limit leak rates.

The aim of the F-Gas regulations is to help the EU to meet its Kyoto Protocol target of reducing emissions of greenhouse gases by 8% from baseline levels by 2008–12. They contain requirements to minimize emissions of fluorinated greenhouse gases and for handlers of F-gases to be qualified and for monitoring of plant and gases (containment, training, labeling, reporting). Chemical companies are making large investments in terms of both time and money in developing new refrigerants that have reduced or negligible environmental effects. Other researchers are looking at the many non-CFC alternatives, including ammonia, propane, butane, carbon dioxide, water, and air, that have been used in the past.

Ammonia is increasingly a common refrigerant in large industrial food cooling and storage plants. It is a cheap, efficient refrigerant whose pungent odor aids leak detection well before toxic exposure or flammable concentrations are reached. The renewed interest in this refrigerant has led to the development of compact, low-charge (i.e. small amounts of ammonia) systems that significantly reduce the possible hazards in the event of leakage. It is expected that ammonia will meet increasing use in large industrial food refrigeration systems. Carbon dioxide is being advocated for retail display cabinets and hydrocarbons, particularly propane and butane or mixtures of both, for domestic refrigerators.

As well as the direct affect of refrigerants on the environment, energy efficiency is increasingly of concern to the food industry. Worldwide, it is estimated that 40% of all food requires refrigeration and 15% of the electricity consumed worldwide is used for refrigeration (Mattarolo, 1990). In the UK, 11% of electricity is consumed by the food industry (DBERR, 2005). However, detailed estimates of what proportion of this is used for refrigeration processes are less clear and often contradictory (James et al., 2009). On the best available data, the energy-saving potential in the top five refrigeration operations (retail, catering, transport, storage, and primary chilling), in terms of the potential to reduce energy consumed, lies between 4300 and 8500 GWh/y in the UK (James et al., 2009).

It is clear that maintenance of food refrigeration systems will reduce energy consumption (James et al., 2009). Repairing door seals and door curtains, ensuring that doors can be closed and cleaning condensers produce significant reductions in energy consumption. In large cold storage sites, it has been shown that energy can be substantially reduced if door protection is improved, pedestrian doors fitted, liquid pressure amplification pumps fitted, defrosts optimized, suction liquid heat exchangers fitted, and other minor issues corrected (James et al., 2009).

In the retail environment, the majority of the refrigeration energy is consumed in chilled and frozen retail display cabinets (James et al., 2009). Laboratory trials have revealed large, up to six-fold, differences in the energy consumption of frozen food display cabinets of similar display areas. In chilled retail display, which accounts for a larger share of the market, similar large differences,

up to five-fold, were measured. A substantial energy saving can therefore be achieved by simply informing and encouraging retailers to replace energy-inefficient cabinets by the best currently available.

New/alternative refrigeration systems/cycles, such as trigeneration, air cycle, sorption-adsorption systems, thermoelectric, Stirling cycle, thermoacoustic and magnetic refrigeration, have the potential to save energy in the future if applied to food refrigeration (Tassou et al., 2009). However, none appears likely to produce a step change reduction in refrigeration energy consumption in the food industry within the next decade.

5.12 Specifying, designing, and commissioning refrigeration systems

In the author's experience, the poor performance of new refrigeration systems used to maintain the cold chain can often be linked to a poor, non-existent, or ambiguous process specification. In older systems, poor performance is often due to a change in use that was not considered in the original specification. There are three stages in obtaining a refrigeration system that works.

1. Determining the process specification, i.e. specifying exactly the condition of the product(s) when they enter and exit the system, and the amounts that have to be processed.

2. Drawing up the engineering specification, i.e. turning processing conditions into terms that a refrigeration engineer can understand, independent of the food process.

3. The procurement and commissioning of the total system, including any services or utilities.

The first task in designing a system is the preparation of a clear specification by the user of how the facility will be used at present and in the foreseeable future. In preparing this specification, the user should consult with all parties concerned; these may be officials enforcing legislation, customers, other departments within the company and engineering consultants or contractors, but the ultimate decisions taken in forming this specification are the user's alone.

The process specification must include, as a minimum, data on the food(s) to be refrigerated, in terms of size, shape, and throughput. The maximum capacity must be catered for and the refrigeration system should also be specified to operate adequately and economically at all other throughputs. The range of temperature requirements for each product must also be clearly stated. If it is intended to minimize loss, it is useful to quantify at an early stage how much extra money can be spent to save a given amount of weight. All the information collected so far, and the decisions taken, will be on existing production. Another question that needs to be asked is, "Will there be any changes in the use of the system in the future?".

The refrigeration system chiller, freezer, storeroom, etc. is one operation in a sequence of operations. It influences the whole production process and interacts with it. An idea must be obtained of how the system will be loaded, unloaded and cleaned, and these operations must always be intimately involved with those of the rest of the production process. There is often a conflict of interest in the usage of a chiller or freezer. In practice, a chiller/freezer can often be used as a marshaling yard for sorting orders, and as a place for storing product not sold. If it is intended that either of these operations is to take place in the chiller/freezer, the design must be made much more flexible in order to cover the conditions needed in a marshaling area or a refrigerated store. In the case of a batch or semi-continuous operation, holding areas may be required at the beginning and end of the process in order to even out flows of material from adjacent processes. The time available for the process will be in part dictated by the space that is available; a slow process will take more space than a fast process, for a given throughput.

Other refrigeration loads, in addition to that caused by the input of heat from the product, also need to be specified. Many of these, such as infiltration through openings, the use of lights, machinery, and people working in the refrigerated space, are all under the control of the user and must be specified so that the heat load given off by them can be incorporated in the final design. Ideally, all the loads should then be summed together on a time basis to produce a load profile. If the refrigeration process is to be incorporated with all other processes within a plant, in order to achieve an economic solution, then the load profile is important. The ambient design conditions, e.g. the temperatures adjacent to the refrigerated equipment and the temperatures of the ambient to which heat will ultimately be rejected, must be specified. If the process is to be integrated with heat reclamation then the temperature of the heat sinks must be specified. Finally, the defrost regime should also be specified. There are times in any process where it is critical that coil defrosting and its accompanying temperature rise do not take place, and that the coil is cleared of frost before commencing the specified part of the process.

Although it is common practice throughout the food industry to leave much of this specification to refrigeration contractors or engineering specialists, the end user should specify all the above requirements. The refrigeration contractors or engineering specialists can give good advice on this. However, since all the above are outside their control, the end user, using their knowledge of how well they can control their overall process, should always take the final decision.

The aim of drawing up an engineering specification is to turn the user requirements into a specification that any refrigeration engineer can then use to design a system. The first step in this process is iterative. A full range of time, temperature, and air velocity options must be assembled for each cooling specification covering the complete range of each product. Each option must then be evaluated against the user requirements. If they are not a fit then another option is selected and the process repeated. If there are no more options available, there are only two alternatives: either standards must be lowered (recognizing in do so that cooling specifications will not be met) or the factory operation must be altered.

A full engineering specification will typically include: the environmental conditions within the refrigerated enclosure, air temperature, air velocity, and humidity; the way the air will move within the refrigerated enclosure; the size of the equipment; the refrigeration load profile; the ambient design conditions; and the defrost requirements. The final phase of the engineering specification should include a schedule for testing the engineering specification prior to handing over the equipment to the owner/operator. This test will be in engineering and not product terms. The specification produced should be the document that forms the basis for quotations and finally the contract between the user and his or her contractor and must be stated in terms that are objectively measurable once the chiller/freezer is completed. Arguments often ensue between contractors and their clients arising from an unclear, ambiguous or unenforceable specification. Such lack of clarity is often expensive to all parties and should be avoided.

5.13 Conclusion

Chilling and freezing are two of the most common methods for preserving foods. Carried out correctly, they can provide a high-quality, nutritious, and safe product for consumption with a long storage life.

The principal factor controlling the safety and quality of a refrigerated (chilled or frozen) food is its temperature. The principle of the refrigerated preservation of foods is to reduce, and maintain, the temperature of the food such that it stops, or significantly reduces, the rate at which detrimental changes occur in the food. In many cases, the time taken to reach the desired temperature is also important. To provide safe, high-quality refrigerated food products, attention must be paid to every aspect of the cold chain from initial chilling or freezing of the raw ingredients, through storage and transport, to retail display.

References

Adams GR (1988) Controlled atmosphere container. In: *Refrigeration for Food and People*. Meeting of IIR Commissions C2, D1, D2/3, E1, Brisbane, Australia, pp. 244–248.

ASHRAE (2006) *ASHRAE Handbook – Refrigeration*. Atlanta, GA: American Society of Heating, Refrigerating and Air-Conditioning Engineers.

Bendall JR (1972) The influence of rate of chilling on the development of rigor and cold shortening. In: *Meat Research Institute Symposium No. 2. Meat Chilling – Why and How?* Bristol: Meat Research Institute, pp. 3.1–3.6.

Berry M, Fletcher J, McClure P, Wilkinson J (2008) Effect of freezing on nutritional and microbiological properties of foods. In: Evans JA (ed) *Frozen Food Science and Technology*. London: Blackwell Publishing, pp. 26–50.

Bøgh-Sørensen L (1980) Product temperatures in chilled cabinets. In: *Proceedings of the 26th European Meeting of Meat Research Workers*, Colorado Springs (USA) n.22. Champaign, IL: American Meat Association.

Brolls EK (1986) Factors affecting retail display cases. In: *Recent Advances and Developments in the Refrigeration of Meat by Chilling*. Meeting of IIR Commission C2, Bristol, Section 9, pp. 405–413.

Brown T, Corry J, James SJ (2005) Humidification of chilled fruit and vegetables on retail display using an ultrasonic fogging system with water/air ozonation. *International Journal of Refrigeration* 27(8): 862–868.

Cairns S (1996) Delivering alternatives: success and failures of home delivery services for food shopping. *Transport Policy* 3: 155–176.

Cano MP (1996) Vegetables. In: Jeremiah O (ed) *Freezing Effects on Food Quality*, New York: Marcel Dekker, pp. 247–298.

Ciobanu A (1976a) Basic principles of food preservation by refrigeration. In: Ciobanu A, Lascu G, Bercescu V, Niculescu L (eds) *Cooling Technology in the Food Industry*. Tunbridge Wells: Abacus Press, pp. 11–22.

Ciobanu A (1976b) Chilling. In: Ciobanu A, Lascu G, Bercescu V, Niculescu L (eds) *Cooling Technology in the Food Industry*. Tunbridge Wells: Abacus Press, pp. 101–137.

Ciobanu A (1976c) Freezing. In: Ciobanu A, Lascu G, Bercescu V, Niculescu L (eds) *Cooling Technology in the Food Industry*. Tunbridge Wells: Abacus Press, pp. 139–222.

Cutting CL, Malton R (1972) Recent observations on UK meat transport. In: *Meat Research Institute Symposium No. 2. Meat Chilling – Why and How?* Bristol: Meat Research Institute, pp. 24.1–24.11.

DBERR (2005) Electricity supply and consumption (DUKES 5.2). Department for Business Enterprise and Regulatory Reform. www.decc.gov.uk/en/content/cms/statistics/statistics.aspx, accessed 8 November 2013.

Earle RL (1983) *Unit Operations in Food Processing*, 2nd edn. New York: Pergamon Press.

Evans JA, Scarcellia S, Swain MVL (2007) Temperature and energy performance of refrigerated retail display and commercial catering cabinets under test conditions. *International Journal of Refrigeration* 30: 398–408.

Fennema OR (1975) *Physical Principles of Food Preservation,*. New York: Marcel Dekker.

Fernández-Martín F, Otero L, Solas MT, Sanz PD (2006) Protein denaturation and structural damage during high-pressure-shift freezing of porcine and bovine muscle. *Journal of Food Science* 65: 1002–1008.

Gac A (2002) Refrigerated transport: what's new? *International Journal of Refrigeration* 25: 501–503.

Gaspar PD, Gonçalves LCC, Pitarma RA (2011) Experimental analysis of the thermal entrainment factor of air curtains in vertical open display cabinets for different ambient air conditions. *Applied Thermal Engineering* 31: 961–969.

Gill CO (1984) Longer shelf life for chilled lamb. In: *23rd New Zealand Meat Industry Research Conference*, Hamilton.

Gill CO, Penney N (1986) Packaging of chilled red meats for shipment to remote markets. In: *Recent Advances and Developments in the Refrigeration of Meat by Chilling*. Meeting of IIR Commission C2, Bristol, Section 10, pp. 521–525.

Gill J (1988) The greenhouse effect. *Food* April: 47, 49, 51.

Gosney WB (1982) *Principles of Refrigeration*. Cambridge: Cambridge University Press.

Hardenburg RE, Alley E, Watada CYW (1986) *The Commercial Storage of Fruits, Vegetables, Florist and Nursery Stocks*. United States Department of Agriculture Handbook No. 66. Washington, DC: United States Department of Agriculture.

Heap RD (1986) Container transport of chilled meat. In: *Recent Advances and Developments in the Refrigeration of Meat by Chilling*. Meeting of IIR Commission C2, Bristol, pp. 505–510.

Heap RD, Mansfield JE (1983) The use of total loss refrigerants in transport of foodstuffs. *Australian Refrigeration, Air Conditioning and Heating* 37(2): 23–26.

Hodgins SG (1990) Current designs and developments in fish chilling and freezing equipment. In: *Chilling and Freezing of*

New Fish Products. Meeting of IIR Commission C2, Aberdeen, Section 6, pp. 363–368.

Hoffmanns W (1994) Chilling, freezing and transport: refrigeration applications using the cryogenic gases liquid nitrogen and carbonic acid. *Fleischwirtschaft* 72(12): 1309–1311.

Hong GP, Flick GJ (1994) Effect of processing variables on microbial quality and shelf-life of blue crabs (*Callinectes sapidus*) meat. *Journal of Muscle Food* 5: 91–102.

Honikel KO (1986) Influence of chilling on biochemical changes and quality of pork. In: *Recent Advances and Developments in the Refrigeration of Meat by Chilling*. Meeting of IIR Commission C2, Bristol, Section 1, pp. 45–53.

Ingram M (1972) Meat chilling – the first reason why. In: *Meat Research Institute Symposium No. 2. Meat Chilling – Why and How?* Bristol: Meat Research Institute, pp. 1.1–1.12.

International Institute of Refrigeration (2000) *Recommendations for Chilled Storage of Perishable Produce*. Paris: International Institute of Refrigeration.

International Institute of Refrigeration (2006) *Recommendations for the Processing and Handling of Frozen Foods*. Paris: International Institute of Refrigeration.

James SJ, James C (2002) *Meat Refrigeration*. Cambridge: Woodhead Publishing.

James SJ, Swain MVL (1986) Retail display conditions for unwrapped chilled foods. In: *Proceedings of the Institute of Refrigeration*, Section 82, pp. 1–7.

James SJ, Gigiel AJ, Hudson WR (1983) The ultra rapid chilling of pork. *Meat Science* 8: 63–78.

James SJ, Swain MJ, Brown T et al. (2009) Improving the energy efficiency of food refrigeration operations. In: *Proceedings of the Institute of Refrigeration*, p. 5.1

Jul M (1982) The intricacies of the freezer chain. *International Journal of Refrigeration* 5: 226–230.

Kerr WL (2005) Frozen food texture. In: Hui YH (ed) *Handbook of Food Science, Technology, and Engineering*, vol 2. Boca Raton, FL: Taylor and Francis Group, pp. 60-1–60-13.

Lu YL, Zhang WH, Yuan P, Xue MD, Qu ZG, Tao WQ (2010) Experimental study of heat transfer intensification by using a novel combined shelf in food refrigerated display cabinets (experimental study of a novel cabinet). *Applied Thermal Engineering* 20: 85–91.

Lyons H, Drew K (1985) A question of degree. *Food* December: 15–17.

Malton R (1971) Some factors affecting temperature of overwrapped trays of meat in retailers display cabinets. In: *Proceedings of the 17th European Meeting of Meat Research Workers*, Bristol, J2.

Mattarolo L (1990) Refrigeration and food processing to ensure the nutrition of the growing world population. In: *Progress in the Science and Technology of Refrigeration in Food Engineering*. Proceedings of meetings of commissions B2, C2, D1, D2–D3,

September 24–28, 1990, Dresden. Paris: Institut International du Froid, pp. 43–54.

McCarthy MA, Matthews RH (1994) Nutritional quality of fruits and vegetables subject to minimal processes. In: Wiley RC (ed) *Minimally Processed Refrigerated Fruits and Vegetables*. London: Chapman and Hall, pp. 313–326.

McGlasson WB, Scott KJ, Mendoza Jr DB (1979) The refrigerated storage of tropical and subtropical products. *International Journal of Refrigeration* 2: 199–206.

McGregor BM (1989) *Tropical Products Transport Handbook*. United States Department of Agriculture Handbook No. 688. Washington, DC: United States Department of Agriculture.

Morelli E, Noel V, Rosset P, Poumeyrol G (2012) Performance and conditions of use of refrigerated display cabinets among producer/vendors of foodstuffs. *Food Control* 26: 363–368.

Morrison CR (1993) Fish and shellfish. In: Mallet CP (ed) *Frozen Food Technology*. London, Blackie Academic and Professional, pp. 196–236.

Moureh J, Letang G, Palvadeau B, Boisson H (2009) Numerical and experimental investigations on the use of mist flow process in refrigerated display cabinets. *International Journal of Refrigeration* 32: 203–219.

Nieboer H (1988) Distribution of dairy products. In: *Cold Chains in Economic Perspective*. Meeting of IIR Commission C2, Wageningen, The Netherlands, pp. 16.1–16.9.

Nunes MCN, Emond JP, Rauth M, Dea S, Chau KV (2009) Environmental conditions encountered during typical consumer retail display affect fruit and vegetable quality and waste. *Post Harvest Biology and Technology* 51: 232–241.

Otero L, Sanz P D (2012) High-pressure shift freezing. In: Sun DW (ed) *Handbook of Frozen Food Processing and Packaging*, 2nd edn. Boca Raton, FL: CRC Press, pp. 667–683.

Paull RE (1999) Effect of temperature and relative humidity on fresh commodity quality. *Postharvest Biology and Technology* 15: 263–277.

Persson PO, Löndahl G (1993) Freezing technology. In: Mallet CP (ed) *Frozen Food Technology*. London, Blackie Academic and Professional, pp. 20–58.

Ponce-Alquicira E (2005) Flavor of frozen foods. In: Hui YH (ed) *Handbook of Food Science, Technology, and Engineering*, vol 2. Boca Raton, FL: Taylor and Francis Group, pp. 60-1–60-7.

Poulsen KP (1977) The freezing process under industrial conditions. In: *Freezing, Frozen Storage and Freeze Drying*. Meeting of IIR Commissions C1, C2, Karlsruhe, Section 6, pp. 347–353.

Reid DS (1994) Basic physical phenomena in the freezing and thawing of plant and animal tissues. In: Mallett L (ed) *Frozen Food Technology*, 2nd edn. Glasgow: Blackie Academic and Professional, pp. 1–19.

Robertson GH, Cipolletti JC, Farkas DF, Secor GE (1976) Methodology for direct contact freezing of vegetables in aqueous freezing media. *Journal of Food Science* 41: 845–851.

Rose SA (1986) Microbiological and temperature observations on pre-packaged ready-to-eat meats retailed from chilled display cabinets. In: *Recent Advances and Developments in the Refrigeration of Meat by Chilling.* Meeting of IIR Commission C2, Bristol, Section 9, pp. 463–469.

Rosenvoid K, Wiklund E (2011) Retail colour display life of chilled lamb as affected by processing conditions and storage temperature. *Meat Science* 88: 354–360.

Sharp AK (1988) Air freight of perishable product. In: *Refrigeration for Food and People.* Meeting of IIR Commissions C2, D1, D2/3, E1, Brisbane, Australia, pp. 219–224.

Smith BK (1986) Liquid nitrogen in-transit refrigeration. In: *Recent Advances and Developments in the Refrigeration of Meat by Chilling.* Meeting of IIR Commission C2, Bristol, pp. 383–390.

Spiess WEL (1979) Impact of freezing rates on product quality of deep-frozen foods. In: *Food Process Engineering, 8th European Food Symposium*, Espo, Finland, pp. 689–694.

Stera AC (1999) Long distance refrigerated transport into the third millennium. In: *20th International Congress of Refrigeration*, IIF/IIR Sydney, Australia, paper 736.

Tassou SA, Lewis J, Ge YT, Hadawey A, Chae I (2009) A review of emerging technologies for food refrigeration applications. *Applied Thermal Engineering* 30: 263.

Taylor AA (1972) Influence of carcass chilling rate on drip in meat. In: *Meat Research Institute Symposium No. 2. Meat Chilling – Why and How?* Bristol: Meat Research Institute, pp. 5.1–5.8.

Toledo RT (2007) Heat transfer. In: *Fundamentals of Food Process Engineering*, 3rd edn. New York: Springer, pp. 223–284.

Trott AR (1989) *Refrigeration and Air-Conditioning.* London: Butterworths.

United Nations Economic Commission for Europe Working Party on the Transport of Perishable Foodstuffs (2012) *ATP Handbook.* Geneva: United Nations.

Willhoft EMA (1991) Continuous monitoring of cryogen consumption during freezing of foodstuffs. In: Bald WB (ed) *Food Freezing: Today and Tomorrow.* London: Springer, pp. 188–200.

Williams DC, Lim MH, Chen AO, Pangborn RM, Whitaker JR (1986) Blanching of vegetables for freezing – which indicator enzyme to choose? *Food Technology* 40: 130–140.

Zheleva I, Kamburova V (2009) Modeling of heating during food processing. In: Costa R, Kristbergsson K (eds) *Predictive Modelling and Risk Assessment.* New York: Springer, pp. 79–99.

Zheng L, Sun DW (2005) Vacuum cooling of foods. In: Sun DW (ed) *Emerging Technologies for Food Processing.* Amsterdam: Elsevier Academic Press, pp. 579–602.

6

Fermentation and Enzyme Technologies in Food Processing

Ali Demirci, Gulten Izmirlioglu, and Duygu Ercan
Department of Agricultural and Biological Engineering, Pennsylvania State University, University Park, Pennsylvania, USA

6.1 Introduction

Fermentation is a microbial biotechnology whereby natural renewable substrates are converted to value-added products such as enzymes, organic acids, alcohols, polymers, and more. Fermentation end-products such as ethanol and lactic acid are proton sinks, whereby NADH is recycled to NAD^+, which allows the cell to continue producing energy via glycolysis by substrate-level phosphorylation. Thus, microorganisms generate many end- or by-products to maintain energy balance.

Today, enhanced production of economically important fermentation products has benefited from targeted genetic engineering techniques to established industrial microbial strains (Campbell-Platt, 1994). The formation of end- or by-products is dependent on microbial strain and the environmental conditions employed. For an optimum fermentation process, a microbial strain should be selected and developed based on the desired product. Strain development technologies include mutation and recombinant DNA technology. Growth of microorganisms and product formation are also affected by temperature, pH, dissolved oxygen, and fermentation medium composition. Therefore, the fermentation media and growth conditions need to be optimized. Moreover, the metabolic pathways should be determined.

Knowledge of biochemical changes in fermented foods can help producers to manipulate the production by changing strains and/or conditions. In addition to growth conditions, types of strains, and media, fermentation modes affect the productivity. Batch, fed-batch, and continuous fermentation modes can be selected for

high productivity. Fed-batch and continuous modes can overcome the substrate limitation during fermentation processes. Higher productivity can also be achieved by cell immobilization, which results in increases in the biomass concentration in the bioreactor, and thus an increased concentration of biocatalysts in the reactor. After the fermentation process, recovery methods of the product from the fermentation medium need to be evaluated and optimized, and are often an economic limitation factor. Indeed, purification of the end-product is often a high-cost step of the process. The microbial end-products can be biomass itself, extracellular products, or intracellular products. Filtration, homogenization, and extraction (liquid and solid) are examples of recovery methods.

Enzymes are products of living organisms and have been used in the industry for many years due to their catalytic activities. Enzyme activity depends on temperature, substrate, pH, inhibitors, etc. and should be optimized for each process. Since enzyme recovery is difficult, the application of enzyme immobilization may reduce process cost. Enzymes can be isolated from plants and mammalian tissues, or can be produced by microorganisms. However, microbial enzymes are preferred due to availability and specificity.

Enzymes are used in the production of over 500 commercial products (Johannes & Zhao, 2006). They have a broad range of applications from food to detergents. Most enzymes are commercially available and used to enhance the product quality in food, detergents, leather, paper, cosmetics, and pharmaceuticals. Commercial enzymes include lipase, amylases, proteases, pectic enzymes, and milk clotting enzymes (rennet). In this

Food Processing: Principles and Applications, Second Edition. Edited by Stephanie Clark, Stephanie Jung, and Buddhi Lamsal.
© 2014 John Wiley & Sons, Ltd. Published 2014 by John Wiley & Sons, Ltd.

chapter, optimization of media and growth conditions and criteria for microbial strain selection and strain development nutrient are summarized. The strains, production conditions, and biochemical changes involved in the production of fermented foods are explained. The disadvantages and advantages of different fermentation modes are discussed. Finally, immobilized fermentation principles and basic recovery methods are also summarized. A brief summary of commercial enzymes and their reactions is provided.

6.2 Fermentation culture requirements

In fermentation processes, environmental conditions, such as temperature, pH, and dissolved oxygen, play a crucial role in microbial growth and product. They need to be carefully evaluated for each fermentation process. Each of these parameters is now briefly discussed.

6.2.1 Culture media

During microbial growth, energy needs are met while the metabolites (by-products) are produced. However, the formation of by-products is usually dependent on the environmental conditions, such as nutrients, pH, and presence of oxygen (Shuler & Kargi, 2008). Optimizing the nutrient requirements of the culture can enhance both growth and production of end/by-products. Nutrients used for fermentation can be categorized as macronutrients and micronutrients. Carbon, nitrogen, oxygen, sulfur, phosphorus, magnesium, and potassium represent major macronutrients. Carbohydrates are the main energy source for the culture. For small-scale fermentation, glucose, sucrose, and fructose are commonly used as a carbon source. However, for industrial-scale fermentation, beet molasses, starch, and corn syrup are the main sources to meet the carbon requirements of microorganisms (Balat et al., 2008). The second major nutrient in a fermentation medium is nitrogen. Ammonia, ammonium salts, proteins, yeast extract, and peptone are some of the nitrogen sources that can be found in a typical fermentation medium (Shuler & Kargi, 2008). Trace elements are also needed for microbial growth and to increase yields and production rates. Iron, zinc, manganese, copper, cobalt, calcium, and sodium represent some examples of trace elements.

Each microorganism has specific requirements for carbon and nitrogen sources, as well as other nutrients such as minerals and vitamins. However, a classic media design approach is used to imitate the elemental composition of the microorganism to be grown. Carbon/nitrogen ratio is used to optimize the amount of carbon and nitrogen in the medium. Magnesium, sulfate, phosphate, and other salts should be equivalent to their presence in the cell (Parekh et al., 2000). In general, a microbial cell is composed of 44–53% carbon for bacteria, yeast, and fungi; 10–14% nitrogen for bacteria, 7–10% for yeasts and fungi; 2–3% phosphorus for bacteria, 0.8–2.6% for yeasts, and 0.4–4.5% for fungi; and 1–4.5% potassium for bacteria and yeasts, 0.2–2.5% for fungi (Ertola et al., 1995).

The composition of media is an important issue when defining the media for a specific product formation. Studies have demonstrated that different end/by-products can be formed in different culture medium and growth conditions with the same microorganism. For example, various value-added products can be generated by *Aspergillus niger* when different culture media are used (Metwally, 1998). It has been reported that glucoamylase production was suppressed when potassium nitrate was used as the nitrogen source compared to ammonium sulfate (Metwally, 1998). On the other hand, Hang and Woodams (1984) reported that the same microorganism, *A. niger*, produced citric acid when a different carbon source was used. In their study, apple pomace was used as the carbon source, and *A. niger* was able to produce 314 g citric acid/kg sugar.

In industrial settings, medium design becomes crucial due to economic feasibility. Moreover, the cost of the downstream process can be affected by the medium composition (Kennedy & Krouse, 1999). Demirci et al. (1997) reported that 30% of ethanol cost is due to fermentation medium costs; in their study, immobilization of cells was used to decrease the cost while increasing productivity. Another study by Kadam and Newman (1997) on ethanol fermentation reported the use of low-cost fermentation medium to reduce cost. Their approach was to reduce the cost by replacing yeast extract with urea, and adding corn steep liquor as a source of vitamins, minerals, and nitrogen. Therefore, medium ingredients play an important role in achieving an economical product.

6.2.2 Temperature

Each microorganism has a specific temperature for optimal growth and production rates (g/L/h). Therefore, microorganisms are categorized into three groups based on their optimum temperature. Psychrophiles prefer a cold environment, less than 20 °C. Most of the psychrophiles are bacteria, *Pseudomonas* and *Vibrio* spp., and Archaea, *Methanogenium*, and *Methanococcus*. Most of the psychrophile bacteria are spoilage microorganisms responsible for spoilage of refrigerated foods. Mesophilic microorganisms require a moderate temperature profile to grow (20–45°C); *Lactobacillus delbrueckii* subsp. *bulgaricus* used in yogurt fermentation and *A. niger* producing amylase enzymes are examples of mesophilic microorganisms. Thermophiles have an optimal temperature greater than 45°C. *Thermomyces lanuginosus* and *Kluyveromyces marxianus* are thermophiles that produce amylase and ethanol, respectively. Above the optimal temperature, growth slows down and thermal death may occur. However, optimum temperature for product formation may differ for microbial growth. Consequently, temperature optimization for a specific product should be determined to obtain maximum product yields.

Optimization of fermentation temperature for most value-added products for various microorganisms and media has been widely reported. Dutta et al. (2004) studied optimization of temperature for extracellular protease production from *Pseudomonas* spp. In their study, three different statistical models were used to optimize culture parameters. As a result, 38°C was determined as the optimum temperature for extracellular protease production, with 58.5 U/mL enzyme activity. In another study, approximately 34% increase in tartaric acid concentration (2.14 kg/m³ after 96 h fermentation) was reported after temperature optimization for tartaric acid production by *Gluconobacter suboxydans* (Chandrasker et al., 1999). Furthermore, *Kluyveromyces marxianus* was chosen to investigate the fermentation conditions in glycerol production from whey permeate (Rapin et al., 1994). These authors reported that at 37°C and pH 7, glycerol yield was 9.5% with a 0.15 L/h specific growth rate, which was the highest yield among examined conditions.

6.2.3 pH

pH is another parameter requiring optimization for high microbial growth and production rates. In general, bacteria have an acceptable range of pH 3–8; for yeast, the range narrows down to pH 3–6, and for molds, optimal pH is 3–7 (Shuler & Kargi, 2008). In fermentation processes, pH can change during microbial growth and product formation due to production of basic compounds or organic acids. Furthermore, the source of nitrogen plays a role in pH fluctuation. Ammonia and nitrate can be given as examples of this; ammonia causes a decrease in pH, while nitrate results in an increase in pH due to

the microbial utilization of hydrogen ions (Shuler & Kargi, 2008).

Consequently, pH optimization and control during fermentation processes is very important for microbial growth and value-added product formation. The pH can be controlled by addition of acid or base during the fermentation (internal buffering and external buffering should be distinguished). pH profiles for products have been studied. Mohawed et al. (1986) optimized the pH value for lipase production by *Aspergillus anthecieus* grown in Dox-Olive oil medium by using different buffers: citrate-phosphate buffer (ph range from 2.6 to 7.0), borate buffer (pH range from 7.6 to 9.2), and phosphate buffer (pH range from 5.8 to 8.8). Highest lipase concentration (156 µg/mL) was reported at pH 6.6 when phosphate buffer was used. Ellaiah et al. (2002) optimized the initial pH (3.0–7.0) for glucoamylase production by a newly isolated *Aspergillus* species, and reported that maximum glucoamylase production (109.0 U/g) was obtained at pH 5. It was also reported that glucoamylase production is highly affected by initial pH level for this particular species. Malik et al. (2010) reported the effect of initial pH (4.5–6.5) for cellulase production by *Trichoderma viride*. Interestingly, at pH 4.5, only 0.2 U/mL/min cellulases were produced, whereas cellulase production reached maximum (1.66 U/mL/min) at an initial pH 5.5. Moreover, further increase in initial pH (>5.5) caused a reduction in cellulase production. Studies have shown that both initial pH value and pH levels during fermentation play a very important role in obtaining and maintaining maximum growth and production rates.

6.2.4 Dissolved oxygen

Dissolved oxygen is an essential substrate for aerobic fermentation. Microorganisms can be categorized as aerobic, anaerobic, and facultative anaerobes (Demirci, 2002). Aerobic microorganisms require oxygen as an electron acceptor during respiration. Strict anaerobes, on the other hand, are killed by oxygen. Facultative anaerobes, such as yeast, will respire when oxygen is present, and will convert to fermentation in the absence of oxygen. Whereas many lactic acid bacteria are indifferent to oxygen, their growth rate is best anaerobically, but most can also grow in the presence of oxygen.

Because solubility of oxygen in the liquid is very low, in aerobic settings, oxygen is continuously added to the fermentation medium as a substrate, and improved by agitation. In some cases, oxygen can be the rate-limiting substrate, for instance, in continuous stirred tank reactors with high cell population (Shuler & Kargi, 2008). Various methods have been reported to overcome this limitation. One is generating oxygen in the reactor by adding hydrogen peroxide and a catalyst system, or using oxygen-catalyzing microorganisms. Also, co-fermentation (pair of oxygen-consuming and -producing microorganisms) can be applied to increase oxygen availability in the reactor. Another approach is utilizing media that promote solubility of oxygen by replacing aqueous solutions with organic solvents (Rols & Goma, 1989). Oxygen vectors, e.g. hydrocarbons and perfluorocarbons, are also introduced as alternatives to improve oxygen solubility (Galaction et al., 2004). However, the growth rate is independent of dissolved oxygen concentration when critical oxygen concentration is exceeded (Shuler & Kargi, 2008). Therefore, the concentration of dissolved oxygen must be sufficient for microbial growth.

Oxygen solubility in culture medium is affected by several factors. Solubility of oxygen decreases when temperature is raised (Demirci, 2002). The presence of other ingredients in the culture medium also has an effect on oxygen solubility. Studies have shown that salts and sugars decrease the solubility of oxygen in the fermentation medium (Demirci, 2002). On the other hand, an increase in the partial pressure of the oxygen can increase the solubility of oxygen linearly.

Oxygen is supplied to the fermentation broth by sparging air for bench-top fermentors, whereas agitation blades and speed are used to break air bubbles into smaller bubbles for industrial-scale fermentors. It is difficult to keep the dissolved oxygen at saturation level while it is being consumed by growing microorganisms. Furthermore, oxygen transfer from gas to the cells is limited by the oxygen transfer rate (OTR), which must be equal to or higher than oxygen uptake rate (OUR) to ensure that the critical oxygen level is achieved. OTR is given as:

$$OTR = k_L a(C^* - C_L)$$

where k_L is the oxygen transfer coefficient (cm/h), a is the gas–liquid interface area (cm^2/cm^3), C^* is saturated dissolved oxygen concentration (mg/L), and C_L is the actual dissolved oxygen concentration in the broth (mg/L). OUR can be given as:

$$OUR = q_{O2}X$$

where q_{O2} is the specific rate of oxygen consumption (mg O_2/g dry weight of cells.h), X is the cell concentration (g dry weight of cells/L).

6.2.5 Strain selection and strain development

Microbial strain selection is important to achieve the desired product at optimum productivity levels in the fermentation process. Microorganisms can be obtained from culture collections all over the world (Table 6.1) or may be isolated from the environment (Hatti-Kaul, 2004). Wild-type strain, that is not mutated, is originally found in nature. The characteristics of microbial strains for industrial fermentation include effective stable productivity of large quantities of a single product and genetic stability. Moreover, the strain should be pure and easy to maintain and cultivate, and it should be grown in inexpensive culture media. In addition to characteristics of strains that are used in industrial production, microorganisms that are used as starter cultures must be neither pathogenic nor capable of producing toxins;

Table 6.1 Centers for culture collections

Agencies	Abbreviation	Web addresses
American Type Culture Collection	ATCC	http://atcc.org
Belgian Co-ordinated Collections of Microorganisms	BCCM	http://belspo.be/bccm
Common Access to Biological Resources and Information	CABRI	http://cabri.org
Czech Collection of Microorganisms	CCM	http://panizzi.shef.ac.uk/msdn/ccm/ccmi.html
Deutsche Sammlung für Mikrorganismen und Zelkulturen	DSMZ	http://dsmz.de
Japan Collection of Microorganisms	JCM	www.jcm.riken.go.jp
Netherlands Culture Collection of Bacteria	NCCB	www.cbs.knaw.nl/NCCB
National Collections of Industrial and Marine Bacteria	NCIMB	www.ncimb.com/culture.html
Northern Regional Research Center	NRRL	http://nrrl.ncaur.usda.gov/

they should also be able to suppress the growth of specific undesirable microorganisms. The starter cultures should be GRAS (Generally Recognized As Safe) organisms, which are safe for human consumption (Geisen & Holzapfel, 1996).

Strain development technologies, such as mutation or recombinant DNA, are applied to enhance metabolic capabilities for industrial applications or to make the strains capable of synthesizing foreign (or heterologous) compounds. Production levels for the wild strains may be too low for economical production. Therefore approaches to increase productivity are typically necessary. Productivity can be increased by strain improvement. Moreover, strain may be improved to reduce oxygen needs and stabilize foaming. For example, Zhang et al. (2009) reported that proteinase A (PrA), encoded by the PEP4 gene, promotes cell growth against insufficient oxygen conditions in steady-state cultivation. A stable head of foam, which is a major consideration in beer quality assessment, was achieved by destruction of the gene responsible for yeast PrA (Wang et al., 2007). The strain may be developed to metabolize inexpensive substrates. Wild strains may produce a mixture of substances, which are chemically related. Mutants can be developed to make the strain able to synthesize one component as the main product. This can simplify product recovery (Parekh et al., 2000; Rowlands, 1984).

Classic strain development depends on random mutation and screening. Mutation is achieved by subjecting the genetic material to physical or chemical agents called mutagens (Parekh et al., 2000). Nitrosoguanidine (NTG), 4-nitroquinolone-1-oxide, methyl methane sulfonate (MMS), ethylmethane sulfonate (EMS), hydroxylamine (HA), γ-ray irradiation, and ultraviolet light (UV) are examples of mutagen agents (Adrio & Demain, 2006; Clarkson et al., 2001). Moreover, single cells, spores, and mycelia have been used for mutagenesis (Adrio & Demain, 2006). Mutation can be classified based on the length of the DNA sequence affected by the mutation. While point mutation affects a single nucleotide, segmental mutation affects several adjacent nucleotides (Graur & Li, 2000). After treatment with mutagens, a diverse population of cells should be screened to identify those with mutations that have desirable traits (Clarkson et al., 2001). Random screening involves the random selection of survivors from the population and testing for their ability for enhanced production of the metabolite of interest by small-scale model fermentation.

Assay of fermentation broth for products of interest and scoring for improved strains is essential for strain development protocols. This process is applied to confirm the statistical superiority of microbial strain compared to control strain used prior to mutation (Parekh et al., 2000). For example, classic mutagenesis was used to improve the yields of α-amylase enzyme production by Bacillus subtilis (Clarkson et al., 2001). Adrio and Demain (2006) reported an increase in the production of penicillin by Penicillium chrysogenum X-1612, which was isolated after X-ray mutagenesis.

In addition to mutation, alteration in DNA can be achieved by the application of genetic recombination. Homologous reciprocal recombination involves the even exchange of homologous sequences between homologous chromosomes. As a result of homologous reciprocal recombination, new combinations of adjacent sequences are produced while retaining both variants. Non-reciprocal recombination involves the uneven replacement of one sequence by another and a process resulting in the loss of one of the variant sequences involved in the recombination (Graur & Li, 2000). Genetic recombination can be achieved by transformation, transduction, or conjugation. A cell takes up isolated DNA molecules from the medium surrounding it in transformation, which allows exchange of genes between non-related organisms. In transduction, the transfer is mediated by bacterial viruses. Another method of genetic recombination is conjugation, which involves the direct transfer of DNA from one cell to another. Conjugation is a natural process in which, by contact between cells, transmissible plasmids are transferred among closely related microbial species (Dale & Park, 2010; Geisen & Holzapfel, 1996).

While strains obtained by transformation of recombinant DNA are subject to genetic engineering laws and regulations, starter cultures, which were obtained by conjugation, are not subject to legal regulations (Geisen & Holzapfel, 1996). Genes can also be transferred artificially by recombinant DNA technology, in which endonuclease and ligase enzymes can be employed. The application of recombinant DNA technology makes the introduction of genes from higher organisms into microbial cells possible. Examples of hosts for such foreign genes include Escherichia coli, Saccharomyces cerevisiae, Kluyveromyces lactis, and other yeasts as well as filamentous fungi such as Aspergillus niger (Stanbury, 2000).

The following aspects should be considered when starter cultures for food production are developed by recombinant DNA technology: acceptor strains must have GRAS status; the DNA to be transferred must be fully identified and have GRAS status; the vector should have food-grade status; the gene integration site should be well identified; and integration in this genomic site should not cause pleiotropic effects (Geisen & Holzapfel, 1996), which means that a single gene can cause multiple effects (Tandon et al., 2005).

Up to now, recombinant DNA technology has been used as an effective method for strain development to produce many products, such as human lysozyme, interleukin-1β, hepatitis B surface antigen, human serum albumin, interferon, insulin, human serum albumin, food-grade β-galactosidase, α-galactosidase, glucoamylase, etc. (Merico et al., 2004; Stanbury, 2000; Zhang et al., 2011). In addition to selection and development of the strain, culture media, growth conditions, and bioreactor design are important to achieve the maximum yield of desired products.

6.3 Fermentation technologies

6.3.1 Batch fermentation

Batch fermentation is carried out in a continuous stirred tank reactor (CSTR) with an initial amount of medium; during fermentation, no medium addition or removal occurs (Sanchez & Cardona, 2008). Fermentation is performed after sterilization of the medium and adjustment of pH by either acid or alkali. After inoculation of yeast or bacteria, product formation takes place by controlling temperature, pH, agitation, and dissolved oxygen, which depend on the characteristics of the cultured microorganism. Because no medium addition occurs, the growth of microorganisms follows four main phases of the growth curve: lag phase, log (exponential) phase, stationary phase, and death phase (Figure 6.1). In the lag phase, cell growth is negligible, because it is an adaptation time for the cells to adjust to the new environment (i.e. substrates

available in the culture medium, metabolic pathway initiation, or synthesizing of specialized enzymes such as amylases and cellulases) (Shuler & Kargi, 2008). The log (exponential) phrase is where microbial growth reaches its maximum rate. The stationary phase is the period during which microbial growth rate equals the death rate. In this phase, cells are still able to produce secondary metabolites. The last period of batch fermentation is the death phase, in which death rate is higher than growth rate due to lack of nutrients or accumulation of inhibitor primary or secondary metabolites, etc.

Malik et al. (2010) reported production of cellulases by *Trichoderma viride* using batch fermentation. Maximum production of enzyme (CMCase 1.57 U/mL/min, FPase 0.921 U/mL/min) was obtained at 30 °C after 72 h of incubation. *Pseudomonas* spp. were investigated for extracellular protease production by using statistical optimization methods in batch fermentation (Dutta et al., 2004). The optimum operating conditions were 38 °C, pH 7.8, and inoculum volume of 1.5% with 58.5 U/mL of protease activity within 24 h of incubation.

In the ethanol industry, batch fermentation has a wide application. Carbohydrate substrates for industrial fermentations can be glucose (corn syrup), sucrose (sugar cane molasses), polysaccharide (starch, cellulose or hemicelluloses) and more. The batch system can be combined with pretreatment processes for ethanol fermentation from starchy and lignocellulosic materials. Hydrolysis of starch (mostly saccharification) and fermentation can be performed simultaneously in simultaneous saccharification fermentation (SSF). Choi et al. (2008) studied ethanol fermentation from sludge containing cassava mash by applying simultaneous saccharification and batch and repeated batch reactor fermentation. They reported that a maximum 85 g/L ethanol concentration was reached after 42 h of batch fermentation.

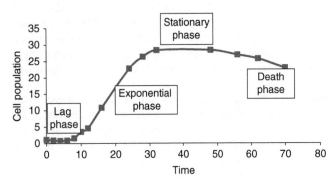

Figure 6.1 A typical batch fermentation curve.

6.3.2 Fed-batch fermentation

In fed-batch processes, one or more nutrients are added to a reactor continuously or intermittently, while fermentation broth is either removed semi-continuously or not removed. By addition of medium, two benefits can be obtained: (1) microbial growth will not be affected due to lack of nutrients, and (2) substrate inhibition will be overcome, if this is a limitation for the process. Ethanol is a metabolic inhibitor for yeast (>10–12% depending on the strain) and a point of concern in fermentation, which will be eliminated in the fed-batch process by slowly decreasing the concentration of end-product during the fermentation process. However, to enhance productivity and ethanol yield, optimization of feeding should be done properly (Sanchez & Cardona, 2008). Fed-batch culture is the most common technology in the ethanol industry in Brazil (Sanchez & Cardona, 2008).

Fed-batch fermentation has been studied to increase the production of cellulase and concentrations of substrate to be consumed by *Trichoderma reesei* (Hendy et al., 1982). In this study, a maximum of 30.4 FPIU/mL was achieved (the highest activity yet reported for this strain) with 47.6 g/L soluble protein. Alfenore et al. (2002) reported a fed-batch fermentation studied to overcome lack of nutrients during ethanol production by *S. cerevisiae*. To overcome nutrient deficiency, vitamin was added during fermentation. Ethanol production increased from 126 g/L to 147 g/L with a maximum productivity of 9.5 g/L/h. Ozmihci and Kargi (2007) also studied fed-batch ethanol fermentation with *S. cerevisiae*, which produced 63 g/L ethanol and 5.3 g/L/h productivity with a 125 g/L sugar feed. Moreover, Sanchez et al. (1999) performed production of enzyme from *Candida rugosa* pilot-plant fed-batch fermentations, and characterized the crude extracellular enzyme preparations in which multiple forms of lipases and esterases were identified.

6.3.3 Continuous fermentation

Continuous fermentation, also known as chemostat or continuous stirred tank reactors (CSTR), involves fresh sterile media fed into a reactor continuously. In addition to feeding the reactor with fresh nutrients, the effluent is removed from the reactor and the volume of the reactor remains constant. The rates of feeding and removal stay equal. To avoid wash-out, which means taking away all the cells from the reactor, growth rate of the microorganism must equal dilution rates (rate of removing cells).

Advantages of continuous fermentation over batch fermentation include low construction costs of bioreactors, lower maintenance and operational requirements, higher yield, and better control of the process (Sanchez & Cardona, 2008). Stability of culture, however, is an issue for continuous fermentation. Even small changes in any parameter (i.e. temperature, dilution rate, substrate concentration of feed, etc.) can decrease yield. Different methods can be used to overcome drawbacks of continuous fermentation. For instance, Sanchez and Cardona (2008) suggest utilization of immobilized cell technology in which cells are captured in the reactor by biofilms, calcium alginate, chrysolite, etc., to enhance yield.

Continuous SSF was studied to produce ethanol from grains by *S. cerevisiae* and yielded 2.75 gal/bushel (Madson & Monceaux, 1995). Kobayashi and Nakamura (2004) studied continuous ethanol fermentation from a starch-containing medium by using recombinant *S. cerevisiae* and reported 7.2 g/L ethanol and 0.23 g/L/h maximum ethanol productivity at $0.026\,h^{-1}$ of the dilution rate in the suspended cell culture; however, ethanol productivity increased approximately 1.5-fold in the immobilized cell culture and reached 3.5 g/L/h productivity at $0.4\,h^{-1}$ dilution rate.

6.3.4 Immobilized cell fermentation

Immobilized cell fermentation can be defined as preventing the free migration of cells in a fermentation medium. The advantage of cell immobilization over suspended cell reactors is to increase biomass concentration or biocatalyst in the reactor, and therefore, enzyme activity as well as fermentation by-products. Long periods of operation in a fermentation plant due to reuse of cells without recovery cost can also be provided by immobilization. In most cases, product recovery can be performed easily in immobilized cell systems compared to suspended cell bioreactors. In continuous fermentation, immobilization can eliminate cell wash-out problems at high dilution rates. Also, better environmental conditions for microbial growth can be provided when cells are immobilized.

According to the physical mechanism of interest, immobilization can be categorized into four groups: immobilization on solid carrier surfaces, entrapment within a porous matrix, aggregation (cell flocculation), and mechanical containment behind a barrier (Kourkoutas et al., 2004; Verbelen et al., 2006).

Immobilization on solid carrier surfaces is simply attachment of cells to a solid support by physical adsorption because of electrostatic forces or covalent binding between cells and the surface (Kourkoutas et al., 2004). One of the important aspects here is thickness of biofilm. Thick biofilms may cause diffusionally limited growth, whereas thin biofilms may result in low biomass concentration caused by low rates of conversion (Shuler & Kargi, 2008). Cell attachment and relocation are possible because there is no barrier between cells and the solution. Waste water treatment, water purification, and enhanced production of value-added products are some of the applications of biofilms (Demirci et al., 2007). Although waste water treatment, microbial leaching, and acetic acid production (also called the "quick vinegar process") using biofilm reactors have been employed in the industry, most biofilm reactors have only been studied at laboratory scale (Bryers, 1994; Cheng et al., 2010). Biofilm reactors have been studied to enhance production of ethanol, organic acids, enzymes, cellulose, pullulan, etc.

Biofilm formation occurs in three steps (Figure 6.2). Attachment, the first step, is the rate at which cells are transported to the surface and is highly dependent on surface properties. In the second stage, colonization, formation of polymer bridges occurs. Colonization is affected by growth conditions, nutrient diffusion, and shear force. In the last step, growth, availability of nutrient plays a crucial role (Demirci et al., 2007).

Solid supports of biofilm reactors should be favorable to microorganisms, inexpensive, easily available, and resistant to mechanical force (Cheng et al., 2010). Plastic composite supports (PCS) are one of the solid supports made from polypropylene, developed at Iowa State University (Pometto et al., 1997).

Entrapment within a porous matrix is one of the most common cell immobilization techniques. Two main methods are employed. First, cells are diffused into a porous matrix until their mobility is prevented by the other growing cells and the matrix. In this case, attachment to the surface is also possible. Some of the materials commonly employed for this method

are sponge, sintered glass, ceramics, silicon carbide, polyurethane foam, and chitosan (Verbelen et al., 2006). Second, entrapment can be obtained *in situ*. In this type of entrapment, hydrocolloidal gels are used with natural gels (Scott, 1987). The gel is formed into droplets after adding the suspended cells to a liquid solution of gelling material (Scott, 1987), and therefore, the polymeric beads are frequently spherical in a range of 0.3–3 mm (Verbelen et al., 2006). Agar, Ca-alginate, gelatin, polystyrene, κ-carrageenan, polyurethane, and polyvinylalcohol are some of the gelling agents that have been used for entrapment. However, the chemical and physical instability of gel, non-regenerability of the beads, diffusion limitations of nutrients, oxygen, and metabolites because of the gel matrix, and high densities in the beads are some of the bottlenecks of the entrapment technique (Verbelen et al., 2006).

Cell flocculation is the aggregation of cells to form a larger unit. The ability to aggregate is common in fungi, molds, and plant cells. Aggregation can be induced by artificial flocculating agents or cross-linkers such as polyelectrolytes and inert powders. Cell aggregation is affected by cell wall composition, pH, dissolved oxygen, agitation, fermentation temperature, microorganism handling and storage conditions, Ca^{2+} concentration, and medium composition.

6.4 Downstream processing

The final step of any fermentation process is downstream processing in which the product is separated from other undesired components of the fermentation broth and further purification is carried out. The products can be the biomass itself, extracellular (present in the fermentation broth) or intracellular (trapped in the cell) products. Since fermentation broth is very complex, recovery of the product of interest requires very comprehensive methods

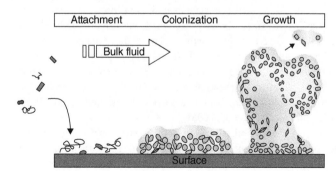

Figure 6.2 Formation of biofilms (reproduced from Cheng et al., 2010, with permission from Springer-Verlag).

of recovery and can account for 50–60% of the total production cost (Cliffe, 1988).

For product recovery, the following characteristics need to be considered during recovery and, in most cases, limit the methods of separation. Commonly, the products are in low concentration in a very complex liquid medium, and temperature sensitive. If products are intracellular, more complicated steps for separation and purification may be needed. Also, physical or chemical properties of the product may be very similar to the other by-products in the fermentation broth, which make downstream processing extremely difficult.

In general, downstream processing is performed in four main steps. In most cases, the first step is the removal of insolubles, in which solid particles such as biomass are separated from the liquid medium. Rupturing the cell wall of biomass and extracting the intracellular compounds is the second step. Removal of solubles from the liquid phase can be considered as the third step, especially for the products present in the medium. The final step is further purification of the product.

6.4.1 Removal of insolubles

In some processes, the product of interest is the biomass itself or the product can be extracellular in the broth; however, in both cases microbial cells must be separated. Typical sizes for cells can be anywhere between 0.5–1 μm for bacteria, 1–7 μm for yeast, 5–15 μm for molds, 20–40 μm for plant cells, and 10–0 μm for suspensions of animal cell (Dutta, 2008).

Removal of insolubles usually involves filtration or centrifugation. Filtration separates the solids from the liquid by forcing the fluid through a filter or solid support on which particles are deposited. This method is fairly straightforward for dilute suspension of large and rigid particles. However, small particles and deformability of the cells can make the process more complicated. There are many types of filters available but plate and frame pressure filters and rotary vacuum filters are most common. Plate and frame pressure filters are employed for small-scale separation of bacteria and fungi from fermentation broth. Rotary vacuum filters are preferred for large-scale commercial processes due to lower labor costs (Perry & Chilton, 1973). Even though filtration is straightforward, for some products pretreatments may be required. Heating for protein denaturation, addition of electrolytes to promote coagulation and flocculation, and addition of filter aids (earths and perlites) to increase

porosity and decrease the compressibility of cakes are some pretreatment examples (Belter et al., 1988).

On the other hand, centrifugation utilizes the density difference between the solids and the surrounding fluid, when solid particles settle down and fluid stands as a clear phase. Even though centrifugation requires more expensive equipment and produces a wet cake compared to filtration, it is still an alternative for separation of small particles, between 100 and 0.1 μm (Shuler & Kargi, 2008). Two basic types of centrifuges are the tubular and the disk centrifuge (Figure 6.3). The tubular centrifuge consists of a cylindrical rotating element in a stationary case. Broth is fed through the bottom and clear liquid collected from top, leaving the solids on the wall of the bowl. The tubular centrifuge can be inexpensive, but is not suitable for large-scale and continuous processes. The disk centrifuge includes a short, wide bowl that turns on a vertical axis. Closely spaced cone-shaped disks decrease the distance of particles to be removed and increase efficiency. This type of centrifuge is utilized for continuous and large-scale processes but is more expensive than the tubular centrifuge (Belter et al., 1988).

6.4.2 Cell disruption

Although some antibiotics, extracellular enzymes, many polysaccharides, and amino acids are secreted into the broth, some products are intracellular, such as lipids, some antibiotics, and most proteins, and trapped in the cell. Cells need to be ruptured to release intracellular products. However, rupturing the cells is usually difficult because of the high osmotic pressure inside the cell and strong cell walls (Dutta, 2008). Techniques used have to be powerful enough to break down the cell wall, but gentle enough not to damage the product.

Physical methods include mechanical disruption by grinding, homogenization, and ultrasonication. Homogenization is effective for mycelia organisms, as well as animal cell or tissues. However, grinding and ultrasonication can be employed for most of the suspension cells. Crushing and orifice type homogenization, on the other hand, are adapted to larger scales (Belter et al., 1988). Chemical methods of cell rupture include treatment with detergent, alkalis, organic solvents, or osmotic shock. Alkali treatment is basically converting cell wall components into detergents. Although alkali treatment is a cheap technique, it is not useful because it is harsh, not specific, and destroys the product (Dutta, 2008). Use of the enzyme lysozyme for digestion of bacterial cell walls is an example of biological methods of cell

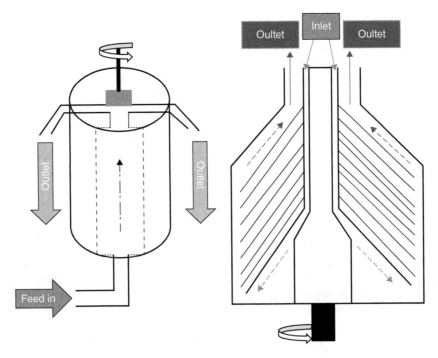

Figure 6.3 Schematics for tubular centrifuge (*left*) and disk centrifuge (*right*).

disruption. It is an effective method due to selectiveness and gentleness, but the high cost of enzymes makes this method impractical for large-scale operations (Shuler & Kargi, 2008).

6.4.3 Removal of solubles

The next step in product recovery is separating intracellular (after cell disruption) and extracellular products from the broth and many methods are available for this purpose. Recovery and purification is a very wide topic, and only a few methods will be explained briefly in this section (Belter et al., 1988; Shuler & Kargi, 2008).

Extraction is the method of separating the solutes of a liquid solution by contact with another insoluble liquid, the solvent (Dutta, 2008). The desired product can be selectively extracted out of the solution into the solvent phase by selecting an appropriate solvent. A good solvent should be easily recovered, have large density difference between two phases, non-toxic, and cheap. The solvent-rich phase is called the extract, whereas the residue of the liquid is called the raffinate (Dutta, 2008).

Adsorption is a technique that relies on the fact that specific substances in solution can be adsorbed selectively by certain solids as a result of physical or chemical interactions between the molecules of the solid and substance adsorbed. Adsorption is a very effective method for separation of dilute substances, because it is very selective while the solute loading on the solid surface is limited (Dutta, 2008). Adsorption can be due to intermolecular forces (van der Waals forces between the molecules and substance), chemical interaction between a solute and ligand, or ionic forces. Adsorption can occur via stage-wise or continuous-contact methods.

Precipitation is a commonly used method for the recovery of proteins or antibiotics (Belter et al., 1988). It can be induced by addition of salts, organic solvents, or heat. The addition of salts precipitates proteins, since protein solubility is reduced by increase in salt concentration in the solution. This method is effective and cheap, but it causes protein denaturation (Belter et al., 1988). Ammonium sulfate is the most commonly used salt but has the disadvantage that it is hard to remove from precipitated protein. Organic solvents precipitate protein at a lower temperature by decreasing the dielectric constant of the solution. Acetone, ethanol, and methanol are the most commonly employed organic solvents. Heating is another way of promoting precipitation of proteins by denaturation. It is used to eliminate unwanted proteins; however, it is difficult to precipitate the unwanted proteins without damaging the desired product.

Chromatography employs a mobile phase and a stationary phase. The mobile phase is the solution containing solutes to be separated and eluent that carries the solution through the stationary phase. The stationary phase can be adsorbent, ion exchange resin, porous solid or gel and is mostly packed in a cylindrical column. A solution is injected at one end of the column and eluent carries the solution through the stationary phase to the other end of the column. Each solute in the original solution moves at a rate proportional to its relative affinity for the stationary phase and is separated at the end of the column (Ogez et al., 1989). Chromatography is employed to purify proteins and peptides from a complex solution at laboratory scale. For large-scale chromatography, a stationary phase must be selected which has the required strength and chemical stability (Dutta, 2008).

Membrane filtration is used to separate small particles using polymeric membranes under pressure. Based on the size of particles, membrane filtration is subcategorized as microfiltration, ultrafiltration, and reverse osmosis. Microfiltration is employed for colloids, fat globules, and cells with a pore size of 0.01–10 µm (Grandison & Finnigan, 1996). Ultrafiltration employs the same principle but is used for smaller particles with a pore size of 10–200 µm. In reverse osmosis, a macromolecule and a solvent are separated by a semi-permeable membrane under pressure (Belter et al., 1988). The pore size (1–10 µm) of the membrane for reverse osmosis enables all suspended and dissolved material to be separated (Dutta, 2008). First-generation membranes were made of cellulose acetate. Nowadays, inorganic and organic membranes are available (Lewis, 1996). Detailed description of membrane filtration can be found in the study of Grandison and Lewis (1996).

6.4.4 Purification

Especially for pharmaceuticals, finely purified final product is required. Therefore, crystallization, ultrafiltration, and various chromatographic methods are employed for final purification. Crystallization operates at low temperature and minimizes the thermal degradation of heat-sensitive materials (Shuler & Kargi, 2008). Drying is the removal of solvent from purified wet product. The physical properties of the product, its heat sensitivity, and desirable final moisture content have to be considered to conduct a drying process. Vacuum-tray driers, freeze driers, rotary drum driers, spray drum driers, and pneumatic conveyor driers are the major types used for purification (Shuler & Kargi, 2008).

6.5 Fermented foods

Fermentation of foods dates back many thousands of years, with grape and barley fermentation for alcoholic beverage production (Campbell-Platt, 1994). Fermentation can improve the sensory characteristics, increase the shelf life, and enhance the nutritional value of food. This section focuses on some of the results of actions of microorganisms, which cause biochemical and physical changes during the production of fermented foods.

6.5.1 Fermented dairy products

Lactic acid bacteria play an important role in the production of fermented dairy products. They ferment lactose and produce lactic acid, which decreases the pH of milk. As a result, pH reaches the isoelectric point of casein, which is the major protein in milk. At the isoelectric point, the total numbers of positive and negative charges on the amino acids are in equilibrium. At that point, casein precipitates and coagulum is formed (Üçüncü, 2005). This causes both physical and biochemical changes in the product. In addition to lactic acid, starter cultures may produce organic molecules, such as acetaldehyde, diacetyl, acetic acid, ethanol, exopolysaccharides, etc., which have effects on textural and flavor profiles of the product (Hutkins, 2006). Therefore, different types of cultures and production conditions create different characteristics for specific products.

6.5.1.1 Yogurt

Yogurt is a dairy product produced by fermentation of lactose in pasteurized milk. The starter cultures of yogurt are *Streptococcus thermophilus* and *Lactobacillus delbrueckii* subsp. *bulgaricus* (Üçüncü, 2005). At the beginning of fermentation, *S. thermophilus* grows faster than *L. delbrueckii* subsp. *bulgaricus* and decreases pH and E_h (oxidation/reduction potential) to levels suitable for the growth of *L. delbrueckii* subsp. *bulgaricus*. Moreover, *S. thermophilus* produces formic acid and CO_2, which are the prime growth factors for *L. delbrueckii* subsp. *bulgaricus* (Gürakan & Altay, 2010). *S. thermophilus* has lower proteolytic activity than *L. delbrueckii* subsp. *bulgaricus*. *L. delbrueckii* subsp. *bulgaricus* produces a proteinase that makes peptides and amino acids available for all organisms, including *S. thermophilus* (Hutkins, 2006).

During manufacturing, *S. thermophilus* and *L. delbrueckii* subsp. *bulgaricus* are prepared separately, because when the conditions are suitable for the growth of *L. delbrueckii* subsp. *bulgaricus*, they can produce more acid than can be tolerated by *S. thermophilus* and suppress the growth of *S. thermophilus*. The growth of streptococci is inhibited at pH of 4.2–4.4, but lactobacilli tolerate pH values in the range of 3.5–3.8 (Lourens-Hattingh & Viljoen, 2001). The fermentation for yogurt production occurs at 40–45 °C for approximately 6 h, until the pH is about 4.4–4.6 or a titratable acidity (as lactic acid) of 0.8–0.9% is reached (Gürakan & Altay, 2010; Hutkins, 2006). Since the growth of *L. delbrueckii* subsp. *bulgaricus* is favored at higher temperature than the growth of *S. thermophilus*, changes in incubation temperature may influence the growth rates of the two organisms, as well as the metabolic products they produce (Hutkins, 2006). For example, the acidity of the product may increase if the growth of *L. delbrueckii* subsp. *bulgaricus* is favored.

Streptococcus thermophilus and *L. delbrueckii* subsp. *bulgaricus* are thermophilic and use a secondary transport system (called LacS) for lactose uptake. Transport of lactose occurs via a symport mechanism, in which the proton gradient serves as the driving force (Hutkins, 2006). Then, *S. thermophilus* and *L. delbrueckii* subsp. *bulgaricus* hydrolyze the intracellular lactose by β-galactosidase and release glucose and galactose. Glucose is subsequently phosphorylated (using ATP as the phosphoryl group donor) by glucokinase (or hexokinase) to form glucose-6-phosphate. Glucose-6-phosphate is used to produce lactic acid via the Embden Meyerhof glycolytic pathway. Most strains of *S. thermophilus* and *L. delbrueckii* subsp. *bulgaricus* do not have an enzyme to metabolize galactose. Therefore, galactose is secreted back into the milk (Gürakan & Altay, 2010; Hutkins, 2006). In addition to lactic acid, both *S. thermophilus* and *L. delbrueckii* subsp. *bulgaricus* can produce acetaldehyde. In one pathway, an aldolase enzyme hydrolyzes the amino acid threonine, and acetaldehyde and glycine are formed directly. In the second pathway, the pyruvate can either be decarboxylated, forming acetaldehyde directly, or converted first to acetyl-CoA and then oxidized to acetaldehyde. Moreover, diacetyl, acetoin, and polysaccharides may also be produced by the yogurt culture (Hutkins, 2006).

6.5.1.2 Cultured buttermilk

Traditionally, cultured buttermilk was made from the fermentation of fluid remaining after cream is churned into butter. Now, low-fat milk is cultured to produce cultured buttermilk. The starter culture for buttermilk usually contains a combination of acid-producing bacteria including *Lactococcus lactis* subsp. *lactis* and/or *L. lactis* subsp. *cremoris* and flavor-producing bacteria, such as *L. lactis* subsp. *lactis* (diacetyl-producing strains), *Leuconostoc mesenteroides* subsp. *cremoris* or *Leuconostoc lactis*, and *Streptococcus lactis* subsp. *diacetylactis* in a ratio of about 5:1 (Hutkins, 2006; Oberman, 1985).

The acid-producing bacteria are homolactic, producing two lactic acid molecules per glucose molecule. The flavor-producing bacteria are heterofermentative and produce lactic acid as well as small amounts of acetic acid, ethanol, and carbon dioxide, which contribute to the flavor. Moreover, these bacteria may ferment citric acid from the Krebs cycle to diacetyl (Hutkins, 2006; Oberman, 1985). Diacetyl is the major flavor compound of cultured buttermilk (Hutkins, 2006). Citrate is transported by a citrate permease (CitP) with an optimum activity between pH 5.0 and 6.0, then acetate and oxaloacetate are produced by hydrolyzation of the intracellular citrate by citrate lyase. The oxaloacetate is decarboxylated by oxaloacetate decarboxylase and produces pyruvate and carbon dioxide, and the acetate is released into the medium. In citrate fermentation, acetaldehyde-TPP is formed from the oxidative decarboxylation of excess pyruvate by thiamine pyrophosphate (TPP)-dependent pyruvate decarboxylase and then, acetaldehyde-TPP condenses with another pyruvate molecule to form α-acetolactate. α-Acetolactate is decarboxylated in the presence of oxygen and forms diacetyl non-enzymatically.

Acetoin and 2,3-butanediol may be formed by the further reduction of diacetyl and these products have no effect on the flavor. Therefore, it is critical to adjust the optimum conditions to obtain desired amount of diacetyl in cultured buttermilk. Oxygen slows down the rate at which diacetyl is reduced to acetoin or 2,3-butanediol. Therefore, it may be useful to introduce air into the final product by agitation. Moreover, after diacetyl is produced, the low pH inhibits the reduction reactions that convert diacetyl to acetoin and 2,3-butanediol. If the temperature during incubation is greater than 24 °C, the citrate fermentors may be inhibited by the growth of the homolactic lactococci, which may cause too much acid production. In contrast, if the incubation temperature is lower than 20 °C, diacetyl formation may not occur due to lack of acid (Hutkins, 2006).

6.5.1.3 Kefir

Kefir is a self-carbonated fermented pasteurized milk drink. Kefir grains consist of bacterial and yeast microorganisms, which grow in a symbyotic relationship. The flavor compounds of plain kefir are lactic and acetic acids, diacetyl, and acetaldehyde, produced by homofermentative and heterofermentative lactic acid bacteria (*Lactobacillus brevis, Lactabacillus fermentum, Lactobacillus kefiri, Lactobacillus plantarum, Leuconostoc mesenteroides* subsp. *cremoris, Streptococcus thermophilus, Lactococcus lactis* subsp. *cremoris*), and ethanol, which is produced when the yeasts (*Candida kefir, Candida maris, Kluyveromyces marxianus, Saccharomyces cerevisiae*, etc.) ferment lactose (Guzel-Seydim et al., 2010; Hutkins, 2006). Acetic acid-producing bacteria, such as *Acetobacter aceti*, can involve kefir grains (Hutkins, 2006). The fermentation takes place at 25 °C for 24 h with agitation. When the pH reaches 4.6, fermentation ends (Guzel-Seydim et al., 2010).

6.5.1.4 Cheese

Cheese making is a very old method for the preservation of unpasteurized milk. When fresh milk becomes sour, the caseins aggregate to form a gel. Whey separation generally occurs when the gelled or clotted milk is kept for some time. These steps are the origin of cheese making. However, for centuries, milk has also been clotted by the addition of specific agents, especially chymosin (also called rennet), an enzyme extracted from the stomach of a calf, and starter cultures (Walstra et al., 2006). There are numerous cheese varieties that follow different production processes. Variations at one or more steps during production of cheese cause different textures and flavors (Gunasekaran & Ak, 2003). Microorganisms have also been used to produce milk-clotting enzymes (rennet). Microbial rennets can be of bacterial or fungal origin. *Bacillus subtilis, Bacillus polymyxa, Bacillus cereus*, and *Escherichia coli* can be used by bacterial rennet producers (Kandarakis et al., 1999; Üçüncü, 2005). Kandarakis et al. (1999) reported that the fermentation produced by chymosin from *E. coli* can be considered a successful alternative to calf rennet used for the manufacture of feta cheese, based on firmness, syncretic and sensory properties. *Mucor pusillus, Mucor miehei*, and *Endothia parasitica* are used to produce fungal origin rennets (Üçüncü, 2005). In acid-coagulated cheese making, acid may be formed from lactose by lactic acid bacteria (Fox et al., 2000; Walstra et al., 2006).

The composition of the starter culture used to make cheese depends on the type of end-product desired. For example, for curds that will be exposed to high temperatures, thermophilic lactic acid bacterial culture should be used to withstand high temperatures during production. Mesophilic cultures, such as *L. lactis* subsp. *lactis* and *L. lactis* subsp. *cremoris*, grow within a temperature range of about 10–40 °C. Thermophilic cultures, mainly *Lactobacillus helveticus, L. delbrueckii* ssp. *bulgaricus* and *S. thermophilus*, grow at a temperature around 42–45 °C. Culture inoculum is about 1% (w/w), which gives an initial cell concentration of about 5×10^6 cells per mL of milk in Cheddar type cheese production. Mozzarella is normally made with a 2% starter and Swiss cheese is made with 0.5% culture (Hutkins, 2006).

Moreover, microorganisms, including adventitious bacteria, yeast, and molds, are present in cheese throughout the curd ripening process and contribute either directly through their metabolic activity or indirectly through the release of enzymes into the cheese matrix through autolysis (Beuvier & Buchin, 2004). Scott et al. (1998) suggested that the primary factors in ripening are (i) storage temperature and humidity, (ii) chemical composition of the curd (fat content, level of amino acids, fatty acids, and other by-products of enzymatic action), and (iii) residual microflora of the curd.

The microflora of cheese and milk have effects on cheese quality. The microflora associated with cheese ripening are extremely diverse. Cheese ripening can be defined as the storage of the cheese at a controlled temperature and humidity until the desired texture and flavor have developed. The microflora are divided into two groups, which are the starter lactic acid bacteria and the secondary microflora (commonly known as non-starter lactic acid bacteria). The secondary microflora, including non-starter lactobacilli, *Pediococcus, Enterococcus*, and *Leuconostoc*, propionic acid bacteria, molds, yeasts, etc., are involved in the ripening process (Beresford & Williams, 2004). Crow et al. (2001) suggested that mesophilic lactobacilli are important in the maturation of cheeses, because they can catabolize citrate and could be involved in proteolysis and other enzymatic processes during cheese ripening (Vidojevic et al., 2007). *Leuconostoc* spp. also have the ability to co-metabolize sugars and citrate. In addition, *Enterococci* spp. metabolize citrate and can form acetaldehyde, acetoin, and diacetyl. Ray (1995) also reported that pediococci can produce diacetyl from glucose. Moreover, non-starter lactic acid bacteria contribute to protein breakdown.

Lactic acid bacteria could be involved in amino acid breakdown by using the α-ketoglutarate dependent transaminase pathway. The resultant α-keto acids are subjected to further enzymatic or chemical reactions to form hydroxy acids, aldehydes, alcohols and carboxylic acids, which are important for cheese flavor (Beresford & Williams, 2004). Lipase and esterase activities are found in non-starter lactobacilli and effect cheese characteristics. Tsakalidou et al. (1993) reported the beneficial effect of enterococci in cheese making as a contribution to the hydrolysis of milk fat by esterases. Methyl ketones and thioesters, which have effects on cheese flavor, can be formed by the conversion of released fatty acids (Beresford & Williams, 2004).

Lactic acid bacteria can protect the cheese against spoilage bacteria by producing lactic acid, H_2O_2, and antibiotics such as nisin. Giraffa (2003) reported that *Enterococcus faecium* is capable of producing a variety of bacteriocins, called enterocins, with activity against *Listeria monocytogenes*, which has an important impact on the safety of cheeses made from raw milk (Abdi et al., 2006). Strains of *L lactis* subsp. *lactis* and *Leuconostoc* spp. metabolize citrate to diacetyl in the presence of a fermentable sugar during production and early ripening. Therefore, diacetyl is a very significant aroma compound for unripened cheeses (Fox et al., 2000). Acetic acid may be the result of the catabolism of lactose, citrate, amino acids, or propionic acid fermentation. Moreover, propionic acid, butyric acid, hexanoic acid, and branched chain volatile fatty acids are important for cheese flavor.

Propionic acid is the end-product of lactic acid fermentation by propionibacteria. Butyric acid can originate from the catabolism of triglycerides and from lactate fermentation by clostridia (Beuvier & Buchin, 2004). Metabolism of fatty acids in cheese by *Penicillium* spp. involves release of fatty acids by the lipolytic systems, oxidation of β-ketoacids, decarboxylation to methyl ketone with one less carbon atom and reduction of methyl ketones to the corresponding secondary alcohol under aerobic conditions (Fox et al., 2000). Therefore, their levels depend on the balance between production and degradation, which depends on the degree of maturity of the cheese (Beuvier & Buchin, 2004). Ketones, especially methyl ketones, have typical odors and are important for the aroma of surface-mold ripened and blue veined cheeses (Curionia & Bossetb, 2002). The concentration of methyl ketones is related to lipolysis. In surface-ripened cheese, these compounds originate from the enzymatic activity of *Penicillium roqueforti*, *Penicillium camemberti* or *Geotrichum candidum* on ketoacids (Curionia & Bossetb, 2002; Fox et al., 2000). Indole and skatole are nitrogen-containing compounds that contribute to the aroma profile of some cheeses (Curionia & Bossetb, 2002).

6.5.2 Meat fermentation

Fermented ground meat (typically beef or pork) products are made from comminuted meat and fat, mixed with salt, curing agents such as $NaNO_2$, KNO_3, and spices, and filled into casings. The fermentation is carried out by lactic acid bacteria and enhances the microbiological stability and organoleptic properties of meat. The fermentation temperature is usually 15–26 °C and the relative humidity is around 90% (Lücke, 1985). The primary organisms which are responsible for the fermentation include species of *Lactobacillus*. *Pediococcus acidilactici*, *P. pentosaceus*, and *P. cerevisiae* are the bacteria used for fast, consistent, large-scale production of fermented meat products. *Lactobacillus sake* and *L. curvatus* are considered psychrotrophic, so they are suitable for fermentations conducted at cool ambient temperatures. Most of the bacteria used as starter cultures for fermented sausage, such as *P. acidilactici*, *L. plantarum*, and *L. sakei*, are able to produce bacteriocins (Hutkins, 2006). The main product of carbohydrate fermentation in meat fermentation is lactic acid. The rate and extent of lactic acid production from glucose, sucrose, or maltose are higher than from lactose, starch, or dextrin. In addition to lactic acid, acetic acid, butyric acid, propionic acid, acetoin, 2,3-butanediol, and/or diacetyl may be produced in minor amounts during fermentation (Lücke, 1985).

The genus of Micrococcaceae is also used in starter culture mixture. They are used in meat fermentation to convert nitrate to nitrite via expression of the enzyme nitrate reductase. This also causes formation of flavor and enhancement of color (Hutkins, 2006). They are the main agents of lipid degradation in the fermentation process. Lipids are the precursor of many odoriferous compounds, such as aldehydes, ketones, and short chain fatty acids. Molds and yeasts, which show lipolytic activity, contribute to the flavor of fermented meat products (Lücke, 1985).

6.5.3 Vegetable fermentation

The production principle of vegetable fermentation is the conversion of sugars to acids by natural lactic microflora, such as *Leuconostoc mesenteroides*, *Weisella kimchii*, *Lactobacillus plantarum*, *Lactobacillus brevis*, *Pediococcus pentosaceus*, etc. (Hutkins, 2006; Vaughn, 1985).

Sauerkraut is made by the fermentation of salted cabbage and includes three stage of microbial succession, which are initiated by heterofermentative *Leuconostoc mesenteroides*, followed by heterofermentative rods such as *Lactobacillus brevis*, and homofermentative rods (*Lactobacillus plantarum*) and cocci (*Pediococcus cerevisiae*). The optimum temperature of sauerkraut fermentation is around 21 °C. In the production of sauerkraut, 2.0–2.5% of salt is added. *L. mesenteroides* is salt tolerant and can grow fast and metabolize sugars via the heterofermentative pathway, yielding lactic and acetic acids, carbon dioxide, and ethanol. The environment created by *Leuconostoc mesenteroides* inhibits the growth of microorganisms except lactic acid bacteria (Battcock & Azam-Ali, 1998; Hutkins, 2006; Vaughn, 1985). Then *Lactobacillus brevis*, *L. plantarum*, and *Pediococcus cerevisiae* begin to grow and produce lactic acid, acetic acid, carbon dioxide, and ethanol (Vaughn, 1985). Moreover, mannitol, diacetyl, acetaldehyde, and other volatile flavor compounds may be produced during fermentation of sauerkraut. Mannitol is produced from fructose via the NADH-dependent enzyme mannitol dehydrogenase by heterofermentative leuconostocs and lactobacilli. The fermentation process completes when the acidity is at about 1.7% (w/v), with a pH of 3.4–3.6 (Hutkins, 2006).

Pickling technology, such as cucumber fermentation, also includes lactic acid fermentation. Unlike sauerkraut fermentation, growth of *Leuconostoc* spp. is inhibited with the addition of salt up to 8% (Vaughn, 1985). The fermentation is initiated by *Pediococcus cerevisiae*, *L. brevis*, and *L. plantarum* (Battcock & Azam-Ali, 1998; Hutkins, 2006; Vaughn, 1985).

In olive fermentation, in addition to lactic acid bacteria, non-lactic organisms (e.g. *Pseudomonas* and Enterobacteriaceae) may affect the fermentation process and the final product consists of lactic acid, but also acetic acid, citric acid, malic acid, carbon dioxide, and ethanol at pH 4.5 and acidity less than 0.6% (Hutkins, 2006).

6.5.4 Vinegar

Vinegar is an aerobic fermentation of various ethanol sources to produce a dilute solution of acetic acid (Adams, 1985). The minimum acetic acid concentration in vinegar should be 4% (w/v). The name of the vinegar indicates its juice source, such as red wine vinegar, apple cider vinegar, malt vinegar, etc. (Hutkins, 2006).

The manufacture of vinegar includes two stages of fermentation. First, ethanol is produced by yeast, such as *S. cerevisiae*, via the fermentation of sugars. Then, the ethanol is oxidized to acetic acid by *Acetobacter*, *Gluconobacter*, and *Gluconoacetobacter* (Adams, 1985; Hutkins, 2006). In the first stage, high sugar concentration and

low pH (~4.5) favor ethanol production and as a result of alcoholic fermentation, anaerobic conditions are created, and acid and ethanol concentrations increase. Then, aerobic conditions are re-established at the surface for the oxidation of ethanol to acetic acid (Adams, 1985). The pathway of acetic acid production includes two main steps. In the first step, acetaldehyde is produced by oxidation of ethanol. Then, the acetaldehyde is oxidized to acetic acid. The acetic acid fermentation is performed by obligate aerobes in the periplasmic space and cytoplasmic membrane (Hutkins, 2006). The conversion of ethanol substrate to acetic acid can be performed by the open vat method, the trickling generator process (quick vinegar process), and aerobic submerged fermentation process (Adams, 1985; Hutkins, 2006).

6.5.5 Alcoholic beverages

6.5.5.1 Wine

Grape wine fermentation is initiated by crushing the grape, producing the juice, removing the pomace and then inoculating with *S. cerevisiae*. The acid and sugar levels of fruit juice and low pH (~4.5) enhance the growth of yeasts and production of ethanol. The ethanol produced restricts the growth of bacteria and fungi. At the beginning of fermentation, several species of yeast, including *Kloeckera*, *Hanseniaspora*, *Candida* and *Metschnikowia*, may grow and produce small amounts of acids, esters, glycerol, and other potential flavor components (Battcock & Azam-Ali, 1998). The optimum temperature is between 10 °C and 18 °C for white wine and between 20 °C and 30 °C for red wine fermentation (Battcock & Azam-Ali, 1998; Hutkins, 2006).

Flavor components, such as 4-methyl-4-mercaptopentan-2-one and 3-mercapto-1-hexanol, can be produced during fermentation and the production of these compounds depends on the strain. Moreover, some common yeast by-products from ethanol fermentations (e.g. glycerol, and acetic acid) affect the flavor of wine. Acetaldehyde is also formed from pyruvate during vinification and is a precursor metabolite for acetate, acetoin, and ethanol synthesis and binds to proteins or individual amino acids to produce aroma compounds. During yeast fermentation, many medium- and long-chain fatty acids, which contribute to the wine aroma, are also formed via the fatty acid synthesis pathway from acetyl-CoA (Styger et al., 2011).

Malolactic fermentation by lactic acid bacteria is desired for high-acid wine. It involves the deacidification of wine via conversion of dicarboxylic L-malic acid to monocarboxylic L-lactic acid and carbon dioxide. *Oenococcus oeni*, *Lactobacillus* spp., *Leuconostoc* spp., and *Pediococcus* spp. carry out the malolactic fermentation (Hutkins, 2006; Styger et al., 2011).

6.5.5.2 Beer

Beer is created by the fermentation of malted barley that is heated with hops to produce the wort for yeast fermentation. The main process stages of beer making are malting, milling, mashing, brewing, cooling, fermentation, maturation, filtering, and packaging. In the fermentation process, yeast ferments the sugars and produces ethanol and CO_2. The strains of *S. cerevisiae* are used for "ale" or "top-fermenting yeast" beers. *Saccharomyces pastorianus* is used for lager beers or "bottom-fermenting yeast." Top-fermenting yeasts form low-density clumps or flocs, which cause the entrapment of CO_2 and rise to the surface. In lager beer fermentations, *S. pastorianus* flocculate, and the flocs settle to the bottom of the fermentation tank. The optimum fermentation temperatures are 18–27 °C for top-fermenting yeast and below 15 °C for bottom-fermenting yeast. The inoculum of yeast population is around 10^6 cells per mL of wort. The fermentation vessel should be sparged with sterile air before inoculation to decrease the lag phase. Moreover, biosynthesis of cell membrane lipids, which are essential for growth of yeast in wort, requires oxygen. Sugar metabolism by yeast occurs via the Embden-Meyerhoff-Parnas (EMP) glycolytic pathway. The ethanol percentage on a weight basis can vary based on the process and is about 3.4% in a regular US beer (Hutkins, 2006).

6.5.6 Bread

Saccharomyces cerevisiae is used for bread production. It should have special characteristics, such as ability to produce good bread flavor, stability and viability during storage. The seed growth medium includes molasses, various ammonium salts such as ammonium phosphate, magnesium sulfate, calcium sulfate, and lesser amounts of the trace minerals (e.g. zinc and iron). The optimum growth conditions are aerobic, 30 °C, pH of 4.0–5.0. Then, yeast is added to flour as yeast cream that contains about 20% total yeast solids. The fermentation process includes the use of glucose via aerobic and/or anaerobic pathways (Hutkins, 2006). Yeast produces CO_2, which causes an elastic, airy texture. Yeast also produces acid, which can affect the flavor of the bread (Hutkins, 2006). In addition to yeast,

Lactobacillus brevis var. *lindneri*, *L. fructivorans*, and *L. farciminis* are used as a starter culture in sour dough rye bread. These cultures also produce acetic acid and some flavor compounds (Hutkins, 2006; Sugihara, 1985).

6.5.7 Tempeh

Tempeh is a popular fermented soybean product that is common in Far East countries such as Indonesia. Tempeh production includes four main steps: soaking, boiling, inoculating of soybean with the fungal organism *Rhizopus microsporus* subsp. *oligosporus*, and incubating at room temperature (Astuti et al., 2000; Hutkins, 2006). The production of tempeh takes about 48 h. During the steeping period, endogenous lactic acid bacteria grow and reduce the pH. This causes inhibition of the growth of undesirable spoilage bacteria and pathogens. *Rhizopus oligosporus* is inoculated at levels from 10^7 to 10^8 spores per kg of beans. The recommended strains of *Rhizopus oligosporus* are NRRL 2710 and DSM 1964, which were isolated from Indonesian tempeh (Hutkins, 2006). In addition to *R. oligosporus*, *R. oryzae*, *R. arhizus*, and *R. stolonifer* are also used in Indonesia (Astuti et al., 2000). Alternatively, "usar," which is produced by inoculating wild *Rhizopus* spores onto the surface of leaves obtained from the indigenous Indonesian *Hibiscus* plant, can be used to inoculate soybeans.

The fermentation process is a solid-state fermentation, which includes soybeans held together by the mold mycelium that grows throughout and in between the individual beans. The fermentation conditions include little excess water and are microaerophilic. The fermentation is considered to be completed when the beans are covered with white mycelium before the mold begins to sporulate. During fermentation, *R. oligosporus* excretes lipases and proteinases. As a result of lipid hydrolysis of oil in soybeans, mono- and diglycerides, free fatty acids, and a small amount of free glycerol are produced. While oxidation of most of the free fatty acids occurs due to *R. oligosporus* acitivity, only about 10% of the released amino acids and peptides are oxidized by *R. oligosporus* (Hutkins, 2006). Astuti et al. (2000) reported that 3.6–27.9% of the total amino acids decrease occurs during fermentation due to consumption of amino acids as a source of nitrogen for growth by strains of *Rhizopus*. *R. oligosporus* also produces polysaccharide-hydrolyzing enzymes. These enzymes hydrolyze pectin, cellulose, and other fiber constituents, and various pentoses (xylose, arabinose) and hexoses (glucose, galactose) are released (Hutkins, 2006).

6.6 Enzyme applications

Enzymes are proteins (MW of 15,000 to several milllion daltons) and catalyze the specific biochemical reactions in biological systems (plants, animal, and microbial) (Aurand et al., 1987; Berk, 1976; Shuler & Kargi, 2008). Many enzymes are conjugated proteins containing specific prosthetic groups. The complete enzyme is called a haloenzyme; the protein part is called the apoenzyme, and the non-protein moiety the prosthetic group (Berk, 1976; deMan, 1996). The prosthetic group is often only loosely attached to the protein and can be separated by dialysis. In this case, the protein moiety is termed the apoenzyme and the prosthetic group the co-enzyme or co-factor (Aurand et al., 1987; Berk, 1976). Nicotinamide adenine dinucleotide (NAD) is an example of a co-enzyme, which binds to enzyme less strongly than prosthetic groups; thus co-enzymes dissociate from the enzyme during or at the termination of a catalytic reaction (Aurand et al., 1987).

At the first stage of enzyme mechanism, the substrate combines with the catalyst into a substrate-enzyme complex. The complex is unstable under reaction conditions and is quickly decomposed. The enzyme is regenerated. The substrate, which emerges from the reaction, can be in the final, changed form or in a highly activated state. When an enzyme reacts with its substrate, only the "active site," which is certain regions of the protein molecule, participates in the process. Special groups of amino acid residues produced by the sequence and particular folding of the enzyme protein form active sites. The portion of the protein molecule outside the active regions provides the backbone and support for the active sites.

The rate of enzyme-catalyzed reaction is proportional to the concentration of the enzyme. The rate also depends on the concentration of the substrate. At low substrate concentration, reaction rate is related to substrate concentration linearly and complies with first-order kinetics (Aurand et al., 1987). Substrate binding is a prerequisite for catalysis. Therefore, enzymes have specificity and susceptibility to inhibition by certain substances. Substrate specificity means that the enzyme is selective towards the substrate. For example, the hydrolytic enzyme arginase catalyzes cleavage of the amino acid L-arginine, into urea and L-ornithine, but will not do so with D-arginine. Enzyme inhibitors slow down the rate of enzymatic reactions. Inhibitors, termed antimetabolites, may block important metabolic pathways. For example, Bowman-Birk trypsin isoinhibitors present in soybean may inhibit the proteolytic action of trypsin (Tan-Wilson et al., 1987). Moreover, the number of active sites available for binding the substrate may decrease due to the effect of a substance which is sufficiently similar in structure to the substrate. For example, if malonic acid is added to succinic acid, it will compete with succinic acid for the active site, which makes the succinic acid dehydrogenase unavailable for converting the succinic acid into fumaric acid (Berk, 1976).

Temperature is a factor of the rate of enzyme-catalyzed reactions. An increase in temperature increases the velocity of the reaction within definite limits (Aurand et al., 1987). Most enzymes are thermolabile and application of 70–80 °C for 2–5 min can destroy their activity (Berk, 1976). Initially the reaction rate increases with increasing temperature to an optimum, but then as the temperature increases further, the reaction rate decreases eventually to zero. Above about 45 °C, thermal denaturation of proteins starts, and the activity of the enzymes gradually decreases (Aurand et al., 1987). The enzyme activity is also greatly affected by changes in pH of its medium, which cause changes in the ionization of the enzyme, substrate, or enzyme-substrate complex. Moreover, low or high pH can cause inactivation due to denaturation of the enzyme protein (Aurand et al., 1987).

Instead of quantification of the enzyme in terms of direct concentration, the quantity is expressed as "activity" or the rate at which the specific reaction proceeds in the presence of specific concentrations of enzyme preparation, under standard conditions (Berk, 1976). The reaction velocity can be measured by static or dynamic methods. In the static method, the substrate and enzyme are mixed and the reaction is stopped at a certain time and the concentration of remaining substrate or of product formed is determined. In the dynamic method, the substrate concentration is continuously monitored for a specific time period and the change in the substrate or product concentration from zero time to time t represents the enzyme activity (Aurand et al., 1987).

The names of enzymes are formed by adding the suffix -ase to the name of the substrate, e.g. lipases, proteinases, amylase, maltase, etc. (Berk, 1976; Eskin, 1990). The Enzyme Commission number (EC number) of the International Union of Biochemists allows systematic classification, which divides the enzymes into six main groups according to the type of reactions involved, as now briefly described.

Oxidoreductase enzymes catalyze the oxidation-reduction reactions. This group includes the enzymes commonly known as dehydrogenases, reductases, oxidases, oxgenases, hydroxylases, and catalases (Berk, 1976). For example, glucose oxidase is used to minimize the Maillard reaction, catalase is used to remove H_2O_2 in the milk industry, and lipoxygenase is used in the baking industry (Eskin, 1990). An example for this reaction is:

$$L\text{-iditol} + NAD \rightarrow L\text{-sorbose} + NADH_2$$
(Boyce & Tipton, 2001)

Transferases are enzymes that catalyze the transfer of various chemical groups (methyl, acetyl, aldehyde, ketone, amine, phosphate residues, etc.) from one substrate to another (Berk, 1976). Transminases and kinases are typical examples of this group (Aurand et al., 1987). An example for this reaction is:

S-adenosyl-L-methionine
  + 3-hexaprenyl-4,5-dihydroxybenzoate →
S-adenosyl-L-homocysteine + 3-hexaprenyl-4-hydroxy
  -5-methoxybenzoate (Boyce & Tipton, 2001)

Hydrolases are responsible for the hydrolytic cleavage of bonds by the addition of a water molecule. For example, amylases are used for the production of glucose syrups from cornstarch, pectinases are used to clarify juices, and proteases are used as meat tenderizers. An example for this reaction is:

$$4\text{-hydroxybenzoyl-CoA} + H_2O \rightarrow$$
$$4 - \text{hydroxybenzoate} + CoA \text{ (Boyce & Tipton, 2001)}$$

Lyases catalyze the non-hydrolytic cleavage of C-C, C-O, and C-N bonds, and other bonds by elimination, leaving double bonds or rings or conversely, adding groups to double bonds. They remove groups from their substrates, leaving double bonds. Examples are carboxylases, aldolases, hydratases, and phenylalanine ammonia lyase (Aurand et al., 1987; Berk, 1976). Pectate lyases have potential applications in the medicine, textile, and food industries to improve the surface properties of natural fibers (Han et al., 2011). An example for this reaction is:

$$2\text{-hydroxyisobutyronitrile} \rightarrow \text{cyanide} + \text{acetone}$$
(Boyce & Tipton, 2001)

Isomerases transform their substrates from one isomeric form to another. Racemases and epimerases are members of this group (Berk, 1976). For example, glucose isomerase is used for the production of high-fructose corn syrups. An example of this reaction is:

$$(2S)\text{-2-methylacyl-CoA} \rightarrow (2R)\text{-2-methylacyl-CoA}$$
(Boyce & Tipton, 2001)

Ligases catalyze reactions that join two molecules together (C-O, C-S, C-N, or C-C bonds), which represents a synthesis (Berk, 1976). For example, aminoacyl-tRNA synthetases catalyze the linking together of two substrate molecules with the breaking of a pyrophosphate bond (Aurand et al., 1987). An example for this reaction is:

$$4\text{-chlorobenzoate} + CoA + ATP \rightarrow 4\text{-chlorobenzoyl-CoA}$$
$$+ AMP + \text{diphosphate (Boyce & Tipton, 2001)}$$

Although there are six main divisions of enzyme, each enzyme group has many subgroups. Enzymes are given a code number, such as EC 3.1.1.2 (carboxylic ester hydrolases) and EC 4.2.1.20 (tryptophan synthase). The EC number contains four elements separated by dots. The first number symbolizes the division of enzyme out of six main enzyme groups. The second and third numbers show the subclass and sub-subclass, respectively. The fourth number indicates the serial number of the enzyme in its sub-subclass (Moss, 1992).

Some of the traditional enzymes are prepared from animal and plant sources. For example, papain from papaya (*Carcia papaya*), bromelain from pineapple (*Ananas sativa*), and ricin from the castor oil plant (*Ricinus communis*) have a high proteolytic effect (Üçüncü, 2005). Rennet, an animal derivative, is extracted from the fourth stomach of a calf or young goat (Carroll, 1994). For isolation of enzyme, the tissue is usually homogenized in the presence of an extraction buffer that often contains an ingredient to protect the enzyme from oxidation and from traces of heavy metal ions. Then, protein impurities are removed by fractional precipitation, ion exchange chromatography, etc. (Belitz et al., 2004).

Nowadays, most industrial enzymes are of microbial origin. Microbial enzymes can be heat stable and have a broader pH optimum. Most of these enzymes are prepared by batch cultivation of highly developed strains of microorganisms. Moreover, gene transfer to carrier microorganisms makes microbial enzyme production possible (deMan, 1996). For example, the gene for prochymosin (rennet) has been cloned in *Escherichia coli*, *Saccharomyces cerevisiae*, *Kluyveromyces marxianus*, *Aspergillus nidulans*, *Aspergillus niger*, and *Trichoderma reesei*. As a result of recombination, enzymatic properties of the recombinant enzymes are indiscernible from those of calf chymosin (rennet). The cheese-making properties of recombinant chymosins have been assessed in many cheese varieties, mostly with very satisfactory results (Fox et al., 2000). Moreover, *Mucor pusillus* and *Aspergillus oryzae* are used to produce proteinase (Belitz et al., 2004).

Enzyme preparations are expensive. Although the enzyme is regenerated in the process and could be used over and over again, its separation from the reaction mixture is not usually economically feasible. Thus immobilized enzymes are employed for enzyme reuse. Immobilization is the attachment of enzymes to insoluble matrices. Therefore, the enzyme can be easily separated from the reaction mixture (Berk, 1976). Immobilized or bound enzymes have been prepared by physically or chemically binding the enzyme to an insoluble support. Immobilized enzymes can be used in batch processes and then recovered by filtration or centrifugation for further use. Applications of this technology facilitate repeated use of the same enzyme preparation in a batch process. Immobilized enzymes can also be used in continuous processes in specially designed reactors, such as continuous feed stirred tank, packed-bed reactor, fluidized bed reactor, and more. Immobilized enzymes are generally more stable to heat than soluble enzymes. Immobilized enzymes have been used in the food industry (Eskin, 1990). For example, an immobilized lactase system has been developed by adsorption of enzyme on porous glass beads and stainless steel, covalent linkage to organic polymers, entrapment within gels or fibers, and microencapsulation within nylon or cellulose microcapsules to produce low-lactose milk (Eskin, 1990). Detailed examples of the application of immobilized enzymes in foods can be found in *Biochemistry of Foods* by Eskin (1990).

6.6.1 Industrially important enzymes

Microbial enzymes have been used in the food industry for centuries. They also had applications in the leather industry, such as using dung for preparation of hides (Underkofler et al., 1958). In the 1930s, enzyme technology was used for the first time in the food industry, to clarify fruit juice (Adler-Nissen, 1987). Since then, enzymes have found uses in the food, detergent, textile, leather, cosmetic, and pharmaceutical industries, due to their benefits, such as specificity, mild conditions, environmentally friendly, and less waste generation (Hasan et al., 2006). In this section, some major industrial enzymes will be presented.

6.6.1.1 Lipases

Lipases (triacylglycerol acylhydrolase, EC 3.1.1.3) act on carboxylic ester bonds, and require no co-factor. Long-chain fatty acids are the natural substrates for lipase (Ghosh et al., 1996). Lipases are of interest for industry because of their natural function of hydrolyzing triglycerides into diglycerides, monoglycerides, fatty acids, and glycerol (Ghosh et al., 1996). Furthermore, substrate specificity, regioselectivity, and enantioselectivity are properties of lipases that attract industrial interest (Hasan et al., 2006). Protein engineering allows the improvement of properties of lipase, including active site, specific activity, stability, specificity, calcium binding, and surfactant compatibility (Svendsen, 2000). Lipases are monomeric proteins with a molecular weight of 19–60 kDa and the characteristics of the enzyme are highly dependent on position of the fatty acids in the glycerol backbone, chain length, degree of unsaturation, and other factors (Aravindan et al., 2007).

Lipase activity is pH dependent but studies have shown that lipases are stable in a pH range of 4–8. On the other hand, the temperature profile of lipases varies based on the source of the enzyme. For example, lipases derived from mesophilic microorganisms are active at 30–35 °C, whereas this value reaches 40–60 °C when the microbial source is a thermophile (Aravindan et al., 2007).

Lipases have applications in several industries and most lipases are commercially available (Houde et al., 2004). The advantages of microbial lipases over plant- or animal-derived enzymes include availability of large amounts of purified lipase, higher stability, activity at elevated temperatures, reduced by-products, cheaper downstream processing, and activity in organic solvents (Hasan et al., 2006).

Infant formulas are a good alternative to breast milk asfar as they have the same properties as human milk. However, the fat in infant formulas is derived from plants and differs from human milk, and binds the calcium and causes constipation. Lipase is used to modify the triglyceride to increase the palmitic acid proportion of the milk and the absorption capability (Houde et al., 2004).

Cocoa butter fat is a high-value product with desired properties such as melting point (37 °C), snap, and gloss. It also provides a cooling sensation and smoothness for chocolate (Hasan et al., 2006; Houde et al., 2004). However, a cheaper substitute for cocoa butter, palm oil, has a melting point of 23 °C, is liquid at room temperature, and is a low-value product. Lipases are used to convert palm oil into cocoa butter substitute by transesterification reactions (Aravindan et al., 2007). Such modification of less expensive fats, such as shea butter, salt fat, and palm oil, can be obtained by transesterification and provides cheap substitutes for cocoa butter (Houde et al., 2004).

In the baking industry, lipases are widely used to improve flavor, texture and softness, as well as to increase loaf volume (van Oort, 2010). Lipase esterifies triglycerides to mono- and diglycerides to enhance the flavor content of bakery products (Aravindan et al., 2007). Studies have shown that a lipid film surrounding the gas cell increases gas cell stability (van Oort, 2010), which can be provided by esterification of triglycerides (Hamer, 1995).

Lipases are commonly used in the dairy industry to hydrolyze milk fat, and current applications of lipases in the dairy industry include cheese ripening, flavor enhancement, manufacturing cheese-like products, and lipolysis of cream and butterfat (Aravindan et al., 2007). Cheese texture is dependent on fat content so lipases that release short-chain fatty acids (C4 and C6) develop the sharp and tangy flavor, whereas release of medium-chain fatty acids (C12 and C14) causes a soapy taste in the product (Hasan et al., 2006). Lipases are also used for enzyme-modified cheeses (EMC) to liberate fatty acids at *sn-1* and *sn-3* positions on the glycerol backbone (Houde et al., 2004). EMC find applications in the food industry to add cheese flavor to salad dressings, dips, soups, sauces, and snacks (Aravindan et al., 2007).

The detergent industry is another example of lipase application for lipid degradation (Gandhi, 1997). In fact, lipases find their most significant application in the detergent industry, approximately 30% of the total enzyme market (Houde et al., 2004). Common commercial uses of lipases for detergents are dish washing, bleaching composition, dry-cleaning solvents, liquid leather cleaner, contact lens cleaning solutions, cleaning of lipid-clogged drains, and so on (Gandhi, 1997). Lipases follow a two-step mechanism: the first step is binding of the enzyme surface and the second step is hydrolysis of carboxyl ester bonds (Galante & Formantici, 2003). Lipases therefore hydrolyze the water-insoluble triglycerides into water-soluble products (mono- and diglycerides, and glycerol) (Galante & Formantici, 2003). Enzymes are preferred in the detergent industry because they can reduce the environmental load of detergent products, enable lower wash temperatures, reduce undesired chemicals in the detergent formulation, leave no harmful residues, and are safe for aquatic life (Hasan et al., 2006).

In the leather industry, fat residues have to be removed during hide and skin processing. Conventional methods involve use of organic solvents and surfactants which are of environmental concern (Hasan et al., 2006). Lipases replace chemical substances to remove the fats and grease. Enzymes are used in the bating step in which non-collagenous, globular proteins are degraded to soften hides before tanning (Galante & Formantici, 2003). Lipases endow a more uniform color and cleaner appearance. They also improve the production of waterproof leather (Hasan et al., 2006). Muthukumaran and Dhar (1982) reported that lipase derived from *Rhizopus nodosus* was used for the production of suede clothing leathers from sheepskins.

Biodiesel is a renewable, environmentally safe, and energy-efficient bioenergy. In the production of biodiesel, plant and animal lipids can be treated with lipase to produce free fatty acids, which are esterified to methanol to produce methyl ester fatty acids and glycerin. Although any fatty acid can be a raw material for the production of biodiesel, waste vegetable oil and animal fats are preferred due to the fact that they are not food-value products (Refaat, 2010). Production of biodiesel from waste oil starts with pretreatment of raw material, which includes filtration to remove dirt, food residues, and non-oil materials. After determining the concentration of fatty acid, transesterification is performed to obtain biodiesel. Korus et al. (1993) reported that some catalysts could be used to improve transesterification such as potassium hydroxide, sodium hydroxide, sodium methoxide, or sodium ethoxide.

Applications of lipases are not limited to the examples given above. They have found applications in cosmetics, for instance in hair waving preparations, antiobesity creams, wax esters in personal care products, retinol, and so on (Hasan et al., 2006). The meat, beverage, paper, and healthy food industries also employ lipases for fat removal and flavor development, aroma improvement, hydrolysis, and transesterification, respectively (Lantto et al., 2010).

6.6.1.2 Amylases

Enzymes which are capable of hydrolyzing the α-1,4-glucosidic linkages of starch are called amylases (Vihinen & Mantsiila, 1989). Although amylases are found in plants and animals, microbial amylases are most common in industry. α-Amylase (EC 3.2.1.1) and glucoamylase (EC 3.2.1.3) are two major amylases (Pandey et al., 2000). Industry has taken advantage of the starch-degrading properties of these enzymes (Bigelis, 1993). Starch is produced mainly in higher plants, and is water insoluble

(Aiyer, 2005). It is composed of two components: amylose and amylopectin. Amylose is a mainly linear polysaccharide which is formed by α-1,4 linked D-glucose residues and some α-1,6-branching points. Amylopectin has a highly branched tree-like structure. The proportion of branches is an important property of the substrate because enzymes hydrolyze different substrates with differing specificities.

Several amylolytic enzymes hydrolyze starch or its degradation products. The actions of these enzymes can be divided into two categories. Endoamylases break down linkages randomly in the interior of the starch molecule. Exoamylases hydrolyze the polysaccharide from the non-reducing end, which produces short end-products. α-Amylase (endo-α-1,4-D-glucan glucohydrolase) hydrolyzes the α-1,4-D-glucosidic linkages in the linear amylase chain, randomly. However, glucoamylase (exo-α-1,4-α-D-glucan glucohydrolase) cleaves the α-1,6-linkages at the branching points of amylopectin as well as α-1,4-linkages (Pandey et al., 2000). β-Amylase (α-1,4-glucan maltohydrolase, EC 3.2.1.2) is originally derived from plants although there are a few microbial strains which can produce it. β-Amylase can cleave non-reducing ends of amylase and amylopectin, which results in incomplete degradation (Pandey et al., 2000).

Amylases are extensively employed for starch processing in which gelatinization, liquefaction, and saccharification occur. Gelatinization is the dissolution of starch granules at above 60 °C in the presence of thermostable α-amylase derived from *Bacillus subtilis* or *B. licheniformis* (Bigelis, 1993; Maarel, 2010). In this step, pH adjustment is important to provide the optimum environment for amylase to function. In addition to pH adjustment, addition of calcium is also necessary to stabilize the enzyme (calcium prevents unfolding of the enzyme) (Maarel, 2010). The next step in starch hydrolysis is liquefaction at which slurry is held at 95–100 °C for 1–2 h. During liquefaction, thermostable amylase breaks down the α-1,4-linkages in both amylose and amylopectin and generates dextrins (Maarel, 2010). The liquefaction process is continued until the desired dextrose equivalent (DE) value is reached (Bigelis, 1993). The DE value is determined as the number of reducing ends relative to a pure glucose of the same concentration. The DE value is 100 for glucose and 0 for starch. In liquefaction, usually, DE of 10–15 is obtained (Bigelis, 1993).

The final step of starch hydrolysis is saccharification. The liquefied slurry is processed at about 60 °C with the addition of glucoamylase. In saccharification, pullanase, glucoamylase, β-amylase, or α-amylase can be employed.

Glucoamylases hydrolyze α-1,4-linkages with additional activity on α-1,6-linkages. Moreover, addition of pullunase in glucoamylase increases the glucose yield about 2%, while a cocktail of pullunase and β-amylase produces maltose (Maarel, 2010). Pullunase addition has more benefits such as shorter saccharification time, reduction of glucoamylase amount used, and increase in dry matter content (Bigelis, 1993). Maltose, glucose, or mixed syrups are end-products of saccharification according to the enzyme used.

Amylases are mainly used in the food processing industry such as for baking, brewing, high-fructose corn syrup, alcoholic beverages, cakes, and fruit juices (Couto & Sanroman, 2006). Amylases produced from *Aspergillus oryzae*, *A. niger*, *A. awamori* or species of *Rhizopus* are used in dough making to degrade starch in the flour into dextrins and provide fermentable sugars for yeast (Souza & Magalhaes, 2010). The addition of amylase therefore reduces the viscosity of dough, increases fermentation rate, and improves taste and crust color by generating sugar (Bigelis, 1993). Volume and anti-firming are the other benefits of using amylases in baking (Hamer, 1995).

High-fructose corn syrup (HFCS) is used as a sweetener in beverages and foods. It has fewer calories and is cheaper than sucrose (Bigelis, 1993). In the United States, corn is the main raw material used to produce glucose syrup by hydrolysis of starch in the presence of amylases (Pomerantz, 1991). After glucose syrups are produced, glucose isomerases (EC 5.3.1.18) can convert glucose syrup into glucose-fructose mixture. Immobilization of glucose isomerase reduces the cost of HFCS and increases the production of glucose-fructose syrup. However, glucose isomerase only converts 50% of the glucose into fructose (Maarel, 2010).

Most alcoholic beverages, such as whiskey, vodka, and brandy, are produced from sugar-containing raw materials. Malted barley, corn, milo, and rye are common raw materials for alcohol fermentation in the United States (Bigelis, 1993). The raw material is cooked to gelatinize the starch for enzymatic degradation, and cooled to room temperature before saccharification by amylase is performed. Fungal amylase, from *Aspergillus* or *Rhizopus* spp., is used for saccharification, because it increases the reaction rate while a complete saccharification is performed. It also produces fewer by-products, for example, maltose, isomaltose, and oligosaccharides, that are not fermentable by yeast (Bigelis, 1993).

Amylases are used in the detergent industry for laundry and dish washing to degrade the starch-containing residues to smaller oligosaccharides and dextrins (Olsen & Faholt, 1998). It is reported that 90% of liquid laundry detergents contain α-amylase (Galante & Formantici, 2003; Souza & Magalhaes, 2010). Amylases are active at lower temperature and alkaline pH (Souza & Magalhaes, 2010). Oxidative-stable amylases are also developed by protein engineering of proteases in *Bacillus licheniformis*. Therefore, among all known microbial amylases, *B. licheniformis* enzyme is the most appropriate because of its thermostability and resistance to proteolytic digestion (Galante & Formantici, 2003).

In the textile industry, amylase is used for desizing. Sizing is defined as coating the yarn before fabric weaving with a layer of removable material to enhance a fast and secure weaving process (Souza & Magalhaes, 2010). However, sizing agents have to be removed before bleaching, dying, or finishing. Sizing agents can be starch, cellulose, gelatin, polyester, etc.; starch is extensively employed because it is cheap, easily available, and can be removed easily (Galante & Formantici, 2003). Amylases are used to remove starch from yarn. Low-temperature (20–40 °C) and high-temperature (90–105 °C) amylases are used. Amylases require calcium to maintain stability at high temperatures but in the textile industry the use of calcium is not practical because free calcium availability is an issue in the presence of cellulose (Galante & Formantici, 2003). However, engineered amylases from *B. licheniformis* have overcome this problem with a wide range of pH and thermal stability (van der Laan, 1995).

Ethanol can be produced synthetically and naturally by microorganisms. Various feedstock and chemically defined media can be used for ethanol fermentation. The most commonly used types of feedstock for ethanol production are corn, sugar cane, and wheat (Balat et al., 2008). Corn is the main raw material for ethanol production in the US, accounting for around 97% of the total ethanol production, whereas in Brazil sugar cane is the main raw material. However, corn is a starchy material and requires starch hydrolysis before fermentation since *S. cerevisiae* cannot ferment polysaccharides.

Amylases have been applied in the fuel industry to convert starch-containing raw material into glucose monomers before fermentation. In the pulp and paper industry, amylases are used to modify the starch for paper coating. Coating provides a smooth and strong paper surface, which improves writing quality. Starch used for paper coating should have low viscosity and high molecular weight (Gupta et al., 2003). α-Amylase is used to partially degrade starch polymers to decrease the viscosity (Souza & Magalhaes, 2010).

6.6.1.3 Pectic enzymes

These enzymes can hydrolyze the long and complicated molecules named pectins that are structural polysaccharides in the plant cell and maintain integrity of the cell wall (Bigelis, 1993). Pectic enzymes are common commercially used enzymes and account for 25% of food enzymes sales worldwide (Jayani et al., 2005). *Aspergillus* species are the most common microbial source for these enzymes but *Coniothyrium diplodiella*, *Sclerotinia libertiana*, *Penicillum* spp., and *Rhizopus* spp. are alternative producers (Bigelis, 1993). Pectic substances are high molecular weight (30,000–300,000 Da), negatively charged complex polysaccharides, with a backbone of galacturonic acid residues linked by α-1,4-linkages (Kashyap et al., 2001). The monomers of galacturonic acid are esterified by methyl groups or galactans, arabinogalactans, arabinans, and rhamnogalacturonans (Bigelis, 1993).

The American Chemical Society has categorized pectic substances into four main groups. The first group, called protopectins, are water-soluble pectic substances composed of pectin or pectic acid. The second group is pectic acid and polymers of galacturonans that contain negligible amounts of methoxyl groups. Pectinic acid, the third group, is the polygalacturonan chain with various amounts of methoxyl groups (0–75%). Pectin is the last group and defines the mixture of differing compositions of galacturonate units esterified with methanol (Jayani et al., 2005; Kashyap et al., 2001; Lea, 1995). As pectic substances are categorized, pectinolytic enzymes are also classified and divided into three groups in general. Protopectinases are enzymes that catalyze the degradation of insoluble propectin to highly polymerized soluble pectin (protopectin + H_2O → pectin). Pectinesterases are the second major group of pectic enymzes that catalyze the de-esterification of pectin by removal of methoxy esters. Pectin methyl esterase (EC 3.1.1.11) is an example of this group of enzymes and hydrolyzes pectin into pectic acid and methanol. The last group are depolymerases which catalyze the hydrolytic cleavage of the α-1,4-glycosidic bonds in the galacturonic acid (Grassin & Coutel, 2010; Jayani et al., 2005). Lyases are subcategorized under depolymerases but differ from hydrolases in terms of cleaving α-1,4-glycosidic bonds by transelimination and produce galacturonide with an unsaturated bond between C4 and C5 at the non-reducing end of the galacturonic acid formed (Kashyap et al., 2001).

Pectic enzymes have been used since the 1930s in manufacturing of wines and fruit juices (Kashyap et al., 2001). Recently, there have been new applications in several industrial processes beside juice and wine, such as textiles, tea, coffee, oil extraction, waste water treatment, etc. In juice and wine making, *Aspergillus niger*-derived pectic enzymes are commonly used. Sparkling clear juices, cloudy juices, and unicellular products are created by pectic enzyme applications in the juice industry. Enzymes for sparkling juices are employed to increase the yield and clarification of juice (Grassin & Coutel, 2010; Kashyap et al., 2001). Filtration time is reduced up to 50% when fruit juices are processed with pectic enzymes (Jayani et al., 2005). Clarification is affected by pH, temperature, enzyme concentration, and enzyme contact time. Lower pH will induce clarification rather than high pH, while elevated temperature will also increase the clarification rate as long as they enzyme is not denatured (Kilara, 1982). Apple, pear, strawberry, raspberry, blackberry, and grape juice are some examples of sparkling clear juices (Grassin & Coutel, 2010).

Another use of pectic enzymes in juice industry is stabilizing the cloud of citrus juices, purees, and nectars. Orange, lemon, mango, apricot, guava, papaya, pineapple, and banana are processed with enzymes to maintain a cloudy texture (Kashyap et al., 2001). Enzyme application differs for each fruit due to its specific requirements and properties. For example, oranges naturally contain pectin esterase and in the presence of calcium, an undesirable sedimentation of cloud particles occurs. Enzymatic treatment increases the stability of cloudiness by degrading the pectin without catalyzing the insoluble pectin that maintains cloud stability (Kashyap et al., 2001). Production of unicellular products is another application of pectic enzymes. In this process, a pulpy product is produced with the addition of enzymes, also called maceration, to use as material for baby foods, puddings, and yogurts. Maceration is transformation of tissue into a suspended cell. A mechanical treatment is performed before enzyme hydrolysis of pectin to enhance product properties (Kashyap et al., 2001).

Application of pectic enzymes is not limited to the fruit juice industry. Pectinases are commercialized to remove sizing agents from cotton with a combination of amylases, hemicellulases, and lipases (Hoondal et al., 2000). Pectinases are also an eco-friendly and economic alternative for degumming (removing gums from fibers) compared to chemical treatments in which pollution and toxicity are of concern (Kapoor et al., 2001). Waste water from vegetable processing plants contains pectin, and pectinases facilitate elimination of those by-products

(Hoondal et al., 2000). Pectinases are also used for production of animal feed (Hoondal et al., 2000), paper making (Reid & Richard, 2004), and extraction of citrus oil (Scott, 1978).

6.6.1.4 Proteases

Proteases are protein-degrading enzymes and catalyze the cleavage of peptide bonds in the proteins. Proteins are linear polymers of amino acids with a general formula of R-$CHNH_2$-$COOH$ in which peptide bonds are formed by the covalent binding of nitrogen atoms of the amino group to the carbon atoms of the preceding carboxyl group with the release of water (Adler-Nissen, 1993).

Proteases are classified according to their catalytic action into endopeptidases and exopeptidases. The EC system subcategorizes the enzymes as 3.4.11–19 for exopeptidases, and 3.4.21–24 for endopeptidases. Endopeptidases act on the polypeptide chain at specifically susceptible peptide bonds, while exopeptides hydrolyze the release of a single amino acid at a time from either the N terminus or C terminus (Adler-Nissen, 1993). Endopeptidases are serine proteases (EC 3.4.21), cysteine proteases (EC 3.4.22), aspartic proteases (EC 3.4.23), and metalloproteases (EC 3.4.24). Serine, cysteine, and aspartic proteases contain serine, cysteine, and aspartic acid, respectively, whereas metalloproteases contain an essential metal atom, usually zinc (Zn), at the active sites (Adler-Nissen, 1993). Exopeptidases, on the other hand, are divided into three major groups: aminopeptidases, carboxypeptidases, and omegapeptidases. Aminopeptidases generate a single amino acid residue, di- or tripeptide, by attacking the N terminus of the polypeptide chain. Carboxypeptides, in contrary, act at the C terminus of the polypeptide and liberate a single amino acid or dipeptide (Rao et al., 1998).

Proteases are found in plants, animals, and microorganisms. Papain, extracted from *Carica papaya*, has been used as a meat tenderizer for a long time. This enzyme is active at pH 5–9 and stable up to 90 °C, but the performance of the enzyme is highly dependent on plant source, climate conditions for growth, and extraction and purification methods. Papain is used for preparation of soluble and flavored protein hydrolyzates in industry (Rao et al., 1998). Bromelain, extracted from pineapples, is a cysteine protease active up to 70 °C in a range of pH 5–9. The other common plant protease is keratinase produced by a botanical plant. Keratinases degrade hair and wool, and find applications in prevention of clogging of waste water systems (Rao et al., 1998).

Well-known animal proteases are trypsin, chymotrypsin, pepsin, and rennin (Boyer, 1971). Trypsin is a digestive enzyme that hydrolyzes food proteins. However, use of trypsin in the food industry is not common because it generates a very bitter taste so it is used for bacterial media formulation and some medical applications (Rao et al., 1998). Chymotrypsin is found in animal pancreases and is expensive, and thus used mainly for analytical applications. Milk protein hydrolyzates are commonly deallergenized by chymotrypsin. Pepsin is an acidic protease and catalyzes the hydrolysis of peptide bonds between two hydrophobic amino acids (Rao et al., 1998). Rennet (rennin, chymosin) cleaves a single peptide bond in κ-casein and produces insoluble para-κ-casein and C-terminal glycopeptides (Law, 2010).

Bacteria, fungi, and viruses are excellent protease sources. Microbial proteases represent about 40% of enzyme sales globally (Godfrey & West, 1996). Bacterial proteases are active at alkaline pH and stable up to $60\,^{\circ}C$ which allows the detergent industry to use them. Fungal proteases also have a wide pH range (4–11) but their reaction rate and heat tolerance are lower compared to bacterial enzymes (Rao et al., 1998).

Proteases have a broad application in the food industry. In the dairy industry, milk-coagulating enzymes (animal rennin, microbial coagulants, engineered chymosin) are extensively used for cheese making. Chymosin has advantages over animal rennin due to its specific activity and availability. The protease-producing GRAS microorganisms are *Mucor michei*, *Bacillus subtilis*, and *Endothia parasitica*. Proteases are used in the baking industry to modify insoluble wheat protein gluten, which determines the characteristics of dough. Enzymatic treatment of dough improves the handling and machining, as well as time reduction and increase in loaf volume. *Aspergillus oryzae* enzymes have been used in the baking industry (Rao et al., 1998). However, proteases can generate a bitter taste, which limits their use in the food industry. Bitterness is relative to the number of hydrophobic amino acids in the hydrolyzate. A combination of endoprotease and amonipeptidase can reduce bitterness (Rao et al., 1998).

Proteases have several applications in the pharmaceutical industry as a digestive aid, in the health food industry for synthesis of aspartame, and in infant formulas for debittering of protein hydrolyzate.

Use of proteases in the detergent industry has grown and the largest application is in household laundry detergent formulation. Proteases remove proteinaceous stains, such as food, blood, and other body secretions (Gupta et al., 2002). Proteases used in detergent industry should be active at a wide range of pH and temperature profiles, and compatible with chelating and oxidizing agents (Rao et al., 1998). *Bacillus* strains are employed for production of proteases used in the detergent industry. In the leather industry, a major problem is removal of skin and hair, which are proteinaceous substances. Therefore, use of proteases having elastolytic and keratinolytic activity is an alternative to chemical treatment in the leather industry in which environmental impacts are of concern. Proteases are employed in the soaking, dehairing, and bating stages of leather processing to destroy undesirable pigments and maintain hair-free leather (Gupta et al., 2002).

6.6.1.5 Oxidoreductases

Oxidoreductases are another group of enzymes that catalyze the oxidation/reduction reaction. Oxidoreductases play a crucial role in foods in terms of taste, texture, shelf life, appearance, and nutritional value. Polyphenol oxidase is a copper-containing enzyme widely found in plants, animals, and humans, and can catalyze phenolic compounds involved in reactions (Hammer, 1993). One of the subclasses of oxidoreductases is peroxidase. Hydrogen peroxide is the main substrate for peroxidases that usually generates colored end-products. Horseradish peroxidase is the well-known enzyme of this group because of its use as an indicator in spectrophotometric and immunoassay techniques (Hammer, 1993). Lactoperoxidase is a hemoprotein that catalyzes the oxidation of thiocyanate and iodine ions in the presence of hydrogen peroxide to produce highly reactive oxidizing agents. Lactoperoxidase is antimicrobial and kills gram-negative and -positive bacteria, and fungi. Catalase, a hemoprotein, catalyzes the decomposition of hydrogen peroxide into oxygen and water (Inamine & Baker, 1989). Oxidation of thiols to their subsequent disulfides is catalyzed by sulfhydryl oxidase using molecular oxygen as the electron acceptor. This enzyme is isolated from *Myrothecium werrucaria*, *Aspergillus sojae*, and *Aspergillus niger* (de la Motte & Wagner, 1987; Katkochin et al., 1986; Mandel, 1956). Lipoxygenase is another subgroup of oxidoreductases which catalyzes the oxidation of polyunsaturated fatty acids by using oxygen; hydroperoxides are produced which are converted into acids, aldehydes or ketones, and other compounds (Matheis & Whitaker, 1987).

Some other applications of oxidoreductases include alcohol oxidase, which catalyzes the oxidation of short-chain linear aliphatic alcohols to aldehydes and H_2O_2. However, alcohol oxidase may not have applications in the alcoholic beverages industry because the optimum pH for this enzyme is 7.5–8. Also, ethanol and methanol have inhibitory effects on the enzyme (Hammer, 1993). Oxidoreductases are also employed in the pasteurization of eggs and cheese by H_2O_2, desugaring of eggs prior to spray drying, preservation of raw milk, and elimination of cooked flavor of UHT (ultra high temperature) pasteurized milk (Szalkucki, 1993).

Because of the limitations of space, only the most important enzymes that have applications in the food industry have been mentioned. However, there are many other enzymes that have been extensively used, including lactases, invertases, cellulases, hemicellulases, dextranases, and lysozyme. More details on enzymes used in food and other industries can be found in Reed (1966), Nagodawithana and Reed (1993), Tucker and Woods (1995), Whitehurst and van Oort (2010), Chen et al. (2003), and Bhat (2000).

6.7 Sustainability

The definition of sustainability varies depending on who defines it. Sustainability can be defined as "… a process of change in which the exploitation of resources, the direction of investments, the orientation of technological development, and institutional change are all in harmony and enhance both current and future potential to meet human needs and aspirations…" according to the Brundtland Report (1987). Therefore, sustainable processes utilize renewable resources and decrease greenhouse gases and other pollutants.

Microbial fermentation has been used for such purposes for many commercial products and enables novel techniques to support sustainable manufacturing. Utilization of non-food value feedstocks, genetically engineered microorganisms, and cost- and energy-efficient fermentation technologies promotes sustainability. In the vitamin production industry, for example, complex vitamin B2 production has come down to a simple fermentation process that has also lowered environmental impact (Gebhard, 2009). Furthermore, production increases 300,000 fold when genetically engineered bacteria (*Bacillus subtilis*) are used (Gavrilescu & Chisti, 2005). Enzymes are commonly used in the chemical industry as catalysts. The specificity of enzymes increases demand compared to conventional chemical catalysts. Enzymes also work under non-toxic and non-corrosive conditions (Gavrilescu & Chisti, 2005).Therefore, more study is needed to develop cost-competitive fermentation and enzyme technologies to reduce material, water and energy consumption and pollutant disposal.

6.8 Concluding remarks and future trends

Fermentation is extensively employed by the food industry to produce value-added products. The success of the fermentation is dependent on the microorganism/strain selection and optimization of growth parameters for the selected microorganism. Fermentation has also been used to enhance the quality and flavor of food. Optimization of fermentation, for a consumer-acceptable fermented food, also should be well considered. On the other hand, enzymes have their own application in the industry due to their catalytic activity.

There is still the need for isolation of new microbial strains and strain development for high yields in the fermentation process. Moreover, genetic and metabolic engineering is an important tool to develop strains, which can produce heterologous products. As a result, products, which are not currently industrially available may become affordable for industrial use. Because some microorganisms do not produce exopolysaccharide, which creates difficulties with attachment to supports, immobilization reactors should be tested for each product to test the effect of immobilization on yield. In addition to yield, the strains and the production parameters of traditional fermented food should be studied. The natural flora improve the flavor and texture of the fermented food. Therefore, microorganisms in traditional foods should be isolated and the functions of these microorganisms should be studied. As a result, the diversity of traditional fermented foods can be protected.

There is always a need to study microbial production and applications of enzymes. The characteristics of enzymes are being improved by protein engineering to increase their potential with respect to their temperature and pH profiles, stability, specificity, etc. Furthermore, novel immobilization techniques can be studied in order to use enzymes more efficiently.

References

Abdi R, Sheikh-Zeinoddin M, Soleimanian-Zad S (2006) Identification of lactic acid bacteria isolated from traditional Iranian lighvan cheese. *Pakistan Journal of Biological Sciences* **9**: 99–103.

Adams RD (1985) Vinegar. In: Wood B (ed) *Microbiology of Fermented Foods*. Philadelphia: Elsevier, pp. 1–49.

Adler-Nissen J (1987) Newer uses of microbial enzymes in food processing. *Tibtech* **5**: 170–174.

Adler-Nissen J (1993) Proteases. In: Tucker G, Woods L (eds) *Enzymes in Food Processing* New York: Chapman and Hall, pp. 191–218.

Adrio JL, Demain A (2006) Genetic improvement of processes yielding microbial products. *FEMS Microbiology Review* **30**: 187–214.

Aiyer P (2005) Amylases and their applications. *African Journal of Biotechnology* **4**: 1525–1529.

Alfenore S, Molina-Jouve C, Guillouet SE, Uribelarrea JL, Goma G, Benbadis L (2002) Improving ethanol production and viability of *Saccharomyces cerevisiae* by a vitamin feeding strategy during fed-batch process. *Applied Microbiol Biotechnology* **60**: 67–72.

Aravindan R, Anbumathi P, Viruthagiri T(2007) Lipase applications in food industry. *Indian Journal of Biotechnology* **6**: 141–158.

Astuti M, Meliala A, Dalais F, Lwahlqvist M (2000) Tempe, a nutritious and healthy food from Indonesia. *Asia Pacific Journal of Clinical Nutrition* **9**(4): 322–332.

Aurand LW, Woods AE, Wells R (1987) Enzymes. In: *Food Composition and Analysis*. New York: Van Nostrand Reinhold, pp. 283–346.

Balat M, Balat H, Oz C (2008) Progress in bioethanol processing. *Progress in Energy and Combustion Science* **34**: 551–573.

Battcock M, Azam-Ali S (1998) *Fermented Fruits and Vegetables. A Global Perspective*. Rome: Food and Agriculture Organization of the United Nations.

Belitz HD, Grosch W, Schieberle P (2004) Enzymes. In: *Food Chemistry*. Berlin: Springer-Verlag, pp. 92– 154.

Belter PA, Cussler EL, Hu WS (1988) *Bioseperations Downstream Processing for Biotechnology*. New York: Wiley-Interscience.

Beresford T, Williams A (2004) *The Microbiology of Cheese Ripening: Cheese, Chemistry, Physics and Microbiology*. New York: Elsevier.

Berk Z (1976) Enzymes. In: *Braverman's Introduction to the Biochemistry of Foods*. Amsterdam: Elsevier, pp. 27–40.

Beuvier E, Buchin S (2004) Raw milk cheeses. In: Fox P, Mc-Sweeney PLH, Cogan TM, Guinee TP (eds) *Cheese: Chemistry, Physics and Microbiology*. London: Elsevier Academic Press, pp. 319–345.

Bhat MK (2000) Cellulases and related enzymes in biotechnology. *Biotechnology Advances* **18**: 355–383.

Bigelis R (1993) Carbohydrases. In: Nagodawithana T, Reed G (eds) *Enzymes* in Food Processing. San Diego, CA: Academic Press, pp. 121–147.

Boyce S, Tipton KF (2001) *Enzyme Classification and Nomenclature*. Encyclopedia of Life Sciences. Nature Publishing Group. www.els.net, accessed 10 November 2013.

Boyer PD (1971) *The Enzymes*, 3rd edn. New York: Academic Press.

Brundtland G (1987) *Our Common Future*. Oxford: Oxford University Press.

Bryers JD (1994) Biofilms and the technological implications of microbial cell adhesion. *Colloids and Surfaces, B, Biointerfaces* **2**: 9–23.

Campbell-Platt G (1994) Fermented foods – a world perspective. *Food Research International* **27**: 253–257.

Carroll R (1994) *Cheesemaking Made Easy: 60 Delicious Varieties*. North Adams: Storey Books.

Chandrasker K, Felse PA, Panda T (1999) Optimization of temperature and initial pH and kinetic analysis of tartaric acid production by *Gluconobacter suboxydans*. *Bioprocess Engineering* **20**: 203–207.

Chen L, Daniel RM, Coolbear T (2003) Detection and impact of protease and lipase activities in milk and milk powders. *International Dairy Journal* **13**: 255–275.

Cheng KC, Demirci A, Catchmark JM (2010) Advances in biofilm reactors for production of value-added products. *Applied Microbiology and Biotechnology* **87**: 445–456.

Choi GW, Moon SK, Kang HW, Min J, Chung BW (2008) Simultaneous saccharification and fermentation of sludge-containing cassava mash for batch and repeated batch production of bioethanol by *Saccharomyces cerevisiae* CHFY0321. *Journal of Chemical Technology and Biotechnology* **84**: 547–553.

Clarkson K, Jones B, Bott R et al. (2001) Enzymes: screening, expression, design and production. In: Bedford M, Partridge G (ed) *Enzymes in Farm Animal Nutrition*. Cambridge, MA: CAB International.

Cliffe K (1988) Downstream processing. In: Scragg A (ed) *Biotechnology for Engineers*. Chichester: Ellis Horwood, pp. 302–321.

Couto SR, Sanroman MA (2006) Application of solid state fermentation to food industry – a review. *Journal of Food Engineering* **76**: 291–302.

Crow V, Curry B, Hayes M (2001) The ecology of non-starter lactic acid bacteria (NSLAB) and their use as adjuncts in New Zealand Cheddar. *International Dairy Journal* **11**: 275–283.

Curionia PMG, Bossetb JO (2002) Key odorants in various cheese types as determined by gas chromatography-olfactometry. *International Dairy Journal* **12**: 959–984.

Dale JW, Park SF (2010) Gene transfer. In: *Molecular Genetics of Bacteria*. Chichester: John Wiley, pp. 178–195.

de la Motte RS, Wagner F (1987) Asperillus niger glucose oxidase. *Biochemistry* **26**: 7363–7371.

deMan JM (1996) Enzymes. In: *Principles of Food Chemistry*. New York: Van Nostrand Reinhold, pp. 373–412.

Demirci A (2002) Aerobic reactions in fermentation processes. In: Heldman DR (ed) *The Encyclopedia of Agricultural and Food Engineering*. New York: Marcel Dekker.

Demirci A, Pometto AL III, Ho K (1997) Ethanol production by *Saccharomyces cerevisiae* in biofilm reactors. *Journal of Industrial Microbiology and Biotechnology* **19**: 299–304.

Demirci A, Pongtharangkul T, Pometto AL III (2007) Applications of biofilm reactors for production of value-added products by microbial fermentation. In: Blaschek PH, Wang HH, Agle EM (eds) *Biofilms in the Food Environment*. Ames, IA: Blackwell Publishing, pp. 167–189.

Dutta R (2008) Downstream processing. In: *Fundamentals of Biochemical Engineering*. India: Ane Books.

Dutta JR, Dutta PK, Banerjee R (2004) Optimization of culture parameters for extracellular protease production from newly isolated *Pseudomonas* sp. using response surface and artificial neural network models. *Process Biochemistry* **39**: 2193–2198.

Ellaiah P, Adinarayana K, Bhavani Y, Padmaja P, Srinivasulu B (2002) Optimization of process parameters for glucoamylase production under solid state fermentation by a newly isolated *Aspergillus* species. *Process Biochemistry* **38**: 615–620.

Ertola RJ, Giulietti A, Castillo F (1995) Design, formulation, optimization of media. In: Asenjo JA, Merchuk JC (eds) *Bioreactor System Design*. New York: Marcel Dekker, pp. 89–137.

Eskin NAM (1990) *Biochemistry of Foods*. San Diego, CA: Academic Press, pp. 467–527.

Fox PF, Guinee TP, Cogan TM, McSweeney P (2000) *Fundamentals of Cheese Science*. Gaithersburg, MA: Aspen Publishers.

Galaction AI, Cascaval D, Oniscu C, Turnea M (2004) Enhancement of oxygen mass transfer in stirred bioreactors using oxygen-vectors.1. Simulated fermentation broths. *Bioprocess and Biosystems Engineering* **26**: 231–238.

Galante YM, Formantici C (2003) Enzyme applications in detergency and in manufacturing industries. *Current Organic Chemistry* **7**: 1399–1422.

Gandhi NN (1997) Applications of lipase. *Journal of the American Oil Chemists' Society* **74**: 621–634.

Gavrilescu M, Chisti Y (2005) Biotechnology – a sustainable alternative for chemical industry. *Biotechnology Advances* **23**: 471–499.

Gebhard R (2009) Sustainable Production of Pharmaceutical Intermediates and API's – *the* challenge of the next decade. www.dsm.com/en_US/downloads/dpp/DSM_Webinar_Sustainable_Production_17Jun09.pdf, accessed 10 November 2013.

Geisen R, Holzapfel WH (1996) Genetically modified starter and protective cultures. *International Journal of Food Microbiology* **3**: 315–324.

Ghosh PK, Saxena R, Gupta R et al. (1996) Microbial lipases: production and applications. *Science Progress* **79**: 119–157.

Giraffa G (2003) Functionality of enterococci in dairy products. *International Journal of Food Microbiology* **88**: 215–222.

Godfrey T, West S (1996) *Industrial Enzymology*, 2nd edn. New York: Macmillan.

Grandison AS, Finnigan T (1996) Microfiltration. In: Grandison A, Lewis J (eds) *Separation and Processes in the Food and Biotechnology Industries – Principles and Applications*. Cambridge: Woodhead Publishing, pp. 141–153.

Grandison A, Lewis J (eds) (1996) *Separation and Processes in the Food and Biotechnology Industries – Principles and Applications*. Cambridge: Woodhead Publishing.

Grassin C, Coutel Y (2010) Enzymes in fruit and vegetable processing and juice extraction. In: Whitehurst R, van Oort M (eds) *Enzymes in Food Technology*, 2nd edn. Ames, IA: Wiley-Blackwell, pp. 103–135.

Graur D, Li HW (2000) *Fundamentals of Molecular Evolution*, 2nd edn. Sunderland, MA: Sinauer Associates.

Gunasekaran S, Ak AM (eds) (2003) *Cheese Rheology and Texture*. Boca Raton, FL: CRC Press.

Gupta R, Beg QK, Lorenz P (2002) Bacterial alkaline proteases: molecular approaches and industrial applications. *Applied Microbiology and Biotechnology* **59**: 15–32.

Gupta R, Gigras P, Mohapatra H et al. (2003) Microbial α-amylases: a biotechnological perspective. *Process Biochemistry* **38**: 1599–1616.

Gürakan GC, Altay N (2010) Yogurt microbiology and biochemistry. In: Yıldız F (ed) *Development and Manufacture of Yogurt and Other Functional Dairy Products*. Boca Raton, FL CRC Press, pp. 97–123.

Guzel-Seydim Z, Kök-Taş T, Greene AK (2010) Kefir and koumiss: microbiology and technology. In: Yıldız F (ed) *Development and Manufacture of Yogurt and Other Functional Dairy Products*. Boca Raton, FL CRC Press, pp. 143–165.

Hamer RJ (1995) Enzymes in the baking industry. In: Tucker G, Woods L (eds) *Enzymes in Food Processing*. New York: Chapman and Hall, pp. 191–218.

Hammer FE (1993) Oxidoreductases. In: TNagodawithana T, Reed G (ed) *Enzymes in Food Processing*, 3rd edn. San Diego, CA: Academic Press , pp. 221–279.

Han Z, Zhao Q, Wang H, Kang Y (2011) Characteristics of the secretory expression of pectate lyase A from *Aspergillus nidulans* in *Escherichia coli*. *African Journal of Microbiology Research* **5**(15): 2155–2159.

Hang YD, Woodams EE (1984) Apple pomace: a potantial substrate for citric acid production by *Aspergillus niger*. *Biotechnology Letters* **6**: 763–764.

Hasan F, Shah AA, Hameed A (2006) Industrial applications of microbial lipases. *Enzyme and Microbial Technology* **39**: 235–251.

Hatti-Kaul R (2004) Enzyme production. In: Doelle HW, Rokem JS, Berovic M (eds) *Encyclopedia of Life Support Systems* (EOLSS). Oxford: UNESCO, EOLSS Publishers. www.eolss.net.

Hendy N, Wilke C, Blanch H (1982) Enhanced cellulase production using solka floc in a fed-batch fermentation. *Biotechnology Letters* **4**: 785–788.

Hoondal GS, Tewari R, Dahiya N, Beg QK (2000) Microbial alkaline pectinases and their applications: a review. *Applied Microbiology and Biotechnology* **59**: 409–418.

Houde A, Kademi A, Leblanc D (2004) Lipases and their industrial applications. *Applied Biuochemistry and Biotechnology* **118**: 155–170.

Hutkins RW (2006) *Microbiology and Technology of Fermented Foods*. Ames, IA: Blackwell Publishing.

Inamine GS, Baker JE (1989) A catalase from tomato fruit. *Phytochemistry* **28**: 345–348.

Jayani RS, Saxena S, Gupta R (2005) Microbial pectinolytic enzymes: a review. *Process Biochemistry* **40**: 2931–2944.

Johannes TW, Zhao H (2006) Directed evolution of enzymes and biosynthetic pathways. *Current Opinion in Microbiology* **9**: 261–267.

Kadam KL, Newman MM (1997) Development of a low cost fermentation medium for ethanol production from biomass. *Applied Microbiology and Biotechnology* **447**: 625–629.

Kandarakis I, Moschopoulou E, Anifantakis E (1999) Use of fermentation produced chymosin from E. coli in the manufacture of Feta cheese. *Milchwissenschaft – Milk Science International* **54**: 24–26.

Kapoor M, Beg QK, Bhushan B et al. (2001) Application of alkaline and thermostable polygalacturonase from *Basillus* spp. MG-cp-2 in degumming of ramie (*Boehmeria nivea*) and sunn hemp (*Crotalaria juncea*) bast fibers. *Process Biochemistry* **36**: 803–807.

Kashyap DR, Vohra PK, Chopra S, Tewari R (2001) Applications of pectinases in the commercial sector: a review. *Bioresource Technology* **77**: 215–227.

Katkochin D, Miller CA, Starnes RL (1986) Microbial sulfhydryl oxidase. *US Patent No.* **4632905**.

Kennedy M, Krouse D (1999) Strategies for improving fermentation medium performance: a review. *Journal of Industrial Microbiology and Biotechnology* **23**: 456–475.

Kilara A (1982) Enzymes and their uses in processed apple industry: a review. *Process Biochemistry* **23**: 35–41.

Kobayashi F, Nakamura Y (2004) Mathematical model of direct ethanol production from starch in immobilized recombinant yeast culture. *Biochemical Engineering Journal* **21**: 93–101.

Korus RA, Hoffman DS, Bam N et al. (1993) Transesterification process to manufacture ethyl ester of rape oil. *Proceedings of the First Biomass Conference of the Americas: Energy, Environment, Agriculture, Industry*, vol. **II**. Golden, CO: National Renewable Energy Laboratory, pp. 815–826.

Kourkoutas Y, Bekatorou A, Banat I et al. (2004) Immobilization technologies and support materials suitable in alcohol beverages production: a review. *Food Microbiology* **21**: 377–397.

Lantto R, Kruus K, Puolanne E et al. (2010) Enzymes in meat processing. In: Whitehurst R, van Oort M (eds) *Enzymes in Food Technology*, 2nd edn. Ames, IA: Wiley-Blackwell, pp. 264–283.

Law BA (2010) Enzymes in dairy product manufacture. In: Whitehurst R, van Oort M (eds) *Enzymes in Food Technology*, 2nd edn. Ames, IA: Wiley-Blackwell, pp. 88–102.

Lea AGH (1995) Enzymes in the production of beverages and fruit juices. In: Tucker G, Woods L (eds) *Enzymes in Food Processing*. New York: Chapman and Hall, pp. 223–250.

Lewis MJ (1996) Ultrafiltration. In: Grandison A, Lewis M (eds) *Separation and Processes in the food and Biotechnology Industries – Principles and Applications*. Cambridge: Woodhead Publishing, pp. 97–122.

Lourens-Hattingh A, Viljoen BC (2001) Yogurt as probiotic carrier food. *International Dairy Journal* **11**: 1–17.

Lücke F (1985) Fermented sausages. In: Wood B (ed) *Microbiology of Fermented Foods*, volume 2. New York: Elsevier, pp. 41–85.

Maarel M (2010) Starch-processing enzymes. In: Whitehurst R, van Oort M (eds) *Enzymes in Food Technology*, 2nd edn. Ames, IA: Wiley-Blackwell

Madson PW, Monceaux DA (1995) Fuel ethanol production. In: Lyons TP, Kelsall DR. Murtagh JE (eds) *The Alcohol Textbook*. Nottingham: Nottingham University Press, pp. 257–268.

Malik SK, Mukhtar H, Farooqi A, Ul-Haq I (2010) Optimization of process parameters for the synthesis of cellulases by *Trichoderma viride*. *Pakistan Journal of Botany* **42**: 4243–4251.

Mandel GR (1956) Properties and surface location of a sulfhydryl oxidizing enzyme in fungus spores. *Journal of Bacteriology* **72**: 230–234.

Matheis G, Whitaker J (1987) A review: enzymatic cross-linking of proteins applicable to foods. *Journal of Food Biochemistry* **11**: 309–327.

Merico A, Capitanio D, Vigentini I, Ranzi B, Compagno C (2004) How physiological and cultural conditions influence heterologous protein production in *Kluyveromyces lactis*. *Journal of Biotechnology* **109**: 139–146.

Metwally M (1998) Glucoamylase production in continuous cultures of *Aspergillus niger* with special emphasis on growth parameters. *World Journal of Microbiology and Biotechnology* **14**: 113–118.

Mohawed SM, Kassim E, El-Shahed A (1986) Studies on the effects of different pH values, vitamins, indoles, giberellic acid on production of lipase by *Aspergillus anthecieus*. *Agricultural Wastes* **17**: 307–312.

Moss GP (1992) *Enzyme Nomenclature*. San Diego, CA: Academic Press.

Muthukumaran N, Dhar S (1982) Comparative studies on the degreasing of skins using acid lipase and solvent with reference to the quality of finished leathers. *Leather Science* **29**: 417–424.

Nagodawithana T, Reed G (1993) *Enzymes in Food Processing*, 3rd edn. San Diego, CA: Academic Press Inc.

Oberman H (1985) Fermented milks. In: Wood B (ed) *Microbiology of Fermented Foods*, volume 1. New York: Elsevier, pp. 186–187.

Ogez JR, Hodgdon J, Beal M, Builder S (1989) Downstream processing of proteins: recent advance. *Biotechnology* 7: 467–488.

Olsen HSO, Faholt P (1998) The role of enzymes in modern detergency. *Journal of Surfactants and Detergents* 1: 555–567.

Ozmihci S, Kargi F (2007) Ethanol fermentation of cheese whey powder solution by repeated fed-batch operation. *Enzyme and Microbial Technology* 41: 169–174.

Pandey A, Poonam N, Soccol C et al. (2000) Advances in microbial amylases. *Biotechnology and Applied Biochemistry* 31: 135–152.

Parekh S, Vinci V, Strobel R (2000) Improvement of microbial strains and fermentation process. *Applied Microbiology and Biotechnology* 54: 287–301.

Perry RH, Chilton C (1973) *Chemical Engineers' Handbook*, 5th edn. New York: McGraw-Hill.

Pomerantz Y (1991) *Functional Properties of Food Components*. San Diego, CA: Academic Press.

Pometto AL III, Demirci A, Johnson K (1997) Immobilization of microorganisms on a support made of synthetic polymer and plant material. *US Patent No.* 5,595,893.

Rao MB, Tanksale A, Ghatge M, Deshpande V (1998) Molecular and biotechnological aspects of microbial proteases. *Microbiology and Molecular Biology Reviews* 62(3): 597–635.

Rapin JD, Marison I, von Stockar U, Reilly P (1994) Glycerol production by yeast fermentation of whey permeate. *Enzyme and Microbial Technology* 16: 143–150.

Ray B (1995) Pediococcus in fermented foods. In: Hui YH, Khachatourians G (eds) *Food Biotechnology: Microorganisms*. New York: Wiley-VCH, pp. 745–795.

Reed G (1966) *Enzymes in Food Processing*. New York: Academic Press.

Refaat AA (2010) Different techniques for the production of biodiesel from waste vegetable oil. *International Journal of Environmental Science and Technology* 7(1): 183–213.

Reid I, Richard M (2004) Purified pectinase lowers cationic demand in peroxide-bleached mechanical pulp. *Enzyme Technology* 34: 499–504.

Rols JL, Goma G (1989) Enhancement of oxygen transfer rates in fermentation using oxygen-vectors. *Biotechnology Advances* 7: 1–14

Rowlands RT (1984) Industrial strain improvement: mutagenesis and random screening procedures. *Enzyme and Microbial Technology* 6: 3–10.

Sanchez A, Ferrer P, Serrano A et al. (1999) Characterization of the lipase and esterase multiple forms in an enzyme preparation from a *Candisa rugosa* pilot-plant scale

fed-batch fermentation. *Enzyme and Microbial Technology* 25: 214–223.

Sanchez OJ, Cardona C (2008) Trends in biotechnological production of fuel ethanol from different feedstocks. *Biosource Technology* 99(13): 5270–5295.

Scott CD (1987) Immobilized cells; a review of recent literature. *Enzyme and Microbial Technology* 9: 66–72.

Scott D (1978) Enzymes, industrial. In: Grayson M, Ekarth D, Othmer K (eds) Encyclopedia of Chemical *Technology*. New York: Wiley, pp. 173–224.

Scott R, Robinson RK, Wilbey RA (1998) *Cheesemaking Practice*, 3rd edn. London: Kluwer Academic/Plenum.

Shuler ML, Kargi F (2008) *Bioprocess Engineering*, 2nd edn. New York: Prentice-Hall.

Souza PM, Magalhaes P (2010) Application of microbial α-amylase in industry – a review. *Brazilian Journal of Microbiology* 41: 850–861.

Stanbury PF (2000) Fermentation technology. In: Walker J, Rapley R (ed) *Molecular Biology and Biotechnology*. Cambridge: Royal Society of Chemistry.

Styger G, Prior B, Bauer F (2011) Wine flavor and aroma. *Journal of Industrial Microbiology and Biotechnology* 38: 1145–1159.

Sugihara TF (1985) Microbiology of breadmaking. In: Wood B (ed) *Microbiology of Fermented Foods*, volume 1. New York: Elsevier, pp. 249–261.

Svendsen A (2000) Lipase protein engineering. *Biochimica et Biophysica Acta* 1543: 223–238.

Szalkucki T (1993) Applications of oxidoreductases. In: Nagodawithana T, Reed G (eds) *Enzymes in Food Processing*, 3rd edn. San Diego, CA: Academic Press, pp. 279–291.

Tandon V, Bano G, Khajuria V, Parihar A, Gupta S (2005) Pleiotropic effects of statins. *Indian Journal of Pharmacology* 37(2): 77–85.

Tan-Wilson AL, Chen J, Duggan M et al. (1987) Soybean Bowman-Birk trypsin isoinhibitors: classification and report of a glycine-rich trypsin inhibitor class 1. *Journal of Agricultural and Food Chemistry* 35: 974–981.

Tsakalidou E, Manolopoulou E, Tsilibari B, Georgalaki M, Kalantzopoulos G (1993) Esterolytic activities from *Enterococcusdurans* and *Enterococcusfaecium* strains isolated from Greek cheese. *Netherlands Milk and Dairy Journal* 47: 145–150.

Tucker GA, Woods L (1995) *Enzymes in Food Processing*. New York: Chapman and Hall.

Üçüncü M (2005) *Süt ve mamülleri teknolojisi*. İzmir: Meta Basım.

Underkofler LA, Barton R, Rennert R (1958) Production of microbial enzymes and their applications. *Applied Microbiology* 6: 212.

van der Laan JM (1995) Novel amylolytic enzymes deduced from the B. licheniformis α-amylase, having improved characteristics. International Patent WO 95/35382.

van Oort M (2010) Enzymes in bread making. In: Whitehurst R, van Oort M (eds) *Enzymes in Food Technology*, 2nd edn. Ames, IA: Wiley-Blackwell, pp. 103–135.

Vaughn RH (1985) The microbiology of vegetable fermentation. In: Wood B (ed) *Microbiology of Fermented Foods*, volume 1. New York: Elsevier, pp. 49–111.

Verbelen PJ, Schutter D, Delvaux F, Vestrepen K (2006) Immobilized yeast cell systems for continuous fermentation applications. *Biotechnology Letters* **28**: 1515–1525.

Vidojevic AT, Vukasinovic M, Veljovic K, Ostojic M, Topisirovic L (2007) Characterization of microflora in homemade semi-hard white Zlatar cheese, *International Journal of Food Microbiology* **114**: 36–42.

Vihinen M, Mantsiila P (1989) Microbial and amylolytic enzymes. *Critical Reviews in Biochemistry and Molecular Biology* **24**: 329–418.

Walstra P, Wouters J, Geurts T (2006) *Dairy Science and Technology*. Boca Raton, FL: CRC Press.

Wang ZZY, He G, Ruan H et al. (2007) Construction of proteinase A deficient transformant of industrial brewing yeast. *European Food Research and Technology* **225**: 831–835.

Whitehurst RJ, van Oort M (2010) *Enzymes in Food Technology*, 2nd edn. Ames, IA: Wiley-Blackwell.

Zhang HHB, Zhang H, Chen Q et al. (2009) Effects of proteinase A on cultivation and viability characteristics of industrial *Saccharomyces cerevisiae* WZ65. *Journal of Zhejiang University Science B* **10**: 769–776.

Zhang W, Wang C, Huang C et al. (2011) Construction and expression of food-grade β-galactosidase gene in *Lactococcus lactis*. *Current Microbiology* **62**: 639–644.

7

Alternative Food Processing Technologies

Hudaa Neetoo[1] and Haiqiang Chen[2]

[1]Faculty of Agriculture, University of Mauritius, Réduit, Mauritius
[2]Department of Animal and Food Sciences, University of Delaware, Newark, Delaware, USA

7.1 Introduction

Thermal and non-thermal technologies are both used in the processing and preservation of foods. Recently, awareness about good nutrition coupled with the increasing demand for fresher tasting food have paved the way for new food processing technologies. These include thermal processing methods such as microwave (MW), radiofrequency (RF), infrared (IR) heating, pressure-assisted thermal sterilization (PATS), and sous-vide processing (SVP), as well as non-thermal methods such as high hydrostatic pressure (HHP) processing, irradiation, ultrasound, pulsed electric field (PEF), and pulsed light (PL) technologies. Significant advances have been made in the research, development, and application of these technologies in food processing (Ohlsson & Bengtsson, 2002; Senorans et al., 2011). This chapter reviews the thermal and non-thermal food processing methods, including the equipment involved, their applications in processing and preservation of foods, and their effects on the nutritional and sensory quality of treated foods.

7.2 Alternative thermal processing technologies

7.2.1 Microwave heating

7.2.1.1 Introduction

Microwaves are a form of electromagnetic radiation characterized with a frequency between 300 MHz to 300 GHz. Microwave (MW) heating is generated by the absorption of microwave by a dielectric material, resulting in the microwaves giving up their energy to the material with a concomitant rise in temperature. Unlike conventional heating, which relies on the slow march of heat from the surface of the material to the interior, heating with MW energy is in effect bulk heating in which the electromagnetic field interacts with the food as a whole. The microwave heating of foods has garnered scientific and consumer interest due to its volumetric (bulk) heating, rapid increase in temperature and relative ease of cleaning (Ahmed & Ramaswamy, 2007). Unlike more traditional forms of thermal processing, such as pasteurization and retorting, which are characterized by a slow thermal diffusion process, the volumetric nature of heat generated by microwaves can substantially reduce total heating time, thereby minimizing the overall severity of the process and leading to a greater retention of the desirable quality attributes of the product (Sumnu & Sahin, 2005). According to Tewari (2007), the time required to come to target process temperature is attained within one-quarter of the time typically reached by conventional heating processes. Microwave technology is also amenable to batch processing and offers the flexibility of being easily turned on or off.

7.2.1.2 Microwave equipment

There are two main mechanisms by which microwaves produce heat in dielectric materials: dipole rotation and ionic polarization. Food materials contain polar molecules such as water. These molecules generally have a random orientation. When an electric field is applied, the molecules orient themselves according to the polarity of the field. Repeated changes in the polarity of the field cause rapid reorientation of the water molecules, resulting

Food Processing: Principles and Applications, Second Edition. Edited by Stephanie Clark, Stephanie Jung, and Buddhi Lamsal.
© 2014 John Wiley & Sons, Ltd. Published 2014 by John Wiley & Sons, Ltd.

in friction with the surrounding medium and hence generation of heat. In addition, when an electric field is applied to food solutions containing ions, the ions move at an accelerated pace due to their inherent charge. The resulting collisions between the ions cause the conversion of kinetic energy of the moving ions into thermal energy.

Microwave ovens come in a variety of designs although the underlying principles are the same. Each MW system typically consists of three basic parts: a MW source, a waveguide, and an applicator. The most widely used source of microwaves for either industrial or commercial applications is the magnetron tube (Metaxas & Meredith, 1983). A magnetron consists of a vacuum tube with a central electron-emitting cathode of highly negative potential surrounded by a structured anode (Regier et al., 2010). The magnetron requires several thousand volts of direct current and converts the power supplied into MW energy, emitting high-frequency radiant energy. The polarity of the emitted radiation changes between negative and positive at high frequencies. The power of the magnetron can range from 300 to 3000 W and for industrial equipment various magnetrons are used to increase the global power output. A waveguide channels the microwaves into the cavity that holds samples for heating. Domestic ovens are designed with reflecting cavity walls that produce several modes of microwaves, thereby maximizing the efficiency of the heating process (Orsat & Raghavan, 2005). Single-mode ovens are also available, which distribute the microwaves into the reactor in a precise way. Typically, MW food processing makes use of two frequencies: 2450 and 915 MHz. Of these two, the 2450 MHz frequency is used for home ovens, while both are used in industrial heating (Regier et al., 2010).

7.2.1.3 Food processing applications of microwaves

There are six major classes of applications for MW processing for foods: pasteurization, sterilization, tempering, dehydration, blanching, and cooking (Fu, 2010). Other uses have also been mentioned, including the use of MW in baking, coagulation, coating, gelatinization, puffing, and roasting. Since MW energy can heat many foods effectively and rapidly, considerable research has focused on the use of MW heating for pasteurization and sterilization applications. MW sterilization operates in the temperature range of 110–130°C while pasteurization is a gentler heat treatment, occurring between 60°C and 82°C. Canumir et al. (2002) demonstrated the suitability of MW pasteurization to inactivate *E. coli* in apple juice. Villamiel et al. (1997) showed that MW pasteurization of

milk by continuous flow achieved a satisfactory level of microbial reduction without excessive damage to sensory quality. In addition, the use of MW heating combined with other methods such as ultraviolet (UV) light, hydrogen peroxide, and γ-irradiation has shown synergistic effects, appearing to be promising food decontamination strategies.

Microwave baking has also been extensively investigated (Icoz et al., 2004; Sumnu & Sahin, 2005; Sumnu et al., 2007). MW technology has made an immense contribution to accelerated baking, leading to an enhanced throughput with negligible additional space required for MW power generators. Accelerated baking, through the combination of MW and conventional or infrared baking, enhances throughput whilst ensuring product quality, appropriate degrees of crust formation, and surface browning (Ohlsson & Bengtsson, 2002). In contrast to conventional baking, MW heating quickly inactivates enzymes such as α-amylase, due to a fast and uniform temperature rise in the whole product, to prevent the starch from extensive digestion, and releasing sufficient carbon dioxide and steam to produce a uniform porous texture (Ohlsson & Bengtsson, 2002). Microwaves can also enhance drying for low-moisture food products (Fu, 2010). The overriding advantage of MW drying and dehydration is that damage to nutrients and sensory quality, typically caused by prolonged drying times and elevated surface temperatures, is avoided (Brewer, 2005).

Although the MW oven is a common household appliance, MW heating has found increasing applications in the food industry in various processing operations. MW processing has been successfully applied on a commercial scale for meat sausage cooking, bacon precooking, and the tempering of products such as beef meats (Farag et al., 2008), fish blocks (Ramaswamy & Tang, 2008), frozen potato purée (Seyhun et al., 2009), and for pasteurizing diverse products including raw meats (Huang & Sites, 2010), beef frankfurters (Huang & Sites, 2007), in-shell eggs (Dev et al., 2008), mashed potatoes (Guan et al., 2004), packaged acidified vegetables (Koskiniemi et al., 2011), orange juice (Cinquanta et al., 2010), and apple cider (Gentry & Roberts, 2005).

7.2.1.4 Effect of microwaves on the sensory and nutritional quality of foods

Milk is a diverse source of vitamins and MW heating has been shown to have a variable effect on its nutrients. Lopez-Fandino et al. (1996) reported an insignificant loss in vitamin A, B_1 (thiamin) and B_2 (riboflavin) although losses of the vitamins E and C, by 17% and 36% respectively, were observed. However, Sierra et al. (1999) compared the heat stability of thiamin and riboflavin in microwave-heated and traditionally heated milk and reported virtually no difference. Valero et al. (2000) compared the chemical and sensory changes in milk pasteurized by microwave and conventional systems during cold storage. Milk heated in the microwave oven or in the conventional system for 15 sec was not distinguished by a sensory panel using a triangle test procedure either after processing or during the storage period at 4.5 °C for up to 15 days.

Cinquanta et al. (2010) investigated the effects of microwave pasteurization on various quality parameters of orange juice such as cloud stability, color, carotenoid compounds, and vitamin C content. The authors observed that the carotenoid content, responsible for the sensory and nutritional quality in fresh juices, decreased by about 13% after MW pasteurization at 70 °C for 1 min. However, the decrease in vitamin C content was minimal with retention ranging from 96.1% to 97%. The authors thus concluded that fine temperature control of the MW oven treatment could result in promising stabilizing treatments. Picouet et al. (2009) examined the effect of MW heating on the various quality parameters of Granny Smith apple purée such as vitamin C stability, total polyphenol content, viscosity, color, and titratable acidity. The authors demonstrated that MW treatment at 652 W for a short duration of 35 sec followed by storage at 5 °C for 15 days resulted in an average loss of vitamin C of the order of 50% although the viscosity and titratable acidity of the product were unaffected during the storage period.

7.2.2 Radiofrequency heating

7.2.2.1 Introduction

Radiofrequency (RF) heating is another form of dielectric heating, which has potential for the rapid heating of solid and semi-solid foods. RF heating refers to the heating of dielectric materials with electromagnetic energy at frequencies between 1 and 300 MHz.

7.2.2.2 Radiofrequency equipment

The equipment needed for RF heating consists of two fundamental components: one responsible for the generation of RF waves (the generator) and one responsible for the application of RF power to the food (the applicator, the main part of which is the electrodes) (Vicente & Castro, 2007). There are basically two groups of units used to produce and apply RF power to foods: the conventional RF heating equipment, where the applicator is part of the RF generator circuit, and a more recent model of RF equipment, whose RF applicator is separate from the RF generator and linked only by a high-power coaxial cable (Rowley, 2001). The design of electrodes in RF heating equipment is one of the most crucial aspects. A number of different electrode configurations are available, depending on factors such as field strength and physical characteristics of the sample such as the moisture content, thickness, geometry, etc. There are three main configurations for the electrodes: throughfield electrodes for thick samples, fringefield electrodes for thin samples, and staggered throughfield electrodes for samples of intermediate thickness (Vicente & Castro, 2007). Electrodes can thus be designed to form unique electric field patterns and uniform heating patterns to suit foods of different geometry (Vicente & Castro, 2007). A schematic of a RF dielectric heating unit is shown in Figure 7.1.

7.2.2.3 Food processing applications of radiofrequency

Radiofrequency has a long history of use in the food processing industry. Food applications described in the literature include blanching, thawing, drying, heating of bread/baking, meat processing, pasteurization, and sterilization. RF postbaking and RF-assisted baking of biscuits, crackers, and snack foods is one of the most accepted and widely used applications of RF heating (Fu, 2010). Other applications include RF drying of grains such as cowpea grains, moisture removal, and "moisture leveling" in finished goods (Jones & Rowley, 1997). RF drying has been described as a "self-leveling" phenomenon, as more energy is dissipated in wetter locations of the samples than drier ones, with the net effect of improvement in quality and consistency of the final products (Jones & Rowley, 1997). RF cooking of pumpable foods has also gained importance because heating is uniform and rapid (Ohlsson, 1999). Another application of RF that has attracted much recent attention is defrosting of frozen meats (Farag et al., 2008, 2009) and seafood (Archer et al., 2008). With regard to sterilization applications, researchers have demonstrated the efficacy of RF energy to inactivate heat-resistant spores to produce shelf-stable foods such as scrambled eggs (Luechapattanaporn et al., 2005) and macaroni and cheese (Wang et al., 2003a,b).

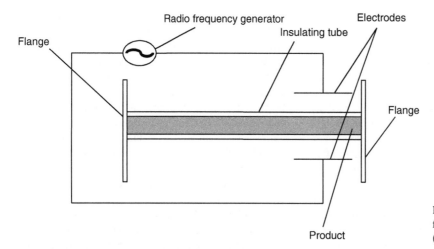

Figure 7.1 Schematic of radio-frequency dielectric heating unit (adapted from Zhao et al., 2000).

7.2.2.4 Effect of radiofrequency on the sensory and nutritional quality of foods

The effects of RF heating on the food constituents, as well as the overall quality, have been rarely documented in the literature. Tang et al. (2005) showed no significant differences in moisture, protein, fat, ash, and sodium chloride content in RF and steam-cooked turkey breast rolls. The contents of water-soluble vitamins (thiamin and riboflavin) were the same in both cases and the texture profile analysis also gave comparable results. However, the researchers did observe that the rate of lipid oxidation was significantly lower in RF-cooked meats compared to their steam-cooked counterparts. Taken together, it can be inferred that product quality of RF-processed foods is maintained if not enhanced, especially by virtue of the rapid volumetric heating characteristics of the technology.

7.2.3 Infrared heating

7.2.3.1 Introduction

Infrared (IR) uses electromagnetic radiation generated from a hot source (quartz lamp, quartz tube or metal rod) resulting from the vibrational and rotational energy of molecules. Thermal energy is thus generated following the absorption of radiating energy. The basic characteristics of infrared heating are the high heat transfer capacity, heat penetration directly into the product, fast process control, and no heating of surrounding air. Compared to conventional heating equipment, the operating and maintenance costs are lower and it is a safer and cleaner process (Vicente & Castro, 2007). Unlike microwave

heating, suitable levels of heating can be achieved at the surface and at the core of the body. The industrial application of this technology is relatively new and IR treatment of food has mainly been limited to experimental or pilot-scale processes.

7.2.3.2 Infrared equipment

The IR heating component consists of a radiator that basically radiates in all directions and a reflector that is responsible for directing the heat radiation to the target location. The maximum energy flux of radiators can range from 50 kW/m^2 (long wave) to 4010 kW/m^2 (ultra short wave). Industrial IR radiators can be categorized as either gas-heated radiators or electrically heated radiators where ohmic heating of a metal within a gas atmosphere (long waves), a ceramic body (long waves) or a quartz tube (medium or short waves) produces the IR radiation (Regier et al., 2010). Reflectors consist of polished metallic surfaces with a low absorption and high emission capacity to reflect the IR. Commercial reflectors can take the form of metallic/gold reflectors, glit twin quartz tube reflectors, or flat metallic/ceramic cassette reflectors (Regier et al., 2010).

7.2.3.3 Food processing applications of infrared

Short- (1 μm) and intermediate- (5 μm) wave IR heating modes have gained wide acceptance and have been applied for the rapid baking, drying, and cooking of foods of even geometry and modest thickness (Vicente & Castro, 2007). Short-wave IR has a penetration depth of several millimeters in many foods and has been successfully applied for the surface pasteurization of bakery products (Tewari, 2007). Long-wave IR (30 μm) has been used for industrial cooking and drying applications, achieving shorter processing times (up to 70% reduction) when compared to convective heating. IR drying of foods is preferred over conventional drying due to the higher drying rate, greater energy savings of up to 50%, and a homogeneous temperature distribution (Vicente & Castro, 2007). The technology has also been used for thawing, surface pasteurization of bread, and decontamination of packaging materials (Skjöldebrand, 2001).

7.2.3.4 Effect of infrared on the sensory and nutritional quality of foods

Baysal (2003) compared the drying characteristics of carrots and garlic dried with hot air, microwaves or IR. Infrared-dried samples displayed the highest rehydration capacity (8.95 g H_2O/g) followed by microwave-dried (8.38 g H_2O)/g) and hot-air dried samples (7.96 g H_2O)/g). The color parameters L, a, and b were determined, where L spans 0–100 ($L = 0$ represents total darkness (black) and $L = 100$ represents total lightness (white)), a runs from −a (green) to +a (red), and b runs from −b (blue) to +b (yellow). L and a values of dehydrated carrot samples were not significantly different. However, the b value of IR-dried carrots was significantly different from the fresh and air-dried products, indicating higher color protection by air drying. It was concluded that the choice of processing methods would ultimately depend on the desired characteristics of the product. The feasibility of IR for drying of shrimps was investigated by Fu and Lien (1998) who demonstrated that the product quality index measured in terms of its thiobarbituric acid (TBA) value – a direct indicator of lipid oxidation – was highest at plate temperature of 357 °C, air temperature of 43 °C, and plate distance of 12.5 cm. IR applied to barley resulted in an improvement in quality parameters such as germination and bulk density (Afzal et al., 1999). Hamanaka et al. (2000) demonstrated that intermittent IR treatment of cereals, i.e. IR heating at 2.0 kW for 30 sec, followed by cooling for 4 h, and treating again for 30 sec with infrared heating, resulted in minimal color changes; continuous treatment longer than 50 sec resulted in the discoloration of the wheat surface (Hamanaka et al. 2000). The irradiation of oysters by IR prior to freezing was found to result in higher moisture retention. Cooking of in-shell eggs using IR not only displayed a more attractive and brighter color, but also reduced the risk of

cracking because the eggs were not in contact with each other (Sakai & Hanzawa, 1994). Finally, the use of IR for bread baking was shown to be highly promising, as IR conditions controlled the crumb thickness, texture, and color of the finished product (Regier et al., 2010).

7.2.4 Ohmic heating

7.2.4.1 Introduction

Ohmic heating, sometimes referred to as joule heating, electrical resistance heating, direct electrical resistance heating, electroheating or electroconductive heating, is the process of passing electric currents through foods or other materials to heat them. The defining characteristics of ohmic heating compared to other electrical methods (MW and RF heating) lie in the frequency and waveforms of the electric field, and the presence of electrodes that contact the material.

7.2.4.2 Ohmic heating equipment

Ohmic heating equipment typically consists of electrodes, a source of power, and a chamber to house the food sample (Figure 7.2). Ohmic heaters may run on either a static (batch) or continuous mode. A typical batch ohmic heater consists of a horizontal cylinder with one electrode placed at each extremity. Continuous ohmic systems call for more flexibility and important considerations in the design of ohmic heaters include electrode configuration,

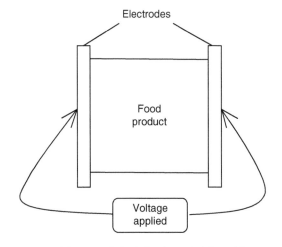

Figure 7.2 Schematic of ohmic heating unit (adapted from Vicente & Castro, 2007).

the distance between electrodes, electrolysis, heater geometry, frequency of AC current, power requirements, current density, applied voltage, product velocity, and velocity profile (Bengtsson et al., 2010). When foods (electrolytes) are subjected to a direct current, electrochemical reactions of reduction and oxidation will occur at the cathode and anode respectively. Under alternating current conditions, cathodes and anodes interchange places according to the frequency. The redox reactions therefore occur alternately at the same electrode site. In order to avoid such electrolytic effects and possible dissolution of the electrode material into the food, electrodes should be coated. This may be achieved by using food-compatible materials such as metal-coated titanium electrodes (Bengtsson et al., 2010). Temperature measurement devices that do not interfere with the electric field produced by the unit should also be used. In addition to the heating device, the unit may also consist of various auxiliary components, including feeding, holding, and cooling equipment, in addition to a tank for storage or direct aseptic filling (Regier et al., 2010).

7.2.4.3 Food processing applications of ohmic heating

Ohmic heating can be applied to a wide variety of foods, including liquids, semi-solid slurries or solid foods accompanied by a suitable carrier liquid. Ohmic heating has been used commercially in the US to pasteurize liquid egg products and in the UK and Japan for processing whole fruits (FDA, 2001). Ohmic heating has been successfully applied to a wide range of commodities such as fruits and vegetables, juices, meats, seafood, soups, crèmes, and pasta dishes (Bengtsson et al., 2010). In addition, it can be used for sterilization purposes to produce high-quality shelf-stable low-acid foods such as ready-prepared meals and high-acid foods such as tomato-based sauces.

In 2001, the FDA reported that "A large number of potential future applications exist for ohmic heating, including its use in blanching, evaporation, dehydration, fermentation, and extraction" (FDA, 2001). Ohmic heating can increase process efficiency in blanching (Wigerstrom, 1976) as well as enhancing the drying of vegetable tissue such as potato and yam by 16% and 43%, respectively (Wang & Sastry, 2000). Ohmic treatment was also found to enhance the extraction of valuable components such as rice bran oil from rice bran by up to 74% (Lakkakula et al., 2004).

7.2.4.4 Effect of ohmic heating on the sensory and nutritional quality of foods

Several researchers have shown that food samples processed by ohmic heating exhibited better quality attributes than their traditionally processed counterparts, with respect to nutrient retention, texture, color, and flavor. Leizerson and Shimoni (2005) observed higher retention of flavor compounds, limonene and myrcene, in ohmically treated orange juice than in conventionally heated juice. Moreover, sensory evaluation showed that ohmically treated and fresh orange juices were indistinguishable. Meat products cooked by ohmic heat were also indistinguishable from traditionally cooked samples with respect to moisture retention and mechanical properties. However, Piette (2004) observed a decrease in textural strength of ohmic-treated sausages and suggested the use of binders to circumvent this problem. Ohmic blanching also appears to be a very promising application, causing minimal deleterious effects on the color of various vegetables and vegetable products. Studies on the vitamin retention of ohmic and conventionally heated strawberry pulps and orange juices have presented similar ascorbic acid degradation kinetics (Vikram et al., 2005).

7.2.5 Sous-vide processing

7.2.5.1 Introduction

Sous-vide is a French term literally meaning "under vacuum." This technology allows food to be thermally processed using vacuum packaging in heat-stable, high-barrier or air-impermeable multilaminate plastics. This form of processing is especially useful for food consisting of partially cooked ingredients alone or combined with raw foods that require low-temperature storage until the food is thoroughly heated immediately prior to serving (Ghazala, 2004). In short, sous-vide is an "assemble-package-pasteurize-cool-store" process (Juneja, 2003). Figure 7.3 provides a simplified flow diagram that outlines the basic steps in sous-vide processing (SVP).

7.2.5.2 Sous-vide processing equipment

Filling of the product is first achieved by dispensing it via a pump from the container into sous-vide pouches or containers on a thermoformer or a conveyor-fed machine (Cole, 1993). The vacuum is carefully controlled to prevent damage to the contents. Usually, sous-vide cooking in food service operations uses a water tank or

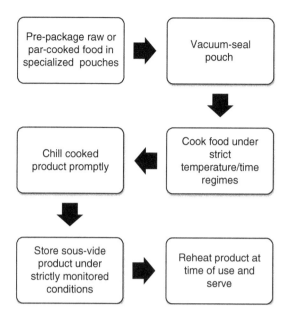

Figure 7.3 Generic flow diagram for a sous-vide processing line.

"bain marie," heated to the required temperature with a paddle to agitate the pouches (Cole, 1993). A slightly different method uses steam combination ovens, which use either convected hot air (dry heat) or low-pressure steam injection with convection heating (wet heat), to achieve temperatures below 100 °C. Following cooking, products are then chilled in an iced water bath system, using paddles to accelerate the cooling effect by ensuring the rapid flow of water around the bags in conjunction with a heat exchanger to maintain a chilled temperature. In addition, products may be cooled in air-blast cabinets, in which the products are placed in trays mounted on trolleys. The products are then stored chilled in standard cold rooms or chill cabinets (Tansey & Gormley, 2005).

Industrial heating equipment for SVP can take several forms including air/vapor, water immersion or steaming water (Schellekens & Martens, 1992).

7.2.5.3 Food processing applications of sous-vide

Sous-vide has been used widely in the processing of various types of raw or par-cooked meat, poultry, fish, and even vegetable-based products to enhance sensory characteristics, improve microbiological safety and extend shelf-life. In North America, food industry and retail food establishments are expected to comply with the principles and practices of their Hazards Analysis

Critical Control Point (HACCP) system, Good Manufacturing Practices (GMPs), sanitation guidelines, to maintain the cold chain from production to consumption and build multiple hurdles into a product for an additional degree of safety. In Europe, recommendations, guidelines, and codes of practice have been developed to ensure the safe production of sous-vide foods (Juneja & Snyder, 2007). The pie chart in Figure 7.4 shows the proportion of published studies on the application of SVP for various food commodities.

Gonzalez-Fandos et al. (2004) demonstrated the capacity of sous-vide cooking to reduce the counts of *Staphylococcus aureus*, *Bacillus cereus*, *Clostridium perfringens*, and *Listeria monocytogenes* on rainbow trout and salmon to below detectable levels and extend the shelf-life to >45 days during storage at 2 °C. Similarly, Shakila et al. (2009) showed an improvement in the microbiological quality of sous-vide fish cakes during chilled storage (3 °C) with an eight-fold increase in its shelf-life compared to their conventionally cooked counterparts. Current research has also focused on spore-forming microorganisms that constitute a safety risk in sous-vide products, such as *Clostridium botulinum*, *C. perfringens*, and *Bacillus* spp. Juneja and Marmer (1996) investigated the growth potential of *C. perfringens* in sous-vide cooked turkey products formulated with 0–3% salt and stored at temperatures of 4–28 °C. Overall, storage at 4 °C and a salt level of 3% proved to be most effective in controlling spore outgrowth. Similarly, Aran (2001) demonstrated that addition of calcium lactate (1.5%) and sodium lactate (3%) completely inhibited *B. cereus* outgrowth in beef goulash samples.

7.2.5.4 Effect of SVP on the sensory and nutritional quality of foods

Sous-vide is an extremely delicate and healthy method of preparing food. Most of the sensory and nutritional benefits are directly related to the fact that the food is placed into an evacuated package, sealed and cooked with mild heat (Ghazala, 2004). The overall effect is to achieve a tight control over heat, oxygen content, and presence of moisture – three primary factors that contribute to a decreased nutritional content of conventionally prepared foods. The presence of the plastic barrier also significantly reduces the extent and rate of oxidation, thereby preserving the qualities of essential polyunsaturated fatty acids (Creed, 1998; Ghazala & Aucoin, 1996). The plastic films also lock in moisture and desirable flavors. Consequently, sensory quality is retained, requiring

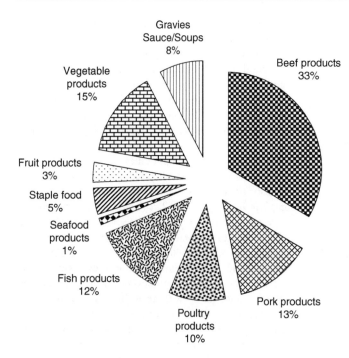

Figure 7.4 Studies from the literature (1989–2011) describing the use of sous-vide cooking for various commodities.

less seasoning and spices (Creed, 1998). The pouch also eliminates mineral and water loss, unlike other traditional cooking methods (Tansey & Gormley, 2005). In addition, vitamins are also protected from destabilization or loss of activity during sous-vide cooking. Research has shown that vitamin C retention was higher after SVP than following pasteurization. In summary, the desirable organoleptic attributes of sous-vide cooked products characterized by their fresh-like textures and vivid flavors, together with their wholesomeness, make them increasingly more appealing to consumers (Tansey & Gormley, 2005). Examples of successful sous-vide products include beefsteaks, meat loaves, fish fillets, and fruits and vegetables.

Unfortunately, there are a few caveats to recognize in SVP. First, only top-quality ingredients should be used and preparation should be accomplished in a clean environment to minimize initial contamination. Second, the time-temperature regime of the cooking step as well as the storage conditions need to be strictly monitored as they would affect the type of microorganisms that may survive and grow in the final product. As a result, these critical factors limit the scope of SVP. The ubiquitous nature of spores of *C. botulinum*, *C. perfringens*, and *B. cereus* in the environment, possible or even probable contamination of ingredients and raw materials and difficulty in maintaining the cold chain at the retail display and in

the domestic refrigerator can all limit the prospects of SVP technology.

7.3 Alternative non-thermal processing technologies

7.3.1 High hydrostatic pressure processing

7.3.1.1 Introduction

The comparative benefits and shortcomings of the various non-thermal processing technologies as well as critical parameters affecting their efficacy are summarized in Table 7.1, together with examples of current and potential products that may become commercially important. Among the array of non-thermal processing technologies, HHP has garnered the most attention since the early 1990s. For the last 20 years, HHP has been explored by food research institutions as well as the food industry with the goal of enhancing the safety, quality, nutritional, and functional properties of a wide variety of foods with minimal deleterious effects on their nutritional and organoleptic characteristics (Jung et al., 2011; Tewari, 2007). During commercial HHP processing, foods are exposed to pressures of the order of 200–1000 MPa for a few minutes using a suitable pressure-transmitting fluid such as water. HHP relies on the isostatic principle, meaning that

Table 7.1 Process considerations, benefits, and shortcomings of alternative non-thermal processing methods

Process	Process considerations	Benefits	Shortcomings	Examples of applications
High hydrostatic pressure	Processing time	Enhances product safety	Equipment is cost-prohibitive	Fruit products
	Treatment temperature	Extends shelf life of product	Phenomenon of "tailing" during microbial inactivation	Yogurts
	Pressure level	Desirable textural changes possible	Changes in sensory quality possible	Smoothies
	Product acidity	Production of "novel" products	Not suitable for foods with air spaces	Condiments
	Water activity	Minimal effect on flavor, nutrients and pigment compounds	Not suitable for dry foods	Salad dressings
	Physiological age of target organisms	Minimal textural loss in high-moisture foods	Refrigeration needed for low-acid foods	Meats and vegetables
	Product composition	Can eliminate spores when combined with high temperature	Elevated temperatures and pressures required for spore inactivation	Sauces
	Vessel size	In-container and bulk processing possible		High-value commodities such as seafood
	Packaging material integrity	Potential for reduction or elimination of chemical preservatives		
	Processing aids	Positive consumer appeal No evidence of toxicity of HHP alone		
Pulsed electric field	Electric field intensity	Effective against vegetative bacteria	Not suitable for non-liquid foods	Fruit juices
	Chamber design	Relatively short processing time	Postprocess recontamination possible	Milk
	Electrodes design	Suitable for pumpable foods	Less effective against enzymes and spores	Whole liquid egg
	Pulse width	Minimal impact on nutrients, flavor or pigment compounds	Adverse electrolytic reactions could occur	Soups
	Treatment time Temperature	No evidence of toxicity	Not currently energy efficient Restricted to foods with low electrical conductivity	Heat-sensitive foods
	Microbial species		Not suitable for product that contain bubbles	
	Microbial load Physiological age of organisms Product acidity Product conductivity Presence of antimicrobials		Scaling up of process difficult	
Ultraviolet light/pulsed UV light	Transmissivity of product	Short processing time	Shadowing effect possible with complex surfaces	Bread

(Continued)

Table 7.1   (*Continued*)

Process	Process considerations	Benefits	Shortcomings	Examples of applications
	Geometric configuration of reactor	Minimal collateral effects on foods	Has low penetration power	Cakes
	Power	Low energy input	Ineffective against spores	Pizza
	Wavelength	Suitable for high- and low-moisture foods	Possible adverse sensory effects at high dosages	Fresh produce
	Physical arrangement of source	Amenable for postpackage processing	Possible adverse chemical effects	Meats
	Product shape/size	Medium cost	Reduced efficacy with high microbial load	Seafood
	Product flow profile		Possible resistance in some microbes	Cheeses
	Radiation path length		Reliability of equipment to be established	Food packages
	Combination with other hurdles			
Ultrasound	Amplitude of ultrasonic waves	Ultrasound effective against vegetative cells	Has little effect on its own	Any food that is heated
	Exposure time	TS and MTS effective against vegetative cells and spores	Challenges with scaling up	
	Microbial species	Reduced process times	Free radicals could damage product quality	
	Volume of food	Amenable to batch and continuous processing	Can induce undesirable textural changes	
	Product composition	Little adaptation required for existing processing plant	Can be damaging to eyes	
	Treatment temperature	Possible modification of food structure and texture	Can cause burns and skin cancer	
		Energy efficient	Depth of penetration affected by solids and air in product	
		Several equipment options	Potential problems with scaling up of plant	
		Effect on enzyme activity		
		Can be combined with other unit operations		
Ionizing radiation	Absorbed dose	Long history of use	High capital cost	Fresh produce
	Water activity	High penetration power	Localized risks from radiation	Herbs and spices
	Freezing	Suitable for sterilization (food and packages)	Hazardous operation	Packaging materials
	Prevailing oxygen	Suitable for postpackage processing	Poor consumer acceptance	Meat and fish
	Microbial load	Suitable for non-microbiological applications (e.g. sprout inhibition)	Changes of flavor due to oxidation	

Table 7.1 *(Continued)*

Process	Process considerations	Benefits	Shortcomings	Examples of applications
	Microbial species	Packaged and frozen foods can be treated	Loss of nutritional value	
	Product composition	Low operating costs	Development of radiation-resistant mutants	
	State of food	Can be scaled up	Microbial toxins could be present	
	Food thickness	Low and medium dose has minimal effect on product quality	Outgrowth of pathogens	
	Particle size	Suitable for low- and high-moisture foods		
	Combination with other hurdles	Diverse applications		

HHP, high hydrostatic pressure; MTS, manothermosonication; TS, thermosonication.

the food product is compressed instantaneously and uniformly from every direction and returns to its original shape when the pressure is removed. This hydrostatic compression is capable of inactivating microorganisms. The mechanisms of inactivation will be covered in greater detail in the later part of this section. HHP can also be combined with heat, where the process temperatures during treatment can vary from subzero temperatures to above the boiling point of water (100 °C) (Caner et al., 2004).

7.3.1.2 High hydrostatic pressure equipment

Generally speaking, HHP equipment consists of a pressurized vessel, a low-pressure pump, an intensifier to generate higher pressures, and system controls (Guan & Hoover, 2005). Additional components include a temperature control device and a product handling system (Mertens, 1995). The construction of vessels may call for one of three common approaches, depending on the vessel operating pressure and diameter. The three cylindrical vessel designs are: a single forged monolithic chamber, a series of concentric tubes shrunk fit on each other to form a multiwall chamber, and a stainless steel core tube compressed by wire winding (Ting, 2010). The monolithic vessel or "monoblock" is commonly fabricated for operating pressures of 400–600 MPa for small vessels with internal diameters not exceeding 15 cm. For higher pressures and larger dimensions, prestressed multilayered (multiwall) or wire-wound vessels are used. HHP vessels require rapid closing and opening systems that allow fast loading and unloading of the product. Small-diameter or lower pressure vessels typically use threaded- or beech-type closures

or in some cases a pin closure (Ting, 2010). At higher pressures and larger diameters, closure loads become so large that a secondary structure such as a yoke or external frame is required to hold the end closures in place (Ting, 2010). The volume of the vessel may vary from less than 1 L (for laboratory-scale applications operating at <1000 MPa) to up to 500 L for processing units operating at 600 MPa.

When operating the unit, a pressure-transmitting fluid such as potable tap water is pumped from the reservoir into the pressure vessel by a pressure intensifier until the target pressure is reached. This is called "indirect compression" as opposed to "direct compression" which uses a piston to compress the hydrostatic medium in the vessel (Mertens, 1995). Temperature control before and during processing can be achieved by pumping a heating/cooling medium around the pressure vessel in the larger or more modern systems (Fonberg-Broczek et al., 1999). This is satisfactory in most applications, where constant temperatures are desired.

Conceptually, HHP systems can either treat unpackaged pumpable liquids or slurries in a semi-continuous bulk mode, or packaged food in a batch mode. A major advantage of the batch over the continuous process is that it reduces the risk of large quantities of product becoming contaminated by the lubricants or wear particles of the machinery (Palou et al., 2007), and potential postprocessing contamination. In addition, the productivity of the batch processes can be increased by operating the pressure vessels in tandem, with no lag in the processing times, so that the system operates sequentially (Palou et al., 2007). Examples of products suitable for batch processing are cooked meats, stews, guacamole, fruits in juice, shellfish, and ready-meal entrées (Jung et al., 2011; Patterson et al., 2006).

7.3.1.3 Food processing applications of high hydrostatic pressure

High hydrostatic pressure processing is applied at a wide range of pressures and temperatures in order to process and preserve foods while retaining their organoleptic and nutritional qualities. HHP used for inactivation of pathogenic and spoilage microorganisms in foods has been the subject of intense study. This section will mainly address the food safety aspects of HHP processing with some mention of other applications such as pressure-assisted freezing, pressure-assisted thawing (Cheftel et al., 2002), pressure-assisted extraction of bioactive compounds (Prasad et al., 2010), and pressure-induced modification of food compounds.

7.3.1.3.1 Microbial inactivation

The mechanism of microbial control and inactivation lies in a combination of processes such as the breakdown of non-covalent bonds in large macromolecules, biochemical effects, effects on the genetic mechanisms of cells, morphological changes and the disruption and permeabilization of the cell membrane (Patterson, 2005). Food microorganisms, such as vegetative bacteria, human infectious viruses, fungi, protozoa and parasites, can be significantly reduced when subjected to high pressure. A range of barotolerances exists among the different microbial groups and even among the different strains of the same species (Patterson, 2005; Vanlint et al., 2012; Whitney et al., 2007).

Several researchers have investigated the efficacy of HHP to inactivate a variety of food-borne pathogens in different commodities to enhance their microbiological safety. The use of Safe Pac™ high-pressure pasteurization to inactivate microorganisms and extend the shelf-life of ready-to-eat products including meats, soups, wet salads, sauces, fruit smoothies, shellfish, and seafood by 200–300% is just one example of the commercial adoption of HHP (Safe Pac, 2010). The synergistic effect of HHP and antimicrobials has also been reported (Jofré et al., 2008). Processing temperature also has an impact on microbial inactivation ratios: treatment temperatures above room temperature, in the range of 45–50 °C, generally bring about dramatic increases in the inactivation rate (Mertens, 1995).

7.3.1.3.2 Other applications

High hydrostatic pressure has also been used in the development of novel "value-added" food products (Torres & Velazquez, 2009). Some of the acquired quality attributes of pressurized products are due to modifications related to the volume reduction that occurs due to the physics of pressure. High pressure induces protein unfolding, a new spatial organization and modified interactions with water molecules, leading to possible changes in food protein functionality such as color, texture, gelation ability, and emulsifying properties. For instance, soy and meat protein extracts can become a gel under pressure, without any heat treatment; these pressure-induced gels have different rheological properties from those of thermal gels. HHP can also induce the modification of dairy proteins such as whey protein concentrate for improved functional properties (Lim et al., 2008; Padiernos et al., 2009). HHP also induces partial unfolding of egg albumen and yolk proteins, producing more elastic gels than heat-induced gels. HHP can also lead to the gelation of polysaccharides.

The effect of high pressure on the structure and physicochemical properties of different starches has been of great interest, undergoing complete gelatinization at 600 MPa (Katopo et al., 2002), and the properties of starch pastes and gels obtained under high pressure differ from those obtained from heat-gelatinized products, opening up opportunities for new applications (Yaldagard et al., 2008). Moreover, research in other areas includes high-pressure freezing applications such as pressure-shift freezing (PSF), high-pressure thawing, and high-pressure non-frozen storage. It is thought that the use of HHP to assist freezing and thawing of foods can result in improved quality and reduced drip loss by generation of small ice crystals (Sanz & Otero, 2005).

7.3.1.4 Effect of high hydrostatic pressure on the sensory and nutritional quality of foods

Covalent bonds of molecules are unaffected by pressure at the level used in food processing. As a result, small molecules that contribute to the color, flavor or nutritional quality of a food are relatively unchanged after exposure to pressure (Pandrangi & Balasubramaniam, 2005). Tomatoes treated at 600 MPa for 1 h at 25 °C underwent minimal change in lycopene and carotenoid content (Butz et al., 2002). Moreover, the antioxidant capacity of pressure-treated carrots and tomatoes was highly comparable to their untreated counterparts. Sancho et al. (1999) also compared the inactivation of multiple vitamins suspended in a model system. Their overall conclusions were that almost all vitamins and nutrients were minimally affected by pressure.

Although it is generally agreed that HHP has minimal effect on the sensory quality of foods, certain researchers have shown that inorganic transition metal ions can be released from their respective compounds under the effects of pressure, and these have been shown to adversely affect the flavor and shelf-life of the treated product due to accelerated lipid oxidation (Cheah & Ledward, 1997). In addition, enzymes are functional proteinaceous macromolecules that are stabilized by ionic as well as hydrophobic forces, and often mediate reactions that dictate the various quality attributes of the product. Since pressure can disrupt the structure and function of enzymes, the organoleptic characteristics of HHP-treated foods may change upon processing (Pandrangi & Balasubramaniam, 2005). HHP can also impart negative structural changes to foods with an open structure, such as queso fresco cheese, affecting its crumbling properties (Hnosko et al., 2012).

7.3.1.5 Pressure-assisted thermal sterilization

Although the application of hydrostatic pressure at a level of 400–800 MPa inactivates most pathogenic and spoilage vegetative bacteria, pressure inactivation of bacterial spores has been more problematic because of the marked resistance of spores to pressure (Black & Hoover, 2011). It is generally accepted that high inactivation ratios of *Bacillus* and *Clostridium* spores cannot be realized with the application of high pressure alone.

Although HHP is often described as a non-thermal processing method, compression of the hydrostatic medium brings about the generation of adiabatic heat (Knoerzer et al., 2010). By capitalizing on the combined effect of high pressures and initial elevated temperatures (hurdle approach), spores can be inactivated by a process called pressure-assisted thermal sterilization (PATS). During PATS, initial chamber temperatures are usually in the range of 70–90 °C and can transiently reach temperatures of 80–130 °C due to internal compression heating while pressures can vary from 600 to 1200 MPa. PATS thus takes advantage of the instantaneous and spatially homogeneous distribution of the adiabatic heat to control the temperature and heat exposure (Black & Hoover, 2011). This can in turn eliminate viable spores, while enhancing product quality (Matser et al., 2004; Wilson & Baker, 1997). According to Rodriguez et al. (2004),

PATS-induced lethality of viable spores is the result of a combined effect of thermally induced protein denaturation, enzyme inactivation and pressure-induced aggregation of proteins.

Reddy et al. (2003, 2006) showed that the application of high pressure at 600 MPa for 30 min at an initial temperature of 75 °C resulted in variable reductions of 0–5.5 log colony-forming units (CFU)/mL, depending on the strain of *C. botulinum*. Kouchma et al. (2005) inactivated spore strips of *C. sporogenes* embedded in egg patties at pressures of 690 MPa and a final product temperature of 110 °C.

A caveat with PATS is the phenomenon of tailing (Margosch et al., 2006). This is thought to occur in response to stabilization of spores by pressure, superdormancy or even the non-linearity of inactivation kinetics reminiscent of studies dealing with large populations of cells (Black & Hoover 2011).

Some researchers have speculated that the high temperature of PATS treatments can affect both covalent and non-covalent bonds and thus these processes require the evaluation of chemical changes of treated foods (Ramirez et al., 2009). Losses of nutrients and formation of hazardous compounds such as acrylamide, heterocyclic amines (HCAs), nitrosamines, polycyclic aromatic hydrocarbons (PAHs), furans, and heterocyclic amines have been reported as concomitants of conventional heating methods. Processing methods that indirectly heat food to very high temperatures, such as HHP, PEF, and ultrasound, may similarly result in the formation of undesirable compounds and significant loss of desirable ones such as vitamins, essential amino acids, and unsaturated fatty acids (Escobedo-Avellaneda et al., 2011; Vazquez Landaverde et al., 2007). Hence, the kinetics of formation of these compounds in PATS-treated food matrices should be evaluated, as this phenomenon is mostly unknown (Segovia Bravo et al., 2011).

In some regions of the world, processed food safety controls are already in place. For example, novel food laws approved in May 1997 in Europe regulate foods or food ingredients in which a production process gives rise to significant modifications in the composition and/or structure of the food or food ingredient (Constable et al., 2007). The safety assessment of these novel foods is based on the application of toxicological tests requiring a case-by-case comparison of the novel and conventional equivalent processed food. If undesirable compounds are found to be present in the traditional or novel product, a detailed risk assessment of these compounds needs to be conducted. This assessment includes identification and characterization of the hazards, an assessment of the human exposure, and final characterization of the toxicological risks. The formation of possible toxicological end-products can place an important constraint to the commercialization of PATS technologies, particularly in countries following the European Union novel food law model. PATS is affected by these regulations because this technology was not used before 15 May 15 1997 (Segovia Bravo et al., 2011). In contrast, a PATS process submitted in 2008 in the US for review by the Food and Drug Administration (FDA) required no such characterization of toxicity risk and was approved in 2009 (Segovia Bravo et al., 2011).

7.3.2 Irradiation

7.3.2.1 Introduction

Ionizing radiation can be used to decontaminate food. During irradiation, molecules absorb energy to form highly labile and reactive ions or free radicals that result in the breakage of a small percentage of chemical bonds (Ohlsson & Bengtsson, 2002). High-energy electrons fracture molecules (typically water molecules) into highly reactive species such as radicals, which can disrupt other cellular macromolecules such as DNA, proteins or structures such as cell membranes. Snyder and Poland (1995) mentioned that cell genetic material is particularly susceptible to radical damage during active replication or transcription.

Several factors affect the efficacy of irradiation to achieve satisfactory microbial inactivation, one of which is the dose of irradiation. In addition, intrinsic or product-related factors, such as prevailing oxygen availability, also affect the bactericidal effect of irradiation in the order of aerobic > hypoxic > anoxic conditions (Lacroix et al., 2002). Reduction of water activity protects bacterial cells against inactivation during irradiation. Freezing has also been shown to increase microbial resistance to irradiation due to reduced availability of reactive water molecules (Thayer, 1994). In addition, other characteristics inherent in the product, such as the state of the food, product composition and matrix, and food thickness and particle size, have a bearing on product decontamination. Although radiation can target the various microbial forms of life, microorganisms can vary considerably in their sensitivity to ionizing radiation in the order of parasites > bacteria > viruses (Niemira, 2003).

7.3.2.2 Irradiation equipment

Irradiation equipment typically consists of a high-energy isotope source to produce γ-rays. γ-Rays originating from cobalt-60 (^{60}Co) are often used commercially. Less commonly, machine sources are also available that produce high-energy electrons and X-rays (Arvanitoyannis & Tserkezou, 2010). Machine sources are electron accelerators, which consist of a heated cathode to supply electrons and an evacuated tube, in which electrons are accelerated by a high-voltage electrostatic field. These three sources of ionizing radiation transfer their energies to materials by ejecting atomic electrons, which can then ionize other atoms through a self-amplifying cascade of collisions. The source of choice for a particular application will depend on a multitude of factors, including the thickness and density of the food, the absorbed dose, and the overall economics of the process (Arvanitoyannis & Tserkezou, 2010). While γ- and X-rays have high penetration power and the potential to treat foods in bulk, high-energy electrons have relatively low penetration power and are only used for thin foods. In order to prevent leakage of radiation beyond the processing area, thick concrete walls and lead shields are often used. For the sake of personnel safety, strict procedures are in place to make sure that access to the irradiation plant is carefully monitored (Fellows, 2000).

7.3.2.3 Food processing applications of irradiation

7.3.2.3.1 Microbial inactivation

Various studies have demonstrated the ability of irradiation to destroy food-borne pathogens and spoilage microflora in various animal-derived and plant-derived products. Irradiation can be used to extend the shelf life of foods by destroying yeasts, molds and viable spoilage microorganisms (radurization) using a dose of 0.4–10 kGy, to reduce viable non-spore forming food-borne pathogens (radicidation) using a dose of 0.1–8 kGy, or sterilize products by killing both vegetative bacteria and spores using dose levels of 10–50 kGy (radappertization) (Fellows, 2000).

Fan and Sokorai (2005) investigated the effects of various doses of irradiation on the quality of fresh-cut iceberg lettuce and determined that 1–2 kGy was the optimum dose to ensure satisfactory inactivation of spoilage bacteria while minimizing sensory changes. Lamb et al. (2002) showed that low-dose γ-irradiation effectively reduced *S. aureus* in ready-to-eat ham and cheese sandwiches and proved to be more effective than refrigeration alone. Other researchers have applied higher doses of up to 10 kGy to frozen poultry or shellfish ($-18\,^{\circ}C$) to destroy *Campylobacter* spp., *Escherichia coli* O157:H7, and *Vibrio* spp. (Crawford & Ruff, 1996). For sterilization purposes, commodities such as herbs and spices are irradiated at a dose of 8–10 kGy to inactivate heat-resistant spore-forming bacteria (Fellows, 2000).

In addition to the application of irradiation alone, combined applications of ionizing radiation and modified atmosphere packaging (MAP) have been investigated. Lambert et al. (2000) showed that the combination of irradiation (1 kGy) and MAP (10–20% O_2, balance N_2) significantly reduced the microbial populations of fresh pork, although to the detriment of organoleptic quality. Similarly, Przybylski et al. (1989) showed that the application of low-dose irradiation and MAP brought about a 4–5-fold extension in the shelf life of fresh catfish fillets.

7.3.2.3.2 Other applications

Certain types of fresh produce, such as strawberries and tomatoes, can be irradiated to extend the shelf life by about 2–3 times when stored at $10\,^{\circ}C$ by slowing down the ripening process (Thomas, 1999). A combination of irradiation and MAP has been shown to deliver a synergistic effect in delaying ripening, lowering the dose required (Fellows, 2000). Irradiation arrests the ripening and maturation of fresh produce by inhibiting hormone production and interrupting the biochemical processes of cell division and growth (Thomas, 1999). Irradiation is also used for disinfestation purposes. The recommended treatment for decontamination of dried spices, herbs, vegetable seasonings, and dry ingredients is irradiation at a dose of 3–30 kGy (ASTM, 1998). Grains and tropical fruits may become infested with insects or larvae (Gupta, 1999). Low dose (<1 kGy) has been shown to be effective for disinfestation of produce and also extends the shelf life by delaying ripening (Gupta, 1999). In addition, irradiation is effective in inhibiting the sprouting of potatoes. In Japan, doses of about 0.15 kGy have been used to control sprouting on potatoes, onions, and garlic (Thomas, 1999).

7.3.2.4 Effect of irradiation on the sensory and nutritional quality of foods

Generally speaking, irradiation results in little adverse chemical change in food. None of the changes known to occur, such as hydroperoxides, have been shown to

be harmful or noxious to humans (Moreira, 2010). Proteins and carbohydrates are relatively stable to radiation doses of up to 10 kGy (Moreira, 2010). However, irradiation applied at particularly high levels can potentially cleave certain amino acids, thereby changing the aroma and flavor of foods. Irradiation can also cause lipids to undergo auto-oxidation reactions, producing hydroperoxides responsible for the development of off-flavors in food. However, selection of appropriate treatment conditions can minimize these objectionable changes in food quality (Olson, 1998).

Nutrients such as vitamins are variably sensitive to irradiation. The extent of vitamin loss depends on the type of vitamins, the dose received, and the type and physical state of the food in question. The general order of sensitivity of water-soluble vitamins to irradiation is: thiamin > ascorbic acid > pyridoxine > riboflavin > folic acid > cobalamin > nicotinic acid (Dock & Floros, 2000). Water-soluble vitamins are more sensitive to irradiation than fat-soluble vitamins. Fat-soluble vitamins rank in the following order of decreasing sensitivity: vitamin E > vitamin A > vitamin K > vitamin D (Kilcast, 1994). A comparative study of the extent of vitamin loss in irradiated and heat-sterilized chicken indicated similar inactivation kinetics. This therefore suggests that irradiation does not produce any special nutritional problems for treated foods (Dock & Floros, 2000; Moreira, 2010) and that the extent of nutritional degradation varies commensurate with various processing factors, similar to other preservation technologies (Moreira, 2010).

7.3.3 Ultraviolet light

7.3.3.1 Introduction

The UV spectrum is customarily divided into three regions: UV-A with a wavelength of 320–400 nm, UV-B with a wavelength of 280–320 nm, and UV-C with a wavelength of 200–280 nm (Sharma, 2010). UV-C possesses germicidal properties and at a dose rate of 1000 J/m^2 or more, bacteria, yeasts, and viruses undergo as much as 4-log reductions. The mechanism of inactivation and cell death is the absorption of UV by DNA and RNA (Sharma, 2010). Because of the very low penetration depth of UV, it is used for surface treatments only.

The fate of microorganisms following exposure to UV depends on a multitude of factors. The range of wavelengths used to irradiate the cells, treatment time, treatment intensity, and target species will affect the lethality of the process. As UV light penetrates the microbial cell, it brings about a number of alterations to the cellular components (Sharma, 2010). However, microbial inactivation is mainly the result of deleterious interactions between UV and nucleic acids. This interaction yields "photoproducts" of which pyrimidine dimers are the most important (Harm & Rupert, 1976). These are formed between two adjacent pyrimidine bases such as thymine and cytosine, on the same DNA strand. Another type of "photoproduct," called a "pyrimidine adduct," is also formed between adjacent bases, but at reduced rates of formation compared to dimers. At sufficiently high UV doses, DNA-protein cross-links are formed while at higher doses still, DNA strand breakages may be induced (Sharma, 2010).

Various processing parameters affect the efficacy of UV light similar to ionizing radiation. Besides the wavelength, other influential factors include the treatment temperature, concentration of hydrogen ions, microbial load, and relative humidity. Spores of bacteria are rather resistant to UV light compared to vegetative cells.

7.3.3.2 Ultraviolet equipment

Ultraviolet light is most commonly produced in low vapor pressure mercury lamps (Sharma, 2010). UV lamps are typically made of special quartz glass that allows 70–90% of UV rays to pass through. Mercury vapor lamps contain an inert gas carrier and a small amount of mercury within the sealed glass tube. When an electric arc is created, it causes the mercury to vaporize and ionize, thereby emitting UV radiation. These form the basic components of a UV-emitting apparatus (Guerrero-Beltràn & Barbosa-Canovàs, 2011). For optimal efficacy, the UV-C emitted should be able to access all parts of the treated food. Some researchers have developed "thin film UV-C disinfection systems" which deliver the (liquid) food via a nozzle in the form of a thin liquid film (Sharma et al., 1996). UV-C lamps are placed to ensure that the light shines towards the center as well as the sides of the disinfection unit. Another approach involves a liquid flowing through a convoluted tube surrounded by UV-C mercury lamps (Anonymous, 1999). While novel thin film devices tend to be used for disinfecting liquids, conveyor belts carrying solid foods past UV sources have also been set up with lamps mounted onto the walls of the enclosures, creating "UV tunnels" (Sharma, 2010). Figure 7.5 shows a commercially available bench-top UV-C test system at the University of Delaware.

7.3.3.3 Food processing applications of ultraviolet

The application of UV-C light technology for food products has been mainly confined to liquid foods and water. Sizer and Balasubramaniam (1999) demonstrated the superficiality of UV-C light when shone into liquid food materials, showing that UV-C light was only able to travel through a distance of ~1 mm underneath the surface of the product (juices) and that 90% of the light energy

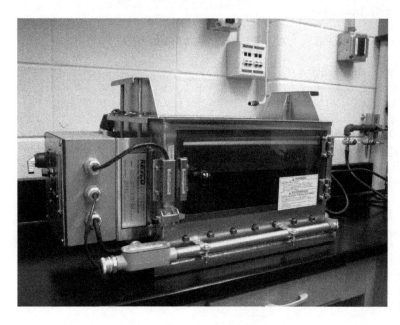

Figure 7.5 Laboratory scale ultraviolet-C test system (courtesy of Reyco Systems, Caldwell, ID). For color details, please see color plate section.

was absorbed. Wright et al. (2000) used a series of chambers to treat apple cider, achieving 3.8-log reduction in *E. coli* O157:H7. Farid et al. (2001) applied UV-C light to orange juice flowing as a thin film to maximize the exposed surface area/volume ratio and observed a two-fold increase in shelf life with no adverse effects on the color or flavor of the product. Guerrero-Beltrán et al. (2009) processed various juices including grape, grapefruit and cranberry juices inoculated with *S. cerevisiae* with UV-C light for a treatment time of 30 min and reported a maximum log reduction of 0.5, 2.4, and 2.4 log CFU/mL, respectively.

The penetration effect of UV-C radiation depends on the type of product, its UV-C absorptivity, soluble solids in the liquid, and suspended matter in the liquid. The larger the amount of soluble solids and the darker the color of the medium, the lower the intensity or penetration of UV-C light will be in the liquid. Guerrero-Beltrán et al. (2009) reported that since grape juice was deep violet, only 0.53-log microbial reductions were obtained after 30 min of exposure to UV light. However, a higher reduction of 1.34-log was observed for *S. cerevisiae* inactivation in apple juice after 30 min of UV light exposure since apple juice is light brown in color and transparent in nature. Even though the color and transparency of cranberry juice (red-brown and clear) and grapefruit juice (pink color and turbid) are entirely different, their response to UV light treatment was similar, suggesting that factors other than color and turbidity, such as the intrinsic composition, acidity and pH, could influence the efficacy of UV light.

In addition to juices, the effect of UV in extending the shelf life of fruits and vegetables has been reported. Lu et al. (1987) studied the use of UV light to extend the shelf life of Walla Walla onions at ambient storage temperature. Gonzalez-Aguilar et al. (2001) demonstrated the ability of UV-C radiation to preserve the postharvest state of ripe mangoes with minimal damage to their quality. UV-C treatments have also been used to control pathogenic as well as spoilage microbes on various meat, fish, and seafood products (Rahman, 2007).

7.3.3.4 Effect of ultraviolet on the sensory and nutritional quality of foods

Application of UV can have a variable impact on the organoleptic characteristics of food (Beaulieu, 2007; Matak et al., 2007; Rossitto et al., 2012). Deleterious effects manifested as surface discoloration, accelerated senescence or sprouting have also been shown to occur (Sharma, 2010).

UV light has the potential to form lipid radicals, superoxide radicals, and hydrogen peroxide (Kolakowska, 2003). Fat-soluble vitamins and pigments can be affected by peroxides produced during extended UV light treatment (Krishnamurthy et al., 2009). Superoxide radicals can further induce carbohydrate cross-linking, protein cross-linking, protein fragmentation, peroxidation of unsaturated fatty acids (Krishnamurthy et al., 2009), and loss of membrane fluidity function that could potentially have a deleterious effect on the textural quality of certain foods. UV light can also degrade vitamins by photodegradation, especially vitamin A, B_2, and C. In high-moisture foods, water molecules absorb UV photons and produce OH^- and H^+ radicals, which in turn interact with other food components (Krishnamurthy et al., 2009). Furthermore, UV radiation may also denature proteins, enzymes, and aromatic amino acids, leading to changes in the composition of the food material. In addition, exposure to UV light can change the flavor profile in certain products. Ohlsson and Bengtsson (2002) reported that during UV light exposure, the oxygen radicals formed induce the production of ozone, leading to the development of off-flavor notes in food products such as sour cream, whipped cream, butter, milk, mayonnaise, and dried vegetable soups. Therefore, UV light not only causes several undesirable chemical reactions, but can bring about deterioration in product quality.

7.3.4 Pulsed light

7.3.4.1 Introduction

Pulsed light (PL) technology is an innovative method for the decontamination and sterilization of foods using very high-power and very short-duration pulses of light emitted by inert gas flash lamps (Palmieri & Cacacea, 2005). This process has shorter treatment times, with concomitantly higher throughput, rendering it very efficient. Exposure to PL can be regarded as a form of UV light, except that it is applied in pulses (rather than statically) and at higher intensity. Several mechanisms have been proposed for the bactericidal action of PL. The UV portion of white light emitted at high power results in microbial inactivation through a photochemical, photothermal, and photophysical route (Krishnamurthy et al., 2010). The lethal effect is thus a combination of action of light pulses on certain cell constituents such as proteins and nucleic acids (photochemical effect), a transient temperature increase caused by heat dissipation of light pulses penetrating the product (photothermal effect) and morphological damage to cells caused by the constant disturbance originating from the high-energy pulses (photophysical effect). Since these light pulses are of short duration and high energy, PL can achieve effective microbial inactivation without any major adverse effect on product properties (Palmieri & Cacacea, 2005).

7.3.4.2 Pulsed light equipment

To generate PL, capacitors release the power as a high-voltage, high-current pulse of electricity to lamps filled with inert gas (Dunn et al., 1995). Inert gases such as xenon or krypton are used to ensure a high conversion efficiency of electrical to light energy. The current ionizes the gas, resulting in a broad-spectrum white light flash, spanning wavelengths of 200 nm (UV) to 1 mm (infrared). White light emitted comprises approximately 25% UV, 45% visible light, and 30% infrared (Dunn, 1996).

Light pulsed at a rate of 1–20 pulses/sec is directed over the material to be treated (Demirci & Krishnamurthy, 2011). Pulsed light systems are equipped with ammeters to measure the lamp current for each flash, which in turn determines the light intensity and spectrum. Silicon photodiode detectors are also used to measure the energy received from the lamp by the sample per unit area during the treatment (a.k.a. "fluence"). Current and fluence are tightly controlled, such that deviations from the set points cause the system to shut down automatically to avoid underprocessing. The flashing frequency can also be adjusted for the processing operation in question. The unit may also comprise one or more inert gas lamps that flash in unison or in tandem (Barbosa-Canovàs et al., 1998).

7.3.4.3 Food processing applications of pulsed light

The influence of surface topography on bacterial inactivation by PL has been shown to be variable. Uesugi et al. (2007) showed greater than 4-log reduction of *Listeria innocua* on rough stainless steel compared to smooth counterparts. Dunn et al. (1995), on the other hand, reported higher microbial reductions of pathogens on simple surfaces than complex surface aspects. Indeed, efficient light absorption depends on the distance through which light is passing, the thickness of the product, and the thickness of the package. Hillegas and Demirci (2003) demonstrated the effect of product thickness on PL-induced microbial inactivation and showed a reduction of 73.9% of *C. sporogenes* spores in clover honey in samples of 2 mm thickness compared to 14.2% in samples that were four-fold thicker. Similarly, Haughton et al. (2011) demonstrated the efficacy of PL to inactivate

Campylobacter on chicken breast and also showed an inverse correlation between the microbial reduction and package thickness. Keklik et al. (2010) demonstrated the optimal efficacy of pulsed UV light to decontaminate boneless chicken breasts for 15 sec when placed at a distance of 5 cm from the quartz window in the pulsed UV light chamber. The authors reported that the treatment resulted in an approximately 2-log reduction in the population of *Salmonella typhimurium*, with minimal impact on the quality and color of the sample tested. Moreover, the applicability of PL to pasteurize bottled beer has also been demonstrated previously (Palmieri & Cacacea, 2005).

The use of PL to enhance the microbiological quality and extend the shelf life of foods has also been examined. For instance, Figueroa-Garcia et al. (2002) showed that the application of PL led to a decrease in coliform and psychrotrophic bacterial counts in catfish fillets. Rice (1994) similarly showed that PL can be used to inactivate mold spores and consequently extend the shelf life of foods, such as sliced white bread and cakes, from a few days to over 2 weeks. The shelf life of cakes, pizza, and bagels was also extended to 11 days during storage at ambient temperature following PL treatment (Ohlsson & Bengtsson, 2002). The use of UV irradiation in conjunction with MAP also extended the shelf life of beef steak by 10 days (Djenane et al., 2003).

7.3.4.4 Effect of pulsed light on the sensory and nutritional quality of foods

Pulsed light treatment of food applied at high doses can negatively affect the sensory and nutritional quality of foods as evidenced by the manifestation of surface discoloration, loss of vitamins and other pigment compounds, as well as textural changes (Demirci & Krishnamurthy, 2011). Gomez-Lopez et al. (2005) conducted a sensory evaluation of PL-treated vegetables and demonstrated that the overall effect depended on the type of product under consideration. White cabbage treated by PL acquired a transient "plastic" off-odor; however, the taste scores, where 1 = fresh, 3 = acceptable, 5 = spoiled, were 2.1 to 2.7 and overall visual quality (OVQ) scores, where 1= excellent, 5 = fair, 9 = extremely poor, were 1.0 to 4.9. These scores were not significantly different from untreated counterparts with taste and OVQ scores of 1.6 to 2.5. Treated iceberg lettuce received higher scores in terms of odor (2.3–3.7), where 1 = no off-odor and 5 = severe off-odor, taste (1.8–2.6) and OVQ (1.5–6.9) compared to their untreated controls. In particular, the extent of leaf edge browning in treated iceberg lettuce (2.9–3.3), where 1 = none and 5 = severe, was lower than the untreated product (3.6–3.7) after 3 and 5 days of chilled storage at 7 °C. This clearly points to the usefulness of pulsed UV in enhancing the organoleptic characteristics of this vegetable (Demirci & Krishnamurthy, 2011).

7.3.5 Pulsed electric field

7.3.5.1 Introduction

Pulsed electric field (PEF) processing involves the application of high-voltage pulses to foods located between a series of electrode pairs. The electrical fields (generally at 20–80 kV/cm) are achieved through capacitors that store electrical energy from DC power supplies (Guan & Hoover, 2005). When a short electric pulse (1–100 μsec) is applied to food, there is a pronounced lethal effect on microorganisms (Ohlsson & Bengtsson, 2002). The precise mechanisms by which the microorganisms are destroyed by electric fields are not fully understood although it is thought that cell inactivation occurs by several mechanisms, including the formation of pores in cell membranes (Toepfl et al., 2007), formation of electrolytic products or highly reactive free radicals, oxidation and reduction reactions within the cell structure that disrupt metabolic processes, disruption of internal organelles and structural changes (Barbosa-Canovas et al., 1999), and production of heat produced by transformation of induced electrical energy.

The antimicrobial efficacy of the PEF process varies as a function of numerous processing parameters, including the electric field strength, number of pulses, pulse duration, pulse shape, processing temperature, and physiological state of the bacteria. Other factors that also influence the degree of inactivation include the temperature of the food, pH, ionic strength, and electrical conductivity (Vega-Mercado et al., 1999). Products of low conductivity include beer (0.143 Siemens/m) and black coffee (0.182 Siemens/m), while products of relatively higher conductivity include vegetable juice cocktail (1.556 Siemens/m), and tomato juice (1.697 Siemens/m). Jeyamkondan et al. (1999) stated that lowering the conductivity of the treated product increases the difference in the ionic concentration between the cytoplasm and the suspending fluid, which can in turn weaken the membrane structure by increasing the flow of ionic substances across the membrane. According to Barbosa-Canovas et al. (1999), foods with high conductivities are difficult

to work with because they generate smaller peak electric field strengths across the treatment chamber.

7.3.5.2 Pulsed electric field equipment

Electrical energy collected at low power levels over an extended period and stored in a capacitor is discharged quasi-instantaneously at very high levels of power. The basic-set up of the PEF unit consists of the following components: a high-voltage DC source also known as a pulse generator, a capacitor bank, and a switch to discharge energy to electrodes around a treatment chamber (Ohlsson & Bengtsson, 2002). The switching mechanism modulates several processing parameters, including the voltage pulse frequency, the duration of the treatment, and the waveform (Ohlsson & Bengtsson, 2002). The voltage and current applied, as well as the strength of the electric field, can be measured using an oscilloscope. Since the process of subjecting food to PEF unavoidably leads to the generation of heat, the treatment chamber is coupled to a cooling system to avoid overheating of the sample (Barbosa-Canovas et al., 1998).

Food samples may be processed in a static chamber (batch mode) or pumped through a chamber (continuous mode). Static PEF treatment chambers typically comprise two electrodes held in position by insulating materials that also enclose the food materials (Vega-Mercado et al., 1999). Uniform electric fields can be achieved using parallel plate electrodes with a gap sufficiently smaller than the electrode surface dimension (Vega-Mercado et al., 1999).

An example of a continuous chamber is a flow-through unit using low flow rates developed at Washington State University (Barbosa-Canovas et al., 1998). The chamber consists of two electrodes, a spacer, and two lids. The parallel-plate stainless steel electrodes have baffled flow channels between the electrodes to eliminate dead corners and ensure uniform treatment (Barbosa-Canovas et al., 1998). The two stainless steel electrodes are separated by a polysulfone spacer. Once treated, the pumpable food is then filled into individual consumer packages or stored using aseptic packaging equipment (Barbosa-Canovas et al., 1998).

7.3.5.3 Food processing applications of pulsed electric field

Several reports have presented the promising PEF-induced inactivation of microorganisms on food matrices. PEF research has mostly focused on the inactivation of microbes suspended in various pumpable non-particulate foods with free flowing characteristics including fruit juices, liquid eggs, milk, and pea soup (Vega-Mercado et al., 1999). Qin et al. (1995) reported a 6-log reduction of *S. cerevisiae* in PEF-treated apple juice after 10 pulses of 35 kV/cm at 22–34 °C. Zhang et al. (1997) showed that the total aerobic counts of reconstituted orange juice were reduced by 3–4-log cycles when treated with a PEF unit operating at <32 kV/cm. Evrendilek et al. (1999) treated fresh apple juice inoculated with *E. coli* O157:H7 and *E. coli* 8739 and reported that PEF achieved a 5-log reduction with a treatment temperature of 35 °C. In orange juice, McDonald (2000) inactivated *Leuconostoc mesenteroides*, *E. coli*, and *L. innocua* by as much as 5-log cycles at 50 kV/cm and 50 °C.

Some authors have also reported the synergistic inactivation upon combination of several hurdles such as PEF, pH, water activity (a_w), salt strength, temperature, and the presence of antimicrobial compounds such as nisin and lysozyme. For instance, Hodgins et al. (2002) studied the combined effect of temperature, acidity and a number of pulses on microbial inactivation in orange juice and reported a 6-log reduction in the natural microbiota following application of 20 pulses of an electric field of 80 kV/cm at a temperature of 44 °C and nisin concentration of 100 IU/mL. Along the same lines, Liang et al. (2002) showed that the application of PEF at moderately high temperatures of 60 °C in combination with the antimicrobial compounds nisin and lysozyme resulted in higher inactivation ratios. It is thought that the presence of antimicrobials sensitizes microbial cells to injury and ultimately loss of viability.

7.3.5.4 Effect of pulsed electric field on the sensory and nutritional quality of foods

The vitamin content, including thiamin (B_1), riboflavin (B_2), cholecalciferol and tocopherol (E), of PEF-treated milk remained unchanged after treatments of up to 400 μsec at field strengths of up to 27 kV/cm (Bendicho et al., 2002). The vitamin C content in milk was only slightly affected. This finding is congruent with those of Grahl and Markl (1996) who reported ~90% retention of ascorbic acid. The authors further demonstrated that vitamin retention was significantly higher than after low-temperature, long-time pasteurization at 63 °C for 30 min (49.7%) or even high-temperature, short-time pasteurization (HTST) at 75 °C for 15 sec. The same authors also reported that the vitamin A content of the PEF-treated milk was not significantly affected. Other

biochemical and chemical properties of milk including pH and titratable acidity (Deeth et al., 2007), fat content (Cortes et al., 2005), fat and protein integrity, cheese production, rennet clotting yield and calcium distribution (Deeth et al., 2007), color, moisture and particle sizes (Deeth et al., 2007) were also reported to be unaffected after PEF treatment at field strengths ranging from 20 to 80 kV/cm. Li et al. (2003) reported that treatment of up to 35 kV/cm for 73 μsec had no effect on bovine milk IgG in a protein-enriched soymilk, suggesting the promising use of PEF in treating products with bioactive components. Flavored milk, yogurt, and yogurt drinks processed by PEF and heat also showed increased shelf life, lower counts of yeasts and molds, and no changes in sensory scores (Yeom et al., 2004).

Besides dairy products, significant research has been conducted to determine the effect of PEF on the sensory and nutritional quality of fruit juices. Hodgins et al. (2002) demonstrated the benefit of PEF over heat treatments in the retention of flavor in fruit juices. The application of combined PEF (80 kV/cm) and thermal treatment of up to 50 °C was reported to have little to no effect on the vitamin C content of orange juice compared to heat-treated samples. PEF-treated apple juice and cider also had higher vitamin C retention (Evrendilek et al., 2000). PEF-processed tomato juice presented better rheological properties following PEF processing, marked by smaller and more uniform-sized particles compared to their heat-treated counterparts. Min and Zhang (2003) observed the overall higher acceptability of PEF-treated tomato juice, due to the reduced extent and rate of browning after treatment and during subsequent storage. Although some researchers reported color fading of PEF-treated apple juice and cider after multiple pulses at 50–466 kV/cm (Ortega-Rivas et al., 1998), other researchers have reported virtually no color loss after 94 μsec at 35 kV/cm.

7.3.6 Ultrasound (US)

7.3.6.1 Introduction

Ultrasonic waves (0.1–20 MHz) have garnered a great deal of interest in recent years in food processing due to their ability to influence the physical, biochemical or microbiological properties of food. This is because US can interact with the food differently depending on the US wave and product characteristics. US is considered both a thermal and a non-thermal treatment by virtue of its mechanism of action (Feng & Yang, 2010).

During the ultrasonication process, regions of compression (positive pressure) and rarefaction (negative pressure) are created. Cavitation or bubbles form and grow in the liquid during rarefaction and collapse during compression, creating micromechanical shocks (shear forces) that disrupt the structural and functional components of cells, ultimately leading to cell lysis (Hoover, 2000). When these cavitation bubbles explode or implode, they generate zones of highly localized extreme pressures ($>10^8$ Pa) and temperatures ($\sim$4000 K) that disrupt the cell structure and inactivate microbes. The mechanism of inactivation can thus be summarized as a combination of thermal and non-thermal effects culminating in damage to cell walls, thinning of membranes (microstreaming) and the production of free radicals that lead to DNA damage (Earnshaw, 1998; Earnshaw et al., 1995).

7.3.6.2 Ultrasound equipment

Different types of apparatus exist for ultrasonic applications, depending on the scale of the process. Some examples of equipment that have been used include whistle reactors, ultrasonic baths, and probe systems (Mason, 1998). A whistle reactor uses a mechanical ultrasonic source causing vibration of a stream of liquid moving past a metal blade. The frequency of vibrations is a function of the liquid flow rate and high enough flow rates can generate US, causing cavitation in the liquid. This set-up is used for food processing applications such as homogenization, emulsification or dispersion (Torley & Bhandari, 2007). Ultrasonic baths have a relatively simple and low-cost set-up, consisting of a metal bath with one or more transducers attached to the walls of the tank. The product to be treated is directly immersed into the bath (Mason, 1990). Another ultrasonic unit makes use of probe systems, comprising a metal horn linked to an ultrasonic transducer, which is made of a piezoelectric material, with the metal horn serving to amplify the vibration generated by the transducer. The need to amplify sound waves is justified since the amplitude of the waves produced by the transducer is too low to deliver any appreciable effect. Probe systems can be directly placed in the fluid-like product with easily controlled power input (Mason, 1990). Technological advances to increase the efficiency and quality of this ultrasonic unit include use of "cup horn" flow cells as well as tube reactors. While "cup horns" incorporate a cooling system to enable better temperature control (Mason, 1990), flow cells and tube reactors allow monitoring of the treatment time during which samples are exposed to US (Mason, 1990).

7.3.6.3 Food processing applications of ultrasound

This section will focus on the use of US in food processing, particularly its application for microbial inactivation, with some mention of other applications.

7.3.6.3.1 Microbial inactivation

Previous research has demonstrated the ability of US technology to pasteurize or even sterilize foods, alone or in combination with other preservation techniques (Earnshaw et al., 1995). The shear forces and rapidly changing pressures created by US waves are effective in destroying vegetative microbial cells, especially when used in conjunction with other preservation treatments including heat, pH modification, and chlorination. The conjunct use of US and heat is called thermosonication (TS) while the combined application of pressure, heat and US is termed manothermosonication (MTS). The effectiveness of TS varies as a function of the intensity, amplitude of sound waves, and treatment time while the effectiveness of MTS also depends on the magnitude of the applied pressure. It is thought that the application of pressure to the food material being ultrasonicated enables cavitation to be maintained at temperatures above the boiling point at atmospheric pressure (Torley & Bhandari, 2007).

Although US can inactivate vegetative cells, insignificant reduction of bacterial spores has been reported even with an extended US treatment (Guan & Hoover, 2005). However, the use of hurdle approaches such as TS and MTS was shown to have greater sporicidal effect, with MTS being more effective than TS (Guan & Hoover, 2005).

Ultrasound can be used in the product sanitization process to improve the efficacy of the washing process (Guan & Hoover, 2005). Seymour (2002) demonstrated that US combined with chlorinated water reduced the population of S. Typhimurium on iceberg lettuce by 1.7 logs compared to 0.7-log reduction by US alone. Ultrasound treatment was also found to be effective in reducing the population of S. Typhimurium on chicken breast skin. Decontamination treatments in a chlorinated solution reduced S. Typhimurium contamination by 0.2–0.9 logs while ultrasonication and chlorinated water delivered 2.4–3.9-log reductions (Lillard, 1993). The combined application of acetic acid and US treatments to clean eggshells has also been investigated with reported similar antimicrobial efficacy (Heath et al., 1980).

7.3.6.3.2 Other applications

Applications of US in food processing operations such as drying, dewatering, filtration, membrane separation, salting, and osmotic dehydration have also been examined (Feng & Yang, 2010). Combining US and air drying has been shown to enhance the rate of drying in various products, including carrot slices (Gallego-Juarez et al., 1999), onion slices (Da-Mota & Palau, 1999), potato cylinders (Bartolome et al., 1969), wheat (Huxsoll & Hall, 1970), corn (Huxsoll & Hall, 1970), and rice (Muralidhara & Ensminger, 1986). Acoustic drying has an overriding advantage compared to conventional drying processes since heat-sensitive foods can be dried faster and at a lower temperature (~50–60 °C) than in conventional hot air driers (~100–115 °C). There has also been considerable research into unit operations such as US-assisted membrane filtration and osmotic dehydration as well as processes such as cheese brining and meat curing.

Emulsification is another application of power US. When a bubble collapses in the vicinity of the phase boundary of two immiscible liquids, the resulting shock wave can provide very efficient mixing of layers. Emulsions generated by US have a number of advantages, including their stability without the addition of surfactant, as well as narrow mean droplet size distribution (Mason et al., 1996). Ultrasound technology has also been used in the making of US knives for "clean cutting" of sticky and brittle food products, including nut, raisins, and other hard fruits (Feng & Yang, 2010). The use of ultrasonication to control enzymatic activity (Feng & Yang, 2010) as well as to extract a range of bioactive compounds has been widely investigated. Several reviews have been published in the past on the use of US to extract plant origin metabolites and flavonoids from foods using a range of solvents (Zhang et al., 2003) and bioactives from herbs (Roldán-Gultiérrez et al., 2008; Vinatoru, 2001). The range of published extraction applications includes herbal extracts (Vinatoru, 2001), almond oils (Riera et al., 2004), soy protein (Moulton & Wang, 1982), and bioactives from plant materials, e.g. flavones (Rostagno et al., 2003) and polyphenolics (Xia et al., 2006).

7.3.6.4 Effect of ultrasound on the sensory and nutritional quality of foods

Several researchers have observed an enhancement in the nutritional value of ultrasonicated foods while others have shown a reduction after sonication and the loss was exacerbated during subsequent storage. Stojanovic and Silva (2007) demonstrated that a certain type of blueberry experienced greater than 60% loss of anthocyanin and phenolics following ultrasound-assisted osmotic dehydration than when treated by osmotic dehydration only. Significant differences in total anthocyanins were found between all three osmotic concentration treatments (osmotically concentrated berries for 12 h (O12), osmotically concentrated berries for 3 h (O3) and osmotically concentrated berries for 3 h with US (O3 + U). Hence, the time of concentration and high-frequency US negatively influenced anthocyanin content in osmoconcentrated berries. Treatments O3, O3 +U and O12 underwent a decrease in anthocyanin content by 20%, 42%, and 59%, respectively, when compared to the untreated sample. The author speculated that this could be due to surface cell rupture caused by cavitation, contributing to the release of anthocyanin and phenolics in berry samples.

Portenlänger and Heusinger (1992) reported that the level of L-ascorbic acid in distilled water was degraded by US, and the authors ascribed this to the generation of H• and OH• radicals. Degradation of astaxanthin by US as well as oxygen-labile nutrients such as ascorbic acid was reported in MTS-treated orange juice (Vercet et al., 2001). In contrast, the level of thiamin and riboflavin in milk was not adversely affected by MTS (Vercet et al., 2001).

Ultrasound-treated milk also showed a higher degree of homogenization, brighter color, and better stability after processing (Banerjee et al., 1996). Other researchers reported that US-treated fruit juices retained better color and flavor with minimal adverse impact on other organoleptic properties. The effect of US applications on the activity of enzymes affecting product quality has also been reported. Vercet et al. (2002) examined the effect of MTS on pectic materials and reported that they were unable to detect any residual activity of pectin methyl esterase (PME) in MTS-treated samples while polygalacturonase (PG) remained fully active. The authors attributed the decrease in PME activity to thermal and mechanical effects of the sonic energy cavitation. The ultrasonic treatment of tomato products was shown to inactivate enzymes and improve its rheological properties. Along the same lines, MTS also inactivated β-subunits of tomato endopolygalacturonase. These units have been shown to be problematic in tomato processing because they could potentially protect other degradative enzymes from thermal denaturation.

Texture of foods can also be variably affected by US, depending on the microstructure of the food. To some extent, the texture of US-treated foods is partly reliant on structural and functional changes of proteins including enzymes during sonication. For liquid food, US was shown to reduce particle sizes. Vercet et al. (2002) demonstrated that yogurt produced from US-treated milk had a stronger and firmer texture than untreated milk. Pohlman et al. (1996, 1997a,b) demonstrated improved meat tenderness following US-assisted cooking. Although there are limited studies investigating the effects of US on the flavor profile of foods, flavor improvement in Mahon cheese (Sánchez et al., 2001a,2001b), generation of off-flavor notes in edible oil (Chemat et al., 2004), and loss of desirable aroma compounds in apple juice (Feng & Lee, 2011) have been reported.

7.4 Sustainability and energy efficiency of processing methods

Energy consumption, energy savings, environmental protection, and waste management in the food industry have been in focus for the last 30 years. These have become critical issues in order to lower production costs, maintain economic growth, and improve sustainability in the food industry (Wang, 2009). This section provides some insight into the environmental impact of alternative and novel technologies.

Novel thermal technologies such as RF, MW, and ohmic heating (OH) methods and non-thermal methods such as PEF and HHP are continuously being developed and evaluated. Many novel technologies can ensure not only energy savings but also water savings, increased reliability, higher product quality, reduced emissions, and improved productivity (Masanet et al., 2008), and consequently, less impact on the environment. However, for many of the aforementioned technologies, such information is still scarce in published literature.

7.4.1 Energy savings

Lung et al. (2006) conducted an enlightening study, which provided estimates on the potential energy savings of PEF and RF drying systems compared to existing technologies. PEF pasteurization was demonstrated to have 100% natural gas savings since thermal input is eliminated. The electricity savings of PEF were estimated to be up to 18%, based on the assumed electricity consumption range of the base technology. For RF drying applications, the estimated natural gas savings range from 73.8 to 147.7 TJ per year, although these savings are masked by the increase in electricity consumption, indispensable to power the RF drying unit. Depending on the natural gas consumption of the base tunnel oven, the primary energy savings of RF drying can range from 0 to 73.8 TJ per year.

For PEF, the preservation of liquid media results in operation costs about 10-fold higher than those of conventional thermal processing. However, with pulse energy dissipation and the simultaneous resistive heating of the suspending medium, capitalizing on the synergistic effects of mild heat temperatures can render PEF more energy efficient (Toepfl et al., 2006). The combined synergistic effects of mild heat treatment temperatures and PEF can provide a shorter treatment time with energy recovery and an energy requirement of less than 40 kJ/kg for a reduction of 6 log cycles of *E. coli*. For instance, an energy input of 40 kJ/kg will lead to a temperature increase of 11 °C in the case of orange juice, showing that with a maximum temperature of 66 °C, the preservation process is still operating at lower maximum temperature and shorter residence times than during conventional heat preservation. Therefore, a reduction of the energy requirements from the original 100 kJ/kg to 40 kJ/kg can be realized, thus rendering PEF energy efficient and able to be easily integrated in existing food processing operations (Heinz et al., 2002). Despite the increase of the delivered electrical power, PEF seems less energy intensive than traditional pasteurization methods, leading to estimated annual savings of 791.2–1055 TJ, while also contributing to reduction of CO_2 emissions (Lelieveld, 2005).

High hydrostatic pressure is one of the most successful developments to date, offering a clean, natural, environment-friendly process. With PATS, instantaneous adiabatic compression during pressurization results in a quick increase in the temperature of food products, which is reversed when the pressure is released, providing rapid heating and cooling conditions and hence shorter processing times (Shao et al., 2008). The combined application of high pressure and heat can be utilized to achieve inactivation of spores of *Clostridium* spp. similar to conventional sterilization. The specific energy input required for sterilization of cans can be reduced from 300 to 270 kJ/kg when applying the PATS treatment (Toepfl et al., 2006). Moreover, if the electricity is generated by an environmentally clean, renewable energy source (e.g. hydroelectric power), then these processes would effectively contribute to reducing the pollution load, helping to preserve the environment. Furthermore, HHP may partly circumvent the use of cooling systems, which often represents 50% of the total electricity consumption (Dalsgaard & Abbots, 2003).

Studies conducted on meat products at the Agri-Food Canada's Food Research and Development Center (FRDC) have indicated that an OH process could result in energy savings of at least 70% (Vicente et al., 2006). In tests at the Louisiana State University Agricultural Center, sweet potato samples were processed using OH prior to freeze drying. OH increased the rate of freeze drying by as much as 25%, which led to significant savings in both processing time and energy use (Lima et al., 2002; Masanet et al., 2008). In addition, since heat is generated in the bulk fluid, less fouling should take place (Bansal & Chen, 2006), thus reducing fouling-related costs.

7.4.2 Reduced gas, effluent emissions, and water savings

The main types of gas emissions into the air from food processing operations are related to heat and power production. Traditional thermal methods require large amounts of fossil fuels and water to generate steam. In fact, it is estimated that the food industry's fossil fuel consumption is ca. 57% (Einstein et al., 2001). Typical boilers may accumulate dirt over time which acts as an insulator, thus reducing the heat transfer rate and wasting energy. In contrast, novel food processing systems powered by electricity may present an environmental advantage. In general, novel processing technologies such as HHP and PEF are considered environment friendly as they may eliminate completely, or at least reduce significantly, the local use of boilers or steam generation systems, and consequently diminish waste water, thus increasing water savings.

Several unit operations such as peeling, blanching, and drying as used by the food industry often require high water use. Ohmic peeling, in contrast, reduces environmental problems associated with lye peeling (e.g. treatment of waste water) because it does not use lye in the process (Wongsa-Ngasri, 2004). Mizrahi (1996) reported that blanching by OH considerably reduced the extent of solid leaching compared to a hot water process and a short blanching time could be used regardless of the shape and size of the product. For example, blanching of mushrooms using OH was reported to maintain a higher solids content than conventional blanching, while reducing the excessive consumption of water (Sensoy & Sastry, 2007).

Ohmic heating, PEF, IR, and RF can also significantly accelerate drying processes when compared to their conventional counterparts (Lima et al., 2002; Nowak & Lewicki, 2004; Wang, 1995), allowing precise control of the process temperature and leading to lower energy costs, reduced gas consumption, and fewer combustion-related emissions.

7.4.3 Generation of solid waste

Food production and consumption generate solid waste that can be classified as "food waste" and "non-food waste." The former is any plant or animal tissue in a raw or cooked state that was intended for human consumption but needs disposal as a result of spoilage, expiration, contamination or excess. The Environmental Protection Agency defines food waste as: "Uneaten food and food preparation wastes from residences and commercial establishments such as grocery stores, restaurants, produce stands, institutional cafeterias and kitchens, and industrial sources such as employee lunchrooms" (EPA, 1997). Innovative processing technologies such as HHP and irradiation can dramatically improve the shelf life of raw and processed products with minimal adverse effects on their sensory quality, thereby decreasing food losses and wastage due to microbial, chemical or enzymatic spoilage. Moreover, these technologies can produce valuable compounds that are attractive to consumers, thus reducing generation of solid waste such as olive mill waste water and other agricultural by-products.

7.5 Conclusion

Conventional thermal processing is a mainstay of the food industry. Alternative processing methods are relatively new in their application but not in their existence. In the quest for better quality, more healthful, minimally processed, and safer foods, these technologies have become the subject of intense research, furthering expansion of the knowledge base in this area, catalyzing research and development and ultimately commercialization of certain innovative food processing techniques. Currently, societal driving forces in the market as well as the technological development of alternative thermal and non-thermal preservation methods have provided a fairly well-grounded platform to deliver safer products that also guarantee higher quality.

Although the various technologies addressed in this chapter are at different stages of development, one may expect the increased "penetration" of more novel products or products processed by innovative methods on the market in the foreseeable future. HHP has been the subject of intense research effort over the last 15–20 years. With HHP, equipment reliability has traditionally been an issue but with advances in instrumentation and experience gained by equipment manufacturers in a number of installations for high-pressure pasteurization, the issue seems to have been resolved. Producers such as Fresherized Foods (formerly Avomex) now pressure process tonnes of guacamole per day. Although the capital cost of HHP equipment remains an issue, the concept of treating a variety of foods in flexible containers that are scalable to larger containers or larger pasteurization high-pressure vessels offers a lot of opportunity to the food industry. Food products pasteurized by high pressure are now commercially available in a number of countries including Japan, France, Spain, North America, and the UK.

For PEF, the technology is restricted to food products that can withstand high electric fields, have low electrical conductivity and do not contain or form bubbles. The particle size of the food in both static and flow treatment modes is a limitation. In addition, there are other major cost drivers that depend on the electrical properties of the food being processed. Intensive research conducted on PEF has brought the technology to the brink of commercial uptake, although PEF energy requirements are too considerable and costly to make it a good choice for most food applications.

With regard to US, although it has interesting potential as a novel preservation method, it still has a long way to go before it can be utilized commercially for this purpose. It does, however, have numerous non-preservation applications, some of which are already being used commercially.

Irradiation is a technology that has been more widely investigated than any other novel preservation method. In the US, electron beam-irradiated beefburgers and ground meat "chubs" have been successfully introduced to the market. Over 5000 US retail stores in 48 states now carry products that have been pasteurized using electron beam irradiation. Though the market for irradiation is still in a nascent stage, the global market is expected to exceed $145 million by 2015 (Global Industry Analysts Inc., 2010).

References

Afzal TM, Abe T, Hikida Y (1999) Energy and quality aspects during combined FIR–convection drying of barley. *Journal of Food Engineering* **42**: 177–182.

Ahmed J, Ramaswamy HS (2007) Microwave pasteurization and sterilization of foods. In: Rahman MS, (ed) Handbook of Food Preservation. Boca Raton, FL: CRC Press, pp. 691–712.

Anonymous (1999) UV light provides an alternative to heat pasteurization of juices. *Food Technology* **53**: 144.

Aran N (2001) The effect of calcium and sodium lactates on growth from spores of *Bacillus cereus* and *Clostridium perfringens* in a sous-vide beef goulash under temperature abuse. *International Journal of Food Microbiologyogy* **63**: 117–123.

Archer M, Edmonds M, George M (2008) Seafood thawing systems. In: Archer M, Edmonds M, George M (eds) Seafood Thawing. Research and Development Department, Campden and Chorleywood Food Research Association, SR **598**, pp. 21–33.

Arvanitoyannis IS, Tserkezou P (2010) Legislation on food irradiation: European Union, United States, Canada and Australia. In: Arvanitoyannis IS (ed) Irradiation of Food Commodities: Techniques, Applications, Detection, Legislation, Safety and Consumer Opinion. London: Elsevier, pp. 3–21.

ASTM (1998) Annual Book of ASTM Standards West Conshohocken, PA: ASTM.

Banerjee R, Chen H, Wu J (1996) Milk protein-based edible film mechanical strength changes due to ultrasound process. *Journal of Food Science* **61**: 824–828.

Bansal B, Chen X (2006) A critical review of milk fouling in heat exchangers. *Comprehensive Reviews in Food Science and Food Safety* **5**: 27–33.

Barbosa-Canovas GV, Pothakamury UR, Palou E, Swanson BG (1998) Light-pulse sterilization of foods and packaging. In: Barbosa-Canovas GV, Pothakamury UR, Palou E, Swanson BG (eds) Nonthermal Preservation of Foods. New York: Marcel Dekker, pp. 139–159.

Barbosa-Canovas GV, Gongora-Nieto MM, Pothakamury UR, Swanson BG (1999) Preservation of Foods with Pulsed Electric Fields. San Diego, CA: Academic Press,.

Bartolome LG, Hoff JE, Purdy KR (1969) Effect of resonant acoustic vibrations on drying rates of potato cylinders. *Food Technology* **23**: 321–324.

Baysal T (2003) Effects of microwave and infrared drying on the quality of carrot and garlic. *European Food Research and Technology* **218**: 68–73.

Beaulieu JC (2007) Effect of UV irradiation on cut cantaloupe: terpenoids and esters. *Journal of Food Science* **72**: 272–281.

Bendicho S, Barbosa-Canovas GV, Martin O (2002) Milk processing by high intensity pulsed electric fields. *Trends in Food Science and Technology* **13**: 195–204.

Bengtsson R, Birdsall E, Feilden S, Bhattiprolu S, Bhale S, Lima M (2010) Ohmic and inductive heating. In: Hui YH (ed) Handbook of Food Science, Technology and Engineering, volume **3**. Boca Raton, FL: CRC Press, pp. 121–127.

Black EP, Hoover DG (2011) Microbiological aspects of high pressure food processing. In: Zhang HQ, Barbosa-Canovas GV, Balasubramaniam VM, Dunne CP, Farkas DF, Yuan JTC (eds) Nonthermal Processing Technologies for Food. Chichester: WileyBlackwell, pp. 51–71.

Brewer M (2005) Microwave processing, nutritional and sensorial quality. In: Helmar S, Regier M (eds) The Microwave Processing of Foods. Cambridge: Woodhead, pp. 76–94.

Butz P, Edenharder R, Garcia AF, Fister H, Merkel C, Tauscher B (2002) Changes in functional properties of vegetables induced by high pressure treatment. *Food Research International* **35**: 295–300.

Caner C, Hernandez RJ, Harte BR (2004) High-pressure processing effects on the mechanical, barrier and mass transfer properties of food packaging flexible structures: a critical review. *Packaging Technology and Science* **17**: 23–29.

Canumir JA, Celis JE, de Bruijn J, Vidal LV (2002) Pasteurisation of apple juice by using microwaves. *Lebensmittel Wissenschaft und Technologie* **35**: 389–392.

Cheah PB, Ledward DA (1997) Catalytic mechanism of lipid oxidation following high pressure treatment in pork fat and meat. *Journal of Food Science* **62**: 1135–1139.

Cheftel JC, Thiebaud M, Dumay E (2002) Pressure-assisted freezing and thawing of foods: a review of recent studies. *High Pressure Research* **22**: 601–611

Chemat F, Grondin I, Shum Cheong Sing A, Smadja J (2004) Deterioration of edibleoils during food processing by ultrasound. *Ultrasonics Sonochemistry* **11**: 13–15.

Cinquanta L, Di Matteo M, Cuccurullo G, Albanese D (2010) Effect on orange juice of batch pasteurization in an improved pilot-scale microwave oven. *Journal of Food Science* **75**: 46–50.

Cole JC (1993) The User's Guide to Cuisine Sous-Vide. Buckinghamshire: JC Group Associates, pp. 63.

Constable A, Jonas D, Cockburn A et al. (2007) History of safe use of novel foods. *Toxicology Letters* **172**: 2–12.

Cortes C, Esteve MJ, Friegola P, Torregosa F (2005) Quality characteristics of horchata (a Spanish vegetable beverage) treated with pulsed electric fields during shelf-life. *Food Chemistry* **9**: 319–325.

Crawford LM, Ruff EH (1996) A review of the safety of cold pasteurization through irradiation. *Food Control* **7**: 87–97.

Creed PG (1998) Sensory and nutritional aspects of sous-vide processed foods. In: Ghazala S (ed) Sous-Vide and Cook-Chill Processing for the Food Industry. Gaithersburg, MA: Aspen Publishers, pp. 57–88.

Dalsgaard H, Abbots A (2003) Improving energy efficiency. In: Mattson B, Sonesson U (eds) Environmentally Friendly Food Processing. Boca Raton, FL: CRC Press, pp. 116–130.

Da-Mota VM, Palau E (1999) Acoustic drying of onion. *Drying Technology* **17**: 855–867.

Deeth HC, Datta N, Ross A, Dam XT (2007) Pulsed electric field technology: effect on milk and fruit juices. In: Tewari G, Juneja VK (eds) Advances in Thermal and Nonthermal Food Preservation. Ames, IA: Blackwell Publishing, pp. 240–279.

Demirci A, Krishnamurthy K (2011) Pulsed ultraviolet light. In: Zhang HQ, Barbosa-Canovas GV, Balasubramaniam VM,

Dunne CP, Farkas DF, Yuan JTC (eds) Nonthermal Processing Technologies for Food. Chichester: Wiley-Blackwell, pp. 249–261.

Dev SRS, Gariepy Y, Raghavan GSV (2008) Dielectric properties of egg components and microwave heating for in-shell pasteurization of eggs. *Journal of Food Engineering* **86**: 207–214.

Djenane D, Sanchez A, Beltran J, Roncales P (2003) Extension of the shelf life of beef steaks packaged in a MAP by treatment with rosemary and display under UV-free lighting. *Meat Science* **64**: 417–426.

Dock LL, Floros JD (2000) Thermal and nonthermal preservation methods. In: Schmidl M, Labuza TP (eds) Essentials of Functional Foods. Gaithersburg, MA: Aspen Publishers, pp. 49–88.

Dunn J (1996) Pulsed light and pulsed electric field for foods and eggs. *Poultry Science* **75**: 1133–1136.

Dunn J, Clark W, Ott T (1995) Pulsed light treatment of food and packaging. *Food Technology* **49**: 95–98.

Earnshaw RG (1998) Ultrasound: a new opportunity for food preservation. In: Povey MJW, Mason T (eds) Ultrasound in Food Processing. London: Blackie Academic and Professional, pp. 183–192.

Earnshaw RJ, Appleyard J, Hurst RM (1995) Understanding physical inactivation processes: combined preservation opportunities using heat, ultrasound and pressure. *International Journal of Food Microbiologyogy* **28**: 197–219.

Einstein D, Worrell E, Khrushch M (2001) Steam systems in industry: energy use and energy efficiency improvement potentials. Lawrence Berkeley National Laboratory. http://escholarship.org/uc/item/3m1781f1, accessed 13 November 2013.

Environmental Protection Agency (EPA) (1997) Terms of Environment: Glossary, Abbreviations and Acronyms. EPA 175-B-97-001. Washington, DC: Environmental Protection Agency.

Escobedo-Avellaneda Z, Moure MP, Chotyakul N, Torres JA, Welti-Chanes J, Lamela CP (2011) Benefits and limitations of food processing by high-pressure technologies: effects on functional compounds and abiotic contaminants. *CyTA Journal of Food* **9**: 351–364.

Evrendilek GA, Zhang QH, Richter ER (1999) Inactivation of *Escherichia coli* O157: H7 and *Escherichia coli* 8739 in apple juice by pulsed electric fields. *Journal of Food Protection* **62**: 793–796.

Evrendilek GA, Jin ZT, Ruhlman KT, Qiu X, Zhang QH, Richter ER (2000) Microbial safety and shelf-life of apple juice and cider processed by bench and pilot-scale PEF systems. *Innovative Food Science and Emerging Technologies* **1**: 77–86.

Fan X, Sokorai KJB (2005) Assessment of radiation sensitivity of fresh-cut vegetables using electrolyte leakage measurement. *Postharvest Biology and Technology* **36**: 191–197.

Farag KW, Cronin DA, Morgan DJ, Lyng JG (2008) Dielectric and thermophysical properties of different beef meat blends over a temperature range of −18 to +10°C. *Meat Science* **79**: 740–747.

Farag KW, Cronin DA, Lyng JG, Duggan E, Morgan DJ (2009) Comparison of conventional and radio frequency defrosting of lean beef meats: effects on water binding characteristics. *Meat Science* **83**: 278–284.

Farid MM, Chen XC, Dost Z (2001) Ultraviolet sterilization of orange juice. In: Welti-Chanes J, Barbosa-Canovas GV, Aguilera JM (eds) Proceedings of the Eighth International Congress on Engineering and Food. Lancaster: Technomic Publishing, pp. 1567–1572.

Fellows PJ (2000) Food Processing Technology: Principles and Practice. Cambridge: CRC Press LLC/Woodhead Publishing, pp. 196–208.

Feng H, Lee H (2011) Effect of power ultrasound on food quality. In: Feng H (ed) Ultrasound Technologies for Food and Bioprocessing. Food Engineering Series. New York: Springer Science and Business Media, pp. 559–582.

Feng H, Yang W (2010) Power ultrasound. In: Hui YH (ed) Handbook of Food Science, Technology and Engineering. Boca Raton, FL: CRC Press, pp. 121–127.

Figueroa-Garcia JE, Silva JL, Kim T, Boeger J, Cover R (2002) Use of pulsed-light to treat raw channel catfish fillets. *Journal of the Mississippi Academy of Science* **47**: 114–120.

Fonberg-Broczek M, Arabas J, Kostrzewa E et al. (1999) High-pressure treatment of fruit, meat and cheese products – equipment, methods and results. In: Oliveira FAR, Oliveira JC (eds) Processing Foods: Quality Optimization and Process Assessment. Boca Raton, FL: CRC Press, pp. 282–298.

Food and Drug Agency (FDA) (2001) *Kinetics of Microbial Inactivation for Alternative Food Processing Technologies: Ohmic and Inductive Heating*. www.fda.gov/~comm/ift–ohm.html, accessed 11 November 2013.

Fu WR, Lien WR (1998) Optimization of far infrared heat dehydration of shrimp using RSM. *Journal of Food Science* **63**: 80–83.

Fu YC (2010) Microwave heating in food processing. In: Hui YH (ed) Handbook of Food Science, Technology and Engineering. Boca Raton, FL: CRC Press, pp. 1–11.

Gallego-Juarez JA, Rodriguez-Corral G, Galvez-Moraleda JC, Yang TS (1999) A new high intensity ultrasonic technology for food dehydration. *Drying Technology* **17**: 597–608.

Gentry TS, Roberts JS (2005) Design and evaluation of a continuous flow microwave pasteurization system for apple cider. *LWT Food Science and Technology* **38**: 227–238.

Ghazala S (2004) Development in cook-chill and sous-vide processing. In: Richardson P (ed) Improving the Thermal Processing of Foods. Cambridge: Woodhead Publishing, pp. 152–173.

Ghazala S, Aucoin EJ (1996) Optimization of pasteurization processes for a sous vide product in rectangular thin profile forms. *Journal of Food Quality* **19**: 203–215.

Global Industry Analysts Inc (2010) *Food Irradiation Trends – A Global Strategic Business Report*, p. 86. www.researchand-markets.com/reports/338681/food_irradiation_trends_global_strategic, accessed 11 November 2013.

Gomez-Lopez VM, Devileghere F, Bonduelle V, Debevere J (2005) Intense light pulses decontamination of minimally processed vegetables and their shelf-life. *International Journal of Food Microbiologyogy* **103**: 79–89.

Gonzalez-Aguilar GA, Wang CY, Buta JG, Krizek DT (2001) Use of UV-C irradiation to prevent decay and maintain postharvest quality of ripe "tommy atkins" mangoes. *International Journal of Food Science and Technology* **36**: 767–773.

Gonzalez-Fandos E, Garcia-Linares MC, Villarino-Rodrıguez A, Garcia-Arias MT, Garcia-Fernandez MC (2004) Evaluation of the microbiological safety and sensory quality of rainbow trout (*Oncorhynchus mykiss*) processed by the sous-vide method. *Food Microbiology* **21**: 193–201.

Grahl T, Markl H (1996) Killing of microorganisms by pulsed electric fields. *Applied Microbiology and Biotechnology* **45**: 148–157.

Guan D, Hoover DG (2005) Novel nonthermal treatments. In: Sapers GM, Gorny JR Yousef AE (eds) Microbiology of Fruits And Vegetables. Boca Raton, FL: CRC Press, pp. 497–523.

Guan D, Cheng M, Wang Y, Tang J (2004) Dielectric properties of mashed potatoes relevant to microwave and radio-frequency pasteurization and sterilization processes. *Journal of Food Science* **69**: 30–37.

Guerrero-Beltran JA, Barbosa-Canovas GV (2011) Ultraviolet-C light processing of liquid food products. In: Zhang HQ, Barbosa-Canovas GV, Balasubramaniam VM, Dunne CP, Farkas DF, Yuan JTC (eds) Nonthermal Processing Technologies for Food. Chichester: Wiley-Blackwell, pp. 262–270.

Guerrero-Beltran J, Barbosa-Canovas GV, Welti-Chanes J (2009) Ultraviolet-C light processing of grape, cranberry and grapefruit juices to inactivate *Saccharomyces cerevisiae*. *Journal of Food Process Engineering* **32**: 916–932.

Gupta SC (1999) Irradiation as an alternative treatment to methyl bromide for insect control. In: Loaharanu P, Thomas P (eds) Irradiation for Food Safety and Quality. Lancaster: Technomic Publishing, pp. 39–49.

Hamanaka D, Dokan S, Yasunga E, Kuroki S, Uchino T, Akimoto K (2000) The Sterilization Effect of Infrared Ray on the Agricultural Products Spoilage Microorganisms. ASABE Paper No. 006090. St Joseph, MI: American Society of Agricultural and Biological Engineering.

Hamanaka D, Uchino T, Inoue A, Kawasaki K, Hori Y, Harm W (2007) Development of the rotating type grain sterilizer using infrared radiation heating. *Journal of the Faculty of Agricultrue Kyushu University* **52**: 107–110.

Harm H, Rupert CS (1976) Analysis of photoenzymatic repair of UV lesions in DNA by single light flashes. XI. Light-induced activation of the yeast photoreactivating enzyme. *Mutation Research* **34**: 75–92.

Haughton PN, Lyng JG, Morgan DJ, Cronin DA, Fanning S, Whyte P (2011) Efficacy of high-intensity pulsed light for the microbiological decontamination of chicken, associated packaging, and contact surfaces. *Foodborne Pathogens and Disease* **8**: 109–117.

Heath JL, Owes SL, Goble JW (1980) Ultrasonic vibration as an aid in the acetic acid method of cleaning eggs. *Poultry Science* **59**: 737–742.

Heinz V, Toepfl S, Knorr D (2002) Impact of temperature on lethality and energy efficiency of apple juice pasteurization by pulsed electric fields treatment. *Innovative Food Science and Emerging Technologies* **4**: 167–175.

Hillegas SL, Demirci A (2003) Inactivation of *Clostridium sporogenes* in clover honey by pulsed UV light treatment. *Agricultural Engineering International* **5**: 1–7.

Hnosko J, San Martin Gonzales MF, Clark S (2012) High-pressure processing inactivates *Listeria innocua* yet compromises Queso Fresco crumbling properties. *Journal of Dairy Science* **95**: 4851–4862.

Hodgins AM, Mittal GS, Griffiths MW (2002) Pasteurization of fresh orange juice using low-energy pulsed electric field. *Journal of Food Science* **67**: 2294–2299.

Hoover DG (2000) Ultrasound. *Journal of Food Science* **1**(suppl): 93–95.

Huang L, Sites J (2007) Automatic control of a microwave heating process for in-package pasteurization of beef frankfurters. *Journal of Food Engineering* **80**: 226–233.

Huang L, Sites J (2010) New automated microwave heating process for cooking and pasteurization of microwaveable foods containing raw meats. *Journal of Food Science* **75**: 110–115.

Huxsoll CC, Hall CW (1970) Effects of sonic irradiation on drying rates of wheat and shelled corn. *Transactions of the American Society of Agricultural Engineers* **13**: 21–24.

Icoz D, Sahin S, Sumnu G (2004) Color and texture development during microwave and conventional baking of breads. *International Journal of Food Properties* **7**: 201–213.

Jeyamkondan S, Jayas DS, Holley RA (1999) Pulsed electric field processing of foods: a review. *Journal of Food Protection* **62**: 1088–1096.

Jofré A, Garriga M, Aymerich T (2008) Inhibition of *Salmonella* sp., *Listeria monocytogenes* and *Staphylococcus aureus* in cooked ham by combining antimicrobials, high hydrostatic pressure and refrigeration. *Meat Science* **78**: 53–59.

Jones PL, Rowley AT (1997) Dielectric dryers. In: Baker CGJ (ed) Industrial Drying of Foods. London: Blackie Academic and Professional, pp. 156–178.

Juneja VK (2003) Sous-vide processed foods: safety hazards and control of microbial risks. In: Novak JS, Sapers GM, Juneja VK (eds) Microbial Safety of Minimally Processed Foods. Boca Raton, FL: CRC Press, pp. 97–124.

Juneja VK, Marmer BS (1996) Growth of *Clostridium perfringens* from spore inocula in sous-vide turkey products. *International Journal of Food Microbiologyogy* **32**: 115–123.

Juneja VK, Snyder OP (2007) Sous vide and cook-chill processing of foods: concept development and microbiological safety. In: Tewari G, Juneja VK (eds) Advances in Thermal and Non-Thermal Food Preservation. Ames, IA: Blackwell Publishing, pp. 145–166.

Jung S, Tonello-Samson C, Lamballerie-Anton MD (2011) High hydrostatic pressure food processing. In: Proctor A (ed) Alternatives to Conventional Food Processing. London: RSC Publishing, pp. 254–306.

Katopo H, Jane JL, Song Y (2002) Effect and mechanism of ultrahigh hydrostatic pressure on the structure and properties of starches. *Carbohydrate Polymers* **47**: 233–244.

Keklik NM, Demirci A, Puri VM (2010) Decontamination of unpackaged and vaccum-packaged boneless chicken breast with pulsed ultraviolet light. *Poultry Science* **89**: 570–581.

Kilcast D (1994) Effect of irradiation on vitamins. *Food Chemistry* **49**: 157–164.

Knoerzer K, Buckow R, Sanguansri P, Versteeg C (2010) Adiabatic compression heating coefficients for high-pressure processing of water, propylene-glycol and mixtures – a combined experimental and numerical approach. *Journal of Food Engineering* **96**: 229–238.

Kolakowska A (2003) Lipid oxidation in food systems. In: Sikorski ZE, Kolakowska A (eds) Chemical and Functional Properties of Food Lipids. New York: CRC Press, pp. 133–168.

Koskiniemi CB, McFeeters RF, Simunovic J, Truong VD (2011) Improvement of heating uniformity in packaged acidified vegetables pasteurized with a 915MHz continuous microwave system. *Journal of Food Engineering* **105**: 149–160.

Kouchma T, Guo B, Patazca E, Parisi B (2005) High pressure-high temperature sterilization from kinetic analysis to process verification. *Journal of Food Process Engineering* **28**: 610–629.

Krishnamurthy K, Irudayaraj J, Demirci A, Yang W (2009) UV pasteurization of food materials. In: Jun S, Irudayaraj JM (eds) Food Processing Operations Modeling: Design and Analysis. Boca Raton, FL: CRC Press, pp. 281–299.

Krishnamurthy K, Tewari J, Irudayaraj J, Demirci A (2010) Microscopic and spectroscopic evaluation of inactivation of *Staphylococcus aureus* by pulsed UV light and infrared heating. *Food Bioprocessing Technology* **3**: 93–104.

Lacroix M, Le TC, Ouattara B et al. (2002) Use of gamma irradiation to produce films from whey, casein, and soy proteins: structure and functional properties. *Radiation Physics and Chemistry* **63**: 827–832.

Lakkakula N, Lima M, Walker T (2004) Rice bran stabilization and rice bran oil extraction using ohmic heating. *Bioresource Technology* **92**: 157–161.

Lamb JL, Gogley JM, Thompson MJ, Solis DR, Sen S (2002) Effect of low-dose gamma irradiation on *Staphylococcus aureus* and product packaging in ready-to-eat ham and cheese sandwiches. *Journal of Food Protection* **65**: 1800–1805.

Lambert Y, Demazeau G, Largeteau A, Bouvier JM (2000) Packaging for high pressure treatments in the food industry. *Packaging Technology and Science* 13: 63–71.

Lelieveld H (2005) PEF – a food industry's view. In: Barbosa-Canovas GV, Tapia MS, Cano MP, Martin-Belloso O, Martinez A (eds) Novel Food Processing Technologies. Boca Raton, FL: CRC Press, pp. 145–157.

Leizerson S, Shimoni E (2005) Effect of ultrahigh-temperature continuous ohmic heating treatment on fresh orange juice. *Journal of Agricultural and Food Chemistry* 53: 3519–3524.

Li SQ, Zhang QH, Lee YZ, Pha TV (2003) Effects of pulsed electric fields and thermal processing on the stability of bovine immunoglobin (IgG) in enriched soymilk. *Journal of Food Science* 68: 1201–1207.

Liang Z, Mittal GS, Griffiths MW (2002) Inactivation of *Salmonella typhimurium* in orange juice containing antimicrobial agents by pulsed electric field. *Journal of Food Protection* 65: 1081–1085.

Lillard HS (1993) Bactericidal effect of chlorine on attached salmonellae with and without sonication. *Journal of Food Protection* 56: 716–717.

Lim SY, Swanson BG, Clark S (2008) High hydrostatic pressure modification of whey protein concentrate for improved functional properties. *Journal of Dairy Science* 91: 1299–1307.

Lima M, Zhong T, Lakkakula NR (2002) Ohmic Heating: A Value-Added Food Processing Tool. A technical report from Louisiana Agriculture Magazine. <www.lsuagcenter.com/en/communications/publications/agmag/Archive/2002/Fall/Ohmic+Heating+A+Valueadded+Food+Processing+Tool.htm, accessed 11 November 2013.

Lopez-Fandino R, Villamiel M, Corzo N, Olano A (1996) Assessment of the thermal treatment of milk during continuous microwave and conventional heating. *Journal of Food Protection* 59: 889–892.

Lu IY, Stevens C, Yakubu P, Loretan PA (1987) Gamma, electron beam and ultraviolet radiation on control of storage rots and quality of Walla Walla onions. *Journal of Food Processing and Preservation* 12: 53–59.

Luechapattanaporn K, Wang Y, Wang J, Tang J, Hallberg LM (2005) Sterilization of scrambled eggs in military polymeric trays by radio frequency energy. *Journal of Food Science* 70: 288–294.

Lung R, Masanet E, McKane A (2006) The role of emerging technologies in improving efficiency: examples from the food processing industry. In: Proceedings of the Industrial Energy Technologies Conference, New Orleans, Louisiana.

Margosch D, Ehrmann MA, Buckow R, Heinz V, Vogel RF, Ganzle MG (2006) High pressure-mediated survival of *Clostridium botulinum* and *Bacillus amyloliquefaciens* endospores at high temperature. *Applied and Environmental Microbiology* 72: 3476–3478.

Masanet E, Worrell E, Graus W, Galitsky C (2008) Energy efficiency improvement and cost saving opportunities for the fruit and vegetable processing industry, an energy star guide for energy and plant managers. Technical report: Environmental Energy Technologies Division. www.energystar.gov/ia/business/industry/Food-Guide.pdf, accessed 13 November 2013.

Mason TJ (1990) A survey of commercially available sources of ultrasound suitable for sonochemistry. In: Mason TJ (ed) Sonochemistry: The Uses of Ultrasound in Chemistry. Cambridge: Royal Society of Chemistry, pp. 60–68.

Mason TJ (1998) Power ultrasound in food processing – the way forward. In: Povey MJW, Mason TJ (eds) Ultrasound in Food Processing. London: Blackie and Academic and Professional, pp. 105–126.

Mason TJ, Paniwnyk L, Lorimer JP (1996) *The Uses of Ultrasound in Food Technology. Ultrasonics Sonochemistry* 3: 253–260.

Matak KE, Sumner SS, Duncan SE et al. (2007) Effects of ultraviolet irradiation on chemical and sensory properties of goat milk. *Journal of Dairy Science* 90: 3178–86.

Matser AM, Krebbers B, van den Berg RW, Bartels PV (2004) Advantages of high-pressure sterilization on quality of food products. *Trends in Food Science and Technology* 15: 79–85.

McDonald CJ (2000) Effects of pulsed electric fields on microorganisms in orange juice using electric field strengths of 30 and 50 kV/cm. *Journal of Food Science* 65: 984–987.

Mertens B (1995) New Methods of Food Preservation. New York: Blackie Academic and Professional, pp. 135.

Metaxas AC, Meredith RJ (1983) Industrial Microwave Heating. IEE Power Engineering Series. London: Peregrinus.

Min Z, Zhang QH (2003) Effects of commercial-scale pulsed electric field processing on flavor and color of tomato juice. *Journal of Food Science* 68: 1600–1606.

Mizrahi S (1996) Leaching of soluble solids during blanching of vegetables by ohmic heating. *Journal of Food Engineering* 29: 153–166.

Moreira RG (2010) Food irradiation using electron-beam accelerators. In: Hui YH (ed) Handbook of Food Science, Technology and Engineering. Boca Raton, FL: CRC Press, 124: 1–8.

Moulton J, Wang C (1982) A pilot plant study of continuous ultrasonic extraction of soybean protein. *Journal of Food Science* 47: 1127–1129.

Muralidhara HS, Ensminger D (1986) Acoustic drying of green rice. *Drying Technology* 4: 137–143.

Niemira B (2003) Irradiation of fresh and minimally processed fruits, vegetables and juices. In: Novak JS, Sapers GM, Juneja VK (eds) Microbial Safety of Minimally Processed Foods. Boca Raton, FL: CRC Press, pp. 279–299.

Nowak D, Lewicki PP (2004) Infrared drying of apple slices. *Innovative Food Science and Emerging Technologies* 5: 353–360.

Ohlsson T (1999) Minimal processing of foods with electric heating. In: Oliviera FAR, Oliviera JC (eds) Processing Foods – Quality Optimization and Process Assessment. New York: CRC Press.

Ohlsson T, Bengtsson N (2002) Minimal processing of foods with nonthermal methods. In: Ohlsson T, Bengtsson N (eds) Minimal Processing Technologies in the Food Industry. Cambridge: Woodhead Publishing, pp. 34–60.

Olson DG (1998) Irradiation of food. *Food Technology* **52**: 56–61.

Orsat V, Raghavan V (2005) Microwave technology for food processing: an overview. In: Helmar S, Regier M (eds) The Microwave Processing of Foods. Cambridge: Woodhead, pp. 105–115.

Ortega-Rivas E, Zarat-Rodriguez E, Barbosa-Canovas GV (1998) Apple juice pasteurization using ultra-filtration and pulsed electric fields. *Food and Bioprocess Technology* **76**: 193–198.

Padiernos CA, Lim SY, Swanson BG, Ross CF, Clark S (2009) High hydrostatic pressure modification of whey protein concentrate for use in low-fat whipping cream improves foaming properties. *Journal of Dairy Science* **92**: 3049–3056.

Palmieri L, Cacacea D (2005) High intensity pulsed light technology. In: Sun DW (ed) Emerging Technologies for Food Processing. California: Elsevier Academic Press, pp. 279–306.

Palou E, Lopez-Malo A, Barbosa-Canovas GV, Swanson BG (2007) High-pressure treatment in food preservation. In: Rahman MS (ed) Handbook of Food Preservation. New York: CRC Press, pp. 815–853.

Pandrangi S, Balasubramaniam VM (2005) High pressure processing of salads and ready-meals. In: Sun DW (ed) Emerging Technologies for Food Processing. California: Elsevier Academic Press, pp. 33–45.

Patterson MF (2005) Microbiology of pressure-treated foods. *Journal of Applied Microbiology* **98**: 1400–1409.

Patterson MF, Ledward DA, Rogers N (2006) High-pressure processing. In: Brennan JG (ed) Food Processing Handbook. Wokingham: Wiley-Vch, pp. 173–197.

Picouet PA, Catellari M, Viclas I, Landl A, Abadias M (2009) Minimal processing of a Granny Smith apple purée by microwave heating. *Innovative Food Science and Emerging Technologies* **10**: 545–550.

Piette L (2004) Ohmic cooking of processed meats and its effects on product quality. *Journal of Food Science* **69**: 71–78.

Pohlman FW, Dickeman ME, Zayas JF, Unruh JA (1996) Effects of ultrasound and convection cooking to different end point temperatures on cooking characteristics, shear force and sensory properties, composition, and microscopic morphology of beef *Longissimus* and *Pectoralis* muscle. *Journal of Animal Science* **75**: 386–401.

Pohlman FW, Dickeman ME, Kropf DH (1997a) Effects of high intensity ultrasound treatment, storage, time and cooking method on shear, sensory, instrumental color and cooking properties of packaged and unpackaged beef *Pectoralis* muscle. *Meat Science* **46**: 89–100.

Pohlman FW, Dickeman, ME, Zayas JF (1997b) The effect of low-intensity ultrasound treatment on shear properties, color stability and shelf-life of vacuum-packed beef *Semitendinosus* and *bicep femoris* muscle. *Meat Science* **45**: 329–337.

Portenländer G, Heusinger H (1992) Chemical reactions induced by ultrasound and γ-rays in aqueous solutions of L-ascorbic acid. *Carbohydrate Research* **232**: 291–301.

Prasad KN, Yang B, Zhao M, Sun J, Wei X, Jiang Y (2010) Effects of high pressure or ultrasonic treatment on extraction yield and antioxidant activity of pericarp tissues of longan fruit. *Journal of Food Biochemistry* **34**: 838–855.

Przybylski LA, Finerty MW, Grodner RM, Gerdes DL (1989) Extension of shelf-life of iced fresh channel catfish fillets using modified atmosphere packaging and low dose irradiation. *Journal of Food Science* **54**: 269–73.

Qin BL, Zhang Q, Barbosa-Canovas GV, Swanson BG, Pedrow PD (1995) Pulsed electric field treatment chamber design for liquid food pasteurization using a finite element method. *Transactions of the American Society of Agricultural Engineers* **38**: 557–562.

Rahman MS (2007) Light energy in food preservation. In: Rahman MS (ed) Handbook of Food Preservation. Boca Raton, FL: CRC Press, pp. 751–757.

Ramaswamy H, Tang J (2008) Microwave and radio frequency heating. *Food Science and Technology International* **14**: 423–427.

Ramirez R, Saraiva J, Lamela CP, Torres JA (2009) Reaction kinetics analysis of chemical changes in pressure-assisted thermal processing. *Food Engineering Review* **1**: 16–30.

Reddy NR, Solomon HM, Tezloff RC, Rhodehamel EJ (2003) Inactivation of *Clostridium botulinum* Type A spores by high pressure processing at elevated temperatures. *Journal of Food Protection* **66**: 1402–1407.

Reddy NR, Tezloff RC, Solomon HM, Larkin JW (2006) Inactivation of *Clostridium botulinum* nonproteolytic type B spores by high pressure processing at moderate to elevated high temperatures. *Innovative Food Science and Emerging Technologies* **7**: 169–175.

Regier M, Rother M, Schuchmann HP (2010) Alternative heating technologies. In: Ortega-Rivas E (ed) Processing Effects on Safety and Quality of Foods. Boca Raton, FL: CRC Press, pp. 188–245.

Rice J (1994) Sterilizing with light and electrical impulses: technological alternative to hydrogen peroxide, heat and irradiation. *Food Processing* **7**: 66.

Riera E, Golás Y, Blanco A, Gallego A, Blasco M, Mulet A (2004) Mass transfer enhancement in supercritical fluids extraction by means of power ultrasound. *Ultrasonics Sonochemistry* **11**: 241–244.

Rodriguez AC, Larkin JW, Dunn J et al. (2004) Model of the inactivation of bacterial spores by moist heat and high pressure. *Journal of Food Science* **69**: 367–374.

Roldán-Gultiérrez JM, Ruiz-Jimenez J, de Castro MDL (2008) Ultrasound-assisted dynamic extraction of valuable compounds from aromatic plants and flowers as compared with steam distillation and superheated liquid extraction. *Talanta* **75**: 1369–1375.

Rossitto PV, Cullor JS, Crook J, Parko J, Sechi P, Cenci-Goga BT (2012) Effects of UV irradiation in a continuous turbulent flow UV reactor on microbiological and sensory characteristics of cow's milk. *Journal of Food Protection* **75**: 2197–2207.

Rostagno A, Palma M, Barroso C (2003) Ultrasound-assisted extraction of soy isoflavones. *Journal of Chromatography* **1012**: 119–128.

Rowley AT (2001) Radiofrequency heating. In: Richardson PS (ed) Thermal Technologies in Food Processing. Cambridge: Woodhead Publishing, pp. 469–493.

Safe Pac Pasteurization (2010) Safe Pac™ expands the food safety universe. Philadelphia, PA. www.safepac.biz, accessed 11 November 2013.

Sakai N, Hanzawa T (1994) Application and advances in far-infrared heating in Japan. *Trends in Food Science and Technology* **5**: 357–362.

Sánchez ES, Simal S, Femenia A, Benedito J, Rosselló C (2001a) Effect of acoustic brining on lipolysis and on sensory characteristics of Mahon cheese. *Journal of Food Science* **66**: 892–896.

Sánchez ES, Simal S, Femenia A, Liull P, Rosselló C (2001b) Proteolysis of Mahon cheese as affected by acoustic-assisted brining. *European Food Research and Technology* **212**: 147–152.

Sancho F, Lambert Y, Demazeau G, Largeteau A, Bouvier JM, Narbonne JF (1999) Effect of ultra-high hydrostatic pressure on hydrosoluble vitamins. *Journal of Food Engineering* **39**: 247–253.

Sanz PD, Otero L (2005) High pressure freezing. In: Sun DW (ed) Emerging Technologies for Food Freezing. California: Elsevier Academic Press, pp. 627–647.

Schellekens W, Martens T (1992) "Sous-Vide" Cooking Part I: Scientific Literature Review. Brussels: Commission of the European Committees Directorate General XII, Research and Development.

Segovia Bravo K, Ramirez R, Durst R et al. (2011) Formation risk of toxic and other unwanted compounds in pressure-assisted thermally processed foods. *Journal of Food Science* **77**: 1–10.

Senorans FJ, Ibanez E, Cifuentes A (2011) New trends in food processing. *Critical Reviews in Food Science and Nutrition* 507–526.

Sensoy I, Sastry S (2007) Ohmic heating of mushrooms. *Journal of Food Process Engineering* **27**: 1–15.

Seyhun N, Sahin S, Ahmed J, Sumnu G (2009) Comparison and modeling of microwave tempering and infrared assisted microwave tempering of frozen potato puree. *Journal of Food Engineering* **92**: 339–344.

Seymour IJ (2002) Ultrasound decontamination of minimally processed fruits and vegetables. *International Journal of Food Science and Technology* **37**: 547–550.

Shakila JR, Jeyasekaran G, Vijayakumar A, Sukumar D (2009) Microbiological quality of sous-vide cook-chill fish cakes during chilled storage (3 °C). *International Journal of Food Science and Technology* **44**: 2120–2126.

Sharma G (2010) Ultraviolet light. In: Hui YH (ed) Handbook of Food Science, Technology and Engineering. Boca Raton, FL: CRC Press, **122**: 1–12.

Sharma G, Peppiatt C, Biguzzi M (1996) A novel thin film photoreactor. *Journal of Chemical Technology and Biotechnology* **65**: 56–64.

Shao YW, Zhu SM, Ramaswamy H, Marcotte M (2008) Compression heating and temperature control for high-pressure destruction of bacterial spores: an experimental method for kinetics evaluation. *Food and Bioprocess Technology* **3**: 71–78.

Sierra I, Vidal-Valverde C, Olano A (1999) The effects of continuous flow microwave treatment and conventional heating on the nutritional value of milk as shown by influence on vitamin B1 retention. *European Food Research and Technology* **209**: 352–354.

Sizer CE, Balasubramaniam VM (1999) New intervention processes for minimally processed juices. *Food Technology* **53**: 64–67.

Snyder OP, Poland DM (1995) Food Irradiation Today. St Paul, MN: Hospitality Institute of Technology and Management Publications.

Skjöldebrand C (2001) Infrared heating. In: Richardson P (ed) Thermal Technologies in Food Processing. New York: CRC Press, pp. 208–227.

Stojanovic J, Silva JL (2007) Influence of osmotic concentration, continuous high frequency ultrasound and dehydration on antioxidants, color and chemical properties of rabbiteye blueberries. *Food Chemistry* **101**: 898–906.

Sumnu G, Sahin S (2005) Recent developments in microwave heating. In: Sun DW (ed) Emerging Technologies for Food Processing. California: Elsevier Academic Press, pp. 419–444.

Sumnu G, Keskin SO, Rakesh V, Datta AK, Sahin S (2007) Transport and related properties of breads baked using various heating modes. *Journal of Food Engineering* **78**: 1382–1387.

Tang X, Cronin DA, Brunton NP (2005) The effect of radio frequency heating on chemical, physical and sensory aspects of quality in turkey breast rolls. *Food Chemistry* **93**: 1–7.

Tansey FS, Gormley TR (2005) Sous vide/freezing technology for ready meals. In: Barbosa-Canovas GV, Tapia MS, Cano MP (eds) Novel Food Processing Technologies. Boca Raton, FL: CRC Press, pp. 477–491.

Tewari G (2007) Microwave and radiofrequency heating. In: Tewari G, Juneja V K (eds) Advances in Thermal and Non-Thermal Food Preservation. Ames, IA: Blackwell Publishing, pp. 91–98.

Thayer DW (1994) Wholesomeness of irradiated foods. *Food Technology* **48**: 132–135.

Thomas P (1999) Control of post-harvest loss of grain, fruits and vegetables by radiation processing. In: Loaharanu P, Thomas P (eds) Irradiation for Food Safety and Quality. Lancaster: Technomic Publishing, pp. 93–102.

Ting E (2010) High pressure processing equipment fundamentals. In: Zhang HQ, Barbosa-Cánovas GV, Balasubramaniam B, Dunne CP, Farkas DF, Yuan JT (eds) Nonthermal Processing Technologies for Food. Oxford: Wiley-Blackwell.

Toepfl S, Mathys A, Heinz V, Knorr D (2006) Potential of high hydrostatic pressure and pulsed electric fields for energy efficient and environmentally friendly food processing. *Food Review International* **22**: 405–423.

Toepfl S, Heinz V, Knorr D (2007) High intensity pulsed electric fields applied for food preservation. *Chemical Engineering and Processing* **46**: 537–546.

Torley PJ, Bhandari BR (2007) Ultrasound in food processing and preservation. In: Rahman MS (ed) Handbook of Food Preservation. Boca Raton, FL: CRC Press, pp. 713–732.

Torres JA, Velazquez G (2009) Hydrostatic pressure processing of foods. In: Jun S, Irudayaraj JM (eds) Food Processing Operations Modeling: Design and Analysis. Boca Raton, FL: CRC Press, pp. 173–205.

Uesugi AR, Woodling SE, Moraru CI (2007) Inactivation kinetics and factors of variability in the pulsed light treatment of *Listeria innocua* cells. *Journal of Food Protection* **70**: 2518–2525.

Valero E, Martinez-Castro I, Sanz J, Villamiel M (2000) Chemical and sensorial changes in milk pasteurised by microwave and conventional systems during cold storage. *Food Chemistry* **70**: 77–81.

Vanlint D, Rutten N, Michiels CW, Aertsen A (2012) Emergence and stability of high-pressure resistance in different foodborne pathogens. *Applied and Environmental Microbiology* **78**: 3234–3241.

Vazquez-Landaverde PA, Qian MC, Torres JA (2007) Kinetic analysis of volatile formation in milk subjected to pressure-assisted thermal treatments. *Journal of Food Science* **72**: 389–398.

Vega-Mercado H, Gongora-Nieto MM, Barbosa-Canovas GV, Swanson BG (1999) Nonthermal preservation of liquid foods using pulsed electric fields. In: Rahman MS (ed) Handbook of Food Preservation. New York: Marcel Dekker, pp. 487–520.

Vercet A, Burgos J, Lopez-Bues P (2001) Manothermosonication of foods and food-resembling system: effect on nutrient content and nonenzymatic browning. *Journal of Agricultural and Food Chemistry* **49**: 483–489.

Vercet A, Sánchez C, Burgos J, Montañés L, Buesa PL (2002) The effects of manothermosonication on tomato pectic enzymes and tomato paste rheological properties. *Journal of Food Engineering* **53**: 273–278.

Vicente A, Castro IA (2007) Novel thermal processing technologies. In: Tewari G, Juneja VK (eds) Advances in Thermal and Nonthermal Food Preservation. Ames, IA: Blackwell, pp. 99–131.

Vicente A, Teixeira J, Castro I (2006) Ohmic heating for food processing. In: Sun D (ed) Thermal Food Processing: New Technologies and Quality Issues. Boca Raton, FL: Taylor and Francis, pp. 459–501.

Villamiel M, Lopez-Fandino R, Olano A (1997) Microwave pasteurization of milk in a continuous flow unit: effects on the cheese-making properties of goat's milk. *Milchwissenschaft* **52**: 29–32.

Vikram VB, Prapulla SG, Ramesh MN (2005) Thermal degradation kinetics of nutrients in orange juice heated by electromagnetic and conventional methods. *Journal of Food Engineering* **69**: 31–40.

Vinatoru M (2001) An overview of the ultrasonically assisted extraction of bioactive principles from herbs. *Ultrasonics Sonochemistry* **8**: 303–313.

Wang L (2009) Energy project management in food processing facilities. In: Wang L (ed) Energy Efficiency and Management in Food Processing Facilities. Boca Raton, FL: CRC Press, pp. 57–83.

Wang W, Sastry S (2000) Effects of thermal and electrothermal pretreatments of hot air drying rate of vegetable tissue. *Journal of Food Process Engineering* **23**: 299–319.

Wang WC (1995) Ohmic heating of foods: physical properties and applications. PhD dissertation, Ohio State University, Columbus, OH, pp. 1–157.

Wang Y, Wig TD, Tang J, Hallberg LM (2003a) Sterilization of foodstuff using radiofrequency heating. *Journal of Food Science* **68**: 539–544.

Wang Y, Wig TD, Tang J, Hallberg LM (2003b) Dielectric properties of food relevant to radiofrequency and microwave pasteurization and sterilization. *Journal of Food Engineering* **57**: 257–268.

Whitney BM, Williams RC, Eifert J, Marcy J (2007) High pressure resistance variation of *Escherichia coli* O157: H7 strains and *Salmonella* serovars in tryptic soy broth, distilled water and fruit juice. *Journal of Food Protection* **70**: 2078–2083.

Wigerstrom K (1976) Passing an electric current of 50–60 Hz through potato pieces during blanching. *US Patent No.* **3,997,678**.

Wilson MJ, Baker R (1997) High temperature/ultrahigh pressure stabilization of low-acid foods. PCT 189601500/1997.

Wongsa-Ngasri MSP (2004) Ohmic heating of biomaterials: peeling and effects of rotating electric field. PhD thesis, Ohio State University, Columbus, OH, pp. 1–192.

Wright JR, Sumner SS, Hackney CR, Pierson MD, Zoecklein BW (2000) Efficacy of ultraviolet light for reducing *Escherichia coli* O157: H7 in unpasteurized apple cider. *Journal of Food Protection* **63**: 563–567.

Xia T, Shi S, Wan X (2006) Impact of ultrasonic-assisted extraction on the chemical and sensory quality of tea infusion. *Journal of Food Engineering* **74**: 557–560.

Yaldagard M, Mortazavi SA, Tabatabaie F (2008) The principles of ultra high-pressure technology and its application in food processing/preservation: a review of microbiological and quality aspects. *African Journal of Biotechnology* **7**: 2739–2767.

Yeom HW, Evrendilek GA, Jin ZT, Zhang QH (2004) Processing of yogurt-based products with pulsed electric fields: microbial, sensory and physical evaluations. *Journal of Food Processing and Preservation* **28**: 161–178.

Zhang QH, Qiu X, Sharma SK (1997) Recent development in pulsed electric field processing. In: New Technologies Yearbook. Washington, DC: National Food Processors Association, pp. 31–42.

Zhang R, Xu Y, Shi Y (2003) The extracting technology of flavonoids compounds. *Food and Machinery* **1**: 21–22.

Zhao Y, Flugstad B, Kolbe E, Park JW, Wells JH (2000) Using capacitive (radiofrequency) dielectric heating in food processing and preservation – a review. *Journal of Food Process Engineering* **23**: 25–55.

8 Nanotechnology for Food: Principles and Selected Applications

Sundaram Gunasekaran

Food and Bioprocess Engineering Laboratory, University of Wisconsin-Madison, Madison, Wisconsin, USA

8.1 Introduction

Nanotechnology deals with "nanoscale" materials and structures, their synthesis, and applications. Though not rigidly defined, the term "nanoscale" is generally considered to represent the size range of 1 nm to 100 nm. Though we realize that "nano" is small, it is helpful to fully appreciate the smallness when it is in the context of a common reference. For example, the upper end of the nanoscale, 100 nm, is about the thickness of a strand of human hair split 100 times. The size scale of different objects spanning millimeters to nanometers is illustrated in Figure 8.1.

The tremendous advantages of nanotechnology stem directly from the unique characteristics exhibited by many materials when reduced to the nanoscale compared to their macroscopic counterparts. Some of the advantages include increased surface area to volume ratio and quantum effects that begin to dominate the behavior of matter. These factors can change and/or enhance material properties such as strength, reactivity, electrical and optical characteristics, etc. A popular and striking example of a unique nanoscale property is that of colloidal gold. While bulk gold that we are familiar with is yellow in color, colloidal suspensions of gold nanoparticles display different colors depending on their size and morphology. And indeed, this optical phenomenon has been exploited in various ways in many nanotechnology applications.

The growth of nanotechnology is very brisk worldwide. It is estimated that by the year 2015, the global impact of nanotechnology in various products will reach one trillion US dollars per year (Roco & Bainbridge, 2001). However, relatively speaking, compared to industries such as healthcare and electronics, the applications of nanotechnology in the food industry are rather limited, although recent indications are that they are showing significant growth. Many new nanotechnological advances are beginning to impact the food industry in aspects ranging from food safety to molecular synthesis of new food products and delivery of bioactive ingredients. Park (2009) estimated that in 2009 over 400 companies were actively pursuing applications of nanotechnology in food and agriculture, with the "nano-food" market reaching several billions of dollars. This trend is expected to continue for the next several years. The major food sectors for nanotechnology applications include food safety, biosecurity, product traceability, and the efficacious delivery of functional foods.

Nanoscale entities comprise all living organisms. For example, cell membranes, hormones, and DNA are all nano-sized entities of life. Foods we consume are composed of molecules such as carbohydrates, proteins and fats, which are nanoscale mergers between sugars, amino acids, and fatty acids (Shrivastava & Dash, 2009). While nanotechnology is potentially useful in all areas of food production and processing, many of the nanotechnology applications are either too expensive or too impractical for the food industry. For this reason, nanoscale techniques are perhaps most cost-effective in the following areas: food formulations, development and delivery of new functional materials, biosensing for detecting pathogens, toxins, allergens and other deleterious food constituents, and active and biodegradable food packaging. In this chapter, nanotechnology applications in two major areas that are relevant to food industry will be discussed: biosensing and packaging.

Food Processing: Principles and Applications, Second Edition. Edited by Stephanie Clark, Stephanie Jung, and Buddhi Lamsal.
© 2014 John Wiley & Sons, Ltd. Published 2014 by John Wiley & Sons, Ltd.

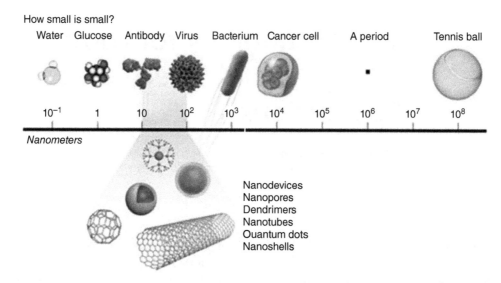

Figure 8.1 Size scale of different objects – macroscale to nanoscale (Anonymous, 2012). (Reproduced with permission from the National Institutes of Health.)

8.2 Biosensing

Biosensors are analytical devices incorporating a biological material (e.g. tissue, microorganisms, organelles, cell receptors, enzymes, antibodies, nucleic acids, natural products, etc.), a biologically derived material (e.g. recombinant antibodies, engineered proteins, aptamers, etc.) or a biomimic (e.g. synthetic catalysts, combinatorial ligands and imprinted polymers) intimately associated with or integrated within a physicochemical transducer or transducing microsystem, which may be optical, electrochemical, thermometric, piezoelectric, magnetic or micromechanical. Four major elements of biosensors are depicted in Figure 8.2.

1. Analyte (or sample), which is the system being sensed.
2. Biointerface, which directly interfaces with the analyte and forms an analyte-bioelement complex. Bioreceptors that are popularly used include nucleic acids, antibodies, enzymes, cells, etc.
3. Electrical interface, which translates the analyte-bioelement complex formation event into a corresponding electrical signal. This is normally achieved using nanostructures, electrodes, etc.
4. Electronic interface, which is a system of electronic components that is used to amplify, filter, and display the biosensor output in a form suitable for the end-user. Of these, the biointerface and electrical interface are the primary focus for innovation in designing biosensor architecture.

The sensing system offers the opportunity for a great deal of innovation, and there have been many over the past several years. Biosensors are usually categorized based on the type of transduction technology employed. Most biosensors currently on the market and/or being developed can be categorized as follows: electrochemical, bioluminescent, piezoelectric, resonant mirror, optoelectronic, and thermistor (Figure 8.3).

Biosensing is one of the major growth opportunities for nanotechnologies. The major users of biosensors are healthcare and environmental monitoring industries (Table 8.1). For example, blood glucose monitoring alone represents about a third of the entire biosensor market. Currently, biosensors are used in some 50 different end-user applications compared to about 30 a few years ago. As far as the food industry is concerned, biosensor development is actively pursued for detecting pathogens, toxins, and other deleterious contaminants, which could compromise the safety and security of our food chain. Biosensors are also being developed for routine food quality evaluation and control. It has been estimated that the global revenue for the biosensors market will continue to exhibit strong growth and will exceed the $14 billion mark by 2016 (Thusu, 2010).

8.2.1 Challenges of biosensing

Tests for detection of biological agents such as pathogens and chemical agents such as toxins are challenged by the following requirements (Golightly et al., 2009).

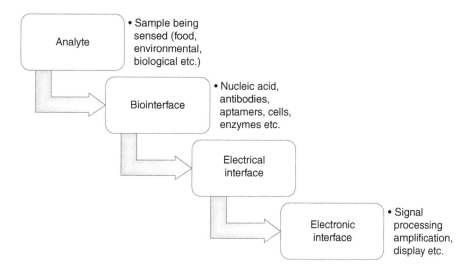

Figure 8.2 Elements and selected components of a typical biosensor.

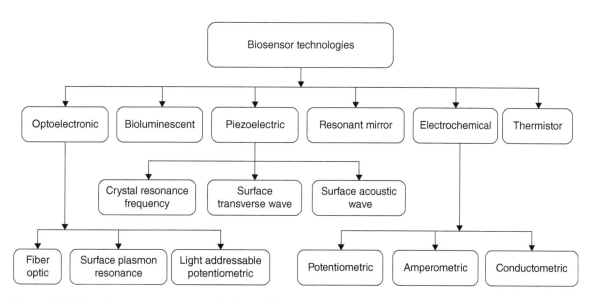

Figure 8.3 Different biosensor transduction technologies (Thusu, 2010).

• The test should detect substances in a very broad range of molecular weights.
• In some cases, the target analyte is not known; this is particularly true for biothreats but may also be the case for chemical warfare agents or even impurities in food or water.
• The test must have an extremely low tolerance for error (both false positives and false negatives) because the cost of a mistake can be enormous.

• Exceptionally high sensitivity is needed because some of the target species are present at very low levels.
• The teset should be able to monitor targets continuously and in real time, or at least within a few minutes.
• The test system should be portable with low use of consumable supplies.
• The test must work in extreme environmental conditions.

Table 8.1 Major biosensor market segments (top) and major targets being sensed (bottom) based on the percent revenues. Data from Thusu (2010)

Market segment	Revenue (%)	Market segment	Revenue (%)
Glucose	31.55	Cryptosporidium	0.32
Toxicity	8.83	Giardia	0.32
Infectious disease	5.80	Bovine spongiform encephalopathy	0.29
Coagulation prothrombin time	4.58	Listeria	0.17
Pregnancy	3.57	Salmonella	0.28
Drug discovery	3.31	Anthrax	0.16
E. coli/Coliform	3.30	Cocaine	0.14
Hematology	2.84	Radixin	0.13
Coagulation activated clotting time	2.60	Ecstasy	0.12
Alcohol sensors	2.30	TNT explosives	0.11
Coagulation	1.72	Opiates	0.09
Drugs of abuse	1.67	2,4-Dinitrotoluene (DNT)	0.07
Cholesterol	1.36	Methamphetamines	0.06
Lactic acid sensor	1.29	Tetryl	0.05
A 1C	0.87	THC (cannabis)	0.05
Peptide sensor	0.69	Nitroglycerin	0.04
E. coli O157:H7	0.60	Pentaerythritol tetranitrate (PETN)	0.03
Microcystins	0.47	Rest of research type	8.47
Severe acute respiratory syndrome (SARS)	0.33	Others	1.45
West Nile virus	0.33		

- The test must handle a wide variety of matrices, from swabs of surfaces (e.g. envelopes) to powders to aqueous solutions to air samples composed predominantly of sand and dust.

The US Centers for Disease Control and Prevention (CDC) estimates that each year roughly one in six Americans (or 48 million people) gets sick, 128,000 are hospitalized, and 3000 die of food-borne diseases (CDC, 2011). Reducing food-borne illness by 10% would keep about 5 million Americans from getting sick each year. Therefore, prompt detection of any pathogen outbreak before it occurs is vitally important. The CDC has identified the top five organisms, among the 31 known pathogens that cause food-borne illnesses, in each of the three classes of pathogens based on the extent of food safety concerns (Table 8.2).

Given the grave threat posed by pathogen contamination to our food chain, the need for simpler and faster pathogen detection methods that will afford real-time and in-line detection cannot be overemphasized. This need has been further exacerbated by the perceived and real threats of bioterrorist attacks. In addition to human costs, the economic losses associated with contaminated foods can be staggering. An annual loss of about

$20 billion in productivity has been estimated (Straub & Chandler, 2003) due to illnesses caused by water-borne pathogens alone. The food- and water-borne pathogens range across a wide spectrum of organisms, including bacteria such as *Escherichia coli* (E. coli), *Salmonella* and *Shigella*, and viruses such as norovirus, rotavirus, and hepatitis A (Hrudey & Hrudey, 2007). The presence of *E. coli* in foodstuffs and drinking water is a chronic worldwide problem. The worldwide food production industry is worth about US $578 billion (Singh et al., 2009). It is estimated that infectious diseases cause about 40% of the approximately 50 million total annual deaths worldwide (Ivnitski et al., 1999).

8.2.2 Bioreceptors

Bioreceptors are recognition elements used in biosensors. During biosensing, bioreceptors come in contact with the target and interact with it in some way. In the case of pathogen detection, bioreceptors are generally biomolecules that have an affinity for epitopes present on the pathogen surface. Depending on the sensing mechanism, a variety of biomolecules can be employed as bioreceptors. They include antibodies, nucleic acid aptamers, carbohydrates,

Table 8.2 Top five pathogens according to different food safety concerns and their percent contribution to total (CDC, 2011)

Pathogen	Domestically acquired food-borne illnesses	Domestically acquired food-borne illnesses resulting in hospitalization	Domestically acquired food-borne illnesses resulting in death
Campylobacter spp.	9	15	6
Clostridium perfringens	10	–	–
E. coli (STEC) O157	–	4	–
Listeria monocytogenes	–	–	19
Norovirus	58	26	11
Salmonella, non-typhoidal	11	35	28
Staphylococcus aureus	3	–	–
Toxoplasma gondii	–	8	24
Total	**91**	**88**	**88**

and antimicrobial peptides (Figure 8.4). Additionally, molecularly imprinted polymers (MIPs) have also been used as biorecognition elements for pathogen detection (Bolisay et al., 2007).

8.2.2.1 Antibodies

Antibodies are the biological entities of choice for a wide range of biosensors, ranging from pathogens to toxins (Iqbal et al., 2000; Leonard et al., 2003). Their high specificity and wide and easy availability make antibodies very popular, and indeed they are the perfect choice for detecting organisms that elicit an antigenic response. Two common types of antibodies are monoclonal and polyclonal; in addition, engineered antibody fragments are also available. The monoclonal antibodies are produced *in vitro* from hybridoma cell lines and consist of an identical, well-defined population of antibodies that bind to a single epitope. Polyclonal antibodies are produced *in vivo* and consist of a suite of antibodies that bind to a number of epitopes on the antigen. While monoclonal antibodies offer greater specificity, polyclonal antibodies have a greater potential for attachment to the antigens on cell surfaces, have higher resistance to pH and salt concentration changes, and are considerably less expensive (Vikesland & Wigginton, 2010). Engineered antibody fragments could offer combined advantages of both monoclonal and polyclonal antibodies: high specificity and low cost (Hudson, 2006). The most common problem associated with the use of antibodies is non-specific binding. Though monoclonal antibodies can help overcome that problem, their cost, short shelf life, and temperature sensitivity pose problems (Jayasena, 1999).

8.2.2.2 Aptamers

Aptamers are single-stranded DNA or RNA ligands which can be selected for different targets from a huge library of molecules containing randomly created sequences. Aptamers have been selected to bind very different targets, from proteins to small organic dyes. In addition to the very important aspect of having an unlimited source of identical affinity recognition molecules available due to the selection process, aptamers can offer advantages over antibodies that make them very promising for analytical applications. The use of aptamers as therapeutic tools is now well established. In contrast, the analytical application of aptamers in diagnostic devices or in systems for environmental and food analysis is still under investigation and the scientific community still needs further research to demonstrate the advancements brought by this new kind of ligands.

Aptamers are small (2–25 kDa) nucleic acid sequences that have high affinity and specificity toward target molecules (Vikesland & Wigginton, 2010). The specificity of aptamers is better than even monoclonal antibodies (Jenison et al., 1994; Win et al., 2006). Aptamers can be synthesized by *in vitro* methods in unlimited quantities, without the need for animal models (Mairal et al., 2008).

Aptamers have been selected to bind very different targets, from proteins to small organic dyes (Tombelli et al., 2007). However, they can suffer from nuclease attack (Tombelli et al., 2007). Aptasensors, the biosensors employing aptamers as recognition elements, have been developed for different food and environmental biosensing applications (Fischer et al., 2007; Homann & Goringer, 1999; So et al., 2008; Tombelli et al., 2007).

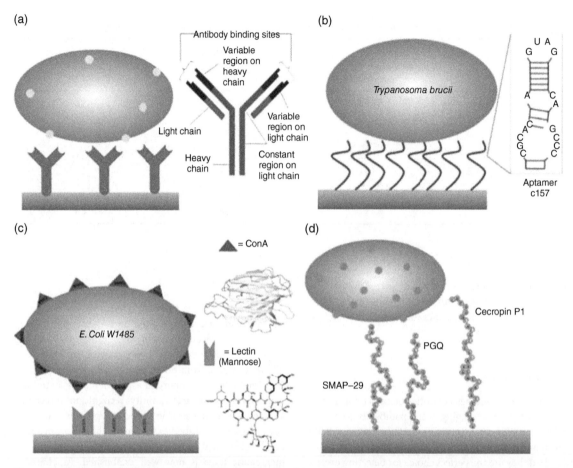

Figure 8.4 Biorecognition elements used for pathogen detection: (a) antibodies; (b) aptamers (cl57 specific for *Trypanosoma brucii*) (Homann & Goringer, 1999); (c) carbohydrates ConA/mannose based sensor for *E. coli* (Shen et al., 2007); and (d) antimicrobial peptides (SMAP-29, PCQ, and cecropin P1) (Arcidiacono et al., 2008). (Reproduced from Vikesland & Wigginton (2010), with permission from the American Chemical Society.)

Aptamers specific to bioterrorism targets, such as anthrax spores, cholera toxin, staphylococcal enterotoxin B, ricin and abrin toxin, have been used to develop different systems for their detection (Bruno & Kiel, 1999, 2002; Kirby et al., 2004; Tang et al., 2007). Bruno and Kiel (1999, 2002) used electrochemiluminescence coupled to magnetic beads for the detection of anthrax spores, cholera toxin and staphylococcal enterotoxin B. Pan et al. (2006) reported an aptamer system to inhibit bacterial action with a reduction of cell invasion of *Salmonella enterica* serovar *typhi* which is an important pathogen for humans and causes typhoid or enteric fever. Stratis-Cullum et al. (2009) used a DNA aptamer for the detection of *Campylobacter jejuni*, a common bacterial cause of food-borne infection in the United States due to mishandling of raw poultry and consumption of undercooked poultry.

8.2.2.3 Carbohydrates

Use of biomolecules other than antibodies and aptamers for biosensing is a recent development. For example, carbohydrates have been used for pathogen detection (de la Fuente & Penades, 2006; Gallego et al., 2004). The advantage of using carbohydrates is that, unlike antibodies, they maintain their activity and structure over changing temperatures and pHs. Also, because carbohydrates are smaller than antibodies, higher densities of carbohydrate-sensing elements per unit surface area are possible (Vikesland & Wigginton, 2010), which can afford carbohydrate-based biosensors better performance than immunoreaction-based sensors (Stine & Pishko, 2005). Though the specificity of carbohydrates is poor (Shen et al., 2007), their broad affinity is an advantage for detecting different organisms (El-Boubbou et al., 2007; Haseley, 2002; Haseley et al., 1999). Carbohydrate binding properties of lectins, which are a group of proteins that strongly bind to specific carbohydrate moieties, are also used in biosensors (Haseley et al., 1999; Milton et al., 1997; Ryu et al., 2010). For example, lectin-based biosensors have been used to detect *E. coli* and other microorganisms (Jelinek, 2004; Shen et al., 2007).

8.2.2.4 Antimicrobial peptides

Antimicrobial peptides (AMPs) are proteins that recognize and semi-selectively bind to microbial surfaces and facilitate pathogen lysis (Zasloff, 2002; Zasloff & Anderson, 2001). Though cell lysis is the primary use of AMPs, their ability to bind microbial surfaces and toxins has been used for biosensing applications to detect pathogens and toxins (Arcidiacono et al., 2008; Kulagina et al., 2005, 2006, 2007; Soares et al., 2009). Kulagina et al. (2005) reported sensitive detection of *E. coli* O157:H7 and *Salmonella typhimurium* bacterial cells or cell fragments in a rapid biosensor assay using AMPs. They immobilized magainin I using biotin-avidin chemistry as well as through direct covalent attachment. The immobilized magainin I reportedly can bind *Salmonella* with detection limits similar to analogous antibody-based assays, while detection limits for *E. coli* were higher than in analogous antibody-based assays.

8.2.3 Conventional biosensing methods

Conventional methods for detecting microorganisms mainly rely on specific microbiological and biochemical identification. Some of these include culturing/colony counting method, immunological method, and polymerase chain reaction (PCR). The culturing/colony counting method is the most reliable and accurate, and hence is still the standard method used in many analyses (Leoni & Legnani, 2001); however, the inordinate amount of time it takes is the main deterrent for its use. For example, some standard methods for the detection of *Listeria monocytogenes* can require up to 1 week (de Boer & Beumer, 1999). Further, when organisms are present in small quantities, as is the case in most foods, these methods require enrichment steps. Other major drawbacks of

the conventional techniques are that they often require expensive instrumentation and/or further analyses.

Immunology-based methods are very popular, as they rely on very specific interactions between antibody and target antigen, which may be a pathogen or another bioentity. There are different platforms under which antibody-antigen interactions can be facilitated. One example is the immunomagnetic method, which uses antibody-coated magnetic beads to capture and extract the targeted pathogen from the bacterial suspension (Gu et al., 2006). However, perhaps the most popular immunological method is the enzyme-linked immunosorbent assay (ELISA). ELISA offers the advantages of high specificity, owing to antibody-antigen interaction, and sensitivity of an easily assayed enzyme by coupling the antibodies and antigens to an enzyme. A common ELISA format is to "sandwich" the target antigen between two antibodies, as illustrated in Figure 8.5.

The invention of PCR signified an important milestone in pathogen detection, as it enabled detection of microorganisms down to single cells (Velusamy et al., 2010). PCR is a nucleic acid amplification technology, i.e. it is based on the isolation, amplification, and quantification of a short DNA sequence including that of target bacteria (Lazcka et al., 2007). Since PCR detects organisms by amplifying the target, rather than the signal, it is less prone to producing false positives. While highly specific and reliable, the PCR method suffers from its inability to discriminate between live and dead cells. Problems such as the sensitivity of the polymerase enzyme to environmental contaminants, difficulties in quantification, the generation of false positives through the detection of naked nucleic acids, non-viable microorganisms, or contamination of samples in the laboratory, etc. may limit the use of PCR for the direct detection of microbial contamination (Toze, 1999). PCR is also rather time consuming (up to 24 h), though it takes much less time than the colony counting method. Recent developments in real-time PCR techniques can provide results within hours (~5 h) (Levi et al., 2003).

8.2.4 Nanomaterials-based biosensing

Once the target entity is recognized and the interaction occurs, for example, via antibody-antigen immunoreaction, this interaction should be transduced into a suitable signal for target identification. In modern biosensors, nanomaterials such as metallic nanoparticles, quantum dots, magnetic nanoparticles, carbon nanotubes, dye-doped nanoparticles, etc. are used in transducing the

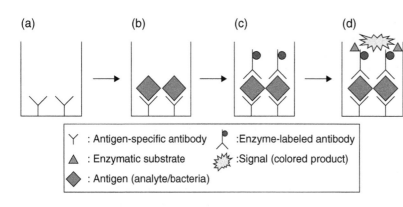

Figure 8.5 Schematic representation of the sandwich-ELISA protocol. (a) immobilization of the antigen-specific antibody on the well; (b) addition of the antigen; (c) addition of the enzyme-labeled antibody; (d) introduction of the enzyme's substrate and apparition of the colored product as signal. (Reproduced from Lazcka et al. (2007), with permission from Elsevier.)

biorecognition reaction to generate optical, magnetic, and electrochemical signals (Gomez-Hens et al., 2008; Medintz et al., 2008). Indeed, biosensors designed to incorporate nanomaterials, called nanobiosensors, are becoming more the norm than the exception. These nanomaterials are attractive as candidates for biosensing because of their small size and correspondingly large surface-to-volume ratio; chemically tailorable physical properties, which directly relate to size, composition, and shape; unusual target binding properties; and overall structural robustness. The characteristics of some of the most commonly used nanomaterials are summarized in Table 8.3. Biosensor development and applications using some of these nanomaterials are described herein.

8.2.4.1 Metallic nanoparticles

Metallic nanoparticles (MNPs) of metals such as gold and silver have many useful optical, optoelectronic and material properties that derive from their nanosize (Hayat, 1991). Gold and silver nanoparticles also have very high light-scattering power (Yguerabide & Yguerabide, 1998). These particles, also called plasmon-resonant particles, are quench resistant and generate very high signal intensities. These properties have long been used in various applications including chemical sensors, spectroscopic enhancers, and nanostructure fabrication (Bassell et al., 1994; Creighton et al., 1979).

Both silver nanoparticles (AgNPs) and gold nanoparticles (AuNPs) have been successfully used in biosensing. Advantages of these nanoparticles include (Daniel & Astruc, 2004; Tansil & Gao, 2006; Wilson, 2008):
- AuNPs can be prepared in a broad range of sizes (2–250 nm) and shapes with a high degree of precision and accuracy
- they are stable during storage

- they can be functionalized with recognition molecules (antibodies, antigens, oligonucleotides, etc.) easily and stably
- their optical properties are not altered by prolonged exposure to light, which allows results of diagnostic tests to be archived directly rather than in electronic form
- attaching biomolecules, such as antibodies and DNA, to these nanoparticles does not affect their optical properties
- even though the material they are made from is prohibitively expensive, the use of MNPs is fairly inexpensive, as they are used in very low concentrations.

Gold nanoparticles have additional benefits, however, and seem to be the more preferred material. A colloidal suspension of AuNPs exhibits an intense red color (absorption peak ~520 nm), due to the interaction of incident light with a collective oscillation of free electrons in the particles, known as localized surface plasmon resonance (Bohren & Huffman, 1983; Wilson, 2008). The extinction cross-section of the particles and the wavelengths at which they absorb and scatter light depend on their size and shape, the dielectric properties (refractive index) of the surrounding medium and their interactions with neighboring particles. The increase in diameter is also accompanied by an increase in the extinction coefficient and a red shift in the plasmon band. The colors of gold and silver nanoparticles of different sizes and shapes are presented in Figure 8.6.

At the typical mean distance of >1000 nm between particles in a sol, there is no significant overlap between the dipoles of neighboring particles. However, when the particles come closer than their diameters and aggregate, the absorption peak shifts towards a longer wavelength and the color of the colloidal AuNPs solution turns purple, as shown in Figure 8.7a (Kreibig & Genzel, 1985; Wilson, 2008). The distance-dependent optical

Table 8.3 Description and properties of some of the commonly used nanomaterials in nanobiosensors (Gilmartin & O'Kennedy, 2012)

Nanomaterial (typical size)	Description	Useful properties	Image
Quantum dots (2–10 nm)	Quantum dots (QDs) are colloidal semi-conducting fluorescent nanoparticles consisting of a semi-conductor material core that has been coated with an additional semi-conductor shell	Photostability and tuneable emission spectra and are used in assays in many modes, such as fluorescence emission and fluorescence quenching, or as energy donors	
Gold nanoparticles (5–110 nm)	Gold nanoparticles (AuNPs) may take the form of spheres, cubes, hexagons, rods or nanoribbons	Ability to resonantly scatter light; chemically highly stable; change color on aggregation from blue to red; excellent conductivity	
Single-walled carbon nanotubes (0.4–3 nm) and multiwalled carbon nanotubes (2–100 nm)	Allotrope of carbon consisting of grapheme sheets rolled up into cylinders; multi-walled carbon nanotubes (MWCNTs) are essentially a number of concentric single-walled carbon nanotubes (SWCNTs)	Exhibit photoluminescence; excellent electrical properties; semi-conductors	
Magnetic nanoparticles (1–100 nm)	Made of compounds of magnetic elements such as iron, nickel and cobalt and can be manipulated using magnetic fields	Used to concentrate particles in assays; excellent conductivity	

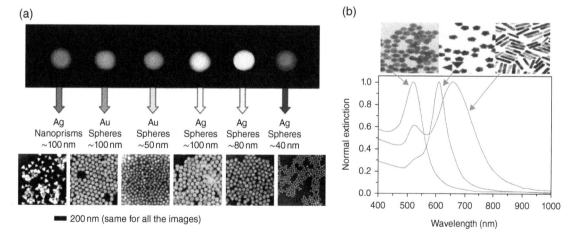

Figure 8.6 (a) Sizes, shapes, and compositions of metal nanoparticles can be systematically varied to produce materials with distinct light-scattering properties. (Reproduced from Rosi & Mirkin (2005), with permission from the American Chemical Society.) (b) TEM micrographs (*top*) and UV–vis spectra (*bottom*) of colloidal gold nanoparticles of different geometries. (Reproduced from Sepulveda et al. (2009), with permission from Elsevier.) For color details, please see color plate section.

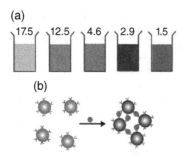

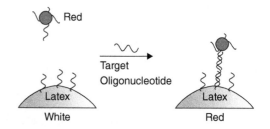

Figure 8.7 (a) Effect of interparticle distance in nanometers on the color of 15 nm AuNPs. (Reproduced from Ung et al. (2001), with permission from the American Chemical Society.) (b) Schematic diagram of distance-dependent sandwich assay for high molecular weight polyvalent antigen (*green circles*) leading to aggregation of antibody-functionalized AuNPs and a red shift in their extinction spectrum. (Reproduced from Wilson (2008), with permission from the Royal Society of Chemistry.) For color details, please see color plate section.

Figure 8.8 Hybridization of oligonucleotide-modified AuNPs to oligonucleotide-modified latex microspheres in the presence of a complementary target oligonucleotide. (Reproduced from Reynolds et al. (2000), with permission from IUPAC.)

properties of AuNPs have been used in a plethora of biomolecular detection methods (Brust et al., 1995; Elghanian et al., 1997; Freeman et al., 1995; Storhoff et al., 2000). Leuvering et al. (1980, 1981, 1983) were the first to develop sol-particle immunoassays based on this color-change principle to detect target molecules in urine and serum. AuNPs have also been conjugated to antibodies and cross-linked to target molecules (Figure 8.7b). At high concentrations of target molecule there is a visible color change from red to blue, but at lower concentrations this change can only be detected with instruments such as a spectrophotometer. Storhoff et al. (2000) proposed a method in which optical properties of AuNPs were controlled by adjusting the length of the DNA linker molecules/target molecules and the average distance between the particles. Optical signals are preferred for many biosensor applications because of their ease of monitoring, high sensitivity, multiplexing potential, and insensitivity to electromagnetic noise; they are also easily monitored using widely available spectroscopic devices (Giljohann et al., 2010).

This principle has been applied to analyses of various substances. For example, Mirkin et al. (1996) reported a DNA analysis method by modifying two sets of AuNPs with different single-stranded DNA probes and mixing them with target DNA. If the target DNA has sequences complementary to both of the two probes, the target cross-links the nanoparticles by hybridization, resulting in aggregation and color change due to a red shift of

the plasmon band. This method of utilizing the optical property change of a colloidal particle dispersion is based on the latex agglutination immunoassay of Gella et al. (1991). Of course, instead of color change when using the AuNPs, the aggregation of antigen- or antibody-modified latex particles cross-linked by the complementary antibody or antigen becomes more turbid. In the presence of a complementary target oligonucleotide, dispersed oligonucleotide-modified AuNPs are cross-linked via hybridization events into aggregated polymeric networks. Reynolds et al. (2000) proposed an alternative assay using gold nanoparticles and latex microspheres, which are linked together by the target DNA strand, generating a white-to-red color change which is easier to discern than red-to-blue color change (Figure 8.8). This method was sensitive (500 pM for single-strand target, 2.5 nM for duplex target), rapid (15 min), straightforward, and inexpensive.

Ko et al. (2010) adapted a similar scheme and developed a colorimetric biosensor for detecting model toxins in foods (Figure 8.9). A toxin analog, 2,4-dinitrophenol-bovine serum albumin (DNP-BSA) was attached to AuNPs and allowed to hybridize via antigen-antibody association with an anti-DNP antibody on latex microspheres, changing the color of latex microspheres to pinkish-red. A model toxin, DNP-glycine (GLY), was detected and quantified via a competition that occurs between the analog-conjugated AuNPs and the toxin molecules for the binding pocket in the anti-DNP antibody. When the AuNPs were displaced from host latex microspheres in the presence of toxin molecules, a visible color change occurred from pinkish-red to white (Figure 8.10). This proof-of-concept self-indicating nanobiosensor detected the model toxin rapidly, and the results were quantifiable.

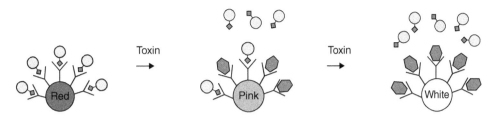

Figure 8.9 Indicator-analyte displacement between analog-conjugated gold nanoparticles (*small circles with diamond tail*) and toxin molecule (*hexagons*). (Reproduced from Ko et al. (2010), with permission from Elsevier.)

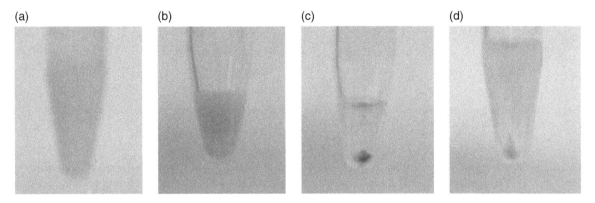

Figure 8.10 Stages of analyte-indicator displacement reaction. (a) Solution of latex microspheres + DNP antibody, (b) addition of AuNPs (conjugated with DNP–BSA), (c) solution in (b) after mild centrifugation shows good binding of latex microspheres and AuNPs, (d) solution (b) after addition of model toxin (DNP-GLY) and light centrifugation (the solution color is still yellow due to the strong yellow color of the DNP-GLY). (Reproduced from Ko et al. (2010), with permission from Elsevier.) For color details, please see color plate section.

Sato et al. (2003) reported a non-cross-linking aggregation method – induced by altering colloidal stability – producing a similar color change. They reported aggregation of DNA-containing colloidal AuNPs due to the lowered stability caused by the hybridization of the surface DNA with the complementary DNA (Mori & Maeda, 2002; Sato et al., 2003). Using various functional groups to anchor the ligands, additional moieties such as oligonucleotides, proteins, and antibodies can impart even greater functionalities; using such gold nanoconjugates has enabled a broad range of biosensor applications. Neeley et al. (2011) reported detection of arsenic, a toxin, and *Salmonella* using a two-photon light-scattering assay. The detection schemes, which involve aggregating AuNPs, are shown in Figure 8.11. For label-free detection of extremely low concentration (100 parts per trillion) of arsenic, aggregation of dithiothreitol (DTT)-functionalized AuNPs, which is mediated by arsenic

present in the sample, was used. And by binding antibody-conjugated AuNPs on *Salmonella*, they were able to detect the presence of *Salmonella* at a 10^3 CFU/mL level. The visible color change accompanying aggregation of AuNPs on the *Salmonella* surface is shown in Figure 8.12. Similar colorimetric changes due to aggregating AuNPs have been reported for detecting several other pathogens (Singh et al., 2009; Wang et al., 2010).

Lim *et al.* (2012) presented a novel approach for colorimetric detection of pathogens and other bioanalytes. Since the extent of aggregation should exceed a certain minimum threshold to produce a visible change, they reported a simple and easy biosensing strategy for controlling the extent of AuNPs aggregation, introducing the concept of a switchable bifunctional linker (SBL), which provides several strategic options to design and enhance the visible detection of wide-ranging targets. An SBL is an element allowing multiple specific bindings

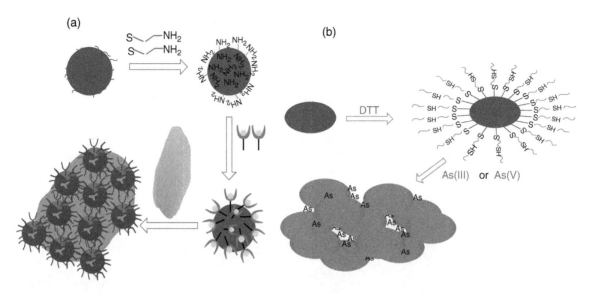

Figure 8.11 (a) Schematic representations of gold nanoparticle-based sensing of (a) Salmonella bacteria and (b) arsenic. (Reproduced from Neeley et al. (2011), with permission from IEEE.)

that can be selectively enabled or disabled to bridge functionalized AuNPs (f-AuNPs). This SBL-based two-step biosensing scheme (Figure 8.13a) is effective for detecting the target with multiple binding sites by performing two functions: bridging two f-AuNPs, and binding to the target selectively, forming a complex with other SBLs, which decreases the number of effective linkers (nLK). The formation of such a linker complex tantamount to switching SLs off, if the number of f-AuNPs that a complex can bridge is less than those that f-AuNPs SBLs can bridge individually. If so, the presence of targets in the sample can effectively cause a change in the extent of aggregation, when the assay is performed in two steps: (i) mixing SBLs with the sample and (ii) adding f-AuNPs. By promptly introducing the sample to the SBLs (Step 1), targets present in the sample cross-link with SBLs to form a linker complex, which reduces the nLK for cross-linking with the f-AuNPs (Step 2).

This strategy to control the extent of aggregation via the SLs provided exceptional sensitivity to NPs-based visible detection of bacteria using immunoreaction. If f-Abs can bridge two f-NPs, they can play the role of SLs. Since the size of f-Abs is too small relative to those of the bacteria and f-NPs, the ability of an f-Ab to bridge two f-NPs is switched off, when an f-Ab binds to an antigen on the cell surface (see Figure 8.13a). Thus, the presence of bacteria should reduce nLK by promptly binding some f-Abs to possible antigens on the cell surface. Moreover,

since the f-NPs are much larger than the interantigen spacing, the f-NPs that attach to the cell may occlude some f-Abs bound to antigens, making those f-Abs unavailable to occupy any binding sites on the f-NPs (Figure 8.13b). When biotinylated anti-*E.coli* polyclonal antibody (b-Ab), labeled with 7–10 biotin molecules, was used as SL to facilitate the aggregation of stAuNPs, the presence of fewer than 100 CFU/mL of *E. coli* was visibly detected in PBS buffer (Figure 8.13c).

8.2.4.2 Quantum dots

Quantum dots (QDs) are semi-conductor nanocrystals composed of atoms from groups II–VI or III–V of the periodic table (e.g. CdSe, CdTe, CdS, ZnSe, PbS, PbTe), and are 1–5 nm in size (Wang et al., 2006). Their small size leads to a quantum confinement effect, affording these nanocrystals unique optical and electronic properties. They have narrow, size-tunable, symmetrical emission spectra and are photochemically stable (Bruchez et al., 1998; Chan & Nie, 1998). The emission wavelengths of QDs scale with particle size (Alivisatos, 2004), shape, and chemical composition (Jamieson et al., 2007), and their spectral properties can be tuned by changing the size of the particle or composition (Figure 8.14). In the visible spectrum region, the emission is usually adjusted by selecting the particle diameter of CdSe QDs, whereas at longer wavelengths the emission

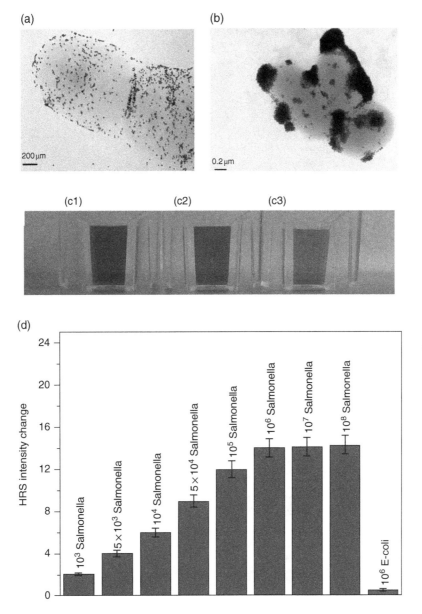

Figure 8.12 TEM images showing aggregation of gold nanoparticles after the addition of (a) 5×10^3 CFU/mL and (b) 8×10^4 CFU/mL *Salmonella*. Photographs showing colorimetric change upon addition of (c1) 2×10^3 *Salmonella*, (c2) 5×10^6 *E. coli*, (c3) 5×10^4 *Salmonella*. (d) Two-photon scattering intensity changes due to the addition of *Salmonella* bacteria to anti-*Salmonella* antibody-conjugated AuNPs. The intensity did not change much when *E. coli* was added. (Reproduced from Neeley et al. (2011), with permission from IEEE.) For color details, please see color plate section.

is adjusted by selecting the particle composition of QDs with different emission colors which can be excited simultaneously at a single wavelength. These features and their binding compatibility with DNA and proteins led QDs to replace fluorophores as biological labeling agents (Medintz et al., 2005). They are also suitable for simultaneous imaging of multiple types of QDs (Han et al., 2001) and hence are used in multiplexed assays

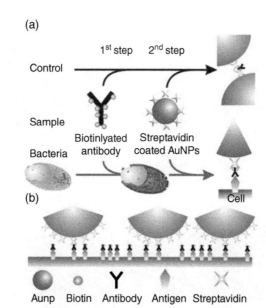

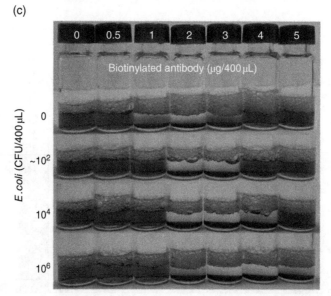

Figure 8.13 Scheme of (a) the two-step assay for visible detection of bacteria using biotinylated antibody (b-Ab) as SLs to control the extent of aggregation of stAuNPs. (b) Relative difference between the size of the cell and that of NPs on switching b-Abs off when they attach on the cell surface. (c) Shift of range exhibiting visible color change in response to the presence of *E. coli*, when tests performed using biotinylated antibody (b-Ab) as SL in 400 μL total sample volume with a fixed stAuNPs concentration (absorption: 0.43 @ 531 nm for 1/10 diluted sample) and. *E. coli* cell loads are typical of several tests, which yielded similar results. (Reproduced from Lim et al. (2012), with permission from Nature Publishing Group.) For color details, please see color plate section.

without the need for multiple excitation sources, which is impractical with conventional fluorescent dyes (Resch-Genger et al., 2008). Presumably, it is not necessary to separate the size-distinguished QD nano-transducers in the area of matrix platform because the unique properties of QDs enable detection of more than six fluorescent signals simultaneously. Since different colors of QD nanotransducers can be conjugated with different reporters, they can be functionally separated on the matrix.

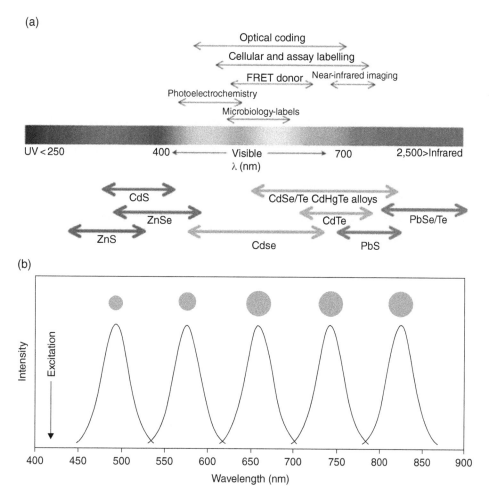

Figure 8.14 (a) Different QDs and their representative areas of biological interest corresponding to the pertinent emission. (Reproduced from Medintz et al. (2005), with permission from Nature Publishing Group.) (b) Size-dependent emission of CdSe QDs. (Reproduced from Wang et al. (2006), with permission from IOP Publishing.) For color details, please see color plate section.

Some of the limitations of QDs include:
- optical blinking, which makes the application of QDs in quantitative assays difficult (van Sark et al., 2002)
- the size of the QDs may affect the function of attached ligand molecules (Rosenthal et al., 2002)
- QDs have to be surface modified to be biocompatible before they are used in live cell or animal experiments. Therefore, QDs are only complementary to conventional organic fluorescent labels and will not take their place completely. They are better suited for applications where good photostability is required (Michalet et al., 2005). Among different types, CdSe/ZnS core-shell QDs are the most commonly used for biological applications (Bakalova et al., 2006). A new class of water-soluble QDs is also used for labeling living tissues *in vivo* (Larson et al., 2003).

Quantum dots have been functionalized with antibodies and aptamers for the fluorescent detection of pathogens as diverse as respiratory syncytial virus (Agrawal et al., 2005; Bentzen et al., 2005), *E. coli* O157:H7 (Hahn et al., 2008), *Bacillus thuringiensis* (Ikanovic et al., 2007), and protozoan parasites such as *Cryptosporidium* and *Giardia* (Zhu et al., 2004). A QD biosensor was shown to detect pathogenic *E. coli* O157:H7 selectively when they were present along with 99% of non-pathogenic DH5R *E. coli*.

8.2.4.3 Surface-enhanced Raman spectroscopy

Surface-enhanced Raman spectroscopy (SERS) is a technique that provides a greatly enhanced Raman signal from Raman-active analyte molecules that have been

(a)

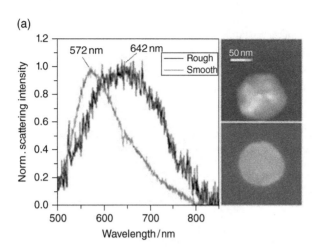

(b)

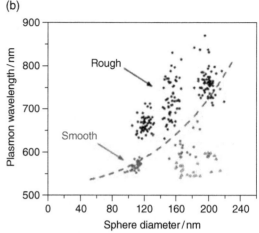

Figure 8.15 (a) Normalized scattering spectra of two Au spheres of same size and shape but different surface roughness, as observed in the scanning electron microscope images: rough (*top*) and smooth (*bottom*). (b) Size dependence of the dipole (*circles*) and quadrupole (*triangles*) plasmon resonance wavelengths in rough (*black*) and smooth (*red*) Au spheres. (Reproduced from Rodriguez-Fernandez et al. (2009), with permission from the Royal Society of Chemistry.) For color details, please see color plate section.

adsorbed onto certain specially prepared metal surfaces. Increases in the intensity of Raman signal of the order of 10^4 to 10^6 have been observed, and for some systems even as high as 10^8 and 10^{14} (Kneipp et al., 1999). SERS is both surface selective and highly sensitive, whereas conventional Raman spectroscopy is neither. SERS selectivity of surface signal results from the presence of surface enhancement mechanisms only at the surface. Thus, the surface signal overwhelms the bulk signal, making bulk subtraction unnecessary.

Theoretically, any metal would be capable of exhibiting surface enhancement; however, SERS is observed primarily for analytes adsorbed onto coinage metal (gold, silver, copper) or alkali metal (lithium, sodium, potassium) surfaces, which provide the strongest enhancement (Moskovits, 1985). The two primary mechanisms of signal enhancement in SERS are electromagnetic and chemical. Of these, the electromagnetic effect is the dominant, while the chemical effect contributes to enhancement by only about an order or two of magnitude (Kambhampati et al., 1998). The electromagnetic enhancement is dependent on the metal surface's roughness features, while the chemical enhancement involves changes to the adsorbate electronic states due to chemisorption of the analyte. The metal roughness features can be developed in a number of ways: oxidation-reduction cycles on electrode surfaces; vapor deposition of metal particles onto substrate; metal spheroid assemblies produced via

lithography; metal colloids; and metal deposition over a deposition mask of polystyrene nanospheres (Jensen et al., 2000; Moskovits, 1985). These nanoscale surface corrugations strongly determine the plasmonic response of AuNPs of size several tens of nanometers. Scattering spectra of individual spheres with a rough surface were found to red-shift and broaden (Figure 8.15) (Rodriguez-Fernandez et al., 2009). Further, rougher spheres display a higher SERS activity, which demonstrates the crucial role of nanoscale surface texturing on the plasmonic response of gold particles. As far as nanotechnology application is concerned, no measurement technique can match SERS in terms of the role of nanostructures (Golightly et al., 2009).

Three key features of Raman in general and SERS in particular are (Golightly et al., 2009):
• SERS is a vibrational spectroscopy that furnishes molecular-level, chemical fingerprint-like information. This enables detection of many different species and also permits multiplexing, whereby multiple species are detected in a single measurement
• the instrumentation meets the needs for point-of-use measurements. Hand-held spectrometers with sufficient spectral resolution and sensitivity are commercially available. Excitation wavelengths can be chosen in the visible or near-infrared spectral regions as needed. Since the phenomenon is based on scattering, it is possible to "point-and-shoot" and collect high-quality spectral data in seconds

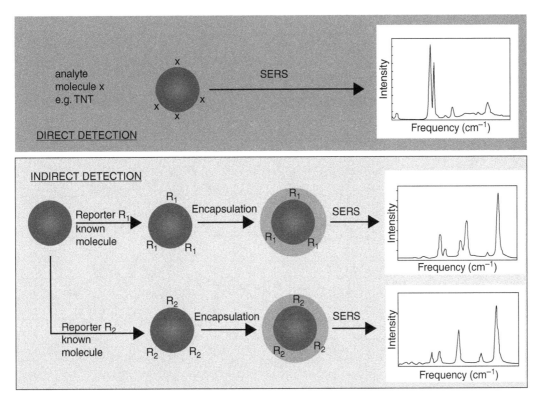

Figure 8.16 SERS detection methods. Direct (*top*) and indirect (*bottom*) using labels. (Reproduced from Golightly et al. (2009), with permission from the American Chemical Society.)

- SERS can be used for both direct and indirect detection. Direct detection refers to the case when the species of interest adsorbs directly to a SERS-active surface (Figure 8.16). Examples of direct detection include bacteria, drug molecules, and the chemical contaminant melamine. SERS is the only optical detection technique capable of analyte detection and identification at nanomolar to picomolar concentration levels (Golightly et al., 2009; Kneipp et al., 1998). Indirect detection, by contrast, makes uses of nanoparticles as SERS-active quantitation labels, similar to the way fluorophores are used. These SERS nanotags (see Figure 8.16) are made by coating a SERS-active particle with an adsorbate ("the reporter") that has a unique spectral fingerprint and then encapsulating it in silica. The resulting spectrum of the tag depends solely on which reporter is used (R_1 versus R_2). Conjugation of antibodies or DNA (or some other molecular recognition motif) to the outer silica surface allows the particles to be used as quantitation labels in ways that are familiar to developers of immunodiagnostic or molecular diagnostic assays. The ability to generate multiple spectrally unique tags enables quantification of multiple analytes in a single measurement.

Holt and Cotton (1989) published the first investigation of a bacterium showing that SERS fingerprints can be used to characterize microorganisms. Subsequently, Efrima and Bronk (1998) published SERS spectra of *E. coli* obtained after mixing the bacterial cell with silver nanoparticles and depositing the cells on a glass slide. The high sensitivity of SERS suggests that only relatively small amounts of bacteria are needed for the identification procedure. However, it is critical not only to detect very low levels of pathogens, but also to differentiate and identify very similar species and strains. It follows that detection of such pathogens would ideally be ultrasensitive, rapid, require no extrinsic labeling steps, and differentiate species and strains of pathogens. SERS is perhaps the only technique that has shown the promise of satisfying all such criteria. The tremendous signal enhancements obtained from SERS, coupled with the large scattering cross-sections of bacterial cells, allow routine acquisition of spectra from individual bacterial cells. Premasri et al. (2005) have shown that SERS spectra can be obtained after deposition of bacteria onto a colloidal Au-based substrate, and they are very distinct for six bacteria species (Figure 8.17). SERS fingerprints are

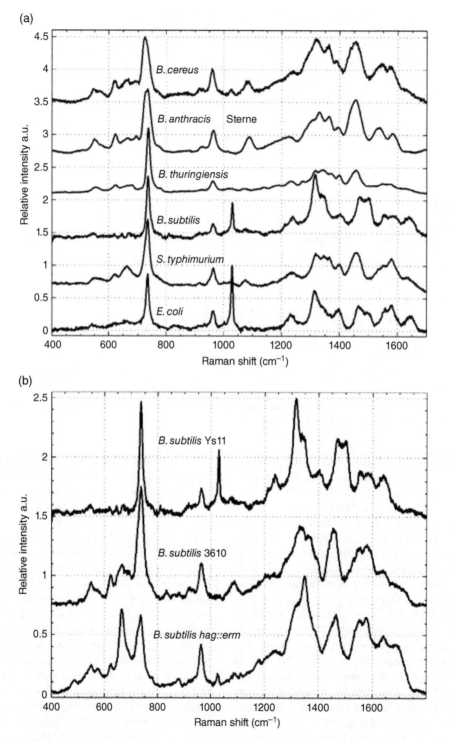

Figure 8.17 (a) SERS spectra of six bacterial species. (b) SERS spectra different strains/mutant of *B. subtilis*. (Reproduced from Premasiri et al. (2005), with permission from the American Chemical Society.)

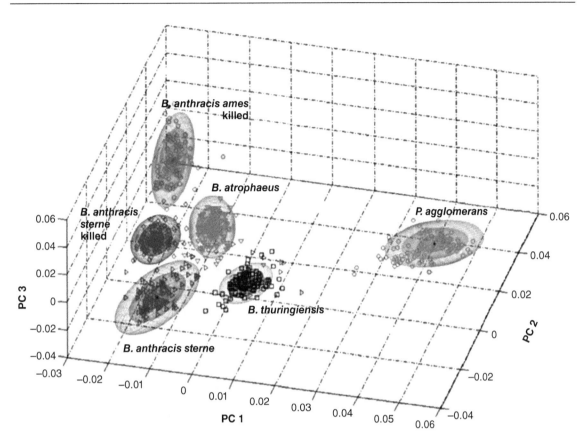

Figure 8.18 Principal component analysis plot showing discrimination of five *Bacillus* spore samples and *P. agglomerans* and killed and live versions of *B.anthracis* based on their SERS emission. (Reproduced from Guicheteau et al. (2008), with permission from Elsevier.)

not only species specific, but also strain and mutant specific. As shown in Figure 8.17b, the SERS spectra of wild-type *B. subtilis* 3610 cells, wild-type *B. subtilis* YS11 cells, and *B. subtilis* 3610 congenic mutant cells lacking flagella are very distinctly different (Premasiri et al., 2005). Further, Guicheteau et al. (2008) demonstrated discrimination of five *Bacillus* species/strains and *Pantoea agglomerans* that had been dried in the presence of colloidal silver and between killed and live versions of *B. anthracis*, after subjecting their SERS spectral data to principal component analysis (Figure 8.18).

The power of SERS technology can be exploited by developing a biochip approach. Grow et al. (2003) described a microscopic technique that uses SERS (μSERS) for label-free identification of microorganisms and toxins. Figure 8.19 is a schematic flow chart of the

μSERS system that can be used to obtain SERS fingerprints of unknown organism(s), compare them against a library of known fingerprints, and identify the organisms.

The μSERS technology was used to detect and identify microbial toxins (Grow et al., 2003). Though toxins are small molecules, toxins as little as 0.02% by weight of the biomolecule-toxin complex have produced SERS fingerprints sufficient for their detection and identification. The fingerprints are also unique for a particular toxin to be identifiable in a mixed toxins sample. As shown in Figure 8.20, in samples containing one or both aflatoxins B1 (312 Da) and G1 (328 Da), each was identified. Similar results were obtained with other toxins ranging from cadaverine (102 Da) to *Staphylococcus* enterotoxin B (28.4 kDa).

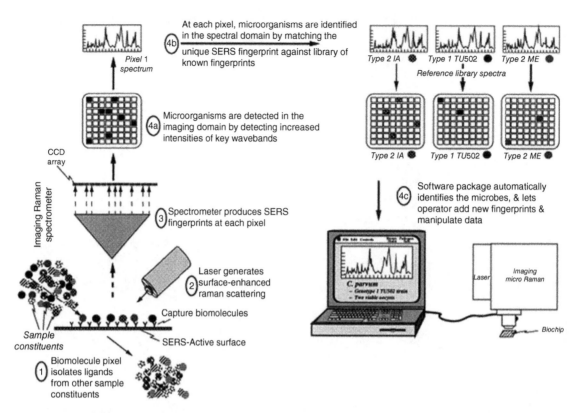

Figure 8.19 Flow diagram of the μSERS process. (Reproduced from Grow et al. (2003), with permission from Elsevier.)

(a)

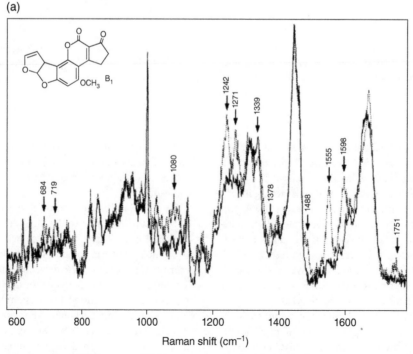

Figure 8.20 Background corrected SERS spectra of RNA polymerase biochips before (—) and after (– – –) incubation in solutions containing aflatoxin B_1 (a) or aflatoxin G_1 (b). Major features of the aflatoxin fingerprints are indicated by arrows. (Reproduced from Grow et al. (2003), with permission from Elsevier.)

(b)

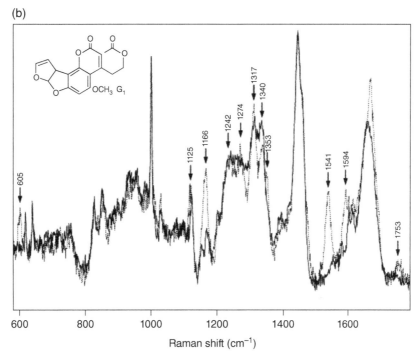

Figure 8.20 (*Continued*).

8.3 Packaging

8.3.1 Use of nanocomposites in food packaging film

As packaging materials, synthetic polymers are fairly inexpensive and offer some superior functional properties to traditional packaging. However, their poor barrier properties (permeability to water vapor, oxygen, carbon dioxide, and other gases and vapors) adversely affect the quality of packaged foods. Therefore, much effort is being focused on improving the barrier properties of food packaging films. The use of synthetic polymers for food materials packaging also contributes to an enormous waste disposal problem. Therefore, environmentally benign packaging alternatives are being sought.

Both natural (polysaccharides and proteins) and synthetic (polylactic acid, PLA) biopolymers are common choices for developing biodegradable food packaging materials. However, biopolymers offer relatively poor mechanical and barrier properties, which currently limit their industrial use. Especially challenging is imparting satisfactory moisture barrier properties due to the inherent hydrophilic nature of biopolymers. Use of polymer blends, applying high barrier coating, and the use of multilayered films containing a high barrier film such as aluminum are among the common approaches. Therefore, composites are formed with organic or inorganic materials such as fillers in polymeric matrices. When these fillers are of nanoscale, the resulting polymeric composites are known as nanocomposites.

Compared to neat polymers and conventional composites, nanocomposites exhibit increased barrier properties, increased mechanical strength, and improved heat resistance, and hence are better food packaging materials (Ray & Okamoto, 2003; Ray et al., 2006; Sorrentino et al., 2007; Thostenson et al., 2005). It is also possible to use thinner, lower gauge nanocomposite film, known as down gauging, because of its superior mechanical properties (Arora & Padua, 2010). However, for improvement in material properties, a nanomaterial must be properly dispersed within the matrix material, and indeed if the nanomaterial is poorly dispersed, the properties of the composite may even be degraded (Gorga & Cohen, 2004). One of the earliest applications of nanocomposites was by the car maker Toyota in early 1990, when they included a very small amount of nanofiller and significantly increased the thermal and mechanical properties of a nylon-6 nanocomposite (Kojima et al., 1993). The properties of nanocomposite materials depend not only on the properties of their individual parents (nanofiller and nylon, in this case), but also on their morphology and interfacial characteristics.

Nanomaterials such as nanoparticles, nanotubes, nanofibers, fullerenes, nanowires, graphene nanosheets, and nanoclays (e.g. montmorillonite and kaolinite) are all used as fillers in preparing food packaging films. For example, carbon nanotubes can be incorporated into polymer structures (liquids, solutions, melts, gels, amorphous and crystalline matrices) to increase their mechanical properties in terms of tensile strength and elasticity (Ruoff & Lorents, 1995). These nanofiller materials may be classified as particles (e.g. carbon black, silica nanoparticle), fibers (nanofibers and carbon nanotubes), and layered materials, which are high aspect ratio (30–1000) plate-like structures (e.g. organosilicate) (Alexandre & Dubois, 2000; Schmidt et al., 2002). This last family of nanocomposites is almost exclusively obtained by the intercalation of the polymer (or a monomer subsequently polymerized) inside the galleries of layered host crystals. There is a wide variety of both synthetic and natural crystalline fillers that are able, under specific conditions, to intercalate a polymer. Depending on the class of nanomaterial used, the composites acquire significantly different properties (Park et al., 2001). The unique combination of characteristics of the nanomaterial suchas size, mechanical properties, and low concentrations necessary to effect change in a polymer matrix, coupled with the advanced characterization and simulation techniques now available, have generated much interest in the field of nanocomposites (Hussain et al., 2006).

Clay and layered silicates are perhaps the most studied nanocomposites, both because they are readily available and their intercalation chemistry is well known (Ogawa & Kuroda, 1997). These nanocomposites exhibit markedly improved mechanical, thermal, optical, and physicochemical properties such as increased moduli and heat resistance and decreased gas permeability and flammability. When layered clay is associated with a polymer, three main types of composites may be obtained (Figure 8.21):

• a phase-separated composite, when the polymer is unable to intercalate between the silicate sheets
• intercalated composite, when one or more extended polymer chains is intercalated between the silicate layers, resulting in a well-ordered multilayer morphology
• an exfoliated or delaminated composite, when the silicate layers are completely and uniformly dispersed in a continuous polymer matrix.

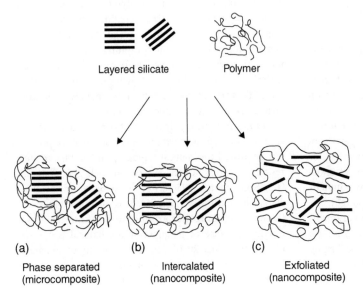

Layered silicate   Polymer

(a)   (b)   (c)
Phase separated   Intercalated   Exfoliated
(microcomposite)   (nanocomposite)   (nanocomposite)

Figure 8.21 Scheme of different types of composite arising from the interaction of layered silicates and polymers: (a) phase-separated microcomposite; (b) intercalated nanocomposite; and (c) exfoliated nanocomposite. (Reproduced from Alexandre & Dubois (2000), with permission from Elsevier.)

8.3.2 Biodegradable nanocomposites

Starch is a popular choice for manufacturing food packaging film due to its biodegradability, inexpensiveness, and wide availability. Biodegradable bionanocomposites prepared from natural biopolymers such as starch and protein exhibit advantages as a food packaging material by providing enhanced organoleptic characteristics such as appearance, odor, and flavor (Zhao et al., 2008). But given its hydrophilic nature, starch is hardly suitable as a food material packaging film, unless the film properties are improved by the addition of inorganic and synthetic polymeric fillers such as montmorillonite clay (Avella et al., 2005; Cyras et al., 2008; Ray & Okamoto, 2003; Yoon & Deng, 2006). Like starch, cellulose also yields poor water barrier films. However, incorporation of carbon nanofibers has been beneficial in improving film properties, including moisture barrier properties (de Azeredo, 2009). Further improvement in film properties has been reported by preparing a nanocomposite film of hydroxypropyl methylcellulose, a cellulose derivative, incorporated with chitosan nanoparticles as fillers (Burdock, 2007; Mattoso et al., 2008, 2009).

A starch/carboxylated multiwall carbon nanotube (CCNT) composite (CCNT-starch) was prepared by covalently grafting a natural polymer starch onto the surfaces of CCNT (Yan et al., 2011). Fourier transform infrared (FTIR) spectroscopy revealed that the covalent bonds between –OH groups of soluble starch and CCNT were formed in CCNT-starch. Transmission electron microscopic (TEM) images showed that CCNTs were covered with the grafted starch (Figure 8.22), which facilitated the dispersion of CCNT-starch in water and chitosan films because of the hydrophilic polysaccharide structure of starch components.

Engineered biopolymer-layered silicate nanocomposites are reported to have markedly improved physical properties, including higher gas barrier properties, tensile strength, and thermal stability (Rhim, 2007; Sorrentino et al., 2007; Zhao et al., 2008). Cabedo et al. (2006) reported improving mechanical, thermal, moisture and gas barrier properties by incorporating chemically treated nanoscale silicate plates in biodegradable blends of amorphous PLA and poly(caprolactone). Extruded starch is a thermoplastic material with low mechanical strength and poor oxygen and moisture barrier properties (Chen & Evans, 2005; Lopez-Rubio et al., 2006; McGlashan & Halley, 2003). However, when this thermoplastic starch is hybridized with nanoclay, the resulting strong interaction improves the tensile strength and water vapor barrier properties (Park et al., 2002).

The film-forming ability of various proteins has been utilized in industrial applications for a long time. Animal proteins (e.g. casein, whey protein, collagen, egg white,

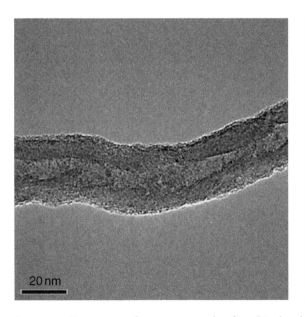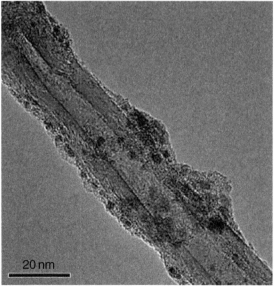

Figure 8.22 Transmission electron micrographs of starch/carboxylated multiwall carbon nanotube (CCNT) (*left*) CCNT–starch composite (*right*). (Reproduced from Yan et al. (2011), with permission from Elsevier.)

and fish myofibrillar protein) (Zhao et al., 2008; Zhou et al., 2009) and plant proteins (e.g. soy protein, corn zein, and wheat gluten) (Brandenburg et al., 1993; Hernandez-Munoz et al., 2003) are popular choices. Protein-based biodegradable films are usually produced by polymerization of proteins under specific conditions. Following denaturation, proteins polymerize into particulate aggregates with various sizes and shapes. Owing to the lack of strong interaction among aggregates, the films produced from these irregular aggregates are weak and brittle. This is the most basic and also the biggest problem for all protein-based films. The improvement in the property of the protein aggregates is crucially important to substantially enhance the functionalities of protein-based films.

Though functionalities of protein films are better than those made from polysaccharides (Miller & Krochta, 1997), concerns regarding their high modulus, high water adsorption, and high gas permeability remain. To reduce the brittleness of the film, a plasticizer, such as glycerol or sorbitol is used. However, using a plasticizer in protein films lowers their strength (McHugh & Krochta, 1994) and increases their water vapor permeability (Ozdemir & Floros, 2008). The biodegradable films are usually hydrophilic and they lose their barrier properties (to moisture, oxygen), or even become solubilized when used with foods with high water activity. Poor moisture barrier properties seriously limit the use of biodegradable films for commercial applications (Bertuzzi et al., 2007)

Whey protein is capable of forming transparent, flexible, colorless, and odorless films that provide good oxygen, aroma, and oil barrier properties. However, films made using whey proteins without any modification have poor moisture barrier properties and relatively low mechanical strength compared with synthetic or other commercial food packaging materials and therefore have limited applications (Cisneros-Zevallos & Krochta, 2003; Mei & Zhao, 2003). Sothornvit et al. (2007) reported the formation of whey protein transparent films, which also acted as an oxygen barrier. Soy protein has been of great interest to researchers for its thermoplastic properties and its potential as a biodegradable plastic. However, because of its poor response against moisture and high rigidity, its biodegradability has not been exploited effectively (Zheng et al., 2009).

Significant efforts have been made to improve the properties of various proteins by applying nanocomposite technology, mainly using nanoclays. Other nanofillers such as titanium nitride, titanium dioxide (TiO_2), zinc oxide (ZnO), nylon, etc. are also incorporated into the protein matrix. Soy protein nanocomposite films showed reduced water vapor permeability, improved elastic modulus and tensile strength compared to their counterparts without fillers (Rhim et al., 2009; Yu et al., 2007). Biodegradable zein films with good mechanical and water vapor barrier properties have been prepared (Lawton, 2002; Shukla & Cheryan, 2001; Yoshino et al., 2002). Nanoparticles of zein used in nanocomposites have also improved the strength of plastic and bioactive food packages, as well as serving as edible carriers for flavor and nutraceutical compounds (Lawton, 2002; Shi et al., 2008). Treating zein with stable silicate complexes (montmorillonite, hectorite, and saponite) has been suggested as a way to improve the barrier properties of zein-based polymers (Shukla & Cheryan, 2001).

Nanotechnology has also helped in preparing oxygen scavengers to counter discoloration, rancidity, off-odor, and texture problems associated with oxygen infiltration into packaged foods. Modification of the surface of nano-sized materials by dispersing agents can provide substrates for the oxidoreductase enzymes. The clay nanoparticles embedded in plastic beer bottles stiffen the packaging, reducing gas permeability and minimizing the loss of carbon dioxide from the beer and the diffusion of oxygen into the bottle, keeping the beer fresher.

Biopolyesters are another important class of biodegradable polymers formed from biological monomers, including polylactic acid (PLA), polyhydroxybutyrate (PHB), and polycaprolactone (PCL) (Sozer & Kokini, 2009). Biopolyesters are biodegradable and biocompatible and can be formed into films or molded into objects (Tharanathan, 2003). However, poor gas barrier properties and brittleness limit biopolyester applications in the food packaging industry. Use of nanoclays as fillers in a biopolyester matrix (e.g. kaolinite nanofillers in PLA film) has been shown to improve thermal stability and mechanical properties of the film without decreasing barrier properties (Cabedo et al., 2006).

8.3.3 Antimicrobial packaging film

In light of growing microbial contamination and food safety issues, much research is being focused on developing food packaging film that is not only biodegradable but also antimicrobial. Antimicrobial packaging systems can be of varied formats: antimicrobial nanoparticle inserts into the package, applying bioactive agents to coat the surface of the packaging material, dispersing bioactive agents in the packaging, forming composites of antimicrobial materials and polymeric matrices (Coma, 2008).

High-moisture foods that are prone to surface spoilage can be protected by contact-packaging impregnated with antimicrobial nanoparticles. Materials such as cinnamon oil and oregano oil have proven useful as antimicrobial agents in paper-based packaging materials (Rodriguez et al., 2007, 2008; Rodriguez-Lafuente et al., 2010; Rojas-Grau et al., 2007). Nanoparticles of silver oxide, zinc oxide, magnesium oxide, and nisin have been used (Coma et al., 2001; Gadang et al., 2008; Jones et al., 2008; Sondi and Salopek-Sondi, 2004). Chitosan nanofibers were fabricated by electrospinning a mixture of cationic chitosan and neutral poly(ethylene oxide) (PEO) at a ratio of 3:1 in aqueous acetic acid, which has antimicrobial properties (Kriegel et al., 2009).

Active food-packaging materials, especially protein-based antimicrobial films, have many desirable characteristics: being environmentally friendly, extending food shelf life, and enhancing food safety. The antimicrobial property is usually imparted by film containing a protein and an antimicrobial agent. Therefore, the active sites of the antimicrobial agent in the film are critical for its antimicrobial activity. When incorporated into the polymer matrix, the activity of an antimicrobial agent might decrease because of its interactions (either covalent or non-covalent) with the protein matrix before and during film formation. Moreover, it is difficult to control the active sites on the antimicrobial molecules such that they reside on the film surface to maximize their effectiveness.

Antimicrobial whey protein films have been produced using different antimicrobial agents such as potassium sorbate (Ozdemir & Floros, 2001; Shen et al., 2010), lysozyme (Bower et al., 2006; Min et al., 2005), sodium lactate, sodium caseinate, and ε-polylysine (Zinoviadou et al., 2010), nisin (Jin et al., 2009), oregano essential oil (Zinoviadou et al., 2009), and chitosan (Fernandez-Saiz et al., 2009; Shen et al., 2010; Ziani et al., 2009). These protein-based antimicrobial films have the ability to kill bacteria, including pathogens. However, the film strength is usually less than 100 MPa. This is mainly determined by the nature of the protein and the processing methods.

To improve antimicrobial activity, silver, magnesium oxide, ZnO or TiO_2 nanoparticles have been incorporated into the film (Morones et al., 2005; Yang et al., 2010; Zhou et al., 2009). Scanning electron microscope pictures of TiO_2 nanoparticles incorporated into whey protein isolate (WPI) film are shown in Figure 8.23. X-ray diffraction (XRD), UV–vis spectra, and fluorescence spectra of the film showed successful incorporation of TiO_2 nanoparticles into the WPI matrix and indicated the interactions between TiO_2 and WPI. Mechanical tests revealed the antiplasticizing effect of TiO_2 nanoparticles on the WPI/TiO_2 film. Small amounts (<1% w/w) of added TiO_2 nanoparticles significantly increased the tensile properties of WPI film, but also decreased the moisture barrier properties; higher amounts (>1% w/w) of TiO_2 improved moisture barrier properties but lowered the tensile properties of the film. Microstructural evaluation confirmed the aggregation and distribution of TiO_2 nanoparticles within the WPI matrix and validated the results of functional properties of the WPI/TiO_2 film. TiO_2 was added to form a nanocomposite with improved antimicrobial properties. Zhou and others (2009) indicated the potential of whey TiO_2 nanocomposites to be used as food-grade, biodegradable packaging materials. Addition of small amounts (<1 wt%) of TiO_2 nanoparticles significantly increased the tensile properties of WPI film (1.69 to 2.38 MPa).

Zinoviadou et al. (2010) demonstrated that addition of ε-polylysine as an antimicrobial into WPI film matrix did not alter the water vapor permeability or water sorption properties of the films, while it induced a plasticizing effect as evidenced by a reduction of peak stress and an increase of elongation at break. This suggests that it is possible to control the pliability of a film by the interaction between whey protein and an antimicrobial agent without/with a very small quantity of plasticizer. Therefore, antimicrobials such as lysozyme, chitosan, and ε-polylysine could act as a plasticizer under controlled conditions, without requiring additional plasticizer.

As an antimicrobial agent, chitosan's action against the growth of pathogenic and spoilage bacteria has been well documented in a range of foods such as bread, strawberries, juices, mayonnaise, milk, and rice cake. It is considered that the polycationic nature of chitosan is crucial. It has been proposed that the positively charged amino groups of the glucosamine units interact with negatively charged components in microbial cell membranes, altering their barrier properties (Liu et al., 2004). In relation to the mode of action of chitosan matrices, a direct relationship between their antimicrobial capacity and the release of glucosamine chains from the film to the medium has recently been demonstrated. Starch-chitosan film has shown activity in inhibiting the growth of *S. aureus* and *Salmonella* (Fernandez-Saiz et al., 2009; Shen et al., 2010).

8.3.4 Nanostructured materials through self-assembled fibrils

It is well known that fibril formation results primarily from the properties of the polypeptide chain that are

Figure 8.23 Whey protein isolate (WPI) film (*left*) and TiO$_2$/WPI blend film containing 0.5% TiO$_2$ nanoparticles (*right*) and their scanning electron micrographs (*bottom row*). (Reproduced from Zhou et al. (2009), with permission from John Wiley & Sons, Ltd.)

common to all peptides and proteins. Polypeptide chains with different sequences, either fully or partially unfolded, can form nanoscale fibrils under specific conditions (pH, ionic strength, temperature, time, denaturants). Hydrophobicity, charge, and secondary structure properties of a protein strongly influence fibril formation and the rate of formation. The hydrophobicity of the side chains determines the aggregation of an unfolded polypeptide chain. The aggregation is affected by the charge that a protein carries. A high net charge either globally or locally may hinder self-association. The aggregation of polypeptide chains can be facilitated by interactions with macromolecules, which exhibit a high compensatory charge. In addition to charge and hydrophobicity, a low propensity to form α-helical structures and a high propensity to form β-sheet structures are also likely to be important

factors facilitating fibril formation. Fibril formation can be preceded by formation of a wide range of aggregates such as unstructured oligomers and structured protofibrils (Chiti & Dobson, 2006).

Fibrils have been observed in food proteins. Langton and Hermansson (1992) reported stiff, short strands (~4 nm) at low pH, and much thicker, longer, and more flexible strands (~10 nm) at high pH in fine-stranded β-lactoglobulin (BLG) and whey proteins gels. Loveday et al. (2012) also reported BLG fibrils formation at pH 2 after heating from 75°C to 120°C. By incubating bovine BLG (0.1%) in 3–5 M urea at 37°C and pH 7, Hamada and Dobson (2002) obtained the fibrils of BLG with a diameter of ~8 to 10 nm. They found that fibril formation by BLG is promoted under conditions where significant accumulation of unfolded proteins occurs, but is inhibited under conditions where higher denaturant concentrations destabilize intermolecular interactions. The addition of preformed fibrils into protein solutions containing urea shows that fibril formation can be accelerated by seeding processes that remove the lag phase. Efficient fibril formation involves a balance between the requirement of a significant population of unfolded or partially unfolded molecules and the need to avoid conditions that strongly destabilize intermolecular interactions. By heating BLG in either water or organic solvents at pH 2, the self-assembled fibrils with distinct structure and morphology were observed (Gosal et al., 2002).

Thus, it is obvious that the formation and rate of formation of protein fibrils can be manipulated by selecting appropriate reaction conditions. Fibrillar gelation of WPI has been observed under high alkali conditions (Mercade-Prieto & Gunasekaran, 2009a, b). These facts indicate that the self-assembled fibril formation and fibril characterization of globular protein could be mediated by controlling the interaction forces, i.e. reaction conditions. Generating such self-assembled supramolecular structures could lead to a class of novel high-performance biomaterials.

It has been recognized that the distribution of inter- versus intramolecular bonding interactions is associated with the transition of proteins from their native globular structures into polymeric supramolecular assemblies. The rigidity of fibrils is described by a common elastic modulus, which is defined predominantly by intermolecular interactions involving the common polypeptide main chain and provides quantitative evidence for the idea that these structures form a generic class of material. The bending rigidities of the different structures vary over nearly four orders of magnitude. Some fibrils, such as those of α-lactalbumin, appear to be very flexible whereas others, such as those formed by the short peptide TTR (105–115) from transthyretin (a blood protein), are extremely stiff (Knowles et al., 2007). The characterization of fibrils can be managed under controlled conditions.

Recently, a very rigid nanostructured film (Young's modulus 5–7 GPa) fabricated from hierarchical self-assembly of proteins has been reported (Knowles et al., 2010). In this, protein is first self-assembled to form amyloid-like fibrils of ~8 to 25 nm in diameter and cast into film by ordering of the fibrils. We could take advantage of the self-assembly of protein fibrils to form a strong, well-ordered nanostructured film. This will create a fundamentally different network structure, which should yield many advantageous properties besides improving the mechanical properties as has been reported.

In addition, using antimicrobial agents such as lysozyme, chitosan, ε-polylysine, etc., we can manipulate the interaction forces between protein fibrils and the antimicrobial agent such that the film we obtain is not only strong, but also highly antimicrobial to effectively inactivate pathogenic bacteria on contact.

Protein molecules have the ability to form a rich variety of natural and artificial structures and materials (Knowles et al., 2007). Protein fibrils can be formed from a range of very different polypeptide sequences. It is assumed that fibril structure is generic, widely accessible, accurate, and stable. It is very appealing for scientists to exploit the stability and accurate self-assembly of unique structures of fibrils and their applications in material science. To be specific, it is feasible to take advantage of the propensity of amyloidogenic protein molecules to create highly specific, non-covalent contacts to fabricate macroscopic materials with controlled nanostructure (Knowles et al., 2007). The hierarchical self-assembly is a powerful tool to create nanostructured materials (Huang et al., 2001; Ikkala & ten Brinke, 2002).

Recently, Knowles et al. (2010) used this technique to produce thin films of fibrillar polypeptide self-assembly in a two-step process, as shown in Figure 8.24. In the first step, protein molecules are assembled into elongated fibrils (Figure 8.24b) under conditions where the formation of intermolecular interactions is favored over intramolecular ones. The resulting nanofibrils are highly stable and rigid by hydrogen bonding. In the second step, the fibrils are cast into thin films (Figure 8.24c). During the casting process, the fibrils align in the film plane (Figure 8.24c). In the presence of plasticizing molecules the fibrils stack with nematic order, resulting in materials that have a hierarchy of length scales: nanometer ordering within the fibrils and micrometer-scale ordering in the stacking of the fibrils. The films obtained by self-assembly of protein fibrils were well ordered and highly rigid, with a Young's modulus of up to 5–7 GPa, which is comparable to the highest values for proteinaceous materials found in nature. This is about 25–500 times stronger than whey protein films (Anker et al., 2001; Ozdemir & Floros, 2008). This success in achieving such strong fabricated nanostructured films suggests that it is possible to create films based on the same nanoscale mechanisms through self-assembly by using high specificity biomolecules while manipulating interaction forces. The major contribution to the rigidity of fibril material stems from a generic inter-backbone hydrogen bonding network that is modulated by variable side-chain interactions (Knowles et al., 2007).

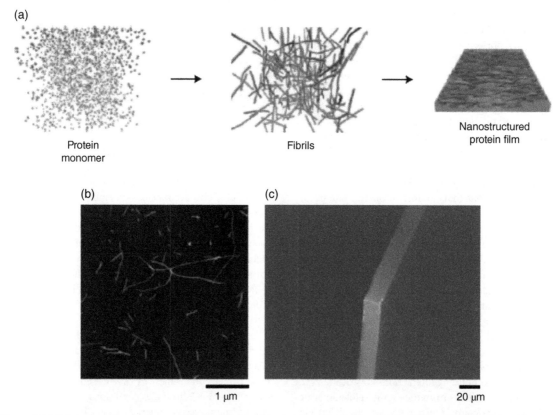

Figure 8.24 Fabrication of nanostructured films through multiscale hierarchical self-assembly. (a) Protein molecules are first assembled into amyloid fibrils, which are then stacked into films. (b) Atomic force micrograph of the component lysozyme fibrils. (c) Scanning electron micrograph of the resulting free-standing protein film. (Reproduced from Knowles et al. (2010), with permission from Nature Publishing Group.)

8.4 Nanotechnology and sustainability

The Brundtland Commission has defined sustainable development as that "which meets the needs of the present without compromising the ability of future generations to meet their own needs"(Brundtland, 1987). Thus, sustainability takes into account various factors pertaining to people, the environment, and the economy (Figure 8.25). As such, it is a function of complex interdependencies that ensure environmental and economic well-being of present and future societies.

The rapid growth of nanotechnology was initially spurred by the opportunities for discovery, characterization, and application of nanoscale materials and phenomena. But with time, the question of sustainability became an important consideration. In fact, even as early as 1999

the "maintenance of industrial sustainability by significant reductions in materials and energy use, reduced sources of pollution, increased opportunities for recycling" was a stated goal of the National Nanotechnology Initiative (Roco et al., 1999). Some 20 years later, sustainability is again a key aspect of Nanotechnology Research Directions for Societal Needs in 2020 (Roco et al., 2010).

During the last several years, nanotechnology solutions have made a significant difference in the areas of water treatment, desalination, and reuse. For example, nanoscale sorbents such as clay, metaloxides, zeolites, carbon fibers, and other polymeric adsorbents have been used to selectively remove anions, cations, and organic solutes from contaminated water. Nanocatalysts and redox-active nanoparticles such as titanium dioxide redox-active zero-valent iron nanoparticles have been used to convert toxic organic solutes and oxyanions into harmless

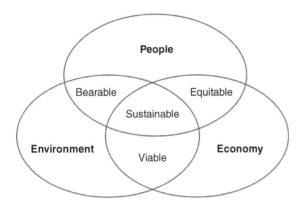

Figure 8.25 Sustainability as intersection of factors pertaining to people, environment, and economy. Based on Brundtland (1987).

by-products. Nanobiocides such as magnesium oxide, silver nanoparticles, and bioactive dendrimers can deactivate bacteria in contaminated water without generating toxic by-products. Carbon nanotube-based filters and reactive membranes are used for water treatment, desalination, and reuse.

In the area of agriculture, food safety and biosecurity, nanotechnology has also contributed significantly in terms of novel formulations of agrochemicals (fertilizers, pesticides, veterinary medicines, etc.), biosensors, nanomaterial-based delivery systems (nanoemulsions, nanocomposites, controlled-release formulations, etc.) and improved biodegradable packaging materials to reduce environmental burden. It is certain that advances in nanotechnology will lead to major improvements in how we grow, process, store, and distribute food.

8.5 Summary

Nanotechnology applications are beginning to make significant impacts in almost every aspect of our lives. Although currently slow, adaptation of nanotechnological advances in the food industry will improve our ability to make food systems safe and secure. The nanotechnological applications discussed in this chapter are only a small subset of a vast range of advances that are taking place, at a rapid pace. Both biosensing and food packaging applications discussed in this chapter have seen robust growth and are considered more advanced in terms of having brought nanotechnology applications to the food industry. However, even as new innovations continue to make

their impact felt, there is an attendant issue that calls for serious consideration: the safety of the nanomaterials themselves. Since not everything is known about the eventual fate of engineered nanomaterials used in our food systems, certain segments of the public, regulatory agencies, and even scientists are rather cautious in espousing nanotechnology. This is clearly an important research need for the future. We need to ensure that, in the name of nanotechnology, we do not adversely affect the very food that we are attempting to improve.

References

Agrawal A, Tripp RA, Anderson LJ, Nie SM (2005) Real-time detection of virus particles and viral protein expression with two-color nanoparticle probes. *Journal of Virology* **79**: 8625–8628.

Alexandre M, Dubois P (2000) Polymer-layered silicate nanocomposites: preparation, properties and uses of a new class of materials. *Materials Science and Engineering R-Reports* **28**: 1–63.

Alivisatos P (2004) The use of nanocrystals in biological detection. *Nature Biotechnology* **22**: 47–52.

Anker M, Stading M, Hermansson AM (2001) Aging of whey protein films and the effect on mechanical and barrier properties. *Journal of Agricultural and Food Chemistry* **49**, 989–995.

Anonymous (2012) *Nanotechnology: The Science of Miniaturization*. www.britannica.com/bps/media–view/73082/1/0/0, accessed 12 November 2013.

Arcidiacono S, Pivarnik P, Mello CM, Senecal A (2008) Cy5 labeled antimicrobial peptides for enhanced detection of Escherichia coli O157: H7. *Biosensors and Bioelectronics* **23**: 1721–1727.

Arora A, Padua GW (2010) Review: nanocomposites in food packaging. *Journal of Food Science* **75**: R43–R49.

Avella M, de Vlieger JJ, Errico ME et al. (2005) Biodegradable starch/clay nanocomposite films for food packaging applications. *Food Chemistry* **93**: 467–474.

Bakalova R, Zhelev Z, Ohba H (2006) Quantum dots open new trends in biosensor evolution. *Sensor Letters* **4**: 452–454.

Bassell GJ, Powers CM, Taneja KL, Singer RH (1994) Single messenger-RNAs visualized by ultrastructural in-situ hybridization are principally localized at actin filament intersections in fibroblasts. *Journal of Cell Biology* **126**: 863–876.

Bentzen L, House F, Utley TJ, Crowe JE, Wright DW (2005) Progression of respiratory syncytial virus infection monitored by fluorescent quantum dot probes. *Nano Letters* **5**: 591–595.

Bertuzzi MA, Armada M, Gottifredi JC (2007) Physicochemical characterization of starch based films. *Journal of Food Engineering* **82**: 17–25.

Bohren CF, Huffman DR (1983) Absorption and Scattering of Light by Small Particles New York: John Wiley and Sons.

Bolisay LD, Culver JN, Kofinas P (2007) Optimization of virus imprinting methods to improve selectivity and reduce non-specific binding. *Biomacromolecules* **8**: 3893–3899.

Bower CK, Avena-Bustillos RJ, Olsen CW, McHugh TH, Bechtel PJ (2006) Characterization of fish-skin gelatin gels and films containing the antimicrobial enzyme lysozyme. *Journal of Food Science* **71**: M141–M145.

Brandenburg AH, Weller CL, Testin RF (1993) Edible films and coatings from soy protein. *Journal of Food Science* **58**: 1086–1089.

Bruchez M, Moronne M, Gin P, Weiss S, Alivisatos AP (1998) Semiconductor nanocrystals as fluorescent biological labels. *Science* **281**: 2013–2016.

Brundtland GH (1987) Our Common Future: From One Earth to One World. World Commission on Environment and Development. Oslo, Norway: United Nations.

Bruno JG, Kiel JL (1999) In vitro selection of DNA aptamers to anthrax spores with electrochemiluminescence detection. *Biosensors and Bioelectronics* **14**: 457–464.

Bruno JG, Kiel JL (2002) Use of magnetic beads in selection and detection of biotoxin aptamers by electrochemiluminescence and enzymatic methods. *Biotechniques* **32**: 178.

Brust M, Bethell D, Schiffrin DJ, Kiely CJ (1995) Novel gold-dithiol nano-networks with nonmetallic electronic properties. *Advanced Materials* **7**: 795.

Burdock GA (2007) Safety assessment of hydroxypropyl methylcellulose as a food ingredient. *Food and Chemical Toxicology* **45**: 2341–2351.

Cabedo L, Feijoo JL, Villanueva MP, Lagaron JM, Gimenez E (2006) Optimization of biodegradable nanocomposites based on aPLA/PCL blends for food packaging applications. *Macromolecular Symposia* **233**: 191–197.

Centers for Disease Control (CDC) (2011) Estimates of Foodborne Illness in the United States. Washington, DC: National Center for Emerging and Zoonotic Infectious Diseases, Division of Foodborne, Waterborne, and Environmental Diseases. www.cdc.gov/foodborneburden/, accessed 12 November 2013.

Chan WCW, Nie SM (1998) Quantum dot bioconjugates for ultrasensitive nonisotopic detection. *Science* **281**: 2016–2018.

Chen BQ, Evans JRG (2005) Thermoplastic starch-clay nanocomposites and their characteristics. *Carbohydrate Polymers* **61**: 455–463.

Chiti F, Dobson CM (2006) Protein misfolding, functional amyloid, and human disease. *Annual Review of Biochemistry* **75**: 333–366.

Cisneros-Zevallos L, Krochta JM (2003) Whey protein coatings for fresh fruits and relative humidity effects. *Journal of Food Science* **68**: 176–181.

Coma V (2008) Bioactive packaging technologies for extended shelf life of meat-based products. *Meat Science* **78**: 90–103.

Coma V, Sebti I, Pardon P et al. (2001) Antimicrobial edible packaging based on cellulosic ethers, fatty acids, and nisin incorporation to inhibit Listeria innocua and Staphylococcus aureus. *Journal of Food Protection* **64**: 470–475.

Creighton JA, Blatchford CG, Albrecht MG (1979) Plasma resonance enhancement of Raman-scattering by pyridine adsorbed on silver or gold sol particles of size comparable to the excitation wavelength. *Journal of the Chemical Society–Faraday Transactions Ii* **75**: 790–798.

Cyras VP, Manfredi LB, Ton-That MT, Vazquez A (2008) Physical and mechanical properties of thermoplastic starch/montmorillonite nanocomposite films. *Carbohydrate Polymers* **73**: 55–63.

Daniel MC, Astruc D (2004) Gold nanoparticles: assembly, supramolecular chemistry, quantum size-related properties, and applications toward biology, catalysis, and nanotechnology. *Chemical Reviews* **104**: 293–346.

de Azeredo HMC (2009) Nanocomposites for food packaging applications. *Food Research International* **42**: 1240–1253.

de Boer E, Beumer RR (1999) Methodology for detection and typing of foodborne microorganisms. *International Journal of Food Microbiology* **50**: 119–130.

de la Fuente JM, Penades S (2006) Glyconanoparticles: types, synthesis and applications in glycoscience, biomedicine and material science. *Biochimica Et Biophysica Acta General Subjects* **1760**: 636–651.

Efrima S, Bronk BV (1998) Silver colloids impregnating or coating bacteria. *Journal of Physical Chemistry B* **102**: 5947–5950.

El-Boubbou K, Gruden C, Huang X (2007) Magnetic glyconanoparticles: a unique tool for rapid pathogen detection, decontamination, and strain differentiation. *Journal of the American Chemical Society* **129**: 13392.

Elghanian R, Storhoff JJ, Mucic RC, Letsinger RL, Mirkin CA (1997) Selective colorimetric detection of polynucleotides based on the distance-dependent optical properties of gold nanoparticles. *Science* **277**: 1078–1081.

Fernandez-Saiz P, Lagaron JM, Ocio MJ (2009) Optimization of the biocide properties of chitosan for its application in the design of active films of interest in the food area. *Food Hydrocolloids* **23**: 913–921.

Fischer N, Tarasow TM, Tok JBH (2007) Aptasensors for biosecurity applications. *Current Opinion in Chemical Biology* **11**: 316–328.

Freeman RG, Grabar KC, Allison KJ et al. (1995) Self-assembled metal colloid monolayers – an approach to SERS substrates. *Science* **267**: 1629–1632.

Gadang VP, Hettiarachchy NS, Johnson MG, Owens C (2008) Evaluation of antibacterial activity of whey protein isolate coating incorporated with nisin, grape seed extract, malic acid, and EDTA on a turkey frankfurter system. *Journal of Food Science* **73**: M389–M394.

Gallego RG, Haseley SR, van Miegem VFL et al. (2004) Identification of carbohydrates binding to lectins by using surface plasmon resonance in combination with HPLC profiling. *Glycobiology* **14**: 373–386.

Gella FJ, Serra J, Gener J (1991) Latex agglutination procedures in immunodiagnosis. *Pure and Applied Chemistry* **63**: 1131–1134.

Giljohann DA, Seferos DS, Daniel WL et al. (2010) Gold nanoparticles for biology and medicine. *Angewandte Chemie – International Edition* **49**: 3280–3294.

Gilmartin N, O'Kennedy R (2012) Nanobiotechnologies for the detection and reduction of pathogens. *Enzyme and Microbial Technology* **50**: 87–95.

Golightly RS, Doering WE, Natan MJ (2009) Surface-enhanced Raman spectroscopy and homeland security: a perfect match? *Acs Nano* **3**: 2859–2869.

Gomez-Hens A, Fernandez-Romero JM, Aguilar-Caballos MP (2008) Nanostructures as analytical tools in bioassays. *Trac-Trends in Analytical Chemistry* **27**: 394–406.

Gorga RE, Cohen RE (2004) Toughness enhancements in poly(methyl methacrylate) by addition of oriented multiwall carbon nanotubes. *Journal of Polymer Science Part B–Polymer Physics* **42**: 2690–2702.

Gosal WS, Clark AH, Pudney PDA, Ross-Murphy SB (2002) Novel amyloid fibrillar networks derived from a globular protein: beta-lactoglobulin. *Langmuir* **18**: 7174–7181.

Grow AE, Wood LL, Claycomb JL, Thompson PA (2003) New biochip technology for label-free detection of pathogens and their toxins. *Journal of Microbiological Methods* **53**: 221–233.

Gu HW, Xu KM, Xu CJ, Xu B (2006) Biofunctional magnetic nanoparticles for protein separation and pathogen detection. *Chemical Communications* **941–949**.

Guicheteau J, Argue L, Emge D et al. (2008) Bacillus spore classification via surface-enhanced Raman spectroscopy and principal component analysis. *Applied Spectroscopy* **62**: 267–272.

Hahn MA, Keng PC, Krauss TD (2008) Flow cytometric analysis to detect pathogens in bacterial cell mixtures using semiconductor quantum dots. *Analytical Chemistry* **80**: 864–872.

Hamada D, Dobson CM (2002) A kinetic study of beta-lactoglobulin amyloid fibril formation promoted by urea. *Protein Science* **11**: 2417–2426.

Han MY, Gao XH, Su JZ, Nie S (2001) Quantum dot-tagged microbeads for multiplexed optical coding of biomolecules. *Nature Biotechnology* **19**: 631–635.

Haseley SR (2002) Carbohydrate recognition: a nascent technology for the detection of bioanalytes. *Analytica Chimica Acta* **457**: 39–45.

Haseley SR, Talaga P, Kamerling JP, Vliegenthart JFG (1999) Characterization of the carbohydrate binding specificity and kinetic parameters of lectins by using surface plasmon resonance. *Analytical Biochemistry* **274**: 203–210.

Hayat MA (ed) (1991) Colloidal Gold: Principles, Methods, and Applications. San Diego, CA: Academic Press.

Hernandez-Munoz P, Kanavouras A, GAVARA R (2003) Development and characterization of biodegradable films made from wheat gluten protein fractions. *Journal of Agricultural and Food Chemistry* **51**: 7647–7654.

Holt RE, Cotton TM (1989) Surface-enhanced resonance Raman and electrochemical investigation of glucose-oxidase catalysis at a silver electrode. *Journal of the American Chemical Society* **111**: 2815–2821.

Homann M, Goringer HU (1999) Combinatorial selection of high affinity RNA ligands to live African trypanosomes. *Nucleic Acids Research* **27**: 2006–2014.

Hrudey SE, Hrudey EJ (2007) Published case studies of water-borne disease outbreaks – evidence of a recurrent threat. *Water Environment Research* **79**: 233–245.

Huang Y, Duan XF, Wei QQ, Lieber CM (2001) Directed assembly of one-dimensional nanostructures into functional networks. *Science* **291**: 630–633.

Hudson PJ (2006) Engineered antibodies for diagnosis and therapy: and "what's hot in NanoBioTechnology?". *Tumor Biology* **27**.

Hussain F, Hojjati M, Okamoto M, Gorga RE (2006) Review article: Polymer-matrix nanocomposites, processing, manufacturing, and application: an overview. *Journal of Composite Materials* **40**: 1511–1575.

Ikanovic M, Rudzinski WE, Bruno JG et al. (2007) Fluorescence assay based on aptamer-quantum dot binding to Bacillus thuringiensis spores. *Journal of Fluorescence* **17**: 193–199.

Ikkala O, ten Brinke G (2002) Functional materials based on self-assembly of polymeric supramolecules. *Science* **295**: 2407–2409.

Iqbal SS, Mayo MW, Bruno JG et al. (2000) A review of molecular recognition technologies for detection of biological threat agents. *Biosensors and Bioelectronics* **15**: 549–578.

Ivnitski D, Abdel-Hamid I, Atanasov P, Wilkins E (1999) Biosensors for detection of pathogenic bacteria. *Biosensors and Bioelectronics* **14**: 599–624.

Jamieson T, Bakhshi R, Petrova D et al. (2007) Biological applications of quantum dots. *Biomaterials* **28**: 4717–4732.

Jayasena SD (1999) Aptamers: an emerging class of molecules that rival antibodies in diagnostics. *Clinical Chemistry* **45**: 1628–1650.

Jelinek R (2004) Carbohydrate biosensors. *Chemical Reviews* **104**: 5987–6015.

Jenison RD, Gill SC, Pardi A, Polisky B (1994) High-resolution molecular discrimination by Rna. *Science* **263**: 1425–1429.

Jensen TR, Malinsky MD, Haynes CL, van Duyne RP (2000) Nanosphere lithography: tunable localized surface plasmon resonance spectra of silver nanoparticles. *Journal of Physical Chemistry B* **104**: 10549–10556.

Jin T, Liu LS, Zhang H, Hicks K (2009) Antimicrobial activity of nisin incorporated in pectin and polylactic acid composite films against Listeria monocytogenes. *International Journal of Food Science and Technology* **44**: 322–329.

Jones N, Ray B, Ranjit KT, Manna AC (2008) Antibacterial activity of ZnO nanoparticle suspensions on a broad spectrum of microorganisms. *FEMS Microbiology Letters* **279**: 71–76.

Kambhampati P, Child CM, Foster MC, Campion A (1998) On the chemical mechanism of surface enhanced Raman scattering: experiment and theory. *Journal of Chemical Physics* **108**: 5013–5026.

Kirby R, Cho EJ, Gehrke B et al. (2004) Aptamer-based sensor arrays for the detection and quantitation of proteins. *Analytical Chemistry* **76**: 4066–4075.

Kneipp K, Kneipp H, Itzkan I, Dasari RR, Feld MS (1999) Surface-enhanced Raman scattering: a new tool for biomedical spectroscopy. *Current Science* **77**: 915–924.

Kneipp K, Kneipp H, Manoharan R et al. (1998) Surface-enhanced Raman scattering (SERS) – a new tool for single molecule detection and identification. *Bioimaging* **6**: 104–110.

Knowles TP, Fitzpatrick AW, Meehan S et al. (2007) Role of intermolecular forces in defining material properties of protein nanofibrils. *Science* **318**: 1900–1903.

Knowles TP, Oppenheim TW, Buell AK et al. (2010) Nanostructured films from hierarchical self-assembly of amyloidogenic proteins. *Nature Nanotechnology* **5**: 204–207.

Ko S, Gunasekaran S, Yu J (2010) Self-indicating nanobiosensor for detection of 2,4-dinitrophenol. *Food Control* **21**: 155–161.

Kojima Y, Usuki A, Kawasumi M et al. (1993) Synthesis of nylon-6-clay hybrid by montmorillonite intercalated with epsilon-caprolactam. *Journal of Polymer Science Part A-Polymer Chemistry* **31**: 983–986.

Kreibig U, Genzel L (1985) Optical-absorption of small metallic particles. *Surface Science* **156**: 678–700.

Kriegel C, Kit KM, McClements DJ, Weiss J (2009) Electrospinning of chitosan-poly(ethylene oxide) blend nanofibers in the presence of micellar surfactant solutions. *Polymer* **50**: 189–200.

Kulagina NV, Lassman ME, Ligler FS, Taitt CR (2005) Antimicrobial peptides for detection of bacteria in biosensor assays. *Analytical Chemistry* **77**: 6504–6508.

Kulagina NV, Shaffer KM, Anderson GP, Ligler FS, Taitt CR (2006) Antimicrobial peptide-based array for Escherichia coli and Salmonella screening. *Analytica Chimica Acta* **575**: 9–15.

Kulagina NV, Shaffer KM, Ligler FS, Taitt CR (2007) Antimicrobial peptides as new recognition molecules for screening challenging species. *Sensors and Actuators B–Chemical* **121**: 150–157.

Langton M, Hermansson AM (1992) Fine-stranded and particulate gels of beta-lactoglobulin and whey-protein at varying Ph. *Food Hydrocolloids* **5**: 523–539.

Larson DR, Zipfel WR, Williams RM et al. (2003) Water-soluble quantum dots for multiphoton fluorescence imaging in vivo. *Science* **300**: 1434–1436.

Lawton JW (2002) Zein: a history of processing and use. *Cereal Chemistry* **79**: 1–18.

Lazcka O, del Campo FJ, Munoz FX (2007) Pathogen detection: a perspective of traditional methods and biosensors. *Biosensors and Bioelectronics* **22**: 1205–1217.

Leonard P, Hearty S, Brennan J et al. (2003) Advances in biosensors for detection of pathogens in food and water. *Enzyme and Microbial Technology* **32**: 3–13.

Leoni E, Legnani PP (2001) Comparison of selective procedures for isolation and enumeration of Legionella species from hot water systems. *Journal of Applied Microbiology* **90**: 27–33.

Leuvering J, Thal P, Vanderwaart M, Schuurs A (1980) Sol particle agglutination immunoassay for human chorionic-gonadotropin. *Fresenius Zeitschrift Fur Analytische Chemie* **301**: 132–132.

Leuvering J, Thal P, Vanderwaart M, Schuurs A (1981) A sol particle agglutination assay for human chorionic-gonadotropin. *Journal of Immunological Methods* **45**: 183–194.

Leuvering J, Thal P, White D, Schuurs A (1983) A homogeneous sol particle immunoassay for total estrogens in urine and serum samples. *Journal of Immunological Methods* **62**: 163–174.

Levi K, Smedley J, Towner KJ (2003) Evaluation of a real-time PCR hybridization assay for rapid detection of Legionella pneumophila in hospital and environmental water samples. *Clinical Microbiology and Infection* **9**: 754–758.

Lim S, Koo OK, You YS et al. (2012) Enhancing nanoparticle-based visible detection by controlling the extent of aggregation. *Scientific Reports*, **2**.

Liu H, Du YM, Wang XH, Sun LP (2004) Chitosan kills bacteria through cell membrane damage. *International Journal of Food Microbiology* **95**: 147–155.

Lopez-Rubio A, Gavara R, Lagaron JA (2006) Bioactive packaging: turning foods into healthier foods through biomaterials. *Trends in Food Science and Technology* **17**: 567–575.

Loveday SM, Wang XL, Rao MA et al. (2012) beta-Lactoglobulin nanofibrils: effect of temperature on fibril formation kinetics, fibril morphology and the rheological properties of fibril dispersions. *Food Hydrocolloids* **27**: 242–249.

Mairal T, Ozalp VC, Sanchez PL et al. (2008) Aptamers: molecular tools for analytical applications. *Analytical and Bioanalytical Chemistry* **390**: 989–1007.

Mattoso LHC, de Moura MR, Aouada FA et al. (2009) Improved barrier and mechanical properties of novel hydroxypropyl methylcellulose edible films with chitosan/tripolyphosphate nanoparticles. *Journal of Food Engineering* **92**: 448–453.

Mattoso LHC, de Moura MR, Avena-Bustillos RJ et al. (2008) Properties of novel hydroxypropyl methylcellulose films containing chitosan nanoparticles. *Journal of Food Science* **73**: N31–N37.

McGlashan SA, Halley PJ (2003) Preparation and characterisation of biodegradable starch-based nanocomposite materials. *Polymer International* **52**: 1767–1773.

McHugh TH, Krochta JM (1994 Milk-protein-based edible films and coatings. *Food Technology* **48**: 97–103.

Medintz IL, Sapsford KE, Bradburne C, Detehanty JB (2008) Sensors for detecting biological agents. *Materials Today* **11**: 38–49.

Medintz IL, Uyeda HT, Goldman ER, Mattoussi H (2005) Quantum dot bioconjugates for imaging, labelling and sensing. *Nature Materials* **4**: 435–446.

Mei Y, Zhao YY (2003) Barrier and mechanical properties of milk protein-based edible films containing nutraceuticals. *Journal of Agricultural and Food Chemistry* **51**: 1914–1918.

Mercade-Prieto R, Gunasekaran S (2009a) Alkali cold gelation of whey proteins. part I: sol–gel-sol(–gel) transitions. *Langmuir* **25**: 5785–5792.

Mercade-Prieto R, Gunasekaran S (2009b) Alkali cold gelation of whey proteins. part II: protein concentration. *Langmuir* **25**: 5793–5801.

Michalet X, Pinaud FF, Bentolila LA et al. (2005) Quantum dots for live cells, in vivo imaging, and diagnostics. *Science* **307**: 538–544.

Miller KS, Krochta JM (1997) Oxygen and aroma barrier properties of edible films: a review. *Trends in Food Science and Technology* **8**: 228–237.

Milton JD, Fernig DG, Rahmoune H, Rhodes JM (1997) The binding characteristics of edible lectins, from peanut and mushroom, to a Gal-GalNAc containing glycoprotein determined on a biosensor device. *Gut* **41**: A144–A144.

Min S, Harris LJ, Han JH, Krochta JM (2005) Listeria monocytogenes inhibition by whey protein films and coatings incorporating lysozyme. *Journal of Food Protection* **68**: 2317–2325.

Mirkin CA, Letsinger RL, Mucic RC, Storhoff JJ (1996) A DNA-based method for rationally assembling nanoparticles into macroscopic materials. *Nature* **382**: 607–609.

Mori T, Maeda M (2002) Stability change of DNA-carrying colloidal particle induced by hybridization with target DNA. *Polymer Journal* **34**: 624–628.

Morones JR, Elechiguerra JL, Camacho A et al. (2005) The bactericidal effect of silver nanoparticles. *Nanotechnology*, **16**: 2346–2353.

Moskovits M (1985) Surface-enhanced spectroscopy. *Reviews of Modern Physics* **57**: 783–826.

Neeley A, Khan SA, Beqa L et al. (2011) Selective detection of chemical and biological toxins using gold-nanoparticle-based two-photon scattering assay. *IEEE Transactions on Nanotechnology* **10**: 26–34.

Ogawa M, Kuroda K (1997) Preparation of inorganic–organic nanocomposites through intercalation of organoammonium ions into layered silicates. *Bulletin of the Chemical Society of Japan* **70**: 2593–2618.

Ozdemir M, Floros JD (2001) Analysis and modeling of potassium sorbate diffusion through edible whey protein films. *Journal of Food Engineering* **47**: 149–155.

Ozdemir M, Floros JD (2008) Optimization of edible whey protein films containing preservatives for water vapor permeability, water solubility and sensory characteristics. *Journal of Food Engineering* **86**: 215–224.

Pan Q, Zhang XL, Wu HY (2006) Aptamers that preferentially bind type IVB pili and inhibit human monocytic-cell invasion by Salmonella enterica serovar typhi. *Antimicrobial Agents and Chemotherapy* **49**: 4052–4060.

Park B (2009) Nanotechnology for food safety. *Cereal Foods World* **54**: 158–162.

Park C, Park O, Lim J, Kim H (2001) The fabrication of syndiotactic polystyrene/organophilic clay nanocomposites and their properties. *Polymer* **42**: 7465–7475.

Park HM, Li XC, Jin CZ et al. (2002) Preparation and properties of biodegradable thermoplastic starch/clay hybrids. *Macromolecular Materials and Engineering* **287**: 553–558.

Premasiri WR, Moir DT, Klempner MS et al. (2005) Characterization of the surface enhanced Raman scattering (SERS) of bacteria. *Journal of Physical Chemistry B* **109**: 312–320.

Ray SS, Okamoto M (2003) Polymer/layered silicate nanocomposites: a review from preparation to processing. *Progress in Polymer Science* **28**: 1539–1641.

Ray S, Quek SY, Easteal A, Chen XD (2006) The potential use of polymer-clay nanocomposites in food packaging. *International Journal of Food Engineering* **2**.

Resch-Genger U, Grabolle M, Cavaliere-Jaricot S, Nitschke R, Nann T (2008) Quantum dots versus organic dyes as fluorescent labels. *Nature Methods* **5**: 763–775.

Reynolds RA, Mirkin CA, Letsinger RL (2000) A gold nanoparticle/latex microsphere-based colorimetric oligonucleotide detection method. *Pure and Applied Chemistry* **72**: 229–235.

Rhim JW (2007) Potential use of biopolymer-based nanocomposite films in food packaging applications. *Food Science and Biotechnology* **16**: 691–709.

Rhim JW, Hong SI, Ha CS (2009) Tensile, water vapor barrier and antimicrobial properties of PLA/nanoclay composite films. *LWT–Food Science and Technology* **42**: 612–617.

Roco M, Bainbridge WS (2001) Societal Implications of Nanoscience and Nanotechnology. Arlington, VA: NSF.

Roco M, Williams RS, Alivisatos P (eds) (1999) Nanotechnology Research Directions. Washington, DC: National Science and Technology Council.

Roco M, Mirkin CA, Hersam MC (eds) (2010) Nanotechnology Research Directions for Societal Needs in 2020: Retrospective and Outlook. Arlington, VA: World Technology Evaluation Center.

Rodriguez A, Batlle R, Nerin C (2007) The use of natural essential oils as antimicrobial solutions in paper packaging. Part II. *Progress in Organic Coatings* **60**: 33–38.

Rodriguez A, Nerin C, Batlle R (2008) New cinnamon-based active paper packaging against Rhizopusstolonifer food spoilage. *Journal of Agricultural and Food Chemistry* **56**: 6364–6369.

Rodriguez-Fernandez J, Funston AM, Perez-Juste J et al. (2009) The effect of surface roughness on the plasmonic response of individual sub-micron gold spheres. *Physical Chemistry Chemical Physics* **11**: 5909–5914.

Rodriguez-Lafuente A, Nerin C, Batlle R (2010) Active paraffin-based paper packaging for extending the shelf life of cherry tomatoes. *Journal of Agricultural and Food Chemistry* **58**: 6780–6786.

Rojas-Grau MA, Avena-Bustillos RJ, Olsen C et al. (2007) Effects of plant essential oils and oil compounds on mechanical, barrier and antimicrobial properties of alginate-apple puree edible films. *Journal of Food Engineering* **81**: 634–641.

Rosenthal SJ, Tomlinson A, Adkins EM et al. (2002) Targeting cell surface receptors with ligand-conjugated nanocrystals. *Journal of the American Chemical Society* **124**: 4586–4594.

Rosi NL, Mirkin CA (2005) Nanostructures in biodiagnostics. *Chemical Reviews* **105**: 1547–1562.

Ruoff RS, Lorents DC (1995) Mechanical and thermal properties of carbon nanotubes. *Carbon* **33**: 925–930.

Ryu G, Dagenais M, Hurley MT, Deshong P (2010) High specificity binding of lectins to carbohydrate-functionalized fiber Bragg gratings: a new model for biosensing applications. *IEEE Journal of Selected Topics in Quantum Electronics* **16**: 647–653.

Sato K, Hosokawa K, Maeda M (2003) Rapid aggregation of gold nanoparticles induced by non-cross-linking DNA hybridization. *Journal of the American Chemical Society* **125**: 8102–8103.

Schmidt D, Shah D, Giannelis EP (2002) New advances in polymer/layered silicate nanocomposites. *Current Opinion in Solid State and Materials Science* **6**: 205–212.

Sepulveda B, Angelome PC, Lechuga LM, Liz-Marzan LM (2009) LSPR-based nanobiosensors. *Nano Today* **4**: 244–251.

Shen XL, Wu JM, Chen YH, Zhao GH (2010) Antimicrobial and physical properties of sweet potato starch films incorporated with potassium sorbate or chitosan. *Food Hydrocolloids* **24**: 285–290.

Shen ZH, Huang MC, Xiao CD et al. (2007) Nonlabeled quartz crystal microbalance biosensor for bacterial detection using carbohydrate and lectin recognitions. *Analytical Chemistry* **79**: 2312–2319.

Shi K, Kokini JL, Huang QR (2008) POLY 19-Engineering zein films with controlled hydrophilicity through alternative solvents and UV/ozone treatment. *Abstracts of Papers of the American Chemical Society* 236.

Shrivastava S, Dash D (2009) Agrifood nanotechnology: a tiny revolution in food and agriculture. *Journal of Nano Research* **6**: 1–14.

Shukla R, Cheryan M (2001) Zein: the industrial protein from corn. *Industrial Crops and Products* **13**: 171–192.

Singh AK, Senapati D, Wang SG et al. (2009) Gold nanorod based selective identification of Escherichia coli bacteria using two-photon RAYLEIGH scattering spectroscopy. *Acs Nano* **3**: 1906–1912.

So HM, Park DW, Jeon EK et al. (2008) Detection and titer estimation of Escherichia coli using aptamer-functionalized single-walled carbon-nanotube field-effect transistors. *Small* **4**: 197–201.

Soares JW, North SH, Doherty LA et al. (2009) ENVR 43-Antimicrobial peptides for detection of bacterial pathogens. *Abstracts of Papers of the American Chemical Society* 238.

Sondi I, Salopek-Sondi B (2004) Silver nanoparticles as antimicrobial agent: a case study on E-coli as a model for Gram-negative bacteria. *Journal of Colloid and Interface Science* **275**: 177–182.

Sorrentino A, Gorrasi G, Vittoria V (2007) Potential perspectives of bio-nanocomposites for food packaging applications. *Trends in Food Science and Technology* **18**: 84–95.

Sothornvit R, Olsen CW, McHugh TH, Krochta JM (2007) Tensile properties of compression-molded whey protein sheets: determination of molding condition and glycerol-content effects and comparison with solution-cast films. *Journal of Food Engineering* **78**: 855–860.

Sozer N, Kokini JL (2009) Nanotechnology and its applications in the food sector. *Trends in Biotechnology* **27**: 82–89.

Stine R, Pishko MV (2005) Comparison of glycosphingolipids and antibodies as receptor molecules for ricin detection. *Analytical Chemistry* **77**: 2882–2888.

Storhoff JJ, Lazarides AA, Mucic RC et al. (2000) What controls the optical properties of DNA-linked gold nanoparticle assemblies? *Journal of the American Chemical Society* **122**: 4640–4650.

Stratis-Cullum DN, McMasters S, Pellegrino PM (2009) Evaluation of relative aptamer binding to Campylobacter jejuni bacteria using affinity probe capillary electrophoresis. *Analytical Letters* **42**: 2389–2402.

Straub TM, Chandler DP (2003) Towards a unified system for detecting waterborne pathogens. *Journal of Microbiological Methods* **53**: 185–197.

Tang JJ, Yu T, Guo L, Xie JW, Shao NS, He ZK (2007) In vitro selection of DNA aptamer against abrin toxin and aptamer-based abrin direct detection. *Biosensors and Bioelectronics* **22**: 2456–2463.

Tansil NC, Gao ZQ (2006) Nanoparticles in biomolecular detection. *Nano Today* **1**: 28–37.

Tharanathan RN (2003) Biodegradable films and composite coatings: past, present and future. *Trends in Food Science and Technology* **14**: 71–78.

Thostenson ET, Li CY, Chou TW (2005) Nanocomposites in context. *Composites Science and Technology* **65**: 491–516.

Thusu R (2010) *Strong Growth Predicted for Biosensors Market.* www.sensorsmag.com/specialty–markets/medical/strong–growth–predicted–biosensors–market–7640, accessed 12 November 2013.

Tombelli S, Minunni M, Mascini M (2007) Aptamers-based assays for diagnostics, environmental and food analysis. *Biomolecular Engineering* **24**: 191–200.

Toze S (1999) PCR and the detection of microbial pathogens in water and wastewater. *Water Research* **33**: 3545–3556.

Ung T, Liz-Marzan LM, Mulvaney P (2001) Optical properties of thin films of Au@SiO2 particles. *Journal of Physical Chemistry B* **105**: 3441–3452.

van Sark W, Frederix, P, Bol AA, Gerritsen HC, Meijerink A (2002) Blueing, bleaching, and blinking of single CdSe/ZnS quantum dots. *Chemphyschem* **3**: 871–879.

Velusamy V, Arshak K, Korostynska O, Oliwa K, Adley C (2010) An overview of foodborne pathogen detection: in the perspective of biosensors. *Biotechnology Advances* **28**: 232–254.

Vikesland PJ, Wigginton KR (2010) Nanomaterial enabled biosensors for pathogen monitoring – a review. *Environmental Science and Technology* **44**: 3656–3669.

Wang F, Tan WB, Zhang Y, Fan X, Wang M (2006) Luminescent nanomaterials for biological labelling. *Nanotechnology* **17**: R1–R13.

Wang SG, Singh AK, Senapati D et al. (2010) Rapid colorimetric identification and targeted photothermal lysis of Salmonella bacteria by using bioconjugated oval-shaped gold nanoparticles. *Chemistry* **16**: 5600–5606.

Wilson R (2008) The use of gold nanoparticles in diagnostics and detection. *Chemical Society Reviews* **37**: 2028–2045.

Win MN, Klein JS, Smolke CD (2006) Codeine-binding RNA aptamers and rapid determination of their binding constants using a direct coupling surface plasmon resonance assay. *Nucleic Acids Research* **34**: 5670–5682.

Yan L, Chang PR, Zheng PW (2011) Preparation and characterization of starch-grafted multiwall carbon nanotube composites. *Carbohydrate Polymers* **84**: 1378–1383.

Yang FM, Li HM, Li F et al. (2010) Effect of nano-packing on preservation quality of fresh strawberry (Fragaria ananassa Duch. cv Fengxiang) during storage at 4 degrees C. *Journal of Food Science* **75**: C236–C240.

Yguerabide J, Yguerabide EE (1998) Light-scattering submicroscopic particles as highly fluorescent analogs and their use as tracer labels in clinical and biological applications – II. Experimental characterization. *Analytical Biochemistry* **262**: 157–176.

Yoon SY, Deng YL (2006) Clay-starch composites and their application in papermaking. *Journal of Applied Polymer Science* **100**: 1032–1038.

Yoshino T, Isobe S, Maekawa T (2002) Influence of preparation conditions on the physical properties of zein films. *Journal of the American Oil Chemists' Society* **79**: 345–349.

Yu JH, Cui GJ, Wei M, Huang J (2007) Facile exfoliation of rectorite nanoplatelets in soy protein matrix and reinforced bionanocomposites thereof. *Journal of Applied Polymer Science* **104**: 3367–3377.

Zasloff M (2002) Antimicrobial peptides of multicellular organisms. *Nature* **415**: 389–395.

Zasloff M, Anderson M (2001) The development of antimicrobial peptides of animal origin as vaginal microbicides: the challenges ahead and the potential. *Aids* **15**: S54–S55.

Zhao RX, Torley P, Halley PJ (2008) Emerging biodegradable materials: starch- and protein-based bio-nanocomposites. *Journal of Materials Science* **43**: 3058–3071.

Zheng H, Ai FJ, Chang PR, Huang J, Dufresne A (2009) Structure and properties of starch nanocrystal-reinforced soy protein plastics. *Polymer Composites* **30**: 474–480.

Zhou JJ, Wang SY, Gunasekaran S (2009) Preparation and characterization of whey protein film incorporated with TiO(2) nanoparticles. *Journal of Food Science* **74**: N50–N56.

Zhu L, Ang S, Liu WT (2004) Quantum dots as a novel immunofluorescent detection system for Cryptosporidium parvum and Giardia lamblia. *Applied and Environmental Microbiology* **70**: 597–598.

Ziani K, Fernandez-Pan I, Royo M, Mate JI (2009) Antifungal activity of films and solutions based on chitosan against typical seed fungi. *Food Hydrocolloids* **23**: 2309–2314.

Zinoviadou KG, Koutsoumanis KP, Biliaderis CG (2009) Physico-chemical properties of whey protein isolate films containing oregano oil and their antimicrobial action against spoilage flora of fresh beef. *Meat Science* **82**: 338–345.

Zinoviadou KG, Koutsoumanis KP, Biliaderis CG (2010) Physical and thermo-mechanical properties of whey protein isolate films containing antimicrobials, and their effect against spoilage flora of fresh beef. *Food Hydrocolloids* **24**: 49–59.

9 Sustainability and Environmental Issues in Food Processing

Fionnuala Murphy,[1] Kevin McDonnell,[1] and Colette C. Fagan[2]
[1]Bioresources Research Centre, School of Biosystems Engineering, University College Dublin, Dublin, Ireland
[2]Department of Food and Nutritional Sciences, University of Reading, Reading, Berkshire, UK

9.1 Introduction

The United Nations has defined sustainable development as development that meets "the needs of the present without compromising the ability of future generations to meet their own needs" (United Nations, 1987). Therefore, a sustainable food chain implies that the entire food chain must be both competitive and resilient if it is to ensure a secure, environmentally sustainable, and healthy supply of food. There are currently a number of challenges to be overcome if we are to develop a truly sustainable food chain. Agricultural production has been estimated to contribute between 17% and 32% of global greenhouse gas emissions; approximately 30–50% of produced food is wasted (Boye & Arcand, 2012), yet widespread malnutrition continues to occur, while food processing accounts for 25% of water consumption worldwide (Baldwin, 2009). An assessment of sustainability should consider environmental, economic, and societal impacts.

When taken holistically, the food chain comprises a number of steps: food production, processing (including transport and distribution), retail, consumption, and end of life (Baldwin, 2009). While it is important to note the environmental impact of food production (i.e. land use change, climate change), the primary focus of this chapter will be food processing (i.e. waste generation, energy use). This chapter will first present motivating factors that are driving the sustainable food processing agenda. A discussion of the key environmental impacts associated with food processing will follow. To achieve better sustainability in food processing, a number of emerging and established technologies and tools will need to be employed; therefore the application of green technologies and sustainability assessment methods in food processing will finally be presented.

9.2 Sustainable food processing drivers

The key drivers in the move towards sustainable food processing are legislative, economic, consumer or corporate based.

9.2.1 Legislative drivers

According to Cooke (Cooke, 2008), environmental legislation relevant to the food processing industry can be considered at three levels:
- at a point in the environment
- at the point of emission
- at the process that creates the emission.

In general, there has been a move towards the third level, that is, controlling the process that emits the pollutant, rather than emissions deemed detrimental to the environment. In Europe, such control effort emerged as the Integrated Pollution Prevention and Control Directive (European Commission, 2008), which has now been superseded by the Industrial Emissions Directive (European Commission, 2010). The objective of such legislation is to minimize pollution from various industrial

Food Processing: Principles and Applications, Second Edition. Edited by Stephanie Clark, Stephanie Jung, and Buddhi Lamsal.
© 2014 John Wiley & Sons, Ltd. Published 2014 by John Wiley & Sons, Ltd.

sources throughout the European Union, including the food and beverage sector. This is achieved through the provision of a "license to operate" to premises that meet certain conditions. Permits must take into account the whole environmental performance of the plant, including aquatic, air and solid waste generation, use of raw materials, energy efficiency, noise, prevention of accidents, and restoration of the site upon closure (European Commission, 2010). Also notable is that emission limit values in the permit must be based on the use of the "best available techniques" (EU Joint Research Centre, 2013); environmental inspections of the facility must be carried out and the public has the right to participate in the decision-making process. Hence, it can be ensured that food processing facilities, through design or upgrade, meet international environmental standards. Therefore, new facilities must be designed at the earliest stage to meet these environment requirements, while older plants may need to be retrofitted with updated technologies.

In the EU, member states are legally bound to optimize the treatment of biowaste. The "waste hierarchy" states that the prevention of waste is the best option, followed by reuse, recycling, and energy recovery. Disposal method, such as landfilling, is defined as the worst environmental option. This has resulted in legislation requiring member states to reduce the amount of biowaste they send to landfill (European Commission, 1999). Therefore, options for waste disposal by food processors are limited, particularly in countries such as Ireland, which were previously heavily reliant on landfill. This has played a role in the rising interest in alternative use and disposal options for waste streams in the food processing sector. This will be further discussed in section 9.3.2.

It should also be noted that there is a move towards linking legislative drivers with economic drivers to improve overall sustainability. For example, the EU Emission Trading System (ETS) is mandatory for food and beverage companies operating combustion installations above a capacity of 20 MW. Examples of facilities which have fallen under phase III of the ETS include those involved in snack food production, distilling and brewing, butter and cheese production, and coffee processing (UK Department of Energy and Climate Change, 2012). The ETS is a "cap and trade" scheme, which has goals of establishing an allowance-trading scheme for emissions, and promoting reductions of greenhouse gases, in particular carbon dioxide. Each EU country is required to submit National Implementation Measures (NIMs) for approval by the European Commission. NIMs determine the levels of free allocation of allowances for installations during the current phase of ETS. These allocations are based on EU community-wide harmonized rules (European Commission, 2011). Through the growth of a carbon market, companies can buy allowances to meet their compliance requirements or sell their surplus allowances. Companies buy allocations through auctions. Two auction platforms are already in place. The European Energy Exchange (EEX) in Leipzig is the common platform for the large majority of countries participating in the EU ETS. The EEX also acts as Germany's auction platform. The second auction platform is ICE Futures Europe (ICE) in London, which acts as the United Kingdom's platform (European Commission, 2013).

9.2.2 Economic drivers

Economic drivers of sustainable food processing can take a number of forms. It has been projected that world marketed energy consumption will grow by 53% between 2008 and 2035 (EIA, 2011). This, coupled with an expectation that energy prices will continue to increase in the long term (EIA, 2011), has become a driver for companies, including food processors, to reduce energy consumption. Selected unit operations in food processing facilities are particularly energy intensive, for example, drying and evaporation. It has been suggested that energy can account for up to 10% of the total production costs for products requiring these unit operations (Marechal & Muller, 2008). While this figure is low in comparison with other industries (oil refining ~60% of total production cost), rising energy costs have forced energy saving and energy management programs to the forefront of many food processing businesses.

Between 30% and 50% of incoming raw materials can end up as waste material during food processing (Henningsson et al., 2004; Schaub & Leonard, 1996). In the UK, food manufacturing generates approximately 2.5 billion kg of food waste annually (Hall & Howe, 2012); in the US, the EPA estimates that it costs approximately $1 billion to dispose of food waste (Kosseva & Steve, 2009). In the EU, landfilling of biowaste is a major concern, with approximately 40% of it ending up in landfill. As mentioned in section 9.2.1, there is a legislative requirement for EU member states to reduce the use of landfill as a waste disposal option. To encourage this, governments have increased the levy on landfill to provide an economic incentive for companies to move to more sustainable methods of waste disposal. Historically, some food waste streams would have been used as animal feed supplements, e.g. whey as pig feed. However, with the recognition of the added value of whey components, the trend is now to utilize it in a much more sustainable manner (see section 9.3.2). Therefore, as a result of the pressure to reduce waste and utilize it more sustainably, there has been an increased focus on using food processing biowaste streams to develop waste-to-energy systems and value-added products, as well as on finding new, more sustainable, waste disposal options. These developments are discussed in more detail in section 9.3.2.

9.2.3 Consumer drivers

In a market economy consumer power and demand can drive market developments and therefore consumer choice and preference could be a driver of sustainability in the food processing industry. A number of studies have found that consumers will differentiate products on the basis of sustainability (Jaffry et al., 2004; Kearney, 2010). However, consumer choice is complex and will be influenced by a multitude of factors including cost, convenience, and habit (Vermeir & Verbeke, 2006). It is estimated that approximately 10% of the population are "socially conscious purchasers" (Sahota et al., 2009). Product claims can range from the use of recyclable material and product life cycle sustainability, to overall company sustainability commitments. The role of certified products, i.e. products verified by a third party to meet a set of standards, in promoting sustainable food processing through consumer-driven demand is growing. Certification can have a regulatory role, such as certification required by the EU or US Food and Drug Administration (FDA), or can occur through voluntary schemes, e.g. Fairtrade or Rainforest Alliance. Schemes such as Fairtrade and Rainforest Alliance focus on the provision of sustainable livelihoods and/or protecting ecosystems and biodiversity in places where products or raw materials originate, while other schemes, e.g. Mobius Loop, indicate that a product or part thereof can be recycled where facilities are available. For voluntary certification schemes to provide maximum benefit to food processors, claims need to be credible, easily understood, and trusted by consumers.

A number of studies have investigated consumer purchasing choices of "green labeled" products. Jaffry et al. (2004) investigated consumer choices for quality and sustainability-labeled seafood products in the UK. They carried out 600 in-home interviews and found that certification denoting sustainability or quality had a significantly positive influence on product choice, with sustainability appearing to have the greatest positive influence on the probability of choice. However, they also found negative attitudes held towards non-governmental certifiers, over governmental certifiers, highlighting the importance of consumer trust in certification schemes. While such studies indicate the potential value to food processors of adopting green labels, it is also critical to ensure that consumers recognize, understand, and value on-package information on production standards (Hoogland et al., 2007).

Research has shown that when food products (chicken, milk, salmon) are labeled with either (a) a certified organic logo combined with details about the animal welfare standards (e.g. outdoor access, painless stunning/killing, organic fodder, etc.), (b) just the certified organic logo, or (c) a statement in which the product was attributed to the world market (e.g. conforms to legal production standards), many consumers failed to realize that the organic logo presented alone (option b) covered all three categories (Hoogland et al., 2007). In fact, they were inclined to underestimate the distinctive advantage of the logo and were more positive towards products that presented both the logo and details of animal welfare.

9.2.4 Corporate performance

Food processing companies are, in general, attempting to address environmental issues. Many companies are now including sustainability as a corporate performance measure (Pirog et al., 2009). There are three "dimensions" of sustainability:

- environmental sustainability, incorporating the management of the effects of human activity so that they do not permanently harm the natural environment
- economic sustainability, which involves managing the financial transactions associated with human activities so that they can be sustained over the long term without incurring unacceptable human hardship
- social/cultural sustainability, i.e. allowing human activity to proceed in such a way that social relationships between people and the many different cultures around the world are not adversely affected or irreversibly degraded.

In order for a business to be truly sustainable, it must succeed in all three categories. This has become known as the "triple bottom line."

Environmental sustainable corporate performance (SCP) is defined as "Good housekeeping through prevention of pollution and waste and efficient use of scarce resources" (Gerbens-Leenes et al., 2003). The challenge for companies can often be how to accurately measure their performance in each of these three criteria. Often companies focus on local-level environmental impacts associated with their operation using a large number of indicators (Gerbens-Leenes et al., 2003). Hence, assessment of sustainability under all three pillars at not only a local but a regional and global scale may be omitted. Overall, the rise in demand for high-quality food, coupled with the fact that the environmentally conscious consumer of the future will consider ecological and ethical criteria in selecting food products, will, in conjunction with legislative, economic, and corporate drivers, play a role in the development of truly sustainable food processing (Roy et al., 2009).

9.3 Environmental impact of food processing

The role of food processing is to convert raw materials (fruit, vegetables, milk, cereals, etc.) into food products fit for human consumption. In addition to raw materials, the major inputs into the system are water and energy. Therefore, the main environmental impacts of the food processing sector are aquatic, atmospheric and solid waste generation, which are influenced by the quantity of resources utilized, waste produced, and transport used in the processing system. Successful management of water and energy resources in food processing is key to cost-effective sustainability practices. Environmental sustainability can be achieved by developing and implementing alternative environmental best-practice technologies and products that maximize the efficient use of resources and achieve cost savings, while minimizing negative human and environmental impacts (Clark, 2010).

9.3.1 Energy

The food industry is a major source of atmospheric emissions, mainly caused by extensive energy use. The food industry consumes a large quantity of energy for heating buildings, in processing and providing processed water, and for refrigeration and the transportation of raw materials and products (Pap et al., 2004). Such energy use contributes to greenhouse gas (GHG) emissions, especially carbon dioxide emissions (Roy et al., 2009). Although GHG emissions are a particular challenge for the food production sector, which relies heavily on transport and fertilizer use, energy reduction and management in food processing have also become increasingly important, as discussed in section 9.2.2, as a result of economic and legislative pressure. Over recent years, the food industry has

made progress in reducing its energy consumption through process optimization and control, energy recovery and recycling systems, and good manufacturing practices. This trend is likely to continue due to the enforcement of legislation in carbon trading systems, e.g. EU ETS (Sellahewa & Martindale, 2010).

Development of energy-efficient equipment designs is, today, standard in the food processing industry. Regeneration is employed in processes such as milk pasteurization to accomplish considerable energy savings. Raw milk entering a plate heat exchange is preheated to ~55 °C by hot pasteurized milk on the other side of the plates; the raw milk then moves from the regeneration section to the next section of the plate heat exchanger, where hot water is used to heat the milk to the required processing temperature of 72 °C. The use of regeneration reduces the amount of hot water required to heat the milk to the pasteurization temperature, thereby reducing the steam/energy requirement of the process. By utilizing regeneration, pasteurization is 95% efficient at recovering heat. Multieffect designs are also standard for processes such as evaporation. Multiple-effect evaporation is evaporation in multiple stages, whereby the vapors generated in one stage serve as a heating source to the next stage. In a multieffect evaporator, for each kg of water evaporated, the quantity of steam consumed is inversely proportional to the number of effects and the quantity of cooling water utilized in the condenser is inversely proportional to the number of effects (Berk, 2009).

Processes such as pasteurization are clearly highly efficient in term of heat recovery. However, in the UK it is still estimated that approximately 2.9 TWh of recoverable heat is wasted annually by the food and beverage processing sector (Law et al., 2013). The successful capture and utilization of such waste heat have the potential to play a critical role in improving the overall energy efficiency of this sector. It is estimated that the recovery of this waste heat could potentially result in cost savings of £70 m and emissions savings of 514,080 tCO_2eq for the UK food and beverage processing industry annually (Law et al., 2013). Meeting such targets will require a number of challenges to be overcome. It is often difficult to exactly match the waste heat with suitable heat sinks, thereby reducing the percent of heat recovered. It is also very difficult to recover waste heat in the range of ambient to 60 °C. Besides the pasteurization and evaporation processes mentioned above, other sources of waste heat include ovens, kilns, dryers, refrigerators, boilers, power plants, and air compressors. Uses for waste heat can be categorized as follows.

- Use within its originating process (e.g. regeneration).
- Use within another process.
- Use of cooling via absorption refrigeration cycle.
- Use for space and water heating (Reay, 2008).

Regardless of its end use, the waste heat source must meet food processing hygiene requirements for the process in which it will be used. Examples of the use of waste heat in other process include:

- recovery of waste heat from two 50 kW compressors, which is used to preheat boiler feed water, resulting in a 3% reduction in boiler energy demand at a biscuit factory (Reay, 2008)
- recovery of waste heat from the fryer section and exhaust stream of a potato crisps manufacturing plant is used to drive an organic Rankine cycle system for power generation and could have the potential to meet the average power requirement as well as 98.58% of the peak power requirement of the manufacturing process (Aneke et al., 2012).

Conventional refrigeration has a high electricity demand. However, the use of waste heat to drive alternative refrigeration cycles can reduce this. In the absorption refrigeration cycle, waste heat can be used to drive a generator where ammonia is evaporated off. The resulting vapor contains some water, so a rectifier and dephlegmator are included to reduce this. This vapor flow condenses into liquid in the condenser, rejecting the heat of condensation. Condensed liquid passes through a valve and low-pressure liquid enters an evaporator to evaporate the heat required for evaporation. The resulting vapor then enters the absorption cycle and is ultimately returned to the generator. While this cycle has the benefit of using waste heat, its coefficient of performance is lower than a conventional mechanical vapor compression cycle. Therefore the absorption cycle can be combined with other refrigeration cycles, such as the ejector refrigeration cycle, to improve the overall coefficient of performance.

In order to improve the sustainability of the food industry significantly, technological advancements in energy efficiency will need to be coupled with increased use of renewable energy such as solar, wind, and bioenergy. Companies, for example Dole (2008) and Kettle Foods (2009), have started to incorporate renewable energy sources, such as biodiesel, wind, and solar energy, to improve their sustainability. Such choices need to be combined with increased efficiency, as highlighted by a recent study of the Australian prune-drying industry, which demonstrated that up to 60% of energy could be saved by optimization and control of the process and utilizing solar energy (Sellahewa & Martindale, 2010). It is also possible to convert food waste stream to energy to reduce energy requirements and cost; this will be discussed in section 9.3.2.

9.3.2 Solid waste

The food industry is a major generator of both solid waste (food, packaging, etc.) and liquid effluents (see section 9.3.3) throughout the process chain. Traditionally, food wastes go to landfill, but this is not now deemed appropriate as it poses a serious environmental concern (see section 9.2.2). When food is wasted, it also contributes to GHG emissions and water usage, as energy and water are used in growing the raw materials, processing the product and in storage and distribution. In a recent life cycle assessment carried out with fresh Australian mangos, it was shown that waste contributed to 53% of the overall GHG emissions during production, distribution, and consumption phases (Ridoutt et al., 2010a). Therefore food waste must be tackled on multiple levels, i.e. processor, distributor/retailer, and consumer.

The major contributors to solid waste from food processing are fruit processing, cocoa/chocolate/confectionery, brewing/distilling, and meat processing (My Dieu, 2009). For example, poultry processing produces 35 kg of biowaste per 1000 birds processed, while distilleries produce 300 kg of biowaste per tonne of final product (My Dieu, 2009). Clearly the composition of the biowaste stream is dependent on the process that generates it, and its composition will, in turn, determine how the waste must be processed. An effective waste management system will aim to reduce the volume of waste produced and recover value components within the waste, followed by treatment and discharge. The minimization of waste production can be achieved through good manufacturing practices, adequate process control, and appropriate equipment maintenance and design. If any of these criteria fail, there can be a decrease in process efficiency and an increase in product losses during production.

Approaches that extend shelf life are also important in reducing overall food waste. Processes such as food irradiation can extend shelf life of products such as potatoes (inhibits sprouting) and strawberries (inhibits growth of spoilage microorganisms). However, irradiation has not received widespread usage. Consumer acceptance of food irradiation varies from country to country, which has also hindered its adoption. It should also be noted that significant efforts have been made by many food processors to reduce food packaging. For example, in the UK, a number of retailers and processors have signed up to the

Waste and Resources Action Program (WRAP) and the Courtauld Commitment 2 (www.wrap.org.uk), whereby they aim, among other things, to reduce packaging by 10%.

For a long time, waste streams were just viewed as by-products of food processing, but today they are often considered as valuable additional revenue streams. Therefore, waste streams are often further processed in order to remove all valuable components prior to discharge. These components are upgraded into value-added products. A good example of this is whey from cheese making. Historically, whey was fed to animals but modern cheese making now encompasses whey processing and extraction of value-added products. In general, fat is separated from the waste stream and can be used for butter manufacture; the reduced fat whey is then subjected to reverse osmosis and membrane filtration to separate out the lactose and whey protein. The resulting water can then be reused within the cheese-making facility. Another example of waste upgrading is presented by Laufenberg et al. (2003), who examined the upgrading of vegetable residues for the production of novel types of products. Plessas et al. (2008) investigated the upgrading of waste orange pulp. They used the pulp for cell growth of kefir and found that bread produced by immobilized kefir on orange pulp had an improved aromatic profile in comparison with bread produced by baker's yeast, and preliminary sensory evaluation of the produced bread was acceptable (Plessas et al., 2008).

Food waste streams can also be used as substrates to generate energy, thus improving sustainability (Sellahewa & Martindale, 2010). These waste-to-energy systems have received much attention in recent years (Banks et al., 2011; Caton et al., 2010; Curry & Pillay, 2012; El-Mashad & Zhang, 2010; Hall & Howe, 2012; Ike et al., 2010; Lai et al., 2009). Options for the conversion of waste to energy include anaerobic digestion (AD), direct combustion, and gasification. The selection of a waste-to-energy conversion process will be dependent on waste composition; in general, waste with a moisture content over 50% will be suitable for biological conversion such as AD, while lower moisture content waste will be more suited to thermochemical conversion (e.g. combustion). In AD, microorganisms under anaerobic conditions break down organic material and produce a biogas that is composed of mainly carbon dioxide and methane. Methane is a high-value fuel, producing 12,000 kcal kg^{-1} with a clean burn (My Dieu, 2009). The four stages of AD are hydrolysis, acidogenesis, acetogenesis, and methanogenesis. In hydrolysis, carbohydrates, lipids, and proteins are broken down into simple sugars, fatty acids, and amino acids, which are subsequently broken down into carbonic acids, volatile fatty acids, and alcohols during acidogenesis. The by-products are hydrogen, carbon dioxide, and ammonia. These products are converted into acetic acid during acetogenesis, with the release of carbon dioxide and water. In the final stage (methanogenesis), acetic acid and hydrogen are converted into methane (50–75%) and carbon dioxide (25–50%). This methane can then be upgraded to a suitable mainline standard.

In gasification, the organic material is exposed to high temperature under low oxygen conditions. This produces a syngas (carbon monoxide, carbon dioxide ,and hydrogen) which, like biogas, can be used as a fuel. Waste oil is another by-product of food processing that can be converted to energy. Kettle Foods, for example, converts 100% of its waste vegetable oil from its production process into biodiesel.

9.3.3 Water and waste water

Ensuring the sustainability of water systems is currently a major concern. The food industry is a large consumer of water across the food production chain, although according to Drastig et al. (2010), agricultural production has accounted for about 90% of global fresh-water consumption during the past century. It is also expected that water consumption for food production will need to increase to meet demands of a 50% larger global population. Milman and Short (2008) stated that 20% of the world's population currently lacks access to clean water and existing levels of access may deteriorate due to worsening urban water infrastructure, climate change, and environmental degradation.

Traditionally water was seen as cheap and plentiful in many parts of the world. This, along with legislation against the use of recycled water in processed foods (Sellahewa & Martindale, 2010), had reduced focus on water use reduction. However, this has recently changed. Increasing water costs and scarcity of supply, in addition to changes in legislation, have lead to increased efforts to improve water use efficiency in the food chain. These efforts will also have the impact of reducing the volume of waste water to be disposed of. Primarily, the focus of developments has been on maximizing water reuse and recycling (Kim & Smith, 2008). Water is key to food processing, as it is used for everything from washing raw material to equipment cleaning programs, as a heat transfer medium, and as a raw material itself. Equipment and process design can include the goal of reduced water requirements. In terms of water system design, three main design options can be considered, as defined by Kim and Smith (2008).

- **Water reuse**: water can be reused between operations. This depends on the water being of sufficient quality. Reused water can be mixed with fresh water if required.
- **Regeneration recycling**: the use of water reclaimed from waste-water treatment, and reused in the same operations.
- **Regeneration reuse**: the use of water reclaimed from waste-water treatment, and reused in a different operation to the one from which it was generated.

Waste-water treatment involves a number of steps. Preliminary treatment involves the removal of large suspended particles through screening and settling. This is followed by primary treatment wherein additional suspended material and material with a high biological oxygen demand (BOD) are removed through the use of coagulation, flocculation, clarification, and aeration. In the secondary treatment stage, systems such as activated sludge utilize microorganisms to reduce the biological and chemical oxygen demand (COD) of the waste along with a reduction in the particulate matter, nutrient, and odor of the waste. This water is non-drinkable but could be used to wash floors. Finally, in the tertiary treatment stage, operations such as membrane separation, pH correction, and ion exchange are used to produce water of potable quality. This reclaimed water could be used as process water but appropriate quality controls need to be in place, including a Hazard Analysis and Critical Control Point (HACCP) plan specifically dealing with the water recycling process. While this process could significantly reduce the volume of water food processing uses, it can be difficult for processors to adopt this approach. There is a significant capital investment required for the implementation of waste-water treatment to drinkable standards and while some companies may already be using technologies employed in tertiary waste-water treatment for other applications, for many there may be a significant technical challenge to be overcome. Finally, consumer acceptance of the use of recycled water as potable water is low and will require a change in consumer perception of the process (Casani et al., 2005). This may be achieved though linking ongoing campaigns aimed at increasing public awareness of the true value of water as well as the environmental impacts related to high consumption of water with the potential for reclaimed water to alleviate some of the pressure on our water resources without compromising public health or food quality.

Combining waste minimization efforts across the food processing chain with environmentally friendly preservation of the food product, the generation of waste can be minimized and the environmental sustainability of the process can be boosted (Pap et al., 2004). In addition, recovery and reuse of by-products are an option to further reduce the final quantity of waste requiring disposal.

9.4 Green technologies: examples in the food processing industry

Green technologies are defined as those that aim to satisfy consumer demands for high-quality products, while optimizing the production process in order to have the least impact on the environment. As discussed in section 9.3, optimization includes the reduced utilization of raw materials, energy, and water, while reducing the generation of process waste and effluent that may contain harmful organic solvents. Green technology also involves reducing the number of processing steps in industrial manufacturing to obtain the same products in fewer processing steps with less energy and waste materials. Life cycle assessment (LCA) is a method for evaluating the environmental impact of a product over its entire life cycle, from raw materials acquisition through processing, to the point of final consumption. Environmental impacts of green technologies are evaluated by LCA by considering land use, raw material consumption, atmospheric emissions, and water-borne pollutants. LCA is discussed in section 9.5.4. Green technologies can be considered as those that use less energy, utilize renewable energy sources and produce fewer waste products with more environmentally friendly disposal options. Processes that utilize "green chemistry," defined by Manley et al. (2008) as the design, development, and implementation of chemical products and processes to reduce or eliminate the use and generation of substances hazardous to human health and the environment, have a role to play in improving the sustainability of food processing.

9.4.1 Separation and extraction technologies

Supercritical fluid extraction is a process whereby a gas or liquid, held at or above its critical temperature and critical pressure, is used as an extracting solvent. Carbon dioxide is commonly used in supercritical fluid extraction as it is inert, non-corrosive, non-flammable, odorless, and tasteless. The advantages of replacing conventional organic solvents with supercritical CO_2 are: a reduction in environmental impact; higher purity extracts can be achieved; reduction in the number of processing steps and the process time; and finally an absence of toxic solvent residues (Shi et al., 2012). Supercritical CO_2 fluid extraction has been used to extract caffeine from coffee (Tello et al., 2011). It was possible to extract 84% of the caffeine using 197 kg CO_2/kg coffee husks. The caffeine is not pure but can be washed in water to achieve 94% purity. Numerous studies have also used supercritical CO_2 fluid extraction to extract flavor compounds (Barton et al., 1992; de Haan et al., 1990; Doneanu & Anitescu, 1998; Khajeh et al., 2004; Oliveira et al., 2009; Yonei et al., 1995). Flavors included those from milk fat, spearmint, peach, and cherry.

Subcritical (also known as superheated) water extraction is another green separation technology. In this case, water under pressure and between 100 °C and 374 °C (critical point) is used as the extractant. The advantage of water is that when heated under pressure, its properties vary significantly. For example, as the temperature rises up to 200 °C, the polarity of the water decreases and its dielectric constant falls. In addition to having the same advantages as supercritical CO_2 extraction, subcritical water extraction has the advantage of the simplicity of using water as a solvent. It has been used to extract essential oils and bioactive compounds such as anthocyanins and total phenolics (Cheigh et al., 2012; Eikani et al., 2007; Hassas-Roudsari et al., 2009; He et al., 2012; Ko et al., 2011; Ozel et al., 2003; Singh & Saldana, 2011).

Mustafa and Turner (2011) also described a pressurized liquid extraction (PLE) process as a green technology, which uses water as a solvent. They described its application for extraction of bioactive compounds and nutraceuticals from plants and herbs whereby it can not only decrease solvent consumption but also improve the yield of the extracted product. PLE is a technique that involves extraction using liquid solvent at elevated temperature and pressure, which enhances the extraction performance compared to those techniques carried out at near room temperature and atmospheric pressure. Extraction performance is improved as PLE enables solvents to be used at temperatures above their atmospheric boiling point, therefore enhancing their solubility and mass transfer properties.

9.4.2 Non-thermal processing

Traditionally, microbial control was achieved in food processing by dehydration, thermal processing or refrigeration. However, novel and emerging technologies can be used for this application without some of the disadvantages of conventional methods, such as high energy requirements. These novel and emerging technologies are usually employed as part of a hurdle technology approach, i.e. in combination with each other or other control methods. Novel processing technologies such as ultrasound, high hydrostatic pressure (HHP), ohmic heating, and pulsed electric field (PEF) processing could all be considered as green technologies if they reduce the energy requirement of a process or replace chemicals used during processing. Ultrasound processing could be used to reduce the time required for processes such as freezing and drying, thereby reducing energy usage. Ohmic heating is a rapid heating technology that should reduce energy usage over conventional thermal processing. Jung et al. (2011) described HHP processing as a green technology in comparison with conventional thermal treatment due to its lower energy requirement and recycling of the pressurization fluid.

Technologies that optimize food processing to limit the unnecessary use of energy could also be considered green technologies. For example, the syneresis control technology described by Fagan et al. (2008) was developed with the goal of optimizing the length of the syneresis process to ensure a consistent high-quality final cheese product. However, it could also have the added benefit of reducing or eliminating the unnecessary heating and stirring of the cheese vats, thereby reducing energy consumption. Regardless of the green technology to be employed, it is vital that an LCA of the system is undertaken to ensure that the correct technology is selected for integration into the process.

9.5 Environmental sustainability assessment methods

This section outlines three different methodological approaches commonly used in estimating the environmental impacts associated with the production of food products based on life cycle thinking. The methods consist of carbon footprint (CF), ecological footprint (EF), and life cycle assessment (LCA). A basic explanation of carbon footprint and ecological footprint is given, along with a brief mention of methodological issues and applications. Life cycle assessment is discussed in more detail later in the chapter. A table with a clear breakdown of methodological issues associated with each of the assessment methods discussed here can be found in the article of Jungbluth et al. (2012).

9.5.1 Carbon footprint (CF)

The process of "carbon footprinting" has become increasingly popular over recent years, as climate change has become a major issue in the public conscience (Weidmann & Minx, 2008). Although there are many definitions of the term "carbon footprint," it is generally referred to as "the total amount of greenhouse gases emitted during a product's production, processing, retailing and consumption" (Plassmann & Edwards-Jones, 2009). CF accounting involves identifying and quantifying emissions data for the product over its entire life cycle. It is important that this is carried out in a methodologically consistent manner in order to ensure comparability of results from different studies. The UK carbon footprint standard was developed for the British Standards Institute in response to this requirement, which outlines specifications for assessing the life cycle GHG emissions of goods and services (jointly referred to as "products") (British Standards Institute, 2011). In addition to this, the International Organization for Standardization is currently developing a draft governing carbon footprinting. The standard, titled "ISO 14067 Carbon footprint of products. Requirements and guidelines for quantification and communication," has been released in a draft format (ISO, 2012). The strength of carbon footprinting is the overall simplicity and ease in calculation of results, easily expressed in a single carbon dioxide emissions value (Weidema et al., 2008). Additionally, these results can be readily identified with and placed in context for the public, thereby influencing consumer choices and decision making, as it is easy to compare between products (Samuel-Fitwi et al., 2012).

Despite the advantages of using carbon footprinting to calculate environmental product information, relying entirely on one indicator can be misleading, as it may result in oversimplification of results. Carbon dioxide emissions contribute to only one of several impact categories and cannot accurately reflect overall environmental system performance. For example, carbon dioxide emissions are mainly related to fuel-based energy use, and food production systems with no energy use are selected as environmentally preferred production systems; non-energy emissions potentially impacting the environment are completely ignored such as land use change and waste management (Samuel-Fitwi et al., 2012). This shows that a carbon footprint might be insufficient for full environmental information and thus the use of a multi-impact assessment tool such as the LCA is recommended for this purpose (Jungbluth et al., 2012).

Examples of food-related carbon footprints include dairy production (Casey & Holden, 2005; Rotz et al., 2010) and conventional food crop cultivation (Hillier et al., 2009; Röös et al., 2010).

9.5.2 Ecological footprint

The ecological footprint concept, first presented by Rees (1992), has evolved into the world's primary measurement of humanity's demands on nature (Wackernagel & Rees, 1996); this parameter is now widely used as an indicator for measuring environmental sustainability (Čuček et al., 2012). The ecological footprint can be defined as "a measure of how much area of biologically productive land and water an individual, population or activity requires to produce all the resources it consumes and to absorb the waste it generates, using prevailing technology and resource management practices" (Global Footprint Network, 2012). The ecological footprint converts the human consumption of natural resources into a normalized measure of land area, referred to as global hectares (gha) (Samuel-Fitwi et al., 2012). Each gha represents the same fraction of the earth's total bioproductivity and is defined as "1 ha of land or water normalized to the world-averaged productivity from all of the biologically-productive land and water, within a given year" (Čuček et al., 2012). The sum of all the available biologically productive area on earth is roughly equal to approximately 120 million square kilometers (Galli et al., 2011).

The Ecological Footprint (EF) standard was developed by the Global Footprint Network and was designed to ensure that footprint assessments are produced consistently and according to community-proposed best practices. It specifies guidelines aimed at ensuring that EF assessments are conducted in an accurate and transparent manner, according to guidelines on issues such as use of source data, derivation of conversion factors, establishment of study boundaries, and communication of findings (Global Footprint Network, 2009). The main strength of the EF concept is that the results are expressed in an intuitive manner, which assists the public to understand complex environmental issues (Čuček et al., 2012). The results are also scientifically robust, making it a reliable instrument to make humanity's dependence on ecosystems clear (Cerutti et al., 2011).

The main drawback of the EF method is that, while it looks at the environmental aspect of sustainability, it does not measure all environmental parameters (Galli et al., 2011). There are also methodological issues related to the EF concept. EF fails to take into account geographic specificity, generalizing the ecosystem properties by assuming a standardized global ecosystem in terms of averaged productivity (Čuček et al., 2012; Samuel-Fitwi et al., 2012). As such, interactions unique to distinct ecosystems are neglected, which results in a simplified EF with equal environmental impacts, regardless of their origin. In this sense, EF fails to capture the variations inherent in different ecosystems, for example, marine, aquatic or forest ecosystems, each unique in their own way (Samuel-Fitwi et al., 2012). Furthermore, data availability is limited, with uncertainty associated with the available data. Converting the data to area units can also be problematic (Čuček et al., 2012). The EF method is applied over a range of scales from assessments of individual products, and from household to regional and country consumption. It is most effective at aggregate levels (Čuček et al., 2012). Examples of EFs related to food include applications in aquaculture (Kautsky et al., 1997), wine (Niccolucci et al., 2008), general food consumption in China (Chen et al., 2010), wheat production (Kissinger & Gottlieb, 2012), and intensive agriculture (Mózner et al., 2012).

The discussion of the CF and EF methodologies highlights the inadequacies of relying on one environmental indicator. The limited nature of CF and EF methodologies in relation to estimating the environmental sustainability of food processing systems shows that they may be insufficient for full environmental information. When it comes to assessing the environmental impacts of agricultural and food processing systems, many studies claim that methods that consider several environmental indicators are more suited to assessing the complexity of these systems (Cerutti et al., 2011). In addition, alternative production systems must be evaluated in a holistic manner in order to identify the possible shifting of impacts from one field/process to another. As such, the use of a multi-impact assessment tool such as the LCA may be more suited to address the complexities of food production systems (Jungbluth et al., 2012). Consequently, the LCA methodology, as applied to food production systems, is discussed in detail in this chapter.

9.5.3 Life cycle assessment

Life cycle assessment (LCA) is a tool created to assess the environmental impact (e.g. use of resources and the environmental consequence of releases) of a product, process, service or system by taking into account each step in its life cycle, from raw material acquisition through production, use, end-of-life treatment, recycling, and final disposal (ISO, 2006a). In this sense, LCA is a "cradle-to-grave" approach, as it begins with the gathering of raw materials from the earth to create the product and ends at the point when all materials are returned to the earth. While the "cradle-to-grave" approach represents a full LCA, partial LCAs can be completed using variants of this approach. Partial LCAs include "cradle-to-gate," from resource extraction to the factory gate, and "gate-to-gate," looking at particular processes in a supply chain. Figure 9.1 illustrates the life cycle stages that can be considered in an LCA (SIAC, 2006). Input and output data such as emissions, waste, energy consumption, and use of resources are also collected for each unit process (Berlin, 2003). Each stage in the life cycle of the product or process is evaluated from the perspective that they are interdependent, meaning that one operation leads to the next one. By including the impacts throughout the product life cycle, an LCA provides a complete view of the environmental aspects of the product or process and a more accurate picture of the true environmental trade-offs in product and process selection (SIAC, 2006). Up until now, LCA has mainly been used to evaluate the environmental impacts of products and industrial services but a growing number of studies incorporating LCA are focusing on the agri-food production sector.

9.5.3.1 LCA method

The LCA process is a systematic, iterative approach and consists of four components (Figure 9.2).
- **Goal and scope definition**: stating the aim, scope, system boundary, and purpose of the LCA study.
- **Inventory analysis**: compiling an inventory of relevant energy and material inputs, and outputs such as products, by-products, wastes, and environmental releases.
- **Impact assessment**: evaluating the potential environmental impacts associated with the identified inputs, outputs, and releases.
- **Interpretation**: interpreting the results to help decision makers make a more informed decision (SIAC, 2006).

9.5.3.2 Standards

The increasing number of LCA analyses carried out in a broad range of study areas prompted the drafting of international standards, ISO 14040 and ISO 14044, in order to

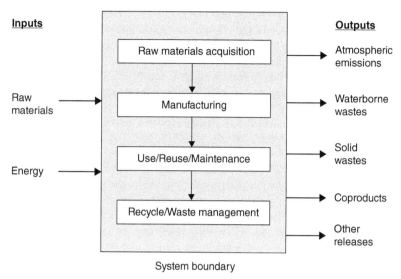

Figure 9.1 Life cycle stages (SAIC, 2006).

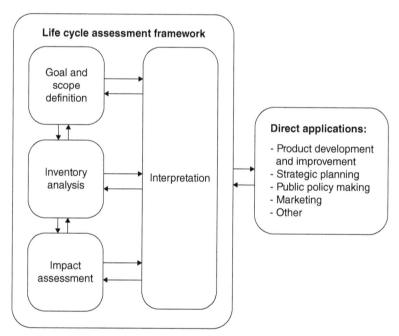

Figure 9.2 Steps of a life cycle assessment (ISO, 2006a).

harmonize the various methodologies that have been used to perform LCAs. The ISO 14040 standard describes the principles and framework for LCA, providing a basic explanation of the standard LCA process (ISO, 2006a). The ISO 14044 standard specifies requirements and provides guidelines for LCA (ISO, 2006b). ISO 14044 outlines the technical requirements, including guidelines on carrying out an LCA, goal and scope definition, inventory analysis, impact assessment, and interpretation. Guidelines are also given on reporting and critical review of the results of an LCA and guidelines for the framework described in ISO 14040. In Britain, a Publicly Available

Specification (PAS) was developed in response to broad community and industry desire for a consistent method for assessing the life cycle GHG emissions of goods and services – PAS 2050:2008 (British Standards Institute, 2008).

9.5.3.3 Goal and scope

The first step in an LCA is potentially the most important and requires the purpose and details of the LCA to be stated in accordance with ISO standards. In this phase, important decisions are made that determine the working plan of the entire LCA (Guinee et al., 2002). The LCA is carried out in line with statements laid out in this section, which defines the purpose of the study and how it is to be carried out. The goal of the study should be clearly stated. The following should be unambiguously stated in the goal: the intended application, the reasons for carrying out the study, the intended audience, whether the results are intended to be used in comparative assertions. The scope of the study should be outlined. The product, process or system is defined and described under the scope. The boundaries of the system are staked out and illustrated by a general input and output flow diagram. All operations that contribute to the life cycle of the product, process, or system are included within the system boundaries. It is essential to determine the functional unit (FU) to which emissions and extractions will then be assigned (Parent & Lavallée, 2011). The FU provides a reference unit to which the inputs and emissions in the inventory are normalized. The FU is often based on the mass of the product under study. However, in food LCA studies, nutritional and economic values of products and land area are also used (Roy et al., 2009). The environmental impacts associated with the FU to be evaluated are identified. Data requirements, assumptions, and limitations of the study should also be stated.

Life cycle assessment requires a number of "value choices" to be made in the goal and scope of the LCA. These choices affect the choosing of the system boundary, functional unit, allocation method, characterization method, etc. Some of these choices will be discussed in more detail further in the chapter.

9.5.3.4 Inventory analysis

The inventory analysis step is the most work-intensive and time-consuming step in an LCA. It involves compiling the required data for the data inventory: all the data relating to the product, process or system being studied, specifically input and output data (Figure 9.3). The qualitative and quantitative data for inclusion in the inventory are collected for each unit process that is included within the system boundary (Guinee et al., 2002). Compiling a data inventory involves identifying and quantifying all input and output flows from the system, including energy, water, and resource usage. Furthermore, releases to the environment from the system are included, such as air emissions (CO_2, CH_4, SO_2, NO_x and CO), water and soil discharges (total suspended solids, biological oxygen demand, chemical oxygen demand, and chlorinated organic compounds) and solid waste generation (municipal solid waste and landfills) (Roy et al., 2009). The collected data are used to quantify the inputs (raw materials, products from other processes, energy, etc.) and outputs (products, emissions, etc.) of unit processes in the LCA. The required data can be measured, calculated or estimated (for product-specific data) and can be obtained from a number of sources including private, government, and university resources (Parent & Lavallée, 2011). LCA databases, which are useful for processes that are not product specific, such as general data on the production of electricity, coal or packaging, are also available and can normally be purchased with LCA software. Data on transport, extraction of raw materials, processing of materials, production of normally used products such as plastic and cardboard, and disposal can normally be found in an LCA database.

The data collected should allow the following questions to be answered:.

• What quantity of energy is required to produce, distribute, and use the product?

• What substances are used during the life phases of the product?

• What are the derivative products, waste, and pollutants released into the environment (water, air, and earth)? (SIAC, 2006)?

9.5.3.5 Impact assessment

The third step involves assessment of the potential human and environmental effects resulting from the various input and output flows to the environment identified in the data inventory. This step relates the energy, water, and raw material usage, along with the environmental releases from the product, process or system, to human and environmental effects (SIAC, 2006). The life cycle impact assessment stage determines which stages and elements of the LCA generate the most impacts and how these impacts may be characterized.

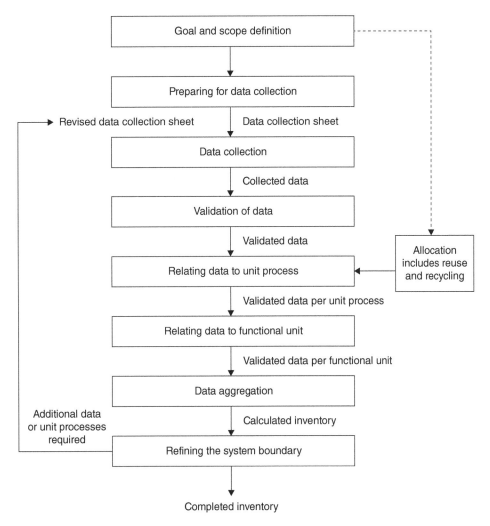

Figure 9.3 Simplified procedures for inventory analysis (ISO, 2006b).

The life cycle impact assessment stage involves the following steps:
• selection of impact categories, category indicators, and characterization models
• assignment of life cycle inventory result to selected impact categories (classification)
• calculation of category indicator results (characterization)

A list of impact categories to be evaluated is defined. There are several impact assessment methods available (detailed under "characterization methods"), each analyzing different impact categories. Common impact assessment categories include global warming, stratospheric ozone destruction, formation of photo-oxidizing (smog) agents, acidification, eutrophication, ecotoxicological impacts, toxicological impact on human beings, use of abiotic resources, use of biotic resources, uses of land. Many of these impacts relate directly to food quality and the concepts of sustainable food security and sustainable agriculture (Parent & Lavallée, 2011).

The results of the inventory analysis, in particular the inventory table, are further processed and assigned to different impact categories, based on the expected types of impacts on the environment. To this end, a list of impact categories is defined, and models for relating the environmental interventions to suitable category indicators for these impact categories are selected (Guinee et al., 2002). This step is known as classification. There are two types of category indicators: mid point and end point. These refer to the distance along the environmental mechanism where the impact is evaluated. According to Bare et al. (2000), mid points are considered to be a point in "the environmental mechanism of a particular impact category, prior to the endpoint, at which characterization factors can be calculated to reflect the relative importance of an emission or extraction in a Life Cycle Inventory." Mid-point indicators include measurements of global warming potential, acidification and eutrophication potential, human and ecotoxicity, etc. End-point indicators reflect issues of concern such as flooding, extinction of species or human lives lost, which occur at the end of the environmental mechanism as a result of global warming, acidification, etc. There is a higher degree of uncertainty related to end-point indicators than mid-point indicators, but end-point results are easier to interpret. Mid-point and end-point indicators are shown in Figure 9.4.

The actual impact assessment results are calculated in the characterization step (Guinee et al., 2002). Characterization involves the assessment of the magnitude of potential impacts of each inventory flow into its

corresponding environmental impact (e.g., modeling the potential impact of carbon dioxide and methane on global warming potential, or the impact of sulfur dioxide and ammonia on acidification potential). Characterization provides a way to directly compare the life cycle inventory (LCI) results within each category. Characterization factors are also known as equivalency factors.

There are a number of characterization methods available for both mid-point and end-point characterization. Mid-point methods include:
- CML 2001 (Guinee et al., 2002)
- TRACI 2.0 (Bare, 2011)
- EDIP (Hauschild & Potting, 2003).

End-point methods include:
- Eco-indicator (Goedkoop & Spriensma, 1999)
- Impact 2002+ (Jolliet et al., 2003)
- ReCiPe (Goedkoop et al., 2009).

Normalization of the results allows the expression of potential impacts in ways that can be compared (e.g. comparing the global warming impact of carbon dioxide and methane for the two options). Valuation is a subjective step in which assessment of the relative importance of environmental burdens identified in the classification, characterization, and normalization stages

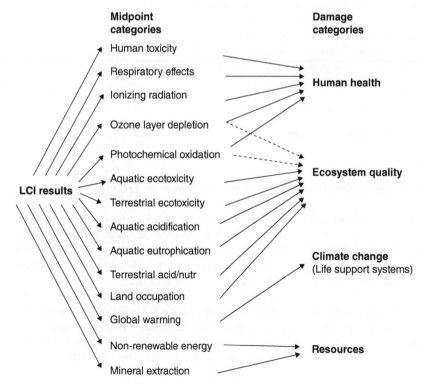

Figure 9.4 Mid-point and end-point indicators (Jolliet et al., 2003).

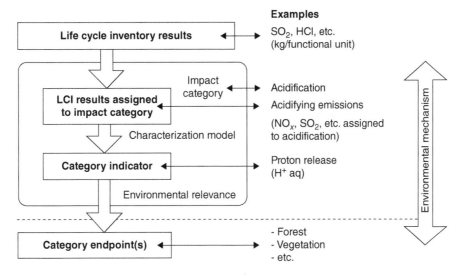

Figure 9.5 Concept of category indicators (ISO, 2006b).

is carried out by weighting them (decided by the stakeholders involved), which allows them to be compared or aggregated (Miettinen & Hämäläinen, 1997). Figure 9.5 outlines the concept of category indicators.

9.5.3.6 Interpretation

The last step in an LCA is the interpretation of the results of the previous three stages in order to draw conclusions that can support a decision or provide a readily understandable result of an LCA. In this step, the data can be analyzed, taking into account assumptions made in the LCA, to assess the quality of the results. During interpretation, the inventory and impact assessment results are discussed together in the case of an LCA, or the inventory only in the case of LCI analysis, and significant environmental issues are identified for conclusions and recommendations consistent with the goal and scope of the study (SIAC, 2006). Interpretation allows for exploration of the areas of the product or process life cycle that could be optimized to reduce environmental impacts and can help identify potential alternative solutions to reduce the environmental impact (Parent & Lavallée, 2011). This may include both quantitative and qualitative measures of improvement, such as changes in product, process and activity design, raw material use, industrial processing, consumer use, and waste management (Roy et al., 2009). Figure 9.6 presents the steps involved in the interpretation phase of an LCA.

9.5.3.7 Life cycle assessment of food products: challenges and applications

Choosing the functional unit (FU) of a food product can be difficult, as food fulfills several functions; it should provide energy and nutrients as well as taste satisfaction. A review by Peacock et al. (2011) found that the FU was not properly identified in many LCA studies, and there was no clear distinction between the FU and the reference flow. The mass of a product is frequently used in LCA studies, but this may not be appropriate for all food products, as it does not take into account the properties of the food. Nutritional values such as protein content and energy content are also commonly used along with economic values (Roy et al., 2009). Other FUs include volume, portions, and pieces of products (Peacock et al., 2011).

The FU is particularly important when comparing the environmental impact of similar food products. According to ISO 14044, in a comparative study, the equivalence of the systems being compared must be determined before the interpreting step in order that the systems can be compared. As a result, the systems must be compared using the same FU (ISO, 2006b).

When analyzing systems that produce multiple products (co-products and by-products), it is very important to allocate the environmental impacts to each of the products in an appropriate manner. Allocation is required in order to determine the quantity of the environmental impacts arising from the entire system, which can be

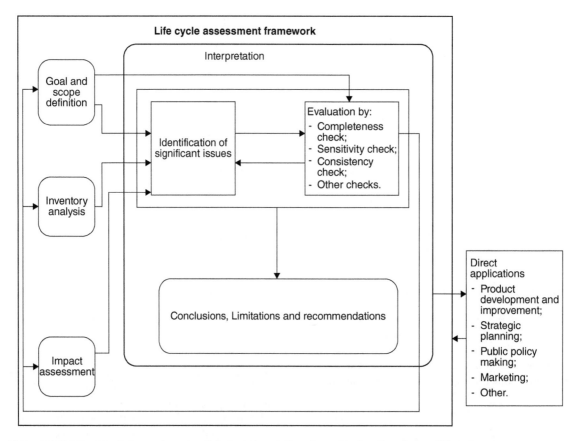

Figure 9.6 Relationships between elements within the interpretation phase with the other phases of life cycle assessment (ISO, 2006b).

apportioned to the individual product in question. Allocation is particularly relevant in food systems where there are often several products produced during a process.

The ISO 14044 standard recommends that allocation should be avoided where possible (ISO, 2006b). This can be achieved by dividing the unit process into subprocesses and collecting the relevant inventory data for the subprocesses, or by system expansion (avoided burden). System expansion involves the expansion of the system boundaries to include the additional functions of the co- or by-products. This can be achieved by identifying an alternative process for the production of the co- or by-product and including it in the expanded boundaries of the original system (Cederberg & Stadig, 2003). If the data inventory exists for this alternative process, it can be subtracted from the inventory of the multi-output process, therefore removing the co- or by-products impact on the final result.

If allocation cannot be avoided then it is necessary to divide the environmental impacts over the different products. The environmental burden can be divided between the products on a mass, energy, physicochemical or economic basis. ISO 14044 has recommended that allocation should reflect underlying physical relationships between inputs and outputs or, in the absence of such knowledge, allocation should reflect other relationships (e.g. economic value). Economic allocation is generally used for food systems, but studies are increasingly using mass and energy contents. Lundie et al. (2007) developed an alternative physicochemical allocation matrix for the dairy industry, which overcomes the issues of mass, process energy, or economic allocations associated with a multiproduct manufacturing plant. This allocation method gives a more realistic evaluation of environmental impact per product.

The use of LCA in the food industry is growing with increasing awareness of the importance of environmental sustainability. LCAs are being applied to a wide range of food products and food production systems. There are several studies that contain comprehensive reviews of LCA studies of food products (Andersson, 2000; de Vries & de Boer, 2010; Milani et al., 2011; Mogensen et al., 2010; Roy et al., 2009). Table 9.1 provides an overview of recent studies. It is clear from this table that LCA studies are being carried out on a wide range of products.

While LCAs have been performed by many food companies, the release of LCA reports to the public is limited due to confidentiality surrounding the production chain. Arla Food, a Danish company, has performed LCA studies on various aspects of its products: dairy products, cheese, and milk packaging (Arla Foods, 2011). Danone (a multinational food product company known as Dannon in the US) has also carried out LCA studies on its products, notably evaluating the energy consumed in producing Evian mineral water and yoghurt (Danone, 2006). More recently, Danone became the first company in Europe to switch to the more environmentally friendly PLA type of packaging for its Activia brand of yoghurt. The decision was based on an LCA study, which determined that switching to this packaging will reduce the packaging carbon footprint by 25% and will use 43% less fossil fuel (Mohan, 2011).

Unilever has implemented LCAs to assist in reducing the environmental impact of its products across its business. Unilever makes use of LCAs in product innovation by comparing new and existing products, in product category analysis and on strategic studies (Unilever, 2011). Examples of Unilever LCA studies include a comparative LCA of butter and margarine consumption (Nilsson et al., 2010) and an estimation of the greenhouse gas emissions associated with Knorr (Milà i Canals et al., 2011).

Life cycle assessments of a food product, process or system are wide and far-reaching. They can identify the processes and materials with the least environmental impact and the optimum combination of energy sources. Weak points (or hotspots) in a system leading to high environmental emissions can be identified using LCAs. As such, LCAs can assist in identifying opportunities to improve the environmental performance of products at various points in their life cycle. An LCA can help in identifying the most environmentally friendly packaging and transport options. It is a problem-solving tool that allows the evaluation of alternative products, processes or systems by comparing like with like. As such, LCA can evaluate the differences between conventional and organic agriculture, and fresh, cold, and preserved foods.

Life cycle assessment allows a system to be environmentally optimized in a holistic manner. This ensures system-wide optimization rather than subsystem optimization. For example, when selecting between two rival products, it may appear that product 1 is better for the environment because it generates fewer GHG emissions than product 2. However, after performing an LCA, it might be determined that the first product actually creates larger environmental impacts when measured across different impact categories and across the three media of air, water, and land. Additionally, product 1 may cause more chemical emissions during the manufacturing stage and as such product 2 may be viewed as producing less environmental harm or impact than the first technology because of its lower chemical emissions (SIAC, 2006). LCAs can be especially useful from a marketing point of view for a food processing company by assisting in implementing an ecolabeling scheme, supporting claims of environmental superiority, or producing an environmental product declaration.

Despite the advantages of using LCAs in assessing environmental sustainability, disadvantages associated with the method do exist. It can be difficult to define the system boundary and goal and scope in a consistent and meaningful manner (Tingström & Karlsson, 2006). When carrying out an LCA, it is unavoidable that assumptions about the system being studied are made, simplifying the system into a more manageable model than in reality. Such assumptions may dilute the accuracy of the results (Ritzén et al., 1996). A large quantity of data is required to carry out an LCA of a product. There are often issues with availability of data, along with quality and reliability of data (Ritzén et al., 1996; Tingström & Karlsson, 2006). Furthermore, aggregation of the inventory data into the standardized environmental impact categories can result in a loss of insight into the emissions from the process or system (Balkema et al., 2002).

The results obtained from the LCA analysis are neither spatially or temporally specific. If using an LCA in process or product development, this can be an issue, as information on temporal and spatial aspects is often required in decision-making processes by the stakeholders involved (Herrchen & Klein, 2000). Finally, LCAs may not be useful in product development, as carrying out a full LCA on a new product is costly, time consuming, and difficult (Ritzén et al., 1996; Tingström & Karlsson, 2006).

Table 9.1 Summary of recent Life Cycle Assessment studies on various food production systems

Ref	Location	Product	By/co-product(s)	Functional Unit	Allocation	GWP	En	OG	AP	EP	LU	H₂O	Tox	OD	POCP	Oth
Meat, fish and dairy products																
Beauchemin (2010)	Canada	Beef	None considered	1 kg of beef produced	Not considered	x										
Pelletier (2010)	USA	Beef	Crop co-products (for feed)	1 kg of beef produced	Gross chemical energy content	x	x		x	x	x					
Pelletier (2008)	Canada	Broiler production	Feed	1 live weight ton of broiler poultry	Gross chemical energy content	x	x		x	x						x
Flysjo (2011)	New Zealand and Sweden	Milk	Surplus calves, meat from culled dairy cows and manure	1 kg of energy corrected milk (ECM) (calculated according to Sjaunja et al. (1990) to correct for fat and protein) provided at the farm gate in NZ and SE, respectively, including all by-products	Economic	x										
Castanheira (2010)	Portugal	Milk	Meat from culled dairy cows	1 ton of raw milk ready to be delivered at the farm gate	Economic	x	x		x	x						x
van Middelaar (2011)	Netherlands	Cheese	Milk	1 kg cheese	Economic	x			x	x	x					
Vasquez-Rowe (2011)	Spain	Hake fillets	None considered	500 g of raw gutted fresh hake fillet reaching the household of an average consumer in the year 2008	Not considered	x			x	x						x
Hospido (2006)	Spain	Canned tuna	None considered	1 ton of raw frozen tuna entering the factory	Economic	x			x	x						x
Pelletier (2010)	USA and Netherlands	Indonesian tilapia	Fish trimmings	1 ton of frozen tilapia fillets from lake-pen cultured fish in Indonesia to market ports in Rotterdam and Chicago	Gross energy content	x			x	x						x
Crops, vegetables and fruit																
Meisterling (2009)	USA	Conventional v organic wheat	None considered	0.67 kg of whole-wheat flour	Not considered	x	x									
Blengini (2009)	Italy	Rice	Residuals	1 kg of refined rice packed and delivered to the supermarket	Economic	x	x		x	x	x				x	x
Beviacqua (2007)	Italy	Pasta	Subproducts	0.5 kg package of a popular brand of durum wheat pasta sold in the Italian market	Not considered	x	x		x	x			x			x
Liu (2010)	China	Chinese pears	None considered	1 ton of pears for direct human consumption, delivered to the point-of-sale	Not considered	x	x									

Reference	Country	Product	Products	Functional unit	Allocation								
Beccali (2010)	Italy	Citrus-based products		1 kg of each final product delivered by the manufacturer to distribution centres in Italy, Central Europe, the United States and Japan (oranges essential oil, oranges natural juice, oranges concentrated juice, lemons essential oil, lemons natural juice, lemons concentrated juice)	Combined mass and economic	x	x	x	x	x	x	x	x
Martinez-Blano (2011)	Spain	Tomatos	Compost	1 ton of commercial tomato	System expansion	x	x	x	x	x	x	x	x
Romero-Gamez (2012)	Spain	Green beans	None considered	1 kg of commercial green beans	Not considered	x	x	x	x	x		x	x
Karakaya (2011)	Turkey	Tomato products	None considered	Several products	Not considered	x	x	x					
Meals													
Davis (2010)	Spain and Sweden	4 protein rich meals (for animals)	Several	one meal served at the table in a household, in two different countries, Sweden and Spain	Economic	x	x	x	x	x	x		
Calderon (2010)	Spain	Ready meals	None considered	1 kg of finished product ready to be consumed	Not considered	x	x	x	x	x			x
Zufia (2008)	Spain	Industrial cooked dish	None considered	2 kg tray of pasteurized tuna with tomato	Not considered	x	x	x	x	x	x		x
Beverages													
Humbert (2009)	Switzerland	Coffee	None considered	provide a 1 dl cup of coffee ready to be drunk at the consumer's home	Not considered	x	x	x					
Pizzigallo (2008)	Italy	Wine	None considered	1 ton of final product	Not considered	x	x						
Gazulla (2010)	Spain	Wine	Pomace, lees, syrup	0.75 l of wine	Economic	x	x	x	x				
Cordella (2008)	Italy	Beer	Roots, spent grains and yeast excess	1 l of beer and the fraction of packaging allocated to such a litre (1/20 of a 20 L steel keg or three 33 cL glass bottles)	System expansion	x	x	x	x	x			x
Water Use													
Ridoutt (2010b)	Australia	Water footprint of skimmed milk powder	*	1 kg of skim milk powder in the temperate South Gippsland region of Victoria	Physiochemical and economic	x	x						

(Continued)

Table 9.1 (*Continued*)

Ref	Location	Product	By/co-product(s)	Functional Unit	Allocation	GWP	En	OG	AP	EP	LU	H$_2$O	Tox	OD	POCP	Oth
Ridoutt (2010a)	Australia	Water footprint of food waste (mango)	Fresh mango	consumption of 1 kg of fresh mango by an Australian household	Economic							x				
Others																
Nilsson (2010)	UK, Germany, France	Margarine and butter	Skim milk, waste for feed etc.	500 g of packaged butter/margarine intended for use as a spread, delivered to the manufacturer's distribution centre in each country	Economic, mass and others	x	x		x	x	x				x	
Avraamides (2008)	Cyprus	Olive oil	Waste and pomace	1 l of extra virgin olive oil extracted	Economic			x								x
Kim (2010)	Republic of Korea	Waste disposal	Animal feeds and composts from food wastes	by-products from dry feeding, wet feeding, composting, and landfilling of 1 ton of food wastes respectively	System expansion	x										
Pasqualino (2011)	Spain	Beverage packaging	None considered	packaging required to contain 1 l of beverage	Not considered	x	x									
Williams (2011)	Sweden	Packaging and food losses (ketchup, bread, milk, cheese, beef)	None considered	1 kg of purchased food	Not considered	x	x		x	x						

GWP – Global Warming Potential, En – Energy use, OG – Other Gases, AP – Acidification Potential, EP – Eutrophication Potential, LU – Land Use, H$_2$O – Water use, OD – Stratospheric ozone depletion, POCP – Photochemical ozone creation potential.

9.5.3.8 Life cycle assessment software and databases

The two main licensed software products for LCA are Simapro, developed by Pré Consultants, and GaBi, developed by PE International (GaBi, 2010). These software products are flexible and can be used to carry out LCAs on any product or systems, including food production and processing systems. The software packages differ based on their user interfaces, and methods of modeling and analyzing the life cycle. Both software packages come with a number of databases containing inventory data which can be used for background for the food system; examples of background data include electricity from the grid and mains water supply. Demo versions of these software products are available with limited access to databases.

There are a number of databases available containing life cycle inventory data for a range of products, processes, and systems. The unit processes from these databases are used in the LCA software in building the system being studied. These databases can provide general data for a wide range of systems but there are a few databases with data specifically concerning food production. The Danish LCA food database contains environmental data on processes in food product chains and on food products at different stages of their production (Nielsen et al., 2003). The Ecoinvent database developed by the Swiss Centre for Life-Cycle Inventories also contains European and Swiss data for agricultural production (Swiss Centre for Life-Cycle Inventories, 2007). The US LCI database developed by the National Renewable Energy Lab contains data for the US (NREL, 2008). Each of these databases is included with Simapro and GaBi when a license is purchased.

9.6 Conclusion

Food safety and quality will remain at top priority for food processors, especially due to increasing numbers of authenticity, adulteration, and safety issues in food products. However, environmental sustainability is becoming increasingly important to companies. The co-management of safety, quality, and sustainability issues will ensure that quality and safety are maintained while ensuring improved sustainability. One of the challenges in achieving this goal will be enabling companies in making informed decisions. Food processors should not only assess the impact of processing but also the wider supply and distribution chains, and end of life of the product or packaging. The use of various sustainability assessment methodologies should assist in achieving this goal.

References

Andersson K (2000) LCA of food products and production systems. *International Journal of Life Cycle Assessment* **5**(4): 239–248.

Aneke M, Agnew B, Underwood C, Wu H, Masheiti S (2012) Power generation from waste heat in a food processing application. *Applied Thermal Engineering* **36**, 171–180.

Arla Foods (2011) *Environment and Climate Log.* Leeds: Arla Foods.

Avraamides M, Fatta D (2008) Resource consumption and emissions from olive oil production: a life cycle inventory case study in Cyprus. *Journal of Cleaner Production* **16**(7): 809–821.

Baldwin CJ (2009) Introduction. In: *Sustainability in the Food Industry.* Ames, IA: Wiley-Blackwell, pp. xiii–xvi.

Balkema AJ, Preisig HA, Otterpohl R, Lambert FJD (2002) Indicators for the sustainability assessment of wastewater treatment systems. *Urban Water* **4**(2): 153–161.

Banks CJ, Salter AM, Heaven S, Riley K (2011) Energetic and environmental benefits of co-digestion of food waste and cattle slurry: a preliminary assessment. *Resources, Conservation and Recycling* **56**(1): 71–79.

Bare J (2011) TRACI 2.0: the tool for the reduction and assessment of chemical and other environmental impacts 2.0) *Clean Technologies and Environmental Policy* 1–10.

Bare J, Hofstetter P, Pennington D, de Haes H (2000) Midpoints versus endpoints: the sacrifices and benefits. *International Journal of Life Cycle Assessment* **5**(6): 319–326.

Barton P, Hughes Jr RE, Hussein MM (1992) Supercritical carbon dioxide extraction of peppermint and spearmint. *Journal of Supercritical Fluids* **5**(3): 157–162.

Beauchemin KA, Henry Janzen H, Little SM, McAllister TA, McGinn SM (2010) Life cycle assessment of greenhouse gas emissions from beef production in western Canada: a case study. *Agricultural Systems* **103**(6): 371–379.

Beccali M, Cellura M, Iudicello M, Mistretta M (2010) Life cycle assessment of Italian citrus-based products. *Sensitivity analysis and improvement scenarios. Journal of Environmental Management* **91**(7): 1415–1428.

Berk Z (2009) *Food Process Engineering and Technology.* Burlington, VA: Academic Press.

Berlin J (2003) Life cycle assessment (LCA): an introduction. In: Mattson B, Sonesson U (eds) *Environmentally-Friendly Food Processing.* Cambridge: Woodhead Publishing Limited.

Bevilacqua M, Braglia M, Carmignani G, Zammori FA (2007) Life cycle assessment of pasta production in Italy. *Journal of Food Quality* **30**(6): 932–952.

Blengini GA, Busto M (2009) The life cycle of rice: LCA of alternative agri-food chain management systems in Vercelli (Italy). *Journal of Environmental Management* **90**(3): 1512–1522.

Boye JI, Arcand Y (2012) *Green Technologies in Food Production and Processing*. New York: Springer.

British Standards Institute (2008) *PAS 2050:2008 – Specification for the Assessment of the Life Cycle Greenhouse Gas Emissions of Goods and Services*. London: British Standards Institute.

British Standards Institute (2011) *PAS 2050:2011 – Specification for the Assessment of the Life Cycle Greenhouse Gas Emissions of Goods and Services*. London: British Standards Institute.

Calderon LA, Iglesias L, Laca A, Herrero M, Diaz M (2010) The utility of Life Cycle Assessment in the ready meal food industry. *Resources, Conservation and Recycling* **54**(12): 1196–1207.

Casani S, Rouhany M, Knøchel S (2005) A discussion paper on challenges and limitations to water reuse and hygiene in the food industry. *Water Research* **39**(6): 1134–1146.

Casey JW, Holden NM (2005) Analysis of greenhouse gas emissions from the average Irish milk production system. *Agricultural Systems* **86**(1): 97–114.

Castanheira ÉG, Dias AC, Arroja L, Amaro R (2010) The environmental performance of milk production on a typical Portuguese dairy farm. *Agricultural Systems*, **103**(7): 498–507.

Caton PA, Carr MA, Kim SS, Beautyman MJ (2010) Energy recovery from waste food by combustion or gasification with the potential for regenerative dehydration: a case study. *Energy Conversion and Management* **51**(6): 1157–1169.

Cederberg C, Stadig M (2003) System expansion and allocation in life cycle assessment of milk and beef production. *International Journal of Life Cycle Assessment* **8**(6): 350–356.

Cerutti AK, Bruun S, Beccaro GL, Bounous G (2011) A review of studies applying environmental impact assessment methods on fruit production systems. *Journal of Environmental Management* **92**(10): 2277–2286.

Cheigh CI, Chung EY, Chung MS (2012) Enhanced extraction of flavanones hesperidin and narirutin from Citrus unshiu peel using subcritical water. *Journal of Food Engineering* **110**(3): 472–477.

Chen D, Gao W, Chen Y, Zhang Q (2010) Ecological footprint analysis of food consumption of rural residents in China in the latest 30 years. *Agriculture and Agricultural Science Procedia* **1**(0): 106–115.

Clark JH (2010) Introduction to green chemistry. In: Proctor A (ed) *Alternatives to Conventional Food Processing*. London: Royal Society of Chemistry.

Cooke P (2008) Legislation and economic issues regarding water and energy management in food processing. In: Klemeš J, Smith R, Kim JK (eds) *Handbook of Water and Energy Management in Food Processing*. Cambridge: Woodhead Publishing, pp. 3–28.

Cordella M, Tugnoli A, Spadoni G, Santarelli F, Zangrando T (2008) LCA of an Italian lager beer. *International Journal of Life Cycle Assessment* **13**(2): 133–139.

Čuček L, Klemeš JJ, Kravanja Z (2012) A review of footprint analysis tools for monitoring impacts on sustainability. *Journal of Cleaner Production* **34**, 9–20.

Curry N, Pillay P (2012) Biogas prediction and design of a food waste to energy system for the urban environment. *Renewable Energy* **41**(0): 200–209.

Danone (2006) *Sustainability Report*. Paris: Danone.

Davis J, Sonesson U, Baumgartner DU, Nemecek T (2010) Environmental impact of four meals with different protein sources: case studies in Spain and Sweden. *Food Research International* **43**(7): 1874–1884.

de Haan AB, de Graauw J, Schaap JE, Badings HT (1990) Extraction of flavors from milk fat with supercritical carbon dioxide. *Journal of Supercritical Fluids* **3**(1): 15–19.

de Vries M, de Boer IJM (2010) Comparing environmental impacts for livestock products: a review of life cycle assessments. *Livestock Science* **128**(1–3): 1–11.

Dole (2008) *Dole Announces Conversion of Farm Equipment to Bio-Diesel Fuel*. www.thefreelibrary.com/Dole+Announces+Conversion+of+Farm+Equipment+to+Bio-Diesel+Fuel.-a0173797625, accessed 13 November 2013.

Doneanu C, Anitescu G (1998) Supercritical carbon dioxide extraction of Angelica archangelica L. root oil. *Journal of Supercritical Fluids* **12**(1): 59–67.

Drastig K, Prochnow A, Kraatz S, Klauss H, Plöchl M (2010) Water footprint analysis for the assessment of milk production in Brandenburg (Germany). *Advances in Geosciences* **27**, 65–70.

Eikani MH, Golmohammad F, Rowshanzamir S (2007) Subcritical water extraction of essential oils from coriander seeds (Coriandrum sativum L.). *Journal of Food Engineering* **80**(2): 735–740.

El-Mashad HM, Zhang R (2010) Biogas production from co-digestion of dairy manure and food waste. *Bioresource Technology* **101**(11): 4021–4028.

Energy Information Administration (EIA) (2011) *International Energy Outlook 2011*. DOE/EIA-0484(2011). Washington, DC: Energy Information Administration.

EU Joint Research Centre (2013) European IPPC Bureau. http://eippcb.jrc.ec.europa.eu/, accessed 19 November 2013.

European Commission (1999) *The Landfill of Waste*. www.central2013.eu/fileadmin/user_upload/Downloads/Document_Centre/OP_Resources/Landfill_Directive_1999_31_EC.pdf, accessed 13 November 2013.

European Commission (2008) *Integrated Pollution Prevention & Control Directive*. http://eur-lex.europa.eu/LexUriServ/LexUriServ.do?uri=CELEX:31996L0061:en:HTML, accessed 19 November 2013.

European Commission (2010) *The Industrial Emissions Directive*. http://ec.europa.eu/environment/air/pollutants/stationary/ied/legislation.htm, accessed 13 November 2013.

European Commission (2011) *Determining Transitional Union-Wide Rules for Harmonised Free Allocation of Emission Allowances Pursuant to Article 10a of Directive 2003/87/Ec of the European Parliament and of the Council (Notified Under Document C(2011) 2772)*. http://eur-lex.europa.eu/LexUriServ/LexUriServ.do?uri=OJ:L:2011:130:FULL:EN:PDF, accessed 19 November 2013.

European Commission (2013) *Auctioning*. http://ec.europa.eu/clima/policies/ets/cap/auctioning/index_en.htm, accessed 19 November 2013.

Fagan CC, Castillo M, O'Donnell CP, O'Callaghan DJ, Payne FA (2008) On-line prediction of cheese making indices using backscatter of near infrared light. *International Dairy Journal* 18(2): 120–128.

Flysjö A, Henriksson M, Cederberg C, Ledgard S, Englund JE (2011) The impact of various parameters on the carbon footprint of milk production in New Zealand and Sweden. *Agricultural Systems* 104(6): 459–469.

GaBi (2010) *GaBi 4 dataset documentation for the software system and databases*. Leinfelden-Echterdingen: University of Stuttgart and PE International GmbH.

Galli A, Wiedmann T, Ercin E, Knoblauch D, Ewing B, Giljum S (2011) Integrating Ecological, Carbon and Water Footprint: Defining the "Footprint Family" and its Application in Tracking Human Pressure on the Planet. One Planet Economy Network. www.oneplaneteconomynetwork.org/resources/programme-documents/WP8_Integrating_Ecological_Carbon_Water_Footprint.pdf, accessed 13 November 2013.

Gazulla C, Raugei M, Fullana-i-Palmer P (2010) Taking a life cycle look at crianza wine production in Spain: where are the bottlenecks? *International Journal of Life Cycle Assessment* 15(4): 330–337.

Gerbens-Leenes PW, Moll HC, Schoot Uiterkamp AJM (2003) Design and development of a measuring method for environmental sustainability in food production systems. *Ecological Economics* 46(2): 231–248.

Global Footprint Network (2009) Ecological Footprint Standards. www.footprintnetwork.org/fr/index.php/blog/af/participate_in_the_2012_ecological_footprint_standards, accessed 13 November 2013.

Global Footprint Network (2012) Glossary. http://www.footprintnetwork.org/en/index.php/GFN/page/glossary/, accessed 13 November 2013.

Goedkoop M, Spriensma R (1999) *The Eco-Indicator 99: A Damage Oriented Method for Life Cycle Impact Assessment*. Amersfoot, Netherlands: PRé Consultants.

Goedkoop M, Heijungs R, Huijbregts M et al. (2009) *ReCiPe 2008. A Life-Cycle Impact Assessment Method Which Comprises Harmonised Category Indicators of the Midpoint and Entry Level*. Amersfoot, Netherlands: PRé Consultants.

Guinee JB, Gorree M, Heijungs R et al. (2002) *Handbook on Life Cycle Assessment. Operational Guide to the ISO Standards. 1: LCA in perspective. 2a: Guide. 2b: Operational Annex. 3: Scientific background*. Dordrecht: Kluwer Academic Publishers.

Hall GM, Howe J (2012) Energy from waste and the food processing industry. *Process Safety and Environmental Protection* 90 (3): 203–212.

Hassas-Roudsari M, Chang PR, Pegg RB, Tyler RT (2009) Antioxidant capacity of bioactives extracted from canola meal by subcritical water, ethanolic and hot water extraction. *Food Chemistry* 114(2): 717–726.

Hauschild M, Potting J (2003) *Spatial Differentiation in Life Cycle Impact Assessment – The EDIP 2003 Methodology*. Lyngby, Denmark: Institute for Product Development, Technical University of Denmark.

He L, Zhang X, Xu H et al. (2012) Subcritical water extraction of phenolic compounds from pomegranate (Punica granatum L.) seed residues and investigation into their antioxidant activities with HPLC-ABTS + assay. *Food and Bioproducts Processing* 90(2): 215–223.

Henningsson S, Hyde K, Smith A, Campbell M (2004) The value of resource efficiency in the food industry: a waste minimisation project in East Anglia, UK. *Journal of Cleaner Production* 12(5): 505–512.

Herrchen M, Klein W (2000) Use of the life-cycle assessment (LCA) toolbox for an environmental evaluation of production processes. *Pure and Applied Chemistry* 72(7): 1247–1252.

Hillier J, Hawes C, Squire G, Hilton A, Wale S, Smith P (2009) The carbon footprints of food crop production. *International Journal of Agricultural Sustainability* 7(2): 107–118.

Hoogland CT, de Boer J, Boersema JJ (2007) Food and sustainability: do consumers recognize, understand and value on-package information on production standards? *Appetite* 49(1): 47–57.

Humbert S, Loerincik Y, Rossi V, Margni M, Jolliet O (2009) Life cycle assessment of spray dried soluble coffee and comparison with alternatives (drip filter and capsule espresso). *Journal of Cleaner Production* 17(15): 1351–1358.

Ike M, Inoue D, Miyano T et al. (2010) Microbial population dynamics during startup of a full-scale anaerobic digester treating industrial food waste in Kyoto eco-energy project. *Bioresource Technology* 101(11): 3952–3957.

ISO (2006a) EN ISO 14040:2006: Environmental management – Life cycle assessment – Principles and framework. www.iso.org/iso/catalogue_detail.htm?csnumber=37456, accessed 13 November 2013.

ISO (2006b) EN ISO 14044:2006: Environmental management – Life cycle assessment – Requirements and guidelines. www.iso.org/iso/catalogue_detail?csnumber=38498, accessed 13 November 2013.

ISO (2012) ISO/TS 14067. Carbon footprint of products. Requirements and guidelines for quantification and communication. www.iso.org/iso/catalogue_detail?csnumber=59521, accessed 13 November 2013.

Jaffry, S, Pickering, H, Ghulam, Y, Whitmarsh, D, Wattage, P (2004) Consumer choices for quality and sustainability labelled seafood products in the UK. *Food Policy*, 29(3): 215–228.

Jolliet O, Margni M, Charles R et al. (2003) IMPACT 2002+: a new life cycle impact assessment methodology. *International Journal of Life Cycle Assessment* 8(6): 324–330.

Jung S, Tonello Samson C, Lamballerie M (2011) High hydrostatic pressure food processing. In: Proctor A (ed) *Alternatives to Conventional Food Processing*. London: Royal Society of Chemistry, pp. 254–306.

Jungbluth N, Büsser S, Frischknecht R, Flury K, Stucki M (2012) Feasibility of environmental product information based on life cycle thinking and recommendations for Switzerland. *Journal of Cleaner Production* 28(0): 187–197.

Karakaya A, Özilgen M (2011) Energy utilization and carbon dioxide emission in the fresh, paste, whole-peeled, diced, and juiced tomato production processes. *Energy* 36(8): 5101–5110.

Kautsky N, Berg H, Folke C, Larsson J, Troell M (1997) Ecological footprint for assessment of resource use and development limitations in shrimp and tilapia aquaculture. *Aquaculture Research* 28, 753–766.

Kearney J (2010) Food consumption trends and drivers. *Philosophical Transactions of the Royal Society B: Biological Sciences* 365(1554): 2793–2807.

Kettle Foods (2009) *Sustainability*. Salem, OR: Kettle Foods.

Khajeh M, Yamini Y, Sefidkon F, Bahramifar N (2004) Comparison of essential oil composition of Carum copticum obtained by supercritical carbon dioxide extraction and hydrodistillation methods. *Food Chemistry* 86(4): 587–591.

Kim JK, Smith R (2008) Methods to minimise water use in food processing. In: Klemeš J, Smith R, Kim JK (eds) *Handbook of Water and Energy Management in Food Processing*. Cambridge: Woodhead Publishing, pp. 113–135.

Kim MH, Kim JW (2010) Comparison through a LCA evaluation analysis of food waste disposal options from the perspective of global warming and resource recovery. *Science of the Total Environment* 408(19): 3998–4006.

Kissinger M, Gottlieb D (2012) From global to place oriented hectares – the case of Israel's wheat ecological footprint and its implications for sustainable resource supply. *Ecological Indicators* 16, 51–57.

Ko,MJ, Cheigh CI, Cho SW, Chung MS (2011) Subcritical water extraction of flavonol quercetin from onion skin. *Journal of Food Engineering* 102(4): 327–333.

Kosseva R, Steve T (2009) Processing of food wastes. In: Taylor S (ed) *Advances in Food and Nutrition Research* 58, 57–136.

Lai CM, Ke GR, Chung MY (2009) Potentials of food wastes for power generation and energy conservation in Taiwan. *Renewable Energy* 34(8): 1913–1915.

Laufenberg G, Kunz,B, Nystroem M (2003) Transformation of vegetable waste into value added products: (A) the upgrading concept; (B) practical implementations. *Bioresource Technology* 87(2): 167–198.

Law R, Harvey A, Reay D (2013) Opportunities for low-grade heat recovery in the UK food processing industry. *Applied Thermal Engineering* 53(2): 188–196.

Liu Y, Langer V, Høgh-Jensen H, Egelyng H (2010) Life cycle assessment of fossil energy use and greenhouse gas emissions in Chinese pear production. *Journal of Cleaner Production* 18(14): 1423–1430.

Lundie S, Dennien G, Morain M, Jones M (2007) Generation of an industry-specific physico-chemical allocation matrix. *Application in the dairy industry and implications for systems analysis. International Journal of Life Cycle Assessment* 12(2): 109–117.

Manley JB, Anastas PT, Cue Jr BW (2008) Frontiers in green chemistry: meeting the grand challenges for sustainability in R&D and manufacturing. *Journal of Cleaner Production* 16(6): 743–750.

Marechal F, Muller D (2008) Energy management methods for the food industry. In: Klemeš J, Smith R, Kim JK (eds) *Handbook of Water and Energy Management in Food Processing*. Cambridge: Woodhead Publishing, pp. 221–255.

Martínez-Blanco J, Muñoz P, Antón A, Rieradevall J (2011) Assessment of tomato Mediterranean production in open-field and standard multi-tunnel greenhouse, with compost or mineral fertilizers, from an agricultural and environmental standpoint. *Journal of Cleaner Production* 19(9–10): 985–997.

Meisterling K, Samaras C, Schweizer V (2009) Decisions to reduce greenhouse gases from agriculture and product transport: LCA case study of organic and conventional wheat. *Journal of Cleaner Production* 17(2): 222–230.

Miettinen,P, Hämäläinen RP (1997) How to benefit from decision analysis in environmental life cycle assessment (LCA). *European Journal of Operational Research* 102(2): 279–294.

Milà i Canals L, Sim S, García-Suárez T et al. (2011) Estimating the greenhouse gas footprint of Knorr. *International Journal of Life Cycle Assessment* 16(1): 50–58.

Milani FX, Nutter D, Thoma G (2011) Invited review: environmental impacts of dairy processing and products: a review. *Journal of Dairy Science* 94(9): 4243–4254.

Milman A, Short A (2008) Incorporating resilience into sustainability indicators: an example for the urban water sector. *Global Environmental Change* 18(4): 758–767.

Mogensen L, Hermansen J, Halberg N, Dalgaard R (2010) Life cycle assessment across the food supply chain. In: Baldwin C (ed) *Sustainability in the Food Industry*. Ames, IA: Wiley-Blackwell.

Mohan,A (2011) Danone first to switch to PLA for yoghurt cup in Germany. www.observatorioplastico.com/detalle_noticia.php?no_id=160079&seccion=reciclado&id_categoria=80001, accessed 13 November 2013.

Mózner Z, Tabi A, Csutora M (2012) Modifying the yield factor based on more efficient use of fertilizer – the environmental impacts of intensive and extensive agricultural practices. *Ecological Indicators* **16**(0): 58–66.

Mustafa A, Turner C (2011) Pressurized liquid extraction as a green approach in food and herbal plants extraction: a review. *Analytica Chimica Acta* **703**(1): 8–18.

My Dieu T (2009) Food processing and food waste. In: Baldwin C (ed) *Sustainability in the Food Industry*. Ames, IA: Wiley-Blackwell, pp. 23–60.

National Renewable Energy Laboratory (NREL) (2008) *US Life-Cycle Inventory Database V 1.6.0*. www.nrel.gov/lci/database/, accessed 19 November 2013.

Niccolucci V, Galli A, Kitzes J, Pulselli RM, Borsa S, Marchettini N (2008) Ecological Footprint analysis applied to the production of two Italian wines. *Agriculture, Ecosystems and Environment* **128**(3): 162–166.

Nielsen P, Nielsen A, Weidema B, Dalgaard R, Halberg N (2003) LCA food data base. www.lcafood.dk.

Nilsson K, Flysjö A, Davis J, Sim S, Unger N, Bell S (2010) Comparative life cycle assessment of margarine and butter consumed in the UK, Germany and France. *International Journal of Life Cycle Assessment* **15**(9): 916–926.

Oliveira AL, Kamimura ES, Rabi JA (2009) Response surface analysis of extract yield and flavour intensity of Brazilian cherry (Eugenia uniflora L.) obtained by supercritical carbon dioxide extraction. *Innovative Food Science and Emerging Technologies* **10**(2): 189–194.

Ozel MZ, Gogus F, Lewis AC (2003) Subcritical water extraction of essential oils from Thymbra spicata. *Food Chemistry* **82**(3): 381–386.

Pap N, Pongrácz E, Myllykoski LRK (2004) Waste minimization and utilization in the food industry: processing of arctic berries, and extraction of valuable compounds from juice – processing by-products. *Proceedings of the Waste Minimization and Resources Use Optimization Conference*, 10 June, University of Oulu, Finland. Oulu: Oulu University Press, pp. 159–168.

Parent G, Lavallée S (2011) LCA potentials and limits within a sustainable agri-food statutory framework. In: Behnassi M, Draggan S, Yaya S (eds) *Global Food Insecurity – Rethinking Agricultural and Rural Development Paradigm and Policy*. Netherlands: Springer, pp .161–171.

Pasqualino J, Meneses M, Castells F (2011) The carbon footprint and energy consumption of beverage packaging selection and disposal. *Journal of Food Engineering* **103**(4): 357–365.

Peacock N, de Camillis C, Pennington D et al. (2011) Towards a harmonised framework methodology for the environmental assessment of food and drink products. *International Journal of Life Cycle Assessment* **16**(3): 189–197.

Pelletier N (2008) Environmental performance in the US broiler poultry sector: life cycle energy use and greenhouse gas, ozone depleting, acidifying and eutrophying emissions. *Agricultural Systems* **98**(2): 67–73.

Pelletier N, Tyedmers P (2010) Life cycle assessment of frozen tilapia fillets from Indonesian lake-based and pond-based intensive aquaculture systems. *Journal of Industrial Ecology* **14**(3): 467–481.

Pelletier N, Pirog R, Rasmussen R (2010) Comparative life cycle environmental impacts of three beef production strategies in the Upper Midwestern United States. *Agricultural Systems* **103**(6): 380–389.

Pirog R, Champion B, Crosby T, Kaplan S, Rasmussen R (2009) Distribution. In: Baldwin C (ed) *Sustainability in the Food Industry*. Ames, IA: Wiley-Blackwell, pp. 61–100.

Pizzigallo ACI, Granai C, Borsa S (2008) The joint use of LCA and emergy evaluation for the analysis of two Italian wine farms. *Journal of Environmental Management* **86**(2): 396–406.

Plassmann K, Edwards-Jones G (2009) Where Does the Carbon Footprint Fall? Developing a Carbon Map of Food Production. London: IIED.

Plessas S, Koliopoulos D, Kourkoutas Y et al. (2008) Upgrading of discarded oranges through fermentation using kefir in food industry. *Food Chemistry* **106**(1): 40–49.

Reay D (2008) Heat recovery in the food industry. In: Klemeš J, Smith R, Kim JK (eds) *Handbook of Water and Energy Management in Food Processing*. Cambridge: Woodhead Publishing, pp. 544–569.

Rees WE (1992) Ecological footprints and appropriated carrying capacity: what urban economics leaves out. *Environment and Urbanization* **4**(2): 121–130.

Ridoutt BG, Juliano P, Sanguansri P, Sellahewa J (2010a) The water footprint of food waste: case study of fresh mango in Australia. *Journal of Cleaner Production* **18**(16–17): 1714–1721.

Ridoutt BG, Williams SRO, Baud S, Fraval S, Marks N (2010b) Short communication: the water footprint of dairy products: case study involving skim milk powder. *Journal of Dairy Science* **93**(11): 5114–5117.

Ritzén S, Hakelius C, Norell M (1996) Life-cycle assessment, implementation and use in industry. In: *Nord-Design '96*. Helsinki, Finland.

Romero-Gámez M, Suárez-Rey EM, Antón A, Castilla N, Soriano T (2012) Environmental impact of screenhouse and open-field cultivation using a life cycle analysis: the case study of green bean production. *Journal of Cleaner Production* **8**, 63–69.

Röös E, Sundberg C, Hansson PA (2010) Uncertainties in the carbon footprint of food products: a case study on table potatoes. *International Journal of Life Cycle Assessment* **15**(5): 478–488.

Rotz CA, Montes F, Chianese DS (2010) The carbon footprint of dairy production systems through partial life cycle assessment. *Journal of Dairy Science* **93**(3): 1266–1282.

Roy P, Nei D, Orikasa T et al. (2009) A review of life cycle assessment (LCA) on some food products. *Journal of Food Engineering* **90**(1): 1–10.

Sahota A, Haumann B, Givens H, Baldwin CJ (2009) Ecolabeling and consumer interest in sustainable products. In: Baldwin C (ed) *Sustainability in the Food Industry*. Ames, IA: Wiley-Blackwell, pp. 159–184.

SAIC (2006) *Life Cycle Assessment: Principles and Practice*. US EPA. 68–C02–067. McLean, VA: SAIC.

Samuel-Fitwi B, Wuertz S, Schroeder JP, Schulz C (2012) Sustainability assessment tools to support aquaculture development. *Journal of Cleaner Production* 32(0): 183–192.

Schaub SM, Leonard JJ (1996) Composting: an alternative waste management option for food processing industries. *Trends in Food Science and Technology* 7(8): 263–268.

Sellahewa JN, Martindale W (2010) The impact of food processing on the sustainability of the food supply chain. *Aspects of Applied Biology*, **102**.

Shi J, Xue SX, Ma Y et al. (2012) Green separation technologies in food processing: supercritical CO2 fluid and subcritical water extraction. In: Boye J, Arcand Y (eds) *Green Technologies in Food Production and Processing*. New York: Springer, pp. 239–272.

Singh PP, Saldana MDA (2011) Subcritical water extraction of phenolic compounds from potato peel. *Food Research International* 44(8): 2452–2458.

Swiss Centre for Life-Cycle Inventories (2007) Ecoinvent data v2.0 ecoinvent reports No. 1-25. Dübendorf: Swiss Centre for Life-Cycle Inventories.

Tello J, Viguera M, Calvo L (2011) Extraction of caffeine from Robusta coffee (Coffea canephora var. Robusta) husks using supercritical carbon dioxide. *Journal of Supercritical Fluids* 59(0): 53–60.

Tingström J, Karlsson R (2006) The relationship between environmental analyses and the dialogue process in product development. *Journal of Cleaner Production* 14(15–16): 1409–1419.

UK Department of Energy and Climate Change (2012) *Modified UK National Implementation Measures for Phase III of the EU Emissions Trading System*. London: Department of Energy and Climate Change

Unilever (2011) *Lifecycle Assessment*. London: Unilever.

United Nations (1987) *Our Common Future*. Report of the World Commission on Environment and Development. New York: United Nations.

van Middelaar CE, Berentsen PBM, Dolman MA, de Boer IJM (2011) Eco-efficiency in the production chain of Dutch semi-hard cheese. *Livestock Science* 139(1–2): 91–99.

Vázquez-Rowe I, Moreira MT, Feijoo G (2011) Life cycle assessment of fresh hake fillets captured by the Galician fleet in the Northern Stock. *Fisheries Research* 110(1): 128–135.

Vermeir I, Verbeke W (2006. Sustainable food consumption: exploring the consumer "attitude – behavioral intention" gap. *Journal of Agricultural and Environmental Ethics* 19(2): 169–194.

Wackernagel M, Rees W (1996) *Our Ecological Footprint: Reducing Human Impact on the Earth*. Gabriola Island, BC: New Society Publishers.

Weidema BP, Thrane M, Christensen P, Schmidt J, Løkke S (2008) Carbon footprint. *Journal of Industrial Ecology* 12(1): 3–6.

Weidmann T, Minx J (2008) A definition of 'carbon footprint'. In: Pertsova CC (ed) *Ecological Economics Research Trends*. New York: Nova Science Publishers, pp .1–11.

Williams H, Wikström F (2011) Environmental impact of packaging and food losses in a life cycle perspective: a comparative analysis of five food items. *Journal of Cleaner Production* 19(1): 43–48.

Yonei Y, Ohinata H, Yoshida R, Shimizu Y, Yokoyama C (1995) Extraction of ginger flavor with liquid or supercritical carbon dioxide. *Journal of Supercritical Fluids* 8(2): 156–161.

Zufia J, Arana L (2008) Life cycle assessment to eco-design food products: industrial cooked dish case study. *Journal of Cleaner Production* 16(17): 1915–1921.

10 Food Safety and Quality Assurance

Tonya C. Schoenfuss[1] and Janet H. Lillemo[2]

[1]Department of Food Science, University of Minnesota, St Paul, Minnesota, USA
[2]Lillemo & Associates, LLC, Plymouth, Minnesota, USA

10.1 Introduction

In order to produce safe, high-quality food products, everyone involved with food products needs to understand their role in the process. There are many definitions for product quality, such as conformance to specifications and requirements, and fitness of a product for use. A more inclusive definition goes beyond the "product" to include everything that is involved from acquiring the product to customer service, to meeting the expectations of the customer (Feigenbaum, 1983). With food products, beyond product performance there is the additional expectation that food was produced in a sanitary manner and will be safe to eat. There are fundamental plant and equipment conditions that must be met, followed by programs that are developed to ensure safety and quality that everyone in the organization must understand and follow. Plant sanitation is just one of the programs under a total quality management system, and in modern food production, being in control of production to produce a safe product through a proactive program is a regulatory, customer, and consumer requirement.

This chapter will focus on the subjects most pertinent to producing food under sanitary conditions, and product quality assurance. Specifically, the sanitary design of plants and equipment, and the prerequisite programs important for food safety during processing are covered as well as quality basics.

10.2 Elements of total quality management

Before the advent of the Industrial Revolution, quality was determined by the expertise of the craftsman making the product. The success of their business rested on their ability to produce products the consumer was pleased with for the price paid. There are sectors where this is still the case, such as small producers of artisan products or producers of extremely high-value products. But as manufacturers grow in scale, the inclusion of processes that define the tasks and responsibilities of people making the product become necessary. Additionally, there is no longer the ability to manually examine every individual product that is produced.

The initiation of a system of scientific management of quality can be traced back to the teachings of people such as Frederick Taylor, who introduced the scientific method when evaluating workplace processes involved in quality and production. He advocated that each task should be standardized, workers trained, and job functions divided to maximize efficiency (Knouse et al., 1993). Walter Shewart is credited with the introduction of the control chart, in 1924, that can be used to monitor the variation of any process or quality factor (Best & Neuhauser, 2006). The control chart is instrumental in understanding the variability of a process, when something is "out of control" or irregular, and for optimizing a process to improve economic efficiency and product quality.

The need to produce supplies for the US war effort in the 1940s led to the development of ways to statistically analyze whether lots of products met specifications. The Military Standard 105D was developed by Harold Dodge and Harry Romig at Bell Labs, and the standard consists of tables that let a processor select a sampling plan based on the size of a sample lot, and the risk a manager is willing to take of accepting a lot when it should have been rejected (ASQ, 2013; Fisher & Nair, 2009). The less risk you are willing to accept, the more samples need to be evaluated within a given lot. Examples of these concepts in use are the acceptance sampling tables codified in the United States Code of Federal Regulations

Food Processing: Principles and Applications, Second Edition. Edited by Stephanie Clark, Stephanie Jung, and Buddhi Lamsal.
© 2014 John Wiley & Sons, Ltd. Published 2014 by John Wiley & Sons, Ltd.

(Title 7: Agriculture Part 43). These are used by the United States Department of Agriculture as part of its grading service (USDA, 2013). The Food and Drug Administration (FDA) uses sampling plans for evaluating microbial hazards, processing compliance and label compliance for net weight and contents. Some standards of identity contain acceptance sampling plans such as Title 21 Part 155 for canned vegetables (FDA, 2013b).

Tools such as control charts and sampling plans go a long way towards improving product quality and consistency, but they do not necessarily lead to continuous improvement. The development of total quality management (TQM), which incorporates improving both the organization and people involved with producing the product, is attributed to the foundational work of both Edward Deming and Joseph Juran (ASQ, 2012). They both recognized that in order to create a quality product, leadership from the top down is critical. Every person involved in the process needs to understand their role in improving the quality of a product, and have the training, tools, and support from top management to continuously improve quality and delight customers. Modern examples of how food companies incorporate these principles can be seen in the applications of the winners of the Malcolm Baldrige Quality Award, administered by the National Institute of Standards and Technology (NIST) under the US Department of Commerce (NIST, 2012). In the manufacturing division, two companies have received the award: Cargill Corporation, twice, for their corn milling and Cargill Kitchen Solutions (formerly Sunny Fresh Foods) divisions, and Nestlé Purina PetCare (NIST, 2012).

The first global quality management system, designed to create a level playing field and enhance international trade, was ISO 9000 (ISO, Geneva, Switzerland). A quality management system specifically for food is ISO 22000, and third-party auditing schemes developed specifically for food production have the additional goal of enhancing food safety. The Global Food Safety Initiative (GFSI) was formed in 2000 by the Consumer Goods Forum (formerly CIES) of large retail conglomerates (such as Wal-Mart, Carrefour and Tesco) with the goal of reducing food safety risks by delivering equivalency between schemes and reducing duplicative audits (and thus reducing costs). They benchmarked third-party audit schemes to establish minimal standards, which were issued in 2007. They reviewed applications from auditing schemes to determine which met the GFSI standards. Once a scheme is GFSI certified, the retailers agree to accept audits from any of the certified schemes (GFSI, 2012). This system has revolutionized food production around the world in that in order to sell a product in any of the major retailers' stores, firms must pass an audit by one of these schemes.

Compared with the past, where certain Good Manufacturing Practice (GMP) audits were sufficient, much greater level of scrutiny of the programs a plant has in place, in addition to GMPs, is required by the GFSI auditing schemes. Hazard Analysis Critical Control Point (HACCP), until recently a voluntary food safety system for most food producers in the US, is mandatory under these schemes. The result is an increased emphasis on hazard analysis and control, and the development of prerequisite programs and the standard operating procedures and documentation that go along with them. While the effect on food safety remains to be seen, it certainly has placed more attention on quality assurance and food safety by food manufacturers.

All the GFSI management schemes have the following basic requirements, with variations for processed foods and agricultural products.

1. Demonstration of management responsibility and commitment
2. Documented procedures and record control
3. Specifications, testing, and product development procedures
4. Sanitary design of plant and equipment
5. Proactive good manufacturing practices/prerequisite programs that include:

   a. Supplier approval (ingredients and packaging)
   b. Water quality
   c. Personnel:
      i. Hygiene practices
      ii. Processing practices
      iii. Training
   d. Calibration of equipment
   e. Management of pests and vermin
   f. Premises and equipment maintenance
   g. Cleaning and sanitation
   h. Control of physical contaminants
   i. Transport and delivery
   j. Waste management and disposal
   k. Allergen control

6. Food safety plan (HACCP)
7. Product identification and traceability (including organic, Kosher and Halal)
8. Market withdrawal and recall procedures
9. Site security/bioterrorism prevention
10. Product information/consumer awareness
11. Complaint management

In addition to the food safety components, some of the third-party auditing schemes under GFSI, and other, include other topics that are of concern to customers such as sustainability, organics, animal welfare, and fair trading practices. Global GAP (Good Agricultural Practices) is for certifying primary agriculture practices that meet food safety and traceability standards specific for the type of agriculture. Part of the auditing process includes evaluation of responsible water use and management of soil fertility (Global GAP, 2013). There are also additional auditing standards under Global GAP that include animal welfare and social practice (worker health and safety). Other organizations specialize in animal welfare such as the Professional Animal Auditor Certification Organization (PAACO) which certifies third-party auditors for dairy cattle and meat animal production. The USDA now governs organic certification in the US and certifies third-party auditors. Organizations such as Fairtrade USA develop standards to certify that the primary producers of a product are being compensated fairly. Generally, the quality function, if not involved with primary production directly, is involved in supplier approval and auditing and so is involved in these additional activities. Additionally, part of TQM is delighting the customer and these additional assurances about the product can be meaningful to them.

In addition to voluntary auditing of food safety, quality, sustainability, and animal welfare programs, companies that manufacture, hold, or ship food are also subject to regulatory inspections by local, state, and federal regulatory agencies. In general, food service and retail establishments are inspected by local and/or state agencies. Local and state regulatory agencies typically inspect against the FDA Food Code or some modification of it. Manufacturing facilities are inspected by state or federal agencies. At times, state agencies will inspect facilities under contract from a federal agency. At the federal level, the FDA Center for Food Safety and Applied Nutrition (CFSAN) has authority over most food products, under regulations outlined in the Code of Federal Regulations (CFR) Title 21. The US Department of Agriculture (USDA) Food Safety and Inspection Service (FSIS) regulates primarily meat, poultry, and egg products under the jurisdiction of 9CFR.

10.3 Hazard Analysis Critical Control Point (HACCP) system

The National Advisory Committee on Microbiological Criteria for Foods (NACMCF) defines HACCP as "a systematic approach to the identification, evaluation, and control of food safety hazards" (NACMCF, 1997). HACCP involves a specific process for defining critical parameters about the food, identifying potential hazards, and determining control measures for those hazards that are reasonably likely to occur in the food processing system. It is a risk-based program that focuses on building food safety prevention into the manufacturing process, rather than relying on end-product testing for detecting food safety issues after they occur.

Until recently, HACCP was required by regulation for only meat, poultry, juice, and seafood. With the implementation of the Food Safety Modernization Act (FSMA), HACCP or a similar preventive control program will be required for all facilities that "manufacture, process, pack or hold food and that are required to register with the US Food and Drug Administration (FDA) under section 415 of the FD&C Act," with a few exceptions for small and very small businesses (FDA, 2013a).

Development of a HACCP plan involves five preliminary steps and seven principles, as shown in Table 10.1. It is imperative, both from a practical level and because of regulatory requirements, that the development of the HACCP plan be led by an individual who is trained in HACCP and its supporting programs. Training is conducted frequently by many independent companies. Other members of the HACCP team should also be trained, although this training may be done internally by the HACCP team leader. A cross-functional approach

Table 10.1 Steps involved in the HACCP process (NACMCF, 1997)

Preliminary steps	HACCP principles
Assemble the HACCP team	Principle 1: Conduct a hazard analysis
Describe the food and its distribution	Principle 2: Determine critical control points (CCPs)
Describe the intended use and consumers of the food	Principle 3: Establish critical limits
Develop a flow diagram which describes the process	Principle 4: Establish monitoring procedures
Verify the flow diagram	Principle 5: Establish corrective actions
	Principle 6: Establish verification procedures
	Principle 7: Establish record-keeping and documentation procedures

with shared responsibility across functions is essential for a fully functioning HACCP plan.

Hazard analysis is the basis on which the HACCP plan is built, but the preliminary steps are necessary to complete a thorough hazard analysis. It is important to have someone knowledgeable in food science and food safety involved in the hazard analysis. Thorough identification of hazards and assessment of their likelihood and severity are essential to developing a plan that ensures the food's safety. Critical control points (CCPs) are steps in the process where control may be applied, and the step is required to prevent, eliminate or reduce hazards to an acceptable level (NACMCF, 1997). One of the "double-edged swords" of HACCP is that corrective actions are spelled out in advance for CCP deviations (situations where the critical limits are not met and therefore the hazard is not controlled). Since the corrective actions are already defined, there is no "management discretion" in determining the disposition of the affected product. This is good in that it ensures that food safety is non-negotiable; it is difficult because it may mean that a significant amount of product must be either destroyed or reworked, and there is no judgment involved in the decision.

Verification involves two parts: validation that the CCPs actually control the hazard for which they were designed (the plan is effective), and verification that the plan is being followed as designed (the plan is properly implemented and maintained). Verification typically involves a review of documentation and observation of those employees responsible for conducting monitoring activities. Validation requires scientific evidence that hazards are being controlled by the preventive controls, including CCPs. From the proposed preventive control rules for FSMA: "Proposed § 117.150(a)(2) would require that… the validation of preventive controls include collecting and evaluating scientific and technical information or, when such information is not available or is insufficient, conducting studies to determine whether the preventive controls, when properly implemented, will effectively control the hazards that are reasonably likely to occur" (FDA, 2013a). Clearly, scientific and technical knowledge of the product and process is necessary to ensure the HACCP plan successfully controls the hazards; outside experts may be helpful in conducting the validation. One final aspect of verification is a periodic review of the entire food safety plan, including both HACCP and its supporting prerequisite programs.

Prerequisite programs are "procedures, including Good Manufacturing Practices, that address operational conditions providing the foundation for the HACCP system" (NACMCF, 1997). Prerequisite programs create an environment in which safe food processing can occur. Some prerequisite program requirements are defined by regulations, such as Current Good Manufacturing Practice (cGMP), which are spelled out in 21CFR §110 (to be replaced by 21CFR §117 when FSMA goes into effect). Even for regulated prerequisite programs, each company must customize programs to be appropriate for their facilities. Together, prerequisite programs and HACCP represent a firm's food safety plan.

10.4 Sanitary processing conditions

10.4.1 Sanitary design and maintenance of plants and equipment

The Federal Food Drug and Cosmetic Act, Chapter IV, Section 402, specifically defines food as adulterated "if it has been prepared, packed, or held under unsanitary conditions whereby it may have become contaminated with filth, or whereby it may have been rendered injurious to health." Maintaining the grounds, facilities, and equipment in a manner that prevents these conditions is not only a good idea, it is a legal and regulatory requirement for manufacturing foods.

There are many checklists (including in the regulations) that give specific requirements for premises and equipment conditions. Any of the GFSI audit schemes include this information. In general, the surrounding grounds must not have standing water, waste accumulation, or discarded equipment that could attract or harbor pests such as insects, rodents, or birds. Exterior walls, doors, and windows must be sealed so pests cannot access the facility. The food processing area must be physically removed from employee welfare areas such as rest rooms, lunch rooms, and locker rooms to prevent inadvertent contamination of the processing area. In the processing area, the floors, ceilings, walls, and any posts or support beams must be constructed in a way that prevents accumulation of food debris or water, and can be easily cleaned and sanitized. In general, wood and other porous materials are not allowed in the production area. Adequate hand washing sinks must be available and must provide warm water, soap, and a method for drying hands. Utensil washing sinks must be separate from hand washing sinks. Sinks must be labeled so their intended use is clear. Only potable water shall come into contact with food equipment. Condensation, dripping water, and inadequate drains can be a major source of contamination, and must be corrected.

Equipment needs to be constructed in a way that allows for easy disassembly for cleaning and sanitizing. There must not be any dead ends, rough welds, or other areas in the equipment that can allow accumulation of food debris and water. Additionally, the ability of equipment to withstand standard cleaning chemicals is important, not only for the food contact surfaces. Equipment can be easily corroded by cleaning and sanitation chemicals and even if not in direct contact with the food, deteriorated equipment can form niche environments for opportunistic food pathogens such as *Listeria monocytogenes* to survive and grow. Formation of biofilms (sticky films of bacteria that adhere to a surface) can contribute to potential cross-contamination of the food product. Additionally, corroded equipment can pose a physical contaminant hazard.

All equipment should be on a routine maintenance schedule. Documenting when equipment requires emergency maintenance can help to determine the proper maintenance frequency for that piece of equipment. User manuals for the equipment may also list a suggested frequency. Proper standard operating procedures (SOPs) for maintaining equipment are important, and the equipment should be formally handed off to the sanitation group after maintenance and before production resumes. It benefits the company to shut down for routine maintenance, as unscheduled repairs can be highly disruptive to the business.

One aspect of maintenance that is sometimes overlooked is equipment calibration. Some equipment can easily be calibrated by facility personnel (e.g. thermometers, pH meters). For other equipment (e.g. load cells, retorts), it is important to have equipment manufacturers calibrate their equipment on a periodic basis, often annually or as required by regulations.

10.4.2 Cleaning and sanitation

Cleaning and sanitation are covered by the regulations in 21CFR 110.35 (21CFR 117.35 after FSMA implementation). Adequate cleaning and sanitation are absolutely essential to maintaining safe, quality food. It is beyond the scope of this book to go into detailed directions for cleaning and sanitation. Proper cleaning and sanitation means that personal protective equipment is used and strict procedures, as outlined by the chemical supplier, are followed by all personnel at all times. Proper cleaning and sanitation will only take place if appropriate cleansers and sanitizers are selected, and specified dwell time, temperature, mechanical action, and concentration are followed. Any deviation can lead to either inadequate level

of chemical action or chemical degradation, or both. For questions as to the best procedures for specific manufacturing equipment and environments, cleaning/sanitation chemical companies and state extension services are good sources of advice.

In general, the processing environment and equipment should be maintained in a clean and sanitary condition. Also, cleaning and sanitation activities should be conducted in a way that does not endanger or contaminate the food, equipment, or environment. Alkaline cleansers are effective in removal of organic soils, while acid cleansers are effective for removal of inorganic soils. Concentrations of cleaning and sanitizing chemicals should be tested for each batch of manually mixed systems, or on a daily basis for automatically mixed systems, to ensure they are within effective and safe levels. Evaluation of water quality on a monthly basis is important, as water quality influences chemical action. Proper dilution is important to the efficacy of some sanitation chemicals. While the tendency may be to think that more is better, in some cases using stronger solutions of chemicals than recommended may render the solution ineffective and/or corrosive.

Periodic housekeeping during production, such as wiping down dust and sweeping floors, will go a long way towards maintaining a clean working environment. There is an old saying in food manufacturing: "if you have time to lean, you have time to clean." All employees should clean their areas when time is available. It is also essential that employees wash hands and replace gloves after handling cleaning tools and before touching food contact surfaces or food products.

Surfaces must be clean prior to sanitizing. Sanitizers do not work if there is organic matter on the surface to be sanitized. This applies to floors and drains as well as food and non-food contact equipment. Pouring sanitizer down a drain full of debris does not effectively control microbial growth in the drain.

Validation and verification of cleaning and sanitation methods of food contact surfaces are important for assuring that the methods are effective. There are many different methods used: visual inspection, ATP swabs, protein swabs, and microbiological swabbing. Use care when choosing which method to use. For example, it is usually unwise to swab for pathogens on food contact surfaces, especially for *Listeria monocytogenes* (*L. mono*), for non-ready-to-eat foods or foods that do not support its growth. There is no acceptable regulatory limit of *L. mono* on food contact surfaces. More information on microbiological testing is given in the upcoming section on environmental testing.

10.4.3 Personnel practices

Personnel practices are regulated in the US as part of the Good Manufacturing Practices, 21CFR 110.10 (to be replaced by 21CFR 117.10 when FSMA is implemented). The purpose of personnel practices is to prevent contamination of the food or the environment by employees.

Employees who are ill or who have open sores, infected wounds, or other sources of microbial contamination that could potentially contaminate food, food contact surfaces, or packaging must be excluded from working in food areas until the health issue is resolved. Minor cuts or injuries should be covered by an appropriate bandage. Bandages used in processing areas are typically colored blue, to be visible in the food, and are metal detectable.

Employees must maintain cleanliness when working in a food environment. Steps to be taken to achieve proper cleanliness include good personal hygiene; protective clothing; proper hand washing and glove use; removing jewelry; wearing hair restraints; refraining from eating, drinking, chewing gum, or smoking in processing areas; storing personal items and clothing in an area separate from the processing area; and avoiding contamination of food with perspiration, medicine, cologne or perfume, lotions, and other cosmetics.

One of the most common findings in warning letters from regulatory agencies is improper hand washing or glove use. Hands must be washed and gloves changed after use of the restroom. Hands must be washed or gloves changed after touching non-food contact surfaces such as cleaning utensils, pens, garbage, pallet jacks, or items that have been on the floor. In addition, hands must be washed or gloves changed after touching one's hair or face, scratching one's arm, etc. Gloves can give workers a false sense of security. If gloves are dirty or damaged, they need to be replaced. Removed disposable gloves should always be discarded, and never reused.

Employees and visitors must be trained in proper personnel practices, and supervisors should reinforce these rules and retrain as necessary.

10.4.4 Transportation and storage

Transportation and storage can play a significant role in the safety and quality of foods. Proper temperature control, pest control, allergen segregation, and lot code traceability aid in maintaining food safety during storage and transportation.

In 2005, Congress passed the Sanitary Food Transportation Act, but rule making was not initiated by the FDA until 2010, and proposed regulations have not yet been published. However, general guidance was issued in 2010 to assist food and transportation companies (FDA, 2010a). Despite the absence of specific regulatory directives, food companies are still responsible for food safety and security during the entire supply chain, including transportation.

Prior to loading a truck carrying raw materials or finished products, the truck should be inspected thoroughly to identify any conditions that could affect the safety or quality of the food. Employees should look for holes in the trailer that could allow access of pests, insects, rodent excreta, dirt, and debris such as pallet wood. Employees should check for unusual odors. For less-than-full load (LTL) trucks, be alert to the presence of chemicals or items already on the truck that could contaminate the food product. The author has rejected trucks containing new tires and pesticides due to the presence of strong odors that could have permeated the pallets of dry bakery mix that were being shipped, as well as a truck with scattered dog food and a water bowl in the back of the truck (the driver's dog was living in the trailer). Record the results of the inspection for future reference. Shipment content, quantities, and associated lot codes should be included on the shipping documentation.

Before receiving a load, inspect the truck for the same issues as noted above. Pallets holding food product should be in good condition, and the load should not have shifted or damaged packages during shipment. The inspector should check to verify that the quantity and lot codes on the shipping documentation match the actual content of the load. Report any discrepancies or sanitation issues to the transportation company prior to accepting the load.

For refrigerated and frozen loads, the actual temperature of the trailer and food should be checked and documented upon receipt and before shipping to ensure the temperature is within the required range. Refrigerated foods should be no more than 5°C (41°F) although the maximum temperature is 7°C (45°F) in refrigerated products governed by the Grade A Pasteurized Milk Ordinance (FDA, 2011a). Foods shipped frozen should still be frozen upon receipt (US Public Health Service and FDA, 2013).

Bulk load and full truckload shipments should be sealed upon loading and the seal numbers recorded on the shipment documentation (e.g. bill of lading). Prior to unloading the shipment, seal numbers should be verified against the numbers on the documentation to determine if the seals have been removed during shipment. The inspector should reject loads with missing or changed seals.

Rail cars should be similarly sealed and inspected. It is common during the warm summer months to fumigate rail cars containing grain products, such as flour, during transportation with pesticides to prevent infestation with storage insects. Trained personnel should remove and discard the fumigant and allow for proper aeration of the rail car prior to unloading. Regulatory jurisdiction of the use of the fumigant chemicals is granted to the Environmental Protection Agency in the Federal Insecticide, Fungicide, and Rodenticide Act and the Federal Food Drug and Cosmetics Act, and modified by the Food Quality Protection Act of 1996 (Public Law 104–170, 1996).

Shippers of bulk tanker loads should provide information on previous loads carried in the tanker, as well as wash tickets to show that the tanker was clean and compatible with the current product prior to loading. Bulk tankers should also be inspected thoroughly to identify unsanitary conditions or evidence of off-odors or previous loads. As with all food safety processes, inspections should be documented.

Ideally, LTL loads should be secured either by seals or a padlock between stops. This is, however, very difficult to enforce, as the driver can affix the seal or lock just before arrival at the dock.

Once a load is received, it should be stored properly to reduce the risk of temperature abuse or cross-contamination. Each pallet should be labeled with the material name and identification code, lot code, receiving date, and allergen content. Storage racks should be off the floor and far enough away from the walls to allow for sanitation and inspection activities. Dangerous chemicals should not be stored over food materials. If a package is damaged, it should be discarded and the pallet restacked. All dry materials should be inspected once per month to identify any issues with storage insects. Materials should be used on a first in-first out basis. Raw materials and products that are on hold should be segregated and secured to prevent inadvertent introduction into the supply chain.

Allergen-containing raw materials and finished products that have the potential to spill should be stored below non-allergen containing materials to reduce the risk of cross-contact. The rule for allergen storage is "same over same, less over more" – that is, only store materials containing the same allergens over other allergen-containing materials, and store materials with more allergens below materials with less. For example, a product with soy and wheat could be stored over a product with soy, wheat and egg, but they could not be inverted because there would be a chance of egg contamination onto the soy/wheat product. Remember that individual allergens need to be treated separately; there is not just one category of "allergens." Individual tree nuts should be stored separately to prevent cross-contact.

10.4.5 Control of physical contaminants

Recalls from foreign materials are unusual, accounting for only 1.3% of all recalls in 2010 through 2012 (FDA Foods and Veterinary Medicine Program, 2013). Most incidents of physical contaminants are detected and removed in the manufacturing process, or are individual incidents involving only one consumer. Whatever the frequency or severity, the presence of physical contaminants in finished products is always a consumer issue, as well as a regulatory non-conformance.

When considering physical contaminants, one must differentiate between natural and unavoidable defects that do not pose a threat to humans and foreign matter introduced from the growing, harvesting, processing, packaging, storage or transportation environment. The latter includes such things as mold, insect or rodent damage, or preharvest infestation, and the FDA has established defect action levels for specific commodities because "it is economically impractical to grow, harvest, or process raw products that are totally free of non-hazardous, naturally occurring, unavoidable defects" (FDA, 2011b). Defect action levels are not average, allowable or acceptable levels; they "represent limits at which FDA will regard the food product 'adulterated'; and subject to enforcement action under Section 402(a)(3) of the Food, Drug, and Cosmetics Act" (FDA, 2011b). Consider fresh, frozen, or canned cherries. Defect action levels are established for rot and insect filth, but not for cherry pits. Cherry pits have the potential to be hazardous, while rot and insect filth are simply aesthetic defects.

The supplier approval program may include a requirement that raw material suppliers include proper foreign material control practices in their manufacturing process. For raw materials that are prone to physical contaminants (such as shells in nuts or pits in cherries), it is common to require suppliers to conduct a formal inspection of the raw material prior to shipment, and to report the absence or presence and level of foreign matter in a Certificate of Analysis (COA). The COA should be compared for compliance to the product specification prior to receipt of the raw material. The receiving facility may also choose to sample and inspect the raw material to verify the content of the COA. Shipments of raw materials with foreign matter levels exceeding the product specification should be rejected.

The FSMA will require that, as part of the HACCP plan development, an assessment is conducted to identify physical hazards from raw materials and processing steps, and control measures are identified to prevent, eliminate, or reduce these hazards to an acceptable level (FDA, 2013a). There are many ways to accomplish this. Good manufacturing practices such as inspection of raw materials upon receipt, preoperational inspections of equipment and the operating environment, restrictions of glass and brittle plastics in production areas, rules for proper utensil selection and usage, and proper equipment maintenance go a long way towards preventing introduction of foreign materials into the product stream.

Foreign material may be detected and removed by a variety of equipment, such as metal detectors, magnets, filters and screens, cyclones or water tanks, destoners, X-ray machines, or other optical scanner or visual control systems (GMA, 2008). Metal detectors are the most common foreign material detection/removal devices. It is critical to place metal detectors as late in the process as possible, preferably after packaging, so as to detect and remove any metal contamination introduced by the production equipment itself. Procedures for validating, calibrating, verifying, and establishing corrective actions, should the metal detector fail to detect test probes, and documenting testing actions, should all be included in the HACCP plan.

10.4.6 Pest control

Pests are any organisms that should not be in food or food plants, and include such things as rodents, insects, birds, weeds, and microorganisms. 21CFR 110.35 c states that no pests shall be allowed in any area of a food plant, and that effective measures shall be taken to exclude them from the premises to protect against contamination. A good pest control plan includes multiple strategies to maintain sanitation, such as inspecting for evidence of pests, identifying them, implementing control procedures, and monitoring the effectiveness of the method. It must be a written plan that allows for monitoring and continuous improvement practices when deficiencies are found. Because of chemicals that may be required to effectively control pests, licensed contractors are often employed to help develop and implement a control system.

First and foremost, the food facility should be designed and maintained so as to not attract pests, and to exclude their entry. The facility grounds and premises should be kept free of waste material and clutter, and a clean weed-free perimeter maintained so that no nesting of rodents can occur on the sides of the building. Access points into the plant such as doors should be self-closing and fly-proofed with methods such as air curtains. The door should have a metal threshold and be flush with the entryway so rodents cannot get under the door. Mice are able to enter through a hole as small as a quarter of an inch. In areas with fork-lift traffic into warehouses, especially in grain facilities, rodent exclusion thresholds can help. When designing facilities, the ability to easily use alternative treatments, such as heat, to kill insects should be incorporated. Warehouses and racking systems should always be installed with an 18 inch perimeter, and with pallets off the floor, to allow for inspection around and underneath for signs of infestation.

In order to monitor and control pests, traps and bait stations are often used. Because of the dangerous chemicals used, bait stations should only be employed outside the food processing facility along the perimeter of the building. They kill rodents, and are used to monitor the level of activity. Chemical sprays are used within facilities to control insects like cockroaches, but the chemicals and the concentrations allowed are regulated (by the Environmental Protection Agency – EPA), so a licensed applicator must be used.

10.4.7 Water quality

Water is used in many ways during food processing, such as:
- for irrigation of crops
- for postharvest cooling of fruits and vegetables
- as a flume for transport of fruits and vegetables
- as an ingredient to rehydrate other ingredients or standardize a formula
- as a heating or cooling medium (either in direct contact with food or indirectly)
- rinsing, cleaning, and sanitation
- for hand washing and other sanitary functions.

The foremost concern with regard to water is its role in food safety. The quality attributes of water (flavor, pH, mineral composition, etc.) and the composition and amount of waste water from a facility are also important to food processors.

Drinking water from municipal sources that is generally considered safe has been a source of contamination and has led to outbreaks such as the 1993 *Cryptosporidium* incident that affected approximately 400,000 people and led to 100 deaths (CDC, 1997). Pathogens such as the hepatitis A virus from fecal contamination, as well as nitrates and environmental pollutants such as arsenic, are a concern in well water. In the US, the Safe Drinking Water Act of 1974 is the law responsible for ensuring safety of water supplies serving 25 people or more, and is administered by the United States EPA; the regulations are found mainly in 40 CFR part 141. In 2008, organisms such as *Salmonella*, *E. coli* O157:H7, *Shigella*, *Campylobacter* and *Legionella* led to 16 disease outbreaks affecting 1672 people and three deaths in the US (CDC, 2013). New concerns are for "emerging contaminants" that are not removed by traditional municipal water treatments and are found in drinking water that, while not at toxic levels, present health concerns (EPA, 2013; Lapworth et al., 2012). These include:

- endocrine-disrupting compounds
- surfactants (from cleaning agents)
- hormones (from the animal industry and urine, especially from women on birth control – estrogen and estradiol)
- pesticides
- phthalates (plasticizers in plastics, used in inks and cosmetics)
- perfluoro-octane sulfonate (PFOS), a chemical used in fabric protectors.

Additionally, food plants contribute some of these chemicals to the aquatic environment in the form of waste water. This is also regulated by the EPA and local agencies. Quite often, plants need to be designed to mitigate the discharge of water that is untreated and need to conduct activities to adjust the pH, and reduce the chemical and biological oxygen demand. Many of the endocrine active compounds found in drinking water are the result of agricultural and food processing activities (Lundgren & Novak, 2009).

10.4.8 Allergens

Approximately 6–8% of children and 2% of adults have food allergies (NIAID, 2003). Allergic reactions to food can be life threatening. According to the FDA, approximately 150 people die from allergic reactions to food each year (FDA, 2010b). Clearly, it is imperative that food and ingredient manufacturers prevent the inadvertent introduction of undeclared allergens in their food.

In the year two report of the Reportable Food Registry (FDA Foods and Veterinary Medicine Program, 2012), the FDA noted that approximately one-third of recalls were due to undeclared allergens. Bakery items, frozen foods, and snack foods were most likely to contain undeclared allergens.

Allergen control needs a multifaceted approach to be successful. Equipment and the facility should be designed to be easily cleaned; this will aid in preventing allergen cross-contact between formulas with incompatible allergens. Air handling systems should be designed to prevent cross-contact from air-borne dust. Formulas should be developed to avoid allergens, if at all possible. Production should be scheduled so that non-allergen containing products are run before allergen-containing products. As described above, storage practices must prevent inadvertent cross-contact. Cleaning procedures should be validated to ensure that they remove allergens if followed diligently.

Labeling should be verified so that all allergens in the product are declared on the label, according to regulatory requirements. Label control includes two important processes: ensuring the ingredient statement and allergen statement on the label are accurate for the current formula of the product, and ensuring the proper, current label is affixed to the packaging.

Finally, consumer reports of allergic reactions should be taken seriously and investigated thoroughly to determine if product should be recalled, and to identify and resolve the root cause of failures in the allergen control system.

10.4.9 Environmental testing

In addition to verifying the effectiveness of cleaning food contact and non-food contact surfaces, the processing environment should also be monitored for environmental pathogens. There have been numerous food safety outbreaks in recent years where environmental pathogens have contaminated ready-to-eat products (FDA, 2013a).

Increased emphasis by regulatory agencies is being placed on environmental testing plans and proper corrective actions when contamination is discovered. Recalls have been initiated from only environmental pathogen contamination, not confirmed product contamination. The FSMA will not specifically require environmental testing as part of the preventive controls section (FDA, 2013a). However, it is essential that facilities, especially those producing ready-to-eat foods, have a robust environmental testing program.

Philosophies on environmental testing have changed over the past 5–10 years. It used to be that facilities were pleased if their environmental swabs came back negative; they recorded the information and filed it away for the next time. Now, emphasis is on using the environmental testing program to learn more about the production environment, to seek out any potential microbial problems, and to solve them. In general, wet clean areas are swabbed for *Listeria* and dry clean areas for *Salmonella*.

The Grocery Manufacturers Association (GMA) has published a guide for controlling *Salmonella* in low-moisture foods, and this document contains excellent advice on developing an environmental program for dry cleaning environments (GMA, 2009). The principles presented can be adapted for wet cleaning environments as well. A four-zone approach is recommended, where Zone 1 represents food contact surfaces, Zone 2 represents non-product surfaces close to Zone 1, Zone 3 is in the processing area but more removed from the product contact surfaces, and Zone 4 is outside the processing area (such as locker rooms, storage rooms, labs). The GMA recommends increased sampling for *Salmonella* in the zone closest to product contact surfaces (Zone 2). Non-product contact surfaces such as Zone 2 or 3 should be tested during normal operating conditions, as swabbing after sanitation does not give good information about the operating environment, only about sanitation techniques. Food contact surfaces (Zone 1) should only be swabbed for indicator organisms such as aerobic plate counts or Enterobacteriaceae, not pathogens, unless there are special circumstances such as suspected product contamination. Before deciding to swab Zone 1, it is wise to determine what actions will be taken should a positive result be found. Swabbing sites should be randomly chosen or chosen on a rotating basis.

Corrective actions in the case of positive results should be spelled out for each zone before swabbing takes place. Root cause analysis, increased cleaning, sanitation, and swabbing are recommended until the environment consistently tests free of contamination. The GMA document includes detailed recommendations on corrective actions.

10.5 Supporting prerequisite programs

10.5.1 Supplier approval

In 2007, the FDA reiterated to food manufacturers that they are responsible for the safety of the ingredients and packaging materials they use in the production of their food products. Manufacturers have a responsibility to their customers, their consumers, their shareholders, and the public at large to ensure they are using the safest, highest quality raw materials available. Many large food companies have hundreds, if not thousands, of raw

materials from an equally large number of supplier facilities. Most companies do not have the personnel or financial resources to inspect every supplier on a frequency that would guarantee that raw materials are consistently safe. Therefore, companies need to develop a risk-based approach to assessing the safety of their suppliers and raw materials.

A necessary component of a risk-based supplier quality system is the risk assessment process. There are many risk assessment tools available, but most include analysis of a combination of severity and likelihood/probability of risk. The Operational Risk Management (ORM) process, developed by the US Air Force for evaluating risk of military operations (Phillips, 1997), can be easily adapted to raw material and supplier risk. A common quality tool, Failure Modes and Effects Analysis (FMEA), can also be used for assessing the risk of suppliers or raw materials. FMEA includes not only severity and occurrence (likelihood), but also a rating of detection (Tague, 2005). Whatever risk assessment method is used, it is beneficial to choose one that provides a numerical ranking to the risks, allowing the user to prioritize actions based on the material or supplier's risk level.

At the raw material level, risk factors to assess include the raw material's inherent food safety risk, the importance of the ingredient to the product's functionality, and the strategic sourcing risk (including the material's availability and market conditions). At the supplier level, risk may include past complaints and how the supplier handled them, or responsiveness to shipping issues (e.g. incomplete orders, wrong products shipped, or incorrect receiving temperature).

Once the overall risk level is determined, supplier approval or periodic assessment activities can be based on the risk level. For example, all extremely high-risk suppliers may be audited by a company auditor, all high-risk suppliers required to be GFSI certified, all medium-risk suppliers required to complete a supplier questionnaire, and all low-risk suppliers required to only provide raw material specifications and other documentation. If a supplier provides multiple raw materials at different risk levels, it would be prudent to assign assessment activities based on the highest risk level.

The contribution of third-party audits and certification, such as GFSI, to this process should not be underestimated. Auditing by company personnel is expensive. Requiring third-party food safety certification allows the company to review the audit report, identify any deficiencies of concern, and address these issues with the supplier without the time and expense of an on-site audit. Do keep in mind, however, that the audit represents only a snapshot in time. Most companies dedicate a significant amount of resources to getting prepared for third-party audits, which are usually scheduled in advance, so these reports represent the best possible situation.

Section 307 of the FSMA sets forth a process for accreditation of third-party auditors to ensure minimum requirements and performance management standards are met; however, since this section is included in Title III of the law, which covers improving the safety of imported food, it is yet unclear whether or not the accreditation requirement will apply to auditors of domestic entities as well as foreign entities (Public Law 111–353, 2011).

It is common to ask all suppliers for basic documentation, regardless of their risk level. Typical documentation requirements include product specifications, allergen information, nutritional values, lot code, shelf life, and storage information, technical and emergency contact information, and Kosher/Halal/organic certificates, if appropriate. Companies usually also require legal documents such as a letter of continuing guarantee, a signed statement saying that the supplier will not change the raw material without advance warning, evidence of liability insurance, and a signed statement agreeing to comply with the company's supplier requirements.

A supplier questionnaire, which collects information about the supplier's food safety programs and practices, may also be required. The questionnaire may identify areas of concern that require further investigation. It is helpful to tell the suppliers that the questionnaire is a pre-audit activity, and that completing it thoroughly may prevent an on-site audit. This increases the probability of thorough, honest responses. A signature should be required stating that the information provided in the questionnaire is true and accurate.

Once a decision is made to approve a new supplier, providing a periodic assessment of the supplier's performance can be beneficial for both the company and the supplier. A balanced scorecard can be developed that focuses on whatever is most important to the company. The scorecard might include complaint rates, documentation status (complete/incomplete), timeliness in responding to issues, and shipping errors. Periodic assessment activities based on supplier/raw material risk should also be conducted to assure current and accurate information is on file.

Unfortunately, all supplier quality programs must include a process to disapprove or delist an existing supplier due to non-compliance or food safety issues. This process should be spelled out thoroughly and communicated both within the company and to suppliers.

When a situation arises, no one should be surprised by the disapproval or delisting process. This process should include a cross-functional discussion of the implications of this action.

10.5.2 Complaint management

Complaints are not necessarily a bad thing. Complaints give insight into where a company's quality and food safety systems fail and into what consumers want and expect from their food products. Companies that analyze and track complaints, and have a robust program for root cause analysis and corrective action, take advantage of the information provided from complaints to improve their products and processes.

Complaint tracking and analysis can vary greatly depending on the size and kind of company. A large international consumer packaged foods company could receive over a million consumer contacts a year. A small wholesale business might only receive one or two complaints a month. Obviously, the approach to addressing complaints in these two companies will be different.

In the consumer foods company, a consumer relations or customer service department will have sole responsibility for responding to consumer calls and emails. The consumer communication is mostly about resolving the consumer's dissatisfaction to increase the probability of future sales. This group will also record critical information about the product and issue to allow trending and statistical analysis of similar complaints by the quality assurance team.

The small wholesale company will work one on one with the customer to solve a complaint issue, often including a visit to the customer's business. These visits are made by a sales representative, a technical service person, or a research and development or quality assurance employee. This type of communication is very different from the large consumer foods company. In this situation, the visit is about solving the specific problem the customer is experiencing, and may involve hands-on laboratory or plant work. The two situations are similar, though, in that both provide an opportunity to satisfy the consumer or customer, retain their business, and learn more about how products are performing in the marketplace.

For serious complaints involving foreign material, illness, injury, or allergic reaction, it is common to involve the company's risk management department or insurance company. These groups are best equipped to handle the liability risk to the company. If complaint analysis identifies trends in serious complaints, the company may need to conduct a market withdrawal or recall of the implicated product.

10.5.3 Market withdrawal and recall plan

If a company identifies issues from complaints, from internal testing, or from regulatory interventions that put the public health at risk, it has a legal and moral responsibility to remedy the situation.

Every company that manufacturers, holds, or ships food should develop a recall manual. Development of this plan in advance of an actual incident is critical. During an emergency, a company does not have time to search for information on how to conduct a recall. As with food safety plans, a recall plan needs to be developed and managed by a cross-functional team. The recall manual should list roles and responsibilities for recall team members, as well as emergency contact information for all key employees and their back-ups, for customers and suppliers, and for local/state/federal regulatory agencies.

The FDA website provides resources that should be included in the recall manual, including detailed lists of recall information to compile, sample press releases, and even sample recall and effectiveness check letters to send to customers. (FDA, 2011c). Another important resource on the FDA website is the contact information for the District Recall Co-ordinators. As soon as a company suspects that it may need to conduct a recall, the District Recall Coordinator should be contacted. They are well equipped to guide companies through the recall process.

The company's internal recall team should conduct mock recall drills at least once a year to ensure that all contact information is current and that everyone understands their role and responsibilities.

Specific actions to include during a food safety incident depend on the severity of the situation. The FDA (2009) classifies recalls according to the following definitions.

• **Class I recall**: a situation in which there is a reasonable probability that the use of or exposure to a violative product will cause serious adverse health consequences or death.

• **Class II recall**: a situation in which use of or exposure to a violative product may cause temporary or medically reversible adverse health consequences or where the probability of serious adverse health consequences is remote.

• **Class III recall**: a situation in which use of or exposure to a violative product is not likely to cause adverse health consequences.

• **Market withdrawal**: occurs when a product has a minor violation that would not be subject to FDA legal action. The firm removes the product from the market or corrects the violation. For example, a product removed from the market due to tampering, without evidence of manufacturing or distribution problems, would be a market withdrawal.

USDA definitions are similar, with the exception that they do not define market withdrawals (USDA, 2011).

• **Class I**: involves a health hazard situation in which there is a reasonable probability that eating the food will cause health problems or death.

• **Class II**: involves a potential health hazard situation in which there is a remote probability of adverse health consequences from eating the food.

• **Class III**: involves a situation in which eating the food will not cause adverse health consequences.

In general, for FDA-regulated recalls, press releases are issued in the case of Class I recalls. For USDA-regulated recalls, press releases are issued for Class I and Class II recalls (USDA, 2011). State requirements vary. The regulatory agency working with individual companies on the recall will guide them.

In 2007, Congress passed the Food and Drug Administration Amendments Act (FDAAA), which established the Reportable Food Registry (RFR) (FDA, 2010c). The purpose of the RFR is to track food safety incidents to identify trends and better allocate limited regulatory resources. Food companies are required to report food safety incidents that meet the definition of a "reportable food," which basically matches the definition of a Class I recall. The RFR is an electronic portal that allows consistent reporting of incidents. The food company is required to submit the report as soon as possible, but not longer than 24 hours after determining that the food is reportable (FDA, 2010c).

The FSMA grants he FDA significant increases in regulatory power. The FDA is now able to mandate recalls and revoke the registration of facilities that do not comply (FDA, 2013a). Prior to the FSMA, the FDA could only recommend that firms recall dangerous product, and seize the product if the firm refused. The FSMA also requires companies to have recall programs. The USDA does not have the authority to mandate recalls, but can seize product in commerce (USDA, 2011).

10.5.4 Traceability

In the case of a recall, one of the most important pieces of information required by the regulatory agencies is the list of consignees – the customers, distribution centers, or stores that received the violative product. The Bioterrorism Act 2002 requires that FDA-regulated companies record and be able to retrieve within 24 hours of request a list of all previous non-transporter sources and subsequent non-transporter recipients of food that the FDA has reasonable suspicion to believe is adulterated (FDA, 2004). In order to fulfill this requirement, firms need to have proper traceability of all incoming raw materials and outgoing finished products. Traceability must include the following elements for the previous source or subsequent recipient (FDA, 2004).

• The contact information for the immediate non-transporter source or recipient,
• including their company name, address, telephone number, fax number and email
• address, if available.
• Specific variety, brand name, quantity and type of package for the raw material or food.
• Date received or released.
• The same contact information as above for the immediate transporter previous source or subsequent recipient.
• Lot code information of raw materials linked to finished products, linked to customer shipments.

During a recall, regulatory agencies expect food companies to account for the entire quantity of the affected lot of raw material received, whether it is still on hand in the warehouse, has been used to manufacture finished product, was discarded, was used for testing, or was subsequently distributed as a raw material. A full mass balance of material coming in, material still on hand, material used, and material going out for these various uses is required. In order to do this, companies must track lot codes throughout the supply chain from initial receipt to subsequent disposition, including which customers received finished product lots containing the raw material in question. The firm has 24 hours from the time of the regulatory agency request to provide the data. This requires a fool-proof system of record keeping.

Electronic lot code tracking systems reduce the amount of time and increase the accuracy and speed of compiling the data. Lot code tracking also facilitates problem solving and quality improvement. As noted above, information from complaints allows trend and root cause analysis. Tying a lot code from a complaint product back to the finished product batch and, if necessary, the raw materials used to make that batch can be very helpful in solving quality and food safety issues.

10.6 Product quality assurance

It should be obvious that in order to be successful, food companies must produce safe food that meets regulatory requirements. However, product quality is also essential for earning repeat business and sustaining growth.

Consumers expect food to taste and look good, to maintain its quality throughout its shelf life, and to perform as expected. Food manufacturers can use a similar approach to quality as they do for food safety: create a risk-based preventive program. The same basic process and steps used in HACCP can be used to create the quality plan.

Quality "hazards" should be identified and once again, they should include biological, chemical, and physical factors. For biological risks, non-pathogenic bacteria, yeast, and molds can affect product quality and reduce shelf life, and should be controlled in the manufacturing process. For example, lactic acid bacteria can produce sour flavors and create performance issues in dairy products and frozen or refrigerated bakery batters. For chemical risks, manufacturing processes must be controlled so that chemical reactions occur at the proper time and at the proper reaction rate. For example, chemical leavening in a cake batter relies on proper batter temperature, mixing time, and batter holding time so that the leavening produces carbon dioxide during the proper stage of baking, making the cake rise once the starch has set so the cake does not fall. Frozen vegetables must be blanched and cooled prior to freezing to inactivate enzymes that could result in off-flavors, colors, or textures. Physical risks may include factors such as excessive sheer, incorrect particle size distribution or striation, dry ingredient bridging, or unwanted physical phase changes. Food scientists and engineers should identify these undesirable effects and determine where and how to control them in the manufacturing process. Again, as with food safety control, the focus should be on building quality into the process (quality assurance) rather than attempting to test finished product for adherence to specifications (quality control).

However, in-process and finished product testing are fundamental for assuring that critical quality attributes have been met. For example, the performance of the product can be evaluated by baking tests (i.e. dry baking mixes) or by viscosity measurements (i.e. yogurt, fillings, etc.). For microbial quality, coliform testing is a standard practice to monitor sanitary quality, but other tests such as total aerobic plate count or yeast and mold counts would be advised. Putting products in abusive storage conditions to evaluate if mold growth or swollen containers occur can help predict if shelf life will be achieved. Ensuring that products meet sensory specifications requires a sensory program in which trained tasters evaluate products post manufacture or at the end of shelf life. These programs are critical to catch sensory defects due to factors such as incorrect dosing of ingredients, improper processing conditions, and ingredient lot variation. These tests are not designed to be "test and release" – they are for monitoring to ensure customer satisfaction.

It is beyond the scope of this chapter to cover all aspects of quality assurance, and the reader is directed to books on the subject for more information (Luning & Marcelis, 2009; Vasconcellos, 2004).

10.7 Conclusion

Producing safe, high-quality food does not happen by accident. It depends on a commitment from all employees at all levels to develop and consistently implement processes to ensure food is being produced in a safe environment with a focus on building safety and quality into the process. As regulatory requirements become more demanding, as science discovers new risks, as industry develops new methods for ensuring safe and quality food, and as consumers become even more selective about what they eat, food manufacturers will need to continually improve their food safety and quality systems. After all, the goal is to keep consumers safe and pleased with the products, and to protect the company's reputation.

References

ASQ (2012) *History of Quality.* http://asq.org/learn-about-quality/history-of-quality/overview/overview.html, accessed 13 November 2013.

ASQ (2013) *Harold F Dodge: Sampling was his forte, simplicity his creed.* http://asq.org/about-asq/who-we-are/bio_dodge.html, accessed 13 November 2013.

Best M, Neuhauser D (2006) Walter A Shewhart, 1924 and the Hawthorne factory. *Quality and Safety in Health Care* **15**: 142–143.

Centers for Disease Control and Prevention (CDC) (1997) Cryptosporidium and Water: A Public Health Handbook. Atlanta, CA: Centers for Disease Control and Prevention.

Centers for Disease Control and Prevention (CDC) (2013) *Waterborne Disease and Outbreak Surveillance System (Wbdoss).* www.cdc.gov/healthywater/surveillance/index.html, accessed 13 November 2013.

Environmental Protection Agency (EPA) (2013) *Water: Contaminants of Emerging Concern.* http://water.epa.gov/scitech/cec/, accessed 13 November 2013.

Feigenbaum AV (1983) Total Quality Control, 3rd edn. New York: McGraw-Hill.

Fisher NI, Nair VN (2009) Quality management and quality practice: perspectives on their history and future. *Applied Stochastic Models in Business and Industry* **25**: 1–28.

Food and Drug Administration (FDA) (2004) What You Need to Know About Establishment and Maintenance of Records – The Public Health Security and Bioterrorism Preparedness and Response Act of 2002. Washington, DC: US Food and Drug Administration.

Food and Drug Administration (FDA) (2007) Letter to Food Manufacturers Regarding Legal Responsibilities for the Safety of Food Ingredients. Washington, DC: US Food and Drug Administration.

Food and Drug Administration (FDA) (2009) *Recalls Background and Definitions.* www.fda.gov/Safety/Recalls/IndustryGuidance/ucm129337.htm, accessed 13 November 2013.

Food and Drug Administration (FDA) (2010a) *Guidance for Industry – Sanitary Transportation of Food.* www.fda.gov/food/guidanceregulation/guidancedocumentsregulatoryinformation/sanitationtransportation/ucm208199.htm, accessed 26 November 2013.

Food and Drug Administration (FDA) (2010b) Food Allergies: What You Need To Know. Silver Spring, MD: US Department of Health and Human Services. www.fda.gov/Food/ResourcesForYou/Consumers/ucm079311.htm, accessed 13 November 2013.

Food and Drug Administration (FDA) (2010c) *Draft Guidance for Industry, Questions and Answers Regarding the Reportable Food Registry as Established by the Food and Drug Administration Amendments Act of 2007 (Edition 2).* www.gpo.gov/fdsys/pkg/FR-2009-06-16/pdf/E9-14048.pdf, accessed 13 November 2013.

Food and Drug Administration (FDA) (2011a) *Grade "A" Pasteurized Milk Ordinance, 2011 revision.* www.fda.gov/downloads/Food/GuidanceRegulation/UCM291757.pdf, accessed 13 November 2013.

Food and Drug Administration (FDA) (2011b) *Defect Levels Handbook – The Food Defect Action Levels.* www.fda.gov/food/guidanceregulation/guidancedocumentsregulatoryinformation/sanitationtransportation/ucm056174.htm, accessed 26 November 2013.

Food and Drug Administration (FDA) (2011c) *Industry Guidance – Information on Recalls of FDA Regulated Products.* www.fda.gov/Safety/Recalls/IndustryGuidance/default.htm, accessed 13 November 2013.

Food and Drug Administration (FDA) (2013a) *FSMA Proposed Rule for Preventive Controls for Human Food.* www.fda.gov/Food/GuidanceRegulation/FSMA/ucm334115.htm, accessed 26 November 2013.

Food and Drug Administration (FDA) (2013b) *Canned Vegetables.* 21 CFR 155.3. www.accessdata.fda.gov/scripts/cdrh/cfdocs/cfcfr/CFRSearch.cfm?fr=155.3, accessed 26 November 2013.

Food and Drug Administration (FDA) Foods and Veterinary Medicine Program (2013c) The Reportable Food Registry: Targeting Inspection Resources and Identifying Patterns of Adulteration. Third Annual Report. Washington, DC: Food and Drug Administration.

Global Food Safety Initiative (GFSI) (2012) www.mygfsi.com/about-gfsi/about-gfsi-main.html, accessed 13 November 2013.

Global GAP Crops Module. www.globalgap.org/uk_en/for-producers/crops/, accessed 21 November 2013.

Grocery Manufacturers Association (GMA) (2008) Food Supply Chain Handbook, version 1.1. Washington, DC: Grocery Manufacturers Association.

Grocery Manufacturers Association (GMA) (2009) Control of Salmonella in Low-Moisture Foods. Washington, DC: Grocery Manufacturers Association.

Knouse SB, Philips Carson P, Carson K (1993) W. Edwards Deming and Frederick Winslow Taylor: a comparison of two leaders who shaped the world's view of management. *International Journal of Public Administration* **16**: 1621–1658.

Lapworth DJ, Baran N, Stuart M, Ward RS (2012) Emerging organic contaminants in groundwater: a review of sources, fate and occurrence. *Environmental Pollution* **163**: 287–303.

Lundgren MS, Novak PJ (2009) Quantification of phytoestrogens in industrial waste streams. *Environmental Toxicology and Chemistry* **28**: 2318–2323.

Luning PA, Marcelis WJ (2009) Food Quality Management: Technological and Managerial Principles and Practices. Wageningen: Academic Press.

National Advisory Committee on Microbiological Criteria for Foods (NACMCF) (1997) Hazard Analysis and Critical Control Point Principles and Application Guidelines. Washington, DC: Food and Drug Administration and US Department of Agriculture.

National Institute of Allergy and Infectious Disease (NIAID) (2003) Report of the Expert Panel on Food Allergy Research. Bethesda, MD: National Institutes of Health.

National Institute of Standards and Technology (NIST) (2012) *Baldrige Performance Excellence Program.* www.nist.gov/baldrige/, accessed 13 November 2013.

Phillips JD (1997) Pocket Guide to USAF Operational Risk Management. Albuquerque, NM: Air Force Safety Center, Department of the Air Force.

Public Law 104–170, August 3, 1996. Food Quality Protection Act. www.epa.gov/pesticides/regulating/laws/fqpa/backgrnd.htm, accessed 13 November 2013.

Tague N (2005) The Quality Toolbox, 2nd edn. Milwaukee, WI: ASQ Quality Press.

US Department of Agriculture (USDA) (2011) *FSIS Food Recalls.* www.fsis.usda.gov/wps/wcm/connect/f00d2ba8-02ee-4b1f-9069-bb41f223aa88/8080.1.pdf?MOD=AJPERES, accessed 26 November 2013.

US Department of Agriculture (USDA) (2013) *Sampling Manual.* www.ams.usda.gov/AMSv1.0/getfile?dDocName=stelprdc5089769, accessed 13 November 2013.

US Public Health Service and Food and Drug Administration (FDA) (2013) *Food Code 2013.* www.fda.gov/Food/GuidanceRegulation/RetailFoodProtection/FoodCode/ucm374275.htm, accessed 26 November 2013.

Vasconcellos JA (2004) Quality Assurance for the Food Industry: A Practical Approach. Boca Raton, FL: CRC Press.

11 Food Packaging

Joongmin Shin[1] and Susan E.M. Selke[2]
[1]Packaging, Engineering and Technology, University of Wisconsin-Stout, Menomonie, Wisconsin, USA
[2]School of Packaging, Michigan State University, East Lansing, Michigan, USA

11.1 Introduction

Food packaging is defined as a co-ordinated system of preparing food for transport, distribution, storage, retailing, and end-use to satisfy the ultimate consumer with optimal cost (Coles et al., 2003). Food packaging is an essential part of modern society; commercially processed food could not be handled and distributed safely and efficiently without packaging. The World Packaging Organization (WPO) estimates that more than 25% of food is wasted because of poor packaging (WPO, 2009). Thus, it is clear that optimal packaging can reduce the large amount of food waste. Moreover, the current consumer demand for convenient and high-quality food products has increased the impact of food packaging. The purpose of this chapter is to provide background knowledge for those who are interested in or may become involved in the development of food packaging and/or processing.

This chapter consists of four major parts. The development of quality food packaging is impossible if the packaging does not perform its required functions. Therefore, the first part of this chapter discusses the important role packaging plays in maintaining food quality and reducing product waste. The second part is about the properties and forms of food packaging materials and systems to facilitate understanding and appreciation of the major packaging materials, including plastic, paper, metal, and glass, which can affect the quality of food and its shelf life. The third part of the chapter explains aseptic packaging, modified atmosphere packaging and active packaging technologies, which have assumed increasing importance in the food industry in recent years. Finally, the last part of the chapter discusses sustainable food packaging issues, including recycling, biodegradable materials, and package design.

11.2 Functions of food packaging

11.2.1 Containment

The term "containment" means, simply, to contain products to enable them to be moved or stored. It is so basic that it is easily overlooked. However, containment is a key factor for all other packaging functions. All products must be contained for delivery from their point of production to their ultimate destination. Even items that consumers consider as "not a packaged product," such as bulk produce, must be packaged for transportation. Without packaging, products are likely to be lost or contaminated by the environment. For this reason, we have actually used packaging for millennia. Early packaging such as animal skins, baskets, or leaves from trees were used to contain liquids, powders, grains, etc. The containment function significantly contributes to protecting and preserving products during their distribution.

11.2.2 Protection/preservation

There are two broad types of damage that fresh and processed foods sustain during storage and transportation. One is physical damage such as shock, vibration, compressive forces, etc. The other is environmental damage that occurs due to exposure to water, light, gases, odors, microorganisms, etc. A good packaging system will protect or reduce these types of damage to the package contents. For example, an essential aroma or flavor in coffee or juice may easily be evaporated or oxidized without optimum barrier packaging. A shelf-stable food in a can or pouch may maintain its stability (especially against microorganisms) as long as the package provides protection. However, in the case of fresh food products, the ideal

Food Processing: Principles and Applications, Second Edition. Edited by Stephanie Clark, Stephanie Jung, and Buddhi Lamsal.
© 2014 John Wiley & Sons, Ltd. Published 2014 by John Wiley & Sons, Ltd.

protection is usually hard to achieve with packaging alone. Since temperature is a major influence on the degradation of food, it is more economical to control temperature through supply chain modification (refrigeration, freezing, etc.). However, packaging can also add a certain level of protection to slow down temperature changes.

11.2.3 Communication

According to the Fair Packaging and Labeling Act (Federal Trade Commission, 1994), food packaging must identify the product, the net quantity of the contents, name/address of business of the manufacturer, packer, or distributor, as well as (usually) nutritional information. The communication function of packaging not only includes the information provided by the written text, but also elements of the packaging design such as package shape, color, recognized symbols or brands. Beyond giving information, the communication function is expected to entice the consumer to purchase the product. Packaging has been regarded as the "silent salesman" (Judd et al., 1989). Consumers may instantly recognize products through appetizing pictures or distinctive brands on packaging, and even simple transparency of the packaging material can attract consumers by allowing them to view the product inside (Selke, 2012).

Another aspect of the communication function is also important. The Universal Product Code (UPC) is widely used to facilitate rapid and accurate checkout in retail stores. Also, most warehouse and distribution centers track and manage their inventory using UPCs. Currently, by using radiofrequency identification (RFID) tags attached to secondary and tertiary packages, manufacturers are able to get better demand signals from customers and markets. An RFID tag can gather data on items automatically without human intervention or data entry. It identifies, categorizes, and manages product and information flow at important inspection and decision points. In addition, RFID tags can be read all at once (e.g. up to 50 per second), while UPC codes can only be read one at a time (Myerson, 2007). RFID technology looks promising in terms of revolutionizing the supply chain. However, it needs to be economical. The cost of individual tags and antenna reading systems is still high. Another problem is the lack of uniformity in global standards in the area of sensor technology. Sensor providers usually provide their own interfaces to communicate with their own tags (Lopez et al., 2011).

11.2.4 Utility

This function of packaging is sometimes termed "convenience." Consumers demand products that fit into their lifestyles and the packaging industry has had to respond to this. Thus, the utility function encompasses all the packaging attributes that provide added value and convenience to the users of the product and/or package. For example, an important social trend is the growing number of mothers in the workforce and smaller households (people living alone and married couples without children). Unquestionably, food products that offer simplification and convenience have grown in popularity with this group; examples include microwavable entrees, steam-in-pouch vegetables, oven-safe meat pouches, pump-action condiments, and so on.

11.3 Packaging systems

We can categorize packaging systems into four groups: primary packaging, secondary packaging, distribution or tertiary packaging, and unit load.

11.3.1 Primary packaging

The first-level package that directly contacts the product is referred to as the "primary package." For example, a beverage can or a jar, a paper envelope for a tea bag, an inner bag in a cereal box, and an individual candy wrap in a pouch are primary packages, and their main function is to contain and preserve the product (Soroka, 2008a). Primary packages must be non-toxic and compatible with the food and should not cause any changes in food attributes such as color changes, undesired chemical reactions, flavor, etc.

11.3.2 Secondary packaging

The secondary package contains two or more primary packages and protects the primary packages from damage during shipment and storage. Secondary packages are also used to prevent dirt and contaminants from soiling the primary packages; they also unitize groups of primary packages. A shrink wrap and a plastic ring connector that bundles two or more cans together to enhance ease of handling are examples of secondary packages.

11.3.3 Tertiary package

The tertiary package is the shipping container, which typically contains a number of the primary or secondary packages. It is also referred to as the "distribution package." A corrugated box is by far the most common form of tertiary package. Its main function is to protect the product during distribution and to provide for efficient handling.

11.3.4 Unit load

A unit load means a group of tertiary packages assembled into a single unit. If the corrugated boxes are placed on a pallet and stretch wrapped for mechanical handling, shipping and storage, the single unit is referred to as a "unit load." The objective is to aid in the automated handling of larger amounts of product. A fork-lift truck or similar equipment is used to transport the unit load.

11.3.5 Consumer/industrial packaging

Packaging systems can also be categorized into consumer and industrial packaging. Consumer packaging means a package that will be delivered to the ultimate consumer in the retail store. Usually, primary and secondary packages fit in this category. Industrial packaging means a package for warehousing and distribution to the retail store. Tertiary packages and unit loads fit in this category.

Not all package systems are actually composed of a set of primary, secondary, and tertiary packages. For example, the packaging system for potato chips usually consists only of a flexible barrier bag and a corrugated shipping container before they are palletized, while mayonnaise jars are sold in a club store as a two-pack consisting of plastic bottles, shrink wrap, corrugated boxes, and pallet. Often, the distinction between consumer and industrial packaging is more clear-cut than between primary, secondary, and tertiary packaging.

11.4 Materials for food packaging

11.4.1 Plastics

Plastics are a special group of polymers that can be formed into a wide variety of shapes using controlled heat and pressure at relatively low temperatures, compared to metals and glass. Plastics are actually a subcategory of polymers, but in packaging the terms tend to be used interchangeably. There are hundreds of identified "species" of synthetic polymers but in practice, only a few polymers are often used for food packaging. The use of plastics has increased more rapidly than any other

material, and plastic is now the second most used material for packaging. Each plastic has its own unique properties, based on its chemical composition. The performance and interaction with a variety of foods are different for each material. Thus, the plastic material for the packaging of a specific food is selected to function well within the parameters of the application. This subsection focuses on the properties and applications of the plastics that are most commonly used for food packaging.

11.4.1.1 Types of plastics and general properties

11.4.1.1.1 Polyethylene (PE)

Polyethylene, polymerized from ethylene, is the plastic most commonly used for food packaging. PE generally has flexibility, good moisture control, oil and chemical resistance, and good impact strength. PE is also an inexpensive plastic, so for applications where its performance is suitable, this plastic is usually the most economical choice. The simplest form of PE is a completely unbranched structure of $-CH_2-$ units. However, some side branches are always formed during polymerization. If the branches are relatively few and short (2–4 carbon atoms), the structure can fold and pack tightly and yields high-density polyethylene (HDPE). Conversely, if there are many long branches, PE becomes low-density polyethylene (LDPE).

Low-density polyethylene is softer and more flexible, and has lower tensile strength than HDPE. Since it has relatively weak intermolecular forces, LDPE has a low melting temperature, 105–115 °C, so it is a useful material for heat sealing. LDPE also has good impact and tear strength. Common applications for LDPE include stretch wraps, shrink wraps, and many types of bags and pouches. LDPE is also used as an adhesive layer for multilayer composite structures, as a coating on paper to provide water protection (such as in milk cartons), in flexible lids for plastic tubs, in squeezable plastic tubes, in soft squeeze bottles, and in a variety of other applications. By far the majority of its use in packaging is in some form of flexible structure, and packaging is the largest market for LDPE.

Like LDPE, HDPE has good oil and grease resistance. It has better barrier properties than LDPE, since permeation occurs almost exclusively through amorphous areas of a polymer, and HDPE has less amorphous area and higher crystallinity than LDPE. It is, therefore, a good water vapor barrier. However, its barrier to oxygen and carbon dioxide, though improved over LDPE, is still very poor. The improved stiffness of HDPE makes it more suitable than LDPE for rigid or semi-rigid packaging applications, such as bottles, tubs, and trays. In particular, blow-molded bottles for food products are the largest single packaging market for HDPE.

Linear LDPE (LLDPE) is a co-polymer of ethylene and a co-monomer that has short "branches" of a uniform length, distributed randomly in the polymer molecule. The density range of LLDPE is the same as that of LDPE. Compared to LDPE of equal density, because of the structure, LLDPE typically has 50–75% higher tensile strength, 50% or greater elongation, and greater stiffness, along with improved impact strength and puncture resistance. LLDPE is more expensive than LDPE but since the superior performance allows the use of significantly less LLDPE in many applications, switching from LDPE to LLDPE often permits significant economic savings. On the other hand, LLDPE has a higher melting temperature and does not heat seal as well as LDPE so LDPE and LLDPE are often blended to get the best mix of performance and cost. New catalysts allow LLDPE to be produced with the equivalent of long-chain branches, improving its heat seal performance, but with added cost.

11.4.1.1.2 Polypropylene (PP)

Polypropylene is polymerized from propylene gas, which is a relatively low-cost feedstock like ethylene (Soroka, 2008a). As with the PE family, PP has good chemical and grease resistance. Barrier properties of PP are similar to those of HDPE; it is a good water vapor barrier but a poor gas barrier. The polypropylene structure includes methyl groups ($-CH_3$) attached to every other carbon in the polymer main chain; consequently PP has a lower density and a higher glass transition temperature (the temperature above which a plastic becomes soft and flexible) and higher melting temperature than PE. At freezing temperatures, unmodified PP is very close to its glass transition temperature, and therefore tends to have serious brittleness problems. On the other hand, PP is suitable for use with products that require moderately high temperatures such as hot filling or reheating (but not cooking) in a microwave oven.

One of the main uses of PP in food packaging is in closures (caps). Particularly for threaded caps, while HDPE is deformed too readily and loses sealing force under stress, PP maintains its original stiffness and performs successfully. PP also has outstanding living hinge properties, which is particularly useful for caps where an integral hinge is part of the design. The use of oriented polypropylene (OPP) film has increased rapidly in recent food

packaging applications because a wide range of properties (such as tensile strength, shrinkage rate, transparency, etc.) can be manipulated by the orientation. OPP film has improved mechanical strength and water barrier properties compared to unoriented (cast) film. OPP film, however, is still not suited for gas barrier applications. Biaxial orientation improves clarity because the variation of crystallized layers in PP is reduced across the thickness of the film (less light refraction).

11.4.1.1.3 Polystyrene (PS)

Polystyrene is a linear addition polymer of styrene resulting in a benzene ring attached to every other carbon in the main polymer chain. It is a material that is brittle and clear and has high surface gloss. The use of PS in food packaging is aesthetically appreciated, but the material cannot generally be used when extended shelf life is required because of its poor water vapor and gas barrier properties.

The brittleness of PS limits its use where good impact resistance is required. In order to reduce the tendency to fracture, oriented polystyrene (OPS) is commonly used. Typical applications include produce and meat trays, lids for drink cups, and inexpensive party glasses.

High-impact PS (HIPS) is a PS co-polymer with polybutadiene (synthetic rubber). Adding the synthetic rubber causes HIPS to become opaque but improves the impact resistance significantly. HIPS is commonly used for disposable cutlery, tubs, and other thermoformed containers.

Polystyrene foam incorporates small bubbles within the plastic, which increase the cushioning properties and insulating ability of PS. PS foam is usually called expanded polystyrene (EPS). While it is not uncommon for this material to be called styrofoam, that name is proprietary for Dow Chemical Company's EPS building insulation and should not be used to describe a packaging material. Foamed PS is commonly used for disposable coffee cups, meat and produce trays, egg cartons, etc.

11.4.1.1.4 Polyvinyl alcohol (PVOH)/ethylene vinyl alcohol (EVOH)

Polyvinyl alcohol is produced by hydrolysis of polyvinyl acetate, PVA. Due to the hydrogen bonding (OH) group in the structure, PVOH can provide an excellent gas barrier when it is totally dry. However, PVOH is readily water soluble and loses its gas barrier properties in humid conditions, which greatly limits its usefulness for food packaging. Also, pure PVOH is difficult to process and cannot be thermoformed or extruded. PVOH is non-toxic and biodegradable once dissolved. This material is generally used in water-soluble pouches such as those for laundry or dishwasher detergent.

Ethylene vinyl alcohol (EVOH) is, in essence, a co-polymer of ethylene and vinyl alcohol. Modification with the ethylene groups decreases the water sensitivity of the material (so it no longer is soluble in water) and greatly improves its process ability. EVOH has high mechanical strength and toughness, good clarity, very high resistance to oil and organic solvents, and excellent gas barrier properties. It is the most widely used packaging plastic for an oxygen barrier. EVOH is expensive and susceptible to moisture so it is usually not used alone. Other films that provide a reasonably good water barrier are generally used to surround and protect EVOH from exposure to moisture.

11.4.1.1.5 Polyester (PET)

Polyethylene terephthalate (PET) is commonly produced by the reaction of ethylene glycol and terephthalic acid and has been one of the fastest growing food packaging plastics for the last several years. While PET is only one member of the general polyester family, the name "polyester" is generally regarded as PET, as it is the most commonly used plastic of the family. The properties of PET are attractive as a food packaging material; it has very high mechanical strength, good chemical resistance, light weight, excellent clarity, and reasonably high barrier properties. PET is also stable over a wide range of temperatures ($-60\,°C$ to $220\,°C$). Thus, under some circumstances PET can be used for "boil-in-the-bag" products which are stored frozen before reheating or in dual-ovenable containers, since it has resistance to higher temperatures than many other plastics. PET is mostly oriented biaxially to improve its mechanical strength and gas barrier properties.

Polyethylene terephthalate recently took over from HDPE as the most widely used plastic in bottles of all types (HDPE still predominates in the overall container category). Its first large-scale use was in bottles for carbonated soft drinks. Its barrier properties and mechanical strength are much higher than those of HDPE, with excellent transparency. It is more expensive than HDPE but offers improved performance, and its cost has decreased as production has increased.

One disadvantage of PET is its low melt strength (the ability to maintain its general shape in molten status). PET flows like a liquid (rather than like a viscous plastic) at its melting temperature, and has a narrow melting temperature range. These characteristics make forming and sealing difficult. Thus, careful control of processing temperatures is important for PET. In order to increase the melt strength of PET, PET co-polymers can be used. The most widely used co-polymer is glycol-modified PET (PETG). This has greatly reduced crystallinity and reasonably good melt strength, which allows it to be thermoformed or extrusion blow molded into clear bottles.

11.4.1.1.6 Polyvinyl chloride (PVC)

Polyvinyl chloride is produced from vinyl chloride monomers. PVC has high toughness and strength, good dimensional stability, good clarity, excellent oil barrier properties, and good heat sealability. Even though it has many beneficial properties, PVC is easily degraded at high temperature. It decomposes and gives off hydrogen chloride (HCl) around its melting temperature. Thus, unmodified PVC is almost impossible to process due to thermal degradation. Most PVC used in packaging is mixed with a large amount of plasticizer to decrease its melting point and hence reduce thermal degradation. Since the incorporation of plasticizer reduces the attractions between neighboring polymer molecules and reduces the melting point, it also has significant impacts on all the material's properties. For example, highly plasticized PVC films have excellent stretch properties and unique "cling," making them ideal for hand wrapping fresh meats, but the films have poor barrier properties. One of the most widespread uses of PVC is in various blister packages (e.g. medical tablets, toothbrushes, etc.) and clamshells (e.g. USB memory cards, batteries, etc.). Due to the ease of thermoforming, excellent transparency, and relatively low cost, PVC is an attractive material for these types of packages.

The high plasticizer content and the presence of residual vinyl chloride monomer have been a concern for use of PVC as a food packaging material. The levels of vinyl chloride monomer (VCM) in PVC food packaging are currently extremely low. The Food and Drug Administration (FDA) proposed limiting the VCM level to between 5 and 10 parts per billion (ppb) (FDA, 2002). So far, no evidence has been presented that PVC itself is a carcinogen, though VCM is known to be one. However, in recent years, many PVC packages, such as water and vegetable oil bottles, have been replaced by PET. PET is rapidly replacing PVC in thermoformed blister packages and clamshells for food products, as well. However, PVC film is still widely used for the stretch wrapping of trays containing fresh red meat and produce.

11.4.1.1.7 Polyvinylidene chloride (PVDC)

Polyvinylidene chloride has one more chlorine atom per monomer unit than PVC. Like PVC, PVDC is also a very heat-sensitive material. It is decomposed and generates HCl at only a few degrees above its melting temperature. PVDC can be modified with various co-monomers, typically in amounts between 6% and 28%. Properties of PVDC depend on the type as well as the amount of the co-monomer. The most noticeable benefit of this plastic is its excellent barrier properties against water vapor, odors/flavors, and gases. Thus, PVDC plastic is commonly used in food and pharmaceuticals as a barrier packaging material. Since it is a relatively expensive material, PVDC is rarely used alone in packaging containers. It is usually used as a component in multilayer structures or applied as a coating layer. One common use of PVDC coating is as a combination oxygen and moisture barrier and heat-seal layer on cellophane. PVDC is also used in co-extruded structures such as plastic cans, for its excellent barrier properties. PVDC co-polymer film is frequently used as a household wrap or a shrink film.

11.4.1.1.8 Polyamides (PA or nylon)

Nylons, or polyamides (PA), are a whole family of synthetic polymers. The term "nylon," formerly a DuPont trade name, is more frequently used in the US. It is formed by condensation polymerization of a diamine and a dibasic acid or by polymerization of certain amino acids. Various chemical structures can be produced but the amide (−CONH−) functional group is always present in the main structure and is largely responsible for the mechanical strength and barrier properties.

Based on their polymerization method, two categories of PA can be identified. One family is made by polymerizing a mixture of diamines and diacids. These can be named using the number of carbons in the straight-chain diacids and straight-chain diamines they are made from. Thus nylon 6,6 is formed from a 6-carbon amine plus a 6-carbon acid. The other family of PAs is formed from only one type of monomer, an amino acid, and is identified by the number of carbons in that amino acid. Nylon 6, for example, is formed from a 6-carbon amino acid. These two types of PA have similar physical and chemical properties. The number of carbons in the structure affects the properties of PA. Longer carbon chains (more CH_2 groups) result in plastics with a lower melting point and increased resistance to water vapor.

A semi-crystalline polyamide, MXD6, was introduced in the 1980s. It is formed by condensing metaxylene diamine and adipic acid (a straight-chain 6-carbon carboxylic acid). MXD6 provides much higher water barrier properties than conventional PAs.

Polyamides in general provide excellent optical clarity, oil and chemical resistance, and mechanical strength over a wide range of temperatures. In packaging applications, the use of PA is often found in the form of film for high-temperature sterilization or hot-filling applications. PAs also act as flavor and gas barriers, but have poor water vapor barrier properties. Thus, for most applications PAs are combined with other materials, such as LDPE and ionomer, to add water vapor barrier and heat sealability properties. These types of materials are used in the vacuum packaging of meats and cheeses.

11.4.1.1.9 Polycarbonate (PC)

Polycarbonate is made from carbonic acid and bisphenol A. The proper name for PC is polybisphenol-A carbonate (as is the case for "polyester," the generic name has come to mean this most used member of the family). The material is a very tough and rigid plastic with excellent clarity. However, it has a relatively high permeability to both water vapor and gases. Thus, it must be coated if good barrier properties are required. Its main packaging uses are large refillable water bottles and refillable milk jugs. PC is also used to a very limited extent in food packaging as a component of multilayer structures to provide transparency and in high strength containers (with high barrier materials). For example, multilayer beer or beverage bottles, containing PC and a thin barrier layer such as EVOH or PVDC, can be used to extend shelf life while still providing transparency (Hernandez, 1997). However, the application is limited by the relatively high cost of PC.

11.4.1.1.10 Ionomers

Ionomers have an unusual structure compared to other plastics. These plastics contain ionic as well as covalent bonds, while other packaging plastics have only covalent bonds in their structure. The plastic is manufactured by neutralization of ethylene-based co-polymers containing acid groups with a base containing a metal (such as sodium, zinc, lithium, etc.). The result is positively charged metal ions and negatively charged ions in the base polymer chain, creating random cross-link like ionic bonds between the polymer chains. This combination of ionic and covalent bonds creates a polymer with excellent toughness as well as transparency. Ionomers also have excellent adhesion properties so they are commonly used in composite structures with film, paper, or aluminum foil to provide an inner layer with excellent heat sealability. They are especially useful in applications where the sealing layer may become contaminated, making it difficult to provide strong heat seals, such as in packaging of processed meats. The excellent impact and puncture resistance of ionomers, even at low temperatures, is also useful for skin packaging of sharp items such as meat cuts containing bone (as well as for inherently sharp products such as knives). Another advantage is ionomers' high infrared absorption, which allows faster heat shrink packaging processes. On the other hand, ionomers are relatively high-cost materials and have relatively poor gas barrier properties.

11.4.1.2 Additives

Additives are auxiliary ingredients intended to modify or enhance a plastic's properties without recognizably changing its chemical structure. Nearly all commercial resins are blended with additives before or during processing into their finished forms. In food packaging, all additives used must comply with the regulations of the appropriate food regulatory authority. For example, the maximum allowable migration of an additive from a packaging material is generally controlled by FDA regulations (FDA, 2005a).

Stabilizers are one of the most common ingredients in plastic resins. Plastics are susceptible to chemical changes during processing, as a result of their exposure to heat, mechanical force (shear), and usually oxygen. Stabilizers minimize the thermal oxidation or other reactions undergone by the polymer.

Plasticizers perform a lubricating function within the plastic material and make it more flexible. They also lower the glass transition and melting temperatures (as discussed in sections 11.4.1.1.6 and 11.4.1.1.7).

Colorants alter the color of plastic materials. Usually pigments (or sometimes dyes) are blended into the plastic as master batch color concentrates at the forming machine. It is possible to obtain virtually any desired color from most plastics, with the appropriate selection of colorants.

Nucleating agents encourage the formation of crystallinity in the polymer. The agents are often used to improve the clarity of PP. By adding a nucleating agent, the crystallite size in the PP can be minimized without any overall decrease in crystallinity, resulting in improved clarity without loss of other performance properties.

There are also additives designed to modify the surface of the plastic. These include slip agents, antislip agents, antiblocking agents, lubricants, mold release agents, antifogging agents, and antistatic agents.

11.4.1.3 Processing and converting of plastics

11.4.1.3.1 Extrusion

Most plastic forming processes begin with melting the plastic in an extruder, except for compression molding and solvent casting techniques. In the extruder, thermoplastics are mixed and softened, which enables shaping into some desired form when they reach an optimum temperature and pressure. Extrusion is used to convert plastic resin into a sheet, film, or tube. There are two main sources of the heat that melts the plastic resin. One is external heating, usually electric heater bands, and the other is friction within the extruder, as the plastic is conveyed. When the melted plastic exits the extruder, it is sent through a die of a desired shape.

In order to convert the molten plastic into film, two processes are commonly used: the cast film process (also called the flat film process) and the blown film process (also called the tubular process). For cast film, the molten plastic is extruded through a slit die, and then it is cooled on a chilled roller (or sometimes in a quenching water bath), as shown in Figure 11.1. Due to the rapid cooling, cast film generally has better clarity than blown film, and the process also results in more uniform thickness. For blown film, the molten plastic is extruded though an annular die. The plastic exits the extruder in the shape of a hollow tube, which is expanded with internal air pressure. As is shown in Figure 11.2, the film is stretched in the longitudinal and circumferential directions during the process which results in biaxial orientation of the film. The blown film process is usually more economical than the cast film process for long runs, and the mechanical

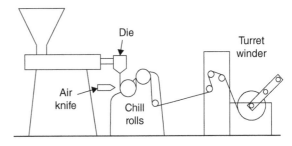

Figure 11.1 Cast film process (from Selke et al., 2004).

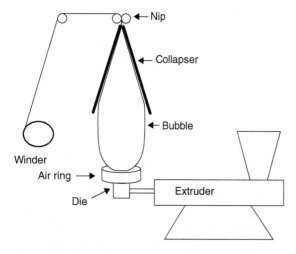

Figure 11.2 Blown film process (from Selke et al., 2004).

properties are often better than cast film, due to the biaxial orientation. Plastic resins for the blown film process must have good melt strength. Thus, not all polymer films can be produced by the blown film process. Blown film tends to have more variation in thickness and less clarity than cast film. Most plastic bags are made using the blown film process.

11.4.1.3.2 Thermoforming

In thermoforming, a plastic sheet heated to its optimum temperature (near its melting temperature) is placed over a mold, and pressure is applied to stretch it into a designed shape. The forming pressure (the pressure used to stretch the sheet) is obtained by pneumatic and/or mechanical means. In the simplest system, a vacuum is drawn through the mold and the forming pressure comes simply from atmospheric pressure pushing the plastic into the mold. In order to get good results from thermoforming, it is best if the plastic sheet is relatively easy to form and the molded shape is relatively simple. Typical thermoplastics used for thermoforming include HIPS, PVC, PP, PA, and PET. Packages made by thermoforming include clamshells, blister packages, and some tubs.

11.4.1.3.3 Injection molding

In an injection molding operation, the molten plastic (from the extruder) is injected into a mold with the desired shape, cooled, and ejected from the mold. Injection molding is widely used for making threaded closures

(caps), tubs, and jars, and for making the initial shapes (preforms) used for injection blow-molded bottles. Injection molding can provide accurate and sophisticated forms as well as a high production rate. However, it is a relatively expensive process and short production runs are not economical. Resins that are commonly used for injection molding include LDPE, HDPE, PP, and PET.

11.4.1.3.4 Blow molding

In the blow molding process, an initial shape (called a parison) is surrounded by a mold with the desired shape, and air is blown into the parison to force it to expand against the wall of the mold. The mold is then opened and the solidified product is ejected. There are two major categories of blow molding: extrusion blow molding and injection blow molding.

Extrusion blow molding is the most used blow molding process. The parison is continuously extruded as a hollow tube. When the parison reaches the proper length, the open halves of the mold are closed around the parison. This tube is usually extruded in a downward direction, and air pressure inflates the hollow tube into the shape of the container. The tubular parison is cut off at the top and pinched shut at the bottom before blowing. Thus, all containers produced by extrusion blow molding have a pinch-off line across the bottom and mold parting lines on the sides. This is a fast and inexpensive process. Extrusion blow molding is commonly used for HDPE 1/2 gallon and 1 gallon milk and water bottles where high barrier properties (against gas and flavors) are not required. The plastic for extrusion blow molding must have sufficient melt strength to maintain that hollow tube shape and permit it to be formed further. Another limitation is rather limited control over the distribution of wall thickness.

Injection blow molding starts with injection molding of a preform. The preform usually has a test tube-like shape and is nearly the same length as the height of the bottle. After injection molding, the hot preform is placed into the container mold and air pressure is used to stretch it into the mold shape. Sometimes, the preform is produced earlier, cooled, and then reheated before the blowing process. After cooling, the mold is opened and the finished container is ejected. This process provides better dimensional accuracy, including uniform wall thickness and a high-quality neck finish (threaded area). Another advantage of this process is that it produces less scrap than extrusion blow molding.

Injection stretch blow molding is similar to injection blow molding but the length of the preform is considerably shorter than the height of the bottle. During the blow molding process, the preform is stretched in both the longitudinal and transverse directions, so the container is biaxially oriented. The finished product has better mechanical strength, better gas and water vapor barrier properties, and better transparency due to the orientation. Careful control over the temperature profile in the parison is extremely important for successful stretch blow molding. Injection stretch blow molding is used to produce PET bottles for carbonated beverages, sports drinks, juices or other bottles, including those used in hot-fill and aseptic processing.

11.4.1.4 Plastic permeability

While glass and metal have almost perfect barrier properties, plastics are permeable in various degrees to gases, water vapor, organic vapors, or other low molecular weight compounds. When the gas or vapor compounds pass through the plastic, they must solubilize at first, diffuse through the material, and finally desorb on the other side. Such vapor or gas mass transfer (or permeability) has a significant impact on the shelf life, quality, and safety of foods. For example, the expected loss of moisture from juice or of CO_2 from carbonated beverages, depending on storage conditions, can be estimated through quantitative evaluation of the package permeability. Similarly, the time required to reach certain atmospheric conditions in the package that favor the growth of aerobic or anaerobic pathogens can also be estimated through calculation of the permeability. These are just two of many examples that could be cited. This section only provides a general theoretical background for permeability and discussion of factors affecting the permeability of gases/vapors through materials.

11.4.1.4.1 Basic theory of permeability

Under steady-state conditions, the permeability coefficient of a non-porous plastic is described by the following equation (Crank, 1975):

$$P = D \times S \qquad (11.1)$$

where P is the permeability coefficient, D is the diffusion coefficient, which is a measure of how fast the permeant compounds are moving through the plastic polymer, and S is the solubility coefficient that shows how much permeant is contained within the plastic.

Calculation of permeability of plastics is based on Fick's first law of diffusion. A gas or vapor will diffuse through a

plastic film at a constant rate if a constant pressure difference is maintained across the plastic:

$$F = -D\frac{\partial c}{\partial x} \qquad (11.2)$$

where F is the flux (vapor mass transfer), D is the diffusion coefficient, c is concentration of permeant in the plastic and x is the distance across the plastic (or thickness). If the flow and diffusion rate are constant, the above equation gives the following equation:

$$F = -D\frac{c_2 - c_1}{L} \qquad (11.3)$$

where c_1 and c_2 are the concentrations of the diffusing compound on the two sides of the plastic film. L refers to the thickness of the film. The flux (F) can be defined as the amount of permeant (Q) passing through a surface of unit area (A) in time (t). Thus, the above equation can be rewritten as follows:

$$F = \frac{Q}{At} \qquad (11.4)$$

$$Q = D\frac{(c_1 - c_2)At}{L} \qquad (11.5)$$

In the case of gas permeation, it is easier to measure the equilibrium vapor pressure (p) rather than the actual concentrations of the permeant in the film. Using Henry's law, the concentrations of the permeants (c) can be expressed as:

$$c = S p \qquad (11.6)$$

where p is the partial pressure and S is the solubility coefficient. Then, by combining Equations 11.5 and 11.6, the following equation is formed:

$$Q = DS\frac{(p_1 - p_2)At}{L} \qquad (11.7)$$

The DS can be replaced with P, the permeability coefficient, based on Equation 11.1. Finally, the permeability coefficient (P) can be rewritten as:

$$P = \frac{QL}{At(p_1 - p_2)} \quad \text{or} \quad \frac{QL}{At\Delta p} \qquad (11.8)$$

The permeation of compounds through a plastic is described by a diffusing model, using Henry's and Fick's

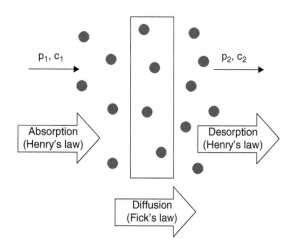

p_1, c_1

p_2, c_2

Absorption
(Henry's law)

Desorption
(Henry's law)

Diffusion
(Fick's law)

● Gas or vapor molecules

Figure 11.3 General mechanism of mass permeation through a plastic film.

laws to obtain the expression above. The mechanism is shown in Figure 11.3.

The permeability coefficient (P) can be used to estimate the shelf life of a product or to determine an appropriate package to provide the desired shelf life. There are various units available to express P. In the SI system, common units for P are:

$$P = \frac{\text{Quantity of permeant} \times \text{thickness}}{\text{area} \times \text{time} \times \text{partial vapor pressure}} = \frac{cm^3 \times cm}{cm^2 \times s \times Pa}$$

Instead of the permeability coefficient, the quantity of permeant flowing per unit area per unit time, such as the oxygen transmission rate (OTR) or water vapor transmission rate (WVTR), is frequently used to express the barrier characteristics of plastic materials. For example, OTR is related to P as follows:

$$P = \frac{QL}{At\Delta p} = OTR\frac{L}{\Delta p} \quad (11.9)$$

Example 11.1 The oxygen transmission rate of PET film with 0.1 cm thickness is 0.41 $cm^3 \cdot cm^{-2} \cdot s^{-1}$. The partial pressure difference (Δp) through the film is 21278 Pa. What is the oxygen permeability coefficient of the film?

Using Equation 11.9:

$$OTRL\frac{L}{\Delta p} = \frac{0.41 cm^3}{cm^2 \times s} \times \frac{0.1 cm}{21278 Pa}$$

$$= 1.926 \times 10^{-6} cm^3 \cdot cm \cdot cm^{-2} \cdot s^{-1} \cdot Pa^{-1}$$

Example 11.2 A food stored in a PET jar with a wall thickness of 0.1 cm and a surface area of 400 cm^2 becomes rancid if it absorbs 3 cm^3 of oxygen. The O_2 permeability coefficient of PET is 1.2×10^{-15} $cm^3 \cdot cm \cdot cm^{-2} \cdot s^{-1} \cdot pa^{-1}$. The oxygen vapor pressure inside the container (Pi) was 0 Pa and outside the container (Po) was 21278 Pa. What is the shelf life of this product?

Using Equation 11.9:

$$P = \frac{QL}{At\Delta p}$$

The t in Equation 11.9 is the shelf life of the product if Q is the maximum allowable amount of gas inside the package. Thus, the expression can be rewritten as:

$$t = \frac{QL}{AP\Delta p}$$

$$= \frac{3 (cm^3) \times 0.1 (cm)}{400 (cm^2) \times 1.2 \times 10^{-15} (cm^3 \cdot cm \cdot cm^{-2} \cdot s^{-1} \cdot pa^{-1}) \times 21278 (pa)}$$

$$= 29373061 s = 400 day$$

Thus, the shelf life of the product in the PET package is 400 days.

11.4.2 Paper and paper-based materials

Paper and paperboard are the most commonly used packaging materials in the world. In the US, over 50% of all packaging is paper based, including food packaging. Paper is produced from plant fibers. More than 95% of paper is made from wood, and the remaining sources are mainly agricultural by-products, such as straw (of wheat, rye, barley, and rice), sugar cane bagasse, cotton, flax, bamboo, corn husks, and so on. Making pulp is the initial stage in making paper or paperboard, and the quality of the paper is closely related to the quality of the pulp. Pulping can be done using mechanical, chemical, or a combination process. Mechanical pulping produces papers that are characterized by relatively high bulk and low strength as well as relatively low cost. Their use in packaging is very limited. Chemical pulping produces stronger and higher quality paper and is also more expensive. Combination processes are intermediate in cost and properties. The pulp produced may be unbleached or bleached to various degrees, and various sizing agents and other additives are used to control functions and appearance.

11.4.2.1 Types of paper and their applications

Different varieties of papers are used in packaging applications. This section will give a brief overview of the major type of papers used for food packaging.

11.4.2.1.1 Kraft paper

Kraft paper is the most used packaging paper and has excellent strength. It is made using the sulfate (kraft) chemical pulping process, and is usually produced from soft wood. Unbleached kraft is the strongest and most economical type of paper. It is used in uncoated form for bags and in the production of corrugated board for boxes, as well as for drums, cans, and other applications. It can be coated or laminated for improved barrier properties and additional strength, or creped for cushioning.

11.4.2.1.2 Bleached paper

Bleached paper is produced using bleached pulps that are relatively white, bright, and soft. Its whiteness enhances print quality and aesthetic appeal. It is generally more expensive and weaker than unbleached paper. This type of paper is used uncoated for fancy bags, envelopes, and labels. However, it is often clay coated for overwraps and labels.

11.4.2.1.3 Greaseproof and glassine

Greaseproof is a dense, opaque, non-porous paper made from highly refined bleached kraft pulp. The prolonged beating during processing results in short fibers. Glassine derives its name from its glassy smooth surface. After the initial paper making process, it is passed through an additional set of calendars (supercalendared) in the presence of steam. The result is a glossy, transparent sheet with good grease and oil resistance (it does not have complete oil barrier but is still fairly resistant to oil). These papers are often used for packaging butter and other fatty foods.

11.4.2.1.4 Waxed paper

Waxed paper is produced by adding paraffin wax to one or both sides of the paper during drying. Many base papers are suitable for waxing, including greaseproof and glassine. The major types are dry waxed, wet waxed, and wax laminated. Dry-waxed paper is produced using a heated roller to allow the wax to soak into the paper. The paper does not have a waxy feel and does not have a continuous wax film on the surface. Wet-waxed paper is produced when the wax is cooled quickly after it is applied, so that the wax remains on the surface of the paper. Wax-laminated paper is bonded with a continuous film of wax which acts as an adhesive, so that it can provide both moisture barrier and a heat-sealable layer.

11.4.2.1.5 Vegetable parchment

Vegetable parchment is produced by adding concentrated sulfuric acid to the surface of the paper to swell and partially dissolve the cellulose fibers. It produces a grease-resistant paper with good wet strength (meaning that it maintains its strength well when it is wet). Vegetable parchment is odorless, tasteless, boilable, and has a fiber-free surface. Labels and inserts on products with high oil or grease content are frequently made from parchment. Parchment can also be treated with mold inhibitors and used to wrap foods such as cheese (Robertson, 2007).

11.4.2.2 Paperboards and their applications

Paper and paperboard can be distinguished by thickness (caliper) and weight of the material. Material is generally termed "paperboard" when its thickness is more than 300 μm and/or its weight exceeds 250 g/m^{-2} (Hanlon et al., 1998). Various types of paperboard are manufactured but paperboard for food packaging generally includes whiteboard, linerboard, foodboard, cartonboard, chipboard, and corrugated board.

11.4.2.2.1 Whiteboard

Whiteboard is made with a bleached pulp liner on one or both sides to improve appearance and printability, and the remaining part is filled with low-grade mechanical pulp. Whiteboard is suitable for contact with food and is often coated with polyethylene or wax for heat sealability. It is used for ice cream, chocolate, and frozen food cartons.

11.4.2.2.2 Linerboard

Linerboard is usually made from softwood kraft paper and is used for the solid faces of corrugated board. Linerboard may have multiple plies. Increasingly, linerboards containing recycled fiber are being used in packaging. The higher quality layer is always placed on top.

11.4.2.2.3 Foodboard

Foodboard is used to produce cartons that are suitable for direct food contact. It is normally made using 100% virgin pulp but recently recycled pulp using an innovative barrier coating with a sustainable coating material is also being used. Foodboard is a sanitary, coated, and water-resistant paperboard. It should be designed to protect against migration of outside contaminants (such as ink or oil) into packaged food. Foodboard can be used for all types of foods, particularly frozen and baked foods.

11.4.2.2.4 Cartonboard (boxboard)

Cartonboard is used to make folding cartons and other types of boxes. Most often, this is a multilayer material made of more than one type of pulp, and often incorporating recycled fibers. To improve its appearance, it may be clay coated or may have a ply of virgin fibers on one or both surfaces.

11.4.2.2.5 Chipboard

Chipboard is the lowest quality and lowest cost paperboard, made from 100% recycled fiber, and is not used in direct contact with foods. Outer cartons for tea and breakfast cereals are some examples. It is also commonly lined with whiteboard to produce a multi-ply board such as cartonboard.

11.4.2.2.6 Corrugated board

Corrugated board has an outer and inner lining of kraft paper with a central corrugating (fluted) material. Corrugated boards resist impact, abrasion, and compression forces so they are commonly used in shipping containers.

11.4.2.3 Paperboard cartons and other containers for food packaging

Folding cartons are made of paperboard, typically between 300 and 1100 μm in thickness. They are creased, scored, cut, and folded into the desired shape. The cartons usually are shipped flat to the product manufacturer (or carton assembler). Paperboard can be coated or laminated when improved function is desired. For example, wax lamination provides moisture resistance, glassine lamination provides oil/grease resistance, and PE lamination provides heat sealing and moisture resistance. Clay and mineral coatings on the exterior provide improved appearance and printing quality.

Molded pulp containers are produced by placing aqueous slurry of cellulosic fibers into a screened mold. Since molded pulp containers are regarded as a sustainable packaging material, they are gaining popularity. Typical applications in food packaging include egg cartons, food trays, and other tray type containers for fruits. Molded pulp containers can be laminated with thermally resistant plastics such as PET to provide functionality as dual-ovenable containers (suitable for use in conventional ovens as well as microwaves).

11.4.3 Metals

11.4.3.1 Types of metal and general properties

Metal is used in packaging in a variety of applications, from rack systems to tuna cans. For food packaging, four types of metal are commonly used: steel, aluminum, tin, and chromium.

Steel and aluminum are commonly used in production of food cans, and are the primary materials for metal packaging. Food cans are most often made of steel, and beverage cans are usually produced from aluminum. Steel tends to oxidize when it is exposed to moisture and oxygen, producing rust. Therefore, tin and chromium are used as protective layers for steel. Tinplate is a composite of tin and steel made by electrolytic coating of bare steel with a thin layer of tin to minimize corrosion. If chromium is used to provide corrosion protection instead of tin, the resulting material is called electrolytic chromium-coated steel (ECCS) or tin-free steel (TFS). ECCS is less resistant to corrosion than tinplate but has better heat resistance and is less expensive.

11.4.3.2 Can forming process

There are two basic styles of cans: three-piece and two-piece. As the name indicates, a three-piece can is made from three pieces (a body blank and two ends), and a two-piece can is made from two pieces (one body and one end).

11.4.3.2.1 Three-piece cans

Hermetically sealed three-piece cans consist of a can body and two endpieces. The process for making three-piece cans starts with rolling steel into a rectangular strip (about 1.8 mm thick). Next, coating is applied, depending on the requirements of the food. Generally, more acidic food has a higher coating weight on the inner side of the strip. Additional coating is applied to improve the surface brightness, to resist corrosion, and to prevent interaction with foods. More details about coating are provided in section 11.4.3.4. In order to make the can body, the steel sheet is cut into a rectangular piece and formed into a cylindrical shape, and its side is seamed. For food packaging, most three-piece cans are made using welded side seams. Soldering was the original method, but has been mostly discontinued due to concerns about lead contamination. Lead solder is no longer permitted for packaging of products sold in the US as well as in many other countries. Next, the body is flanged, and one end of the can is attached using a double seaming process. The double seam forms a hermetic seal by interlocking the end cover and body of the can with a rubbery sealing compound between them. After filling, the second end is attached in the same manner. The structure and main components of a double seam are shown in Figure 11.4.

Cans for food packaging are often subjected to external and internal pressure during processing and storage. The can body may be rippled or beaded to increase its strength.

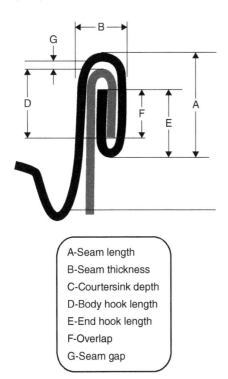

A-Seam length
B-Seam thickness
C-Courtersink depth
D-Body hook length
E-End hook length
F-Overlap
G-Seam gap

Figure 11.4 Structure and main components for double seaming.

11.4.3.2.2 Two-piece can

A two-piece can does not have a side seam. It is composed of the body and one cover for the top. There are two main methods of producing two-piece cans: draw and iron (DI) and draw and redraw (DRD). In the DRD process, a metal blank is punched into a die (drawn) to form a shallow can shape. The diameter of the cup produced in the initial draw is then further reduced by a similar redraw process. Some very short cans do not require this second draw so are not really DRD cans – these are referred to as shallow draw cans. The wall and base thickness, as well as the surface area, are the same as those of the original blank. The DRD process is commonly used for TFS cans. After the body is shaped, the can is trimmed, often beaded, and after filling the end is double seamed onto the can body.

In the DI process, typical for aluminum beverage cans, a circular disk shape blank is cut and drawn to a shallow cup. The cup is usually redrawn once, and then is passed through a series of ironing dies that extend and thin the walls. Thus, the base of the can ends up thicker than the walls. DI cans are usually used for carbonated beverages. The ends of DI cans are often necked (narrowed) to reduce the size of the end, as this improves the ability to stack the cans, and lowers the overall cost (by saving metal).

11.4.3.3 Metal foil and containers

Aluminum foil is the most commonly produced metal foil. It is manufactured by passing aluminum sheet between a series of rollers under pressure. Pure aluminum (purity >99.4%) is passed through rollers to reduce the thickness to less than 150 µm and then annealed to provide dead-folding properties. Foil is widely used for wraps (9 µm), bottle caps (50 µm), and trays for ready-to-eat meals (50–100 µm). Aluminum foil has excellent barrier properties against gases and water vapor. Thus, it is also used as the barrier material in laminated films for packages, such as those in retort pouches.

Collapsible aluminum tubes can be used for the packaging of viscous products. The collapsible tube allows the user to apply precise amounts of the products when required because the tubes permanently collapse as they are squeezed and prevent air being drawn into the container again. These types of tube applications for foods are rare in the US but more common in Europe. Typical applications include condiments packages such as mustard, mayonnaise, and ketchup.

Metallized films are also used in food packaging applications for their excellent barrier properties. The principle of metallization is to use a vapor deposition process to deposit an extremely thin layer of metal on another substrate (film). The typical thickness of the aluminum layer in metallized film is 400–500 Å. Oriented polypropylene (OPP) is the material most often used for metallized film applications. Nylon and PET are also common film substrates.

11.4.3.4 Coating

One of the major problems associated with metal packaging is corrosion. The inside of a can is normally coated to prevent interaction between the can and its contents. The outside of a can is generally also coated to provide protection from the environment. Coatings used in cans need to provide an inert barrier (must not impart flavor to the product), must usually resist physical deformation during fabrication, be flexible, spread evenly and completely cover the surface of the metal, and the coating must adhere well to the metal and be non-toxic (for food packaging).

Two methods are used for the application of protective coatings to metal containers: roller coating and spraying. The roller coating process is used for external coating of cylindrical can bodies and spraying is used if physical contact is difficult, such as for the inside surface of can bodies. Typical can coatings are polymers applied in a liquid state and then dried after application by solvent removal, oxidation or heat-induced polymerization. The commonly used coating materials for food packaging include the following (Robertson, 2007).

- **Epoxy-phenolic compounds**: these are used for all types of steel and cans. They are resistant to acids and have good heat resistance and flexibility. They are used for beer, soft drinks, meat, fish, fruits, vegetables, and so on. They are especially suitable for acidic products and have excellent properties as a basecoat under acrylic and vinyl enamels.
- **Vinyl compounds**: vinyl compounds have good adhesion and flexibility, and are resistant to acid and alkaline products. However, they are not suitable for high-temperature processes such as retorting of food. They are used for canned beers, wines, fruit juices and carbonated beverages and as clear exterior coatings.
- **Phenolic lacquers**: phenolic compounds are inexpensive and resistant to acid and sulfide compounds. They are used for acid fruits, fish, meats, soups, and vegetables.
- **Polybutadiene lacquers**: polybutadiene compounds have good adhesion, chemical resistance, and high heat resistance. They are used for beer and soft drinks, soups, and vegetables (if zinc oxide is added to the coating).

- **Acrylic lacquers**: acrylic lacquers are expensive coating materials. They take heat processing well and provide an excellent white coat. They are used both internally and externally for fruits and vegetables.
- **Epoxy amine lacquers**: epoxy amine lacquers are also expensive. They have excellent adhesion, heat and abrasion resistance, and flexibility, and no off-flavor. They are used for beer and soft drinks, dairy products, and fish.
- **Alkyd lacquers**: alkyd lacquers are low cost and used mostly as an exterior varnish over inks (due to flavor and color problems inside the can).

11.4.4 Glass

Glass is defined as "an amorphous inorganic product of fusion that has been cooled to a rigid condition without crystallizing" (ASTM, 2003). For food packaging, bottles or jars are the types of glass packaging most often used, bottles being the primary use. In the US, 75% of all glass food containers are bottles.

Glass is made primarily of silica, derived from sand or sandstone. For most glass, silica is combined with other raw materials in various proportions. For example, soda-lime glass, the glass typically used for food packaging, contains silica (68–73%), limestone (10–13%), soda ash (12–15%), and alumina (1.5–2%). Glass is inert to a wide variety of food and non-food products, very rigid and strong against pressure, transparent, and nonpermeable (excellent barrier properties). However, glass has disadvantages due to its heavy weight and fragility. For food packaging, the fragility has caused some safety concerns such as the possibility of the presence of chipped glass in food products. Glass for food packaging has declined over the last three decades, with glass losing market share to metal cans and, increasingly, to plastics. However, it still plays an important role in packaging.

11.4.4.1 Forming of glass

The glass making process begins with weighing out and mixing of the raw materials and introduction of the raw material to the glass melting furnace, which is maintained at approximately 1500 °C. Cullet, broken or recycled glass, is also an important ingredient in glass production. In the melting furnace, the solid materials are converted to liquid, homogenized, and refined (getting the bubbles out). At the end of the furnace, a lump of molten glass, called a "gob," is transferred to the glass forming process.

For food packaging, glass can be formed using the blow-and-blow process, wide-mouth-press-and-blow process, or narrow-neck-press-and-blow process. In the blow-and-blow process, compressed air blows the gob into the blank mold of the forming machine and creates the shape of the parison. Then, the completed parison is transferred into the blow mold where air blows the parison to form a final shape. In the wide-mouth-press-and-blow process, a metal plunger is used to form the gob into the parison shape, instead of using air blowing. As in the blow-and-blow process, the compressed air blows the container into its final shape. In the narrow-neck-press-and-blow process, the overall process is similar to the wide-mouth-press but a much smaller metal plunger is used to make the parison shape. Less than 38 mm of finish diameter is regarded as narrow mouth and over 38 mm is called wide mouth. (Glass Packaging Institute, 2012). The press-and-blow process provides increased productivity, less weight, and more uniform wall thickness compared to the blow-and-blow process. Beer or beverage bottles are common applications for the narrow-neck-press-and-blow process.

Once the finished container is formed, it is transferred to a large oven known as a lehr for the annealing process. The function of annealing is to reheat and gradually cool the container in order to relieve the residual thermal stress. Surface coatings are often applied to the glass container for strengthening and lubricating the surface. Hot-end coatings are applied before the container enters the annealing oven (when the glass is still hot due to the previous forming process). Hot end coatings consist of tin chloride (which reacts to form tin oxide) or organo-tin. These compounds are applied in vapor form and leave a rough high-friction surface on the glass container, which provides a good adhesive surface for the cold-end coatings. They also supply hardness, fill in minor cracks, and compress the glass surface. After the glass containers are cooled, a cold-end coating is applied to increase lubricity and minimize the scratching of surfaces. These coatings typically consist of lubricants such as waxes, polyethylene, polyvinyl alcohol, and silicone. Since cold-end coatings make the glass surface more slippery, it is important to check the compatibility of the cold-end treatment with adhesives used in labeling.

11.5 Other packaging types

11.5.1 Aseptic packaging

Aseptic packaging is the filling of a commercially sterilized product into a sterilized container under aseptic conditions and then sealing it hermetically to prevent contamination. Therefore, the sterility can be maintained throughout the handling and distribution process. The aseptic packaging permits shorter heat exposure for the food than typical thermal sterilization processes such as canning and retorting; therefore it generally results in superior food quality compared to typical thermal sterilization. Because of the separate package and food product sterilization, the selection of material and container design can be more flexible than in traditional thermal sterilization.

The required microbial count reduction (referred as log reduction) for the sterilization of a food contact packaging material is determined by the type of product. For non-sterile acidic products (pH <4.5), a minimum 4-log reduction in bacterial spores is required. For sterile, neutral, low-acid products (pH >4.5), a 6-log reduction is required. In addition, if sterility against *Clostridium botulinum* endospores is required, a 12-log reduction is required, and the process is called commercial sterilization. The sterilization methods for aseptic packaging materials include irradiation (such as ultraviolet rays, infrared rays, and ionizing radiation), heat (such as saturated steam, superheated steam, hot air, hot air and steam), and chemical treatment (such as hydrogen peroxide, ethylene oxide, peracetic acid). These processes can be used either individually or in combination.

The type of aseptic packaging material used is influenced by the nature of the product, the cost of both product and packaging, and the preference of consumers. The most widely used aseptic package is the paperboard laminated carton. The typical structure of this material consists of unbleached or bleached paperboard, polyethylene, and aluminum foil. The laminated structure is impermeable to liquid, gas, and light. The detailed structure of a typical paperboard carton (which is produced by Tetra Pak) is shown in Figure 11.5.

Aseptic packages can also be in the form of cans, bottles, pouches, trays, or cups. Can type aseptic packages are the same basic types of metal cans as in the regular canning process: tinplate, ECCS, and aluminum. Under an aseptic packaging process, the heating time for sterilization and food quality are the same for both large and small containers, while the traditional canning process has longer heating times for larger containers. Bottle type aseptic packages are produced from plastics as an economical alternative to glass for non-returnable containers. HDPE, PP, and PET are the most commonly used materials for this type of package. Pouch type aseptic packages are made from similar laminated paperboard structures. Also, multilayer films containing LLDPE and

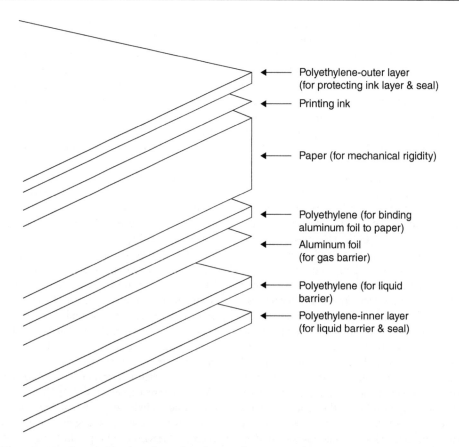

Figure 11.5 The structure of a typical laminated paperboard carton for aseptic packaging.

EVOH are used. Cup type aseptic packages are produced from HIPS, PP, or multilayer film. Multilayer films are the choice when high barrier properties are required. PVDC or EVOH is typically used as the barrier layer.

Maintaining the package integrity during distribution and handling is one of the most important issues in aseptic food packaging. Thus, various integrity tests are used commercially. Typically, electrolytic testing, dye penetration or vacuum leak tests are used for quality control during manufacturing. However, these traditional methods are destructive and therefore it is impossible to test and reject all faulty packages in the processing line. Thus, non-destructive test methods, such as gas leak detectors and ultrasound techniques, are gaining more attention from industry.

11.5.2 Modified atmosphere packaging

Modified atmosphere packaging (MAP) is based on modifying or altering the atmosphere inside the package to prolong shelf life and maintain quality of products. The modification of the atmosphere can be achieved actively or passively. In active type MAP, the optimum gaseous environment is obtained by flushing a controlled mixture of gases in a package ("gas flush"). The passive type of MAP modifies the optimum gaseous environment in a package by a combination of the food's respiration and the metabolism of microorganisms associated with the food and the permeability of the packaging. With the optimized gaseous atmosphere, degradation reactions in foods such as enzyme activity, oxidation, moisture loss, and postharvest metabolic activities as well as the growth of microorganisms are delayed. The three main gases used for MAP are nitrogen (N_2), carbon dioxide (CO_2), and oxygen (O_2). They are used either alone or, in most cases, in combination. Mixtures of carbon dioxide (CO) and argon (Ar) are also utilized commercially. Examples of gas mixtures that are used for fresh and processed foods are shown in Table 11.1.

Table 11.1 Examples of optimum headspace gas mixtures (%) and storage temperature for fresh and processed food products

Product	Temperature (°C)	Oxygen (%)	Carbon dioxide (%)	Nitrogen (%)
Snack	23	0	20–30	70–80
Bread	23	0	60–70	30–40
Cake	23	0	60	40
Cheese (hard)	4	0	60	40
Cheese (soft)	4	0	30	70
Pasta	4	0	80	20
Pizza	5	0–10	40–60	40–60
White fish	0–2	30	40	30
Oily fish	0–2	0	60	40
Shrimp	0–2	30	40	30
Fresh red meat	0–2	40–80	20	0
Cooked/cured meat	1–3	0	30	70
Pork	0–2	40–80	20	0
Poultry	0–2	0	20–100	
Sausage	4	0	80	20
Apples	0–3	1–3	0–3	
Banana	12–15	2–5	2–5	
Broccoli	0–5	5–10	1–2	
Lettuce	0–5	2–3	5–6	
Tomato	7–12	4	4	

Adapted from: FDA (2001), Brody (2000), Parry (1993).

11.5.2.0.1 Nitrogen (N₂)

Nitrogen is the most commonly used gas in MAP. It is an inert gas with no odor or taste. Nitrogen also has low solubility in water (0.009 g/kg at 20 °C). Nitrogen does not directly provide any microbial retardation but it delays aerobic microbial growth and oxidation by replacing oxygen.

11.5.2.0.2 Carbon dioxide (CO₂)

Carbon dioxide is colorless with a slightly pungent odor. The most important function of this gas is related to its bacteriostatic and fungistatic properties. Usually, a higher concentration is more effective against microorganisms. However, color changes and acid tastes have been reported by several researchers due to exposure to high CO_2. Carbon dioxide has high solubility in water (1.69 g/kg at 20 °C). Thus, packages containing moist foods with CO_2 in their headspace may collapse.

11.5.2.0.3 Oxygen (O₂)

Oxygen is a colorless, odorless, and highly reactive gas. It promotes food deteriorative reactions such as fat oxidation, browning reactions, and aerobic microbial growth. For this reason, the desired oxygen content in the headspace of most foods is often extremely low. However, oxygen is still needed for the retention of color in red meat and for fruit/vegetable respiration.

11.5.2.0.4 Carbon monoxide (CO)

Carbon monoxide is a colorless, tasteless, and odorless gas. When carbon monoxide combines with myoglobin, a bright red pigment (carboxymyoglobin) is formed, which is more stable than normal red meat pigment (oxymyoglobin). The use of carbon monoxide is not for quality issues but for its visual effect. Since carbon monoxide is a toxic and highly flammable gas, its commercial use is limited. In the US, carbon monoxide is currently used for red meat but the level of its use is limited to 0.4% (FDA, 2005a).

11.5.2.0.5 Argon (Ar)

Argon is a chemically inert, colorless, tasteless, and odorless gas which is denser and heavier than air. Compared to nitrogen, argon is a more effective gas for use in flushing out air. However, due to the high cost of argon, nitrogen is mostly used for flushing.

11.5.2.1 Modified atmosphere packaging for meat

For red meat, oxygen is necessary to maintain the red color of oxymyoglobin in unprocessed meats. On the other hand, a low oxygen level is desired to prevent the growth of microorganisms and oxidative rancidity of fat. Typically, the shelf life of fresh red meat is extended by packaging it in an atmosphere of 20% CO_2, 60–80% O_2, and up to 20% N_2. However, off-odors and rancidity have been reported in meats stored at high O_2 concentrations (Taylor, 1985). Thus, maintaining a low temperature (0–2 °C) is desired when high-O_2 MAP is used.

Poultry has low myoglobin content so it does not need oxygen to maintain its color. A higher CO_2 concentration (20–100%) is possible to extend the shelf life. However, a few studies report color change with high CO_2 concentration in poultry products (Dawson, 2004).

11.5.2.2 Modified atmosphere packaging for seafood

Seafood such as fish and shellfish is highly perishable owing to its high water activity (a_w), pH, and the presence of autolytic enzymes, which cause rapid development of undesirable odors. A low oxygen atmosphere can delay the occurrence of undesirable odor and growth of aerobic microorganisms. For example, packaging tilapia fillets in 75% CO_2:25% N_2 extended their shelf life up to 80 days at 4 °C (Sivertsvik et al., 2002). However, fresh fish may be contaminated with the anaerobic C. *botulinum* either as a result of being present in the microbiota of the fish ecosystem or due to postcatch contamination during processing. A low oxygen atmosphere poses a potential threat for a packaged fish to become toxic prior to spoilage. To assure the safety of MAP fish products, the product must be maintained at or below 3 °C at all times. Since chilled storage control is a critical factor to determine the shelf life, some companies use time temperature indicators (TTI) to monitor for temperature abuse.

11.5.2.3 Modified atmosphere packaging for fresh fruits and vegetables

Fresh fruits and vegetables keep consuming O_2 and emitting CO_2 even after harvest. The purpose of MAP for fresh fruits and vegetables is to minimize the respiration and senescence without causing suffocation and damage to metabolic activity that rapidly reduces their shelf life. However, a low oxygen and high CO_2 atmosphere that develops inside the package due to respiration of the product may result in the accumulation of ethanol and acetaldehyde, and fermentation could start. Thus, the package material needs to be somewhat permeable to oxygen and CO_2 to allow the transfer of the gases from outside and inside. The change in gas composition during storage depends on the permeability of the container to water vapor and gas, storage temperature, and the mass of the food.

11.5.3 Active packaging in food processing

Active packaging is an important and rapidly growing area. There are several different definitions that can be found in the literature for active packaging. This type of packaging usually involves an interaction between the packaging components and the food product beyond the inert passive barrier function of the packaging material (Labuza & Breene, 1989; Soroka, 2008b). The major active packaging technologies include oxygen scavengers, moisture absorbers, antimicrobial agent releasers, ethylene scavengers, flavor/odor absorbers, and temperature control packaging. In order to apply this technology, the major deteriorative factor(s) for food products should be understood. For example, the shelf life of a packaged food is affected by numerous factors such as acidity (pH), water activity (a_w), respiration rate, oxidation, microbial spoilage, temperature, etc. By carefully considering all of these factors, active packaging can be developed and applied to maintain the quality of the product and/or to extend its shelf life. Active agents are contained in sachets or incorporated directly into packaging containers.

11.5.3.1 Oxygen scavengers

Oxygen scavengers are the most commercially applied technology in the active packaging market. By scavenging oxygen molecules in the package headspace, oxidative damage to food components such as oil, flavors, vitamins, color, etc. can be prevented. In addition, reduced O_2 concentration in the package retards the growth of aerobic bacteria and mold.

The basic principle of oxygen scavenging is related to oxidation of the scavenging agents (either metallic or non-metallic based materials) to consume oxygen. The most common O_2 scavengers used in the food industry are sachets with iron powder, which are highly permeable to oxygen. By using iron powder, the O_2 concentration in the headspace can be reduced to less than 0.01% while vacuum or gas flushing typically achieves 0.3–3.0% residual O_2 levels (Robertson, 2007). The metal-based oxygen scavengers normally cannot pass the metal detectors on the packaging lines and cannot provide transparent packages if they are directly incorporated into packages. Non-metallic O_2 scavengers, such as ascorbic acid or glucose oxidase, can be used as an alternative choice. However, the use of non-metallic O_2 scavengers is not widespread.

O_2 scavengers can be incorporated into plastic film/sheet when there are market concerns about accidental consumption of sachets or when an O_2 scavenger needs to be used for liquid foods. In this case, the O_2 scavenger-impregnated layer is typically sandwiched between film layers. The outside layer provides high oxygen protection and the inner layer prevents direct contact between the scavenger-containing matrix layer and the food.

11.5.3.2 Moisture absorbers

Moisture in packages is a major cause of food deterioration such as microbial growth-related spoilage and product softening. Moisture-absorbing sachets are used in food packaging for humidity control. Several desiccants such as silica gel, calcium oxide, and activated clays and minerals are typically contained in sachets. Drip-absorbent pads and sheets are also used to absorb liquid in high a_w foods such as meats, fish, poultry, fruits, and vegetables. A superabsorbent polymer is sandwiched by a microporous non-woven plastic film such as polyethylene or polypropylene. Polyacrylate salts, carboxymethyl cellulose (CMC), and starch co-polymers are typically used as the absorbent polymer.

11.5.3.3 Antimicrobial agent releasers

Typically, surface contamination during food handling and transportation is one of the most common sources of food-borne illness, and lower amounts of antimicrobial agents are required to control the surface microbial contamination/growth if they are incorporated into or onto packaging material rather than directly added into the food itself. There are two mechanisms for antimicrobial action: one is controlling microbial growth by slow and controlled release (or migration) of the antimicrobial agents over the product shelf life. The other type is controlling microbial growth by contact without release of the antimicrobial agents, which are immobilized on the surface of the package. Despite the large number of experimental studies on antimicrobial packaging, the

technology has not been widely used in food markets yet, because of cost or regulatory constraints. An example of a commercial application of an antimicrobial packaging is silver ion-based film. This film is an effective antimicrobial agent with very low human toxicity and it has been used in food containers. Other examples of antimicrobials used (or studied) include nisin, pediocin, organic acids, grapefruit seed extract, cinnamon, and horseradish (Han, 2003).

11.5.3.4 Ethylene scavengers

Ethylene works as a plant growth regulator which accelerates the respiration rate and senescence of fruits and vegetables. Removing ethylene from the environment can extend the shelf life of horticultural products. One of the most common ethylene scavengers is made from potassium permanganate ($KMnO_4$) immobilized on an inert mineral substrate such as alumina or silica gel. Activated carbon-based scavengers with various metal catalysts are also used as effective ethylene scavengers. In recent years, packaging films and bags have been commercialized based on the reputed ability of certain finely dispersed minerals to absorb ethylene (such as clays, zeolites, coral, ceramics, etc.) (Rooney, 2005).

11.5.3.5 Flavor and odor absorbers

Undesirable flavor scalping by the packaging material can result in the loss of desirable food flavors, while foods can pick up undesirable odors or flavors from the package or the surrounding environment. For example, fruit juices in PET bottles or water in HDPE bottles can absorb unwanted odors that result from aldehydes such as hexanal and heptanal originating in oxidation of the plastics. Another example is that an unpleasant "confinement odor" can accumulate inside PE bags in a distribution center. One type of flavor/odor-absorbing technology is based on a molecular sieve with pore sizes of approximately 5 nm. Cyclodextrin is a common example of an odor absorber. Wood (2011) reported that cyclodextrin, grafted to a PE layer, was applied to military ration pouches to absorb accumulated odor from food decomposition. Synthetic aluminosilicate zeolite, which has a highly porous structure, has been incorporated in packaging materials, especially papers, to absorb odorous gases such as aldehyde (Day, 2003).

On the other hand, absorbers can also be used in a flavor/aroma-releasing system. Aroma compounds can be released inside the package to enhance the flavor/aroma of the product or outside the package to attract consumers to the products on a retail store shelf (Lagaron & Lopez-Rubio, 2009).

11.5.3.6 Temperature-controlled packages

Temperature-controlled active packages generally include self-heating and self-cooling systems. The fundamental concepts for self-heating are not new. An exothermic chemical reaction between CaO (quicklime) and a water-based solution generates heat. The challenging part of this system is optimizing the reaction and the thermal design of the container to provide an efficient, safe, and cost-effective package. Self-heating packages are commercially available for coffee, tea, and ready-to-eat meals. Self-cooling cans use endothermic chemical reactions; the latent heat of evaporating water is often used to produce the cooling effect. One example is dissolution of ammonium nitrate and chloride in water to cool the product (Day, 2008).

11.6 Sustainable food packaging

11.6.1 Recycling of food packaging

Food packaging has the largest demand for packaging industries, whether it is paper, plastics, glass, or metal. Finding or improving ways to reduce landfilled waste is important to meet current demands for environmentally friendly packaging (Arvanitoyannis & Kasaverti, 2008). Recycling can be defined as diverting materials from the solid waste stream for use as raw materials in the manufacture of new products. The overall recycling rate of packaging material in the US is 40%, and is far behind the rate in Europe, which is about 59% (Fischer & Davisen, 2010).

11.6.1.1 Recycling of paper and paperboard

Packaging materials are the largest sector in which recycled paper is used in the US, as shown in Figure 11.6. The recycling rate for paper and paperboard has been increasing during the last decade. In 2010, 37.7 million tonnes of paper and paperboard packaging waste were generated and 71.3% of the used material was recycled (EPA, 2010). Even if most types of paper are recyclable, recycled paper is not suitable for most food contact packaging applications because the recycling processes may allow contaminants to be present in the recycled paper products. In many applications, recycled

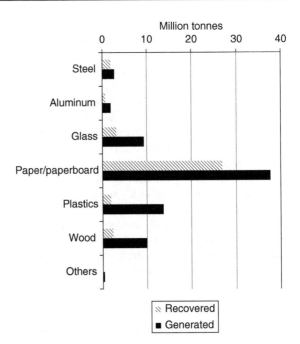

Figure 11.6 Generation and recovery of packaging materials in the United States (EPA 2010).

paper is not used in direct contact with food, so migration is of little or no concern. Another issue is related to paper properties: recycled paper normally has a damaged and weak fiber structure. In order to obtain suitable physical and mechanical characteristics, mixing virgin and recovered fibers in different proportions is often required.

11.6.1.2 Recycling of glass

Glass is the next most recycled packaging material, in terms of weight recovered; 9.36 million tonnes of glass packaging waste were generated in the US in 2010, and 33.4% of the used material was recycled (EPA, 2010). Unlike paper-based material, the properties of glass are not affected by recycling. Cullet can be reprocessed into glass containers with physical properties identical to the original material. Glass recycling also reduces energy consumption: addition of 10% cullet reduces energy consumption by 2.5% (Dainelli, 2008). One of the major problems for food packaging applications originating from recycling is the presence of contaminants. Contaminants such as metals, papers, plastics, organic substances, ceramics, and heat-resistant glass (such as borosilicates), if not removed, can cause problems in making new food containers. Combination with colored glass is another problem, especially in the manufacture of colorless glass. This can result in off-color containers that are not acceptable for demanding applications such as food packaging. Mixed color cullet may be used for alternative applications such as abrasive paper, water filtration media, construction materials, and so on. Efficiently removing or sorting these contaminants or colored cullet requires a high level of investment. In addition, cullet is heavy and therefore expensive to ship. Collection within an acceptable transport distance is a critical point in glass recycling.

11.6.1.3 Recycling of aluminum

Aluminum is a metal largely used for industrial and consumer goods applications; 1.9 million tonnes of total aluminum packaging waste were generated in the US in 2010, and 35.8% of the used material was recycled (EPA, 2010). Like glass, recycled aluminum does not lose its physical properties and is safe for food packaging. Using recycled aluminum can save 75–90% of energy compared with its production from natural sources (Dainelli, 2008). Collection of aluminum for recycling is through a combination of deposit, curbside, and drop-off programs. During the recycling process, aluminum is easily separated from other metals. Since it is lighter and not magnetic, iron or other ferrous metals can be separated using a magnetic separator or flotation. Internal coatings, printed ink, and any other organic contaminants are destroyed during the recycling process.

11.6.1.4 Recycling of plastics

Plastics are recycled the least of the major packaging materials in the US; 13.68 million tonnes of plastic packaging waste was generated in the US in 2010, and 13.5% of the used material was recycled (EPA, 2010). During recycling, plastics can undergo several types of reactions such as chain scission, cross-linking, oxidation, and hydrolysis. Thus, the overall physical performance of the recycled plastics may decrease significantly. In addition, all recycled plastics cannot be mixed together due to their chemical incompatibility. In most applications, achieving good performance properties requires separation of the recycled plastics by resin type. Therefore, recycling of plastic packaging wastes is more difficult and costly than that of some other packaging materials. Energy recovery, such as incineration, may be the preferred option for multilayer plastics which cannot be separated by type, while cleaning and reprocessing into pellets for use in new

plastic products is often the best choice for homogeneous plastics. Another option in some cases is depolymerization to monomers, purification, and repolymerization.

Since the potential migration of contaminants to food or other products is often a concern, most recycled plastics are used for non-food applications, including packaging containers, film and sheet, and also non-packaging applications such as fiber and carpet. In order to obtain food contact-grade plastic, the recycled plastics need to be processed and used in a way that effectively removes the potential for such contamination.

One option, as mentioned above, is chemical recycling, in which after depolymerization, the monomers are purified and repolymerized. Polymers produced in this way are identical to those produced from ordinary raw materials. Chemical recycling can be used for PET and PA, but is often not economical.

Another option for recycled plastic in food packaging is using it as a component in a multilayer structure. The recycled plastic is sandwiched inside, and a "barrier" layer of virgin plastic separates the recycled plastic from the product.

Another widely used option for PET is intensive cleaning to remove most potential contaminants. A number of companies have received "non-objection" letters from the US FDA for plastics recycling processes that have been demonstrated to remove potential contaminants from recycled PET streams to a degree that makes them acceptable for food contact (FDA, 2008). One example is the Superclean process, which consists of a series of processes that can remove the volatile contaminants as well as increase the viscosity of the recycled PET so that it is suitable for injection blow molding (Franz & Welle, 2002).

Most "non-objection" letters for recycled plastics have been issued for PET recycling processes. HDPE is both more susceptible to sorption of contaminants and more difficult to clean. While there have been a few processes approved for use of recycled HDPE in food packaging, there is little commercial use. On the other hand, recycled PET is used to a considerable extent in food packaging applications.

11.6.2 Biodegradable and compostable food packaging

Due to growing concerns about waste disposal problems and the environmental effects of petroleum-based plastics, natural biopolymers derived from renewable sources that are biodegradable appear to be a good alternative to conventional plastics. In addition, in recent times, oil prices have increased markedly. These facts have caused increased interest in non-petroleum based biodegradable polymers. The biodegradation takes place through the action of enzymes and/or biochemical deterioration associated with living organisms, and the biodegradability depends not on the raw material sources to produce the polymer but rather on the chemical structure of the polymer and the environmental conditions because chemical structure, such as the chemical linkage, pending groups, etc., is related to susceptibility to degradation, and environmental condition is related to living organisms' activities.

The common challenges in using biodegradable packaging materials are ensuring durability to maintain their mechanical and/or barrier properties during the product's shelf life, and then, ideally, the ability to biodegrade quickly on disposal. Ideally, the materials need to function in a similar way to conventional packaging in filling and sealing equipment with equivalent costs. Biobased biodegradable polymers can be classified into three main categories according to their origin and method of production (van Tuil et al., 2000).

- Polymers directly extracted/removed from biomaterials (for example, starch, cellulose, casein, etc.).
- Polymers produced by classic chemical synthesis from monomers produced from biomaterials (for example, polylactide polymerized from lactic acid monomers).
- Polymers produced directly by microorganisms (for example, polyhydroxyalkanoates).

A schematic presentation of these three categories is given in Figure 11.7.

11.6.2.1 Biodegradable polymers from agricultural crops

Polymers produced from starch are examples for this class. Starch is a widely available, environmentally friendly material with low cost. Corn is currently the most commonly used source of starch for bioplastics. However, potato, wheat, rice, barley, and oats can also be used as

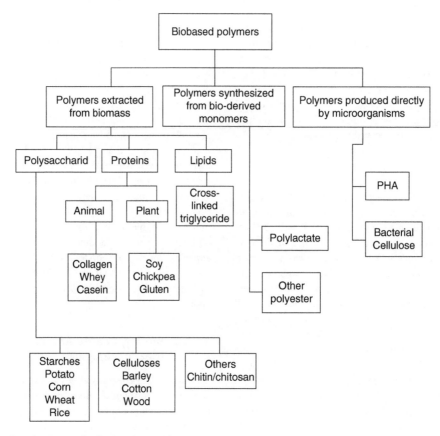

Figure 11.7 Biobased polymers for food packaging (from van Tuil et al., 2000).

starch sources (Liu, 2006). Without modification, starch films are hydrophilic and have relatively poor mechanical strength. They cannot be used for packaging applications. The brittleness of starch-based bioplastics can be decreased by using biodegradable plasticizers, including glycerol and other low molecular weight polyhydroxy compounds, polyethers, etc. (van Tuil et al., 2000).

Cellulose is the most widespread polysaccharide resource produced by plants. Cellulose is composed of glucose monomers that form a linear polymer with very long macromolecular chains. Cellulose is highly crystalline, brittle, infusible, and insoluble in all organic solvents (Chandra et al., 2007). These properties make cellulose impossible to process without modification. Cellophane films are produced by chemical modification of cellulose to render it soluble, and then regeneration of the cellulose after it is formed into film.

Another common practice is to use cellulose derivatives (cellulosic plastics) to improve the properties. The derivatives, such as ethers, esters and acetals, are produced by the reaction of one or more hydroxyl groups in the repeating unit, and can impart good film-forming properties. Cellulosic plastics are used for wrapping films, and for injection molded or blow molded containers. Tenite® (Eastman, USA), Bioceta® (Mazzucchelli, Italy), Fasal® (IFA, Austria), and Natureflex® (UCB, Germany) are trade names of some cellulose-based polymers.

11.6.2.2 Biodegradable polymers synthesized from bio-derived monomers

Polylactic acid (PLA) is a biodegradable, thermoplastic polyester that is derived from lactic acid. Lactic acid can be produced economically by microbial fermentation of glucose obtained from biomaterials such as corn or wheat starch, lactose in whey, or sucrose. PLA is now produced on a comparatively large scale. PLA is usually obtained from ring-opening condensation of lactide, which is a lactic acid dimer. Two forms of the monomeric acid (D- or L-lactic acid) exist. The properties of PLAs can be varied by adjusting the relative amounts of the two lactic acid isomers (D- or L-) in the polymer. For example, 100% L-lactic acid or 100% D-lactic acid forms highly crystalline PLA whereas amorphous PLA is obtained from DL-co-polymers across a wide composition range. The L-isomer is predominant in most PLA resins as most lactic acid obtained from biological sources is the L form; 90%L/10%D PLA is a common formulation. The formation is able to crystallize but melts more easily than 100%L and is suitable for production of packaging films.

Different companies have commercialized PLA with various ratios of D/L lactide, with trade names including Natureworks ®, Galacid®, Heplon®, etc.

Polylactic acids generally have reasonable moisture and oxygen barrier properties and are suitable for various plastic package forming processes such as blown and cast film, injection molding, and vacuum forming. PLA is currently utilized in wraps for bakery and confectionery products, paperboard coatings for cartons, disposable foodservice tableware items, containers for fresh produce, and water bottles. The rate of degradation of PLA depends on the degree of crystallinity. Increasing the amount of D-isomer in predominantly L-PLA tends to suppress crystallinity and therefore increase the rate of biodegradation. There has been some research on enhancing the biodegradability of PLA by grafting with chitosan (Luckachan & Pillai, 2006).

11.6.2.3 Biodegradable polymers produced directly by microorganisms

Poly(β-hydroxyalcanoate)s (PHAs) are natural polyesters which are produced by bacteria from sugars or lipids. They are actually "grown" inside the cellular structure and then harvested. The use of PHAs is currently limited due to their high production costs. The performance properties are similar to conventional plastics, but the polymers are completely biodegradable. Thus, PHAs have potential as biodegradable alternative materials for conventional bulk commodity plastics (Foster et al., 2001). One of the family of PHAs, polyhydroxybutyrate (PHB), is the most commonly produced and researched bioplastic. PHB has high thermal resistance and water barrier properties. However, a narrow processability window and low impact resistance have hampered widespread use of PHB for packaging application. Blends of PHB with other polymers may improve its properties. For example, poly(ethylene oxide), poly(vinyl butyral), poly(vinyl acetate), poly(vinylphenol), cellulose acetate butyrate, chitin, and chitosan have been studied as blend materials with PHB. Another common method to improve the processability of PHB is to induce the microorganisms to form a co-polymer rather than the homopolymer. Polyhydroxybutyrate-valerate, PHBV, is the most common of these.

11.6.2.4 Synthetic biodegradable polymers

In addition to the biodegradable plastics based on natural substrates, there are synthetic biodegradable plastics produced from petrochemical feedstocks that have groups which are susceptible to hydrolytic microbial attack. Poly-caprolactone (PCL) is a semi-crystalline aliphatic polyester which has a relatively low melting point (60 °C). It is completely biodegradable in marine, sewage, sludge, soil, and compost ecosystems (Khatiwala et al. 2008).

Polyvinyl alcohol (PVOH) is another synthetic biodegradable polymer which is completely soluble in water. The combination of starch and PVOH as a biodegradable packaging material has been studied since 1970. Currently, it is used to produce starch-based loose fillers as a substitute for expanded PS. Other types of synthetic biodegradable polymers include polyesters, polyamides, polyurethanes and polyureas, poly(amide-enamine)s, polyanhydrides (Chandra et al., 1998; Nair & Laurencin, 2007).

Synthetic polymers can be manipulated for a wide range of properties to obtain required mechanical properties (flexibility, toughness, etc.) as well as the degree of degradation. Rather than food packaging, the application of synthetic biodegradable polymers has been gaining more attention in the biomedical area such as tissue engineering scaffolds, orthopedic fixation devices, etc. (Gunatillake at al., 2006).

References

Arvanitoyannis IS, Kasaverti A (2008) Consumer attitude to food packaging and the market for environmentally compatible products. In: Chiellini E (ed) Environmentally Compatible Food Packaging. Boca Raton, FL: CRC Press, pp. 161–179.

ASTM (2003) Standard Terminology of Glass and Glass Products West Conshohocken, PA: ASTM International.

Brody AL (2000) Packaging: part IV – controlled/modified atmosphere/vacuum food packaging. In: Francis FJ (ed) Wiley Encyclopedia of Food Science and Technology, 2nd edn, vol. 3. New York: Wiley, pp. 1830–1839,

Chandra R, Rustgi R (1998) Biodegradable polymers. *Progress in Polymer Science* 23: 1273–1335.

Chandra R, Bura R, Mabee WE et al. (2007) Substrate pretreatment: the key to effective enzymatic hydrolysis of lignocellulosics. *Advanced Biochemistry Engineering and Biotechnology* 108: 67–93.

Coles R, McDowell D, Kirwan MJ (2003) Introduction. In: Coles R, McDowell D, Kirwan, M (eds) Food Packaging Technology. Boca Raton, FL: CRC Press, pp. 1–31.

Crank J (1975) Mathematics of Diffusion. London: Oxford University Press, pp. 1–11.

Dainelli D (2008) Recycling of food packaging materials: an overview. In: Chiellini E (ed) Environmentally Compatible Food Packaging. Boca Raton, FL: CRC Press, pp. 294–325.

Dawson PL (2004) Poultry packaging. In: Mead GC (ed) Poultry Meat Processing and Quality. Cambridge: Woodhead Publishing, pp. 135–159.

Day BPF (2003) Active packaging. In: Coles R, McDowell D, Kirwan, M (eds) Food Packaging Technology. Boca Raton, FL: CRC Press, pp. 282–302.

Day BPF (2008) Active packaging of food. In: Kerry J, Butler P (eds) Smart Packaging Technologies for Fast Moving Consumer Goods. Chichester: John Wiley, pp. 1–17.

Environmental Protection Agency (EPA) (2010) Municipal Solid Waste Generation, Recycling, and Disposal in the United States: Facts and Figures for 2009. www.epa.gov/epawaste/nonhaz/municipal/pubs/msw_2010_rev_factsheet.pdf, accessed 14 November 2013.

Federal Trade Commission (FTC) (1994) The Fair Packaging and Labeling Act. www.ftc.gov/os/statutes/fpla/fplact.html, accessed 14 November 2013.

Fischer C, Davisen C (2010) Europe as a Recycling Society: The European Recycling Map. ETC/SCP Working Paper. Denmark: European Topic Centre on Sustainable Consumption and Production (ETC/SCP).

Food and Drug Administration (FDA) (2001) *Microbiological Safety of Controlled and Modified Atmosphere Packaging for Fresh and Fresh-Cut Produce, Analysis and Evaluation of Preventive Control Measures for the Control and Reduction/Elimination of Microbial Hazards on Fresh and Fresh-Cut Produce.* www.fda.gov/Food/FoodScienceResearch/SafePracticesforFoodProcesses/ucm090977.htm, accessed 18 November 2013.

Food and Drug Administration (FDA) (2002) Guidance For Industry, Preparation of Food Contact Notifications and Food Additive Petitions for Food Contact Substances: Chemistry Recommendations Final Guidance. Center for Food Safety and Applied Nutrition, Office of Food Additive Safety. www.fda.gov/food/guidanceregulation/guidancedocumentsregulatoryinformation/ingredientsadditivesgraspackaging/ucm081825.htm, accessed 20 November 2013.

Food and Drug Administration (FDA) (2005a) *Guidance for Industry: Submitting Requests under 21 CFR 170.39 Threshold of Regulation for Substances Used in Food-Contact Articles.* www.fda.gov/food/guidanceregulation/guidancedocuments-regulatoryinformation/ingredientsadditivesgraspackaging/ucm081833.htm, accessed 20 November 2013.

Food and Drug Administration (FDA) (2005b) GRAS letter. www.accessdata.fda.gov/scripts/fcn/gras_notices/grn000167.pdf, accessed 14 November 2013.

Food and Drug Administration (FDA) (2008) No Objection Letter for Recycled Plastics #117. www.fda.gov/food/ingredientspackaginglabeling/packagingfcs/recycledplastics/ucm155232, accessed 20 November 2013.

Foster LJR, Saufi A, Holden PJ (2001) Environmental concentrations of polyhydroxyalkanoates and their potential as bioindicators of pollution. *Biotechnology Letters* 23: 893.

Franz R, Welle F (2002) Post-consumer poly(ethylene terephthalate) for direct food contact application – challenge text of an inline recycling process. *Food Additives and Contaminants* 19(5): 502–511.

Glass Packaging Institute (2012) *Section 3.4 Forming Process.* http://gpi.org/glassresources/education/manufacturing/section-34-forming-process.html, accessed 14 November 2013.

Gunatillake P, Mayadunne R, Adhikari R (2006) Recent developments in biodegradable synthetic polymers. *Biotechnology Annual Review* 12: 301–347.

Han JH (2003) Antimicrobial food packaging. In: Ahvenainen R (ed) Novel Food Packaging Techniques. Cambridge: Woodhead Publishing, pp. 50–70.

Hanlon JH, Kelsey RJ, Forcinio HE (1998) Paper and paperboard. In: Hanlon JH, Kelsey RJ, Forcinio HE (eds) Handbook of Package Engineering. Boca Raton, FL: CRC Press, pp. 31–57.

Hernandez RJ (1997) Food packaging materials, barrier properties and selection. In: Valentas KJ, Rotstein E, Singh RP (eds) Handbook of Food Engineering Practice. Boca Raton, FL: CRC Press, pp. 291–360.

Judd D, Aalders B, Melis T (1989) The Silent Salesman. Singapore: Octogram Design.

Khatiwala VK, Nilanshu S, Aggarwal S, Mandal UK (2008) Biodegradation of poly (ε-caprolactone) (PCL) film by Alcaligenes faecalis. *Journal of Polymers and the Environment* 16: 61.

Labuza TP, Breene WM (1989) Applications of "active packaging" for improvement of shelf life and nutritional quality of fresh and extended shelf life of foods. *Journal of Food Processing and Preservation* 13: 1–69.

Lagaron JM, Lopez-Rubio A (2009) Latest developments and future trends in food packaging and biopackaging. In: Passos ML, Ribeiro CP (eds) Innovation in Food Engineering: New Techniques and Products. Boca Raton, FL: CRC Prezzss, pp. 485–510.

Liu L (2006) Bioplastics in Food Packaging: Innovative Technologies for Biodegradable Packaging. San Jose, CA: Packaging Engineering Department, San Jose University. www.iopp.org/files/public/SanJoseLiuCompetitionFeb06.pdf, accessed 14 November 2013.

Lopez TS, Ranasighe DC, Patkai B, Farlane DM (2011) Taxonomy, technology and applications of smart objects. *Information of Systems Frontiers* 13: 281–300.

Luckachan GE, Pillai CK (2006) Chitosan/oligo L-lactide graft copolymers: effect of hydrophobic side chains on the physico-chemical properties and biodegradability. *Carbohydrate Polymers* 24: 254–266.

Myerson JM (2007) RFID in the Supply Chain: A Guide to Selection and Implementation. Boca Raton, FL: Taylor and Francis Group.

Nair LS, Laurencin CT (2007) Biodegradable polymers as biomaterials. *Progress in Polymer Science* 32: 762–798.

Parry RT (1993) Introduction, In: Parry RT (ed) Principles and Applications of MAP of Foods. New York: Blackie Academic and Professional, pp. 1–18.

Robertson GL (2007) Food Packaging: Principles and Practice, 2nd edn. Boca Raton, FL: Taylor and Francis.

Rooney ML (ed) (1995) Active Food Packaging. London: Chapman and Hall.

Selke S (2012) Green packaging. In: Boye J, Arcand Y (eds) Green Technologies in Food Production and Processing. New York: Springer, pp. 443–470.

Selke S, Culter JD, Hernandez RJ (2004) Extrusion, film and sheet. In: Plastics Packaging, 2nd edn. Cincinnati, OH: Hanser, pp. 193–223.

Sivertsvik M, Jeksrud WK, Rosnes T (2002) A review of modified atmosphere packaging of fish and fishery products – significance of microbial growth, activities and safety. *International Journal of Food Science and Technology* 37: 107–127.

Soroka W (2008a) Packaging functions. In: Soroka W (ed) Fundamentals of Packaging Technology, Naperville, IL: Institute of Packaging Professionals, pp. 29–48.

Soroka W (2008b) Packaging terms. In: Soroka W (ed) Illustrated Glossary of Packaging Terms. Naperville, IL: Institute of Packaging Professionals.

Taylor AA (1985) Packaging fresh meat. In: Lawrie RI (ed) Developments in Meat Science. Edinburgh: Elsevier, pp. 89–113.

van Tuil R, Fowler P, Lowther M, Weber CJ (2000) Properties of biobased packaging materials. In: Weber CJ (ed) Biobased Packaging Materials for the Food Industry – Status and Perspectives. Food Biopack Project, EU Directorate 12, pp. 13–44. www.plastice.org/fileadmin/files/Book_on_biopolymers__Eng_.pdf, accessed 20 November 2013.

Wood W (2011) Amphoteric grafted barrier material (# 20WO200511613605). Cellresin Technologies. www.patentlens.net/patentlens/patents.html?patnums=EP_1753817_B1&language=&patnum=EP_1753817_B1&language=en&query=&stemming=&pid=p0, accessed 14 November 2013.

World Packaging Organization (WPO) (2009) *Packaging is the Answer to World Hunger.* www.google.com/url?sa=t&rct=j&q=&esrc=s&frm=1&source=web&cd=2&ved=0CEAQFjAB&url=http%3A%2F%2Fwww.worldpackaging.org%2Fi4a%2Fdoclibrary%2Fgetfile.cfm%3Fdoc_id%3D12&ei=XqiGUsO-ILTs2AWRo4HgAw&usg=AFQjCNGOz68KZ4HaxEkOBXSmcVF8inXSsA, accessed 20 November 2013.

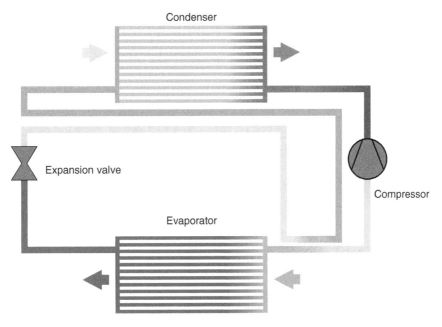

Figure 5.1 A mechanical vapor compression refrigeration system.

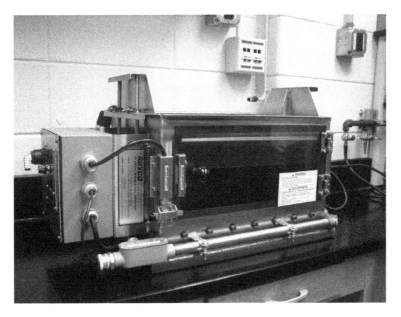

Figure 7.5 Laboratory scale ultraviolet-C test system (courtesy of Reyco Systems, Caldwell, ID).

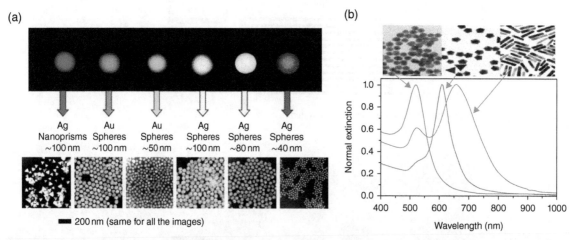

Figure 8.6 (a) Sizes, shapes, and compositions of metal nanoparticles can be systematically varied to produce materials with distinct light-scattering properties. (Reproduced from Rosi & Mirkin (2005), with permission from the American Chemical Society.) (b) TEM micrographs (*top*) and UV–vis spectra (*bottom*) of colloidal gold nanoparticles of different geometries. (Reproduced from Sepulveda et al. (2009), with permission from Elsevier.)

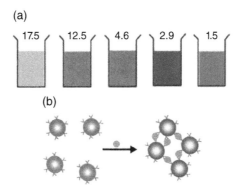

Figure 8.7 (a) Effect of interparticle distance in nanometers on the color of 15 nm AuNPs. (Reproduced from Ung et al. (2001), with permission from the American Chemical Society.) (b) Schematic diagram of distance-dependent sandwich assay for high molecular weight polyvalent antigen (*green circles*) leading to aggregation of antibody-functionalized AuNPs and a red shift in their extinction spectrum. (Reproduced from Wilson (2008), with permission from the Royal Society of Chemistry.)

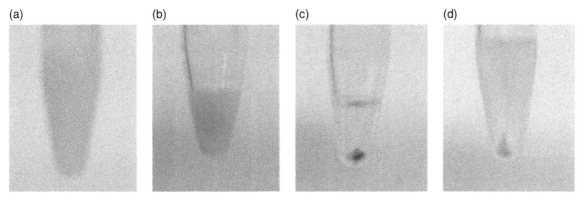

Figure 8.10 Stages of analyte-indicator displacement reaction. (a) Solution of latex microspheres + DNP antibody, (b) addition of AuNPs (conjugated with DNP–BSA), (c) solution in (b) after mild centrifugation shows good binding of latex microspheres and AuNPs, (d) solution (b) after addition of model toxin (DNP-GLY) and light centrifugation (the solution color is still yellow due to the strong yellow color of the DNP-GLY). (Reproduced from Ko et al. (2010), with permission from Elsevier.)

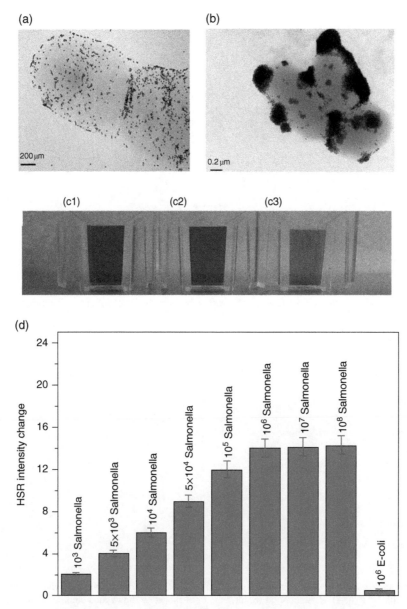

Figure 8.12 TEM images showing aggregation of gold nanoparticles after the addition of (a) 5 × 103 CFU/mL and (b) 8 × 104 CFU/mL *Salmonella*. Photographs showing colorimetric change upon addition of (c1) 2 × 103 *Salmonella*, (c2) 5 × 106 *E. coli*, (c3) 5 × 104 *Salmonella*. (d) Two-photon scattering intensity changes due to the addition of *Salmonella* bacteria to anti-*Salmonella* antibody-conjugated AuNPs. The intensity did not change much when *E. coli* was added. (Reproduced from Neeley et al. (2011), with permission from IEEE.)

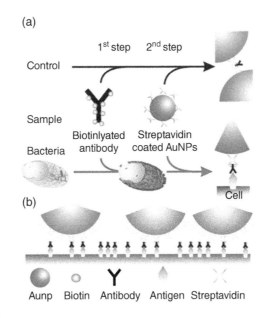

(a)

(b)

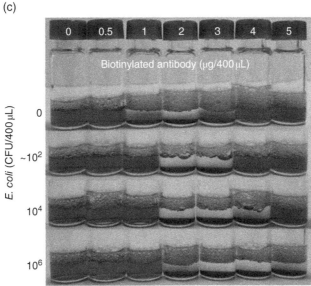

(c)

Figure 8.13 Scheme of (a) the two-step assay for visible detection of bacteria using biotinylated antibody (b-Ab) as SLs to control the extent of aggregation of stAuNPs. (b) Relative difference between the size of the cell and that of NPs on switching b-Abs off when they attach on the cell surface. (c) Shift of range exhibiting visible color change in response to the presence of *E. coli*, when tests performed using biotinylated antibody (b-Ab) as SL in 400 μL total sample volume with a fixed stAuNPs concentration (absorption: 0.43 @ 531 nm for 1/10 diluted sample) and. *E.coli* cell loads are typical of several tests, which yielded similar results. (Reproduced from Lim et al. (2012), with permission from Nature Publishing Group.)

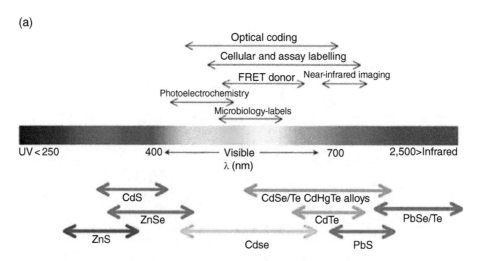

Figure 8.14a Different QDs and their representative areas of biological interest corresponding to the pertinent emission. (Reproduced from Medintz et al. (2005), with permission from Nature Publishing Group.)

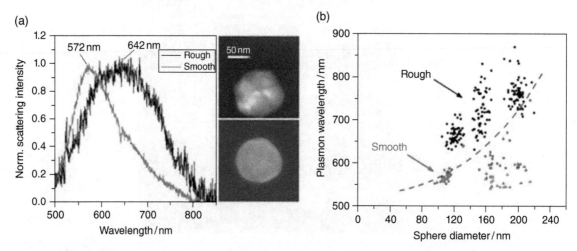

Figure 8.15 (a) Normalized scattering spectra of two Au spheres of same size and shape but different surface roughness, as observed in the scanning electron microscope images: rough (*top*) and smooth (*bottom*). (b) Size dependence of the dipole (*circles*) and quadrupole (*triangles*) plasmon resonance wavelengths in rough (*black*) and smooth (*red*) Au spheres. (Reproduced from Rodriguez-Fernandez et al. (2009), with permission from the Royal Society of Chemistry.)

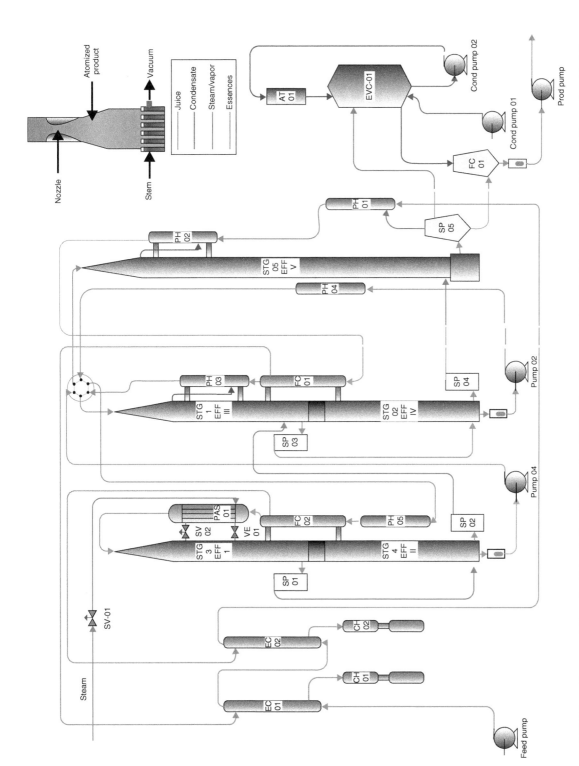

Figure 15.5 Schematic representation of a pilot-scale five-effect, five-stage TASTE. Top right corner, schematic of the cross-section of the top of stages 1 and 3. CH, chiller; EC, essence cooler; EFF, effect; FC, essence recovery; PH, preheater; SP, separator; STG, stage.

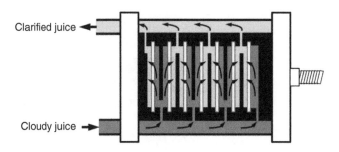

Figure 15.7 Schematic representation of the cross-section of a press filter.

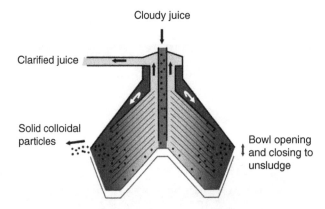

Figure 15.8 Schematic representation of a continuous centrifuge for juice clarification.

Figure 22.5 A variety of dried fish, abalone, shellfish, sea cucumber, and small fish in Hong Kong. Photo credit: Gleyn Bledsoe.

12 Food Laws and Regulations

Barbara Rasco

School of Food Science, Washington State University, Pullman, Washington, USA

12.1 Introduction

Food regulation in the US will be dramatically transformed by the Food Safety Modernization Act (FSMA) signed into law January 2011 (Public Law 111-353). The Food and Drug Administration (FDA) regulates about 80% of the food we eat, and requirements for a food protection plan with suitable preventive controls to guard food from intentional and unintentional contamination advancing global food safety are now a requirement for all food, domestically produced or imported, in the US. Provisions of the new law will be adopted either formally or through market-induced changes to business practices globally and at a substantial cost. Many are hopeful this new law will improve food safety, but it is likely to come at the cost of further consolidation throughout the industry as regulatory compliance costs drive marginal producers, particularly of "high-risk foods" and potentially fresh produce and aquatic foods, out of the market due to the high cost of regulatory compliance associated with the provisions of this new law. The federal government has consolidated its control over food safety programs within the US over the past six decades, most recently through the expansion of federal jurisdiction under this and other recent Acts, and the regulatory strategies by which the agencies choose to implement them.[1]

Changes in the regulations and practice of the United States Department of Agriculture Food Safety and Inspection Service (USDAFSIS), which subsumed state programs in recent decades, have further expanded federal authority. With 10–15% of all foods consumed in the US now imported, an increased focus is being placed upon the safety and quality of imported foods and ingredients along with the strategies to regulate their safety. Import product safety is prominent in the FSMA and in USDA programs, and with this will come the evolution of additional testing and certification programs. Many companies outsource testing services and inspections to third parties due to a lack of in-house expertise or because requirements of their retail customers specify that third-party audits or certifications be conducted to one of a variety of different standards. The proliferation of audit programs with the inherent conflict of interest that inevitably accompanies them will be exacerbated under FSMA and will increase food costs.

As a food technologist employed in the private sector, your responsibility will include assisting the company to develop strategies to support business operations by successfully implementing and managing an effective food protection program compliant with myriad government and external certification requirements that could lead to reduced litigation exposure. User fees, costs of testing, and third-party inspections will continue to increase the cost of food for consumers across the world, exacerbating the global food security crisis. Resources are limited in both economically developed and developing countries, and unfortunately, resources dedicated to overly stringent food safety and compliance programs are taken directly from food production and processing that could be used to make more food for more people, following the law of diminishing returns.

This chapter will outline some of the legal issues of which a food technologist should be cognizant and which will

---

[1] LM Lewis (2011) Informal guidance and FDA. *Food and Drug Law Journal* 66(4): 507–550 provides a historical overview of how the FDA develops regulations and guidance documents and raises recent examples of how the FDA has attempted to pre-empt state law and expand the scope of statutory authority through the issuance of informal guidance. Lewis notes that the FDA is very unlikely to change its position or incorporate any substantive stakeholder feedback into a guidance or a final rule. He further states that there is essentially no recourse to challenge FDA guidance in court, through a process known as judicial review.

---

likely impact his/her professional life. This chapter is not an exposition on civics and will not cover in detail information on food safety plans and programs such as HACCP, GAP, cGMPs, SSOPs, product labeling, etc, which are covered in other chapters. Instead, the objective is to provide an overview of newer legal trends and requirements, with an emphasis on issues that impact international trade.

12.2 The regulatory status of food ingredients and additives

Many useful substances are added to processed foods. The FDA determines the regulatory status of substances added to food, including any food ingredient or other substance that might become a component of the food, such as from a food contact surface or from packaging, under Section 409 of the Food Drug and Cosmetic Act. Companies are legally responsible for the safety of a new food or an additive. Common ingredients added to foods such as food acids, salt, and sugar are "generally recognized as safe" (GRAS) because of a long history of safe use as a food additive. However, if a food additive is not "generally recognized as safe by experts or previously approved informally by the federal government for the described use" the FDA will review safety and issue a decision (premarket approval) on if and how the additive can be used. One class of additives that receives special scrutiny is artificial colorants; these are subjected to pretesting and safety clearance by the FDA. The Center for Food Safety and Applied Nutrition reviews petitions for food and color additives along with notices submitted for the use of materials added to food that may be potentially harmful, such as food or ingredients made using modern genetic technologies.[2] Any new food additives that

are not GRAS are required to undergo premarket approval regardless of the type of method used to produce them.

12.3 Adulteration and misbranding

It is illegal to sell foods that are either adulterated or misbranded, and the FDA has a number of regulatory tools to force removal of foods that pose a risk to public health, are not wholesome, or mislead consumers about the quality, quantity or marketable features of the food.[3] A food is adulterated if it bears or contains any added poisonous or deleterious substance that may render it injurious to health, pathogenic microorganisms included. Under the Food Drug and Cosmetic Act (Section 402), a food is also deemed adulterated if:

1. any valuable constituent has been omitted in whole or in part
2. any substance has been substituted in whole or in part
3. damage or inferiority has been concealed in any manner, or
4. any substance has been added so as to increase its bulk or weight, or reduce its quality or strength, or make it appear better or of greater value than it is.

Food is adulterated if it is unwholesome or otherwise unfit for consumption.

A food is misbranded under the Act (Section 403) if its labeling is false or misleading in any particular aspect. To determine if a label is misleading, the FDA considers representations made about the product on the label, and the extent to which the labeling fails to reveal material facts in light of such representations made or suggested in the labeling or accompanying material. Its review emphasizes safety (and may not pay as much attention to issues that would be called "economic adulteration"),

---

[2] These are commonly called "genetically modified" (GMO) foods. Genetically modified plants or animals have had their DNA altered in some way to obtain a desired trait such as higher productivity, different fatty acid profile, or pesticide resistance. To date, the FDA has treated genetic techniques as a process and as long as the foods themselves do not pose a food safety risk, such as enhanced allergenicity, unknown toxicity or unusual functionality from a change in the composition of the food, there has been a tendency to grant approval. Any ingredient that would be a "major new dietary component" of food would receive an extensive premarket review prior to approval; one example is sucrose polyesters used as fat replacers. Numerous ingredients such as oil, starch, and sweeteners derived from genetically engineered plants (e.g. corn, soybeans) are on the market. However, no premarket approval has been granted for food from a genetically modified animal. The regulatory status

of a GMO Atlantic salmon, that grows more rapidly because of the incorporation of a king (Chinook) salmon growth hormone gene, has languished for 15 years after hopes in 2010 that FDA approval was pending following a series of high-level hearings. Foods from cloned animals have received approval because the animals do not contain foreign genes.
[3] An omission in the Food Safety Modernization Act is failure to define "economic adulteration" or "economically motivated adulteration" but it does retain the definition under the Food Drug and Cosmetic Act. This is a problem because economic fraud has been the basis for a number of recent high-profile food adulteration incidents that have resulted in numerous illnesses and deaths, the most notorious being the series of melamine contaminations of plant proteins, wheat, and milk powder in China (2007–2010).

Box 12.1 General food labeling requirements in the US (www.fda.gov)

1. General Food Labeling Requirements
   - Name of Food
   - Net Quantity of Contents Statements
   - Ingredient Lists
     - Food Allergen Labeling
   - Nutrition Labeling
     - General
     - Nutrient Declaration
     - Products with Separately Packaged Ingredients/Assortments of Foods
     - Label Formats/Graphics
       - Label Formats
       - Serving Size
       - Exemptions/Special Labeling Provisions
2. Claims
   - Nutrient Content Claims
   - Health Claims
   - Qualified Health Claims
   - Structure/Function Claims

and label reviews are conducted with an eye towards determining consequences that may result if a consumer uses the food as per the instructions provided or under conditions of customary or usual use. The omission of certain material facts from the label or labeling of a food is a form of misbranding. Product information, claims, package inserts, and accompanying literature are covered by similar provision to protect against consumer deception.[4]

Food must be labeled truthfully and provide information that will not mislead consumers. A summary of information to be provided on a food label is presented in Box 12.1. A misbranded food has one or more of the following features.

1. A false or misleading label. If the label is false or misleading in any particular, or if its advertising is false or misleading in a material aspect or if it is offered for sale under the name of another food.

2. Imitation of another food. If it is an imitation, then the label must contain in type, print size, and prominence the word "imitation" immediately before the name of the food. A food is an imitation if it lacks or is deficient in one or more characteristics considered to be important for the food, for example, low fat or vitamin content.

3. Misleading container. Containers cannot be formed, made, or filled to be misleading.

4. Improper package form. The package must contain a label with (1) the name and address of business of the manufacturer, packer or distributor, and (2) an accurate statement of the quantity of the contents in terms of weight, measure, or numerical count.

5. Lack of prominence of information on the label. If any word, statement or other information is required to appear on the label, this must be conspicuous and readable, and likely to be read and understood by ordinary individuals under customary conditions of purchase and use.

6. Lack of representation as to definition and standard of identity. If the food has a standard of identity it must conform to the compositional standards, ingredient statement, and be labeled in accordance with the standard.

7. Lack of representation as to standards of quality and fill of container. Requirements for foods with a standard of identity must meet requirements for fill or be marked that they do not meet these requirements.

8. Lack of labels for foods with no standard of identity. Foods and beverages must contain the common or usual name of the food/beverage and the common name of ingredients. Ingredients must be listed in decreasing order. In a beverage claiming to be a vegetable or fruit juice, the quantity of juice (total percentage) must be labeled prominently.

9. Lack of representation for special dietary use. Label must contain information about vitamins, minerals and dietary properties necessary to inform purchasers about the value of the product for such uses.

10. List of artificial flavoring, artificial coloring, or chemical preservatives missing or misleading. Foods with these components must have them listed on the label. There are special requirements for artificial coloring in butter, cheese, and ice cream. Labeling of pesticide chemicals on raw agricultural products (produce of the soil) is exempt (see 11).

11. List of pesticide chemicals on raw agricultural commodities missing or misleading. If a food is a raw agricultural commodity that is the produce of the soil, bearing or containing a pesticide chemical applied after harvest, the shipping container must be labeled with the common or usual name of the pesticide chemical(s) and its function.

12. Color additive list missing or misleading. Color additive packaging and labeling must conform to requirements. Certain colors, for example yellow 5, must be separately labeled.

---

[4] The Federal Trade Commission regulates food advertising and applies the same legal standards as the FDA when determining whether advertising is "misleading" to a reasonable consumer under the circumstances, and whether it is "material", in the sense that it will likely affect a consumer's conduct or decision to purchase a product.

13. Misleading nutrition information. Label must contain nutritional information as follows.

a. Serving size in amount customarily consumed. This must be expressed in a common household measure, and by weight.

b. Number of servings per container.

c. Total calories and calories from fat on a per serving basis.

d. The amount of the following nutrients: macronutrients [(total fat, saturated fat, trans fat (transesterified fatty acids)) total carbohydrate, complex carbohydrate (optional), sugars, total dietary fiber (required), soluble and insoluble fiber (optional) and total protein], cholesterol, sodium, vitamins (A, C), and minerals (calcium, iron). Labeling of other constituents is permissible. The labeling is factual, is not misleading or not otherwise prohibited by regulation. If a food has been fortified with added nutrients, this should be noted. The FDA may permit certain nutrients to be highlighted on the label if this will help consumers maintain healthy dietary practices.

e. Bulk foods and popular unpackaged fruits vegetable and seafood items should have nutritional information provided for consumers at point of retail sale.

14. Misleading nutrition levels and health-related claims. Express or implied claims on the label that characterize the level of any nutrient or the relationship of any nutrient with a disease or health-related condition must comply with FDA guidelines.

a. May not state the absence of a nutrient unless the nutrient is usually present in the food (or in a food which substitutes for the food in question) and the nutrient level or health-related claims are permitted by the FDA on the basis of a finding that such a statement would assist consumers to maintain healthy dietary practices. Claims for foods containing cholesterol must not contain fat or saturated fat in an amount that could increase the risk of disease or a diet-related health-related condition unless the cholesterol level is substantially less than normally present in the food (or a substitute), or cholesterol is not normally present in the food and a statement would assist consumers to maintain healthy dietary practices. The level of saturated fat and total fat must be prominent and in immediate proximity to the cholesterol claim on the label.

b. May not state that a food is high in dietary fiber unless the food is also low in total fat. Total fat must be disclosed on the label prominently and in close proximity to fiber claim on the label.

c. Claims may not be made if they are prohibited by regulation or if the FDA determines that the claim is misleading in the light of the level of another nutrient in the food.

d. If a nutrient claim is made for a food containing a nutrient at a level that increases the risk of a diet-related disease or a health-related condition to persons in the general population, the food shall prominently place the statement "See nutrition information for ---- content" in immediate proximity to the nutrient claim.

e. Claims for which there are no regulations may be made if a scientific body of the US government has published an authoritative statement regarding the nutrient; notice and claims approval provisions apply.[5]

---

[5] Dietary supplements, from a safety standpoint, are regulated in the same ways as foods and not like over-the-counter drugs, which might be more appropriate because these items are consumed, not as a food is consumed, but instead in small quantities as a pill, capsule, or as a small volume in liquid or solid form. However, labeling of nutrient supplements has significantly more leeway and more liberal labeling provisions than those permitted for food and are outlined in the Dietary Supplement Health and Education Act and its accompanying regulations. Nutrient label claims for dietary supplements composed of vitamins, minerals, or other similar nutritional substances and more provisions are allowed. For example, a supplement can have claims highlighting a benefit of consumption for alleviating a classic nutrient deficiency (e.g. iron deficient anemia), role of nutrient or dietary ingredients on the structure or function of the human body, characterization of a documented mechanism by which the nutrient acts, or describes the general well-being from consumption are permissible if such statements are truthful and not misleading. However, the statement: "This statement has not been evaluated by the Food and Drug Administration. This product is not intended to diagnose, treat, cure, or prevent any disease" must be prominently displayed on packages upon which nutrient claims are made. See *Pearson v Shalala* (1999) (164 F.3d 650, 334 U.S. App. D.C. 71) in which the appellate court declared that FDA regulations for evaluating health claims for dietary supplements violated first amendment rights of commercial free speech when the agency refused to authorize supplement health claims accompanied by a "reasonable disclaimer." The court also invalidated the manner in which the FDA conducted reviews for supplement labeling. Because of this ruling, for supplements, but not for foods, the FDA has the burden of proving that a claim is inherently misleading and that this cannot be "neutralized" by including a disclaimer on the product packaging. A common disclaimer is:" These statements have not been evaluated by the Food and Drug Administration. This product is not intended to diagnose, treat, cure or prevent any disease."

f. An implied claim made within a brand name may be made following a petition to the FDA for its use; claim cannot be misleading.

15. Lack of proper allergen labeling. The common or usual name of each of the major food allergens the product contains is required on the product label or in the ingredient statement if the ingredient name includes the name of the allergenic food source. Allergen labeling is required for foods containing: tree nuts, peanuts, fish, crustaceans, milk, eggs, wheat, and soybeans or ingredients obtained from them. Flavorings, colorings or incidental additives that contain a major food allergen are subject to labeling; some highly refined foods, such as oils, may be exempt from allergen labeling because the levels of allergenic components would be very low. There are specific provisions for "gluten free" labeling.

12.4 The global food trade: risk from adulterated and misbranded foods

Economic fraud poses one of the greatest risks to our food supply. Adulteration is rampant in developing countries, with at least 30 identified adulterants added to milk to improve shelf life or increase apparent protein, solids or fat content. The melamine scandal in China made international headlines in 2007 after the death of domestic animals in the USA, Europe, South Africa, Canada and throughout Asia resulting from addition of melamine to increase the apparent protein content of plant protein ingredients that were used in animal feed. In 2008, milk powder containing melamine led to a global recall of milk powder and thousands of food products containing milk powder from China. Hundreds of thousands of children were affected in China alone, and untold numbers of others in countries in which this melamine-tainted milk powder was sold and consumed. Many of these children will likely suffer from chronic kidney problems from this exposure.

This disastrous failure of our food safety systems finally focused worldwide attention on the problem of intentional contamination in which the motivation is economic fraud. Several provisions of the FSMA are directed at economic fraud, in the sense that this is a form of intentional contamination that can potentially cause serious adverse health consequences or death. However, addressing this form of insidious and clever adulteration in a preventive measure program is difficult. The FDA recognizes this dilemma and has recharacterized itself as a global agency that must prepare itself to "regulate in an environment in which product safety and quality know no borders."

The US Government Accounting Office, in a review of FDA efforts in this area, offers a definition of economic adulteration: *economic adulteration* is fraudulent, intentional substitution or addition of a substance in a product for the purpose of increasing the apparent value of the product or reducing the cost of its production, for economic gain (GAO, 2011). Examples of economic adulteration include the dilution of products with increased quantities of an already present substance (e.g. increasing inactive ingredients of a drug with a resulting reduction in the strength of the finished product, or watering down milk or juice), as well as the addition or substitution of substances in order to mask dilution. This definition is the same as the working definition of "economically motivated adulteration" that the FDA developed in 2009, but which is likely to be included in regulation or guidance.

The FDA should build criminal cases against companies engaged in economic fraud and prosecute such cases vigorously (GAO, 2011). This would provide a strategy for keeping adulterated ingredients out of the food supply. Industry self-policing with market sanctions is critical for curtailing economic fraud and often brings fraud to the attention of governmental authorities.

Traceability will be the key to managing the impact of economic fraud along a complex value chain, although the technologies that are available to address this problem remain inadequate. Provisions for country of origin labeling will not be effective for monitoring the source of ingredients in a food that changes hands several times during processing and distribution if the initial source of the food is not disclosed. Similarly, ingredients are sourced from numerous regions and countries, depending upon season, demand, and price, making tracking difficult further down the supply chain. Provisions in the FSMA require companies to track the immediate previous source of an ingredient and then track the ingredient into the production lot(s) in which it is used. This information will help the company, as well as the FDA, to locate foods affected by a recall if ingredients are deemed to be unsafe or if other problems arise with a product. Still, because multiple parties are involved along the supply chain for the manufacture and distribution of foods containing multiple components from multiple sources, and with increasing numbers of transfers of components between parties in different countries, opportunities for adulteration are widespread for each party further down the distribution chain. The FDA lacks the expertise in forensic investigation and analytical capacity to scratch the surface of this problem, but passing the burden to companies is irresponsible, since companies, particularly smaller ones, often have few resources for this activity. Regardless, traceability programs need to be implemented and continually improved since effective tracking reduces the public health and market impact of an adulteration incident. The FDA is seeking assistance from the private sector regarding traceability systems and will likely be issuing guidance in this area.

The FDA has required registration of food facilities for the past 10 years under provisions of the Bioterrorism Act of 2002 and has incorporated these requirements into the FSMA, as will be discussed in detail later in this chapter. The registration requirements cover both domestic and foreign producers and directly affects over 700,000 entities as of this writing. Foods regulated by the FDA cannot be sold unless the firm is registered. If food fails to meet certain regulatory requirements, the agency can pull the registration, meaning that it will not be possible for the food to be sold. This registration provision was devised to provide a means of tracking food to its source and is tied into traceability provisions under the law.

A major complication in the current regulatory scheme is that the US does not have the legal authority to inspect foreign operations or to accredit third parties in foreign countries unless this is voluntarily granted by the foreign entity. Despite the lack of jurisdiction, Congress has placed an obligation upon the agency to inspect foreign facilities and establish overseas office(s) as a means of increasing oversight of imported foods. Some of this enhanced oversight will be through third-party certifiers. Unfortunately, corruption within the third-party inspection process and accreditation process is common and the burden to protect our food supply from adulteration, economic or otherwise, will fall upon the importing country and border control programs to prevent entry of adulterated food.

12.5 US Department of Agriculture programs

The USDA is engaged in numerous activities that promote agriculture and, since the 1960s, an expanded mission for food security. The USDA engages in market support and market and economic research, promotion of international agricultural trade and market grades and standards, plant and animal health, food safety, and research on many aspects of agriculture, with a recently expanded emphasis on resource conservation, including water and soil management and climate change.

12.5.1 Food safety: the Food Safety and Inspection Service

Food safety within the USDA is within the purview of the Food Safety and Inspection Service (FSIS). The FSIS inspects and monitors all meat, poultry, and liquid egg products sold in interstate and foreign commerce. The Federal Meat, Poultry and Egg Products Inspection Acts ensure compliance with mandatory US food safety standards. The Federal Meat Inspection Act provides for inspection of all meat products, with the exception of custom-processed meat covered by state programs, and reinspection of imported meat products. The Poultry Products Inspection Act similarly gives the FSIS authority for all poultry products sold in interstate commerce. The Egg Products Inspection Act provides for inspection of domestically produced liquid eggs and reinspection of imported liquid eggs. (Eggs in the shell are inspected by the FDA.) For processed eggs, eggs are examined before and after breaking. The Humane Methods of Livestock Slaughter Act mandates that livestock slaughter must be conducted humanely.

Procedures for obtaining meat or poultry inspection are straightforward and outlined in the Federal Grant of Inspection Guide found on the USDA website (USDA, 2012). A company applies for a grant of inspection and with this must show that the facility can meet performance standards for the products in question, develop a sanitation plan (sanitation SOPs), and Hazard Analysis Critical Control Point (HACCP) program that meets FSIS criteria, have approval of a potable water source (a report from a municipal water authority showing water is potable) and a sewage discharge system that prevents back-up of sewage into areas where food is processed, handled or stored. Model plans are provided for different facilities, including small and very small processing facilities. Agency label review is mandatory and guidance is provided. Once a grant of inspection is made, a facility will be provided with an establishment number and mark that is affixed to the FSIS inspected and passed product. The FSIS can withdraw inspection and this will prevent sale of food until the facility is back in compliance.

Withdrawing inspection of a facility by the FSIS can occur under the following circumstances.

1. An establishment produced and shipped adulterated product.

2. An establishment did not have or maintain a HACCP plan.

3. An establishment did not have or maintain sanitation SOPs.

4. An establishment did not maintain sanitary conditions.

5. An establishment did not collect and analyze samples of *Escherichia coli* biotype I and record results as required.

6. An establishment did not comply with *Salmonella* spp. performance standards.

7. An establishment did not slaughter or handle livestock humanely.

8. An establishment operator, officer, employee or agent assaulted, threatened to assault, intimidated or interfered with an FSIS program employee.

9. A recipient of inspection or anyone responsibly connected to the recipient is unfit to engage in any business requiring inspection.

12.5.2 Food safety: the Animal and Plant Health Inspection Service

The Animal and Plant Health Inspection Service (APHIS) is charged with protecting the health of agricultural animals and plants, regulating genetically engineered organisms, administering the Animal Welfare Act and managing certain wildlife activities. One of the most important roles the APHIS plays is to guard against the introduction of agricultural insect pests and diseases into domestic food production. The APHIS plays a role in import inspections and manages quarantine programs for animals and plants, including emergency protocols to manage and eradicate an outbreak conducted in conjunction with state and international partners. The agency develops sanitary and phytosanitary standards to promote international trade and prevent unjustified trade restrictions or non-tariff trade barriers.

12.5.3 International trade

12.5.3.1 Marketing and regulatory programs

The USDA has a number of marketing and regulatory programs to promote and facilitate domestic and international marketing of US agricultural products. USDA employees are actively involved in setting national and international market and trade standards for agricultural products. These include health and care of animals and plants, import requirements and harmonization of international standards to encourage global trade, and a number of market promotion programs.

12.5.3.1.1 Agricultural Marketing Service

The Agricultural Marketing Service (AMS) provides quality grade standards, grading programs, certification, audit and accreditation programs, and laboratory analysis and approval, and equipment review as part of a voluntary program to promote trade and promote quality to customers domestically and internationally. Companies pay for these services.

The USDA quality grade marks are seen in the marketplace for beef, lamb, chicken, turkey, dairy products, and eggs. The grading service is used by wholesalers. Depending on how wholesalers want to market their products, they may include a quality grade mark on fresh or processed fruits or vegetables. Quality grades make business transactions easier since product attributes are clearly defined and understood by market actors. A number of countries have adopted US grade standards for commerce.

Quality standards are based on measurable attributes that describe the value and utility of the product. For example, beef quality standards are based on attributes such as marbling (the amount of fat interspersed with lean meat), color, firmness, texture, and age of the animal, for each grade. Quality grade standards describe a range of attributes for which there is a standard for a product, and the number of grades varies by commodity. These include yield, size, level of defects, factors associated with functional properties, cosmetic appearance, and sensory characteristics. There are eight grades for beef and three each for chickens, eggs, and turkeys. Thirty eight grades exist for cotton, and more than 312 for fruit, vegetables, and specialty product standards (www.usda.ams.gov). The AMS also provides audit and accreditation programs based on International Organization for Standardization (ISO) standards and/or HACCP principles and guidelines, and verifies these documented programs through their provision of independent, third-party audits. AMS audit and accreditation programs are voluntary and paid through hourly user-fees (www.usda.gov).

12.5.3.1.2 Country of origin labeling (COOL)

Country of origin labeling (7 CFR Part 60 *et seq*) is a retail-level requirement that provides consumers with notification about the geographic origin of foods they buy. Manufacturers provide country of origin on packaged foods under this mandatory program and voluntarily on other items. This label is required for muscle cut and ground meats: beef, veal, pork, lamb, goat, and chicken; wild and farm-raised fish and shellfish; fresh and frozen fruits and vegetables; peanuts, pecans, and macadamia nuts; and ginseng. For meat, poultry and seafood items, a "United States country of origin" label can only be used for animals born, raised, and slaughtered in the US, and for aquatic foods harvested from US waters or by a US-flagged vessel. For peanuts, pecans, ginseng, and macadamia nuts, these must be exclusively produced in the US in order to receive a US country of origin designation. Any of the food items covered under this regulation must retain their original country designation unless they have undergone "substantial transformation[6]" following import into the US. As an example, in the case of seafood, the food would be labeled: "From [country X], processed in the United States." Alternatively, the food could be labeled: "Product of country X and the United States." Aquatic foods must also have a designation as to whether they are cultivated or wild harvested.

A food can have multiple countries of origin if the raw materials from a number of countries have commingled in a production lot; for example, a label on ground beef might read: "Product of country X, Y and Z." However, to reduce consumer confusion, "or" and "and/or" in the

country of origin designation declaration are not allowed; for example, retailers would not be allowed to put: "Product of the US, Canada, and/or Mexico."

Processed foods are exempt under COOL, but may require country of origin labeling under the Tariff Act of 1930 (Tariff Act). One example is for frozen mixed vegetables containing items from a number of countries blended together in the US which must have the separate countries listed on the retail package. Another example where origin of the raw material is required is a container of roast nuts consisting of peanuts of foreign origin roasted in the US; similarly canned salmon made from imported Chilean salmon. An example of where such a label is not required is for a candy bar or trail mix containing peanuts because of the presence of other "substantive food ingredients," and processing to the extent that the character of the peanuts is transformed into a completely different food item.

12.5.3.2 Organic foods

The USDA manages one of the most widely recognized organic programs in the world. The USDA and Washington, Oregon and California states are international leaders in the development of organic food requirements and over a period of years have led efforts to build a comprehensive organic certification program for foods, consumer products, and fabrics. The USDA defines organic as a:

> labeling term signifying that the food or other agricultural product has been produced using approved methods that integrate cultural, biological, and mechanical practices that foster cycling of resources, promote ecological balance, and conserve biodiversity. Use of synthetic fertilizers, prohibited pesticides, sewage sludge, irradiation, and genetic engineering is prohibited.

Development of rigorous organic standards arose in response to consumer pressure in the US in the 1990s to label foods containing ingredients from genetically modified organisms and in response to scares associated with pesticide and agricultural chemical use. The National Organics Program regulates all products certified in the US to the USDA standards through organic certifiers. Agricultural products in international markets can be certified to USDA standards and the USDA seal is found on a variety of foods around the world. The USDA conducts investigations and enforcement activities to ensure compliance with products on which its organic seal or

---

[6] Substantial transformation is defined by Customs and Border Protection (CBP) and is a designation assigned for collections of duties and tariffs, a food item is substantially transformed if it has been processed. CBP does not consider blanching, cutting, freezing, and combining and packaging different vegetables to exclude them from CBP marking requirements.

Here is an example of a definition of processed foods for country of origin labeling (see 7 CFR Sec. 65.220). A processed food item means a retail item derived from a covered commodity that has undergone specific processing resulting in a change in the character of the covered commodity, or that has been combined with at least one other covered commodity or other substantive food component (e.g. chocolate, breading, tomato sauce), except that the addition of a component (such as water, salt, or sugar) that enhances or represents a further step in the preparation of the product for consumption, would not in itself result in a processed food item. Specific processing that results in a change in the character of the covered commodity includes cooking (e.g. frying, broiling, grilling, boiling, steaming, baking, roasting), curing (e.g. salt curing, sugar curing, drying), smoking (hot or cold), and restructuring (e.g. emulsifying and extruding). Examples of items excluded include teriyaki flavored pork loin, roasted peanuts, breaded chicken tenders, and fruit medley.

logo is used. The USDA seal can be affixed to organic products if the agricultural product or food is certified organic and contains 95% or more organic content. Agricultural chemical residue testing programs are an important component of organic certification and one of its most costly features.

Natural products are not necessarily organic, and are defined by the USDA FSIS for products under its jurisdiction as: "meat, poultry or egg products that have been minimally processed and contain no artificial ingredients." "Natural" does not imply any farm practice standards. The FDA has not defined "natural" for products that it regulates. Organic products do not necessarily meet environmental sustainability or fair trade criteria, all of which, to this point, are voluntary privately managed certification programs. Other organic certification programs are common in European markets.

12.6 Environmental Protection Agency programs

The United States Environmental Protection Agency (EPA) regulates many aspects of agricultural and food production, including fertilizer and pesticide/herbicide usage; discharge of solid, liquid and gaseous materials into water and air; management of solid waste and transport storage and disposal of hazardous materials; and environmental management of concentrated animal feeding establishments. Many of these programs are managed by complicated permitting processes, along with compliance with industry-specific discharge requirements. State and local authorities also play a role in setting requirements for food processing operations for air and water discharges, including run-off, shoreline and wetland management programs, and various zoning/building requirements. A list of laws directed at or including agricultural and food operations is presented in Box 12.2.

The EPA has no compunction about levying egregious civil penalties[7] for minor labeling infractions, late

paperwork filings, and for small but environmentally insignificant discharges of pollutants. Discharge permit requirements by the EPA or state departments of ecology or environment along with the cost and delays associated with obtaining these permits have caused many companies to curtail or abandon their US operations, jeopardizing our ability as a nation to provide sufficient food to feed ourselves. Violations of federal laws often lead to prosecution under similar state laws and vice versa and liability in multiple jurisdictions for the same activity.

12.7 The Food Safety Modernization Act

In the past decade, the jurisdiction of the US FDA has been greatly expanded through new provisions of the FSMA and the incorporation within it of some of the earlier provisions in PL 107-188 The Public Health Security and Bioterrorism Preparedness and Response Act of 2002 (the Bioterrorism Act), requiring registration of food processing facilities, prior notice to the agency for imported foods and ingredients entering the US, new provisions for traceability, and delegation to the FDA of a new authority to administratively detain food.

From a food processing perspective, provisions in the FSMA for Preventative Control Plan (similar to HACCP)-based food protection programs, and stricter traceability requirements are the parts of the law likely to impact food technologists to the greatest extent from

---

[7] The EPA is primarily funded by user fees and civil penalties. This provides an incentive to maximize fees. Under various mandates, the EPA can assess civil penalties on a daily basis or on a per container basis. Incidents from the past year coming to the author's attention include the following. Hundreds of boxes of a pesticide chemical in a warehouse were considered by an EPA inspector to be out of compliance because a required label on the primary package had not additionally been placed upon the outside carton. This inspector treated each box as a separate violation and levied a fine of around 100 million dollars. The company's labeling was not

in error and none of the allegedly out of compliance containers had left company custody. The company challenged the fine and it was reversed because the company was not only in compliance but had also passed a number of recent inspections in which the same labeling scheme had been used. The inspector was educated as to the difference between a primary and secondary container and hopefully reassigned. A disconcerting aspect of this incident was that if the fine had been 1 million dollars, the company would have paid it just to prevent damage to its reputation.

Reporting requirements for minor discharges have to be made within a short period, normally 48 h or less regardless of staffing, size of the company, or hours of operations, making a minor 5 gallon (18.9 L) ammonia discharge on a weekend from a refrigeration system into a $50,000 proposition if paperwork is not filed in a timely fashion by 9 am Monday morning.

Good faith efforts on the part of companies in their attempt to interpret ambiguous and confusing permitting requirements are not recognized. One small firm was fined tens of thousands of dollars for failing to properly transfer a water discharge permit after closely following instructions of district and regional offices on how to conduct the transfer.

Box 12.2 Federal environmental laws affecting agriculture

Air Quality
   Clean Air Act – control of criteria pollutants (carbon monoxide, lead, nitrogen oxides, ozone, sulfur dioxide, particulate solids), development of state implementation plans to reduce pollutant levels
   National Emission Standards for Hazardous Air Pollutants (NESHAP) (pesticide active ingredient production industry; industrial, commercial and institutional boilers and process heaters)
Chemical use
   Toxic Substance Control Act – premanufacture or import notice to EPA for new chemicals
   • Insecticide, Fungicide and Rodenticide Act – use of pesticide agents, food residues
   • Emergency Planning and Community Right to Know Act – community-based emergency notification, planning, mandatory reports of releases
Solid waste
   Solid Waste Disposal Act – minimum requirements for solid waste disposal; source reduction
   Resource Conservation and Recovery Act (RCRA) – wastes that are toxic, flammable, corrosive, or reactive may be considered hazardous and require special waste disposal provisions
Water quality
   Safe Drinking Water Act – state development of safe well-head programs, watershed protection, and surface and ground water source protection
   Clean Water Act (Federal Water Pollution Control Act); National Pollutant Discharge Elimination System (NPDES) permit required for "point source" discharges
   States have similar discharge permit programs with requirements for individual discharge permits

---

the passage of this new law. Specific provisions in the FMSA are outlined in Box 12.3.[8]

The FSMA incorporates many of the provisions of the Bioterrorism Act. The most important of these for a food technologist is the addition to conventional HACCP-based food safety plans that require a food producer to address intentional adulteration risk (vulnerability assessment) along with the development of mitigation strategies to control intentional contamination risks, including those from biological, chemical, and radiological hazards.

The FSMA is the greatest expansion of federal authority over food production and sales since implementation of the Food Drug and Cosmetic Act of 1938 (FDCA). The FDCA provided direct federal authority over foods by statute and provided enhanced enforcement authority and control over the food industry by prohibiting the sale of adulterated or misbranded foods through an enforcement system based upon periodic inspections of food facilities, federal checks for compliance with good manufacturing practices, and records-based reviews. The FDCA followed about 30 years after the adoption of the Federal Meat Inspection Act (1906) and Poultry Products Inspection Act (1906)

that mandated in-plant inspection programs, including continuous presence of USDA employees in meat and poultry plants to conduct veterinary inspections of animals prior to slaughter, carcass inspections after slaughter, and later adoption of microbial sampling and HACCP-based food safety programs.

The strategies used by both the FDA and FSIS are converging. The FDA was first to adopt HACCP-based programs with the first programs being with seafood products (1995) (Procedures for the Safe and Sanitary Processing and Importing of Fish and Fishery Products, 60 Federal Register 65096, December 18, 1995), followed by FSIS HACCP programs in 1996 (Pathogen Reduction: Hazard Analysis and Critical Control Point (HACCP) System, 61 Federal Register 38806, July 25, 1996). Meat and poultry products have been labeled as passing inspection since the beginning of in-plant meat and poultry inspection programs, with slaughter and processing facilities for meat- or poultry-containing foods receiving an establishment mark that permits them to sell the products they make. Facilities out of compliance will lose the right to use their mark, and will not be able to sell food until the facility is back in compliance and use of the mark restored by the FSIS.

The FDA, through the new FSMA law, has adopted a similar strategy by requiring registration of food processing and warehousing facilities domestically and internationally as previously mentioned. If the FDA believes that a food has "a reasonable probability of causing serious adverse health

---

[8] Agency regulations are the means by which a federal government agency implements a statute such as the Food Safety Modernization Act. As of this writing, few of the regulations pertaining to the FSMA have been issued, so exactly how the FDA plans on implementing this new statute is not known at this time.

Box 12.3 Provisions of the Food Safety Modernization Act (www.fda.gov/Food/GuidanceRegulation/ FSMA/ucm247548.htm)

TITLE I – IMPROVING CAPACITY TO PREVENT FOOD SAFETY PROBLEMS
Sec. 101. Inspections of records.
Sec. 102. Registration of food facilities.
Sec. 103. Hazard analysis and risk-based preventive controls.
Sec. 104. Performance standards.
Sec. 105. Standards for produce safety.
Sec. 106. Protection against intentional adulteration.
Sec. 107. Authority to collect fees.
Sec. 108. National agriculture and food defense strategy.
Sec. 109. Food and Agriculture Coordinating Councils.
Sec. 110. Building domestic capacity.
Sec. 111. Sanitary transportation of food.
Sec. 112. Food allergy and anaphylaxis management.
Sec. 113. New dietary ingredients.
Sec. 114. Requirement for guidance relating to post harvest processing of raw oysters.
Sec. 115. Port shopping.
Sec. 116. Alcohol-related facilities.

TITLE II – IMPROVING CAPACITY TO DETECT AND RESPOND TO FOOD SAFETY PROBLEMS
Sec. 201. Targeting of inspection resources for domestic facilities, foreign facilities, and ports of entry; annual report.
Sec. 202. Laboratory accreditation for analyses of foods.
Sec. 203. Integrated consortium of laboratory networks.
Sec. 204. Enhancing tracking and tracing of food and recordkeeping.
Sec. 205. Surveillance.
Sec. 206. Mandatory recall authority.
Sec. 207. Administrative detention of food.
Sec. 208. Decontamination and disposal standards and plans.
Sec. 209. Improving the training of state, local, territorial, and tribal food safety officials.
Sec. 210. Enhancing food safety.
Sec. 211. Improving the reportable food registry.

TITLE III – IMPROVING THE SAFETY OF IMPORTED FOOD
Sec. 301. Foreign supplier verification program.
Sec. 302. Voluntary qualified importer program.
Sec. 303. Authority to require import certifications for food.
Sec. 304. Prior notice of imported food shipments.
Sec. 305. Building capacity of foreign governments with respect to food safety.
Sec. 306. Inspection of foreign food facilities.
Sec. 307. Accreditation of third-party auditors.
Sec. 308. Foreign offices of the Food and Drug Administration.
Sec. 309. Smuggled food.

TITLE IV – MISCELLANEOUS PROVISIONS
Sec. 401. Funding for food safety.
Sec. 402. Employee protections.
Sec. 403. Jurisdiction; authorities.
Sec. 404. Compliance with international agreements.
Sec. 405. Determination of budgetary effects.

consequences or death to humans or animals," a registration can be "suspended" and food cannot be sold. Registration can be restored when the FDA is satisfied that the facility is back in compliance.

The FSMA provides for mandatory recalls without having hard evidence that the food is contaminated but instead, only that there is a "reasonable probability" that the food is contaminated. The mandatory HACCP-based food protection program requires that allergens and decomposition that could make a food unsafe be addressed in addition to the more conventional biological, chemical, and physical hazards. Relatively rigid record-keeping requirements are envisioned for high-risk foods. There are provisions in the FSMA for new rules on fresh produce safety.[9] A new employee protection provision is introduced to protect whistleblowers and those co-operating with the FDA in a food safety investigation from employer retaliation. Substantial new costs are anticipated for laboratory testing, which after 2015 will have to be conducted in accredited laboratories, provisions for third-party audits for imported foods, costs of compliance with new import requirements and agency fees for reinspection are new added costs. Details of the major provisions in the FSMA are listed in Box 12.3 and elaborated upon in the following text.

SEC. 101. Inspections of records. The FDA can request, in writing, that a company provide records about any food that it reasonably believes is adulterated or "any other article of food that [it] reasonably believes is likely to be affected in a similar manner." The FDA would use the information it finds to determine if "there is a reasonable probability that the use of or exposure to an article of food, and any other article of food that the Secretary reasonably believes is likely to be affected in a similar manner, will cause serious adverse health consequences or death to humans or animals." This applies to "all records relating to the manufacture, processing, packing, distribution, receipt, holding, or importation of any potentially affect for in paper or electronic formats wherever they may be located."

SEC. 102. Registration of food facilities. All domestic and foreign food facilities must register and renew their registration every even year. The purpose is to identify both animal and human food facilities that are using high-risk products and those that have not been inspected. Only registered facilities can sell food in the US. The registration stipulates that the FDA can inspect the facility. Registration is suspended if the FDA "determines that food manufactured, processed, packed, received, or held by a registered facility has a reasonable probability of causing serious adverse health consequences or death to humans or animals."[10] The company can request an informal hearing not later than 2 business days after suspension about the actions that it needs to take to have its registration reinstated and to propose a corrective action plan to correct conditions. Reinstatement is solely at the agency's discretion.

SEC. 103. Hazard analysis and risk-based preventive controls. The individual in charge of a food facility that processes, packs, or holds foods must identify and implement preventive measures to "significantly minimize or prevent the occurrence" of certain hazards and also ensure that food is not adulterated or misbranded. This is an expansion of HACCP as it is currently implemented in the industry which addresses food safety hazards that are "reasonably likely to occur" and have the potential to cause serious injury or death, with preventive measures to eliminate the hazard or reduce it to an acceptable level.

Under the FSMA, a facility is now required to:
1. identify and evaluate known or reasonably foreseeable hazards that may be associated with the facility, including:

   a. biological, chemical, physical, and radiological hazards, natural toxins, pesticides, drug residues, decomposition, parasites, allergens, and unapproved food and color additives; and

   b. hazards that occur naturally, or may be unintentionally introduced; and

2. identify and evaluate hazards that may be intentionally introduced, including by acts of terrorism; and
3. develop a written analysis of the hazards.

Preventive controls include measures to significantly minimize or prevent the identified hazards and assure that the food is not adulterated or misbranded. These need to be monitored as part of a written plan, either as a critical control point or control point, if most appropriate, or as part of good manufacturing practices (GMPs) or as part of a sanitation program (SOP). If preventive controls fail, any affected food has to be segregated and not sold, and activities taken to correct the problem so that it is not likely to reoccur. The preventive measures have

---

[9] Along with politically motivated inclusions, somewhat unrelated, are provisions for postproduction handling of oysters, allergens in foods served in schools, and anabolic steroid-containing dietary ingredients.

[10] A registration may be suspended if it created, caused, or was otherwise responsible for such reasonable probability; knew of, or had reason to know of such reasonable probability; and packed, received, or held such food.

to be verified to show that they are adequate to control the hazards identified, with recommendations for environmental and product testing programs. A written plan with monitoring and corrective actions is required. Facilities must conduct a new hazard analysis at least every 3 years or when there is a significant change in the operation of the facility, including type and sourcing of ingredients that could either introduce a new hazard or remove an old one. Records must be retained for 2 years for review by the FDA on either a written or oral request and include monitoring records, documentation of non-conformance critical to food safety along with corrective actions, verification activities including test results. Needless to say, care must be taken as part of this new HACCP-based program to ensure the completeness of records; equally critical is protection of intellectual property, specifically formulations and proprietary processes, so that these are not released to the agency without appropriate actions taken to protect confidential information.

SEC. 105. Standards for produce safety. Science-based minimum standards for the safe production and harvesting of raw fruits and vegetables to minimize the risk of serious adverse health consequences or death are being established with proposed rules for produce safety slated for 2014.

SEC. 106. Protection against intentional adulteration. The FDA will conduct a vulnerability assessment of the overall food system considering Department of Homeland Security biological, chemical, radiological, or other terrorism risk assessments to better understand the uncertainties, risks, costs, and benefits associated with guarding against intentional adulteration of food at vulnerable points, and determine the types of science-based mitigation strategies or measures that are necessary to protect against the intentional adulteration of food. A series of regulations were proposed in 2013 to protect against the intentional adulteration of food, specifying how a person shall assess such risks and implement mitigation strategies intended to protect against the intentional adulteration of food. These regulations would apply only to foods for which there is a high risk of intentional contamination, and could cause serious adverse health consequences or death to humans or animals and shall include those foods with clearly identified vulnerabilities such as a short shelf life or susceptibility to intentional contamination at critical control points, and are in bulk or batch form (such as milk), prior to being packaged for the final consumer. It is uncertain given the lack of technical information available whether the target date for these regulations is realistic.

SEC. 107. Authority to collect fees. The FDA will institute new fees for food facility reinspection (domestic and foreign), certain recall activities and import activities, such as the voluntary qualified importer program, and is authorized to recover the full cost including administrative fees associated with these activities. The total fee assessed will depend on the number of hours the FDA spends directly on the reinspection-related activities or food recall activities associated with a recall order.

SEC 108. National agriculture and food defense strategy. Part of this program is for the agency to work with the private sector to develop business recovery plans to rapidly resume agriculture, food production, and international trade following a food safety disaster. This would involve conducting exercises of the plans with the goal of long-term recovery results; rapidly removing and effectively disposing of contaminated agriculture and food products, and infected plants and animals; and decontaminating and restoring areas affected by an agriculture or food emergency.

SEC. 111. Sanitary transportation of food. The FDA will conduct a study of food transportation systems, including air transport, and evaluate the unique needs of safe and sanitary food delivery to rural communities. Guidance will be provided on how to properly sanitize containers and vehicles used for food transportation to minimize cross-contamination.

SEC. 113. New dietary ingredients. This particular provision relates only to the safety of dietary ingredients and supplements containing anabolic steroids or their analogs.

SEC. 114. Requirement for guidance relating to post-harvest processing of raw oysters. Certain individuals within the agency have targeted the consumption of raw molluscan shellfish for years, with the intent to ban sales. This provision reflects a compromise relating to postharvest processing for raw oysters and controls that could improve product safety. It applies to both domestic and foreign producers. Depuration (purification) is likely the most appropriate control and is not excluded in the text of the statute.

SEC. 115. Port shopping. Port shopping is when an importer attempts to select a port that will provide less scrutiny to incoming shipments. Prior notice provisions will make port shopping more difficult, and with this provision will provide notice to other ports that a certain shipment has been denied entry so that it will not be admitted at another location.

SEC. 201. Targeting of inspection resources for domestic facilities, foreign facilities, and ports of entry; annual report. The emphasis of inspections will be upon high-risk facilities and products based upon the following factors.

1. Known safety risks of the food.
2. Compliance history of a facility including recalls, food-borne illness outbreaks, and food safety violations.
3. Rigor and effectiveness of the facility's hazard analysis and preventive controls plan.
4. For imports, these additional criteria apply: known risks with the imported food, known food safety risks of the country or regions of origin and countries through which the food may be transported, compliance history of the importer, rigor of the importer's food safety program including compliance with the foreign supplier verification program and their participation in the voluntary qualified importer program, current physical inspection rate, number of line items for the food imported, and FDA experience in the exporting country.

Domestic facilities will be inspected at least every 5 years initially and every 3 years thereafter, and for lower risk foods, not less than every 7 years initially and every 5 years thereafter. Inspection of foreign facilities will increase from 600 during the first year (2012) to more than 19,000 by 2018. Reliance will be placed upon state and local inspectors and those of foreign governments and third parties since the FDA will not have the resources to conduct this task. How the FDA will co-ordinate these activities to assure consistency is not clear at this time.

SEC. 202. Laboratory accreditation for analyses of foods. By 2015, any laboratory testing that is to be conducted with regard to a food safety investigation[11] will have to be conducted in an "accredited laboratory" approved by an "accrediting body." This expensive and tedious process, applicable not just to the laboratory but for any sampling program and analytical testing protocol conducted within it, will substantially increase the cost of laboratory testing and may place it out of reach of many small businesses. The process will not necessarily increase the reliability of the testing, just its expense, and will undermine the ability to establish in-house laboratories within small and medium-sized food businesses because of the accreditation expense. The FDA would establish a registry of accreditation bodies that will determine if a laboratory has "demonstrated capability to conduct one or more sampling and analytical testing methodologies for food." Foreign laboratories would also require accreditation and will be required to meet the same standards as domestic laboratories. The FDA "shall develop model standards that a laboratory shall meet to be accredited by a recognized accreditation body for a specified sampling or analytical testing methodology and included in the registry" and will review accreditation bodies every 5 years.

SEC. 204. Enhancing tracking and tracing of food and record keeping. The requirement for one-step-forward one-step-back traceability remains a technical challenge. In response to this requirement the FDA will be spearheading a pilot project "in co-ordination with the food industry to explore and evaluate methods to rapidly and effectively identify recipients of food to prevent or mitigate a food-borne illness outbreak and to address credible threats of serious adverse health consequences or death to humans or animals as a result of such food being adulterated or misbranded," with studies on the processed food sector, raw fruits and vegetables, foods that have been the subject of significant outbreaks in the past 5 years and a diversity of other foods, including imports. The objectives are to build a tracking system that the agency can effectively implement and to develop and demonstrate methods and appropriate technologies for rapid and effective tracking of foods that is practicable for facilities of varying sizes, including small businesses.

SEC. 206. Mandatory recall authority. The FDA has mandatory recall authority, although many question whether this provision is necessary for domestic food producers, and consider the checks upon agency action through involvement of the individual states or the Justice Department to have been perfectly adequate to ensure that the US food supply remains safe. It is rare for a company not to initiate a food recall voluntarily once it is put on notice that its food may pose a risk to consumers.

A company may be notified of a "reasonable probability" that its food is adulterated or misbranded from an

---

[11] (A) by or on behalf of an owner or consignee: (i) in response to a specific testing requirement under this Act or implementing regulations, when applied to address an identified or suspected food safety problem; and (ii) as required by the Secretary, as the Secretary deems appropriate, to address an identified or suspected food safety problem; or (B) on behalf of an owner or consignee: (i) in support of admission of an article of food under section 801(a); and (ii) under an Import Alert that requires successful consecutive tests. (2) Results of testing. The results of any such testing shall be sent directly to the Food and Drug Administration, except the Secretary may by regulation exempt test results from such submission requirement if the Secretary determines that such results do not contribute to the protection of public health. Test results required to be submitted may be submitted to the Food and Drug Administration through electronic means.

internal audit or testing, information received from or by its customers, or from the Reportable Food Registry. The registry is described in a later section. A recall is mandated when "the use of or exposure to such article will cause serious adverse health consequences or death to humans or animals." A market withdrawal under the FDCA is conducted when food is mislabeled, poses a minor food safety risk or has a quality problem so that the company decides to remove the food from commerce. Under a voluntary recall, the FDA would determine how much product must be accounted for, monitor the progress of the recall, and make a determination when the recall can be terminated. Public notifications are made, including notices of recalls of various products, on the agency website, which forms a historical record.

Under the FSMA mandatory recall provision, the FDA can require a company to immediately cease distribution of the food and immediately notify others in the supply chain to also cease distribution of the food. The agency can amend the order at any time. The FDA would determine the scope and timetable, geographic area of the recall, and when the recall would terminate. Public notification occurs by press releases, alerts, and public notices at the retail and consumer levels describing the product, possible risks, and "information for consumers about similar articles of food that are not affected by the recall."

SEC. 207. Administrative detention of food. The FDA may order the detention of any article of food that is found during an inspection, examination, or investigation in a secure facility if it has reason to believe that the article of food is adulterated or misbranded. This was written into the Act to ensure that rejected product from other countries did not enter into the US. Food that is regulated exclusively by the USDA under the Federal Meat Inspection Act, the Poultry Products Inspection Act, or the Egg Products Inspection Act is not subject to administrative detention. All other food is subject to this regulation, when it enters interstate commerce. Import detention applies to food offered for import into the US and that may be subject to refusal of admission. Perishable food can be detained for 7 days with recommendations to the Department of Justice for seizure within 4 days and other food, 20 days. Detention can be appealed. Administrative detention is likely to be less costly to the agency than other enforcement actions. The cost to the private sector is another matter, and the fact that 65% of seizure actions are challenged and half of these reversed indicates that the process is far from foolproof.

SEC. 208. Decontamination and disposal standards and plans. The EPA, in co-ordination with Health and Human Services, Homeland Security, and the Department of Agriculture, shall provide support and technical assistance to state, local, and tribal governments to prepare for, assess, decontaminate, and recover from an agriculture or food emergency. This will involve conducting annual exercises and the development of specific standards and protocols and model plans to clean up, clear, and conduct recovery activities following decontamination of people, equipment and facilities, and disposal of specific threat agents and foreign animal diseases, and disposal of contaminated animals and crops.

SEC 211. Reportable Food Registry. Retail food stores are required to provide a means for notifying customers at the register and at other prominent locations in the store providing targeted recall information for consumers.

SEC. 301. Foreign supplier verification program. Importers are now required to perform risk-based foreign supplier verification activities to show that their foods are neither adulterated nor misbranded. This requires verification that the processors they represent have appropriate risk-based preventive controls in place and that imported foods are as safe as those produced in the US. Verification activities may include monitoring records for shipments, lot-by-lot certification of compliance, annual on-site inspections, checking the hazard analysis and risk-based preventive control plan of the foreign supplier, and periodically testing and sampling shipments. Importer records must be held for 2 years.

SEC. 302. Voluntary qualified importer program. The Voluntary Qualified Importer Program provides importers with expedited entry of food into the US if certain provisions are met. An importer must request participation and may be accepted to participate for a 3-year period, subject to renewal. Various factors are considered.

1. The known safety risks of the food to be imported.
2. The compliance history of foreign suppliers used by the importer, as appropriate.
3. The capability of the regulatory system of the country of export to ensure compliance with United States food safety standards for a designated food.
4. The compliance of the importer with the other requirements (section 805).
5. The record keeping, testing, inspections and audits of facilities, traceability of articles of food, temperature controls, and sourcing practices of the importer.
6. The potential risk for intentional adulteration of the food.

7. Any other factor that the Secretary determines appropriate.

SEC. 303. Authority to require import certifications for food. Imported food must be accompanied by a certificate or other assurance that the food meets applicable requirements of the FSMA in the form of shipment-specific certificates, a listing of certified facilities that manufacture, process, pack, or hold such food, or in some other manner. Certification requirements track the criteria listed above in SEC. 302. An agency or a representative of the government of the country from which the article of food at issue originated could issue a certificate as could other entities recognized by the FDA as reliable certifiers. If the FDA determines that the food safety programs, systems, and standards in a foreign region, country, or territory are inadequate to ensure that an article of food is as safe as a similar article of food that is manufactured, processed, packed, or held in the US, it shall identify such inadequacies and establish a process by which the foreign region, country, or territory may inform the Secretary of improvements made to the food safety program, system, or standard and demonstrate that those controls are adequate to ensure that an article of food is safe.

SEC. 304. Prior notice of imported food shipments. Prior notice provides the FDA with the opportunity to schedule inspections of food before it arrives in port. These must be submitted electronically. Much detailed information is required on these notices; for example, there is a requirement to identify each article of food in a shipment, not just to provide a general description of the cargo. Failure to comply with prior notice provisions, including reporting the name of any country to which food or animal feed within the shipment has been denied entry, could lead to bars to further shipments.

SEC. 306. Inspection of foreign food facilities. The FDA intends to enter into agreements with foreign governments to facilitate their inspection of registered foreign facilities, directing resources to high-risk foods. Food will be "refused admission into the United States if it is from a foreign factory, warehouse, or other establishment of which the owner, operator, or agent in charge, or the government of the foreign country, refuses to permit entry of United States inspectors or other individuals duly designated by the Secretary, upon request, to inspect such factory, warehouse, or other establishment."

SEC. 307. Accreditation of third-party auditors. The FDA is developing a process for accreditation of third-party food safety auditors (audit agents) because they lack the resources to conduct the mandated inspections. An "audit agent" is an individual who is an employee or agent of an accredited third-party auditor and, although not individually accredited, qualified to conduct food safety audits on behalf of an accredited third-party auditor. This auditor is required to be recognized by an "accreditation body" which is an authority that performs accreditation of third-party auditors and will most likely be an agency of a foreign government. A third-party auditor may be a single individual or an employer of others who would conduct audits to determine if a firm is in compliance with the FSMA and applicable industry standards (a consultative audit) or if a food can be certified for export (regulatory audit) if the firm is in compliance with the provisions of this Act, and the results of which determine whether food manufactured, processed, packed, or held the entity is eligible to receive a food certification.

SEC. 308. Foreign offices of the Food and Drug Administration. The FDA is intent upon establishing foreign offices to provide assistance to the appropriate governmental entities to improve the safety of exported food, including directly conducting risk-based inspections of foreign food facilities and supporting foreign government inspection programs.

SEC. 402. Employee protections. Companies that produce, distribute, transport or sell food may not discharge an employee or otherwise discriminate against an employee with respect to compensation, terms, conditions, or privileges of employment because the employee, whether at the employee's initiative or in the ordinary course of employment:

1. provided, caused to be provided, or is about to provide or cause to be provided to the employer, the Federal Government, or the attorney general of a State information relating to any violation of, or any act or omission the employee reasonably believes to be a violation of any provision of this Act;

2. testified or is about to testify in a proceeding concerning such violation;

3. assisted or participated or is about to assist or participate in such a proceeding; or

4. objected to, or refused to participate in, any activity, policy, practice, or assigned task that the employee (or other such person) reasonably believed to be in violation of any provision of this Act. A person who believes that he or she has been discharged or otherwise discriminated against may file a complaint with the Secretary of Labor alleging unfair discharge or discrimination and identifying the person responsible and an investigation will ensue with a finding in 60 days and a final order in 120 days.

An employer must show by clear and convincing evidence that it would have otherwise taken the same unfavorable personnel action against the employee. If the finding is in favor of the employee, relief necessary to make the employee whole can be granted, including injunctive relief and compensatory damages, including: reinstatement with the same seniority status that the employee would have had, but for the discharge or discrimination; the amount of back pay, with interest; and compensation for any special damages sustained as a result of the discharge or discrimination, including litigation costs, expert witness fees, and reasonable attorney's fees. In contrast, sanctions for a bad faith claim are reasonable attorneys' fee, not exceeding $1000, to be paid by the complainant. The final order is not subject to judicial review.

12.8 Summary

The food regulatory environment, with its safety, record-keeping and labeling requirements, is becoming increasingly complicated and increasingly global. The international market impact of potential new products must be considered now more than ever, before the development cycle begins. Managing a global supply chain and the associated regulatory requirements are issues that had limited impact on food technologists a few years ago but which now play a major role in their professional lives. Having an understanding of the Food Safety Modernization Act and examples of some of the common labeling requirements and agricultural marketing programs as outlined here should provide a basis upon which to build a stronger understanding of the basic legal requirements surrounding the manufacture and sale of foods. Most nations are moving to science-based risk assessment, coupled with HACCP-based preventive controls for food safety, with varying degrees of rigidity regarding monitoring, records, and testing.

References

Citations are included in the body of the text. The other documents listed here are general references besides statutes and regulations.

General Accounting Office (GAO) (2011) *Food and Drug Administration Better Coordination Could Enhance Efforts to Address Economic Adulteration and Protect the Public Health.* Report to Congressional Requesters October 2011. GAO-12-46. Washington, DC: United States Government Accountability Office.

Kellam J, Guarni ET (2000) *International Food Law.* Washington, DC: Food and Drug Law Institute.

Rasco BA, Bledsoe GE (2005) *Bioterrorism and Food Safety.* New York: CRC Press.

US Department of Agriculture (2012) *Federal Grant of Inspection Guide.* www.fsis.usda.gov/wps/portal/fsis/topics/inspection/apply-for-a-federal-grant-of-inspection, accessed 15 November 2013.

13 Crops – Cereals

Kent D. Rausch and Vijay Singh

Department of Agricultural and Biological Engineering, University of Illinois, Urbana, Illinois, USA

13.1 Introduction

Cereal grains, such as corn, wheat, rice, and sorghum, are grown and processed into a wide variety of ingredients for animal and human food uses. The focus of this chapter will be on corn and wheat, but other cereals have significance in global processing and human nutrition. For example, rice is grown throughout Asia and the Pacific Rim and is primarily consumed with the endosperm intact. Rice production requires large amounts of water during a relatively long growing season and therefore is grown in warm, moist climates. In contrast, grain sorghum (or milo) is processed to recover its endosperm with reduced particle size to be used in many foods throughout Africa and arid regions. Relative to other cereal crops, sorghum species are more frugal with water utilization at higher temperatures (above approximately 30 °C), making them better suited to hot and dry growing conditions such as the western plains of the US and sub-Saharan Africa. Many processing techniques used with grain sorghum have strong similarities to those used for processing corn. In this chapter, however, we chose to focus on corn and wheat processing technologies, especially wet processing of these two cereals for food, feed, and industrial purposes.

13.2 Industrial corn processing for food uses

Corn, known as maize outside the US, is a high-yielding and genetically diverse crop grown for grain in many regions of the world. In the US, corn was grown as an ingredient in animal diets, but use for direct human consumption has increased during the past several decades. The US produced 32.1% of the world's corn production for the period 2012–2013 and was the largest global producer (Capehart et al., 2013). Corn has been used for centuries as a food product. Archeological evidence suggests corn was cultivated 7000 years ago in Mexico before being cultivated elsewhere in North and South America (Farnham et al., 2003).

Industrial processing of corn started in the 1800s to optimize production of starch from corn kernels to be used in human food products as well as for industrial starches (e.g. adhesives, paper, and cardboard). Later, other corn components (oil, fiber, and protein) were also recovered as co-products. In 2010, US corn producers produced 12.45 billion bushels of corn, averaging 153 bushels per acre (NCGA, 2011). US corn usage totaled 11.3 billion bushels in 2012, consisting of feed and residual (4.45 billion bushels), exports (0.95 billion bushels) and food, seed, and industrial uses (3.47 billion bushels). Corn used for fuel ethanol production was 3.5 billion bushels (30.7% of corn use); corn used for syrups, sweeteners, and starch products was 1.0 billion bushels (8.9% of use in 2012); corn used for cereals, snacks, and beverage alcohol totaled 0.337 billion bushels, while seed use was 0.025 billion bushels (0.2%) (Capehart et al., 2013). It should be noted that these figures are for overall corn use, and do not reflect that industrial processes co-produce animal food ingredients (distillers dried grains with solubles, corn germ meal, corn gluten feed, and corn gluten meal) as 30–40% of their output. Sweetcorn grown for freezing and canning purposes is a very small fraction of the overall corn crop.

There are four major segments of the corn processing industry: corn wet milling, corn dry milling, dry grind ethanol, and masa. Wet milling, dry milling, and masa are used for producing food products from corn for human consumption and will be discussed in this chapter. Dry grind is the fastest growing segment of the corn processing industry and is primarily used for fuel ethanol production.

Food Processing: Principles and Applications, Second Edition. Edited by Stephanie Clark, Stephanie Jung, and Buddhi Lamsal.
© 2014 John Wiley & Sons, Ltd. Published 2014 by John Wiley & Sons, Ltd.

13.2.1 Corn kernel composition

Structurally, the corn kernel has four main parts: tip cap, pericarp, germ, and endosperm. Watson (2003) gave the percentage component parts and the composition of these parts of dent corn kernels. The tip cap is the location where the kernel is attached to the corn cob and where most of the water enters the kernel during process steps such as tempering and steeping. The pericarp is formed of plant tissues that surround the kernel and protect it during growth and maturation in the field as well as during harvest, storage, and handling. The germ comprises 83% of the fat and 26% of the protein in the corn kernel. The solubles in corn kernels are mostly proteins in the corn germ and monomeric sugars that leach out from the corn kernel when soaked in water. Most of the phytic acid in the corn kernel is in the germ fraction. The endosperm consists of starch, protein, and endosperm fiber and forms about 82% of the kernel weight. Two kinds of fiber are derived from two regions of the kernel: pericarp and endosperm. Pericarp fiber constitutes 50% of the fiber in the corn kernel. Endosperm fiber is the fiber fraction composed of cellular material from the endosperm interior. Depending on processing technique, different corn structural components will end up in various final products (Figure 13.1).

Because of its genetic diversity and relative ease of creating new hybrids, corn can have various colors, endosperm hardness, germ sizes, and overall kernel composition. Major classifications of commercially grown corn include regular dent, waxy, high amylose, flint, high oil, popcorn, and sweetcorn. In the US, regular dent is the primary crop. The endosperm of dent corn contains soft (characterized by spherical starch granules) and hard (characterized by polygonal granules) regions. As the corn kernel matures and begins to dry in the field, the softer regions of the endosperm lose moisture and shrink, causing the top end of the kernel (the portion on the outer part of the ear) to form a dimple or dent. Regular dent corn is a relatively broad classification; some dent varieties are well suited for wet milling but not for dry milling, and vice versa. Waxy (nearly 100% amylopectin starch in the kernel) and high-amylose (30–70% amylose starch) types are typically grown under contract and are channeled to wet milling facilities that produce specialty starch products. Flint corn is more common in South America and has primarily hard or vitreous endosperm.

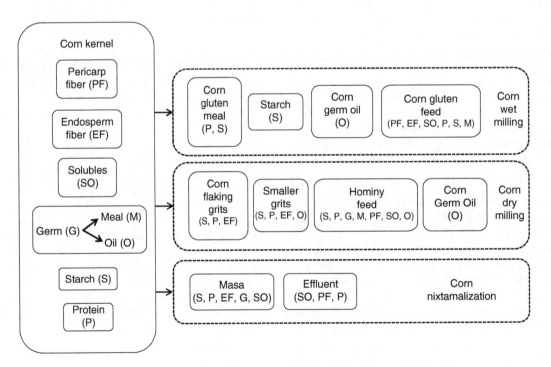

Figure 13.1 Fractionation of corn components into different co-products based on processing technologies.

Due to the hard endosperm, this type of corn does not form a dent at the top of the kernel as with other softer varieties. Flint corn is difficult to steep if wet milled and can be too hard to be desirable for dry milling, but is less susceptible to breakage during storage, handling, and transportation. High-oil corn has oil contents from 6% to as high as 12% by increasing the size of the germ or concentration of the oil in the germ. This crop is typically grown under contract for animal diets near large animal feeding operations. Popcorn is a niche snack product, which has small, hard kernels that will pop when heated properly. Sweetcorn hybrids have high sugar contents and are harvested prior to kernel maturity and sold with kernels still attached to the ear, or processed into canned or frozen products. Even though highly visible to the average consumer, popcorn and sweetcorn consumption forms less than 1% of the total US corn crop. In the US and in most of the rest of the world, regular dent corn is used for industrial processing due to its availability as a commodity crop and its high starch content.

13.2.2 The corn wet milling process

In the wet milling process, corn is fractionated into individual components of starch, protein, fiber, germ, and solubles in an aqueous medium (Blanchard, 1992; Johnson & May, 2003). The industrial corn wet milling industry started when the Thomas Kingsford starch factory was established in 1848 in Oswego, New York. By 1879, about 140 corn wet milling companies had been established to produce starch as food ingredients and as a laundry aid (Peckham, 2000). In the 19th century, corn syrups were beginning to be used in the candy, baking, brewing, and vinegar industries (Peckham, 2000). Purified starch, the primary co-product of wet milling, can be modified for food and industrial uses, hydrolyzed to produce syrups and sweeteners, or converted to produce ethanol, lysine, polylactic acid, and other biochemicals. Present-day corn wet milling fractions are used in more than 1000 commercial products. Currently there are eight wet milling companies with 27 plants operating in the US; the major US wet milling companies include Archer Daniels Midland, Cargill, Corn Products International, Grain Processing Corporation, Roquette America, and Tate and Lyle Ingredients America. Most of the corn wet milling plants are located in the US Midwest region, which is also the highest corn-producing region in the country.

The first step in the corn wet milling process involves hydration of corn kernels in 0.1–0.2% SO_2 at 48–52 °C.

This process of hydration is called steeping and is conducted for 24–36 h (Figure 13.2). During steeping, the SO_2 reacts with the disulfide bonds in the endosperm, allowing subsequent separation of the starch granules from the other corn components. After steeping, corn is ground using attrition mills (specific to corn), which open the kernels and release germ. Germ is recovered by density separation using hydrocyclones. The remaining slurry is passed over a set of screens to recover fiber (pericarp and endosperm) and washed to recover residual starch and protein from fiber. After fiber recovery, starch and protein slurries are processed through a system of centrifuges to separate starch and protein (called gluten) and to recover a concentrated protein (corn gluten meal, CGM) fraction. Starch is further washed using hydrocyclones to remove residual protein and achieve 99.5% db purity. The starch is further processed to produce many different food and industrial products. Fiber, solubles (concentrated steepwater) and sometimes germ meal are mixed together to produce corn gluten feed (CGF). Wet milled germ is used for the extraction of corn oil for human food uses. CGM and CGF are used as ingredients in animal diets. The terms CGM and CGF are misnomers, even though they are used commercially as official names for these coproducts; the corn kernel does not contain gluten protein.

From the corn wet milling process, starch and corn oil are two major products produced for human consumption.

13.2.2.1 Starch from corn wet milling

The primary objective of corn wet milling is the recovery of high-purity starch to be further processed into many products. Starch accounts for 60–75% of the weight of the corn kernel, and is used in several food and industrial applications. Starch is a polymer of six-carbon sugar (D-glucopyranose) molecules linked by α1,4 and α1,6 glycosidic bonds. There are two polymers: amylose and amylopectin. Amylose is a smaller linear polymer of glucose molecules attached by α1,4 bonds. Amylopectin is a larger branched glucose polymer using α1,4 bonds and α1,6 bonds; the branching of amylopectin is due to α1,6 bonds. The molecular size of the starch polymers (degree of polymerization) is 1500 to 6000 (from 1500 to 6000 glucose molecules) for amylose and 300,000 to 3,000,000 for amylopectin. Dent corn typically has 23–25% amylose and 75–77% amylopectin. Waxy corn has less than 1% amylose and more than 99% amylopectin. High-amylose corn has 55–70% amylose and 30–45% amylopectin.

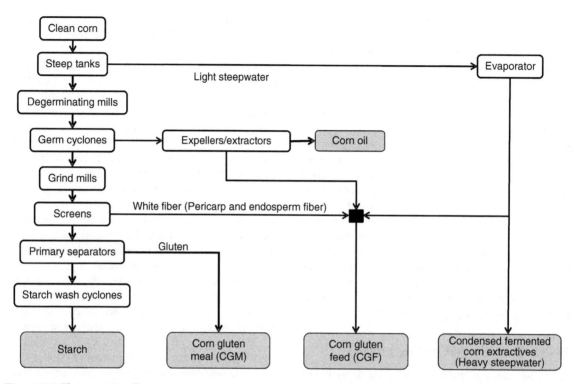

Figure 13.2 The corn wet milling process.

Depending upon the final use, starch can be used unmodified or modified (chemically or physically) to impart certain functional characteristics. Dry, unmodified corn starches are used as thickening and gelling agents, dry mixes for gravies and puddings, batters, and other food applications. Chemically modified starches (cross-linked, substituted, acid-thinned, oxidized, and dextrinized) are produced for food applications that require retaining viscosity, improving stability, gelling, adhesion or emulsification properties. Chemical modification improves the visual appearance of final food products, stabilizes texture, makes food product preparation more convenient, or improves shelf stability of the food product. Modified starches have many applications and uses in a range of food and industrial products relative to unmodified counterparts. Cross-linked starches have improved shear and temperature stability in food processing systems; substituted starches have improved stability during temperature cycling (e.g. freeze-thaw stability) and reduced gelatinization temperatures; acid-thinned starches can have reduced viscosity in high solids systems; oxidized starches have adhesive characteristics needed for batters and coatings; dextrins can be used in adhesives and coatings.

Starch can also be hydrolyzed into glucose or short polymers using acid, enzyme or a combination to produce products that are classified based on their dextrose equivalent (DE). DE is a measure of the total reducing sugars in a product. Hydrolyzed starch food products consist of maltodextrins (5–20 DE), low-DE corn syrups (22–30 DE), and high-DE corn syrups (above 60 DE). A low-DE syrup can be used in applications that require body, cohesiveness, foam stability, prevention of sucrose crystallization, and viscosity whereas a high-DE syrup is used when browning, fermentability, enhanced flavor, lowering of freezing point, increased osmotic pressure, and sweetness are required (Table 13.1) (Hull, 2010).

13.2.2.2 Oil from corn wet milling

Corn oil ranks third in the world production of edible plant oils, with an annual production of 2.4 million tons (Moreau, 2008). The corn kernel contains typically 4.0% oil in commercial corn hybrids. Most of this oil (≈85%) is located in the corn germ (corn germ oil). A small amount of oil (6–7%) is also located in the fiber fraction (corn fiber oil). On a commercial scale, corn oil is recovered

Table 13.1 Uses of corn starch hydrolyzates

Hydrolyzed products	DE	Uses
Glucose syrup	28–32	Coffee whitener; spray drying; meat products
	42	Caramels and toffees; chewing gums; fudge and fondants; gums and jellies; muesli bars; nougat; adhesives; bakery glaze; caramel color; glaze; ice cream; toys
	63	Baking; brewing; fermented drinks and liqueurs; jams; sauces; ketchup and mayonnaise; soft confectionery; soft drinks; Tabasco
	95	Antibiotics; baking; brewing; building industry; caramel color, carrier for soluble favor and colors; cattle lick blocks; cider, confectionery; fruit preparations; ice cream; industrial fermentation; preservatives; well drilling
Maltose/high maltose syrup		Confectionery; brewing
Dextrose monohydrate		Antibiotics; bakery; browning agent; carrier for flavors
High-fructose corn syrup	42 and 55	Bakery products; breakfast cereals; confectionery; diabetic foods and drinks; fruit preparations, ice creams and frozen desserts; jams, preserves, fruit fillings and topping syrups, sauces, ketchups, mayonnaise and pickles; soft drinks
Maltodextrins	<20	Brewing; bulking agent; fat and calorie reduction; carriers; cereal coatings, confectionery; dairy fillings, dry mixes; encapsulation; film forming; frozen foods; geriatric, invalid, baby food; ice cream; margarine; meat analogs; personal hygiene; pharmaceuticals; snacks; sport drinks; spreads

DE, dextrose equivalent.

from the germ which has been separated during the corn wet milling process. Corn wet-milled germ is composed of 40–50% oil by weight. This oil is usually removed by either hexane extraction or mechanical prepressing followed by hexane extraction (Moreau, 2011). Crude corn oil contains 96% triacylglycerols, 0.3–1.7% free fatty acids, 1% phytosterolsm and 1% phospholipids (Moreau et al., 1999b; Orthoefer & Sinram, 1987). Currently all refined corn oil for human use is produced from corn germ; however, oil can also be recovered from corn fiber (corn fiber oil) or from the entire corn kernel (corn kernel oil). Processes have been developed to recover corn from the whole corn kernel (Hojilla-Evangelista et al., 1992) or from the fiber fraction (Singh et al., 1999, 2005). Modifications in conventional corn wet milling (Johnston & Singh, 2001) have been implemented that could potentially allow recovery of pericarp and endosperm fiber fractions for recovery of corn fiber oil and other components for food applications (Doner et al., 2001).

Depending upon the solvent used (i.e. ethanol, methylene chloride, chloroform/methanol), corn kernel oil was shown to contain higher levels of three phytosterol lipid classes (Moreau et al., 1996) as well as the yellow pigments lutein and xanthophylls (Moreau et al., 2007), compared to hexane-extracted corn germ oil. Corn fiber oil is rich in phytosterol esters and can be extracted from different corn fiber fractions (Moreau et al., 1996, 1998). Corn fiber oil has been found to contain the highest levels of natural phytosterols and phytostanols of any known plant extracts (Hicks & Moreau, 2001). In clinical studies, phytosterols have been shown to reduce serum cholesterol (Hendriks et al., 1999; Liu, 2007; Miettinen et al., 1995; Moreau et al., 1999a). Major food uses of corn oil are in cooking, salad oil, margarines, and spreads (Moreau, 2011).

13.2.3 The corn dry milling process

In the dry milling process, corn is dry fractionated into grits (endosperm), germ, pericarp fiber, and flour (Duensing et al., 2003). The most common commercial-scale dry milling process consists of a tempering-degerming milling process (Figure 13.3). Cleaned corn is tempered (with steam or hot water) to increase its moisture content from 15% to about 22% (weight basis, wb). Tempering facilitates the removal of germ and bran (pericarp). The tempering moisture is rapidly absorbed by the pericarp and germ, causing differential swelling between these tissues and the endosperm; tempering is terminated before moisture is evenly distributed throughout the kernel, facilitating separation of kernel components. The moistened pericarp and germ become more resilient and resist breakage during subsequent milling steps. Tempered corn is passed through a degerminator. Several commercial degerminator systems, such as Beall, Buhler, Satake, and Ocrim, use a combination of abrasion, shear and crushing forces to break the tempered kernel

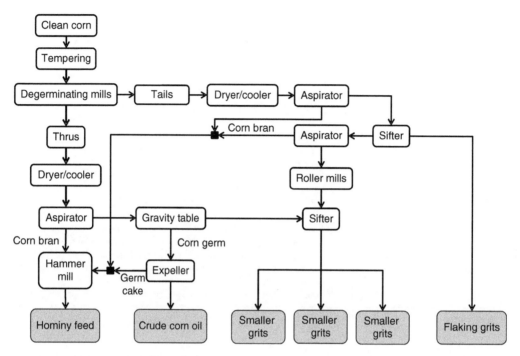

Figure 13.3 The corn dry milling process.

into desired endosperm, germ, and pericarp fractions. The degermination step produces two fractions (tails and thrus) (Duensing et al., 2003). Tails are primarily large pieces of endosperm (grits) that exit the tail end of the degerminator. Thrus primarily consist of germ, bran, and smaller endosperm pieces and exit through screens at the sides of the degerminator. The highest valued co-products from the dry milling process are the grits streams, classified by size and fat content. For highest value and longest shelf life, grits must have fat contents less than 1% crude fat. The tails and thrus fractions are subjected to drying, cooling, aspiration, density separation, and sizing to produce grits (flaking and other coarse grits), flour and other co-products (bran and hominy feed). In 2001, there were a total of 48 corn dry milling plants in the US; 12 of these plants had capacities of more than 12,000 bushels/day (Duensing et al., 2003).

13.2.3.1 Flaking and smaller grits

Flaking and smaller grits are mainly used in food applications. Flaking grits are used to make corn flakes and are valued for their large size; each corn flake is made from a single grit (Fast, 2003). Brewers grits, corn meal, and corn cones are smaller endosperm grit materials that

are used in a wide variety of snack, baking, and breakfast products. Typical composition of degermed corn products, granulation and uses are listed in Table 13.2 (Duensing et al., 2003). Dry-milled germ, bran, low-grade flour (standard meal), and broken corn are mixed together to make hominy feed, which is sold mainly as an ingredient for ruminant animal diets.

13.2.3.2 Dry milled corn oil

In dry milling, all corn components are dry fractionated; therefore, separation of germ from endosperm is not complete, resulting in small pieces of endosperm and pericarp adhering to germ. Consequently, the oil content in dry-milled germ is about 18–22% (Johnston et al., 2005). Starch content of dry-milled germ is twice that of wet-milled germ and dry-milled germ value is half that of wet-milled germ (Johnston et al., 2005). Moreau et al. (1999b) compared the composition of laboratory wet- and dry-milled germ. They found that dry-milled germ had lower diacylglycerols and γ-tocopherols but had higher phytosterol compounds compared to wet-milled germ. The composition of free fatty acids and triacylglycerols was similar (Moreau et al., 1999b). Dry millers are unable to sell their germ to corn oil

Table 13.2 Composition of corn dry milled co-products and typical uses (adapted from Duensing et al., 2003)

Component	Flaking grits	Brewers' grits	Corn meal	Corn cones	Break flour
Moisture (%)	13.8	11.7	12.0	11.5	12.0
Protein (%)	7.5	7.7	7.0	8.0	6.0
Fat (%)	0.4	0.7	0.7	0.6	2.2
Crude fiber (%)	0.3	0.4	0.5	0.4	0.6
Ash (%)	0.2	0.3	0.4	0.3	0.6
Carbohydrates (%)	77.8	79.2	79.4	79.2	78.6
Granulation (US Sieve)*	−3.5 + 6.0	−12 + 30	−30 + 40	−40 + 80	−60
Uses	RTE breakfast cereals,	Brewing adjuncts	Extruded and sheeted snacks	Extruded and sheeted snacks	Breading, batters and prepared mixes,fortified foods

* Granulation: "−3.5 + 6.0" means material passing through US Sieve 3.5 and material retained on US Sieve 6.0.
RTE, ready to eat.

extraction facilities due to low oil and high starch contents of germ. Some dry millers will mechanically expel and/or solvent extract their germ for producing corn oil. Crude corn oil that has been expelled from dry-milled germ can be sold into specialty niche food markets since expelling without solvents has supposed advantages over hexane-extracted oil.

13.2.3.3 Dry-milled corn bran

Bran can be aspirated from the thrus process stream (see Figure 13.3) based on its particle size and density differences. This bran is high in total dietary fiber (TDF) (Burge & Duensing, 1989) and can be cleaned to remove adhering endosperm pieces and processed further (ground) and used to increase the fiber content of food products. Coarse ground bran is used in ready-to-eat cereals and extruded products and finely ground bran is used in high-fiber baked products (Duensing et al., 2003).

13.2.4 The corn masa process

Corn masa processing or nixtamalization is a process of cooking corn in a 1% lime (wb) solution to form eventually a unique dough called masa (Serna-Saldivar, 2010a). Cooking is done typically at 85–110 °C for 15–60 min to loosen pericarp from the kernel. After cooking, corn is steeped for 8–16 h at temperatures above 68 °C (Figure 13.4). Cooked/steeped corn is washed with water to remove pericarp and lime to recover a product called nejayote. The washed corn (nixtamal) is stone ground to produce the masa dough. This dough is used to

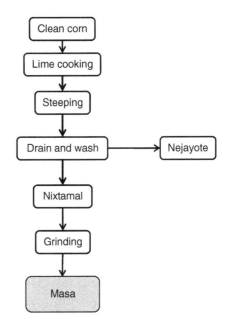

Figure 13.4 The corn masa process.

make a variety of corn-based products including tortillas, snack chips, and other food items. Masa can be pressed to form sheets, cut and baked to produce table tortillas or fried to produce tortilla chips or taco shells. Masa can also be extruded and fried to produce corn chips. Process parameters (cooking and steeping times and temperatures, lime concentrations, and agitation) and equipment used for cooking, steeping, and grinding (direct steam injected or jacketed kettles, cooker, volcanic or synthetic stones with different groove designs, cutters, mills) is varied depending upon the final product and cultural preferences. The masa dough can be used directly or dried and ground to produce shelf-stable packaged flour. Several sources in the literature (Maya-Cortes et al., 2010; Rooney & Serna-Saldivar, 2003; Serna-Saldivar, 2010b; Serna-Saldivar et al., 1990) report without providing specific detail that nixtamalization has beneficial effects on the nutritional value of nixtamal proteins by improving the essential amino acid profile and leads to assimilation of calcium ions (originating from the lime i the cooking step) into the masa material. However, the washing process that recovers the nejayote also removes protein and other valuable nutrients.

13.3 Industrial wheat processing for food uses

For thousands of years, wheat has been ground into flour to make bread of various types all over the world. During the 20th century, the flour production process was developed to remove wheat kernel pericarp (bran) and germ, resulting in a lighter colored flour that was more shelf stable and gave bread a whiter appearance. As protein research progressed, scientists identified the protein that gave quality wheat bread a desirable volume and texture during mass production of bread: wheat gluten. It was discovered that quality gluten, or vital gluten, would improve the bread loaf characteristics of many bread formulations, since gluten was elastic and would allow the bread dough to rise uniformly during the proofing stage and remain stable during baking. Vital gluten became a commodity in demand for commercial bread making. Wheat starch was co-produced with vital wheat gluten and markets for this co-product were developed.

13.3.1 Wheat kernel composition

For commercial processing, wheat is classified by its hardness (soft, hard, and durum), its protein content and its color (white and red). Soft, hard, and durum varieties have protein content ranges of 8.5–9.5%, 11.5–15%, and 14–15%, respectively (Delcour & Hoseney, 2010b). Soft varieties are used in baking of cakes, biscuits, cookies, and pastries; hard wheat is used for bread making; durum wheat is used to produce semolina and pasta. The wheat kernel (35 mg) is approximately one-tenth the weight of a corn kernel. The major components of the kernel are the endosperm, germ, and pericarp, comprising 90%, 2–3.5%

and 5% of the kernel, respectively. Other components of lesser importance are the brush hairs at the kernel tip and the crease, which runs along the longitudinal axis of the kernel. The starch granules present in the endosperm have two size groupings: large granules (about 40 micron diameter) called A starch and small granules (2–8 microns) called B starch.

13.3.2 The wheat flour process

Wheat flour production has some similarities to corn dry milling, but due to the differences in process objectives and in wheat and corn kernel morphologies, there are distinct differences in process strategies and technologies.

The flour production process begins with whole, cleaned kernels of wheat being subjected to a tempering process (Figure 13.5). Unlike corn tempering, the objective of wheat tempering is to increase the kernel moisture uniformly so the bran layer does not shatter during further processing and to prepare the endosperm for milling. In contrast to corn dry milling, moisture is allowed sufficient time (5–24 h) to fully penetrate the wheat kernel. The amount of moisture and the length of tempering time are increased as kernel hardness increases. Wheat bran layers do not provide a barrier to moisture transfer as in the corn kernel. Tempered kernels are sent to break rolls, which crush and shear the kernels to separate bran and germ components from the endosperm. Break rolls, arranged in pairs, are corrugated and rotate at differential speeds to generate the shear needed. The shearing action avoids breaking the bran component into smaller fragments and making subsequent separation more difficult (Delcour & Hoseney, 2010a).

Following the break rolls, the fractured kernel components are sifted through a series of screening cloths to classify the particles by size. The smallest fragments, those passing through a 10XX cloth (132 micron openings), are considered to be flour. The largest particles, classified as bran, are abraded in the scalper reels in an attempt to remove endosperm particles from pericarp. The small particles are sent along with other small particles from the classifier to the reduction rolls. These rolls do not have corrugations and are used to reduce particle size. The material leaving the reduction rolls is classified again, separating material into reduction flour, shorts (small endosperm particles with bran attached), bran and a low-grade flour called red dog. Germ is typically recovered with the shorts fraction. Shorts, bran, red dog and sometimes germ are blended to form the millfeed

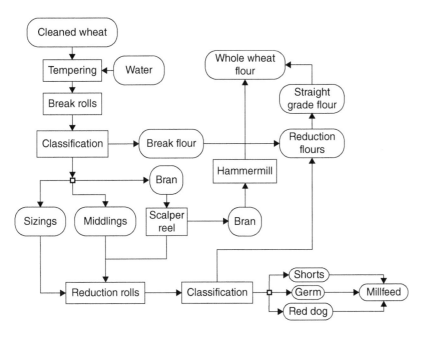

Figure 13.5 The wheat flour milling process.

co-product. Germ may be recovered separately, dried or toasted to deactivate enzymes that oxidize the fat and sold separately for specialty food ingredients or oil extraction. Germ yields are much lower than those for corn, typically 0.5–1.0% of total mill production.

Several grades of flour are produced. For whole wheat flour, all kernel components are present in the flour (bran, germ, and endosperm), but this type of flour will have shortened shelf life due to the higher fat content unless preservatives are added. Most flours have had the bran and germ removed to improve shelf life and lighten the appearance of bakery products. Flours are blended based on their ash content since this is an indication of baking characteristics (Delcour & Hoseney, 2010a). A flour miller will adjust flour composition based on customer requirements and market prices. Straight grade flour includes all flour-sized particles from the endosperm. Long patent flour originates from the break rolls, is one of the highest quality flours produced and is typically 65% of total products made by the mill. Short patent flour has the lowest ash content of all the flours and is about 45% of total mill production. Low-grade or red dog flour is about 7% of production and has a high ash content and dark color; usually, red dog is not used for bread making but is blended with the millfeed co-product, which consists of endosperm particles that have bran attached along with wheat bran and germ.

13.3.3 The wet wheat milling process

Wheat or wheat flour can be wet milled to produce vital wheat gluten and wheat starch. Wet wheat milling is practiced in many countries primarily for the vital gluten to be used in baking and meat products, but also to recover starch in regions and conditions where corn starch cannot be economically produced.

Many plants begin with wheat flour, rather than producing flour within the facility or processing whole kernels. There are many versions of the wet milling process for wheat, but the most common is the Martin process (Serna-Saldivar, 2010c), shown in Figure 13.6. Unlike corn endosperm, which requires a lengthy acidic steeping process for starch-protein separation, in the wheat endosperm, starch and gluten particles are separated by addition of water and agitation. Water is mixed with flour to form a dough or batter and allowed to rest. The addition of water, mixing and rest initiates agglomeration of the gluten protein particles, which makes separation possible. The mixture is sent to an extractor, which is a rotating screen that allows starch granules to pass through and gluten to remain on top. The extractor sends the wet gluten to dewatering rolls and a ring dryer to produce dry, vital wheat gluten. The gluten protein must be dewatered and dried in such a manner as to preserve its functionality in baking and other products. If the functionality is lost,

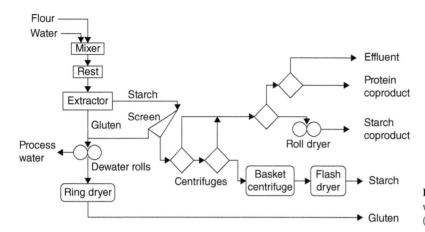

Figure 13.6 The Martin process for wet milling of wheat flour (Rausch, 2002).

the non-vital gluten is sold as a lower valued co-product for use in animal diets, aquaculture, and pet food.

The starch stream is further refined to recover gluten particles, centrifuged and dried to make a high-value starch product as well as other co-products (see Figure 13.6). The A starch fraction can be recovered using hydrocyclone technology and can be used, modified or converted in much the same way as corn starch can be utilized. The challenges in wet milling of wheat flour have been to recover small granule starch (B starch) in an economical fashion and reduce the amount of water and water-soluble components of wheat being discharged into the waste treatment stream. Due to their small diameter, B starch granules are more difficult to recover and yet have significant implications if discharged as a waste stream. Many developments have occurred to recover this fraction as a separate co-product that can be used in production of glucose as well as beverage and fuel ethanol.

13.4 Sustainability of corn and wheat processing

Processing of corn and wheat in industrialized countries and regions requires significant inputs of resources. These resources are transferred to the processing facility using a well-developed infrastructure to provide water, electricity, natural gas, coal, and petroleum. Processes using large amounts of water, specifically the corn wet milling and wheat wet milling processes, must properly treat waste water to comply with regulations and reduce the environmental footprint of the discharge from the facility. Because processing of cereal grains such as corn and

wheat is carried out at a large scale (commonly 1000–10,000 tonnes per day), these processes must be operated in a sustainable manner.

Corn wet milling has improved its use of water so that the amount of water per bushel has decreased significantly, cutting costs for the processor but also reducing environmental implications. The process is still water intensive, using 10–12 gal per bushel (Johnson & May, 2003), with water use strongly correlated to energy use at the plant since much of the water must be evaporated, resulting in intensive energy use as well. However, challenges in water use remain because water usage is largely controlled by the amount needed to properly steep the corn. Johnson and May (2003) reported that wet milling operations require 1.48 million kilocalories per metric tonne corn, with about 20% of the total for electricity and the remainder for steam generation. Galitsky et al. (2003) reported that corn wet milling was the largest consumer of energy in the US food industry sector.

Wet processes used for the production of wheat gluten and starch have improved in their consumption of water and energy, but there are few publications quantifying specific water and energy use other than reports from European facilities. The Martin process, which is the oldest commercial process used to produce wheat starch and gluten, uses more than 10–15 parts fresh water per part wheat flour, resulting in large volumes of process water with low solids contents. This process water contains low levels of starch, protein and water-soluble components such as arabinoxylans and β-glucans. While these components could be of commercial value, typically the process water is considered effluent and treated as waste water. Other processes, such as the batter, Fesca and Raiso processes, were developed in part because they had potential to use less water during gluten and starch production. Improvements in the Martin and batter processes reduced water consumption to 5–7 parts fresh water per part flour (Maningat et al., 2009; Zwitserloot, 1989). With the introduction of modern centrifuges into wet wheat processing, ratios were reduced to 4–5:1 (Meyer et al., 1999).

In the context of a growing global population, the sustainability of these processes should also be concerned with process design innovations and the co-products made at a processing facility. Processes historically designed to produce pure starch have implicitly placed a lower value on cereal protein; conversely, processes designed to recover protein have typically recovered starch and other carbohydrates as an afterthought. Process improvements tend to focus on commodities from a business standpoint, or on regulatory compliance. As the global population grows, all cereal processes must continue to increase in their respective sophistication. Not only should water and energy resources be used sustainably, but the components of cereal grains should also.

References

Blanchard P (1992) Wet milling. In: *Technology of Corn Wet Milling and Associated Processes*. Amsterdam: Elsevier.

Burge RM, Duensing WJ (1989) Processing and dietary fiber ingredient applications of corn bran. *Cereal Foods World* **34**: 535–538.

Capehart T, Allen E, Bond J (2013) *Feed Outlook*. Report No. FDS-13a. Washington, DC: Economic Research Service, United States Department of Agriculture.

Delcour JA, Hoseney RC (2010a) Dry milling. In: *Principles of Cereal Science and Technology*, 3rd edn. St Paul, MN: AACC International, pp. 121–137.

Delcour JA, Hoseney RC (2010b) Structure of cereals. In: *Principles of Cereal Science and Technology*, 3rd edn. St Paul, MN: AACC International, pp. 1–22.

Doner LW, Johnston DB, Singh V (2001) Analysis and properties of arabinoxylans from discrete corn wet-milling fiber fractions. *Journal of Agricultural and Food Chemistry* **49**: 1266–1269.

Duensing WJ, Roskens AB, Alexander RJ (2003) Corn dry milling: processes, products and applications. In: White PJ, Johnson LA (eds) *Corn Chemistry and Technology*. St Paul, MN: American Association of Cereal Chemists, pp. 407–447.

Farnham DE, Benson GO, Pearce RB (2003) Corn perspective and culture. In: White PJ, Johnson LA (eds) *Corn Chemistry and Technology*. St Paul, MN: AACC International, pp. 1–33.

Fast RB (2003) Manufacturing technology of ready-to-eat cereals. In: Fast RB, Caldwell EF (eds) *Breakfast Cereals and How They are Made*. St Paul, MN: AACC International.

Galitsky C, Worrell E, Ruth M (2003) *Energy Efficiency Improvement and Cost Saving Opportunities for the Corn Wet Milling Industry: An Energy Star Guide for Energy and Plant Managers*. Report No. LBNL-52307: Berkeley, CA: Lawrence Berkeley National Laboratory.

Hendriks HFJ, Weststrate JA, van Vliet T, Meijer GW (1999) Spreads enriched with three different levels of vegetable oil sterols and the degree of cholesterol lowering in normocholesterolaemic and mildly hypercholesterolaemic subjects. *European Journal of Clinical Nutrition* **53**: 319–327.

Hicks KB, Moreau RA (2001) Phytosterols and phytostanols: functional food cholesterol busters. *Food Technology* **55**: 63–67.

Hojilla-Evangelista MP, Johnson LA, Myers DJ (1992) Sequential extraction processing of flaked whole corn: alternative corn fractionation technology for ethanol production. *Cereal Chemistry* **69**: 643–647.

Hull P (2010) Syrup applications: an overview. In: *Glucose Syrups: Technology and Applications*. Chichester: Wiley-Blackwell.

Johnson LA, May JB (2003) Wet milling: the basis for corn biorefineries. In: White PJ, Johnson LA (eds) *Corn Chemistry and Technology*. St Paul, MN: AACC International, pp. 449–495.

Johnston DB, McAloon AJ, Moreau RA, Hicks KB, Singh V (2005) Composition and economic comparison of germ fractions from modified corn processing technologies. *Journal of the American Oil Chemists' Society* **82**: 603–608.

Johnston DB, Singh V (2001) Use of proteases to reduce steep time and so2 requirements in a corn wet-milling process. *Cereal Chemistry* **78**: 405–411.

Liu RH (2007) Whole grain phytochemicals and health. *Journal of Cereal Science* **46**: 207–219.

Maningat CC, Seib PA, Bassi SD, Woo KS (2009) Wheat starch: production, properties, modification and uses. In: BeMiller JN, Whistler RL (eds) *Starch: Chemistry and Technology*, 3rd edn. Amsterdam: Elsevier, pp. 442–510.

Maya-Cortes DC, Cardenas JDF, Garnica-Romo MG et al. (2010) Whole-grain corn tortilla prepared using an ecological nixtamalisation process and its impact on the nutritional value. *International Journal of Food Science and Technology* **45**: 23–28.

Meyer D, Hanneforth U, Bergthaller W (1999) Quality of wheat for wafer making and starch production. In: Mugnozza GT, Porceddu E, Pagnotta MA (eds) *Genetics and Breeding for Crop Quality and Resistance*. Amsterdam: Kluwer Academic Publishers, pp. 347–355.

Miettinen TA, Puska P, Gylling H, Vanhanen H, Vartiainen E (1995) Reduction of serum-cholesterol with sitostanol-ester margarine in a mildly hypercholesterolemic population. *New England Journal of Medicine* **333**: 1308–1312.

Moreau RA (2008) Corn kernel oil and corn fiber oil. In: Moreau RA, Kamal-Eldin A (eds) *Gourmet and Health-Promoting Speciality Oils*. Urbana, IL: American Oil Chemists' Society, pp. 409–431.

Moreau RA (2011) Corn oil. In: Gunstone FD (ed) *Vegetable Oils in Food Technology: Composition, Properties and Uses*, 2nd edn. Chichester: Wiley-Blackwell, pp. 273–289.

Moreau RA, Powell MJ, Hicks KB (1996) Extraction and quantitative analysis of oil from commercial corn fiber. *Journal of Agricultural and Food Chemistry* **44**: 2149–2154.

Moreau RA, Hicks KB, Nicolosi RJ, Norton RA (1998) Corn fiber oil – its preparation, composition, and use. US Patent No. 5,843,499.

Moreau RA, Norton RA, Hicks KB (1999a) Phytosterols and phytostanols lower cholesterol. *INFORM – International News on Fats, Oils and Related Materials* **10**: 572–577.

Moreau RA, Singh V, Eckhoff SR, Powell MJ, Hicks KB, Norton RA (1999b) Comparison of yield and composition of oil extracted from corn fiber and corn bran. *Cereal Chemistry* **76**: 449–451.

Moreau RA, Johnston DB, Hicks KB (2007) A comparison of the levels of lutein and zeaxanthin in corn germ oil, corn fiber oil and corn kernel oil. *Journal of the American Oil Chemists' Society* **84**: 1039–1044.

National Corn Growers Association (NCGA) (2011) Corn production. In: *World of Corn*. Washington, DC: National Corn Growers Association, p. 13.

Orthoefer FT, Sinram RD (1987) Corn oil: composition, processing, and utilization. In: Watson SA, Ramstad PE (eds) *Corn: Chemistry and Technology*. St Paul, MN: American Association of Cereal Chemists, pp. 535–552.

Peckham BW (2000) The first hundred years of corn refining in the United States. In: *Corn Annual*. Washington, DC: Corn Refiners Association, pp. 18–21.

Rausch KD (2002) Front end to backpipe: membrane technology in the starch processing industry. *Starch/Starke* **54**: 273–284.

Rooney L, Serna-Saldivar SO (2003) Food use of whole corn and dry milled fractions. In: White PJ, Johnson LA (eds) *Corn Chemistry and Technology*. St Paul, MN: AACC International, pp. 495–535.

Serna-Saldivar SO (2010a) *Cereal Grains: Properties, Processing, and Nutritional Attributes*. New York: CRC Press.

Serna-Saldivar SO (2010b) Milling of maize into lime-cooked products. In: *Cereal Grains: Properties, Processing, and Nutritional Attributes*. New York: CRC Press, pp. 239–257.

Serna-Saldivar SO (2010c) Wet milling operations. In: *Cereal Grains: Properties, Processing, and Nutritional Attributes*. New York: CRC Press, pp. 225–238.

Serna-Saldivar SO, Gomez MH, Rooney LW (1990) Technology, chemistry and nutritive value of alkaline cooked corn products. In: Pomeranz Y (ed) *Advances in Cereal Science and Technology*. St Paul, MN: American Association of Cereal Chemists, pp. 243–295.

Singh V, Moreau RA, Doner LW, Eckhoff SR, Hicks KB (1999) Recovery of fiber in the corn dry-grind ethanol process: a feedstock for valuable coproducts. *Cereal Chemistry* **76**: 868–872.

Singh V, Johnston DB, Naidu K, Rausch KD, Belyea RL, Tumbleson ME (2005) Comparison of modified dry-grind corn processes for fermentation characteristics and DDGS composition. *Cereal Chemistry* **82**: 187–190.

Watson SA (2003) Description, development, structure and composition of the corn kernel. n: White PJ, Johnson LA (eds) *Corn Chemistry and Technology*. St Paul, MN: AACC International, pp. 69–106.

Zwitserloot WRM (1989) Production of wheat starch and gluten: historical review and development into a new approach. In: Pomeranz Y (ed) *Wheat is Unique*. St Paul, MN: AACC International, pp. 509–519.

14 Crops – Legumes

George Amponsah Annor,[1] Zhen Ma,[2] and Joyce Irene Boye[2]

[1]Department of Nutrition and Food Science, University of Ghana, Legon-Accra, Ghana
[2]Food Research and Development Centre, Agriculture and Agri-Food Canada, Québec, Canada

14.1 Introduction

Legumes belong to the family Leguminosae and consist of oilseeds such as soybeans, peanuts, alfafa, clover, mesquite, and pulses, including the dry grains of peas, chickpeas, lentils, peas, beans, and lupins. Production and use of legumes date back to ancient cultures in Asia, the Middle East, South America, and North Africa. They are cultivated throughout the world for their seeds, harvested and marketed as primary products. Grain legumes are grouped into pulses and oilseeds. The pulses are different from the leguminous oilseeds, which are primarily utilized for oil (Schneider, 2002). There are about 1300 species of legumes, with only about 20 commonly consumed by humans (Reyes-Moreno & Paredes-Lopez, 1993). Notable amongst legume species are chickpeas (*Cicer arietinum*), pigeon pea (*Cajanus cajan*), lentil (*Lens culinaris*), mung bean (*Vigna radiata*), soybean (*Glycine max*), winged bean (*Psophocarpus tetragonoloba*), cowpea (*Vigna unguiculata*), pea (*Pisum sativum*), groundnut (*Arachis hypogaea*), and black gram (*Vigna mungo*), to mention but a few. Some of the most important legumes in the world are peas, beans, peanuts, soybeans, and chickpeas (Reyes-Moreno et al., 2000).

Canada is the leading producer of peas in the world, with about 3,379,400 metric tonnes (MT) produced in 2009 (FAO, 2009). In 2010, the US was the leading global producer of soybeans and the second and third top producer of peas and lentils, respectively (Table 14.1). In the same year, the US was also the fourth leading producer of dry beans and peanuts. Canada again tops the production of lentils and is the seventh and ninth leading producer of soybeans and chickpeas. Similarly, production of dry beans and chickpeas in Mexico is high and the country ranks fifth and eighth in global production, respectively. Brazil was the leading producer of dry beans in 2009. The leading producer of chickpeas in 2009 was India with 7,060,000 MT. Cutting-edge research in plant breeding and agronomic practices in the last several decades has allowed suitable varieties and cultivars for the North American climate to be identified, which has resulted in marked increases in legume production in the region. Although a large percentage of the legumes produced in North America are exported, there is growing interest in expanding domestic consumption, due to increased awareness of their health benefits.

Legumes have a special place in the diet of humans, because they contain nearly 2–3 times more protein than cereals (Reyes-Moreno & Paredes-Lopez, 1993). Cowpeas, for example, contain about 25% protein (Annor et al., 2010). Legumes are also excellent sources of complex carbohydrates and have been reported as beneficial for cardiovascular diseases and diabetes by some researchers (Hu, 2003; Jacobs & Gallaher, 2004), probably due to the large amounts of water-soluble fiber and a large content of phenolics (Enujiugha, 2010). Legumes are also a good source of vitamins (thiamine, riboflavin, niacin, vitamin B_6, and folic acid) and certain minerals (Ca, Fe, Cu, Zn, P, K, and Mg), and are an excellent source of polyunsaturated fatty acids (linoleic and linolenic acids) (Augustin et al., 1989). Indeed, several studies suggest that increased consumption of legumes may provide protection against diseases such as cancer, diabetes, osteoporosis, and cardiovascular diseases, among others (Hu, 2003; Pihlanto & Korhonen, 2003; Tharanathan & Mahadevamma, 2003). Legumes further offer a practical avenue for diet diversification as consumers look for greater balance between plant and animal food sources. With growing concerns about the impact of agricultural practices on the environment,

---

Food Processing: Principles and Applications, Second Edition. Edited by Stephanie Clark, Stephanie Jung, and Buddhi Lamsal.
© 2014 John Wiley & Sons, Ltd. Published 2014 by John Wiley & Sons, Ltd.

Table 14.1 Production of legumes in North America and top 20 global production ranking

Legume	USA		Canada		Mexico	
	Rank	Production (MT)	Rank	Production (MT)	Rank	Production (MT)
Beans, dry	4	1442470	15	253700	5	1156250
Chickpea	15	87952	9	128300	8	131895
Groundnut, with shell	4	1885510				
Lentil	3	392675	1	1947100		
Pea, dry	2	645050	1	2862400		
Soybean	1	90605500	7	4345300		

MT, metric tonnes.
Source: http://faostat.fao.org/site/339/default.aspx.

addition of legumes in crop rotation cycles can have beneficial impacts as they have the capacity to fix nitrogen in soils, thereby reducing the need for chemical fertilizers.

Peanuts are the most commonly consumed (by humans) and convenient of the legumes as they form part of the mainstream diet and can be easily obtained and consumed as roasted seeds or in the form of peanut butter. In the US, for example, peanuts and peanut butter comprise over two-thirds of all nut consumption (www. peanut-institute.org/peanut-facts/history-of-peanuts.asp, accessed 18 November 2013). Soybeans and pulse legumes, on the other hand, are more alien to the North American diets. Factors that have limited their consumption in North America include the longer time required for their preparation, the possible gastrointestinal (GI) discomfort due to the presence of indigestible carbohydrates which ferment in the GI tract causing gas and bloating, and their typical beany flavor. Extensive research in breeding, food quality, and processing has helped to overcome some of these limitations, increasing the acceptability of legumes in the North American diet and facilitating their use in food formulation.

Although legumes are rich in proteins, the quality of their protein is not nutritionally adequate. This is because they lack sulfur-containing amino acids such as methionine and cysteine. These limiting amino acids are, however, complemented by the use of legume cereal blends in diets. Cereals, being rich in sulfur-containing amino acids, complement the legume proteins, hence improving the quality of the protein. Other factors such as low protein digestibility, presence of antinutritional factors such as trypsin inhibitors, lectins, phytates, polyphenols, and flatulence factors make some legume seeds underutilized (Enujiugha, 2005; Mubarak, 2005;

Ragab et al., 2010). Most of these antinutritional factors can, however, be reduced or eliminated by various processing techniques. In this chapter, we discuss different processing technologies applied to legumes, which are grouped as traditional and modern processing technologies.

14.2 Technologies involved in legume processing

The processing of legumes can be conveniently grouped into traditional and modern, depending on the complexity of the processing steps involved and the types of equipment used. Traditional methods include simple technologies and simple equipment that can be used at household level, whereas modern processing includes much more sophisticated processes and equipment at industrial level.

Legumes can be cooked and consumed as fresh beans or after drying, which is done to extend their shelf life. The term "pulse," for example, specifically refers to the dried grains of pea, chickpea, bean, lentil, and lupin, which distinguishes them from the fresh beans. Cooking of legumes inactivates antinutritional factors such as trypsin and amylase inhibitors, thus improving their nutritional quality. As is done in Asia and other parts of the world, in North America, fresh soybeans in the pod, peas, green beans, sugar snap beans, and string beans can be cooked and eaten as a side dish or with salads. Peanuts and soybeans are also roasted whole and consumed as is or used to prepare peanut butter and soybean butter. Increasingly, peas have been subjected to similar processing and are roasted for consumption as a snack or further processed to obtain pea butter.

The majority of soybean and pulses produced in North America are dried post harvest. For pulses, the seeds can be subsequently dehulled and split, which reduces cooking time. Details on the techniques used for primary processing of legumes (e.g. harvesting, cleaning, sorting, dehulling, splitting) are described elsewhere (Erskine et al., 2009; Snyder & Kwon, 1987; Subuola et al., 2012; Tiwari et al., 2011).

Dry legume seeds (e.g. soybeans, alfalfa, clover, pea, beans, chickpeas, lentils) are sometimes allowed to germinate after soaking and sold in the sprouted form. Sprouted legumes are of interest nutritionally, as the germination process helps to increase protein digestibility and mineral bioavailability, and in some instances can reduce the concentration of tannins, phytic acid, and indigestible carbohydrates (Boye et al., 2012). Sprouting will be discussed later in this chapter.

Another technique used to preserve and extend the shelf life of legumes, particularly pulses, is canning, which will be discussed later in this chapter. Whole seeds are first soaked, blanched, and cooked and then packaged in cans with a variety of sauces, which eases their use in food preparation. A wide variety of canned legumes can be found in North American supermarkets, with the most popular perhaps being canned baked beans.

Another growing market for both household use and in the food service sector is the frozen precooked legumes category. Frozen vegetables are perceived by some consumers to have higher nutritional value than canned foods. Rickman et al. (2007) point out that this perception may not always be true, as the effects of processing, storage and cooking are highly variable by commodity. Nevertheless, there is a growing market for precooked frozen legumes, which offer convenience when they can be quickly warmed on stove-top or in the microwave prior to consumption. The export market for frozen vegetables in North America was valued at US$292 M in 2011 (www.icongrouponline.com, accessed 18 November 2013). Individually quick-frozen (IQF) vegetables may be classified as ready-to-use, reheat-and-serve foods. They are first blanched/cooked prior to quick-freezing, which helps to preserve physical and nutritional quality.

Infrared heating is another technique applied to whole pulse seeds to decrease the time required for cooking. The process is sometimes called micronization. A company located in Canada, Infraready Products Ltd. (www.infrareadyproducts.com, accessed 18 November 2013), uses this technique to precook pulses (i.e., peas,

beans, chickpeas and lentils). Described benefits of micronization include shorter cooking times, increased water and moisture absorption and retention, decreased microbial and enzymatic activity, increased shelf life, softer texture and flavor enhancement due to the addition of toasted notes to finished food products. Pulses can also be fully precooked in water and then dehydrated and sold as is or ground into flour, which will be discussed later in this chapter.

A variety of food products can be processed directly from dried legume seeds. In traditional markets, soybeans are processed into soymilk, tofu, yuba, miso, natto, sufu, and tempeh (Keshun, 1997). With the growing migrant population and the increasing trend towards exotic foods, these traditional products are now available in North America. Similarly, whole pulses are typically used to prepare soups, sauces, fried and baked products in places like India, Africa, and South America; these food products are now being made from pulses in North America. Research is further exploring novel uses and promising application areas of legumes for the North American market (Figure 14.1).

14.3 Traditional processing technologies

The traditional processing of legumes is labor intensive and is mostly done by women, especially in developing countries in Asia and Africa. The major traditional techniques used in the processing of legumes are soaking, dehulling, milling, boiling/cooking, roasting, pounding and grinding, frying, steaming, germination, fermentation, and popping, among others. Irrespective of the type of food that is prepared from legumes, they are taken through at least one of these processes. In this section, we discuss what some of these technologies are and the principles behind them.

14.3.1 Soaking

Legumes are primarily soaked in water and/or salt solutions (0.25–1%) to soften the cotyledon, which then hastens cooking (Silva et al., 1981). Soaking involves adding water and/or salt solution to the legumes and discarding the water after a period of time or cooking with the soak water. Sodium chloride, acetic acid, and sodium bicarbonate solutions have been used in the soaking of legumes (Huma et al., 2008). Different soaking times have also been reported (Huma et al., 2008; Xu & Chang, 2008), but in most cases, the soaking is done overnight. Soaking

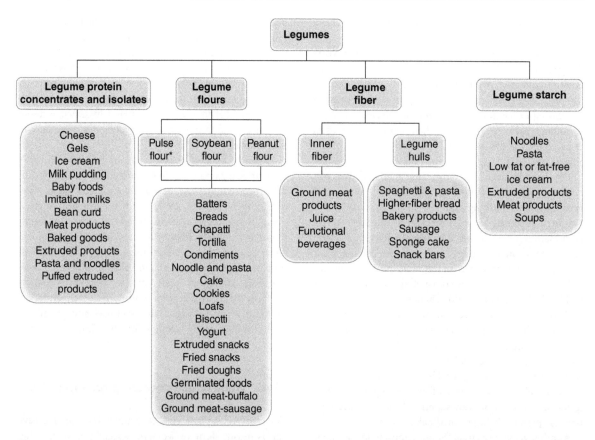

Figure 14.1 Potential and current food applications of legume flours and fractions. *Pulse flours include flours prepared from yellow pea, green gram, cowpea, navy bean, faba bean, field pea, lupin, lentils, great northern bean, pinto bean, red bean, white bean, black bean, winged bean, and pigeon pea.

of legumes can be done in either warm water or water at ambient temperature. Beside its primary role of shortening the cooking times of legumes, soaking has been reported to significantly reduce the phytate and phytic acid contents of legumes (Toledo & Canniatti-Brazaca, 2008). This was observed when legumes were not cooked with the soaking water. The flatulence factors in legumes are also reduced by soaking, as a result of the leaching out of stachyose and raffinose (Shimelis & Rakshit, 2005). These oligosaccharides are used as substrates by microorganisms in the large intestines, resulting in the production of carbon dioxide, leading to flatulence and intestinal discomfort. The addition of sodium bicarbonate to the soak water results in significant reduction in stachyose and raffinose. Soaking also increases the protein digestibility of legumes (Toledo & Canniatti-Brazaca, 2008), as confirmed in chickpeas, lentils, and different types of legumes (Martín-Cabrejas et al., 2009).

Soaking further results in the reduction of the mineral contents of legumes, due to the loss in the soaking water, especially when the water is discarded; however, their bioavailability is increased after soaking (Martín-Cabrejas et al., 2009). The increase in the bioavailability of minerals may be attributed to the reduction in antinutritional factors during soaking. These antinutritional factors are known to bind to the minerals in legumes, making them unavailable to the human body. Generally, soaking and cooking legumes without the soaking water reduces their carbohydrate contents (Martín-Cabrejas et al., 2009).

14.3.2 Cooking

The cooking of legumes has been practiced for years. It is one of the most common processing techniques applied to legumes, and involves boiling the legume seeds in water till they are soft. Traditionally, determination of the required softness of cooked legumes is done by pressing the legumes with the thumb. Several changes occur during the cooking of legumes apart from softening: gelatinization of starch, denaturation of proteins, and browning of the seeds (Enujiugha, 2005: Onigbinde & Onobun, 1993). Besides reducing the antinutritive factors in legumes, cooking reduces the amounts of stachyose and raffinose. The longer cooking times required have, however, been an obstacle to legumesuse. To reduce the legume cooking times, potash has been used traditionally to help soften the legume cotyledons. Sodium bicarbonate, trisodium phosphate, and ammonium carbonates have also been exploited to reduce the cooking times of legumes (Bueno et al., 1980).

14.3.3 Fermentation

Solid-state fermentation (SSF) is one of the alternative technologies for processing a great variety of legumes and/or cereals with the aim of improving their nutritional quality and obtaining edible products with palatable sensory characteristics (Reyes-Moreno et al., 2004). The advantages of fermentation include the development of flavor, texture, taste, reduction in product volume and the increase in product stability and shelf life through the preservation of the foods (Steinkraus, 2002). Legumes have been fermented into a variety of products. Notable among these products are tempeh and soy sauce, which are particularly popular in Asian countries.

Tempeh is a traditional fermented food produced with different strains of *Rhizopus* species (*R. oligosporus, R. stolonifer, R. oryzae, R. arrhizus*) fermenting boiled and dehulled soybeans (Annor et al., 2010: Keuth et al., 1993). It has a pleasant mushroom-like aroma and a nutty flavor, making it an excellent option for meat, fish and poultry products (Pride, 1984). During the fermentation process, stachyose and raffinose are broken down to digestible sugars. The fermentation process also improves the flavor, nutritional and functional properties of the product (Bavia et al., 2012). Even though soybeans are the main legumes used for the preparation of tempeh, other substrates have been used, e.g. cowpeas (Annor et al., 2010) and chickpeas (Reyes-Moreno et al., 1993). The process variables for the production of tempeh from chickpeas were optimized by Reyes-Moreno and colleagues (1993). According to their study, the optimum combination of process variables for production of optimized chickpea tempeh flour through the SSF process was incubating at 34.9 °C for 51.3 h. The chickpea tempeh was prepared by soaking the chickpeas at 25 °C for 16 h in

four volumes of acetic acid solution (pH 3.1). The seeds were then drained and their seed coats removed. The cotyledons were then cooked for 30 min and then cooled to 25 °C, packed in polyethylene bags and fermented with a suspension of *R. oligosporus* spores.

Soy sauce is a traditional Asian fermented soybean product that has gained international acceptance as a condiment or seasoning sauce due to its distinctive flavor. A combination of soybeans and wheat is normally used as the raw material. The soybeans and wheat are first fermented with *Aspergillus oryza* and then yeast and lactic acid bacteria are added later, after the addition of brine solution (8%) (Zhao et al., 2013). Two types of fermentation for soy sauce can be used: low-salt solid state and high-salt liquid state. The fermentation times for these two different types of fermentations differ significantly; while the former takes about 1 month, the latter takes as long as 6 months. Different concentrations of brine are also used for the different fermentation types. For the low-salt solid state fermentation, about 8% is used, while about 17% is used for the high-salt liquid state fermentation. The different fermentation types result in different tastes and flavors, with the high-salt liquid state fermented soy sauces having an edge over the taste and flavor of the low-salt solid state fermented product. Soy sauce has been reported to have antihypertensive properties, due to the presence of angiotensin I-converting enzyme, which was found to decrease blood pressure in hypertensive rats (Kinoshita et al., 1993).

The list of fermented legume foods is endless, with soybean arguably being the most fermented legume. Some of the products are *dawadawa*, from the African locust bean (*Parkia tilicoidea*), natto from soybeans, tempe *kedelee* from soybeans, *oncomhitam* from peanuts, *ketjap* (soy sauce) from black soybeans and *channakiwaries* prepared from bengal gram (*Cicerarietinum L.*) and black gram (*Vignamungo*) flour.

14.3.4 Dehulling

Dehulling involves the removal of the hulls of grain seeds, in this case legume seeds. It is one of the basic processing steps in legume processing. Dehulling can be done traditionally with mortar and pestle, which makes the process laborious and time consuming (Ehiwe & Reichert, 1987). The dehulling of legumes results in reduction of fiber and tannin content, and, most importantly, affects the appearance, texture, cooking quality, digestibility, and palatability of the grains (Deshpande et al., 1982). It has been demonstrated that there are marked differences in the dehulling efficiency of legumes (Reichert, 1984). Soybeans (*Glycine max*), faba beans (*Vicia faba equine* L.) and field peas (*Pisum sativum* L.) have better dehulling efficiencies (about 70%) compared to the others. These dehulling efficiencies were attributed to the resistance of seed splitting during dehulling and also to fact that the seed coat of these legumes is loosely bound to their cotyledons (Ehiwe & Reichert, 1987).

14.3.5 Germination/sprouting

Germination is one of the most common and effective legume processing methods, with the aim of improving nutritional quality. It can be defined as the transformation of seeds (herein referred to as legumes) from their dormant state to a metabolically active state, involving the mobilization of stored reserves of these seeds. As a result, there is a rapid increase in respiration, synthesis of proteins and nucleic acids, and the elongation and division of cells (Kadlec et al., 2008). Germination is normally preceded by soaking. During germination, the degradation of stored carbohydrates in the seeds by enzymes takes place. This results in significant changes in the physicochemical characteristics of the legumes, including the modification of antioxidant activities (López-Amorós et al., 2006). The process of germinating legumes, as is traditionally practiced, involves soaking seeds in water for 24 h at room temperature, draining and then spreading them on a damp cloth for about 48 h. In some cases where traditional germination is done on a large scale, large wet baskets are used.

14.3.6 Puffing

Puffing is one of the traditional technologies used to process legumes. It is commonly applied to chickpeas and peas, resulting in a light and crispy product. Puffed legumes are commonly eaten as snack foods, though they can also be milled into flour and used for other purposes. Traditionally, puffed legumes are prepared by soaking the legumes in water for about 15–20 min, followed by draining the water. The wet grains are then tempered in a closed vessel for about 4 h, after which they are cooked in sand, heated to about 200 °C for about a minute. This normally results in the expansion of the grains, leading to the splitting of the husk of the legumes. Puffed legumes are known to retain all the nutrients and also result in improved protein and carbohydrate digestibility (Baskaran et al., 1999).

14.4 Modern processing technologies

Modern processing technologies for legumes involve the use of sophisticated equipment and result in the mass production of products. Some of these technologies include extrusion cooking, high-pressure cooking, air classification, agglomeration, and canning. In this section, some of these technologies as applied to the processing of legumes and their effects on the nutritional and physical characteristics of legumes are discussed.

14.4.1 Extrusion cooking

Extrusion cooking is a high-temperature, short-time process that can be applied to foods to modify and/or improve their quality attributes. It consists of the thermomechanical cooking of foods at high temperatures, pressure and shear, generated inside a screw-barrel assembly (Battacharya & Prakash, 1994). Extrusion has been applied to legumes for the production of ready-to-eat products. Attempts to use extrusion cooking as a means to decontaminate aflatoxin in peanuts (Grehaigne et al., 1983; Saalia & Phillips, 2011) and canavanine in jack beans (Tepal et al., 1994) have been mentioned.

Extrusion has several effects on the nutritional properties of the resulting extruded products. Improvements in the protein and starch digestibility of extruded faba and kidney beans were reported by Alonso et al. (2000). According to Phillips (1989), the conditions used in extrusion cooking result in physical and chemical transformations such as protein cross-linking (Stanley, 1989), isopeptide bonding (Burges & Stanley 1976), or amino acid racemization, that directly influence the nutritional composition of extruded products. Extrusion cooking has also been effectively used in the production of textured vegetable protein (TVP) and textured soy protein (TSP), used extensively as food ingredients.

The extruder consists of a sturdy screw or screws rotating inside a smooth or grooved cylindrical barrel. The barrel can be heated externally for certain applications. For the production of extruded legume products, legume flour, which is conditioned to a moisture content of about 20–25% with live steam, is fed into the extruder. As the flour-water mixture goes through the barrel, it is heated rapidly by friction and external heat; coupled with high pressures, temperatures as high as 150–180 °C are attained. The legume flour-water mixture then goes through a process called thermoplastic "melting," also known as thermoplastic extrusion. The intense heat and pressure conditions applied to the product result in the denaturation of the soy proteins and puffing of the mixture. The extrudate is then cut and dried.

14.4.2 High-pressure cooking

High-pressure cooking involves the application of hydrostatic pressure of several hundred MPa to foods for the purpose of sterilization, protein denaturation, and control of enzyme and chemical reactions, amongst others (Estrada-Giron et al., 2005). It basically involves the cooking of food in a high-pressure cooker. High-pressure cooking, also commonly referred to as high hydrostatic pressure (HHP) cooking, is gaining worldwide interest, especially in Japan, the US and Europe, because of its advantages over most processing methods. In the US, consumers can purchase HHP-processed sauces, oysters, and guacamole (Estrada-Giron et al., 2005). HHP results in significant inactivation of microorganisms (Knorr 1993), improved food quality and retention of ingredients in the products (Cheftel, 1991). HHP can also be used in the modification of texture, whipping, emulsification and dough-forming properties of foods (Hoover, 1989). HHP has been applied to the processing of legumes, especially soybeans. HHP-produced tofu was found to have a much longer shelf life due to the significant reduction in its microbial population (Prestamo et al., 2000). The solubilization of protein from whole soybean grains, subjected to pressure of up to 700 MPa, has also been reported (Omi et al., 1996). The activity of lipoxygenase, an enzyme which is responsible for the off-flavors produced in soybeans, has been found to be sensitive to high pressures (Ludikhuyze et al., 1998).

14.4.3 Canning

Canning is a heat sterilization process applied to foods to ensure they are commercially sterile (i.e. the products are free from microorganisms capable of growing in the food at normal non-refrigerated temperatures). Properly sealed and heated canned foods should remain stable and indefinitely unspoiled in the absence of refrigeration. The effectiveness of the canning process is determined by the type of food, pH, container size and consistency or bulkiness of the food, but heating of food for longer than necessary is undesirable, as the nutritional and eating quality of foods are affected negatively by prolonged heating (Brock et al., 1994). Canning, like many other food processes, is applied to legumes to improve their

shelf life. Canning of legumes is mainly composed of two processes: soaking/blanching and thermal processing/heat sterilization. Soaking is done before canning to remove foreign material, facilitate cleaning, aid in can filling through uniform expansion, ensure product tenderness, and improve color. Soaking also results in the reduction of antinutritive factors in the legumes, due to their leaching out (Uebersax et al., 1987). Blanching inactivates enzymes, which might produce off-flavors, but also softens the product and removes gases to reduce strain on can seams during retorting (Beckett, 1996). Afoakwa et al. (2006) optimized the preprocessing conditions for the canning of Bambara groundnuts. They concluded that soaking time of 12 h, blanching time of 5 min and sodium hexametaphosphate salt concentration of 0.5% gave the best quality canned product from Bambara groundnut with acceptable quality characteristics.

Conditions for heat sterilization of low-acid foods are defined to ensure that all spores of *Clostridium botulinum* are destroyed and to prevent the spoilage of the product by heat-resistant, non-pathogenic organisms. Sterilization should normally be performed at 121 °C for at least 3 min (Beckett, 1996). In the case of legumes, additional sterilization would also provide adequate softening of the seeds (van Loggerenberg, 2004). Canning has significant effects on the nutritional properties of legumes. Wang et al. (1988) found that canning decreased the protein content of drained beans, with the exception of one cultivar. Canning results in nitrogenous components, such as amino acids and small chain polypeptides, leaching from bean tissue into brine (Drumm et al., 1990); crude protein also leaches into the canning medium (Lu & Chang, 1996). Canning of legumes also causes mineral losses. Iron, magnesium, manganese, potassium, and zinc losses occur during soaking, blanching and/or thermal processing, but phosphorus and copper levels remain the same in canned beans. The sodium and chloride levels increase in canned beans, due to the sodium chloride (NaCl) added to the filling medium of cans (Lopez & Williams, 1988).

14.4.4 Air classification

Air classification is a technique for the dry separation of particles from finely ground powders and flours, according to their size, shape, and density. Air classification is typically applied to pulse products and is a relatively simple technique that allows expansion of product offerings. It has been proven as an effective method for the production of starch-rich and protein-rich fractions of meals (King & Dietz, 1987). Air classification has been carried out on pea (*P. sativum*), faba bean (*V. faba*), mung bean (*Vigna radiata*), green lentil (*Lens culinaris*), navy bean (*Phaseolus vulgaris*), baby lima bean (*Phaseolus lunatus*), and cowpea (*Vigna unguiculata*) (Tyler et al., 1981). The first process of air classification involves milling. Tyler et al. (1981) used a pin mill when they studied the air classification of legumes. In their study, dehulled legume seeds were milled with a pin mill and then fractioned into starch fraction and protein concentrate using an air classifier. These fractions were remilled several times and air classified again to increase the quality and also the yield of the protein concentrates. Air classification, aside from separating the starch and protein fractions, results in enrichment of the fractions.

14.4.5 Agglomeration

Agglomeration, in general, can be defined as a process during which primary particles are joined together so that bigger porous secondary particles (conglomerates) are formed (Palzer, 2005). According to this definition, even caking of hygroscopic raw materials during storage can be regarded as a kind of undesired agglomeration. Agglomeration is a physical phenomenon and can be described as the sticking of particulate solids, which is caused by short-range physical or chemical forces among the particles themselves due to physical or chemical modifications of the surface of the solid. This phenomenon is triggered by specific processing conditions, or binders and substances, which adhere chemically or physically on the solid surfaces to form a bridge between particles (Pietsch, 2003). The basic principle of the process of agglomeration is that powdery flour is dispersed in a humid atmosphere to wet the surface of the flour particles, resulting in the particles adhering to each other. This process has been used in the preparation of cous-cous in North Africa. Agglomeration enhances the swelling and dispersion properties of legume flours.

14.5 Ingredients from legumes

14.5.1 Oil

Due to their high oil content, soybeans and peanuts have served as important raw materials for oil production. Soybean oil production in the US in 2011 was 8.4 million MT (www.soystats.com, accessed 18 November 2013), whereas peanut oil production is more limited and was 89 thousand MT in 2012 (www.indexmundi.com,

accessed 18 November 2013). Even though Mexico and Canada rank seventh and 15th in global production of soybean oil, respectively, there is hardly any peanut oil production in Canada.

Hydraulic pressing, expeller, solvent extraction and combinations of these techniques are the main processes used for vegetable oil extraction. With a few exceptions, the process for extracting oil from soybeans and peanuts is very similar (Figure 14.2). Typically, the oilseed is first cleaned to remove foreign materials, dried if needed to facilitate dehulling and cracking, which allows the seeds to be broken into smaller pieces, followed by conditioning and flaking. The flaking process involves passage of the broken pieces between rolls, which ruptures the oil cells and expands the surface area for solvent penetration and subsequent oil extraction.

Hexane is the most common solvent used for oil extraction. The miscella obtained after solvent extraction contains a mixture of oil and solvent, which is passed through a solvent recovery system to remove solvent. The crude oil remaining is subjected to further refining to obtain edible oil and other derived products. The defatted flakes are passed through a desolventizer/toaster to remove residual solvent. The meal obtained is rich in protein, and can be used as is or further processed downstream to extract protein. Lui (1997) provides further detail on soybean oil extraction processes. Concerns about trace remnants of hexane in solvent-extracted defatted oilseed meals have spurred research on alternative defatting techniques such as the use of other solvents, mechanical extraction, enzyme-assisted aqueous extraction, and supercritical extraction (Russin et al., 2011).

14.5.2 Flours

Whole, dehulled and defatted legume seeds, flours, and flakes can be milled to obtain a variety of flour products. Subuola et al. (2012) and Tiwari et al. (2011) provide an overview of the methods used to obtain flours from soybean, pulses, and peanuts. The major component found in legume flours is carbohydrate, which can range from 25% to 68%. Depending on whether flours are defatted or not, protein content can range from 17% to 56%. Table 14.2 presents the composition of legume flours prepared using a variety of processing techniques.

Legume flours are of interest in food processing due to both their nutritional and functional properties. Functional properties that aid in food formulation and processing include protein solubility, water holding

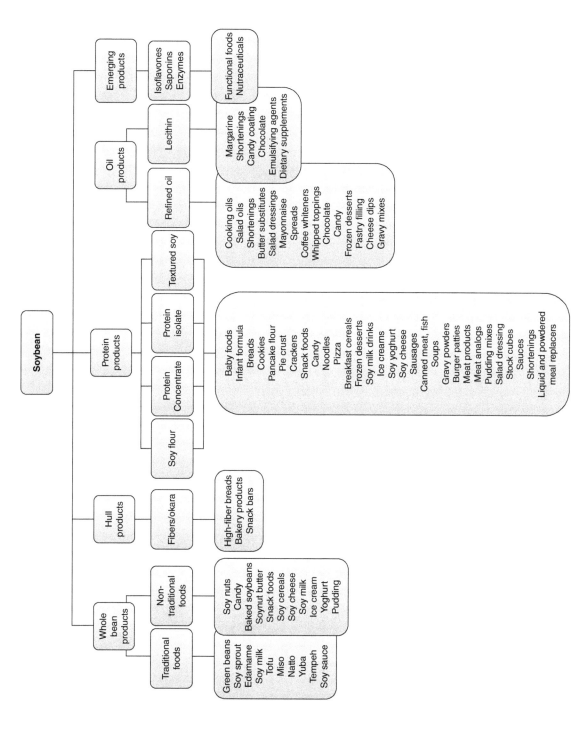

Figure 14.2 Soybean foods and ingredients. Reprinted from L'Hocine & Boye (2007), with permission from Taylor & Francis.

Table 14.2 Chemical composition of legume flours and legume protein concentrates/isolates

	Processing techniques	Protein (%)	Fat (%)	Carbohydrates (%)	Moisture (%)	Ash (%)	Starch (%)	Crude fiber (%)	References
Legume flours									
Full-fat Desi chickpea flour	Centrifugal mill (5-mesh sieve with 1.5 mm pore size)	22.3	5.16	39.19	db	3.37			(Mondor et al., 2010)
Full-fat Kabuli chickpea flour	Centrifugal mill (5-mesh sieve with 1.5 mm pore size)	18.9	6.70	71.24	db	3.16			(Mondor et al., 2010)
Kabuli chickpea flour	Centrifugal grinding mill and dehulling	16.71	7.34	61.14	12.06	2.76			(Boye et al., 2010a)
Desi chickpea flour	Centrifugal grinding mill and dehulling	20.52	5.23	61.94	9.26	3.04			(Boye et al., 2010a)
Defatted chickpea flour	Grinding and sieving (80-mesh sieve), defatted (10% w/v hexane) and air dried	17.2	0.3			2.8		1.1	(Paredes-López et al., 2006)
Chickpea flour	Grinding with a coffee mil, sieving through a 60-mesh sieve	21.37	7.17	58.92	7.40	2.98		2.16	(Bencini, 2006)
Chickpea flour	NR	22.9	6.4		10.4	2.83	40.4		(Marconi et al., 2000)
Bengal gram flour (*Cicer arietinum*)	Dehulling, grinding and sieving (40- or 60-mesh)	21.2	5.6	66.1	3.2	2.6		1.8	(Nagmani & Prakash, 1997)
Whole pea flour	Grinding and passing through a 4 mm screen	23.7			13.9	3.28	52.7	5.5	(Maaroufi et al., 2000)
Pigeon pea flour	Boiling, dehulling and dry milling	22.4	2.63	59.4	5.24	5.76		3.82	(Oshodi & Ekperigin, 1989)
Pea flour	NR (Fernández-Quintela et al., 1997)	23.93	3.12	59.39		2.58		8.77	
Yellow pea flour	Centrifugal grinding mill and dehulling	21.09	2.01	60.29	14.19	2.42			(Boye et al., 2010a)
Field pea flour	Commercial	25.0	1.1		db	2.7	55.7	1.9	(Sosulski & McCurdy, 1987)

Sample	Processing								Reference
Green lentil flour	Centrifugal grinding mill and dehulling	23.03	0.82	63.08	10.68	2.39			(Boye et al., 2010a)
Red lentil flour	Centrifugal grinding mill and dehulling	25.88	0.53	63.10	9.27	2.34			(Boye et al., 2010a)
Lentil flour	Grinding of whole flours	20.6	2.15	56.1	11.2	2.80		6.83	(de Almeida Costa et al., 2006)
Lentil flour	Dehulling and grinding into powder by passing through a 0.4 mm screen	32.38–33.39	1.95–2.10	47.04–51.49	7.98–10.37	2.70–3.78		2.43–4.13	(Suliman et al., 2006)
Lentil flour	Dehulling, grinding and sieving (40- or 60-mesh)	26.0	0.8	65.2	5.0	2.3		0.7	(Nagmani & Prakash, 1997)
Common bean flour	Grounding the whole flours	20.9	2.49	54.3	9.93	3.80		8.55	(de Almeida Costa et al., 2006)
Black gram flour (*Phaseolus mungo*)	Dehulling, grinding and sieving (40- or 60-mesh)	23.2	1.4	67.9	3.3	3.3		0.8	(Nagmani & Prakash, 1997)
Green gram flour (*Phaseolus aureus*)	Dehulling, grinding and sieving (40- or 60-mesh)	25.6	1.3	67.7	1.3	3.3		0.8	(Nagmani & Prakash, 1997)
Common bean flour	NR	20.8	2.6		10.4	3.68	37.9		(Marconi et al., 2000)
Defatted peanut flour	Dehulling, flaking and defatting (using butane and propane)	55.88	1.50	25.14	8.12	4.85			(Wu et al., 2009)
Defatted soybean flour	Grinding flakes to pass through 100-mesh or finer, and defatting	50.5	1.5	34.2		5.8		3.2	(Wolf, 1970)
Soybean flour	Commercial	48.2	0.9		db	5.8	2.4	4.2	(Sosulski & McCurdy, 1987)
Ranges		16.71–55.88	0.3–7.34	25.14–67.9	1.3–14.19	2.34–5.76	2.4–55.7	0.8–8.77	
Legume protein concentrates									
Soybean protein concentrate	Aqueous alcohol washing, IEP and leaching	66.2	0.3		6.7				(Wolf, 1970)

(*Continued*)

Table 14.2 (Continued)

	Processing techniques	Protein (%)	Fat (%)	Carbohydrates (%)	Moisture (%)	Ash (%)	Starch (%)	Crude fiber (%)	References
Full-fat Desi chickpea protein concentrate (IEP)	Alkaline extraction/IEP	78.6	11.37	6.88	db	3.18			(Mondor et al., 2010)
Full-fat Kabuli chickpea protein concentrate (IEP)	Alkaline extraction/IEP	69.9	21.55	5.39	db	3.16			(Mondor et al., 2010)
Defatted Desi chickpea protein concentrate (IEF)	Alkaline extraction/IEP	86.9	3.71	5.92	db	3.47			(Mondor et al., 2010)
Defatted Kabuli chickpea protein concentrate (IEF)	Alkaline extraction/IEP	85.6	10.44	0.59	db	3.37			(Mondor et al., 2010)
Peanut protein concentrate (IPPPC)	Acid extraction/IEP	72.35	1.13	17.97	1.48	3.05			(Wu et al., 2009)
Peanut protein concentrate (AAPPC)	Aqueous alcohol precipitation	69.54	0.70	16.06	1.51	2.06			(Wu et al., 2009)
Peanut protein concentrate (IAPPC)	IEP and alcohol precipitation	71.49	0.84	16.46	1.49	2.03			(Wu et al., 2009)
Ranges		66.2–86.9	0.3–21.55	0.59–17.97	1.48–6.7	2.03–3.47			
Legume protein isolates									
Micelle chickpea protein isolate	Micellization (NaCl extraction and ultrafiltration) from defatted flour	87.8	1.8			2.3		0.2	(Paredes-López et al., 2006)
Isoelectric chickpea protein isolate	IEP (alkaline extraction) from defatted flour	84.8	1.9			2.7		0.2	(Paredes-López et al., 2006)

Product	Method								Reference
Peanut protein isolate	IEP (pH 4.5) and alcohol precipitation	96.65	0.20	0.36	1.61	2.22			(Wu et al., 2009)
Soybean protein isolate	Alkaline extraction from defatted flour or flask, IEP, centrifugation/filtration	92.8	<0.1		4.7				(Wolf, 1970)
Soybean protein isolate	Commercial	82.3	0.4		db	4.0	1.8	0.6	(Sosulski & McCurdy, 1987)
Field pea protein isolate	Alkaline extraction and IEP	80.3	1.7		db	4.4	2.7	1.3	(Sosulski & McCurdy, 1987)
Faba bean protein isolate	Acid extraction and IEP	86.3	2.0		db	3.9	1.8	0.6	(Sosulski & McCurdy, 1987)
Pea protein isolate	Dehulling, alkaline extraction, centrifugation and lyophylizing	84.09	3.32	6.57		7.88		5.01	(Fernández-Quintela et al., 1997)
Faba bean protein isolate	Dehulling, alkaline extraction, centrifugation and lyophylizing	81.24	3.83	8.47		7.89		6.90	(Fernández-Quintela et al., 1997)
Soybean protein isolate	Dehulling, defatting, alkaline extraction, centrifugation and lyophylizing	82.16	1.46	5.64		7.73		3.17	(Fernández-Quintela et al., 1997)
Ranges		80.3–96.65	0.1–3.83	0.36–8.47	1.61–4.7	2.22–7.89	1.8–2.7	0.2–6.90	

db, dry basis; IEP, isoelectric precipitation; NR, not reported.

Table 14.3 Functional properties of legume flours, protein concentrates, and isolates

	Protein content (%)	PS (%)	WHC	FAC	LGC	BD (g/mL)	FE	FC	FS	EC	EA (%)	ES	References
Legume flours													
Red kidney bean flour	23.32		2.25 (g/g)	1.52 (g/g)	10 (%)	0.556		45.7 (mL/100 mL)	41.2 (mL/100 mL)	55.0 (mL/100 mL)		52.4 (mL/100 mL)	(Siddiq et al., 2010)
Small red kidney flour	20.93		2.65 (g/g)	1.23 (g/g)	10 (%)	0.526		38.2 (mL/100 mL)	43.3 (mL/100 mL)	60.5 (mL/100 mL)		62.3 (mL/100 mL)	(Siddiq et al., 2010)
Cranberry flour	23.62		2.41 (g/g)	1.48 (g/g)	12 (%)	0.539		49.6 (mL/100 mL)	54.9 (mL/100 mL)	53.4 (mL/100 mL)		52.4 (mL/100 mL)	(Siddiq et al., 2010)
Black bean flour	23.24		2.23 (g/g)	1.34 (g/g)	12 (%)	0.515		37.4 (mL/100 mL)	39.4 (mL/100 mL)	45.6 (mL/100 mL)		48.2 (mL/100 mL)	(Siddiq et al., 2010)
Yam bean flour	20.43		131.9 (%)	0.6 (mL/g)	14.3 (%)			40.2 (%)		50.7 (%)			(Obatolu et al., 2007)
Green gram flour	NR		1226 (g/kg)	900 (g/kg)		0.69		16 (%)		48 (mL oil/g of sample)	54.0	51.8 (%)	(Ghavidel & Prakash, 2006)
Bengal gram flour	NR		1362 (g/kg)	788 (g/kg)		0.73		12 (%)		185 (mL oil/g of sample)	51.6	49.4 (%)	(Ghavidel & Prakash, 2006)
Pigeon pea flour	22.4		138 (%)	89.7 (%)	12% (w/v)			68 (%)	20 (%)	49.4 (%)			(Oshodi & Ekperigin, 1989)
Cowpea flour	NR		1285 (g/kg)	993 (g/kg)		0.65		40 (%)		69 (mL oil/g of sample)	51.9	50.1 (%)	(Ghavidel & Prakash, 2006)
Field pea flour	25.0	80.3 (%)	0.78 (g /g)	0.41 (g oil/g sample)						34.6 (mL oil/0/1 g sample)			(Sosulski & McCurdy, 2006)
Lentil flour	NR		974 (g/kg)	857 (g/kg)	8.0 (%)	0.85		22 (%)		58 (mL oil/g of sample)	50.5	48.1 (%)	(Ghavidel & Prakash, 2006)
Lentil flour	NR		3.20 (mL/g)	0.95 (mL/g)		0.91		40.0 (%)			47.4		(Aguilera et al., 2009)
Desi chickpea flour	20.6–24.3		1.34–1.39 (g/g)	1.05–1.17 (g/g)	10–14 (%)						59.6–68.8	76.6–81.3 (%)	(Kaur and Singh, 2005)
Kabuli chickpea flour	26.7		1.33 (g/g)	1.24 (g/g)	10 (%)						58.2	82.1 (%)	(Kaur & Singh, 2005)
Chickpea flour	NR		2.10 (mL/g)	1.10 (mL/g)	8.0 (%)	0.71		24.0 (%)			22.9		(Aguilera et al., 2009)
Faba bean flour	29.2	85.9 (%)	0.72 (g /g)	0.47 (g oil/g sample)						34.6 (mL oil/0.1 g sample)			(Sosulski & McCurdy, 2006)
Peanut flour	52.73		1.67 (mL/g)	2.67 (mL/g)				0.06 (mL/g)		87.08 (mL/g)			(Yu et al., 2007)
Soybean flour	48.2	20.6 (%)	1.75 (g/g)	0.56 (g oil/g sample)						37.2 (mL oil/0.1 g sample)			(Sosulski & McCurdy, 2006)

Legume protein isolates

Field pea protein isolate	80.3	38.1 (%)	2.52 (g /g)	0.98 (g oil/g sample)			36.6 (mL oil/0.1 g sample)		(Sosulski & McCurdy, 2006)
Faba bean protein isolate	86.3	40.0 (%)	2.16 (g /g)	1.78 (g oil/g sample)			38.6 (mL oil/0.1 g sample)		(Sosulski & McCurdy, 2006)
Micelle chickpea protein isolate	87.8	72.5 (%)	4.9 (mL/g)	2.0 (mL/g protein)	43.3 (%)	59.2(%)	63.7	94.3 (%)	(Paredes-López et al., 2006)
Isoelectric chickpea protein isolate	84.8	60.4 (%)	2.4 (mL/g)	1.7 (mL/g protein)	47.5 (%)	66.6(%)	72.9	85.0 (%)	(Paredes-López et al., 2006)
Soybean protein isolate	82.3	30.6 (%)	2.65 (g /g)	1.03 (g oil/g sample)			45.1 (mL oil/0.1 g sample)		(Sosulski & McCurdy, 2006)
Soybean protein isolate	NR	21.2 (%)	5.7 (mL/g)	1.9 (mL/g protein)	41.8 (%)	53.2(%)	50.8	99.7 (%)	(Paredes-López et al., 2006)
Soybean protein isolate	90.2	22.2 (%)	584 (%)	144 (%)			75.1		(Naczk et al., 1986)
Cowpea protein isolate	95.7		2.20 (mL/g)	1.10 (mL/g)	6 (%)		50		(Ragab et al., 2004)
Legume protein concentrates									
Soybean protein concentrate	69.6	31.5 (%)	445 (%)	157 (%)			59.4		(Naczk et al., 1986)
Peanut protein concentrate	77.82		1.11 (mL/g)	0.90 (mL/g)	0.02 (mL/g)		87.50 (mL/g)		(Yu et al., 2007)
P. angularis	79.6		5.05 (g/g)	4.38 (g/g)	80.4–140.1 (%, pH 2–10)		54.7–57.0 (pH 2–10)	93.2–96.7 (%, pH 2–10)	(Chau et al., 1997)
P. calcaratus	78.0		5.28 (g/g)	4.71 (g/g)	80.2–130.0 (%, pH 2–10)		54.5–57.7 (pH 2–10)	94.5–97.3 (%, pH 2–10)	(Chau et al., 1997)
D. lablab	85.0		5.08 (g/g)	4.77 (g/g)	60.5–140.2 (%, pH 2–10)		53.0–57.9 (pH 2–10)	94.9–97.1 (%, pH 2–10)	(Chau et al., 1997)
Soybean protein concentrate	78.7		3.46 (g/g)	3.06 (g/g)	50.8–100.2 (%, pH 2–10)		54.5–58.1 (pH 2–10)	94.6–97.8 (%, pH 2–10)	(Chau et al., 1997)
Pea protein concentrate	55.5		153.0 (%, V/W)	113.0 (%, V/W)	0.45		22.4		(Conc & Blend, 1981)

BD, bulk density; EA, emulsifying activity; EC, emulsifying capacity; ES, emulsifying stability; FAC, fat absorption capacity; FC, foaming capacity; FE, foaming expansion; FS, foaming stability; LGC, least gelation concentration; NR, not reported; PS, protein solubility; WHC, water-holding capacity.

and fat absorption capacity and gelling, foaming, and emulsifying. A comparison of the functional properties of different legume flours is provided in Table 14.3. Depending on the final particle size, ground flakes and seeds of oilseeds and pulses can be classified as medium, fine or coarse. Final particle size distribution of the flour can have an impact on ingredient functionality and thus should be kept in mind during processing.

Table 14.4 presents a list of some commercially available legume flour products in North America. One of the interesting products on the list is a fermented soybean powder prepared with non-genetically modified soybean, which is fermented with *L. acidophilus*, *Bifidobacterium* spp., *L. bulgaricus* and *S. thermophilus* prior to drying. Due to the reported health benefits of probiotic bacteria (e.g. reducing risks of gastrointestinal tract disease, colorectal cancer and constipation, as well as boosting the immune system) (Ouwehand et al., 2002), there is growing interest in North America in formulating foods containing probiotics. As shown in Table 14.4, the legume flour products presented are touted as being gluten free, and their high-protein and high-fiber characteristics are highlighted. Suggested application areas include bakery products, cereals, nutritional bars, pastas, soups, sauces, processed meat products, batters and breadings, confections, frostings, and fillings.

14.5.3 Protein concentrates and isolates

Proteins are essential components of food. The quality of many cereal proteins is improved by complementation with legume proteins due to the high concentration of the amino acid lysine in legumes, which is often limited in cereals. Among legumes, soybeans have a Protein Digestibility Amino Acid Score (PDCAAS) of 100%, making them equivalent to dairy and meat in their ability to meet the essential amino acid requirements of humans. Pulses have lower PDCAAS scores but when used in mixed diets, they help to improve the protein quality score of food (Table 14.5) (Boye et al., 2012). Attempts to balance animal and plant sources of protein in the diet have created a market for plant-based proteins.

Several processes have been developed to extract proteins from legume flours and flakes (Figures 14.3–14.5). The process for protein extraction is typically the same for all legumes and can be done either through dry processing or wet processing. Dry processing involves air classification after milling of the flours as described above. However, protein content of the flours is much lower than for wet extracted ingredients. Protein contents reported for enriched flour fractions obtained using air classification range from 40% to 62% (Aguilera et al., 1984; Elkowicz & Sosulski, 1982; Gujska & Khan, 1991).

Wet processing (see Figure 14.5) provides flours with higher protein contents than dry processing. The process typically involves pretreatment of the oilseed or pulse to remove fiber and fat and decrease the particle size of the seed for efficient extraction. The next step is an alkaline extraction to solubilize protein, followed by filtration to remove insoluble carbohydrate material. Protein extraction can also be done with water or under acidic conditions (Boye et al., 2010a, 2010b). After filtration, the liquid extract is subjected to isoelectric precipitation or membrane filtration. At the isoelectric point, there is no net charge on the proteins, which allows them to precipitate out of solution. Membrane separation takes advantage of the differences in the molecular weight of the proteins to separate them from other soluble components in the extract. During isoelectric precipitation, the precipitate obtained is removed by centrifugation or filtration, washed and dried to obtain the concentrate or isolate, depending on protein purity. For membrane separation, the retentate, which contains the desired protein material, may be subjected to further diafiltration to remove salts, followed by drying (Boye et al., 2010a, b).

Washing of flours with alcohol can alternatively be used to obtain protein concentrates. In this instance, aqueous alcohol is used to remove alcohol-soluble carbohydrates and other alcohol-soluble compounds (e.g. flavors) from the defatted materials, leaving behind a higher protein-containing meal (protein concentrate) that is subsequently desolventized and dried.

Protein concentrates generally contain lower amounts of protein (65%, dry weight basis, dwb) than protein isolates, which typically have >85% protein dwb (Boye et al., 2010b). This classification is loosely interpreted by scientists as can be seen from Table 14.3 which presents the composition and functional properties of a variety of legume protein isolates and concentrates. Functional characteristics exhibited by legume protein concentrate and isolates are similar to those of the flours and include water- and fat-holding capacity, gelling, emulsifying, and thickening. As shown in Table 14.3, specific properties vary depending on the type of legume and product. Table 14.6 further lists some protein products prepared from various legumes that are commercially available in North America.

Table 14.4 Examples of commercial legume flour products in North America

Company	Products	Composition (source of products)	Characteristics indicated	Suggested applications
Best Cooking Pulses, Inc.[1] (Canada)	Pea flour Chickpea flour Lentil flour Beans flour	100% flours are obtained by milling whole pulse seeds	Gluten free, low in fat, high in protein, fiber and micronutrients, and can improve the nutritional quality of cooked and baked goods	Baked goods, cereals, extruded snacks, nutrition bars, pasta, soups and sauces, and processed meat products
Diefenbaker Seed Processor Ltd.[2] (Canada)	Chickpea flour Yellow pea flour	Chickpea flour is made from 100% pure Canadian chana dahl (yellow gram); yellow pea flour is derived from 100% Canadian yellow split peas which are ground to a superfine powder	Gluten-free products that can be used in many vegetarian and ethnic homes	Chickpea flour can be used to prepare onion bhajias, traditional potato and vegetable pakoras, desserts and battered dishes; yellow pea flour is traditionally used to thicken soups and stews
Parrheim Foods Inc.[3] (Canada)	Fiesta flour	Fiesta flour is a finely ground flour from yellow field pea (contains 67% starch and a minimum of 22% protein)	Gluten-free products that can be used in a wide variety of applications	Extruded snacks, batters (as a water-binding agent), and breading (either in the breading or di mix); baked goods and sauces (as a water control agent)
Golden Peanut Company[4] (USA)	Peanut flour	Natural roasted partially defatted flour made from high oleic US grade peanuts (available in either 12% or 28% fat levels in various roasted colors)	Gluten free, GMO free, peanut flour can be used to add texture, peanut flavor, aroma or protein to different food products	Confections, nutritional bars, baked goods, seasoning blends, frosting and fillings, sauces and dressings, baking mixes, and peanut spreads
Thebes Trade International[5] (Canada)	Textured soy flour	Flours are obtained by milling the whole soybean seeds	The flours can be used as raw materials for frozen, fast and vegetable foods	Can be used in a broad range of instant snacks, ready and convenience meals (meat and non-meat), and functional health foods, meat substitute, vegetarian foods
Now Foods Inc.[6] (USA)	NOW™ fermented soy powder	Prepared with non-GMO soybeans, and cultured with *L. acidophilus*, *Bifidobacterium* spp., *L. bulgaricus* & *S. thermophilus*	This product combines the nutritive value of soybeans with the benefits of fermentation; it offers a broader nutrient profile than traditional soybean products with higher content of isoflavones	Can be consumed as a dietary supplement by mixing 2 tablespoons (32 g) daily with water or beverage

[1] Website for Best cooking Pulses Inc.: www.bestcookingpulses.com/
[2] Website for Diefenbaker seed processor Ltd.: www.dspdirect.ca/
[3] Website for Parrheim Foods Inc.: www.parrheimfoods.com/
[4] Website for Golden Peanut Company: www.goldenpeanut.com/
[5] Website for Thebes trade international: http://nova2000.en.gongchang.com/
[6] Website for Now Foods Inc.: www.nowfoods.com/

Table 14.5 Protein Digestibility-Corrected Amino Acid Score (PDCAAS) for some legumes

Food	Food processing	PDCAAS (%) reported	PDCAAS (%) recalculated using reference pattern for 1–2 yr child, and LAA[1]	PDCAAS (%) recalculated using reference pattern for 3–10 yr child, and LAA[2]	References
Black beans	Raw, ground	72	69, Met + Cys	75, Met + Cys	(Sarwar, 1997)
Chickpea	Defatted flour from seeds soaked, decorticated, and dried	44	59, Met + Cys	64, Met + Cys	(Tavano et al. 2008)[3]
Cowpea, var. Bechuana white	Whole grain flour, raw	80 (not given)	80 (not given)	87 (Lys)	(Anyango et al., 2011)
Kidney bean, red, Canadian	Raw	28	37 Met + Cys	40 Met + Cys	(Khattab et al., 2009)
Lentil (*Lens culinaris*, cv. Medik)	Soaked 18 h; drained; autoclaved 10 min at 121 °C; cooled and freeze dried. Finely ground	52	(AA data not given)		(Sarwar & Peace, 1986)
Pea (organic cultivation, 2002)	Cooked (and freeze-dried)	75	82 Met + Cys		(Jorgensen et al., 2008)
Peanut	Roasted in electric oven for 30 min at 140 °C and then ground	70	65, Lys	70, Lys	(Fernandes et al., 2010)
Soybean	Meal, raw	80	88, Met + Cys	96, Met + Cys	(Sarwar, 1997)
Cowpea, var. Bechuana white	Whole grain flour, raw	80 (not given)	80 (not given)	87 (Lys)	(Anyango et al., 2011)

Source: Boye et al. (2012).

PDCAAS recalculated using LAA and reference pattern for 1–2 yr child. Neither the AAS nor PDCAAS was truncated. A few studies had digestibility values for individual amino acids as well as for protein. In such cases, the protein digestibility value was used. Trp and His were not determined in some studies. In all the *in vivo* digestibility studies, the diet fed to animals was not the individual food item but included corn starch, sucrose, oil, vitamins, minerals, cellulose, etc.

[1] PDCAAS recalculated using LAA and reference pattern for 1–2 yr child (WHO, 2007).

[2] Diets included sucrose, fat, vitamins and minerals, fiber, choline bitartrate, and corn starch.

[3] LAA, L-aspartic acid; Lys, lysine; Met, methionine.

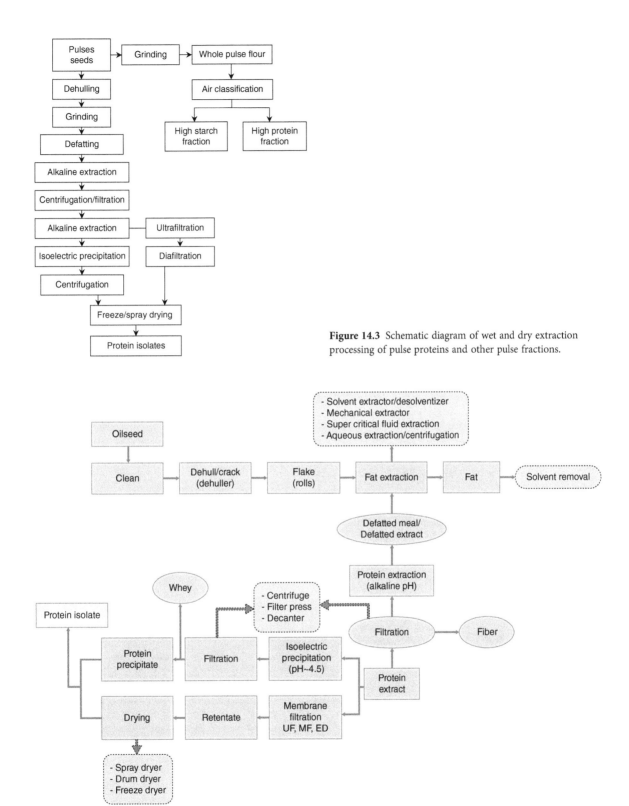

Figure 14.3 Schematic diagram of wet and dry extraction processing of pulse proteins and other pulse fractions.

Figure 14.4 Schematic flow diagram for the extraction of fat, fiber, and protein from oilseed legumes. ED, electrodialysis; MF, microfiltration; UF, ultrafiltration.

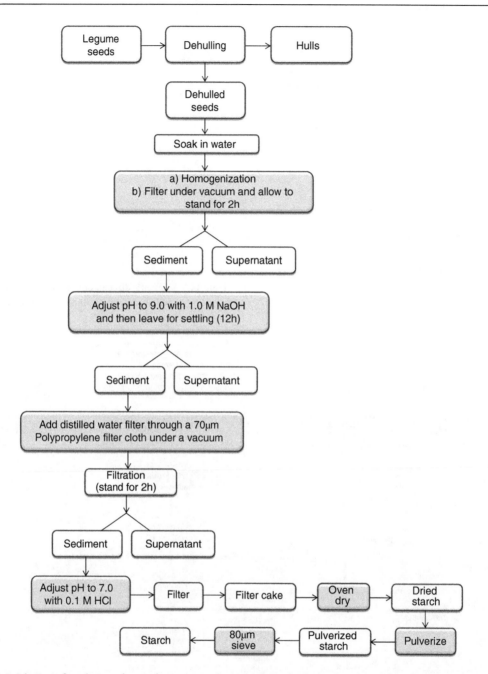

Figure 14.5 Schematic flow diagram for starch preparation by wet milling of pulse seeds. Adapted from Hoover et al. (2010).

14.5.4 Starch flours and concentrates

The composition of starch in various legume flours and fractions is provided in Table 14.2. Whereas very little starch is found in soybeans (trace) (Keshun, 1997) and peanut (3–6%) (Isleib et al., 2004), starch represents a major component of pulse legumes (38–56%) (Maaroufi et al., 2000; Marconi et al., 2000; Sosulski & McCurdy, 1987). Hoover et al. (2010) provide a comprehensive review on the composition, molecular structure, processing, and properties of pulse starches.

Table 14.6 Examples of commercial legume protein products produced in North America

Company	Products	Composition (source of products)	Characteristics indicated	Suggested applications
Nutri-Pea Limited[1] (Canada)	Propulse™ Propulse N™	A natural food-grade pea protein isolate (with protein content of 82%)	Offers a high level of functionality and nutrition (with an excellent amino acid profile, absent in gluten, lactose, cholesterol and antinutrients), can be used for protein enrichment	Beverage dry mix, nutritional bars, meal replacement beverages, baby food formulations, vegetarian applications, pasta, meat and seafood products, breads, dressings
Herbal Extracts Plus[2] (USA)	Soy capsule	100% soybean standardized extract (10% isoflavones)	Each capsule contains 600 mg of the extracted materials, and provides protein-rich soybean in the diet which may aid in lowering the risk of heart disease	Take one to two pea fiber capsules, two to three times each day at mealtimes
Norben Company Inc.[3] (USA)	Pea protein Soy protein concentrates Soy protein isolates	Protein extracts from pea and soybeans	Could be used as supplements in various food products; offers both nutritional and functional characteristics	Pasta, beverage, breads, dressings
Parrheim Foods Inc.[4] (Canada)	Prestige protein	Prestige protein is pea protein concentrate derived from field pea which contains 50% protein	Excellent emulsification capacity, oil and water absorption and foaming capacity; excellent amino acid profile; gluten free and has a low allergenicity	Specialty feeds, aquaculture feeds, pet food and food recipes
Parrheim Foods Inc.[4] (Canada)	Propel protein	Propel protein is concentrated from the yellow field pea and contains 44% protein	GMO free with very low allergenicity	Aquaculture, poultry and young swine diets due to its lysine content
Parrheim Foods Inc.[4] (Canada)	Fababean protein Progress protein Great Northern bean protein	Fababean protein is prepared from faba beans; Progress protein is produced from the yellow or green field pea; Great Northern bean protein is produced from great northern beans	Non-GMO, low allergenic, gluten free, functional and natural; excellent and valuable ingredients in various food applications	Different food recipes
Now Foods Inc.[5] (USA)	NOW™ Sports Pea protein	100% pure non-GMO pea protein isolates	Dietary supplement which is high in branched chain amino acids (each scoop contains over 4800 mg branched chain amino acids and over 2000 mg of L-arginine); has high solubility and is easy to digest	Mixe one scoop (33 g) of pea protein with 8 oz. of cold water, juice, or beverage and blend

[1] Website for Nutri-pea Ltd.: www.nutripea.com/index.htm
[2] Website for Herbal Extracts Plus: www.herbalextractsplus.com/
[3] Website for Norben Company Inc.: www.norbencompany.com/
[4] Website for Parrheim Foods Inc.: www.parrheimfoods.com/
[5] Website for Now Foods Inc.: www.nowfoods.com/

As with proteins, pulse starches can be processed using either dry (air classification) or wet processing techniques. Figure 14.5 presents the process for wet milling of starch. Similar to the protein extraction process, starch purity is higher when processed by wet milling. The properties of legume starch vary depending on the ratios of amylose and amylopectin, the two major components of starch that determine starch functionality in food applications. Starches are of interest in food formulation as they provide texture through their rheological and swelling properties. High-amylose starches retrograde to a greater extent than high-amylopectin starches, resulting in increased degrees of crystallinity, syneresis, and gel firmness.

Table 14.7 provides a list of some of the important properties of legume starches. Although commercial applications have been limited due to the poor swelling power and dispersibility and high gelatinization temperature and water exudation of pulse starches (Hoover et al., 2010), there is technological and nutritional interest in pulse starches as they provide unique functional characteristics. From the nutritional perspective, there is interest in the high resistant starch (RS) content of some pulse flours, as RS has been linked with health benefits such as reduced risk of colon cancer and diabetes, and providing a substrate for growth of probioctic organisms (Boye & Ma, 2012; Hoover et al., 2010). A list of some of the commercial legume starch products available in North America, their characteristics and suggested applications is provided in Table 14.8 and Figure 14.1.

14.5.5 Fiber products

Legumes are a good source of dietary fiber (both soluble and insoluble). Insoluble fibers found in legumes include cellulose, hemicellulose and lignin, whereas soluble fibers comprise oligosaccharides such as stachyose, raffinose, verbascose, and pectin. Increased fiber consumption can reduce the risk of many diseases including heart disease, diabetes, obesity, and some forms of cancer (Marlett et al., 2002). Fibers also provide physicochemical functionality to foods through their ability to bind and hold liquids such as water and fat, their swelling properties, and their impact on product viscosity. In the recent past, there has been growing interest in the use of fibers in food formulation due to their beneficial properties.

Legume hulls are removed as one of the first steps in processing, and this provides a by-product that is rich in fiber and which can be used as an ingredient by food processors. As particle size of the final fiber product can influence its physicochemical properties (e.g. dispersibility, water absorption) and nutritional quality (e.g. role in colonic function, transit time, etc.) (Tosh & Yada, 2010), the technique used for milling needs to be carefully considered. The particle size of the hull fiber can be controlled by judicious selection of milling procedures and screens (Ngoddy et al., 1986). Additionally the composition of protein, fat, minerals, and carbohydrates in the final product will also affect functionality.

During wet extraction of proteins and starch, the fiber fraction is separated, and this by-product can be dried and milled for use as a value-added product (see Figures 14.1, 14.2). Perhaps the best known legume fiber product is okara, the by-product obtained during soymilk production. To prepare soymilk, soybeans are soaked in water, cooked, ground and filtered to obtain the smooth-textured soymilk. This leaves behind a residue, the okara, which contains the bulk of the soybean fiber. In addition to fiber, soy okara contains protein (32%), oil (15%), and ash (3%) (Espinosa-Martos & Rupérez, 2009), which makes it a promising nutritive ingredient that could be incorporated in functional high-fiber foods.

Legume fibers as prebiotics for the growing probiotic food market are particularly promising. Prebiotics are carbohydrate sources of food for probiotics. In addition to the indigestible oligosaccharides found in legumes, legume starches with high amounts of amylose, which resists enzymatic hydrolysis (also classified as resistant starch), can serve as prebiotics (Conway, 2001; Niba, 2002; Wollowski et al., 2001). Various researchers are studying the resistant starch characteristics of different legume fibers in order to identify market applications (Bravo et al., 1998; Cairns et al., 1996; Tovar & Melito, 1996). Table 14.9 provides a list of some of the commercial legume fiber products available, their characteristics, and suggested applications.

14.5.6 Nutraceutical products

In addition to the major components, legumes contain minor components that may have health benefits. Isoflavones from soybeans have been of particular interest due to their reported beneficial effects on menopausal symptoms. Isoflavones are naturally occurring polyphenolic compounds belonging to the phytoestrogen class. Twelve different types have been found, which include daidzin, genistin, glycitin, acetyldaidzin, acetylgenistin,

Table 14.7 Properties of legume starches

Legumes	Processing techniques	Yield (%)	Total amylose (%)	SP or SF[1]	Pasting temp. (°C)	T_o (°C)[2]	T_P (°C)[2]	T_c (°C)[2]	Crystallinity (%)	ΔH	References
Kabuli chickpea	Soaking with H_2O containing 0.2% sodium hydrogen sulfite, dehulling, grinding, filtering with 100-mesh sieve and centrifuging, then washing sediment repeatedly with water	37.94	31.8	11.61 SP (g/g)	73.4	59.4	68.6	77.8	13.12	1.198 (J/g)	(Miao et al., 2009)
Desi chickpea	Same as the above	29.65	35.24	13.3 SP (g/g)	70.7	62.2	67.0	72.0	12.0	1.87 (J/g)	(Miao et al., 2009)
Black bean	Steeping in water containing 0.01% sodium metabisulfite, homogenizing and passing through a 202 screen, residue was homogenized and passed through a 70 μm screen, filtrate was left standing for sedimentation. Sediment was suspended in 0.2 NaOH and left standing and the final sediment was suspended in water and passed through a 70 μm screen, neutralized and then dried	16.37–21.80	37.17–39.32	8.2–17.7 SF (v/v)		61.0–65.7	70.9–74.9	81.2–86.7	32.1–32.7	17.8–20.1 $\Delta H / AP^3$ (mJ/mg)	(Zhou et al., 2004)
Pinto bean	Same as above	25.01–28.25	31.34–31.93	9.9–10.4 SF (v/v)		63.3–64.5	70.9–76.5	85.1–88.8	33.0–33.4	17.9–18.8 $\Delta H / AP^3$ (mJ/mg)	(Zhou et al., 2004)
Lentil	Same as above	27.44–34.07	30.51–32.29	16.0–18.4 SF (v/v)		60.0–60.1	66.0–66.6	76.4–77.5	31.7–32.3	15.5–16.6 $\Delta H / AP^3$ (mJ/mg)	(Zhou et al., 2004)

(Continued)

Table 14.7 (*Continued*)

Legumes	Processing techniques	Yield (%)	Total amylose (%)	SP or SF[1]	Pasting temp. (°C)	T_o (°C)[2]	T_p (°C)[2]	T_c (°C)[2]	Crystallinity (%)	ΔH	References
Smooth pea	Same as above	19.40–28.90	34.73–35.09	16.2–16.6 SF (v/v)		61.1–63.9	67.7–70.6	77.3–80.1	30.0–30.3	14.6–16.3 ΔH/AP[3] (ml/mg)	(Zhou et al., 2004)
Wrinkled pea	Same as above	21.60	78.42	3.4 SF (v/v)	NR	NR	NR	NR	17.7	NR	(Zhou et al., 2004)
Chickpea	Steeping in water containing 0.16% sodium hydrogen sulfite, grinding and screening through 100-mesh, centrifuging the filtrating slurry after removing supernatant, drying	29.0–35.2	28.6–34.3	11.4–13.6 SP (g/g)	75.4–76.7	61.5–64.8	66.7–69.0	71.3–73.8	NR	7.6–8.7 (J/g)	(Singh et al., 2005)
Soybean	Blending in 0.3% sodium metabisulfite, filtering through a screen of 106 μm mesh and centrifuging, the sediment was washed with 10% toluene in NaCl and left standing; sediment was then washed with water and filtered	NR	11.8–16.2	NR	71.6–83.8	52.0–53.5	56.5–57.9	NR	NR	12.3–12.9 (J/g)	(Stevenson et al., 2006)
Soybean	Same as above	NR	16.5–19.8	NR	68.9–82.4	51.2–51.8	55.2–55.8	NR	NR	10.7–12.7 (J/g)	(Stevenson et al., 2007)
Range		16.37–37.94	11.8–78.42	8.2–16.6	68.9–83.8	51.2–65.7	55.2–76.5	71.3–88.8	12.0–33.4		

[1] SP represents swelling power which is the ratio of the wet weight of the sedimented gel to its dry weight; SF represents swelling factor which is the ratio of the volume of sedimented gel to the volume of the dry starch granules with a density of 1.4 g/mL.
[2] T_o, T_p, T_c indicate the onset, peak and end temperature of gelatinization, respectively.
[3] Gelatinization enthalpy (ml/mg)/amylopectin content.
NR, not reported.

Table 14.8 Examples of commercial legume starch products in North America

Company	Products	Composition (source of products)	Characteristics indicated	Suggested applications
Nutri-Pea Limited[1] (Canada)	Accu-Gel™	A food-grade, native pulse starch	Its superior gelling properties allow it to be used at a 20–30% lower usage level. It offers good body and mouthfeel without altering flavor	Oriental noodles, french fry batters and coatings, surimi, extrusion applications, low-cost meat formulations, fat-free sour cream, vegetarian applications, canned products, fruit fillings, and sauces
Parrheim Foods Inc.[3] (Canada)	Starlite™	A pea starch concentrate derived from yellow field pea, which contains 85% starch (db) with a blend of 6% protein	Starlite expands well during extrusion. The use of Starlite in various food applications can increase resistant starch and slowly digestible starch content of the finished product; it can also improve moistness and water retention	Snack foods and breakfast cereals, noodle, pasta and baking recipes, making of bean-thread vermicelli
Parrheim Foods Inc.[2] (Canada)	Probond Starch	A blend of pea starch and pea protein	Can be used as a binding agent; produces a solid pellet and improves pelleting and extruding functions	Uses range from nutritional petfood recipes to the most rugged industrial binding application
Now Foods Inc.[3] (USA)	NOW™ PHASE 2™ starch neutralizer	A natural white kidney bean starch extract, also contains cellulose, cellulose powder and gum acacia	Product shown in non-clinical studies to help reduce the breakdown and absorption of complex carbohydrates, by limiting the action of α-amylase	Could be consumed as a dietary supplement by taking before any meal containing complex carbohydrates or starches

[1] Website for Nutri-pea Ltd. : http://www.nutripea.com/index.htm;
[2] Website for Parrheim Foods Inc.: http://www.parrheimfoods.com/
[3] Website for Now Foods Inc.: http://www.nowfoods.com/

acetylglycitin, malonyldaidzin, malonylgenistin, and malonylglycitin. Some companies have developed processes to extract soybean isoflavones for nutraceutical and functional food applications, as shown in Table 14.10. Other microcomponents in legumes of growing interest are saponins, lectins, amylase, and protease inhibitors due to their reported benefits in decreasing risks of a variety of diseases including cardiovascular disease, diabetes, obesity, and cancer (Campos-Vega et al., 2010).

14.6 Novel applications

The use of legumes in food product formulations in North America has been made easier with the advent of technologies that have allowed novel convenient ingredients to be developed. Soybeans led this growth, with the expansion of novel product formulation in the last two decades. Figure 14.2 lists some of the soybean foods and products available today. A similar growth is occurring with pulse

Table 14.9 Examples of commercial fiber products prepared from legume

Company	Products	Composition (source of products)	Characteristics indicated	Suggested applications
Herbal Extracts Plus[1] (USA)	Pea fiber capsule	100% pea fiber botanical powder	Each capsule contains approximately 600 mg of powdered pea fiber material; the bulking action of pea fiber may help to provide many healthful benefits	One to two pea fiber capsules, two to three times each day at mealtimes
Euringus Ingredients[2] (France)	Organic pea fiber	Contains insoluble dietary fiber from pea	Bringing many health benefits (such as reducing risk of colon cancer and coronary heart disease and cholesterol-lowering effects) as well as functionality (such as texture improver, fat replacer, and texture modifier) when supplementing organic pea fiber products into various food applications	Bread products (8–12% substitution); fiber additive (substitute for wheat, oat); meat products (sausage, beefburgers); meat products filler (by replacing starch); cookies and muffins (up to 25% substitution for flours); energy, health and wellness bars; noodles (3–5% substitution for wheat flours)
Best Cooking Pulses Inc.[3] (Canada)	Pea fiber	100% pea fiber	Product is approved by Health Canada's Bureau of Nutritional Science, the FDA and the USDA	Health drinks, baked goods, cereals, extruded snacks, nutrition bars
Parrheim Foods Inc.[4] (Canada)	Exlite fiber	Concentrated from yellow field pea	A functional fiber (soluble and insoluble), high in fiber and mineral content (Fe and Ca); product made without any chemical extraction or modification; finely ground and tasty	Breads, rolls, muffins, cookies, breakfast cereals, pasta products, snack foods, and specialty health foods and beverages; Exlite can also be used to substitute (up to 25%) for wheat flour in cookies, cakes, and muffins
Nutri-Pea Limited[5] (Canada)	Uptake 80™	A food-grade vegetable fiber	Offering both nutrition (fiber enrichment) and functionality (bulking agent and texture modifier) due to its high level of soluble and insoluble fiber, with high water-binding capacity	Low-fat and fat-free applications, hamburgers, veggie burgers and wieners, sauces and fillings, nutritional bars, cookies and brownies, processed fish products
Nutri-Pea Limited[5] (Canada)	Centu-Tex™	A food-grade vegetable fiber	Offering similar benefits as for Uptake 80™ with added benefit of a higher water- and fat-binding capacity	Low-fat and fat-free applications, hamburgers, veggie burgers and wieners, sauces and fillings, nutritional bars, cookies and brownies, processed fish products

(Continued)

Table 14.9 (*Continued*)

Company	Products	Composition (source of products)	Characteristics indicated	Suggested applications
Nutri-Pea Limited[5] (Canada)	Centara III™ Centara IV™ Centara 5™	A natural food-grade vegetable fiber manufactured from the hulls of Canadian yellow peas	Offering an excellent means of insoluble fiber fortification without significant alteration in color, flavor or odor. Centara IV and 5 are progressively whiter in color and are particularly suited for color-sensitive applications	Nutritional bars, white breads, bagels, tortillas, pasta, vegetarian applications, cookies and crackers
Golden Peanut Company[6] (USA)	AgForm 100 AgForm ES AG Granules Granules ES	Made from peanut hull and fiber	An excellent source of cellulose and crude fiber, peanut hulls are high in liquid absorbency, and have chemical inertness and biodegradability	Peanut hull and fiber products are primarily used in animal feed, as pesticide and fertilizer carriers, and as industrial absorbents; other uses include fiber ingredient, cellulosic products, fiber filler products, extender products, composite products, and inert carrier products

[1] Website for Herbal extracts plus: www.herbalextractsplus.com/
[2] Website for Euringus Ingredients: www.euringus.com/
[3] Website for Best cooking Pulses Inc.: www.bestcookingpulses.com/
[4] Website for Parrheim Foods Inc.: www.parrheimfoods.com/
[5] Website for Nutri-pea Limited: www.nutripea.com/index.htm
[6] Website for Golden Peanut Company: www.goldenpeanut.com

legumes as pulse flours, protein, starch and fiber fractions become increasingly available. A list of some of the potential and current applications of legume flours and fractions including pulses is presented in Figure 14.1.

14.7 Conclusion

Socio-economic, environmental and population challenges will likely continue to exert pressures on food prices and food supply in the coming decades, which will have impacts on global food security and the environment. Legumes are important sources of protein, carbohydrate and oil and other critical micronutrients. Maximizing their benefits as integral components of the global food supply requires the following: (a) agricultural technologies to enhance their production; (b) primary processing technologies to minimize losses during harvest and post harvest; and (c) effective techniques for their secondary and tertiary processing to expand convenience and ease of use. Currently, legumes are used as whole foods and as ingredients in food formulation. Growth in the sector has resulted from robust research over many years. Whereas much progress has been made with soybeans, there remains room for innovation, especially with regard to pulse legumes. For oilseed legumes, specifically soybeans and peanuts, research on ways to reduce allergenicity in order to expand their consumption will be useful. Legume processing is one of the most important activities in the food industry and results in a large variety of products with unique properties. The processing of legumes makes them acceptable to consumers. Both traditional and modern methods of processing legumes result in products that are suited for different purposes so both avenues should be exploited.

Table 14.10 Examples of commercial products prepared from micronutrients in soybean

Company	Products	Composition (source of products)	Characteristics	Suggested applications
Now Foods Inc.[1] (USA)	NOW[TM] Extra Strength Soy Isoflavones Vcaps[TM]	Soybean extracts and other ingredients including rice flour, cellulose, silica and magnesium stearate	Product extracted through a proprietary process that results in the highest natural levels of genistein	Could be used as a dietary supplement by taking one Vcaps one to three times daily
SISU[2] (Canada)	Soy Isoflavones	Soybean extract, standardized to contain 20% total isoflavones (genistein, daidzein and glycitein), also contains microcrystalline cellulose, magnesium stearate	Does not have estrogenic effect; product can alleviate menopause symptoms such as hot flashes; contains antioxidant that reduces free radicals that cause tissue damage	Suitable for vegans and contains no ingredients that are a source of gluten
Webber Naturals[®3] (Canada)	Webber Naturals® Soy Isoflavone Complex	Each capsule contains 50 mg soybean (*Glycine max*) (bean extract) and 20 mg isoflavones (40%), as well as non-medicinal ingredients such as gelatin capsule (gelatin, purified water), microcrystalline cellulose, vegetable-grade magnesium stearate (lubricant)	Contains isoflavones as phytoestrogens (plant source estrogens) that help to reduce or eliminate menopausal hot flashes	Could be consumed by taking one capsule two to three times daily, or as directed by a physician
Natural Factors[®4] (Canada)	Soy Isoflavones Complex Capsules	Each capsule contains 50 mg soybean isoflavone extract, total isoflavones of 13.8 mg AIE (aglycone isoflavone equivalents), and genistein compounds of 2.2 mg AIE	Product contains naturally balanced isoflavones, genistein and daidzein. These well-known flavonoids are complemented by other healthful natural compounds found in soybeans and soy foods. Provides support for menopausal symptoms and may slow bone density and inhibit bone reabsorption	Recommended to consume six capsules daily

[1] Website for Now Foods Inc.: www.nowfoods.com
[2] Website for Sisu: www.sisu.com/sisu
[3] Website for Webber Naturals: http://webbernaturals.com/caen
[4] Website for Natural Factors: http://naturalfactors.com/caen

References

Admassu Shimelis E, Kumar Rakshit S (2005) Antinutritional factors and in vitro protein digestibility of improved haricot bean (*Phaseolus vulgaris* L.) varieties grown in Ethiopia. *International Journal of Food Sciences and Nutrition* **56**(6): 377–387.

Afoakwa EO, Budu AS, Merson AB (2007) Application of response surface methodology for optimizing the preprocessing conditions of bambara groundnut (*Voandzei subterranea*) during canning. *International Journal of Food Engineering* 2(5).

Aguilera J, Crisafulli E, Lusas E, Uebersax M, Zabik M (1984) Air classification and extrusion of navy bean fractions. *Journal of Food Science* 49(2): 543–546.

Aguilera Y, Esteban RM, Benítez V, Mollá E, Martín-Cabrejas M. A (2009) Starch, functional properties, and microstructural characteristics in chickpea and lentil as affected by thermal processing. *Journal of Agricultural and Food Chemistry* 57 (22): 10682–10688.

Alonso R, Grant G, Dewey P, Marzo F (2000) Nutritional assessment in vitro and in vivo of raw and extruded peas (*Pisum sativum* L.). *Journal of Agricultural and Food Chemistry* 48(6): 2286–2290.

Annor GA, Sakyi-Dawson E, Ssaalia FK et al. (2010) Response surface methodology for studying the quality characteristics of cowpea (*Vigna unguiculata*)-based tempeh. *Journal of Food Process Engineering* 33(4): 606–625.

Anyango JO, de Kock HL, Taylor J (2011) Impact of cowpea addition on the protein digestibility corrected amino acid score and other protein quality parameters of traditional African foods made from non-tannin and tannin sorghum. *Food Chemistry* 124(3): 775–780.

Augustin J, Klein B (1989) Nutrient composition of raw, cooked, canned, and sprouted legumes. In: Mathews RH (ed) *Legumes: Chemistry, Technology and Human Nutrition*, 2nd edn. Rome: FAO.

Baskaran V, Malleshi N, Shankara R, Lokesh B (1999) Acceptability of supplementary foods based on popped cereals and legumes suitable for rural mothers and children. *Plant Foods for Human Nutrition* 53(3): 237–247.

Bavia ACF, Silva CE, Ferreira MP, Leite RS, Mandarino JMG, Carrão-Panizzi MC (2012) Chemical composition of tempeh from soybean cultivars specially developed for human consumption. *Ciência e Tecnologia De Alimentos* 32(3): 613–620.

Beckett ST (1996) *Physico-chemical Aspects of Food Processing*. New York: Springer.

Bencini MC (1986) Functional properties of drum-dried chickpea (*Cicer arietinum* L.) flours. *Journal of Food Science* 51(6): 1518–1521.

Bhattacharya S, Prakash M (1994) Extrusion of blends of rice and chick pea flours: a response surface analysis. *Journal of Food Engineering* 21(3): 315–330.

Boye J, Ma Z (2012) Finger on the pulse. *Food Science and Technology* 26(2): 20–24.

Boye J, Zare F, Pletch A (2010a) Pulse proteins: processing, characterization, functional properties and applications in food and feed. *Food Research International* 43(2): 414–431.

Boye J, Aksay S, Roufik S, Ribéreau S, Mondor M, Farnworth E (2010b) Comparison of the functional properties of pea, chickpea and lentil protein concentrates processed using ultrafiltration and isoelectric precipitation techniques. *Food Research International* 43(2): 537–546.

Boye J, Wijesinha-Bettoni R, Burlingame B (2012) Protein quality evaluation twenty years after the introduction of the protein digestibility corrected amino acid score method. *British Journal of Nutrition* 108(S2): S183–S211.

Bravo L, Siddhuraju P, Saura-Calixto F (1998) Effect of various processing methods on the in vitro starch digestibility and resistant starch content of Indian pulses. *Journal of Agricultural and Food Chemistry* 46(11): 4667–4674.

Brock TD, Madigan MT, Martinko JM, Parker J (1994) *Biology of Microorganisms*, 7th edn. New Jersey: Prentice-Hall.

Burgess L, Stanley D (1976) A possible mechanism for thermal texturization of soybean protein. *Canadian Institute of Food Science and Technology Journal* 9(4): 228–231.

Cairns P, Morris V, Botham R, Ring S (1996) Physicochemical studies on resistant starch *in vitro* and *in vivo*. *Journal of Cereal Science* 23(3): 265–275.

Campos-Vega R, Loarca-Piña G, Oomah BD (2010) Minor components of pulses and their potential impact on human health. *Food Research International* 43(2): 461–482.

Caro Bueno E, Narasimha H, Desikachar H (1980) Studies on the improvement of cooking quality of kidney beans (*Phaseolus vulgaris*). *Journal of Food Science and Technology* 17(5): 235–237.

Chau C, Cheung PC, Wong Y (1997) Functional properties of protein concentrates from three Chinese indigenous legume seeds. *Journal of Agricultural and Food Chemistry* 45(7): 2500–2503.

Cheftel J (1991) Applications des hautes pressions en technologie alimentaire. *Industries Alimentaires et Agricoles* 108(3): 141–153.

Conc WWL, Blend P (1981) Preparation and properties of spray-dried pea protein concentrate-cheese whey blends. *Cereal Chemistry* 58(4): 249–255.

Conway PL (2001) Prebiotics and human health: the state-of-the-art and future perspectives. *Food and Nutrition Research* 45: 13–21.

de Almeida Costa GE, da Silva Queiroz-Monici K, Pissini Machado Reis SM, de Oliveira AC (2006) Chemical composition, dietary fibre and resistant starch contents of raw and cooked pea, common bean, chickpea and lentil legumes. *Food Chemistry* 94(3): 327–330.

Deshpande S, Sathe S, Salunkhe D, Cornforth DP (1982) Effects of dehulling on phytic acid, polyphenols, and enzyme inhibitors of dry beans (phaseolus vulgaris L.). *Journal of Food Science* 47(6): 1846–1850.

Drumm TD, Gray JI, Hosfield GL, Uebersax MA (1990) Lipid, saccharide, protein, phenolic acid and saponin contents of four market classes of edible dry beans as influenced by soaking and canning. *Journal of the Science of Food and Agriculture* 51(4): 425–435.

Ehiwe A, Reichert R (1987) Variability in dehulling quality of cowpea, pigeon pea, and mung bean cultivars determined with the tangential abrasive dehulling device. *Cereal Chemistry* **64**(2): 89–90.

Elkowicz K, Sosulski F (1982) Antinutritive factors in eleven legumes and their air-classified protein and starch fractions. *Journal of Food Science* **47**(4): 1301–1304.

Enujiugha V (2005) Quality dynamics in the processing of underutilized legumes and oilseeds. In: *Crops: Growth, Quality and Biotechnology*. Helsinki: WFL, pp. 732–746.

Enujiugha V (2010) The antioxidant and free radical-scavenging capacity of phenolics from African locust bean seeds (*Parkia biglobosa*). *Advances in Food Sciences* **32**(2): 88–93.

Erskine W (2009) *The Lentil: Botany, Production and Uses*. Wallingford, Oxfordshire: CABI.

Espinosa-Martos I, Rupérez P (2009) Indigestible fraction of okara from soybean: composition, physicochemical properties and in vitro fermentability by pure cultures of lactobacillus acidophilus and bifidobacterium bifidum. *European Food Research and Technology* **228**(5): 685–693.

Estrada-Girón Y, Swanson B, Barbosa-Cánovas G (2005) Advances in the use of high hydrostatic pressure for processing cereal grains and legumes. *Trends in Food Science and Technology* **16**(5): 194–203.

Fernandes DC, Freitas JB, Czeder LP, Naves MMV (2010) Nutritional composition and protein value of the baru (*Dipteryx alata* vog.) almond from the Brazilian savanna. *Journal of the Science of Food and Agriculture* **90**(10): 1650–1655.

Fernández-Quintela A, Macarulla M, del Barrio A, Martínez J (1997) Composition and functional properties of protein isolates obtained from commercial legumes grown in northern Spain. *Plant Foods for Human Nutrition* **51**(4): 331–341.

Food and Agriculture Organization (FAO) (2009) *Consumption: Crops Primary Equivalent*. Rome: Food and Agriculture Organization of the United Nations. http://faostat.fao.org/site/339/default.aspx, accessed 18 November 2013.

Ghavidel RA, Prakash J (2006) Effect of germination and dehulling on functional properties of legume flours. *Journal of the Science of Food and Agriculture* **86**(8): 1189–1195.

Grehaigne B, Chouvel H, Pina M, Graille J, Cheftel J (1983) Extrusion-cooking of aflatoxin-containing peanut meal with and without addition of ammonium hydroxide. *Lebensmittel Wissenschaft Technologie* **16**(6): 317–322.

Gujska E, Khan K (1991) Functional properties of extrudates from high starch fractions of navy and pinto beans and corn meal blended with legume high protein fractions. *Journal of Food Science* **56**(2): 431–435.

Hoover DG (1989) Biological effects of high hydrostatic pressure on food microorganisms. *Food Technology* **43**: 99–107.

Hoover R, Hughes T, Chung H, Liu Q (2010) Composition, molecular structure, properties, and modification of pulse starches: a review. *Food Research International* **43**(2): 399–413.

Hu FB (2003) Plant-based foods and prevention of cardiovascular disease: an overview. *American Journal of Clinical Nutrition* **78**(3): 544S–551S.

Huma N, Anjum M, Sehar S, Khan MI, Hussain S (2008) Effect of soaking and cooking on nutritional quality and safety of legumes. *Nutrition and Food Science* **38**(6): 570–577.

Isleib TG, Pattee HE, Giesbrecht FG (2004) Oil, sugar, and starch characteristics in peanut breeding lines selected for low and high oil content and their combining ability. *Journal of Agricultural and Food Chemistry* **52**(10): 3165–3168.

Jacobs DR, Gallaher DD (2004) Whole grain intake and cardiovascular disease: a review. *Current Atherosclerosis Reports* **6**: 415–423.

Jørgensen H, Brandt K, Lauridsen C (2008) Year rather than farming system influences protein utilization and energy value of vegetables when measured in a rat model. *Nutrition Research* **28**(12): 866–878.

Kadlec P, Dostalova J, Bernaskova J, Skulinova M (2008) Degradation of alpha-galactosides during the germination of grain legume seeds. *Czech Journal of Food Sciences* **26**(2): 99.

Kaur M, Singh N (2005) Studies on functional, thermal and pasting properties of flours from different chickpea (*Cicer arietinum* L.) cultivars. *Food Chemistry* **91**(3): 403–411.

Keshun L (1997) *Soybeans: Chemistry, Technology, and Utilization*. London: Chapman and Hall.

Keuth S, Bisping B (1993) Formation of vitamins by pure cultures of tempe moulds and bacteria during the tempe solid substrate fermentation. *Journal of Applied Microbiology* **75**(5): 427–434.

Khattab R, Arntfield S, Nyachoti C (2009) Nutritional quality of legume seeds as affected by some physical treatments, part 1: Protein quality evaluation. *LWT-Food Science and Technology* **42**(6): 1107–1112.

King R, Dietz H (1987) Air classification of rapeseed meal. *Cereal Chemistry* **64**(6): 411–413.

Kinoshita E, Ozawa Y, Aishima T (1997) Novel tartaric acid isoflavone derivatives that play key roles in differentiating Japanese soy sauces. *Journal of Agricultural and Food Chemistry* **45**(10): 3753–3759.

Knorr D (1993) Effects of high-hydrostatic-pressure processes on food safety and quality. *Food Technology* **47**(6): 156–161.

L'Hocine L, Boye JI (2007) Allergenicity of soybean: new developments in identification of allergenic proteins, cross-reactivities and hypoallergenization technologies. *Critical Reviews in Food Science and Nutrition* **47**(2): 127–143.

Lopez A, Williams H (1988) Essential elements in dry and canned kidney beans (Phaseolus vulgaris L.). *Journal of Food Protection* **51**.

López-Amorós M, Hernández T, Estrella I (2006) Effect of germination on legume phenolic compounds and their antioxidant activity. *Journal of Food Composition and Analysis* **19**(4): 277–283.

Lu W, Chang K (1996) Correlations between chemical composition and canning quality attributes of navy bean (*Phaseolus vulgaris* L.). *Cereal Chemistry* **73**(6): 785–787.

Ludikhuyze L, van den Broeck I, Weemaes C, Hendrickx M (1998) Effect of combined pressure and temperature on soybean lipoxygenase. 2. modeling inactivation kinetics under static and dynamic conditions. *Journal of Agricultural and Food Chemistry* **46**(10): 4081–4086.

Maaroufi C, Melcion J, de Monredon F, Giboulot B, Guibert D, Le Guen M (2000) Fractionation of pea flour with pilot scale sieving. I. physical and chemical characteristics of pea seed fractions. *Animal Feed Science and Technology* **85**(1): 61–78.

Marconi E, Ruggeri S, Cappelloni M, Leonardi D, Carnovale E (2000) Physicochemical, nutritional, and microstructural characteristics of chickpeas (*Cicer arietinum* L.) and common beans (*Phaseolus vulgaris* L.) following microwave cooking. *Journal of Agricultural and Food Chemistry* **48**(12): 5986–5994.

Marlett JA, McBurney MI, Slavin JL (2002) Position of the American Dietetic Association: health implications of dietary fiber. *Journal of the American Dietetic Association* **102**(7): 993–1000.

Martín-Cabrejas MA, Aguilera Y, Pedrosa MM et al. (2009) The impact of dehydration process on antinutrients and protein digestibility of some legume flours. *Food Chemistry* **114**(3): 1063–1068.

Miao M, Zhang T, Jiang B (2009) Characterisations of kabuli and desi chickpea starches cultivated in china. *Food Chemistry* **113**(4): 1025–1032.

Mondor M, Aksay S, Drolet H, Roufik S, Farnworth E, Boye JI (2009) Influence of processing on composition and antinutritional factors of chickpea protein concentrates produced by isoelectric precipitation and ultrafiltration. *Innovative Food Science and Emerging Technologies* **10**(3): 342–347.

Mubarak A (2005) Nutritional composition and antinutritional factors of mung bean seeds (*Phaseolus aureus*) as affected by some home traditional processes. *Food Chemistry* **89**(4): 489–495.

Naczk M, Rubin L, Shahidi F (1986) Functional properties and phytate content of pea protein preparations. *Journal of Food Science* **51**(5): 1245–1247.

Nagmani B, Prakash J (1997) Functional properties of thermally treated legume flours. *International Journal of Food Sciences and Nutrition* **48**(3): 205–214.

Ngoddy P, Enwere N, Onvorah V (1986) Cowpea flour performance in akara and moin-moin preparations. *Tropical Science* **26**(2): 101–119.

Niba LL (2002) Resistant starch: a potential functional food ingredient. *Nutrition and Food Science* **32**(2): 62–67.

Obatolu V, Fasoyiro S, Ogunsunmi L (2007) Processing and functional properties of yam beans (*Sphenostylis stenocarpa*). *Journal of Food Processing and Preservation* **31**(2): 240–249.

Omi Y, Kato T, Ishida K, Kato H, Matsuda T (1996) Pressure-induced release of basic 7S globulin from cotyledon dermal tissue of soybean seeds. *Journal of Agricultural and Food Chemistry* **44**(12): 3763–3767.

Onigbinde A, Onobun V (1993) Effect of pH on some cooking properties of cowpea (*V. unguiculata*). *Food Chemistry* **47**(2): 125–127.

Oshodi A, Ekperigin M (1989) Functional properties of pigeon pea (*Cajanus cajan*) flour. *Food Chemistry* **34**(3): 187–191.

Ouwehand AC, Salminen S, Isolauri E (2002) Probiotics: an overview of beneficial effects. *Antonie Van Leeuwenhoek* **82**(1–4): 279–289.

Palzer S (2005) The effect of glass transition on the desired and undesired agglomeration of amorphous food powders. *Chemical Engineering Science* **60**(14): 3959–3968.

Paredes-López O, Ordorica-Falomir C, Olivares-Vázquez M (1991) Chickpea protein isolates: physicochemical, functional and nutritional characterization. *Journal of Food Science* **56**(3): 726–729.

Phillips R (1989) Effect of extrusion cooking on the nutritional quality of plant proteins. In: Phillips RD, Finley JW (eds) *Protein Quality and the Effects of Processing.* New York: Marcel Dekker, pp. 219–246.

Pietsch W (2003) An interdisciplinary approach to size enlargement by agglomeration. *Powder Technology* **130**(1): 8–13.

Pihlanto A, Korhonen H (2003) Bioactive peptides and proteins. *Advances in Food and Nutrition Research* **47**: 175–276.

Prestamo G, Lesmes M, Otero L, Arroyo G (2000) Soybean vegetable protein (tofu) preserved with high pressure. *Journal of Agricultural and Food Chemistry* **48**(7): 2943–2947.

Pride C (1984) *Tempeh Cookery.* Summertown, TN: Book Publishing Company.

Ragab DM, Babiker EE, Eltinay AH (2004) Fractionation, solubility and functional properties of cowpea (*Vigna unguiculata*) proteins as affected by pH and/or salt concentration. *Food Chemistry* **84**(2): 207–212.

Ragab H, Kijora C, Ati KA, Danier J (2010) Effect of traditional processing on the nutritional value of some legumes seeds produced in Sudan for poultry feeding. *International Journal of Poultry Science* **9**(2): 198–204.

Reichert R, Oomah B, Youngs C (1984) Factors affecting the efficiency of abrasive-type dehulling of grain legumes investigated with a new intermediate-sized, batch dehuller. *Journal of Food Science* **49**(1): 267–272.

Reyes-Moreno C, Paredes-López O, Gonzalez E (1993) Hard-to-cook phenomenon in common beans – a review. *Critical Reviews in Food Science and Nutrition* **33**(3): 227–286.

Reyes-Moreno C, Okamura-Esparza J, Armienta-Rodelo E, Gomez-Garza R, Milán-Carrillo J (2000) Hard-to-cook phenomenon in chickpeas (*Cicer arietinum* L): effect of accelerated storage on quality. *Plant Foods for Human Nutrition* **55**(3): 229–241.

Reyes-Moreno C, Cuevas-Rodríguez E, Milán-Carrillo J, Cárdenas-Valenzuela O, Barrón-Hoyos J (2004) Solid state fermentation process for producing chickpea (*Cicer arietinum* L) tempeh flour. Physicochemical and nutritional characteristics of the product. *Journal of the Science of Food and Agriculture* **84**(3): 271–278.

Rickman JC, Barrett DM, Bruhn CM (2007) Nutritional comparison of fresh, frozen and canned fruits and vegetables. part 1. vitamins C and B and phenolic compounds. *Journal of the Science of Food and Agriculture* **87**(6): 930–944.

Russin TA, Boye JI, Arcand Y, Rajamohamed SH (2011) Alternative techniques for defatting soy: a practical review. *Food and Bioprocess Technology* **4**(2): 200–223.

Saalia FK, Phillips RD (2011) Degradation of aflatoxins by extrusion cooking: effects on nutritional quality of extrudates. *LWT-Food Science and Technology* **44**(6): 1496–1501.

Sarwar G (1997) The protein digestibility-corrected amino acid score method overestimates quality of proteins containing antinutritional factors and of poorly digestible proteins supplemented with limiting amino acids in rats. *Journal of Nutrition* **127**(5): 758–764.

Sarwar G, Peace RW (1986) Comparisons between true digestibility of total nitrogen and limiting amino acids in vegetable proteins fed to rats. *Journal of Nutrition* **116**(7): 1172–1184.

Schneider AV (2002) Overview of the market and consumption of pulses in Europe. *British Journal of Nutrition* **88**: S243–S250.

Siddiq M, Ravi R, Harte J, Dolan K (2010) Physical and functional characteristics of selected dry bean (Phaseolus vulgaris L.) flours. *LWT-Food Science and Technology* **43**(2): 232–237.

Silva C, Bates R, Deng J (1981) Influence of soaking and cooking upon the softening and eating quality of black beans (*Phaseolus vulgaris*). *Journal of Food Science* **46**(6): 1716–1720.

Singh B, Nagi H, Sekhon K, Singh N (2005) Studies on the functional characteristics of flour/starch from wrinkled peas (*Pisum sativum*). *International Journal of Food Properties* **8**(1): 35–48.

Singh N, Singh Sandhu K, Kaur M (2004) Characterization of starches separated from Indian chickpea (*Cicer arietinum* L.) cultivars. *Journal of Food Engineering* **63**(4): 441–449.

Snyder HE, Kwon T (1987) *Soybean Utilization*. New York: Van Nostrand Reinhold.

Sosulski F, McCurdy A (1987) Functionality of flours, protein fractions and isolates from field peas and faba bean. *Journal of Food Science* **52**(4): 1010–1014.

Stanley D (1989) Protein reactions during extrusion processing. In: Mercier C, Linko P, Harper JM (eds) *Extrusion Cooking*. St Paul, MN: American Association of Cereal Chemists, pp. 321–341.

Steinkraus K (2002) Fermentations in world food processing. *Comprehensive Reviews in Food Science and Food Safety* **1**(1): 23–32.

Stevenson DG, Doorenbos RK, Jane J, Inglett GE (2006) Structures and functional properties of starch from seeds of three soybean (*Glycine max* (L.) merr.) varieties∗. *Starch-Stärke* **58**(10): 509–519.

Stevenson DG, Jane J, Inglett GE (2007) Structures and physicochemical properties of starch from immature seeds of soybean varieties (Glycine max(L.) merr.) exhibiting normal, low-linolenic or low-saturated fatty acid oil profiles at maturity. *Carbohydrate Polymers* **70**(2): 149–159.

Subuola F, Widodo Y, Kehinde T (2012) Processing and utilization of legumes in the tropics. In: Eissa A (ed) *Trends in Vital Food and Control Engineering*. Rijeka, Croatia: InTech, p. 71.

Suliman MA, El Tinay AH, Elkhalifa AEO, Babiker EE, Elkhalil EA (2006) Solubility as influenced by pH and NaCl concentration and functional properties of lentil proteins isolate. *Pakistan Journal of Nutrition* **5**(6): 589–593.

Tavano OL, da Silva S Jr, Demonte A, Neves VA (2008) Nutritional responses of rats to diets based on chickpea (*Cicer arietinum* L.) seed meal or its protein fractions. *Journal of Agricultural and Food Chemistry* **56**(22): 11006–11010.

Tepal JA, Castellanos R, Larios A, Tejada I (1994) Detoxification of jack beans (*Canavalia ensiformis*): I. Extrusion and canavanine elimination. *Journal of the Science of Food and Agriculture* **66**(3): 373–379.

Tharanathan,R, Mahadevamma S (2003) Grain legumes – a boon to human nutrition. *Trends in Food Science and Technology* **14**(12): 507–518.

Tiwari BK, Gowen A, McKenna B (2011) *Pulse Foods: Processing, Quality and Nutraceutical Applications*. New York: Academic Press.

Toledo TCF, Canniatti-Brazaca SG (2008) Chemical and nutritional evaluation of carioca beans (*Phaseolus vulgaris* L.) cooked by different methods. *Ciência e Tecnologia De Alimentos* **28**(2): 355–360.

Tosh SM, Yada S (2010) Dietary fibres in pulse seeds and fractions: characterization, functional attributes, and applications. *Food Research International* **43**(2): 450–460.

Tovar J, Melito C (1996) Steam-cooking and dry heating produce resistant starch in legumes. *Journal of Agricultural and Food Chemistry* **44**(9): 2642–2645.

Tyler R, Youngs C, Sosulski F (1981) Air classification of legumes [beans, lentils, peas]. I. separation efficiency, yield, and composition of the starch and protein fractions. *Cereal Chemistry* **58**: 144–148.

Uebersax M, Ruengsakulrach S, Srisuma N (1987) Aspects of calcium and water hardness associated with dry bean processing. *Michigan Dry Bean Digest* **12**: 8–10.

van Loggerenberg M (2004) Development and Application of a Small-Scale Canning Procedure for the Evaluation of Small White Beans (Phaseoulus Vulgaris). PhD thesis, University of the Free State.

Wang CCR, Chang SK (1988) Effect of selected canning methods on trypsin inhibitor activity, sterilization value, and firmness of canned navy beans. *Journal of Agricultural and Food Chemistry* **36**(5): 1015–1018.

Wolf WJ (1970) Soybean proteins. their functional, chemical, and physical properties. *Journal of Agricultural and Food Chemistry* **18**(6): 969–976.

Wollowski I, Rechkemmer G, Pool-Zobel BL (2001) Protective role of probiotics and prebiotics in colon cancer. *American Journal of Clinical Nutrition* **73**(2): 451s–455s.

Wu H, Wang Q, Ma T, Ren J (2009) Comparative studies on the functional properties of various protein concentrate preparations of peanut protein. *Food Research International* **42**(3): 343–348.

Xu B, Chang S (2008) Total phenolic content and antioxidant properties of eclipse black beans (*Phaseolus vulgaris* L.) as affected by processing methods. *Journal of Food Science* **73** (2): H19–H27.

Yu J, Ahmedna M, Goktepe I (2007) Peanut protein concentrate: production and functional properties as affected by processing. *Food Chemistry* **103**(1): 121–129.

Zhao G, Yao Y, Wang,X, Hou L, Wang C, Cao X (2012) Functional properties of soy sauce and metabolism genes of strains for fermentation. *International Journal of Food Science and Technology* **48**(5): 903–909.

Zhou Y, Hoover R, Liu Q (2004) Relationship between α-amylase degradation and the structure and physicochemical properties of legume starches. *Carbohydrate Polymers* **57**(3): 2.

15 Processing of Fruit and Vegetable Beverages

José I. Reyes-De-Corcuera,[1] Renée M. Goodrich-Schneider,[2]
Sheryl A. Barringer,[3] and Miguel A. Landeros-Urbina[4]

[1]Department of Food Science and Technology, University of Georgia, Athens, Georgia, USA

[2]Food Science and Human Nutrition Department, University of Florida, Gainesville, Florida, USA

[3]Department of Food Science and Technology, The Ohio State University, Columbus, Ohio, USA

[4]Coca-Cola FEMSA, Mexico City, Mexico

15.1 Introduction

15.1.1 Classification and regulations

In 2011, world fruit production was 638 million metric tonnes (MT). China, India, Brazil, USA, Italy, Indonesia, and Mexico are the largest producers, accounting for 51% of the world production (FAO, 2010). An important portion of fruit production is processed into juice, but data on processed juice processing are readily available for only a few countries. Table 15.1 summarizes the production of the major fruits used for processing of juices and nectars in the world; the third column of Table 15.1 shows the 2010 sum of the production of fruit by the five largest producers of each of the fruits. These fruits were selected because they are typically used for juice processing. Juices are defined as mechanically extracted juices from fruits or vegetables to which no water or other exogenous substances are added. Juices are commercialized as single-strength or concentrates, as 100% from a particular fruit or, less often but increasingly commonly, as blends (CODEX STAN 247-2005).

Not-from-concentrate (NFC) juices (sometimes labeled as "premium") typically retain more of a fresh-like quality compared to reconstituted juices that undergo longer thermal treatments and handling steps that affect color and flavor. Soluble solid content (SSC) and titratable acidity (TA) of both NFC and reconstituted juice must fall within the range of what results from extracting the juice from a mature fruit. These ranges are often regulated and regulations differ slightly in different parts of the world.

Adjustment of reconstituted juice quality is achieved by proper dilution. NFC products are adjusted by blending juices extracted from fruits with different levels of maturity. Storage and transportation of NFC are more costly than for reconstituted juices; therefore, consumers also have to pay a premium. Table 15.2 shows the total imports and exports for main fruit juices in the US. Figure 15.1 shows the average annual consumption of orange and apple juice, as well as the total citrus and non-citrus juice consumption in the US in the last three decades. Orange juice is the most consumed fruit juice (56–62%) but its consumption has declined compared to apple, which has increased from 18% to 26% in the last three decades. Both juices combined account for 80% of the total fruit juice consumed in the US.

"Nectar" typically refers to beverages produced by dilution of fruit pastes or juices with or without the addition of sweeteners. Nectars are commercialized as from a single fruit or as blends. In some countries, e.g. the UK, a fruit beverage labeled "nectar" must contain at least 25–50% fruit juice, depending on the specific fruit (SI 2003/1564). Fruit drinks are beverages with a small content of juice.

In the US, any beverage containing fruit juice must be labeled with the percent juice content. Beverages containing less than 1% juice must also be labeled as such (21 CFR

Food Processing: Principles and Applications, Second Edition. Edited by Stephanie Clark, Stephanie Jung, and Buddhi Lamsal.
© 2014 John Wiley & Sons, Ltd. Published 2014 by John Wiley & Sons, Ltd.

Table 15.1 Production of fruit for processing into commercial juices and nectars in 2010 with data from FAO (2010)

Fruit/vegetable	Growing regions	Total fruit production from the 5 world largest producers (MMT)
Juices		
Orange	Subtropical/tropical	41.9
Apple	Temperate/subtropical	44.4
Grapefruit	Subtropical/tropical	5.0
Mandarin/tangerine	Subtropical/tropical	14.6
Lemon/lime	Subtropical/tropical	8.2
Pineapple	Tropical	9.7
Grape	Temperate	34.6
Pear	Temperate/subtropical	17.9
Tomato	Temperate/subtropical	85.4
Pomegranate	Semi-arid, mild temperate, subtropical	Not reported
Cranberry	Temperate, moist	0.4
Coconut water or milk	Tropical	52.0
Nectars		
Mango/mangosteen/guavas	Tropical	26.7
Guava	Tropical/subtropical	Included in Mango
Peach/nectarine	Temperate/subtropical	15.1
Apricot	Temperate/subtropical	1.7
Passion fruit	Tropical/subtropical	Not reported
Papaya	Tropical/subtropical	8.6
Guanávana	Tropical	Not reported
Strawberry	Subtropical	2.3
Banana	Tropical	65.8
Tamarind	Tropical	Not reported
Plum	Temperate	7.5

MMT, million metric tonnes.

Table 15.2 Volume of US imports and exports of selected juices to and from the rest of the world, with data from USDA-ERS for 2010 (ERS, 2010)

Fruit juice	Volume in millions of liters SSE	
	Imports	Exports
Apple	2182	33
Grape	184	55
Grapefruit	2	53
Lemon	98	
Lime	49	
Orange	1150	583
Pineapple	255	
Wine	932	386

SSE, single strength equivalent.

101.30). Several countries that do not have their own definitions or standards typically adopt those of the Codex Alimentarius (CODEX STAN 247-2005), the EU, or the US standards. In addition to standards of identity, juice and nectar labeling and trade are regulated in some countries and in some cases on the basis of country of origin. This is quite controversial, because to maintain quality year around, often fruit juices from different origins need to be blended. In the case of blends of juices from different fruits, it becomes impossible to establish a single country of origin. Beyond labeling, processing of fruit juices is regulated, in particular to ensure safety. In the US, the Food and Drug Administration (FDA) has mandated that Hazard Analysis and Critical Control Point (HACCP) plans be implemented in all juice and fruit paste processing plants (21 CFR 120; Goodrich et al., 2008).

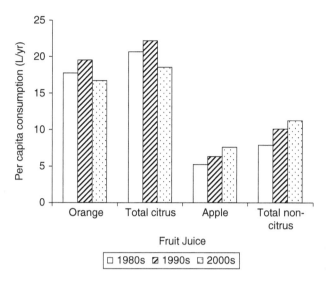

Fruit Juice

□ 1980s ▨ 1990s □ 2000s

Figure 15.1 Fruit juice consumption per capita in the USA with data from the USDA-ERS 2009 *Fruit and Tree Nut Yearbook.*

15.1.2 General processing operations

All juice processing lines have several unit operations in common. A general flow process diagram (FPD) for juice production is shown in Figure 15.2. Fruit composition, geometry, and other physical properties dictate the method of juice or nectar extraction and defect removal, as well as the need for peeling prior to extraction or the inclusion of de-aeration and other secondary operations. Relative to other goods, the profitability of most processed foods, including fruit and vegetable beverages, is based on large volumes with small profit margins. Also, consumers are increasingly informed and concerned about nutrition and health with the general perception that the closer to fresh, the better a product is. Therefore, it is critical to minimize the extent of processing and product losses in the system.

Contrary to common belief, juices are often pasteurized more than once, subjecting the product to thermal abuse. A typical example of this is depicted in Figure 15.2. For example, in the production of juice blends, juices and nectars are pasteurized after extraction and again after blending or ingredient addition and prior to packaging. In the case of orange juice, often the juice is stored for a long period of time between the first pasteurization and packaging. In such cases, a second pasteurization is also needed before packaging. In many cases, blending occurs at locations different from where juices or fruit pastes are produced. To minimize thermal abuse, two major strategies are employed: aseptic processing and non-thermal processing. Both represent an additional cost and are not always economically viable.

15.2 Juices

15.2.1 Citrus

15.2.1.1 Growing regions, cultivars, world production, major producers, and processors

Citrus cultivars commercially used for the production of beverages include sweet orange (*Citrus sinensis*), which is the most abundant, grapefruit (*Citrus paradisi*), mandarin (*Citrus reticulata*), lemon (*Citrus limon*), Persian lime (*Citrus latifolia*) and Key or Mexican lime (*Citrus aurantifolia*). The earliest reports of citriculture in China, where most likely citrus originated, date from 2200 BC (Zhaoling, 1986). In about 400–300 BC, after spreading throughout Asia, citrus was introduced to Europe. In the 1500s citrus was introduced to the Americas. Between the 1700s and late 1800s citrus spread to western US and Mexico.

The major citrus production is located within the tropical and subtropical regions of the world; that is, between latitudes 40°S and 40°N, also called the world citrus belt. However, in subtropical regions, between 20° and 40°N and S, defined seasons with cool nights and alternate rain and drought result in longer maturation periods, higher sugar accumulation, and better fruit quality. Figure 15.1 shows the consumption of orange, citrus, apple, and non-citrus juices in the US in the last three decades. In 2010, US consumers drank on average 13 L/yr of orange juice, accounting for 50% of the total juice consumption in that country (USDA-ERS, 2011).

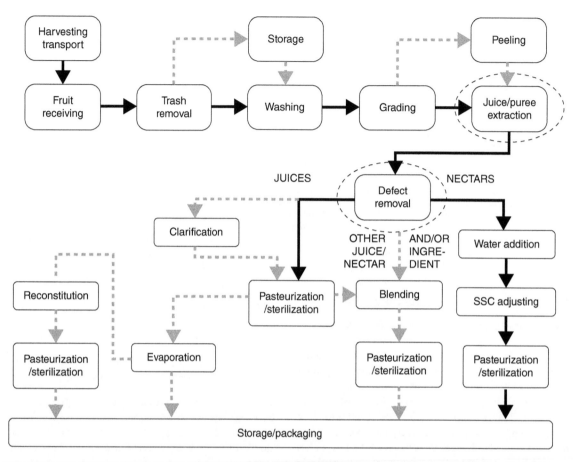

Figure 15.2 General flow process diagram for fruit juice and nectar processing. Solid lines represent operations that are followed most of the time. Dashed lines represent alternative operations often found in industry and that are specific only to the processing of certain fruit or vegetables. Dashed ovals indicate unit operations that differ the most in the processing of different types of fruit.

15.2.1.2 Harvest and handling for processing

Each citrus cultivar has a different harvesting period. Several orange varieties are used to ensure almost year round availability for processing. In the northern hemisphere, oranges are harvested between October and June, in the southern hemisphere between June and February. The most commonly utilized varieties of oranges for processing are Valencia, Hamlin, Pineapple, and Ambersweet. Hamlin oranges are harvested from mid-fall to mid-winter (October to January in the northern hemisphere (NH)), Pineapple and Ambersweet are harvested mid-winter to early spring (January to March in the NH) and Valencia, the most abundant variety, is harvested during the spring and early summer (March to June in the NH). There are two types of grapefruits, white and colored (pink or red).

The best known white varieties are Duncan and Marsh, and the colored varieties are Redblush and Star Ruby. Mature grapefruit can be harvested from mid-fall to mid-spring. Mature fruit can remain on the tree for that period of time. Because of its bitterness, the market for white grapefruit juice has been largely replaced by red grapefruit juice. Tangerines or mandarins are mostly used for the fresh market but in the US, up to 10% of tangerine juice can be added to orange juice without having to label it. This practice is mainly done to improve color of juice from early season varieties. In contrast to orange and grapefruit, where juice is the main product, lemon and lime juices are the by-products, while peel oil is the most valuable and main product.

Although mechanical harvesting machines are commercially used, in the US, most oranges or grapefruits are harvested by hand. The harvested fruit is dumped into trucks that carry approximately 20 MT to the processing plants. Fruit is unloaded by tilting the trucks on concrete or hydraulic ramps and is pregraded while it is conveyed into large vertical storage bins. Storage bins are designed with alternating baffles at an angle that prevent the fruit from bruising during bin loading and also to prevent fruit in the bottom of the bin from being crushed by the weight of the fruit on the top. Representative samples of each truck load are taken to assess fruit quality. Fruit from bins is washed, culls are removed often by hand and the fruit is sized before extraction.

15.2.1.3 Process description (Figure 15.3)

15.2.1.3.1 Juice extraction and finishing

There are two main juice extractor manufacturers, John Bean Technologies Corporation (JBT) and Brown International Corporation LLC. Detailed information

on these extractors is given in the websites of the two companies:

• www.jbtfoodtech.com/solutions/equipment/citrus-juice-extractor.aspx
• www.brown-intl.com/products/#Extraction.

Briefly, in Brown extractors, fruit is loaded into a carousel made of a set of hemispherical cups that hold the fruit while a blade cuts it into two pieces. The cups spread apart while retaining half of an orange each. Then, a spinning reamer presses the half-orange against the cup to extract the juice. Pressure and gap between reamer and cup are among the critical adjustment parameters in Brown extractors. Each extractor can handle 4–14 tonne/h of fruit.

John Bean Technologies extractors are typically configured with three, five or eight cups, with five cups being the most common. Cups are made of a hard cast stainless steel alloy and are formed by an array of rigid "fingers" that join at the bottom of the cup with an orifice at the center. Fruit is fed onto the lower cups and then pushed downwards with the upper cup. As the fruit is pushed, the peel

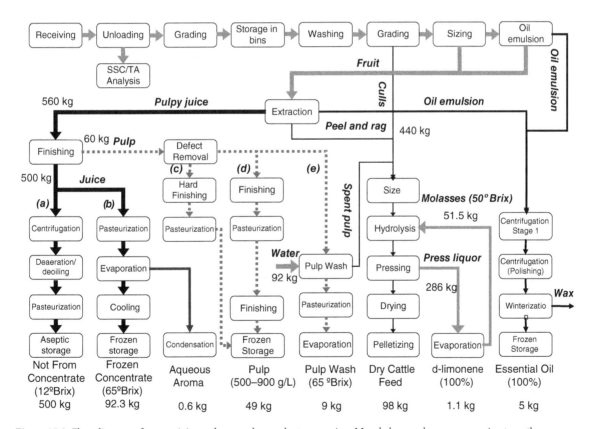

Figure 15.3 Flow diagram of orange juice and orange by-product processing. Mass balance values are approximate as they vary depending on processing conditions. Reproduced from Reyes-de-Corcuera et al. (2012), with permission from John Wiley & Sons, Inc.

detaches from the juice segments. The peel falls off and out of the extractor and the juice, with the fruit core and segment membranes, is then pushed inside the strainer tube. The juice flows through the strainer tube holes, acting as a prefinisher. The fruit core, segment membrane, and peel particles are ejected out the bottom of the tube. Synchronization of all moving parts is critical to effective and efficient operation of JBT extractors. Also, the size of the holes in the prefinisher is critical to the quality of the juice and the pulp recovered from this extractor. JBT extractors have a capacity of up to 600 fruit per minute or about 10 tonne/h (Reyes-de-Corcuera et al., 2012).

Extracted juice flows by gravity to screw or paddle finishers that decrease pulp content to about 20%. Juice fed into the finisher is conveyed and pushed by the screw or the paddle against a mesh drum. Pulp is retained inside the drum, conveyed to one end of the finisher, and recovered as a by-product (see Pulp Recovery section). The juice flows through the mesh and is then pumped to a pasteurizer or an evaporator. Mesh size and gap between the screw (or paddle) and the sieving screen determine the extent of separation. Tight finishing (i.e. smaller gap between the screw or the paddle and the sieve) increases yield and produces a drier pulp but reduces juice quality because pulp particles are extruded through the finisher mesh and bitter compounds formed in the pulp are released into the juice. Some processors use a cyclone between extractors and finishers to remove defects such as embryonic seeds. Defect removal is particularly important when pulp is intended to be sold as a by-product.

15.2.1.3.2 Essential oil

In JBT extractors, as the peel is separated from the rest of the orange, it is shredded and pressed and thus, the oil glands are burst. A manifold of water nozzles sprays the peel, forming an oil-in-water emulsion that drains through the back of the extractor. Brown oil extractors (BOEs) are placed before extractors. Whole oranges are conveyed into rollers with thousands of small, sharp blades that puncture the fruit peel. The puncturing rollers are immersed in a shallow vat with water, where the oil emulsion is formed. Oil emulsion is recovered and decanted to recover an oil-rich phase that is then centrifuged to recover 100% cold-pressed oil. Part of this cold-pressed oil is reincorporated to concentrate and another part is sold to flavor companies that, by physical means only, separate and concentrate flavor fractions that are then added back to juice before packaging. Addition of the so-called flavor packs produces more uniform quality

products. Because all the components of the flavor pack are from orange, after pack addition the final concentration of oil in juice is typically below 0.04% and only a small fraction of that amount is responsible for the aroma of the juice, so it is labeled 100% orange juice. This has recurrently produced controversy among some consumers, who perceive that addition of such oils is "artificial" and argue that the same concentrations of flavors would not be present in freshly squeezed orange juices. The rest of the cold-pressed oil is sold to other industries.

15.2.1.3.3 Pulp recovery

Pulp refers to burst juice vesicles. The market for orange pulp has increased in recent years. Some pulp is added back to juice to produce what is marketed as "home style" or "country style." A very large portion of pulp is sold to the Asian and European markets to provide mouthfeel and the perception of a fresher, more natural product to other fruit-based beverages. Figure 15.4 shows the schematic representation of a pulp recovery system. After separation of defects in a hydroclone, pulp is recovered at approximately 500 g/L. This measurement of pulp concentration is on wet basis and is based on an empirical test commonly used in the industry (Kimball, 1999). An instrument based on nuclear magnetic resonance (NMR) is also used, mostly by large processors. Pulp is then pasteurized and finished to increase pulp concentration to approximately 900 g/L, packaged, and frozen.

Some processing plants pasteurize 900 g/L or more concentrated pulp for aseptic filling. Pulp at such high concentrations behaves like a paste that displays wall slippage (Payne, 2011). At flow rates encountered in the industry, flow is laminar and heat transfer from tubular heat exchangers occurs mostly by conduction. Therefore, to maximize temperature uniformity, small pipe diameters and static mixers are used. The main issue of this approach is the enormous pressure drop and associated pumping costs. For that reason, some manufacturers prefer to produce frozen pulp. The size and integrity of pulp vesicles are of great importance to processors. Both parameters affect the ability of pulp to float and consumer sensory perception. Monitoring and controlling the extent of pulp vesicle degradation, i.e. tearing and size reduction, is critical as it affects quality of the end-product and the total amount of pulp that needs to be added back to the juice or to any beverage to produce the right balance of floating and sinking pulp. One must keep in mind that unit operations that involve high shear stress affect pulp quality.

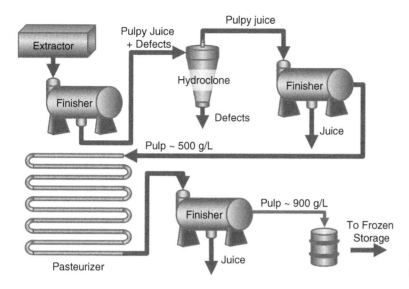

Figure 15.4 Schematic representation of typical pulp recovery system.

15.2.1.3.4 Frozen concentrate orange juice

Frozen concentrate orange juice (FCOJ) was developed in the mid-1940s to provide a good source of vitamin C to US allies after World War II. The main rationale behind concentration was to reduce the cost of transportation. By 1970, it represented 78% of the US orange juice market. Today the market share for FCOJ is only about 8%, due to strong marketing efforts for NFC and reconstituted orange juice. A detailed historical description of the evolution of the orange juice market has been published recently (Morris, 2010).

Orange juice is concentrated from 12 °Bx to 65 °Bx in thermally accelerated short-time evaporators (TASTEs). TASTEs are 5–7 effect falling film evaporators. The capacity of these evaporators ranges from 14,000 to 91,000 kg/h of evaporated water. At the last stages of evaporation (40–45 °Bx), juice is often homogenized to reduce viscosity and facilitate flow of product. Figure 15.5 shows the schematic representation of a five-effect, five-stage TASTE in mixed configuration. In this configuration, single-strength juice is fed to the third effect, which operates at a lower temperature than the first and second effects. Hence, aroma compounds that are recovered there experience less thermal degradation. Because during evaporation the juice is stripped from all aroma volatile compounds, essence oil and aqueous aroma must be added back to FCOJ. Approximately 0.01% cold-pressed oil is added to the concentrate and it is chilled to –9 °C. The headspace of FCOJ tanks is flushed with nitrogen to remove oxygen and to minimize oxidative reactions that affect vitamin C and flavor during storage. FCOJ tanks are kept cold in refrigerated rooms. Large processing plants have over 76,000 m³ storage capacities for 65 °Bx concentrate. Concentrate can also be stored in 200 L drums. Concentrate is marketed as reconstituted juice (12°Bx) packaged in half-gallon or 1 gallon containers, as 40°Bx, distributed to food service businesses, or as 40°Bx, 8 oz cans, to be sold at grocery stores and reconstituted by the consumer.

15.2.1.3.5 Pasteurization

Microbial inactivation

Most disease outbreaks in orange juice have been caused by *Salmonella* spp. (Danyluk et al., 2012). Several of these outbreaks were associated with unpasteurized orange juice. Therefore *Salmonella* is the recommended pathogen of interest for the purpose of pasteurization. *Escherichia coli* O157:H7 can grow in fruit juices with pH greater than 4. Therefore, although no *E. coli* O157:H7 outbreaks have been reported, in late season fruit and under poor sanitation conditions, this bacterium is a potential hazard. *Alicyclobacillus* spp. have been found in orange juice. This spoilage microorganism does not typically affect refrigerated juices because vegetative cells are killed during pasteurization and spores do not germinate at low temperatures. However, it can affect the quality of shelf-stable products and its presence impacts the ability to export. Lactic acid bacteria such as *Lactobacillus* and *Leuconostoc* have also been reported to affect orange juice quality (Rushing et al., 1956).

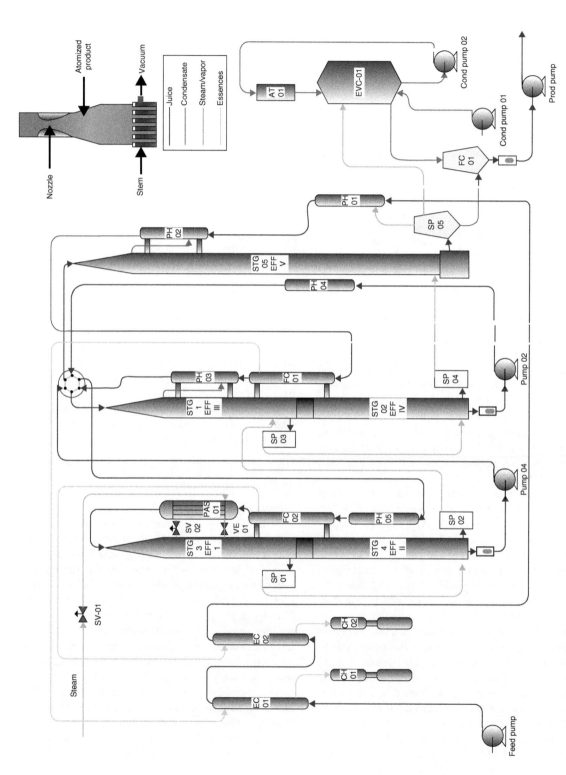

Figure 15.5 Schematic representation of a pilot-scale five-effect, five-stage TASTE. Top right corner, schematic of the cross-section of the top of stages 1 and 3. CH, chiller; EC, essence cooler; EFF, effect; FC, essence recovery; PH, preheater; SP, separator; STG, stage. For color details, please see color plate section.

Enzyme inactivation

Orange juice pasteurization aims not only to kill pathogenic and spoilage microorganisms, but also to inactivate pectinmethylesterase (PME). PME catalyzes the de-esterification of methoxy groups of pectin, which in turn, and especially in the presence of divalent cations, form soft gels that favor the precipitation of the juice cloud. Orange juice pasteurization conditions are adjusted to achieve PME inactivation because PME is more heat resistant than the pathogens of interest. The time-temperature profile for pasteurization is typically 85–90 °C for 10–15 sec. Pulpy orange juice is pasteurized in tubular pasteurizers. Plate heat exchangers are occasionally used when pulp content is low because pulp fouls conventional plate heat exchangers. Because PME is mostly bound to pulp particles, physical removal of pulp by finishing or centrifugation decreases the PME activity in the juice.

Of all non-thermal pasteurization and enzyme inactivation technologies that have been widely researched in orange juice, to the best of our knowledge, only high hydrostatic pressure is used commercially. Challenges associated with early designs that directly pressurized juices included keeping the product aseptic after depressurization, the cost of operation and maintenance, and safety concerns. New systems that pressurize juice filled and sealed bottles with water appear promising. A main challenge is the inactivation of endogenous enzymes that often require higher pressures and longer processing times than the pathogen of interest. Also, because of the batch nature of the process, this technology is only viable for small quantities of premium products.

15.2.1.4 Product quality

Flavor is arguably the most important quality aspect of any juice, but it is quite complex, as it depends on the physical properties of the juice such as viscosity and pulp presence, as well as on the composition of non-volatile and volatile compounds. Routinely measured quality parameters in orange juice are soluble solid content (SSC), SCC to titratable acidity (TA), expressed as percent citric acid or Bx/acid ratio (BAR), color, cloud, pulp content, and vitamin C content. The concentration of oil in juice is also determined by titration, as percent d-limonene. The maximal appropriate concentration is 0.04%. Above that concentration, most consumers experience a burning sensation on the lips and mouth. Of that 0.04%, only about 5% contains the aroma active volatiles. Compounds responsible for characteristic orange aroma include terpenes, aldehydes, ketones, alcohols, esters, and organic acids. Aldehydes level is used as an indicator of relative quality, with decanal and octanal as the most abundant. Detailed reviews of aroma compounds have been published elsewhere (Maarse & Visscher, 1989). Flavor may be negatively impacted by limonin, a bitter compound present in the seeds. Limonin is also formed after extraction in juice from Navel oranges. This juice needs to be debittered, typically by adsorption with resins.

Thermal pasteurization of orange juice affects mostly vitamin C, α-carotene, β-carotene, β-cryptoxanthin, and vitamin A. Other vitamins, as well as carbohydrates, lipids, amino acids and minerals, are not significantly affected by processing. The presence of dissolved oxygen also affects flavor, color, and vitamin C content. For that reason, prior to pasteurization, orange juice is often de-aerated. Single-strength and concentrate juices are stored in tanks with the overhead space filled with nitrogen gas to minimize the reincorporation of oxygen. For the same reason, orange juice is bottled leaving a minimal headspace volume. Some processors also displace headspace oxygen with nitrogen to maximize shelf life. Oxygen permeability of packaging materials is critical.

15.2.1.5 Sustainability

In the last 20 years, the orange juice industry in the US has experienced a strong consolidation. Most small companies were acquired by larger ones or closed in the face of competition based on economy of scale. Fourteen processors are members of the Florida Citrus Processors Association. To remain competitive, most plants have systems for by-product recovery, water recycling, and/or reduction of emissions of volatile organic compounds. By-products not mentioned above include dried peel, also called "dried pulp," which should not be confused with juice vesicles. Pulp is shredded, hydrolyzed, pressed, and dried in rotary drums. Press liquor that is recovered after peel pressing can be concentrated using the waste heat from the drier and a multiple effect waste heat evaporator (WHE). A fraction of the molasses is recirculated to the pressed peel to increase the percent dry matter and improve the efficiency of the drying operation. Citrus peel molasses are also fermented to produce food-grade alcohol. D-limonene from the peel is recovered as well. It finds markets as a solvent in the electronic component manufacture, for paints and other household products. The dried peel is pelletized and sold as cattle feed. Peel driers are operated by burning fuel by direct or indirect exposure to the combustion gas. The economic viability of this operation is directly related to the oil prices and alternative cattle feed prices, which is in many cases tied to the supply of corn. A very small portion of orange peel is also used for marmalades or dried spice.

From an environmental perspective, citrus processing plants have proactively integrated the reuse of water to minimize waste water. Modern plants operate steam with almost full return of condensate to the boiler. Similarly, cooling towers are operated at high efficiency. Condensate water from the FCOJ evaporator is used to wash incoming fruit or other parts of the plant. Waste water is treated to decrease the biological oxygen demand to a level where it can be used for agricultural field irrigation. The WHE and the cold storage of very large volumes or NFC or FCOJ account for most of the energy consumption in a citrus processing plant. Some processing facilities have co-generation plants that reduce electrical consumption from the grid. Citrus by-products have been comprehensively discussed elsewhere (Braddock, 1999).

15.2.2 Apple

15.2.2.1 Growing regions, world production, and major producers

Apple is the second most consumed fruit juice in the world, with total export value of single-strength and concentrated juice of US$3936 million in 2009 (FAO, 2010). The apple species, *Malus pumila*, originated in southwestern Asia and Europe and was brought to the Americas by European colonists.

The major apple production is located within the temperate regions of the world; that is, between latitudes 35° and 50°N, and between latitudes 30° and 45°S. China, the US, Turkey, Poland, and Iran are the largest apple producers, accounting for over 70% of the world production in 2009. In the US, apple acreage is a little over one half that of orange, with Washington, Michigan, New York, and California as the major producers. In 2010, of the utilized US production, 14.5% was processed into juice . There are hundreds of apple cultivars but some of the most common include Red Delicious, Golden Delicious, McIntosh, Rome, Beauty, Granny Smith, Fuji, and Braeburn. In China, the most popular varieties are Fuji (45%), New Red Star (12%), Qinguan (10%), and Guoguang (10%) (O'Rourke, 2003). Dynamic development of new cultivars and changing preferences are occurring around the world. In 2011, 394,400 MT of apples were processed into juice (AMS, 2011). In 2010 Americans consumed 8 L/yr of apple juice, accounting for 31% of the total juice consumption.

15.2.2.2 Harvest and handling for processing

Apples are harvested by hand and the large majority of apples are harvested for the fresh market. Mostly culls, in particular small apples or apples not suitable for other processed products such as slices or sauces, are used for juice processing. However, with the increased demand for apple juice, some apples are being harvested for juice processing. It is crucial to maintain fruit wholesomeness and ensure fruit maturity prior to processing to avoid contamination. In particular, postharvest spoilage microorganisms may contaminate the fruit during storage. Of particular concern are *Alicyclobacillus* spp., a group of acid-tolerant, spore-forming bacteria that resist thermal processing and whose spores germinate during storage of shelf-stable products. Although not posing a health threat, *Alicyclobacillus* spp. produce guaiacol, which has an unpleasant medicinal off-flavor.

15.2.2.3 Process description

Several books and book chapters on apple juice processing have been published in the past decades (Binnig & Possmann, 1993; Downing, 1989; Lea, 1995; Root & Barrett, 2005). Figure 15.6 shows a flow diagram of apple juice and apple by-product processing, with approximate mass balance.

15.2.2.3.1 Washing and sorting

Apples are typically dumped into tanks where they soak to loosen soil to facilitate downstream removal with rotating brushes. Washing is critical to remove molds and bacteria that may spoil the juice. Patulin, a mycotoxin, is of particular concern. Hence, fruit with signs of spoilages are either removed or cut to remove spoiled portions.

15.2.2.3.2 Maceration

Whole apples are then macerated using a hammer mill or a disintegrator to produce a slurry, called "mash" or "pulp." The particle size and consistency of the mash depend on fruit variety and maturity and affect juice extraction yield. Juice is extracted by pressing the mash with a hydraulic, pneumatic, screw, or basket press. Some presses require press aids that minimize slippage and increase yield. Enzyme treatment with cocktails of PME, polygalacturonase and pectinlyase and cellulase are used to hydrolyze pectin and fruit cell wall, facilitating juice release during pressing and increasing yield. The catalytic activity and mechanism of action of these enzymes have been studied extensively. Enzyme treatment requires heating because it is most effective at around 50 °C and requires some reaction time. However, press throughput

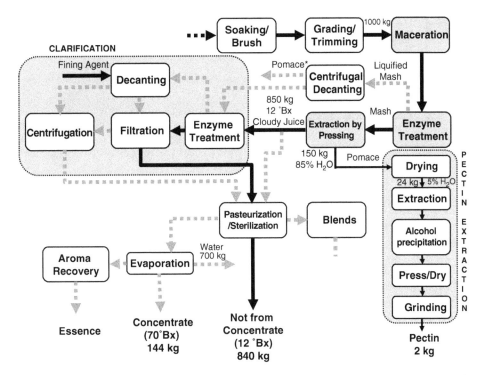

Figure 15.6 Flow diagram of apple juice and apple by-product processing. Mass balance values are approximate as they vary depending on processing conditions.

can be increased by up to 40% and juice yield increases by about 20% (Root & Barrett, 2005). Small amounts of enzyme are required (~100 mL/t), making this approach cost-effective. At 50 °C some of the enzymes in the cocktail denature and lose activity. However, treatment at lower temperatures (~40 °C) favors microbial growth, and even lower temperatures result in decreased enzyme activity. Therefore, enzyme cocktail formulations require a careful balance of kinetic activity and stability. A complex enzyme cocktail with increased cellulase, hemicellulase, oligomerase and other enzyme activities is used for mash liquefaction. Addition of cellulases can increase SSC by almost 5%. Research and mining for heat-stable enzymes or for enzymes that are highly active at low temperature is under way. Enzyme catalysis at high pressure and temperature has also been proposed.

Another important consideration is the downstream use of the product and by-product. If pomace (solid residue after juice extraction) is to be used for pectin production, excessive pectin hydrolysis produces short galacturonic acid oligomers, losing the desired functional properties of the polymer. Residual activity also prevents gel formation when concentrate is used to produce jellies.

15.2.2.3.3 Juice extraction

Apple juice is extracted by pressing the mash in a belt, hydraulic, pneumatic, other type of press or a combination of presses. Typically, hydraulic presses are more efficient (3–5%) than their pneumatic counterparts. For liquefied mash, a decanting centrifuge is used. Juice yield is affected by pressing conditions (temperature, pressure, presence, and type of pressing aid) and by the quality of the mash, which in turn depends on the quality of the apples, particle size after maceration and extent of enzyme treatment. Typical juice yields range between 70% and 95%. Cellulose or rice hulls are often added (1–2%) as press aids that help to uniformly distribute pressure across apple particles that constitute the pulp and mash. Press aids can increase extraction yields by up to 10%.

15.2.2.3.4 Clarification

Most apple juice is clarified, although the market for cloudy, "country style" apple juice appears to be growing with the demand for more fresh-like, minimally processed food products. Most processors currently use pectinase

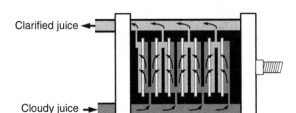

Figure 15.7 Schematic representation of the cross-section of a press filter. For color details, please see color plate section.

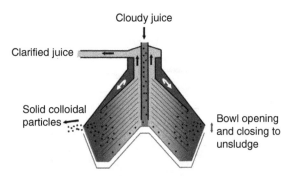

Figure 15.8 Schematic representation of a continuous centrifuge for juice clarification. For color details, please see color plate section.

cocktails to clarify apple juice. Cleavage of pectin decreases juice viscosity, which facilitates filtration. Similar to maceration, a pectinase formulation is used for clarification and the conditions are also either 10–20 °C for 8–10 h or 45–55 °C for 1–2 h. The amount of enzyme formulation is adjusted by each processor but is in the range of 20–30 mL/m^3 for liquid formulations. Although in many processes enzymes are a large portion of the processing costs, pectinases are relatively inexpensive. Unlike with citrus, addition of enzymes to apple and other clarified juices does not violate the standards of identity. Fining, that is, further removal of suspended colloidal particles by decanting, is facilitated by addition of positively charged gelatin or bentonite during or after enzyme clarification. The decanted juice is further clarified by filtration and/or centrifugation. There are different types of filters, including press (Figure 15.7), vacuum, and rotary filters. Filter aids commonly used are diatomaceous or cellulose. Continuous disk centrifuges (Figure 15.8) are typically used.

Because apples of many different varieties are used for juice production, blending is commonly done to produce uniform quality product.

15.2.2.3.5 Concentration

Apple juice concentration is done mostly by evaporation, although freeze concentration can be done. Apple juice is typically concentrated to 70°Bx in 4–5 effects. High-temperature, short-time (HTST) falling film evaporators that operate under high vacuum at 90–100 °C are commonly used. Like orange, apple juice evaporators are operated in mixed configuration. Single-strength juice is fed to the second or third stage to separate volatiles at a lower temperature than if fed at the first stage. Aromas are recovered and concentrated by distillation. High temperatures are used to pasteurize the juice during evaporation to ensure sufficient microbial kill and avoid

concentrate spoilage. Plate evaporators and rising film evaporators have also been used for apple juice concentration.

15.2.2.4 Relevant processing conditions

15.2.2.4.1 Pasteurization

Microbial inactivation

Yeasts, molds, and bacteria are commonly found in apple juice. For the purpose of pasteurization, the US FDA recommends that *E. coli* O157:H7 and *Cryptosporidium parvum*, whichever is the most tolerant microorganism for any given pasteurization treatment, be considered as the pertinent microorganism for apple juice. As in citrus juices, *Alicyclobacillus* spp., though not pathogenic, is of concern to the apple juice industry. This is particularly important because of the large production of shelf-stable apple juice. At room temperature, *Alicyclobacillus* spores that are not destroyed during thermal treatment germinate and produce a strong medicinal off-flavor that is characteristic of guaiacol. Combined thermal and pressure treatments were reported to synergistically kill *Alicyclobacillus* spores in apple juice concentrate (Lee et al., 2006). Yeasts and molds can spoil juice by fermenting it but the main concern in apple juice is patulin, a mycotoxin produced by several species of mold. These molds are present in bruised apples and cannot be removed during washing. FDA guidance establishes a maximum level of 50 μg/L in foods. However, in 1993 the FDA found that one in five samples of apple juice contained patulin concentrations above 50 μg/L.

Pasteurization also inactivates polyphenol oxidase (PPO), the enzyme responsible for the browning (oxidation) of apple juice. Clarified apple juice is typically pasteurized at 95 °C for 10–30 sec or at 85 °C for 15–120 sec (Lea, 1990).

15.2.2.5 Product quality

The composition of apple phenolics depends largely on variety, maturity, storage, and extraction conditions. Concentrations of phenolics change drastically during processing, mainly by the action of PPO. This enzyme is mostly bound to the cell membrane. When juice is allowed to oxidize with pulp, total polyphenols decrease in the clarified juice due to adsorption of colored compounds onto pulp particles. Total polyphenol content also decreases. In contrast, when clarified juice is allowed to oxidize, discoloration increases until PPO is inactivated by the products of phenolic oxidation. Addition of ascorbic acid or SO_2 to inhibit the enzyme and reduce oxidation products can be done to control discoloration (Lea, 1990).

The overall aroma compound concentration of apples is estimated at around 200 ppm. Early characterization of aroma active compounds identified 56 volatile compounds, out of which only three (1-hexanal, trans-2-hexanal, and 2-methyl butyrate) had "apple-like" characteristics (Flath et al., 1967). Hexanal and trans-2-hexanal are formed after disruption of the cell structure during processing. A more recent study of apples of selected cultivars at selected levels of maturity reported 36 aroma active compounds, of which 24 were common to all the extracts analyzed (Mehinagic et al., 2006). Butyl acetate, 2-methylbutyl acetate, hexyl acetate, and hexyl hexanoate have been associated with the overall aroma of fresh (Young et al., 1996) or stored apples (Lopez et al., 2000; Plotto et al., 1999).

Apple juice quality is assessed by SSC and TA, but in contrast to orange juice, where TA is reported as citric acid (triprotic), for apple juice TA is reported as malic acid (biprotic). Also in contrast to citrus juices, most apple juice is clarified. Therefore, turbidity is assessed. Because pectin contributes to the stability of the cloud, the so-called "alcohol precipitation" method is used to determine residual pectin content. Acidified ethanol is added to the juice after enzyme treatment to precipitate water-soluble pectin. Adequate clarification results in little to no precipitation.

Like any other juice, apple flavor is affected by pasteurization temperature and time. However, thermal processing has a less pronounced effect on the acceptability of apple juice compared to orange. Therefore, shelf-stable apple juice is the most common. Non-thermal processing of apple juice has been widely researched. Higher retention of volatile aroma compounds has been reported for non-thermal technologies (Aguilar-Rosas et al., 2007).

A major concern in the international marketing of juice concentrates is the content of pesticides that are not volatile and therefore appear in concentrations above the limits approved by some countries or are not allowed at all in certain countries.

15.2.2.6 Major processors and markets

The largest exporters of apple juice concentrate are Poland, Austria, Hungary, US, and Ukraine. Poland accounts for 28.1%, Austria for 10.5% and the rest, under 10%. The largest exporters of single-strength apple juice are China, Germany, Italy, Poland, and Austria. China accounts for 51.9%, and all others under 10% of the world exports (FAO, 2010). The largest importers of apple juice concentrate are US, UK, Russian Federation, Japan, and Germany, The US accounts for 36.5% of imports. The largest importers of single-strength apple juice are Germany, Netherlands, France, Austria, and US, with Germany accounting for 33.2% of the imports (FAO, 2010).

15.2.3 Tomato

15.2.3.1 Growing regions, world production, and major producers

Tomatoes are grown worldwide for the fresh market, the processing market, and increasingly the fresh-cut market. The fresh-cut market is usually included in fresh tomato statistics. Tomato (*Lycopersicon esculentum*) is the second most important vegetable crop in the world, followed by potatoes (FAO, 2010). In 2003, the US production of fresh-market tomatoes was over 1.6 million tonnes, with Florida and California responsible for 43% and 28%, respectively, of that amount (Sargent et al., 2005). In comparison, the 2003 US figure for processing tomato production was 9.8 million tonnes, with that figure rising to 14.0 million tonnes in 2009 (ERS, 2010). California produces about 95% of all US tomatoes used for processing into products such as sauce, paste, canned products, and juice. Worldwide production of tomato totaled 158,368,530 MT in 2009, an increase of 3.7% from the previous year. The top tomato producer in 2009 was China, which accounted for about 25% of the world production, followed by the US, Turkey, India, Egypt, and Italy (NASS, 2010).

15.2.3.2 Harvest and handling for processing

Tomatoes for processing can be mechanically harvested, a process that has been utilized since the mid-1960s and that is widely recognized as one of the reasons for the success of the California processed tomato industry. The tomatoes are transported to the processing plant as quickly as possible, and unloaded as soon as practical. Fruit quality can deteriorate rapidly even if the trucks are shaded. Tomatoes are off-loaded onto belts or conveyors, or sometimes into water-filled flumes. The major handling step at this part of the process is grading. In the US, voluntary quality standards may be adopted or grading may follow that outlined in the marketing order pertaining to processing tomatoes that are destined for paste and juice (the Processing Tomato Advisory Board). In any grading schedule, the purpose is to establish quality standards for processing tomatoes and to conduct a grading program to assure the orderly marketing of uniform quality processing tomatoes. Tomatoes are graded primarily on the basis of color and defects (worms, weather damage, mechanical harvesting damage, mold, and decay). For the production of juice and paste products, the SSC is important and is considered in the grading process.

15.2.3.3 Process description

Figure 15.9 outlines the major processing steps for processed tomato juice and paste (concentrate). After grading, tomatoes are washed to remove dirt, insects, mold, and other possible contaminants. Proper washing results in products that are of better quality (microbial and hygiene) and can be enhanced by several practices: agitation of the fruit to physically remove the soil, warm water spray or dip (although water is more often used to cool the product), application of a surfactant to help remove soil, and

transport through a water flume. The water flume is used to minimize damage to the fruit during handling and also to separate the physical debris that might have been picked up during the mechanical harvesting process. A final rinse is generally accomplished through a spray nozzle system.

15.2.3.3.1 Sorting

Optical color sorting is used to remove green and pink tomatoes from the fruit stream. Color is an important attribute of tomato juice (see discussion below) and while green tomatoes do not adversely affect the color of juice in small numbers, pink tomatoes significantly affect the red color in even small numbers. For juice or paste, tomato size is unimportant. A final sorting may take place where human sorters are utilized. Any unacceptable tomatoes and debris are removed and discarded at this step.

15.2.3.3.2 Break

A relatively unique operation in the production of tomato juice and paste is the so-called break process. Tomatoes can be processed into juice by either a hot break or cold break method, although the hot break is most common for juices that are further processed into concentrate. Briefly, the hot break method involves chopping the tomatoes and heating the resultant mixture to at least 85°C to inactivate pectolytic enzymes that would, if not inactivated, result in the loss of desirable high viscosity in the tomato product. As most tomato juice is converted

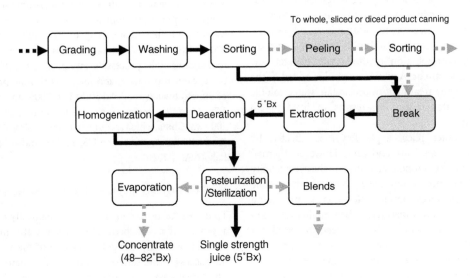

Figure 15.9 Flow diagram of tomato juice production.

into tomato paste, the retention of viscosity is a desirable quality attribute. Most hot break processes are conducted at 92–99 °C for fast inactivation of enzymes.

The cold break process is favored by tomato juice processors who do not necessarily require a high viscosity product or who might be using or selling the juice as an ingredient in another beverage. In this case, the tomatoes are chopped and mildly heated to temperatures around 60–66 °C, the optimum range for enzymatic activity. The resultant tissue breakdown leads to higher yields, as well as slightly lower viscosity of the resulting juice. In cold break PME activity resulted in the fast formation of methanol and increased TA (Anthon & Barrett, 2012).

15.2.3.3.3 Juice extraction

After the break process (hot or cold), the seeds and skins are removed through a finishing or pulping step that also serves as the juice extraction method. Either paddle or screw-type finishers can be used, although screw finishers generally result in better retention of bioactive compounds such as lycopene and ascorbic acid due to less inherent aeration compared to paddle finishers. Juice is de-aerated to remove oxygen incorporated during extraction that would otherwise rapidly oxidize bioactive components. Oxidation is exacerbated by the high temperatures resulting from hot break. Vacuum de-aeration is the type most commonly employed for the removal of entrained and dissolved oxygen, as with many other juice and beverage products.

15.2.3.3.4 Homogenization

Tomato juice can be homogenized to prevent separation of solids from the serum, and this can also cause a slight increase in viscosity. Homogenization is accomplished by forcing the juice through small orifices at high pressures; instead of the fat globule reduction that occurs in dairy homogenization, a shearing of pulp and solids particles takes place, thus leading to increased stability of the juice. High-shear in-line mixers have replaced some homogenizers in the juice industry because of their lower capital and operation costs.

15.2.3.3.5 Packaging

Prior to filling into a consumer package, salt is usually added as an ingredient, along with ascorbic acid, which is added to achieve 120% of the Referenced Daily Intake

(CFR 21 104.20) in a 240 mL (8 oz) serving. The exact level of these ingredients depends on the consumer market for which the product is intended. As neither ingredient contributes to the soluble solids of the juice, the juice can still be labeled as 100% juice, which is defined as all of the soluble solids (mostly sugars) being derived from the fruit/vegetable and the product meets the minimum SSC level as defined by the FDA. Tomato juice is commonly packaged in cans that are hot filled and heat treated, bottles that are hot filled, or aseptically packaged in barrier-layer packages; these products are all considered shelf stable with a 1–2-year shelf life. Alternatively, the juice can be pasteurized and filled into plastic (usually PET) bottles and sold as a refrigerated product with a somewhat more limited shelf life of 1–2 months.

15.2.3.3.6 Concentration

If the final product is not juice, the juice is next concentrated to paste. Concentration occurs in forced circulation, multiple-effect, vacuum evaporators. Typically, three- or four-effect evaporators are used, and most modern equipment now uses four effects. The temperature is raised as the juice goes to each successive effect. A typical range is 48–82 °C. Vapor is collected from later effects and used to heat the product in previous effects, thus conserving energy. The reduced pressure (~350–700 mmHg) lowers the temperature, minimizing color and flavor loss. The paste is concentrated to a final solids content of at least 24% natural tomato soluble solids (NTSS) to meet the USDA definition of paste.

Commercial paste is available in a range of solids contents, finishes, and Bostwick consistencies. The larger the screen size used for extraction, the coarser the particles and the larger the finish. Bostwick measurements may range from 2.5 to 8 cm (tested at 12% NTSS). The paste is heated in a tube-in-tube or scraped surface heat exchanger, held for a few minutes to pasteurize the product, then cooled and filled into sterile containers, in an aseptic filler. A typical process might heat to 109 °C, then hold for 2.25 min or heat to 96 °C and hold for 3 min. Aseptically processed products must be cooled before filling, both to maintain high quality and because many aseptic packages will not withstand temperatures above 38 °C. An aseptic bag-in-drum or bag-in-crate filler is used to fill the paste into bags previously steam sterilized. Paste is typically sold in 55 gallon drums or 300 gallon bag-in-box containers.

15.2.3.4 Relevant processing conditions

Tomato juice quality is highly dependent on utilizing the proper time-temperature regimes throughout the entire production process. These processes affect the microbial stability of the product, as well as the enzymes inherent in the tomato fruit. These processes in turn have a profound effect on product quality, including nutritional and sensory quality.

15.2.3.4.1 Microbial inactivation

Canned tomato juice is a traditional product, and canning protocols have been well known since the 1920s. Individual tomatoes range in pH from 4.05 to 4.65, although in general, the pH of tomato juice without the addition of some sort of acid is in the range of 4.1–4.35, which is quite close to that of a non-acid food product. In some cases small amounts of citric or ascorbic acid are added to minimize the risk of spore-forming microorganisms causing quality or safety issues with juice. A typical heating regime for a non-acidified tomato juice is heating the juice to 121 °C for 45 sec, cooling to 93 °C in order to fill the package, then sealing the package and agitating, maintaining that temperature for 3 min, in order to achieve overall commercial sterility. This process is designed to destroy the vegetative cells and the spores of *Bacillus* spp., which are the most common spoilage organisms of tomato juice. Specifically, *B. coagulans* is responsible for the common type of tomato and tomato juice spoilage termed "flat-sour" which is not evident by package swelling or obvious spoilage, but has an uncharacteristic acidity due to the production of lactic acid.

Tomato juice packaged in plastic bottles and sold in a refrigerated case undergoes a less rigorous prepackaging pasteurization regime, as the enzymatic treatment has already occurred during the break process and the product has a limited shelf life. Spore-forming microorganisms are generally not an issue with refrigerated products over their limited shelf life as long as proper refrigeration is maintained.

15.2.3.4.2 Enzyme inactivation

Pectolytic enzymes are the focus of the previously discussed break process. The two major enzymes are polygalacturonase (PG) and pectinmethylesterase (PME). Enzyme activity of PG and PME is slowed or stopped through the high temperatures that occur in the hot break process as described above, and enhanced under the cold break process as the pectolytic enzyme activity is at its maximum at the cold break temperatures of 60–66 °C. Although both processes have been utilized for several decades, researchers continue to elucidate the exact mechanism of pectin breakdown and subsequent physical parameters as a result of these processes (Chong et al., 2009; Lin et al., 2005).

15.2.3.5 Product quality

Tomato juice quality can be assessed through voluntary grading programs, for example, US Grade A Fancy, but this discussion will focus on the accepted technical measures of tomato juice quality. The soluble solid content of tomato juice, if reconstituted, must reach the minimum of 5.0 °Bx upon reconstitution and packaging of the juice (CFR 21-101.30). Additionally color, viscosity, flavor, and defect level are factors that are included in the quality assessment.

The nutritional quality of tomato juice is thought to be due to high levels of the antioxidant lycopene, which is also responsible in large part for the red color of the tomato juice. Processing and storage techniques have developed with the purpose of maintaining as high levels of lycopene as possible. The ascorbic acid content, whether endogenous or added, also contributes to antioxidant activity of tomato juice (Jacob et al., 2008), which enjoys a relatively healthy image, especially in low-sodium forms.

15.2.3.6 Major processors, markets, and sustainability

A global trade organization for tomato processing supports the industry worldwide (World Processing Tomato Council; www.wptc.to/) with the California processing industry the major representative in the US. Major tomato processers worldwide include Hunts and Contadina, and increasing numbers of processors in China.

Waste water disposal is a major issue in the tomato processing industry, due to both the amounts of water used in fluming operations and the strict environmental laws in the major US tomato processing area.

15.2.4 Carrot

15.2.4.1 Growing regions, world production, and major producers

Carrots, a member of the parsley family (Umbelliferae), are thought to have originated in central or western Asia but are grown worldwide. Carrot cultivation in the US was first practiced several centuries ago by early settlers in the Virginia region (AGMRC, 2011); by 2004 the US was the third largest carrot-producing country. Russia produced slightly more carrots but the production from both countries was dwarfed by that of China during the 2004–2009 period. Total US carrot production for the fresh and processing markets was valued at more than $627 million in 2010 (NASS, 2010). The US carrot market is overwhelmingly fresh, representing 95% of the value of the crop in 2010. California, Michigan, and Texas were the top fresh carrot-producing states, with California representing the majority of fresh market production (84%). Processed carrot products, including canned, frozen, dehydrated and juiced, represented just 5% of the total value of the carrot market in 2010; Washington, Wisconsin, Minnesota, and California were the leading producers of carrots for the processed market (AGMRC, 2011). Carrot juice has traditionally been a home-processed product and is sold in many food service outlets and juice bars, particularly those targeting the "health and wellness" consumer (Sloan, 2012). While there are several large, global carrot juice producers and packagers, carrot juice is still a relatively small portion of the overall processed carrot category; except for proprietary sales figures there is little data available for US or global processed carrot juice consumption.

15.2.4.2 Harvest and handling for processing

Carrot production in the US is highly mechanized and carrots grown for processing are usually selected for yield, processing utility, and maturity at a consistent time. The Chantenay variety is most often used for processing, due to its larger relative size and high soluble solids. Some fresh varieties are processed as culls from fresh or fresh-cut operations. Carrots destined for both fresh and processed markets are mechanically harvested (AGMRC, 2011), usually from August to late fall in the western US.

15.2.4.3 Process description

Carrots are delivered in bulk to the processing facilities, either directly from field harvest or as culls from fresh pack or fresh-cut operations. Figure 15.10 represents a simplified summary of typical carrot juice processing operations. Carrots, topped by the removal of excess leafy greens, are rinsed to remove excess dirt, washed with cold water and sent to the comminuter unpeeled. The carrots are run through a size-reduction device, often a hammer mill, to achieve a finely ground start. The mash may be

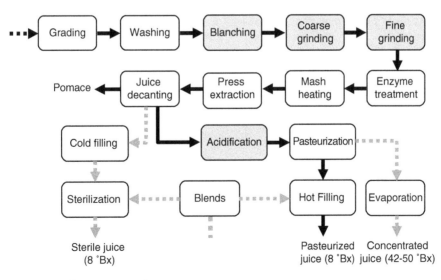

Figure 15.10 Flow diagram of carrot juice production.

heat-treated to inactivate native enzymes (see Enzyme Inactivation section) or treated with enzymes to enhance yield, then the carrots are pressed to extract the juice from the mash. The type of press most often used is a continuous belt press, but a bladder press is occasionally used in small batch operations. After pressing, the juice may undergo centrifugation to remove small particles of pulp, or the juice may be decanted in a vessel to achieve removal of the larger particles. Carrot juice can also be concentrated to 42–50°Bx, sent to a pasteurizer or sterilization unit for bulk storage or filling into consumer packages, or canned and retorted under sufficient conditions to assure shelf stability.

15.2.4.4 Relevant processing conditions

15.2.4.4.1 Microbial inactivation

Carrot juice is classified as a low-acid product and has traditionally been processed and filled into cans and bottles in a manner that will prevent the growth of *Clostridium botulium*, a spore-forming bacterium that can produce a highly potent toxin at pH greater than 4.6, anaerobic environment, and non-refrigerated temperatures. Specific processes have been developed to destroy the spores of this organism. Chilled carrot juice is not required to undergo this rigorous thermal process, but due to a recent outbreak of temperature-abused chilled carrot juice, there are still safety issues associated with the refrigerated product. Processors use a variety of hurdle processes such as acidification, pasteurization, and carrot surface treatment to ensure the safety of this product and warn consumers with adequate labeling.

15.2.4.4.2 Enzyme inactivation

Carrots processed as frozen, canned or dehydrated products are blanched at approximately 90 °C for 3–5 min to inactivate enzymes that might otherwise result in degradation of the preserved product. The enzyme of most interest in carrot juice is PME which, as in the case of orange juice, causes precipitation of the cloud (Reiter et al., 2003). This is in contrast to tomato juice, where the pectolytic enzyme inactivation is primarily used to manipulate juice viscosity.

15.2.4.5 Product quality

The standard SSC of carrot juice, as designated by the FDA, is 8 °Bx. This measure of soluble solids, comprising almost all sugars in the case of carrot juice, represents the minimum level to which a carrot juice product could be diluted with water. Carrot juice processed as NFC may have slightly higher or lower SSC, depending on the maturity and type of carrot. There are no federal grade standards for carrot juice due to its relatively small market, but consumer testing has determined that color, fresh carrot taste, sweetness, and low bitterness are important attributes in producing a desirable product.

15.2.4.6 Major processors, markets, and sustainability

Two major carrot operations in California dominate the production of carrot juice: Bolthouse Farms and Grimmway Farms. As both plants operate in California, there are substantial environmental issues that impact the production of carrot juice. These include the ever-present tension of water use for agriculture in areas of high urban pressures, as well as the need to thoroughly treat and minimize any waste water produced in the plant.

Organic carrot production has increased in recent years, and organic carrot juice can be found in some smaller outlets. The sustainability of organic food production and products is perceived by some consumers as being superior to that of conventional products.

15.3 Nectars

As described in the Classification and Regulations section, nectars are prepared by addition of water, sugars, and other ingredients to a fruit paste, concentrate, or purée. Hence, essentially, nectars can be made from any fruit, including the ones that are typically sold as juices. Even though in some countries, nectars produced by diluting juices are popular because of their lower cost and local preference for such beverages, most of the nectars sold as such in the market come from fruit from which a juice could not be produced because of the high viscosity and low SSC relative to single-strength juices. Some of the most common nectars found in the market are apricot, guava, mango, peach, pineapple, and strawberry. Mango, peach, and other stone fruit are processed in a very similar way. Therefore, in this section we only describe mango processing.

15.3.1 Mango

15.3.1.1 Growing regions, world production, and major producers

Mango (*Mangifera indica*) is a tropical climacteric fruit. It can grow at altitudes from sea level to 1200 m and in frost-free subtropical regions. India is the largest producer of mangoes with 13.5 million MT, followed by China, Thailand, Indonesia, Pakistan, and Mexico, all producing more than 1.5 million MT. The world total production was almost 33 million MT in 2009. There are hundreds of varieties. The most important varieties in India include Alphonso Toatpuri, Kesar, and in Mexico and Brazil, Tommy Atkins and Palmer. In 2009, world exports of mango juice were 90,582 MT, with a value of $49.0 million (FAO, 2010).

15.3.1.2 Harvest and handling for processing

Mangoes are harvested during the summer, unripe but mature, 75–135 days after blooming. In regions closer to the equator, harvesting periods are more extended. Mangoes are typically harvested by hand, avoiding caustic sap getting in contact with the fruit peel, as it can produce undesirable dark spots and localized softening. This is more of a problem for fruit directed to the fresh market

than for processing. Fruit is then ripened in controlled temperature rooms. Ripening can be achieved in two days (Wu et al., 1993).

15.3.1.3 Process description

Mango purée is best produced from varieties that are not fibrous and produce a smooth purée. Because of the large diversity of cultivars with different quality attributes including color and flavor, mango purées are often blended. Depending on the cultivar, peel represents between 6.5% and 24.4% of the fresh fruit weight and the flesh accounts for 32–85%. Detailed information for each cultivar was reported by Wu et al. (1993). Often, purées are produced by processors different from those who produce nectars. Nectars are then produced by major brands or co-packed by large juice processing plants. A flow process diagram of mango nectar production is shown in Figure 15.11.

15.3.1.3.1 Washing and sorting

Mangoes are dumped into a conveyor, where they are rinsed with a manifold of water spray nozzles and culls are removed by hand. The fruit is then brush-washed and rinsed on the conveyor.

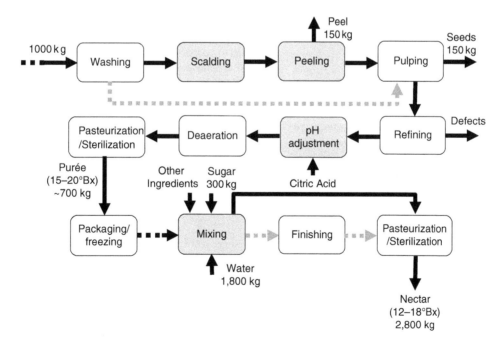

Figure 15.11 Flow diagram of mango nectar production.

15.3.1.3.2 Peeling

Varieties with thin skin and low content of polyphenols may be processed without peeling but other varieties require peeling to avoid introduction of off-flavors. Where labor costs allow, mangoes are peeled raw by hand using knives. Mangoes can be scalded (steamed) for 2–3 min, cooled down in water, and the peel is removed by hand. This scalding step not only facilitates peel detachment but achieves microbial reduction. Lye peeling is done for thin-skinned mangoes. The fruit is scored (i.e. an incision is made) with stainless steel brushes and immersed in hot sodium hydroxide solution (~20%) with a surfactant. Then, the peel is removed by water washing and abrasion on a rotary rod.

15.3.1.3.3 Pulping

Mangoes are pulped in a paddle pulper or destoner where the seeds are separated. Disintegrated flesh is separated from peel residues and fibrous material and other defects in centrifugal separators or finishers. This process is also called refining. Because of differences in cultivars and levels of maturity, mango pulp pH varies from batch to batch and needs to be lowered to ensure microbial stability. To minimize browning and other oxidative reactions, mango purée is often de-aerated prior to pasteurization.

Some processes include enzymatic liquefaction to reduce viscosity and increase yield but require additional thermal inactivation of the added enzymes.

15.3.1.4 Relevant processing conditions

15.3.1.4.1 Microbial inactivation

Pulp is then pasteurized at 90 °C for 1 min and stored frozen. Because of its high viscosity, pasteurization is typically done in scraped surface heat exchangers. Aseptic pulp systems are also available. During the production of nectar, after water, sugar and ingredient addition, nectar is again pasteurized at 95 °C for 1 min and packaged aseptically in plastic-lined carton containers for retail sale. Alternatively, mango nectar can be heated to 80 °C, filled in cans held at the same temperature for 10 min, and then cooled down.

15.3.1.4.2 Enzyme inactivation

Polyphenol oxidase is the most relevant enzyme affecting the quality of mango pureé and nectar, as it causes product browning. Peroxidase, a more thermally stable enzyme, may have some residual activity without impacting quality.

15.3.1.5 Product quality

The Codex general standard for fruit juices and nectars stipulates that mango nectars should have at least 25% v/v content of mango pulp (CODEX STAN 247-2005). Soluble solid content is between 12 and 18°Bx, pH around 3.4, and TA between 0.2% and 0.3% as citric acid. Color and aroma are the most relevant fruit-derived quality attributes. Of processed mangoes, Alphonso has the richest aroma profile; however, other cultivars such as Tommy Atkins and Totapuri are commonly used. The effects of pH, high hydrostatic processing and anti-browning agents as alternative non-thermal processes have been studied (Guerrero-Beltran et al., 2005, 2006).

15.3.1.6 Major processors, markets, and sustainability

According to the Food and Agriculture Organization (FAO, 2010), in 2009, the largest exporters of mango purée or juice were Egypt, China, the Philippines, Jordan and Senegal, and the largest importers are China, Jordan, Libya, Maldives, and Senegal. However, the largest producers of mango purée are India, Mexico, Colombia, Egypt, and Thailand. About 30% of mango fruit is peel or seed and in most cases it is waste. There have been some studies to recover lipids and antioxidants from seed kernels (Puravankara, 2000), or to make flour from pulp and peel (Noor Aziah et al., 2012). Mangoes grow in the wild in many poor tropical countries and opportunities exist to develop local and export markets in countries like Mali or Haiti, where producers and processing plants can hire many people for harvesting and sorting and help boost such economies.

15.4 Clean-in-place

Equipment sanitation is a critical part of the economical sustainability of any food processing operation. Failure to adequately sanitize a processing line may result in product contamination, costly product recalls, and consumers getting sick. Also, a large portion of waste water comes from equipment sanitation. Prior to clean-in-place (CIP), processing plants were periodically disassembled and each piece of pipe and equipment was sanitized by hand. Pipes were immersed in cleaning solutions and sanitized with chlorine solutions at concentrations that occasionally led to stainless steel corrosion. CIP became possible as a result of process automation and control and the design of equipment with washing ports and configurations accessible to cleaning solutions and rinsing.

Installation of a CIP requires the additional capital investment of CIP tanks, heat exchanger, special valves (such as double-seat and double butterfly), instrumentation, and programmable logic controllers (PLCs). A CIP also almost doubles the length of pipe required to supply and return the CIP solutions. However, the payback is twofold: significantly lower labor costs and enhanced product safety and quality. CIP solutions are prepared by pumping solution concentrates (caustic or sanitizer) to CIP tanks. After the product has been processed, all pieces of equipment, tanks, and pipes with which the product was in contact must be sanitized.

In a typical CIP system (Figure 15.12), the first step is to rinse at about 50 °C. A plate heat exchanger is typically used to heat in a single pass the first rinse solution, which is fed from the main treated water supply. Then, a hot (~85 °C) caustic solution, typically 2% sodium hydroxide, is used to dissolve oily materials. Solubility of food residues increases with temperature. Heat also contributes largely to microbial kill. The caustic solution is heated to the set point by pumping it through the heat exchanger and circulating to the caustic solution tank until the temperature at the exit of the heat exchanger reaches the desired level. The set point is often determined by a threshold temperature at the return of the CIP.

The following steps are a temperate rinse, a cold rinse, and then a cold sanitation using peracetic acid and other sanitizers. Sanitation is often done cold, and the concentration is low, otherwise polymer (Buna or EPDM) gaskets used in pipe unions and equipment seals can be chemically degraded. However, today some processors prefer hot sanitizers and use Teflon or Viton gaskets. In this stage, microbial kill is chemically driven. Finally, a last rinse, typically used to remove sanitizer residues, is done.

Critical to the effectiveness of any CIP is adequate mechanical shear between the rinses and CIP solutions and the wall of pipes and other pieces of equipment. Mean flow velocities of 1.5–2.1 m/sec are recommended in pipes. Impingement or shear stress in tank surfaces is not uniform and largely depends on the size of the tank, the flow rate of solutions, and the design and location of the spray ball or nozzles used. A very important factor is the duration of the CIP. In theory, there are an infinite number of combinations of CIP solution formulations, temperatures, and flow rate that can achieve adequate sanitation. Heuristic rules are used as starting points, and each processor adjusts them to their specific need. One must bear in mind that even identical systems in which the same product is processed may need different CIP conditions. For example, a juice processing plant

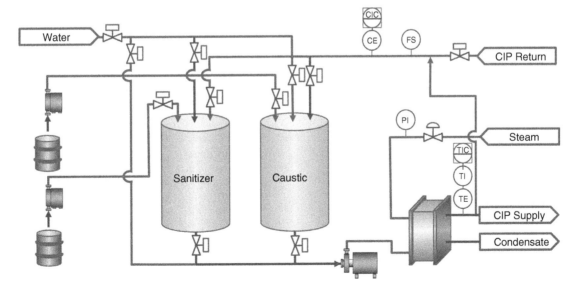

Figure 15.12 Schematic representation of a two-tank clean-in-place (CIP) system.

located at a high elevation cannot use a temperature as high as one at sea level without producing a steam bath. Also, the microbial populations in a particular location may be different, i.e. be more or less tolerant to heat or sanitizer, from another.

Water from the first rinse contains large loads of dissolved and suspended solids. Small juice processors have to drain it, pay fees, and in some cases fines to the municipality to treat the effluent. Larger processing facilities send the waste water to their own water treatment plant to reduce the organic matter content to acceptable levels so that it can then be discarded to the municipality or used for irrigation if located near agricultural land. In small to medium size juice processing plants, a two-tank CIP system like the one shown in Figure 15.12 is common. However, large processors use a third tank in which water from the last rinse is recovered and used for the first rinse. This makes the operation more economically and environmentally sustainable. Other processors use an additional buffer tank with water for the first rinse and even a fifth tank with hot water.

Juices and juice concentrates are often transported from processors to packers in tanker trucks. Proper sanitation is crucial. Tankers with only one port for CIP nozzles may not be properly sanitized, especially when the port is not located near the middle of the tanker. Validation of tanker sanitation is, in many cases, still needed.

In summary, effective CIP requires considering the flow rate, temperature, chemical concentration, and time of each step. To ensure the efficacy of routine CIP, processors use rapid methods based on swabs of selected pieces of equipment and determine by bioluminescence the amount of residual adenosine triphosphate (ATP). Although ATP determinations are non-specific, this type of test has proven very valuable and affordable. Finally, as alternatives to conventional caustic and acid sanitizers, the use of ozone or electrochemically activated water (EAW) has been proposed. Despite its outstanding oxidizing power, ozone does not have a surfactant effect, so for most juices, a caustic wash is required. Also, all gaskets must be EPDM, Viton or Teflon. With regard to EAW systems, more research needs to be done to assess their efficacy and their impact on processing equipment.

15.5 Conclusion

In summary, although processed fruit and vegetable juices have been on the market for decades, new technologies continue to be developed to improve quality retention, nutrition value, and shelf life. Thermal processes that ensure product safety and acceptable product shelf life will continue to be predominant. However, as the cost of non-thermal technologies decreases, a greater number of products are expected to become available, still at a premium price. Consumers, processors, and governments have become increasingly aware of the impact of industrial operations in the environment, especially as population in urban areas continues to grow. Technologies that minimize energy consumption (e.g. carbon footprint) and maximize efficient use of water are becoming increasingly relevant to ensure the sustainability of the fruit beverage industry.

References

Agricultural Marketing Resource Center (AGMRC) (2011) Carrots. www.agmrc.org/commodities__products/vegetables/carrots/, accessed 15 December 2013.

Agricultural Marketing Service (AMS) (2011) *National Apple Processing Report*. USDA, Agricultural Marketing Service, Fruit and Vegetable Programs. www.ams.usda.gov/mnreports/fvwaplproc.pdf, accessed 15 August 2012.

Aguilar-Rosas SF, Ballinas-Casarrubias ML, Nevarez-Moorillon GV, Martin-Belloso O, Ortega-Rivas E (2007) Thermal and pulsed electric fields pasteurization of apple juice: effects on physicochemical properties and flavour compounds. *Journal of Food Engineering* **83**: 41–46.

Anthon GE, Barrett DM (2012) Pectin methylesterase activity and other factors affecting pH and titratable acidity in processing tomatoes. *Food Chemistry* **132**: 915–920.

Binnig R, Possmann P (1993) Apple juice. In: Nagy S, Chen CS, Shaw PE (eds) *Fruit Juice Processing Technology*. Auburndale, FL: Agscience, Inc.

Braddock RJ (1999) *Handbook of Citrus By-Products and Processing Technology*. New York: John Wiley.

Chong HH, Simsek S, Reuhs BL (2009) Analysis of cell-wall pectin from hot and cold break tomato preparations. *Food Research International* **42**: 770–772.

Danyluk MD, Goodrich-Schneider RM, Schneider KR, Harris LJ, Worobo RW (2012) *Outbreaks of Foodborne Disease Associated with Fruit and Vegetable Juices, 1922–2010*. IFAS Extension FSHN12–04. http://edis.ifas.ufl.edu/fs188, accessed 19 November 2013.

Downing DL (ed) (1989) *Processed Apple Products*. New York: Van Nostrand Reinhold.

ERS (2010) *Fruit and Tree Nut Yearbook*. USDA Economics, Statistics, and Market Information System. http://usda01.library.cornell.edu/usda/ers/89022/FTS2011.pdf, accessed 19 November 2013.

Flath RA, Black DR, Guadagni DG et al. (1967) Identification and organoleptic evaluation of compounds in Delicious apple essence. *Journal of Agricultural and Food Chemistry* **15**: 29–35.

Food and Agriculture Organization (FAO) (2010) FAOSTAT. http://faostat3.fao.org/home/index.html, accessed 19 November 2013.

Goodrich RM, Schneider KR, Parish ME (2008) *The Juice HACCP Program: An Overview*. IFAS Extension, 1–4. http://edis.ifas.ufl.edu/fs124, accessed 19 November 2013.

Guerrero-Beltran JA, Swanson BG, Barbosa-Canovas GV (2005) High hydrostatic pressure processing of mango puree containing antibrowning agents. *Food Science and Technology International* **11**: 261–267.

Guerrero-Beltran JA, Barbosa-Canovas GV, Moraga-Ballesteros G, Moraga-Ballesteros MJ, Swanson BG (2006) Effect of pH and ascorbic acid on high hydrostatic pressure-processed mango puree. *Journal of Food Processing and Preservation* **30**: 582–596.

Jacob K, Periago MJ, Bohm V, Berruezo GR (2008) Influence of lycopene and vitamin C from tomato juice on biomarkers of oxidative stress and inflammation. *British Journal of Nutrition* **99**: 137–146.

Kimball DA (1999) *Citrus Processing. A Complete Guide*. Gaithersburg, MD: Aspen Publishers.

Lea AGH (1990) Apple juice. In: Hicks D (ed) *Production and Packaging of Non-Carbonated Fruit Juices and Fruit Beverages*. New York: Van Nostrand Reinhold.

Lea AGH (1995) Apple juice. In: Ashurst PR (ed) *Production and Packaging of Non-Carbonated Fruit Juices and Fruit Beverages*. London: Blackie Academic and Professional.

Lee SY, Chung HJ, Kang DY (2006) Combined treatment of high pressure and heat on killing spores of Alicyclobacillus acidoterrestris in apple juice concentrate. *Journal of Food Protection* **69**: 1056–1060.

Lin HJ, Qin XM, Aizawa K, Inakuma T, Yamauchi R, Kato K (2005) Chemical properties of water-soluble pectins in hot- and cold-break tomato pastes. *Food Chemistry* **93**: 409–415.

Lopez ML, Lavilla MT, Recasens I, Graell J, Vendrell M (2000) Changes in aroma quality of 'Golden Delicious' apples after storage at different oxygen and carbon dioxide concentrations. *Journal of the Science of Food and Agriculture* **80**: 311–324.

Maarse H, Visscher CA (1989) *Citrus Fruit. Volatile Compounds in Food. Quantitative and Qualitative Data*. Zeist, Netherlands: TNVO-CIVO Food Analysis Institute.

Mehinagic E, Royer G, Symoneaux R, Jourjon F, Prost C (2006) Characterization of odor-active volatiles in apples: influence of cultivars and maturity stage. *Journal of Agricultural and Food Chemistry* **54**: 2678–2687.

Morris RA (2010) *The U.S. Orange and Grapefruit Juice Markets: History, Development, Growth and Change*. IFAS Extension, 1–14. http://edis.ifas.ufl.edu/fe834, accessed 19 November 2013.

National Agricultural Statistics Services (NASS) (2010) Vegetables. http://usda01.library.cornell.edu/usda/nass/Vege//2010s/2010/Vege-10-01-2010.txt, accessed 15 December 2013.

Noor Aziah AA, Wong L, Rajeev B, Cheng L (2012) Evaluation of processed green and ripe mango peel and pulp flours (Mangifera indica var. Chokanan) in terms of chemical composition, antioxidant compounds and functional properties. *Journal of the Science of Food and Agriculture* **92**: 557–563.

O'Rourke D (2003) World production, trade,consumption and economic outlook for apples. In: Ferree DC, Warrington IJ (eds) *Apples. Botany, Production and Uses*. Wallingford, Oxfordshire: CABI Publishing.

Payne EM (2011) *Rheological Properties of High and Low Concentrated Orange Pulp*. MSc thesis, University of Florida.

Plotto A, Mcdaniel MR, Mattheis JP (1999) Characterization of 'Gala' apple aroma and flavor: differences between controlled atmosphere and air storage. *Journal of the American Society for Horticultural Science* **124**: 416–423.

Reiter M, Neidhart S, Carle R (2003) Sedimentation behaviour and turbidity of carrot juices in relation to the characteristics of their cloud particles. *Journal of the Science of Food and Agriculture* **83**: 745–751.

Reyes-de-Corcuera JI, Goodrich-Schneider RM, Braddock RJ (2012) Oranges. In: Siddiq M (ed) *Tropical and Subtropical Fruit Processing and Packaging*. Ames, IA: John Wiley, pp. 399–417.

Root WH, Barrett DM (2005) Apples and apple processing. In: Barrett DM, Somogyi L, Ramaswamy H (eds) *Processing Fruits. Science and Technology*, 2nd edn. Boca Raton, FL: CRC Press.

Rushing NB, Veldhuis MK, Senn J (1956) Growth rates of Lactobacillus and Leuconostoc species in orange juice as affected by pH and juice concentration. *Applied Microbiology* **4**: 97–100.

Sargent SA, Brecht J, Olczyk T (2005) Handling Florida vegetables series — round and Roma tomato types. Electronic Database Information System, SS-VEC-928. University of Florida, Gainesville, FL.

Sloan AE (2012) Top 10 functional food trends. *Food Technology* **66**(4): 24–41.

USDA-ERS (2011) Table F–14, Selected fruit juices: Per capita use, 1980/81to date. In: *Fruit and Tree Nut Yearbook*. http://usda01.library.cornell.edu/usda/ers/89022/2011/FTS2011.pdf, accessed 19 November 2013.

Wu JSB, Chen H, Fang T (1993) Mango juice. In: Nagy S, Chen CS, Shaw PE (eds) *Fruit Juice Processing Technology*. Auburndale, FL: Agscience, Inc.

Young H, Gilbert JM, Murray SH, Ball RD (1996) Causal effects of aroma compounds on Royal Gala apple flavours. *Journal of the Science of Food and Agriculture* **71**: 329–336.

Zhaoling H (1986) China. In: Wardowski W, Nagy S, Grierson W (eds) *Fresh Citrus Fruits.* New York: Van Nostrand Reinhold.

Further reading

Ashurst PR (ed) (1995) *Production and Packaging of Non-Carbonated Fruit Juices and Fruit Beverages.* London: Blackie Academic and Professional.

Barrett DM, Somogyi L, Ramaswamy H (eds) *Processing Fruits. Science and Technology*, 2nd edn. Boca Raton, FL: CRC Press.

Hicks D (ed) *Production and Packaging of Non-Carbonated Fruit Juices and Fruit Beverages.* New York: Van Nostrand Reinhold.

Nagy S, Chen CS, Shaw PE (eds) *Fruit Juice Processing Technology.* Auburndale, FL: Agscience, Inc.

Puravankara D, Boghra V, Sharma RS (2000) Effect of antioxidant principles isolated from mango (Mangifera indica L) seed kernels on oxidative stability of buffalo ghee (butter-fat). *Journal of the Science of Food and Agriculture* **80**: 522–526.

16 Fruits and Vegetables – Processing Technologies and Applications

Nutsuda Sumonsiri and Sheryl A. Barringer
Department of Food Science and Technology, The Ohio State University, Columbus, Ohio, USA

16.1 Raw materials

16.1.1 Chemical composition

The chemical composition of fruits and vegetables depends on many factors including variety, weather, cultivation, degree of maturity at harvesting, and storage conditions. Table 16.1 shows the composition of some raw fruits and vegetables. The most important component in fruits and vegetables is water, which ranges from 70% to 95% and 10% to 20% of the weight in wet crops and dry crops, respectively (Haard, 1984). Due to the high quantity of water in fruits and vegetables, a decrease in the moisture content causes significant wilting and weight loss, and causes both undesirable and desirable changes during processing (Brecht et al., 2007).

The majority of the solids in most fruits and vegetables is carbohydrates, including starches, sugars, cellulose, hemicelluloses, and pectins (Haard, 1984; FAO, 1995). In most fruits and vegetables, carbohydrates represent approximately 75% of the dry weight. Starches are normally found in intercellular plastids or starch granules to provide energy and reserve energy in plants, seeds, and tubers. In some crops, such as potato, sweet potato, and cereals, the starch concentration can be up to 74% of the dry weight (Brecht et al., 2007; FAO, 1995; Hucl & Chibbar, 1996). The sugars found in edible plants include fructose, glucose, and sucrose. Some fruits and vegetables have a sugar content of more than 20%, such as ripe banana, while other fruits and vegetables do not contain significant amounts of sugar, such as avocado, which contains a sugar content of approximately 4% of the

dry weight (Richings et al., 2000). Some sugar alcohols, including but not limited to sorbitol, mannitol, and xylitol, are also found in fruit tissues (Brecht et al., 2007). Cellulose, hemicelluloses, pectin, and lignin are cell wall constituents that produce supporting structures in plant tissues. They do not provide any energy because they cannot be digested by humans in the upper digestive tract. However, they are considered a good source of beneficial dietary fiber, and are broken down in the colon to provide prebiotic benefits.

After harvesting, fruits and vegetables still undergo active biological processes, such as respiration, ripening in fruits, and senescence. In some fruits and vegetables, these activities cause significant changes in the quality so the postharvest storage conditions and processing steps need to be carefully conducted to prevent these changes. For example, the level of sugar in potatoes increases up to 5–10 times the original sugar concentration at harvest if they are stored below 10 °C after harvesting. The high sugar content in these potatoes can cause Maillard browning reactions during further processing steps, especially drying and frying. In ripe sweet corn, the opposite reaction is of concern. During storage, the level of sugars decreases and starch is produced, causing losses in flavor and texture (Cottrell et al., 1993).

In most fruits and vegetables, lipids are found in cytoplasmic membranes and in the endosperm (Brecht et al., 2007; Haard, 1984). The amount of lipid typically varies from 0.1% to 1% of the fresh weight; however, some fruits and vegetables have high amounts of storage lipids, such as avocado (up to 15.5%), oilseeds (up to 18.5%), and nuts (up to 65.2%) (Haard, 1984; USDA Agricultural Research

Food Processing: Principles and Applications, Second Edition. Edited by Stephanie Clark, Stephanie Jung, and Buddhi Lamsal.
© 2014 John Wiley & Sons, Ltd. Published 2014 by John Wiley & Sons, Ltd.

Table 16.1 Composition of some fruits and vegetables (per 100 g of raw fruits and vegetables with skin)

Fruit/vegetable	Water (g)	Carbohydrate (g)	Protein (g)	Fat (g)	Vitamin C (mg)	Thiamin (mg)	Potassium (mg)
Apple	85.56	13.81	0.26	0.17	4.6	0.017	107
Tomato (ripe)	94.52	3.89	0.88	0.20	13.7	0.037	237
Potato	83.29	12.44	2.57	0.10	11.4	0.021	413
Green beans	90.32	6.97	1.83	0.22	12.2	0.082	211
Apricot	86.35	11.12	1.40	0.39	10.0	0.030	259
Strawberries	90.95	7.68	0.67	0.30	58.8	0.024	153
Carrot	88.29	9.58	0.93	0.24	5.9	0.066	320
Avocado	73.23	8.53	2.00	14.66	10.0	0.067	485

Source: USDA National Agricultural Statistics Service (2011).

Service, 2011). The lipids found in fruits and vegetables are phospholipids, glycolipids, and triacylglycerols. Other lipid-like substances, such as carotenoids, are also found in many crops, for example, orange, pineapple, cantaloupe, carrot, and tomato. Waxes are found on the surfaces of leaves, fruits, and seeds (Brecht et al., 2007), and can decrease moisture loss during storage and processing. Waxes on fruits and vegetables also slow down the rate of sodium hydroxide diffusion during lye peeling (Floros & Chinnan, 1989).

In most fruit, proteins and other nitrogenous compounds can be found at less than 2% of the fresh weight (Haard, 1984) while vegetables contain 1.0–5.5% of these compounds (FAO, 1995). Some crops, such as vegetables in the family Leguminosae, accumulate storage proteins and contain up to 40% protein (Haard, 1984). The legume seeds, including peas, beans, lentils, peanuts, and soybean, are good sources of all the essential amino acids, except methionine and cysteine (Brecht et al., 2007). The presence of amino compounds and reducing sugars causes Maillard browning reactions, resulting in changes in color and flavor during processing, especially in dehydration and concentration processes.

Fruits and vegetables are a significant source of vitamins. The vitamin content depends on variety and conditions during growth, postharvest storage, and processing. Edible plants contribute approximately 50% of vitamin A, 58% of thiamin, 26% of riboflavin, 47% of niacin, and 94% of vitamin C intake in the US diet (Brecht et al., 2007). In general, plants do not have vitamin A, but they contain β-carotene, which is converted into vitamin A after consumption. Orange and yellow crops, and green leafy vegetables, such as squash, sweet potatoes, carrots, and spinach, are an excellent source of β-carotene. Vitamin C can be found in many fruits and vegetables,

for example, strawberry, oranges, guava, and kale. Most vitamins, especially C, A, and B, are sensitive to changes in pH, air, light, and heat; therefore, there is usually at least 30% loss of these vitamins during processing (Karmas & Harris, 1988).

Minerals in fruits and vegetables are in the form of salts of organic or inorganic acids or complex organic combinations. There are more mineral substances found in vegetables than in fruits. The mineral content in edible plants varies from less than 0.1% to 5% of the fresh weight. The minerals found in edible plants are calcium, iron, potassium, magnesium, sulfur, phosphorus, and nitrogen (Brecht et al., 2007). Fruits and vegetables that are rich in minerals include cabbage, tomatoes, carrots, raspberries, cherries, peaches, and strawberries (FAO, 1995). Minerals are more stable than vitamins during processing; however, they can be increased or decreased due to different processing steps and conditions. For example, calcium, potassium, and sodium levels are frequently increased during processing of canned vegetables from hard water uptake or addition of minerals during processing (Rickman et al., 2007).

Organic acids, such as citric acid, malic acid, and tartaric acid, are present in many fruits, such as oranges, lemon, apples, and grapes, and provide tartness and decrease the tendency for microbial spoilage. During storage and ripening, the organic acids of many fruits, such as apples, pears, and oranges, decrease; therefore, degree of acidity and sugar content are used to predict ripeness for harvesting. Organic acids are present at up to 50 mEq acid per 100 g of tissue in acidic crops (Spanos & Wrolstad, 1992).

Some acids, such as chlorogenic acid, affect browning of fruits and fruit juices, particularly the enzymatic browning reaction, which is catalyzed by the presence of oxygen and PPO (polyphenol oxidase), resulting in the formation of brown quinones. Many phenolic acids in plants, such as coumaric, caffeic, and benzoic acids in apple, pear, and grape, also have a large effect on flavor and color of fruit and vegetable products, especially fruit juices and wines (Spanos & Wrolstad, 1992). Moreover, benzoic acid can serve as an antifungal agent in some fruits, such as cranberry (Brecht et al., 2007).

16.1.2 Pigments

16.1.2.1 Chlorophyll

16.1.2.1.1 Location and function in the plant

Chlorophylls, magnesium complexes derived from porphin, are green, oil-soluble pigments found in the chloroplasts of green plants, photosynthetic bacteria, and algae. In most green plants, chlorophyll a and b are in the ratio of 3:1. Chlorophyll c and d are commonly found in marine algae and red algae, respectively (von Elbe & Schwartz, 1996). In the plant, chlorophyll plays an important role in producing carbohydrates from CO_2, water, and light using photosynthesis.

16.1.2.1.2 Effect on color changes

The color of chlorophyll is affected by heat, pH, and light. During heating, chlorophyll is converted to the olive green or brown color of pheophytin, pyropheophytin, pheophorbide, and pyropheophorbide since two hydrogen ions easily replace the magnesium atom in chlorophyll (Figure 16.1) (von Elbe & Schwartz, 1996). The heat stability of chlorophyll is affected by pH. Chlorophyll is only heat stable at pH above 9.0 (von Elbe & Schwartz, 1996). At neutral conditions, the color of chlorophyll is rapidly changed by thermal treatment. Acid conditions (pH below 5.8) change the color of chlorophyll to olive green, even under cold conditions (Maharaj & Sankat, 1996).

In healthy plant tissues, light cannot destroy chlorophyll since it is protected by carotenoids and other lipids that have weak linkages with chlorophyll (von Elbe & Schwartz, 1996). However, the presence of light increases the degradation of chlorophyll in plant leaves during senescence or when cells are damaged during processing (Kar & Choudhuri, 1987; von Elbe & Schwartz, 1996).

16.1.2.1.3 Reaction with additives

Since chlorophylls are stable in basic conditions, sodium carbonate or other alkali could theoretically protect the bright green color in some vegetables. However, alkali

is not commercially used because the texture of the vegetable is softened, unpleasant flavors are developed, and some vitamins, such as vitamin C and thiamine, are destroyed under alkaline conditions and heat treatment (FAO, 1995). A patented process using Zn^{2+} or Cu^{2+} salts can retain the bright green color of vegetables. These ions substitute for the Mg^{2+} atom in chlorophyll and form zinc or copper complexes of chlorophyll derivatives, which are bright green and stable to heat. The chlorophyll derivatives are allowed for use in canned foods, candy, soups, and dairy products in most European countries. In the US, Zn^{2+} is allowed only in vegetables and Cu^{2+} is allowed in confections (LaBorde & von Elbe, 1996).

16.1.2.2 Anthocyanins

16.1.2.2.1 Location and function in the plant

Anthocyanins are water-soluble pigments in a subgroup of flavonoids that provide a wide range of colors in fruits and vegetables, including purple, violet, magenta, blue, and red. The pigments can be found in berries, grapes, eggplant, and cherries. Anthocyanins have a wide structural diversity with as many as 250 different structures in plants, depending on the cultivar and maturity. The structure of anthocyanin is based on 2-phenylbenzopyrylium of flavylium salt. The factors affecting the different structures of anthocyanins are the number, types, and sites of sugars attached to the molecules, number and types of aliphatic or aromatic groups attached to the sugars, and number of hydroxyl and/or methoxy groups. When the attached sugar in an anthocyanin is hydrolyzed, the non-sugar form of anthocyanin is called an anthocyanidin. There are six common anthocyanidins found in food: pelargonidin, cynaidin, delphinidin, peonidin, petunidin, and malvidin (von Elbe & Schwartz, 1996).

16.1.2.2.2 Effect on color changes

The color of anthocyanins is sensitive to pH, enzymes, and metal ions during extraction from tissues of fruits and vegetables, as well as during processing and storage of the product. There are four possible forms of anthocyanins in an aqueous medium or foods with different pH, including the blue quinonoidal base, the red flavylium cation, the colorless carbinol pseudobase, and the colorless chalcone (Figure 16.2). These structures play an important role in color changes of anthocyanins at various pH. For example, the color of some anthocyanins turns from red to blue with the increase of pH in the

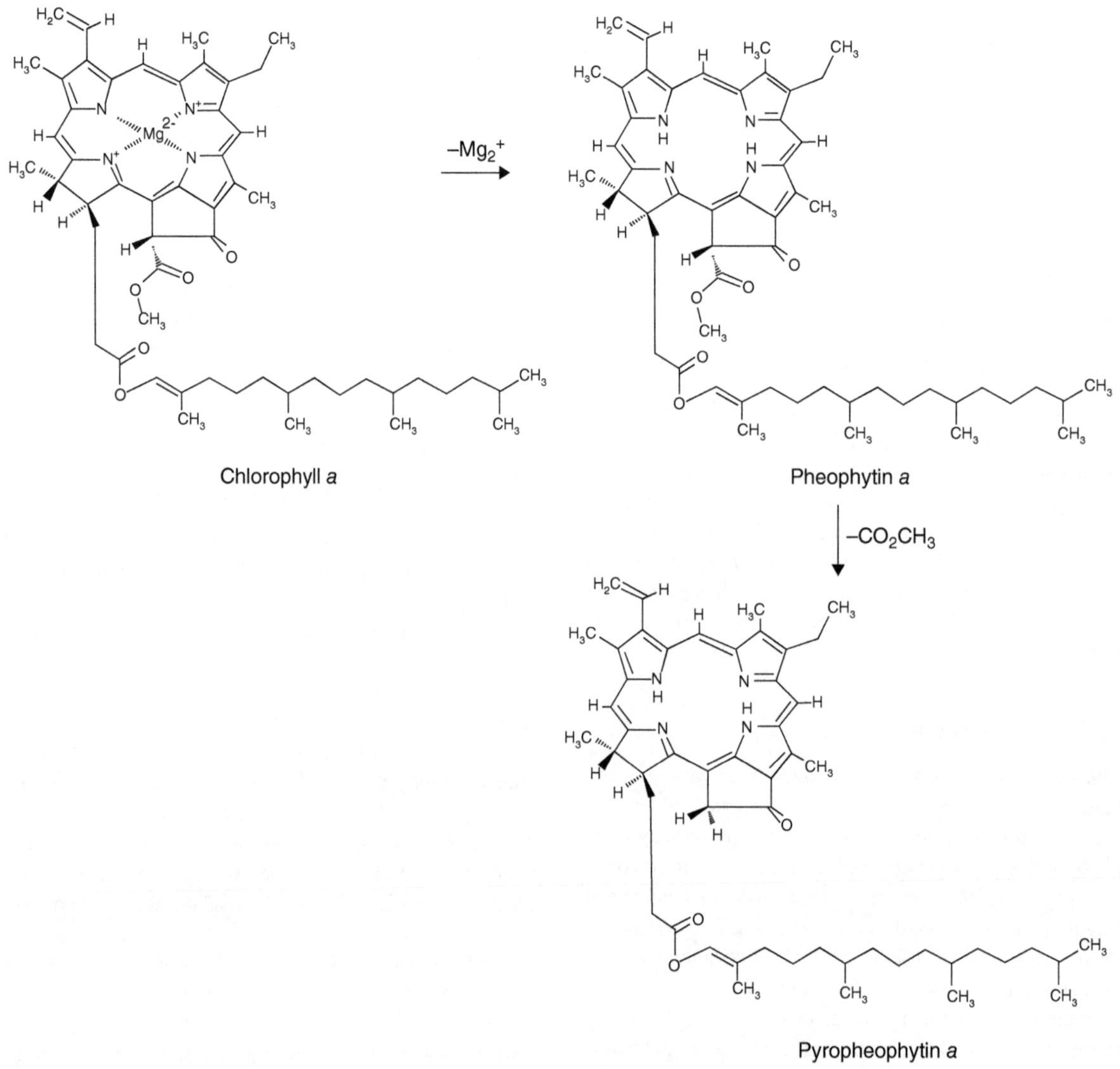

Figure 16.1 The degradation of chlorophyll *a*.

0–6 pH range since the red flavylium dominates at low pH and the blue quinonoid dominates as pH is increased (von Elbe & Schwartz, 1996).

There are two groups of enzymes that can affect the color of anthocyanins: glycosidases and polyphenol oxidases. Glycosidases can hydrolyze glycosidic linkages and produce sugars and the aglycone, resulting in decreased solubility and color loss of anthocyanins. Polyphenol oxidases can oxidize diphenols in the presence of oxygen and produce brown *o*-benzoquinone, which is usually undesirable, especially in fruit juice processing (Yokotsuka & Singleton, 1997).

16.1.2.2.3 Reaction with additives

Anthocyanins are decolorized in sulfur dioxide solution, which can be either reversible or irreversible, depending on the concentration of sulfur dioxide. Some fruits containing anthocyanins are held in sulfur dioxide solution at low concentration (500–2000 ppm) to reduce microbial spoilage during storage. This process causes color loss of the fruits; however, a desulfuring process through washing can restore the color before processing. During the production of maraschino and candied cherries, a high concentration of sulfur dioxide (0.8–1.5%) is used to bleach the color

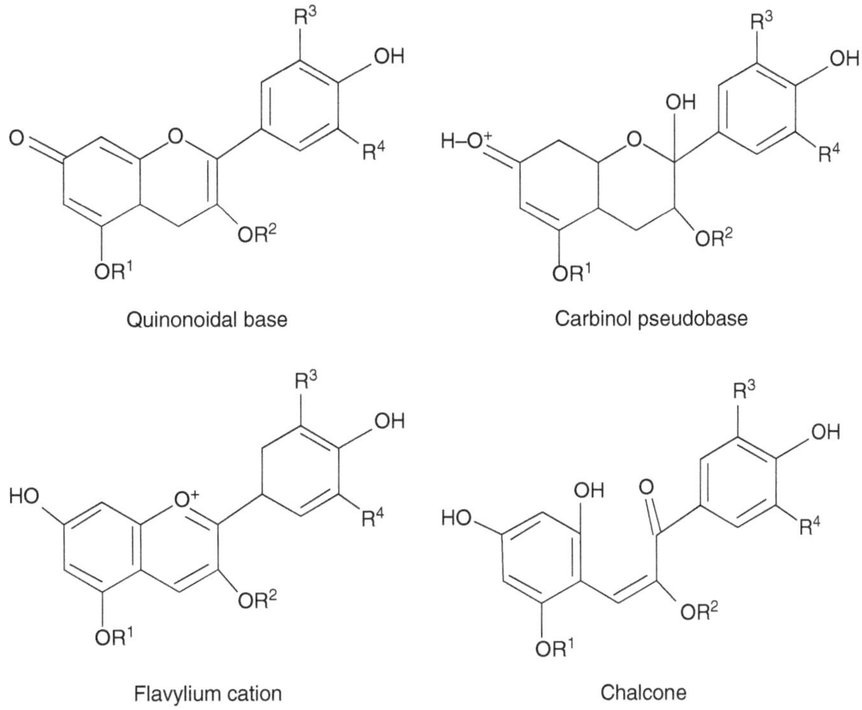

Quinonoidal base

Carbinol pseudobase

Flavylium cation

Chalcone

Figure 16.2 Four possible structures of anthocyanins in an aqueous medium or foods.

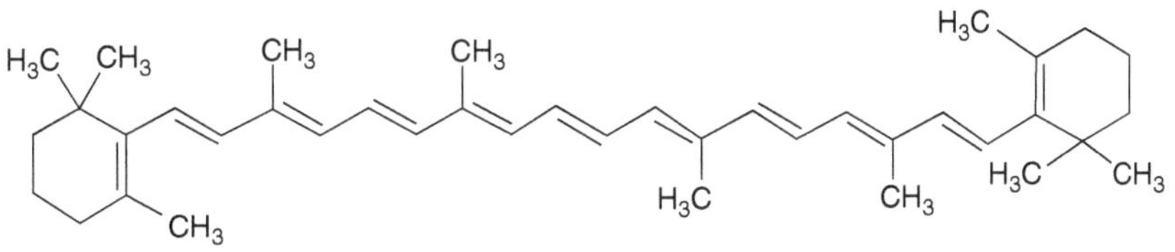

Figure 16.3 Structure of β-carotene.

of anthocyanins and this effect is irreversible (von Elbe & Schwartz, 1996; Wrolstad, 2009). However, at very low concentration (30 ppm), sulfur dioxide can be used to prevent enzymatic browning in sour cherry juice (Unten et al., 1997). In frozen strawberry, sucrose is also added to stabilize the color of anthocyanins and reduce the browning reaction by functioning as a diluent and interrupting the condensation reactions, as well as serving as an inhibitor of polyphenol oxidase (Wrolstad et al., 1990).

Metal complexes of anthocyanin are commonly found in plants, especially in flowers (Takeda, 2006; von Elbe & Schwartz, 1996). The anthocyanins can react with metal ions, such as Al, Sn, and Fe, and turn violet or blue; therefore, the inside of a metal can has to be lacquered to protect the color of fruits and vegetables containing anthocyanins.

16.1.2.3 Carotenoids

16.1.2.3.1 Location and function in the plant

Carotenoids are fat-soluble pigments that have a wide range of colors, from yellow through orange to red, such as the orange carotene in carrot, peach, apricot, and citrus fruits, the red lycopene in tomato and watermelon, and the yellow xanthophylls in corn and squash. The general structural backbone of carotenoids is a symmetrical molecule of isoprene units, which are covalently linked in a head-to-tail or a tail-to-tail fashion. The most common carotenoid in the plant is β-carotene (Figure 16.3) (von Elbe & Schwartz, 1996).

The carotenoids are often found along with chlorophylls in the chloroplasts and play an important role in photosynthesis and photoprotection in plants. In the human diet, some carotenoids, such as β-carotene, α-carotene, γ-carotene, and cryptoxanthin, can be converted into vitamin A. β-Carotene has the greatest provitamin A activity since it has two β-ionone rings, which is a retinoid structure (see Figure 16.3). Fruits and vegetables containing provitamin A carotenoids can provide 30–100% of the vitamin A required for the human diet (von Elbe & Schwartz, 1996).

16.1.2.3.2 Effect on color changes

Carotenoids are stable to pH, heat, and water leaching. They can be degraded by oxidation due to the presence of light, oxygen, or lipoxygenase, resulting in loss of both color and vitamin A activity. In the degradation of carotenoids due to enzyme activity, carotenoids react with peroxides, which are produced when lipoxygenase catalyzes oxidation of unsaturated fatty acids. Since carotenoids can be easily oxidized due to several conjugated double bonds in their structure (see Figure 16.3), they are considered antioxidants, which can quench singlet oxygen and prevent damage to the cell from oxidation.

16.1.2.3.3 Reaction with additives

L-ascorbate is usually used to prevent the oxidation of carotenoids and preserve the color in beverages and margarine (Delgado-Vargas et al., 2000). Green tea polyphenols can also be added to beverages and margarine to prevent discoloration of carotenoids since the phenolic hydroxyl group stabilizes the peroxide radical and prevents the oxidation of β-carotene (Unten et al., 1997).

16.1.2.4 Betalains

16.1.2.4.1 Location and function in the plant

Betalains are water-soluble pigments found in vacuoles of plant cells that produce similar colors to carotenoids and anthocyanins. The color range of betalains varies from the red of betacyanins to the yellow of betaxanthins. These pigments can be found in only 10 families of the order Centrospermae, including red beets, cacti, Swiss chard, and amaranth. The basic structure of betalains (Figure 16.4) is formed by the condensation of amine with betalamic acid (von Elbe & Schwartz, 1996). Betalains are believed to play an important role in preventing damage from UV radiation in plants that can tolerate extremely

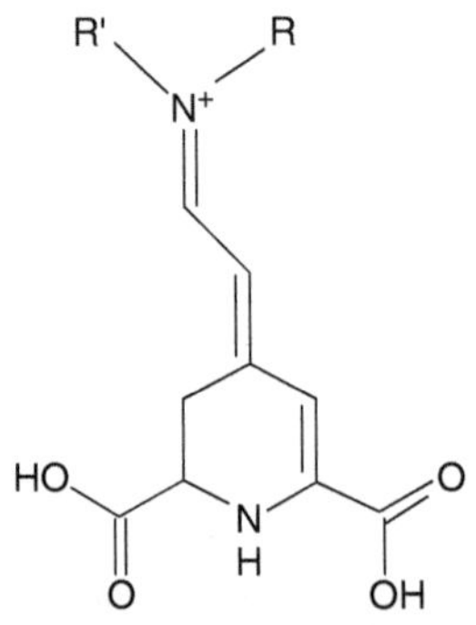

Figure 16.4 Basic structure of betalains.

dry environments with high salinity, such as the ice plant (Ibdah et al., 2002), and protecting the plant tissue from pathogens in vegetables that grow underground, such as the red beet (Stafford, 1994).

16.1.2.4.2 Effect on color changes

The color of betalains is unstable to changes in pH. Under alkaline conditions, betacyanins dominate and produce a bluish red or bluish violet color. Under acidic condition, the color of betalains is yellow due to the formation of 14,15-dehydrobetanin (Mabry et al., 1967).

16.1.2.4.3 Reaction with additives

The presence of metal cations, such as Cu^{2+}, Al^{3+}, Fe^{2+}, and Fe^{3+}, increases the degradation rate of betalains due to the formation of metal-pigment complexes (Herbach et al., 2006). Ethylenediamine tetra-acetic acid (EDTA) can form complexes with metal ions and therefore increase the stability of betalains (Pasch & von Elbe, 1979).

16.1.3 Enzymes

16.1.3.1 Enzymes affecting color

There are two enzymes that have a significant effect on the color of fruits and vegetables: lipoxygenase (EC 1.13.11) and polyphenol oxidase (EC 1.10.3.1). Lipoxygenase can be found in plants, animal tissues, and mushrooms (Oliw, 2002). It initially catalyzes the oxidation of unsaturated fatty acids and produces hydroperoxides and free radicals, which are responsible for bleaching chlorophylls and carotenoids in unblanched stored vegetables, resulting in loss of color.

Polyphenol oxidase is an enzyme catalyzing the oxidation of phenolic compounds, which include anthocyanins as well as several of the flavor compounds, found in plants, animals, and some microorganisms (Ruenroengklin et al., 2009). The products from the reaction, usually *o*-benzoquinone, are not stable, and are further oxidized and polymerized, and finally produce the brown color of melanins. This is responsible for the undesirable brown color in cut fruits and vegetables, such as peaches, potatoes, apples, lettuce, and bananas, and especially in tropical fruit juices. However, some brown and black colors from these reactions, found in coffee, raisins, and prunes, are desirable (Whitaker, 1996).

16.1.3.2 Enzymes affecting texture

The texture of fruits and vegetables is significantly affected by two enzymes, which work together: pectin methylesterase (EC 3.1.1.11) and polygalacturonase (EC 3.2.1.15). Pectin methylesterase converts pectin polymers into pectic acid, which polygalacturonase can hydrolyze. These enzymes, therefore, cause a significant decrease in texture, or softening of fruits and vegetables, especially those rich in pectin, such as tomatoes, apples, bananas, and avocado. In the processing of many tomato products, including ketchup, pizza, and spaghetti sauces, a high viscosity is desired; therefore, the hot break process, which is a rapid high heat treatment at above 82 °C for 15 sec, is applied after grinding to inactivate the enzymes. During the production of tomato juice, a low viscosity is desired so the cold break process uses a temperature less than 66 °C to speed up the activity of these enzymes and decrease viscosity of the product (Madhavi & Salunkhe, 1998).

16.1.3.3 Enzymes affecting flavor

Enzymes in fruits and vegetables that can cause great changes in flavor are peroxidase (EC 1.11.1.7) and lipoxygenase during the oxidation of unsaturated fatty acids and other compounds. These enzymes can cause the formation of aromas, such as *cis*-3-hexenal, *cis*-3-hexenol, *trans*-2-hexenol, and 2-methylbutanal, in lettuce, corn, broccoli, green beans, cauliflower, and peas (Azarnia et al., 2011; Deza-Durand & Petersen, 2011). Therefore, vegetables need to be blanched prior to freezing or drying to stabilize their flavor. Peroxidase is very resistant to heat. The thermal inactivation of this enzyme is in the temperature range of 70–95 °C, so it is usually used as an indicator of proper thermal treatment of the product (Morales-Blancas et al., 2002).

16.1.3.4 Effect of temperature, pH, and water activity on enzyme activity

Temperature, pH, and water activity affect enzyme activity, which causes changes in the product quality. Temperature has a significant effect on the stability of enzymes and velocity of the reaction. As temperature increases, the enzyme activity increases. However, as temperature increases beyond 60–70 °C, enzymes become unstable and there is reduced activity due to the denaturation of the enzyme (Whitaker, 1996).

pH also has a great effect on enzyme activity. Each enzyme has a particular optimal pH depending on its type, source, and other conditions, such as temperature, substrate concentration and accessibility, enzyme concentration, and the presence of inhibitors. For example, the optimal pH for lipoxygenase in soybean is 7–9, while that for polygalacturonase in tomatoes is 4 (Whitaker, 1996).

Water activity has a significant effect on enzyme activity since most enzymes are in aqueous media. Some water is required for enzyme activity to mobilize and solubilize the reactants. As the moisture content increases, enzyme activity increases since there is enough water to dissolve the substrate and increase the diffusion at the active site. Enzyme activity decreases at low water activity; for example, phospholipase has no activity on lecithin at water activity below 0.35 (Whitaker, 1996).

16.2 Basic processing

Basic processing of fruits and vegetables usually starts with grading, washing, cooling, and peeling, depending on the characteristics of the crops and the final products. Figure 16.5 shows the production figures for the commercial fruits and vegetables discussed in this chapter.

16.2.1 Grading

One of the initial steps fruits and vegetables go through is grading, to determine the price paid to the farmer. This is done at the processing facility or at a centralized station before going to the processing facility. Individual companies may set their own grading standards, use the voluntary USDA grading standards, or use locally determined standards, such as those of the Processing Tomato Advisory Board in California. The farmer is paid based on the percentage of fruits and vegetables in each category. Typically, companies hire USDA graders or hold an annual grading school to train their graders.

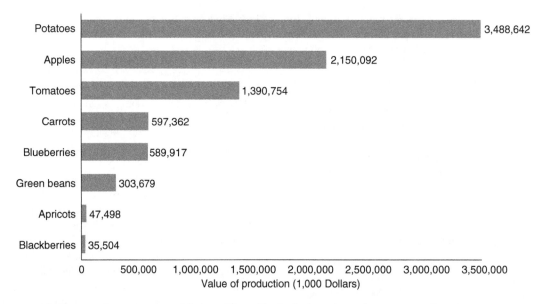

Figure 16.5 Production of some commercial fruits and vegetables for fresh market and processing in the US in 2010. Source: USDA National Agricultural Statistics Service (2011).

16.2.1.1 Tomato grading

The USDA divides tomatoes for processing into categories, the highest being A, followed by B, C, and culls (USDA, 1983a). Grading is done on the basis of color and percentage of defects. Color can be determined visually by estimation of what percentage of the surface is red, with an electronic colorimeter on a composite raw juice sample, or with a portable colorimeter on a whole tomato. Defects include worms, worm damage, freeze damage, stems, mechanical damage, anthracnose and other disease, mold, and decay. The allowable percentage of extraneous matter may also be specified. Extraneous matter includes stems, vines, dirt, stones, and trash.

Tomatoes for canning whole, sliced or diced are graded on the basis of color, firmness, defects, and size. Solids content is unimportant, unlike in tomatoes for juice or paste. Graders must be trained to evaluate and score color and firmness. Color should be a uniform red across the entire surface of the tomato. Color is graded using USDA-issued plastic color comparators, the Munsell colorimeter or the Agtron colorimeter, or the tomato is ground into juice and used in a colorimeter with a correlation equation to convert it to the Munsell scale. Firmness, or character, is important to be sure the tomato will go through canning and remain intact. Soft, watery tomatoes or tomatoes possessing large seed cavities give

an unattractive appearance and therefore receive a lower grade. Size is not a grading characteristic *per se*, but all tomatoes must be above a minimum agreed upon size.

The Processing Tomato Advisory Board inspects all tomatoes for processing in California. Their standards are similar to those of the USDA, but more geared for the paste industry. They inspect fruit for color, soluble solids, and damage. A load of tomatoes may be rejected for any of the following reasons: >2% of fruit is affected by worm or insect damage, >8% is affected by mold, >4% is green, or >3% contains material other than tomatoes, such as extraneous material, dirt, and detached stems (California Department of Food and Agriculture, 2001).

16.2.1.2 Potato grading

Potatoes for French fry processing are divided into two categories during grading: US no. 1 processing and US no. 2 processing, based on the size of the potatoes (USDA, 1983b). The potatoes in US no. 1 processing must be more than 5 cm in diameter or 4 oz in weight. The whole potatoes in US no. 2 processing must be more than 3.8 cm in diameter or the usable pieces, which are the portion of potato after trimming, must be more than 4 oz in weight. Potatoes in both categories have to be free from serious damage, freezing injury, blackheart, late blight tuber rot, insects, worms, southern bacterial wilt, bacterial ring rot, soft rot and wet breakdown, loose sprouts, dirt, and foreign material. The number of potatoes in different size categories may also be specified.

Besides the size of potatoes and defects, which are the basic requirements, specific gravity and fry color are optional tests during grading of potatoes for French fries. For specific gravity, potatoes are randomly taken from the lot and at least three corrected readings are averaged. The specific gravity is determined by the weights of the potatoes in air (5000 g) and in water at a specific temperature using the USDA-approved equipment. The reading from each test is corrected for temperature variations using correction factors provided by the USDA. In the test for fry color, 20 potatoes are randomly chosen and then sliced into 3.23 cm^2 strips and fried in oil for at least 3 min at 176.67 °C or 2.5 min at 190.56 °C. The fry color can be determined by using the Munsell Color Standards for Frozen French Fried Potatoes.

Potatoes for chipping are also divided into two categories during grading: US no. 1 and 2, based on the size of potatoes (USDA, 1978). The US no. 1 potatoes for chipping should be more than 4.8 cm in diameter while US no. 2 potatoes for chipping should be more than 4.4 cm in diameter. Potatoes in both categories have to be free from serious damage, freezing, blackheart, late blight tuber rot, nuts of nut sedge, southern bacterial wilt, bacterial ring rot, and tuber moth injury, soft rot, and wet breakdown. Similar to potatoes for French fry processing, specific gravity and fry color are optional tests for potatoes for chipping. The test for specific gravity is the same but the fry color is determined by frying at least 40 potatoe slices (1.3 mm thick) in oil for at least 1 min and 40 sec at 185 °C. The color of fried potato chips should be more than a reading of 25 on a photoelectric colorimeter approved by the USDA, such as Agtron M-30A or M-300A.

16.2.2 Washing

Washing is a critical control step in producing fruit and vegetable products with a low microbial count. After harvesting, fruits and vegetables are washed to eliminate soil, dirt, surface microorganisms, mold, insects, *Drosophila* eggs, fungicide, insecticide, and other pesticide residues (FAO, 1995). The efficiency of the washing process will determine microbial counts in the final product. Spoiled fruits and vegetables should be removed before washing to minimize the contamination of washing tools, equipment, and produce during washing. The washing process should reduce the surface microorganisms by a six-fold, or 1-log, reduction and there should not be any molds and yeast in the water from the final wash. Lye or surfactants may be added to the water to improve the efficiency of dirt removal; however, surfactants have been shown to promote infiltration of some bacteria into fruits and vegetables by reducing the surface tension at the pores (Bartz, 1999), which jeopardizes food safety. The washing step also serves to cool fruits and vegetables. Since some of them are harvested on hot summer days, washing removes the field heat, slowing respiration and therefore quality loss.

Several methods can be used to increase the efficiency of the washing step. Agitation increases the efficiency of soil removal. The warmer the water spray or dip, up to 90 °C, the lower the microbial count (Adsule et al., 1982), although warm water is not typically used because of economic concerns. Immersion or spraying is usually used with the application of detergents, 1.5% HCl solution, warm water (approximately 50 °C), or high water pressure (for spray or shower washing). For washing vegetables, detergents or sanitizers can be used in equipment designed for the shape, size, and fragility of the vegetables, such as a flotation cleaner for peas and small vegetables, and a rotary washer with overhead spray for fragile vegetables.

Chlorine is frequently added to the wash water at 100–150 ppm. Chlorine will not significantly reduce microbial counts on fruits and vegetables itself because the residence time is too short. However, it is effective at keeping down the number of microorganisms present in the flume water (Heil et al., 1984). When there is a large amount of organic material in the water, such as occurs in dirty water, chlorine is used up rapidly, so it must be continuously monitored. Other potential alternative sanitizers, such as peroxyacetic acid and chlorine dioxide, are also suggested for some fruits and vegetables such as lettuce leaves and apples (Wisniewsky et al., 2000).

16.2.2.1 Tomato washing

Tomatoes are typically transported in a water flume to minimize damage to the fruit. Therefore, tomato washing can be a separate step in a water tank or it can be built into the flume system (Figure 16.6). The final rinse step uses pressurized spray nozzles at the end of the soaking process. Flume water may be used in a counterflow system, so that the final rinse is with fresh water, while the initial wash is done with used water. In either system, the first flume frequently inoculates rather than washes the

Figure 16.6 Tomato washing in a tank before transporting to the lye peeler.

tomatoes because all the dirt in the truck is washed into the flume water (Heil et al., 1984). When the water is reused, high microbial counts on the fruit may result if careful controls are not kept.

16.2.2.2 Potato washing

A washing step is necessary before processing potatoes in order to remove stones, soil, debris, and pesticide residues (Schorneck, 1961; Zohair, 2001). Potatoes are usually dumped into a vat with an adjustable conveyor system that contains washing water. A water tank is also used to separate stones from the tubers, since the stones settle to the bottom. The tank also removes debris floating on top (Figure 16.7).

Several solutions are reported to be efficient for eliminating the pesticide residues in potatoes during washing, including citric acid, acetic acid, hydrogen peroxide, sodium chloride, and sodium carbonate, at a concentration between 5% and 10% in wash water (Soliman, 2001; Zohair, 2001). The acidic solutions are more efficient than neutral and alkaline solutions in removing pesticides, especially organochloride compounds.

16.2.3 Cooling

Cooling is used to remove the field heat from fresh fruits and vegetables before further processing. This reduces water loss, slows down respiration and ripening (for fruits), and minimizes microbial growth. The cooling

(a)

(b)

Figure 16.7 (a) A destoner used for washing potatoes. Stones settle to the bottom. (b) Debris on top is removed with a screen.

conditions for different fruits and vegetables depend on the type, maturity, and cultivar. The most common cooling methods are water cooling, vacuum cooling, and air cooling.

16.2.3.1 Water cooling

In water cooling or hydro cooling, fruits and vegetables are immersed in cold water, which is usually in the bulk bins or flumes for transporting fruits and vegetables from the truck to the next processing step. This cooling method can be used for stem vegetables, leafy vegetables, and small fruits, such as peas, carrots, asparagus, tomatoes, melons, and peaches. However, some fruits and vegetables, such as strawberries, cannot be cooled by cold water since more water on the surface increases the risk of microbial spoilage (Aked, 2002). Water cooling produces uniform cooling and there is no weight loss from dehydration; however, it produces a lot of waste water and there can be a high risk of microbial contamination in the cooling water. Sanitation, such as chlorination, of the water is often used to prevent contamination of the fruits and vegetables.

16.2.3.2 Vacuum cooling

Vacuum cooling is one of the most rapid cooling methods, providing uniform cooling using a vacuum chamber. The pressure around the fruits and vegetables is decreased, decreasing the boiling point of water. The heat in the fruits and vegetables is absorbed by the surface water as it evaporates. This method is used for cooling fruits and vegetables that have a large surface area to volume ratio, such as spinach, cabbage, lettuce, and other leafy vegetables. Vacuum cooling can cause up to 3% moisture loss in the product but water sprayed on the surface of fruits and vegetables before cooling can help reduce the loss of water during cooling (Aked, 2002).

16.2.3.3 Air cooling

Air cooling cools fruits and vegetables by heat transfer from the product to cold air circulating at −1 °C to 16 °C with a relative humidity (RH) of 85–90%. This step can also be done with room temperature air. Air cooling is efficient for cooling tomatoes, apples, and cherries. This cooling method requires an intermediate investment cost and the system is easy to control; however, air cooling takes more time when compared to other methods. The rate of cooling can be improved by using forced air, where the cold air is forced with a pressure gradient into the chamber or container (Aked, 2002).

16.2.3.4 Tomato cooling

After harvesting, tomatoes need to be cooled to remove field heat, increase shelf life, and preserve quality. Ripe tomatoes for fresh consumption are cooled down within 12 h of harvest to 7–10 °C at RH 90% using forced air cooling. Tomatoes for processing are flume cooled to room temperature (Narayanasamy, 2006).

16.2.3.5 Apple cooling

The postharvest cooling of apples removes field heat, prolongs shelf life and reduces respiration. Within 12 h of harvest, apples are cooled down by either forced air cooling or hydro cooling at −1 °C to 4.5 °C with RH of 90–95% (Narayanasamy, 2006).

16.2.4 Peeling

Peeling is a critical step in the processing of many fruits and vegetables to remove undesirable parts which are either inedible or difficult to digest, and to enhance the physical appearance of the product. Efficient peeling methods remove minimal skin to produce a clean and undamaged surface. They should also use minimal energy and labor, and have low operating costs. The main methods for peeling fruits and vegetables are lye peeling, steam peeling, and mechanical peeling.

16.2.4.1 Lye peeling

Lye or caustic peeling applies a solution of lye (sodium hydroxide) at 10–20% at 100–120 °C for 2–6 min. During this process, the lye hydrolyzes the pectin, loosening the skin, and a high-pressure water spray with rubber disks or a perforated mesh cage is then used to remove the skin. The average product loss during this peeling method is 17% (Fellows, 2000). Lye peeling can be used in the peeling of peaches, nectarines, apricots, pears, tomatoes, potatoes, apples, carrots, sweet potatoes, and onions.

Lye peeling produces waste water that contains a high organic load and high pH. Time in the lye, temperature of the bath, and concentration are the three major controllable factors that determine peeling efficiency. Increasing any of these factors increases the extent of peel removal. Time and temperature are linearly correlated, while time and concentration are correlated exponentially; therefore, longer time in the lye at higher lye concentration and higher temperature increases peel removal (Bayindirli, 1994).

16.2.4.2 Steam peeling

Steam peeling is the application of high-pressure steam at 1500 kPa in a pressure vessel to peel fruits and vegetables (Fellows, 2000). It can be used in the peeling of beets, potatoes, tomatoes, carrots, and onions. In steam peeling, peel removal is possible because of rupture of the cells just underneath the peel. Due to the high temperature and

(a)

(b)

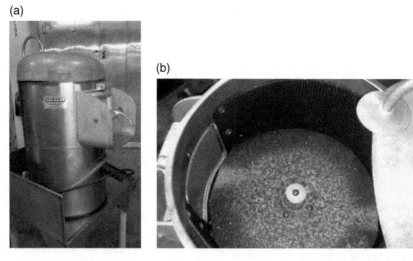

Figure 16.8 (a) Abrasion peeler for potatoes and sweet potatoes. (b) Rotating cylinder with an abrasive surface along the inner wall to remove the skin as the cylinder rotates.

pressure, the temperature of the water inside these cells exceeds the boiling point, but remains in a liquid state. When the pressure in the chamber is released, the water changes to steam, bursting the cells. Time, temperature, and pressure are the most critical factors to control to optimize the peeling process. The higher the temperature and pressure, the shorter the time required, and the more complete the peel removal. At higher temperatures, there is also less mushiness in the fruit due to cooking. The process uses relatively little water and produces little waste effluent; however, the peeling is less complete than in lye peeling. The waste peels that are produced can be used as fertilizer or animal feed or processed into other products, such as lycopene extract from tomato peeling (Knoblich et al., 2005).

16.2.4.3 Mechanical peeling

Mechanical peeling is mainly used for peeling fruits, such as apples, pears, pineapples, oranges, and other citrus fruits. Some vegetables can also be peeled by mechanical peeling, such as carrots, potatoes, and sweet potatoes. The most common mechanical peeling method uses either cutting tools (knife peeling) or an abrasive peeler. In knife peeling, the skin of fruits and vegetables is removed by either pressing stationary blades against the surface of fruits and vegetables, which are rotated, or rotating blades against the surface of stationary fruits and vegetables. In abrasion peeling of carrots, stiff brushes in a trough are used. In abrasion peeling of potatoes and sweet potatoes

(Figure 16.8), the peelers have carborundum rollers or a rotating cylinder with an abrasive surface along the inner wall, which removes the skin of the product as the roller or cylinder rotates. The skin is then washed away. This method operates with low energy and capital costs and no heat damage occurs; however, the average product loss can be up to 25% (Fellows, 2000).

16.2.4.4 Tomato peeling

Tomatoes are typically peeled before further processing. The Food and Drug Administration (FDA) standard of identity does allow for canned, unpeeled tomatoes if the processor so desires. This is not common in the market, though there are some unpeeled salsas. This is probably because the peel is very tough and undesirable to the consumer; in addition, unpeeled tomatoes would show many blemishes that are hidden from the consumer by peeling. Some easy-peel varieties have been bred that may be suitable for canning with the peel on, since the peel is less tough. However, these varieties also have less resistance to insect and microbial attack on the plant and so are not typically used by growers.

There are two commonly used peeling methods for tomatoes: steam and lye. In California, most peeling is done by steam, while in the Midwest United States and in Canada, peeling is done with a hot lye solution. In steam peeling, the tomatoes are placed on a moving belt one layer deep and pass through a steam box in a semi-continuous process. Steam peeling is done at 165.5–186.2 kPa, which equals about 127°C, for 25–40 sec.

In lye peeling, the tomatoes pass on a conveyor belt under jets of hot lye (sodium hydroxide) or through a lye tank in a continuous operation. The tomatoes go through a solution of 12–18% lye at 85–100 °C for 30 sec, followed by holding for 30–60 sec at room temperature to allow the lye to react. The lye dissolves the cuticular wax and hydrolyzes the pectin. The hydrolysis of the pectin in the middle lamella causes the cells to separate from each other, or rupture, causing the peel to come off (Bayindirli, 1994).

Lye peeling typically produces a higher yield of well-peeled tomatoes than steam peeling, but disposal of the lye waste water can be difficult. Steam gives a higher total tomato yield, but removes much less of the peel than lye. A 65% peel removal is considered good for steam peeling, while peel removal with lye is close to 100%. For this reason, lye is used exclusively in the Midwest United States, where peeled tomatoes are the most important tomato product.

After either steam or lye peeling, the tomatoes pass through a series of rubber disks or through a rotating drum under high-pressure water sprays to remove the adhering peel. Fruits with irregular shape and wrinkled skin are difficult to peel and result in excessive loss during the peeling step. Thus varieties prone to these characteristics are undesirable. Overpeeling is undesirable because it lowers the yield, results in higher waste, and strips the fruit of the red, lycopene-rich layer immediately underneath the peel, exposing the less attractive yellow vascular bundles.

16.2.4.5 Potato peeling

In potato processing, peeling is a critical step before further processing. Potatoes can be peeled by the application of steam, lye, or mechanical methods. Peeling can also be done by hand or using drum-shaped mechanized knife peelers but these methods are time consuming and the loss of product is high. Abrasion or steam peelers are commonly used before canning, freezing, or dehydration. In steam peeling, potatoes are loaded in a pressure vessel, which is slowly rotated at 5–6 rpm, and steam is quickly flashed to the product and immediately removed after steaming to avoid overheating (Saravacos & Kostaropoulos, 2002). In lye peeling, potatoes go through an immersion (Draper) lye scalder or a lye-spray scalder. In the immersion lye scalder, the peeling is achieved in a long tank with a capacity of approximately 20 tonnes/h. The lye-spray scalder consists of a conveyor belt, which can move the product slowly. The hot lye solution is sprayed on the potatoes and the skin is finally removed with sprays of wash water.

In lye peeling, the tissue of potatoes is damaged and a heat ring is formed inside the potato; therefore, this type of peeling is not used for potato chips. Potatoes for chipping are peeled using an abrasive peeler. The peeling drum has carborundum on the inner surface to remove potato skin by abrasion and a water spraying unit is used to wash the potato skin from the drum. In an abrasive peeler with 100 kg/h capacity, the loss of product was approximately 6% and the peeling efficiency was 78% (Singh & Shukla, 1995).

16.2.5 Blanching

Blanching is the application of heat at 85–95 °C for a few minutes, depending on the product, to inactivate enzymes in some fruits and vegetables before processing. The other reasons for blanching are reducing surface microbial load, removing intercellular gases, preheating the materials, softening the product, and stopping respiration and maturation. Enzymes that cause quality loss and changes in fruits and vegetables include lipoxygenase, polygalacturonase, and polyphenol oxidase. Peroxidase is a heat-resistant enzyme found in most vegetables, which is frequently used as a marker enzyme to indicate whether the products are correctly blanched. There are two common blanching methods used commercially: steam and hot water. The fruits and vegetables are rapidly heated by steam or hot water immersion to a preset temperature and then, depending on the next step, may be cooled to ambient temperature. The blanching time depends on several factors, including size and type of product, heating method, and blanching temperature.

16.2.5.1 Green bean blanching

Green beans are commercially blanched at 98–100 °C by hot water for 3.5 min before further processing, usually in the production of frozen green beans, in order to inactivate enzymes, especially lipoxygenase which causes off-flavors from lipid oxidation (Lee & Smith, 1988). Unblanched frozen green beans develop off-flavors after 4 weeks of storage while blanched products can be stored up to 9 months without significant off-flavors (Katsaboxalis & Papanicolaou, 1984; Stone & Young, 1985).

16.2.5.2 Apple blanching

Many fruits are not blanched during processing because of texture damage. However, apples are generally blanched before canning, freezing, or dehydration to inactivate the enzymatic browning activity of polyphenol oxidase and soften the texture, especially in the production of apple sauce and apple pie filling. During blanching, apples are either passed through a steam tunnel or immersed in hot water at 77–93 °C for 3–10 min (Arthey, 1997).

16.2.6 Size reduction

Size reduction decreases the average size of the pieces of the fruit or vegetable by cutting. Beside consumer demand and standard of identity, size reduction of fruits and vegetables before thermal processing decreases the thickness of the products, increasing the efficiency and rate of heat transfer during freezing, drying, and heating. In fruit and vegetable processing, methods of size reduction include chopping, cutting, slicing, and dicing, depending on the specific requirement of the processing technology. The equipment for size reduction includes slicers, dicers, shredders, and bowl choppers, depending on the preferred size of the final products.

16.2.6.1 Tomato size reduction

During the production of salsa, tomatoes are diced by rotating blades, which slice the fruits first and then cut them into strips. The second rotating blades then cut the strips into cubes or dices (Figure 16.9). In the production of chilled sandwiches, tomatoes are sliced using high-speed slicers, which hold the fruit against the slicer blades by centrifugal force.

16.2.6.2 Potato size reduction

Potatoes are usually cut in one plane to produce slices for making potato chips. In the production of French fries, potatoes are cut in two planes to produce strips. Potato slices for drying and frying are often rinsed in water before further processing. The rinsing stops polyphenol oxidase activity and minimizes black specks from burnt starch, as well as separating the slices from each other and from the equipment. The period of rinsing after slicing can be 10 to over 120 sec (Wicklund & Ivers, 1981).

16.2.7 Freezing

Freezing reduces the temperature of fruits and vegetables to below the freezing point of the product. This lowers water activity, slows down enzymatic reactions, and stops microbial growth. Ice crystal formation during freezing can damage texture. Many fruits and vegetables have to be blanched before freezing in order to reduce microbial load and stop enzymatic activity, especially enzymatic browning by polyphenol oxidase (PPO) and lipid

(a)

Figure 16.9 (a) Tomato dicer. (b) Rotating blades inside the dicer. The first rotating blades slice and then cut tomato into strips. The second rotating blades cut the strips into cubes.

oxidation by lipoxygenase. Lipoxygenase is frequently the target of blanching in frozen vegetables since it causes the destruction of product quality, including producing off-flavors, undesirable odors, loss of carotene, chlorophylls, and essential fatty acids. However, it is difficult to analyze for lipoxygenase activity so peroxidase is analyzed as the target of blanching instead. Commercial frozen fruits and vegetables include strawberries, raspberries, spinach, peas, green beans, and potatoes.

There are two main types of freezers: mechanical and cryogenic. Mechanical refrigerators use cooled fluid, cooled air or cooled surfaces to remove heat from the product. Spiral blast freezers are the most common mechanical refrigerator in commercial freezing. In this freezer, cooled air at $-30\,^{\circ}$C to $-40\,^{\circ}$C is circulated over fruits and vegetables using a high velocity at 1.5–6.0 m/sec. The products are placed on a continuous spiral conveyor belt that passes through the freezing tower.

Plate freezers are also used in freezing fruits and vegetables. In this type of freezer, the products are placed in layers on plates, which are cooled to $-40\,^{\circ}$C. Cryogenic freezers use liquid or solid carbon dioxide or liquid nitrogen, which directly contacts the product (Fellows, 2000). Plate freezers are used for prepackaged products in rectangular containers, while cryogenic freezing is not commonly used except for juices.

16.2.7.1 Apricot freezing

Apricots are frozen as peeled halves. Before freezing, either blanching or dipping the product in ascorbic acid solution is necessary to minimize browning due to polyphenol oxidase. The product is then packed in sugar or syrup at a 3:1 or 4:1 ratio of fruit to sugar prior to the freezing process. For freezing apricots, air-blast freezers are used and then the apricots are vacuum packaged to minimize color change. The storage of frozen apricots should be below $-18\,^{\circ}$C (Reid & Barrett, 2005).

16.2.7.2 Berry freezing

Many types of berries can be frozen either in syrup or as individual fruits. Before freezing, the berries are usually packed or dipped in 30–50% syrup as a barrier to oxygen and to protect flavor, color, and nutritional quality during freezing (FAO, 2005). The appropriate freezers are air-blast or cryogenic freezers. During freezing, the temperature of the freezer should be below $-26\,^{\circ}$C, so that it takes less than 48 h for the center of the container to reach $0\,^{\circ}$C,

and freezing should be maintained until the center temperature reaches $-17\,^{\circ}$C. The storage of frozen berries should be below $-18\,^{\circ}$C (Reid & Barrett, 2005).

16.2.8 Dehydration

Dehydration increases the shelf life of fruits and vegetables by decreasing water activity, which is the available water for microbial growth and enzyme activity. Therefore, sufficient water removal can prevent microbial growth and undesirable enzyme activity. Dehydration removes water by evaporation, such as sun drying, tunnel drying, and freeze drying. Sun and solar drying are widely used to dehydrate fruits and vegetables around the world due to their simplicity and low cost. However, this method causes microbial contamination and some changes in quality, such as color and texture. It also requires a large amount of land and a hot, dry climate. Tunnel driers are commonly used for dehydrating fruits and vegetables in a short time. In tunnel driers, the products are placed on trays stacking on trucks, which continuously move through a tunnel. Freeze drying is a method developed for minimizing quality changes. This method removes water in fruits and vegetables by sublimation at low temperature along with low pressure to preserve the quality of the product; therefore, water in the products is changed into ice first and then changed directly from the solid into the gaseous state without becoming liquid.

16.2.8.1 Apple dehydration

Apples, with an original water content of 85%, are usually dehydrated at 55–75 $^{\circ}$C for 5–6 h in tunnel driers until the product has a moisture content of approximately 20%. To prevent both non-enzymatic and enzymatic browning of apples during drying and storage, apple pieces are treated in 8000 ppm sulfur dioxide for 1 min (FAO, 1995). Dipping in ascorbic acid (0.7%) can also be used to minimize enzymatic browning and increase vitamin C content in the final product.

16.2.8.2 Potato dehydration

After grading and washing, potatoes must be cooked completely prior to cutting and drying to preserve the flavor, color, and nutrition of the product. Dried uncooked potatoes are not acceptable due to their dark color and poor flavor. Water or steam cooking can be used to cook potatoes for approximately 20–40 min

(Shaw & Booth, 1983). During production of potato flakes, some antioxidants, such as butylated hydroxyanisole, are added to protect carotenoids (Baardseth, 1989).

16.2.9 Canning process

The canning industry uses thermal processing to ensure microbial safety and shelf life extension of food products. Retorting is a process that relies on the transfer of heat to guarantee the safety of canned food. In this process, cans are filled with the food product and then sealed hermetically before retorting. Wet heat and pressure are applied within the retort to sterilize both the container and food product. This heat sterilization is essential in canning, especially for low-acid foods, which have a pH greater than 4.6 and a water activity greater than 0.85, such as papaya, bananas, melons, corn, green beans, and peas. These foods provide appropriate conditions for some spore-forming microorganisms and anaerobic microorganisms, such as *Bacillus coagulans* and *Clostridium botulinum*, to grow. Therefore low = acid food requires more severe heat treatment, such as 121.1 °C for 25 min for canning small carrots (FAO, 1995), than acid or acidified food (pH below 4.6 and water activity below 0.85), which can be canned at 100 °C.

Many fruits are acid and have a pH below 4.6, such as apricots, grapefruit, pineapples, tomatoes, and peaches. A thermal process at temperatures at or below 100 °C is used to destroy vegetative cells of spoilage microorganisms and inactivate enzymes. Hot-filling, a process of heating the juice with a heat exchanger to a fill temperature of 88–95 °C, then filling the juice into a container, is also sufficient for acidic beverages, such as cherry, cranberry, and apple juice. The maximum pH of the fruit juice to be hot-filled is 4 and the shelf life of the product is between 9 and 12 months (Mclellan & Padilla-Zakour, 2005).

16.2.9.1 Tomato canning

Tomato products can be hot filled or processed in a retort as needed to minimize spoilage. Because tomatoes are a high-acid food with a pH of 4.0–4.5, they do not have to be sterilized. However, most tomato products undergo a retort process to ensure an adequate shelf life of 24–30 months. The continuous rotary retort is most commonly used for tomato products. This retort provides agitation of the product and can handle large quantities in a continuous process. Because tomatoes are a high-acid food, a continuous rotary retort set at 104 °C for 30–40 min is common. Exact processing conditions depend on the

product being packed, the size of the can, and the type and brand of retort used. The key is for the internal temperature of the tomatoes to reach at least 88 °C.

16.2.9.2 Potato canning

Since potatoes are a low-acid food, the required processing temperature is 115.6–121.1 °C for 27–50 min, depending on the temperature, type of retort, and can size (FAO, 1995; Mishra & Sinha, 2011). Some commercial retorts include static retort, continuous retort, and hydrostatic pressure sterilization. Before canning, potatoes should be half-cooked using steam or boiling water to inactivate the enzymes and prevent discoloration.

16.2.10 Minimal processing

Many fresh fruits and vegetables are minimally processed to keep them fresh, prevent quality loss, and prolong shelf life. The shelf life of minimally processed fruits and vegetables is at least 4–7 days at 5 °C. Commercial minimally processed products are ready-to-eat prepeeled, sliced, grated, or shredded fruits and vegetables, such as precut lettuce, grated carrot, and shredded Chinese cabbage for salad mixes. Minimal processing of fresh fruits and vegetables includes strict hygiene and good manufacturing practices, careful cleaning and washing before and after peeling, mild additives in washing, gentle peeling, cutting, slicing, or shredding, and a low temperature (usually below 5 °C) during processing (Laurila & Ahvenainen, 2002).

In the production of minimally processed fruits and vegetables, packaging is an important factor that helps prolong their shelf life. Modified-atmosphere packaging (MAP) is one of the packaging methods used to reduce the respiration activity of the produce since it balances the levels of CO_2 and concentration of O_2 (generally at 2–5% for both gases) inside the package by using appropriate permeable packaging materials and/or a specific gas mixture in the package (Laurila & Ahvenainen, 2002).

16.2.10.1 Minimally processed potatoes

The minimal processing of potatoes is usually for prepeeled and sliced potatoes. The processing temperature is at 4–5 °C. The main issue is browning so citric acid with ascorbic acid at a maximum of 0.5% for both, combined with 2% (w/v) calcium chloride, 4-hexyl resorcinol or sodium benzoate, is used during washing. After washing, the produce should be packaged immediately in a gas mixture of 20% CO_2 and 80% N_2 in a package made of 80 μm nylon-polyethylene, which has an oxygen permeability of $70 \, cm^3 \, m^{-2}$ per 24 h, 101.3 kPa, 23 °C, RH 0%. The produce should be stored in the dark at 4–5 °C. The shelf life of prepeeled potatoes is 7–8 days and that of sliced potatoes is 3–4 days at 5 °C (Laurila & Ahvenainen, 2002).

16.2.10.2 Minimally processed carrots

Minimally processed grated carrots are processed at 4–5 °C. The optimal grate size for carrots is 3–5 mm. After grating, the product should be lightly sprayed with water and then mildly centrifuged to remove loose water. After centrifugation, grated carrots are packaging immediately with normal air in a package made of oriented polypropylene or polyethylene-ethylene vinyl acetate-oriented polypropylene, which has an oxygen permeability between 1200 and $5800 \, cm^3 \, m^{-2}$ per 24 h, 101.3 kPa, 23 °C, RH 0 %. The produce should be stored in the dark at 0–5 °C. The shelf life of grated carrots is 7–8 days at 5 °C (Laurila & Ahvenainen, 2002).

16.2.11 Sustainability

The major waste streams produced in fruit and vegetable processing are solid waste, from trimming and peeling, and waste water from washing, fluming, peeling, can washing, cooling, and clean-up. Table 16.2 shows the amount of waste water and solid waste produced in some fruit and vegetable industries. Waste management is beneficial not only to the environment, but also to the business, including reducing costs for waste treatment, decreasing resource consumption, and improving the reputation of the business. There are several methods of decreasing waste generation and improving waste management, such as recycling and reuse of materials, improving raw material management, and employee training.

Waste water from the fruit and vegetable industry can be categorized into low-polluted (wash water) and high-polluted (process water) waste. Low-polluted waste has a biological oxygen demand (BOD) of less than 200 mg/L while high-polluted waste has a BOD around 20,000 mg/L. Aerobic systems, such as aerated lagoons, are adequate to treat low-polluted waste. For high-polluted water, anaerobic biological treatment can be used with 70–95% efficiency (Tran, 2009).

Solid waste can be recycled or reused depending on its composition and properties. For example, pectin is recovered from peeled fruit skins and used in other food products. Waste materials containing cellulose can be fermented to sugar for production of organic acids, ethanol, and edible oils. Other fruit and vegetable solids are commonly used as fertilizers or animal feed.

Table 16.2 Amount of waste water and solid waste produced in some fruit and vegetable industries

Product	Water waste (m^3/metric tonne production)	Solid waste (kg/metric tonne production)
Carrots	12.0	200
Canned corn	4.5	40
Canned peas	20.0	40
Frozen potato	11.0	40
Canned peaches	13.0	180
Pears	12.0	200

Source: Monspart-Sényi (2006), Sogi and Siddiq (2010).

References

Adsule PG, Amba D, Onkarayya H (1982) Effects of hot water dipping on tomatoes. *Indian Food Packer* **36**(5): 34–37.

Aked J (2002) Maintaining the post-harvest quality of fruits and vegetables. In: Jongen WMF (ed) *Fruit and Vegetable Processing: Improving Quality*. Boca Raton, FL: CRC Press, pp. 119–149.

Arthey D (1997) Fruit and vegetable products. In: Ranken MD, Kill RC, Baker CGJ (eds) *Food Industries Manual*, 24th edn. London: Blackie Academic and Professional, pp. 139–171.

Azarnia S, Boye JI, Warkentin T, Malcolmson L (2011) Changes in volatile flavor compounds in field pea cultivars as affected by storage conditions. *International Journal of Food Science and Technology* **46**(11): 2408–2419.

Baardseth P (1989) Effect of selected antioxidants on the stability of dehydrated mashed potatoes. *Food Additives and Contaminants* **6**(2): 201–207.

Bartz JA (1999) Washing fresh fruits and vegetables: lessons from treatment of tomatoes and potatoes with water. *Dairy, Food and Environmental Sanitation* **19**(12): 853–864.

Bayindirli L (1994) Mathematical analysis of lye peeling in tomatoes. *Journal of Food Engineering* **23**(2): 225–231.

Brecht JK, Ritenour MA, Haard NF, Chism GW (2007) Postharvest physiology of edible plant tissues. In: Damodaran S, Parkin KL, Fennema OR (eds) *Food Chemistry*, 4th edn. Boca Raton, FL: CRC Press, pp. 975–1050.

California Department of Food and Agriculture (2001) *California Processing Tomato Inspection Program.* West Sacramento, CA: California Department of Food and Agriculture, Marketing Branch.

Cottrell JE, Duffus CM, Paterson L, Mackay GR, Allison MJ, Bain H (1993) The effect of storage temperature on reducing sugar concentration and the activities of three amylolytic enzymes in tubers of the cultivated potato, *Solanum tuberosum* L. *Potato Research* **36**(2): 107–117.

Delgado-Vargas F, Jiménez AR, Paredes-López O (2000) Natural pigments: carotenoids, anthocyanins, and betalains – characteristics, biosynthesis, processing, and stability. *Critical Review of Food Science and Nutrition* **40**(3): 173–289.

Deza-Durand KM, Petersen MA (2011) The effect of cutting direction on aroma compounds and respiration rate of fresh-cut iceberg lettuce (*Lactuca sativa* L.). *Postharvest Biology and Technology* **61**(1): 83–90.

Fellows P (2000) *Food Processing Technology: Principles and Practice*, 2nd edn. Boca Raton, FL: CRC Press.

Floros JD, Chinnan MS (1989) Determining the diffusivity of sodium hydroxide through tomato and capsicum skins. *Journal of Food Engineering* **9**(2): 129–141.

Food and Agriculture Organization (FAO) (1995) *Fruit and Vegetable Processing.* Bulletin no. 119. Rome: Food and Agriculture Organization.

Food and Agriculture Organization (FAO) (2005) *Freezing of Fruits and Vegetables.* Bulletin no. 158. Rome: Food and Agriculture Organization.

Haard NF (1984) Postharvest physiology and biochemistry of fruits and vegetables. *Journal of Chemical Education* **61**(4): 277–283.

Heil JR, SLeonard S, Patino H (1984) Microbiological evaluation of commercial fluming of tomatoes. *Food Technology* **38**(4): 121–126.

Herbach KM, Stintzing FC, Carle R (2006) Betalain stability and degradation – structural and chromatic aspects. *Journal of Food Science* **71**(4): R41–R50.

Hucl P, Chibbar RN (1996) Variation for starch concentration in spring wheat and its repeatability relative to protein concentration. *Cereal Chemistry* **73**(6): 756–758.

Ibdah M, Krins A, Seidlitz HK, Heller W, Strack D, Vogt T (2002) Spectral dependence of flavonol and betacyanin accumulation in *Mesembryanthemum crystallinum* under enhanced ultraviolet radiation. *Plant, Cell and Environment* **25**(9): 1145–1154.

Karmas E, Harris RS (1988) *Nutritional Evaluation of Food Processing*, 3rd edn. New York: Van Nostrand Reinhold.

Kar RK, Choudhuri MA (1987) Possible mechanisms of light-induced chlorophyll degradation in senescing leaves of *Hydrilla verticillata*. *Physiologia Plantarum* **70**(4): 729–734.

Katsaboxalis KZ, Papanicolaou DN (1984) *The Consequences of Varying Degrees of Blanching on the Quality of Frozen Green*

Beans. London: Proceedings of the Seminar of European Cooperation in Scientific and Technical Research, pp. 684–690.

Knoblich M, Anderson B, Latshaw D (2005) Analyses of tomato peel and seed byproducts and their use as a source of carotenoids. *Journal of the Science of Food and Agriculture* **85**(7): 1166–1170.

LaBorde LF, von Elbe JH (1996) Methods of improving the color of containerized green vegetables. US Patent no. 5482727.

Laurila E, Ahvenainen R (2002) Minimal processing of fresh fruits and vegetables. In: Jongen WMF (ed) *Fruit and Vegetable Processing: Improving Quality.* Boca Raton, FL: CRC Press, pp. 288–309.

Lee CY, Smith NL (1988) Enzyme activity and quality of frozen green beans as affected by blanching and storage. *Journal of Food Quality* **11**(4): 279–287.

Mabry TJ, Wyler H, Parikh I, Dreiding AS (1967) The conversion of betanidin and betanin to neobetanidin derivatives. *Tetrahedron* **23**(7): 3111–3127.

Madhavi DL, Salunkhe DK (1998) Tomato. In: Salunkhe DK, Kadam SS (eds) *Handbook of Vegetable Science and Technology: Production, Composition, Storage, and Processing.* New York: Marcel Dekker, pp. 171–202.

Maharaj V, Sankat CK (1996) Quality changes in dehydrated dasheen leaves: effects of blanching pre-treatments and drying conditions. *Food Research International* **29**(5-6): 563–568.

Mclellan MR, Padilla-Zakour OI (2005) Juice processing. In: Barrett DM, Somogyi LP, Ramaswamy HS (eds) *Processing Fruits: Science and Technology.* Boca Raton, FL: CRC Press, pp. 73–96.

Mishra DK, Sinha NK (2011) Principles of vegetable canning. In: Sinha NK (ed) *Handbook of Vegetables and Vegetable Processing.* Danvers, MA: Blackwell Publishing, pp. 243–258.

Monspart-Sényi J (2006) Fruit processing waste management. In: Hui YU (ed) *Handbook of Fruits and Fruit Processing.* Danvers, MA: Blackwell Publishing, pp. 171–188.

Morales-Blancas EF, Chandia VE, Cisneros-Zevallos L (2002) Thermal inactivation kinetics of peroxidase and lipoxygenase from broccoli, green asparagus and carrots. *Journal of Food Science* **67**(1): 146–154.

Narayanasamy P (2006) *Postharvest Pathogens and Disease Management.* Hoboken, NJ: John Wiley, pp. 213–219.

Oliw EH (2002) Plant and fungal lipoxygenases. *Prostglandins and Other Lipid Mediators* **68–69**: 313–323.

Pasch JH, von Elbe JH (1979) Betanine stability in buffered solution containing organic acids, metal cations, antioxidants, or sequeatrants. *Journal of Food Science* **44**(1): 71–75.

Reid DS, Barrett DM (2005) Fruit freezing. In: Barrett DM, Somogyi LP, Ramaswamy HS (eds) *Processing Fruits: Science and Technology.* Boca Raton, FL: CRC Press, pp. 161–172.

Richings EW, Cripps RF, Cowan AK (2000) Factors affecting 'Hass' avocado fruit size: carbohydrate, abscisic acid and isoprenoid metabolism in normal and phenotypically small fruit. *Physiologia Plantarum* **109**(1): 81–89.

Rickman JC, Bruhn CM, Barrett DM (2007) Nutritional comparison of fresh, frozen, and canned fruits and vegetables II. Vitamin A and carotenoids, vitamin E, minerals and fiber – a review. *Journal of the Science of Food and Agriculture* **87**(7): 1185–1196.

Ruenroengklin N, Sun J, Shi J, Xue SJ, Jiang Y (2009) Role of endogenous and exogenous phenolics in litchi anthocyanin degradation caused by polyphenol oxidase. *Journal of Food Chemistry* **115**(4): 1253–1256.

Saravacos GD, Kostaropoulos AE (2002) *Handbook of Food Processing Equipment.* New York: Kluwer Academic/Plenum Publishers.

Schoeneck O (1961) Wash tank and stone separator. *US Patent no.* **2990064**.

Shaw R, Booth R (1983) *Simple Processing of Dehydrated Potatoes and Potato Starch.* Lima, Peru: International Potato Center, pp. 14–26.

Singh KK, Shukla BD (1995) Abrasive peeling of potatoes. *Journal of Food Engineering* **26**(4): 431–442.

Sogi DS, Siddiq M (2010) Waste management and utilization in vegetable processing. In Sinha NK (ed) *Handbook of Vegetables and Vegetable Processing.* Danvers, MA: Blackwell Publishing, pp. 423–442.

Soliman KM (2001) Changes in concentration of pesticide residues in potatoes during washing and home preparation. *Food and Chemical Toxicology* **39**(8): 887–891.

Spanos GA, Wrolstad RE (1992) Phenolics of apple, pear, and white grape juices and their changes with processing and storage – a review. *Journal of Agricultural and Food Chemistry* **40**(9): 1478–1487.

Stafford HA (1994) Anthocyanins and betalains: evolution of the mutually exclusive pathways. *Plant Science* **101**(2): 91–98.

Stone MB, Young CM (1985) Effects of cultivars, blanching techniques, and cooking methods on quality of frozen green beans as measured by physical and sensory attributes. *Journal of Food Quality* **7**(4): 255–265.

Takeda K (2006) Blue metal complex pigments involved in blue flower color. *Proceedings of the Japan Academy, Series B* **82**(4): 142–154.

Tran MDT (2009) Food processing and food waste. In: Baldwin CJ (ed) *Sustainability in the Food Industry.* Danvers, MA: Blackwell Publishing, pp. 23–60.

Unten L, Koketsu M, Kim M (1997) Antidiscoloring activity of green tea polyphenols on β-carotene. *Journal of Agricultural and Food Chemistry* **45**(6): 2009–2012.

US Department of Agriculture (USDA) (1978) *United States Standards for Grades of Potatoes for Chipping.* Washington, DC: Fruit and Vegetable Division, Agricultural Marketing Service, USDA.

US Department of Agriculture (USDA) (1983a) *United States Standards for Grades of Tomatoes for Processing.* Washington, DC: Fruit and Vegetable Division, Agricultural Marketing Service, USDA.

US Department of Agriculture (USDA) (1983b) *United States Standards for Grades of Potatoes for Processing.* Washington, DC: Fruit and Vegetable Division, Agricultural Marketing Service, USDA.

US Department of Agriculture (USDA) Agricultural Research Service (2011) *USDA National Nutrient Database for Standard Reference, Release 24.* Nutrient Data Laboratory home page. http://ndb.nal.usda.gov/, accessed 20 November 2013.

von Elbe JH, Schwartz SJ (1996) Colorants. In: Fennema OR (ed) *Food Chemistry*, 3rd edn. New York: Marcel Dekker, pp. 651–722.

Whitaker JR (1996) Enzymes. In: Fennema OR (ed) *Food Chemistry*, 3rd edn. New York: Marcel Dekker, pp. 431–530.

Wicklund PA, Ivers JT (1981) Process of making potato chips. US Patent no. 4277510.

Wisniewsky MA, Glatz BA, Gleason ML, Reitmeier CA (2000) Reduction of *Escherichia coli* O157:H7 counts on whole fresh apples by treatment with sanitizers. *Journal of Food Protection* **63**(6): 703–708.

Wrolstad RE (2009) Marashino cherry: a laboratory-lecture unit. *Journal of Food Science and Education* **8**(1): 20–28.

Wrolstad RE, Skrede G, Lea P, Enersen G (1990) Influence of sugar on anthocyanin pigment stability in frozen strawberries. *Journal of Food Science* **55**(4): 1064–1065.

Yokotsuka K, Singleton VL (1997) Disappearance of anthocyanins as grape juice is prepared and oxidized with PPO and PPO substrates. *American Journal of Enology and Viticulture* **48**(1): 13–25.

Zohair A (2001) Behavior of some organophosphorus and organochlorine pesticides in potatoes during soaking in different solutions. *Food and Chemical Toxicology* **39**(7): 751–755.

17 Milk and Ice Cream Processing

Maneesha S. Mohan,[1] Jonathan Hopkinson,[2] and Federico Harte[1]

[1]Department of Food Science and Technology, University of Tennessee, Knoxville, Tennessee, USA
[2]Danisco USA, New Century, Kansas, USA

17.1 Introduction

Milk products have been consumed by humans since prehistoric times. Archeological records show that humans began milking goats *ca.* 9000 years ago (Evershed et al., 2008) and manufacturing cheese at least *ca.* 7000 years ago (Salque et al., 2012). Milk has become an important constituent of our diet for a number of reasons, including but not limited to its high nutritional quality. The annual production/consumption of milk in the world was 749 million metric tonnes and the average per capita consumption of milk was 107.3 kg/person/year in 2011 (IDF, 2012). According to the Food and Agriculture Organization (FAO), the largest producers of milk in the world in 2011 were USA and India (FAO, 2011). The annual US milk production increased from 56 million metric tonnes in 1979 (USDA-NASS, 1980) to 87 million metric tonnes in 2010 (USDA-NASS, 2011). California and Wisconsin are the largest milk-producing states in the US. The availability of fluid milk and cream (half and half, light, heavy, sour cream, and eggnog) in the US in 2009 was 92 kg/person, and for all dairy products about 276 kg (including fluid milk cream, cheese, frozen dairy products, evaporated and condensed milks and dry dairy products) (USDA-ERS, 2011). In order to sustain the growing consumption, the US dairy industry has grown considerably over the years. Along with the amount processed, the variety of dairy products on the market has also increased.

In the present chapter, we focus on fresh milk products including fluid milks and ice cream. Other milk products on the market, including fermented milk products (e.g. yogurt, kefir, and kumiss) and fermented-coagulated products (e.g. cheese), are handled in the next chapter.

17.1.1 Milk – definition and composition

According to the US Code of Federal Regulations (21 CFR Ch. 1, Subpart B, 131.110):

> "Milk is the lacteal secretion, practically free from colostrum, obtained by the complete milking of one or more healthy cows. Milk that is in final package form for beverage use shall have been pasteurized or ultra-pasteurized, and shall contain not less than 8.25% milk solids not fat and not less than 3.25% milk fat. Milk may have been adjusted by separating part of the milk fat therefrom, or by adding thereto cream, concentrated milk, dry whole milk, skim milk, concentrated skim milk, or nonfat dry milk. Milk may be homogenized."

Milk intended for interstate shipment and sale requires pasteurization, while intrastate shipment and sale of milk are regulated by states.

The major constituents of milk are shown in Table 17.1. However, bovine milk composition varies with breed, type of milking (time, intervals, procedure, and completeness), season, age of cow, stage of lactation, variations in individual cow, health and condition of cattle, and nutrition provided to the cattle (Jenkins & McGuire, 2006; Laben, 1963; Legates, 1960; Loganathan & Thompson, 1968).

17.2 Physical and chemical properties of milk constituents

17.2.1 Fat

Milk fat is secreted as a fat globule surrounded by a milk fat globule membrane (MFGM), which maintains the integrity of the globule and maintains fat as an emulsion in the surrounding aqueous phase. The interfacial tension between milk fat globule and milk serum is about

*Department of Health and Human Services

Food Processing: Principles and Applications, Second Edition. Edited by Stephanie Clark, Stephanie Jung, and Buddhi Lamsal.
© 2014 John Wiley & Sons, Ltd. Published 2014 by John Wiley & Sons, Ltd.

Table 17.1 Major constituents of cow milk

Major constituents	Fractions (% of major constituent)	Minor fractions
Water (87%)		
Lipids (4%)	Triacylglycerides (95.8%)	
	Di- and monoglycerides (2.4%)	
	Phospholipids (1.1%)	
	Sterols (0.5%)	
	Free fatty acids (0.3%)	
Solids not fat (8.9%)		
Proteins (3.3%)	Casein (78.3%)	α_{S1}, α_{S2}, β, κ, γ caseins
	Whey proteins (17.0%)	α-Lactalbumin
		β-Lactoglobulin
		Immunoglobulins (Ig) – IgG and IgM
		Proteose peptone
		Bovine serum albumin
	Membrane proteins (2%)	
	Enzymes	Lipases, proteinasaes, phosphatases, nucleases, lactoperoxidase
Carbohydrates (5%)	Lactose (95.8%)	
	Glucose (2.0%)	
	Others (2.2%)	Galactose
		Oligosaccharides
Organic and inorganic ions (0.7%)	Chloride (27.9%)	
	Citrates* (25.8%)	
	Phosphates* (23.5%)	
	Potassium (22.2%)	
	Calcium* (18.7%)	
	Sodium (9.2%)	
	Magnesium* (1.9%)	
Minor constituents	Organic acid	
	Vitamins	Vitamin A
		Vitamin B complex: thiamine, riboflavin, niacin, pyridoxine, pantothenic acid, biotin, folic acid, cyanocobalamin
		Vitamin C
		Vitamin D
		Vitamin E
		Vitamin K
	Non-protein nitrogen	Amino acids
		Nitrogenous compounds: ammonia, urea, creatine, creatinine, uric acid, nitrogen
		Nucleic acids

* Part of it is soluble and part of it is present as a colloidal dispersion.
Sources: Flynn & Cashman (1992), Walstra et al. (2006), Chandan (2006).

$2\,Nm^{-1}\,s^{-1}$ and that between non-globular liquid fat and milk serum is about $15\,Nm^{-1}\,s^{-1}$ (Fox & McSweeney, 1998a). This reduction in the interfacial tension by the MFGM stabilizes the fat emulsion in the serum phase. Milk contains a high concentration of saturated fatty acids (62%), followed by lower concentrations of monounsaturated (29%) and polyunsaturated fatty acids (4%). Among saturated fatty acids, short- and medium-chain fatty acids (C4:0 to C18:0) are in abundance. Among short-chain fatty acids, butyric acid (C4:0) is in highest concentration, at 3.8%. Oleic acid is the dominant unsaturated fatty acid residue (27.42%). Milk fat exists mainly as triacylglycerides and other fractions of lipids are mono- and diglycerides, free fatty acids, and compound lipids (phospholipids and cholesterols) (Chandan, 2006).

Most of the fat globules in bovine milk have a size distribution of 0.1–20 μm (Fox & McSweeney, 1998a). The thermal conductivity of cream is about $0.36 \, W.m^{-1}{}^{\circ}C^{-1}$ (Sweat & Parmelee, 1978). The specific heat of milk fat is about $2177 \, kJkg^{-1} \, K^{-1}$, which is temperature dependent. The electrical conductivity is less than $10^{-12} \, S/cm$ and the dielectric constant is about 3.1, depending on frequency (Walstra & Jenness, 1984). The refractive index of milk fat is in the range of 1.3440–1.4552 at $40\,^{\circ}C$ (Fox & McSweeney, 1998a). Milk fat is liquid above $40\,^{\circ}C$ and solid below $-40\,^{\circ}C$. In between these temperatures, it exists as a mixture of crystals and oil. Milk fat has a specific gravity of 0.900, which is less than the rest of milk, with specific gravity close to 1.036 (Mulder & Walstra, 1974). This difference in specific gravity enables the separation of fat from skim milk by centrifugal and gravitational forces.

17.2.2 Protein

The protein in milk are primarily caseins (ca. 80%) and whey proteins (ca. 20%). Casein is defined as those proteins in milk that precipitate at or above pH = 4.5. The caseins are phosphoproteins that form a complex quaternary structure (casein micelle), but lack tertiary structure. The five major casein fractions in bovine milk are α_{S1}, α_{S2}, β, γ, and κ fractions. Fragments of β-casein, called γ-casein and proteose peptone, are formed by hydrolysis. The α_{S1}-, α_{S2}-, β-, γ-, and κ-caseins are in the approximate ratio of 4:1:3.5:1.5 in milk (Swaisgood, 2003). The casein micelles contain in excess of 20,000 individual protein molecules hydrated with 3–4 g of water per g of protein (Dalgleish, 2011). In these micelles, α_S- and β-caseins are extensively phosphorylated and are bound with calcium phosphate. The α_S- and β-caseins are highly calcium sensitive (precipitates in the presence of 0.25 M calcium chloride), while κ-casein (does not precipitate in the presence of 0.25 M calcium chloride) is calcium insensitive (Fox, 2003). The κ-casein stabilizes the particles by extending the caseinomacropeptide moiety, containing sufficient amounts of hydrophilic amino acids, into the serum. This forms a "hairy" layer on the micelle surface that provides steric and electrostatic stabilization (de Kruif & Zhulina, 1996). The α_S- and β-caseins form the internal structure of the micelle, owing to the presence of patches of hydrophobic residues, which enable hydrophobic interactions with other hydrophobic proteins.

The internal structure of casein micelles is still being debated. A number of models have been suggested by researchers depicting the internal structure of the casein micelles, including the submicelle (Walstra, 1990), dual binding (Horne, 1998), and nanocluster (Holt et al., 2003) models. They all emphasize the importance of hydrophobic bonds and calcium bridges in holding the micelle together. All these models have been reviewed by Dalgleish (2011).

The size of casein micelles ranges from 80 to 400 nm, with an average of 200 nm (de Kruif, 1998). Its molecular weight is $3.7 \times 10^8 \, Da$. The casein micelles, owing to their colloidal dimensions, are capable of scattering light and are responsible for the white color in milk. The micelles carry an overall negative charge at neutral pH, equivalent to zeta potential of about $-20 \, mV$ (Dalgleish, 2011). Caseins precipitate close to pH 4.6 (isoelectric pH) and at high temperatures of $140\,^{\circ}C$ for 15–20 min at normal pH of milk.

Whey proteins are globular proteins consisting mainly of α-lactalbumin and β-lactoglobulin, forming approximately 20% of proteins in bovine milk. The globular structure is mainly attributed to the disulfide bonds present in them (Smith & Campbell, 2007). Whey proteins have a net negative charge at pH 6.6 (normal milk pH). β-Lactoglobulin is the major whey protein in bovine milk (50% of total whey proteins). It does not occur in human milk and is one of the primary allergenic protein in bovine milk for human infants. Whey proteins are rich in sulfur amino acids and have high biological value. α-Lactalbumin is a calcium-binding metalloprotein and constitutes 20% of whey proteins.

17.2.3 Salts

Milk salts include organic and inorganic salts. The inorganic salts include Na^+, K^+, Ca^{2+}, Mg^{2+}, Cl^-, $CO_3{}^{2-}$, $SO_4{}^{2-}$, and $PO_4{}^{3-}$; organic salts include amines, citrate, carboxylic acid, and phosphoric esters. Salts in milk can be present in the dissolved and undissolved (colloidal) form (as given in Table 17.1). The dissolved salts play an important role in milk protein stability. A part of the colloidal salts is associated as counter ions to the negatively charged casein micelles. The rest of the undissolved salts consist of calcium and phosphate that are integral to the structure of the casein micelles. Salts are the principal buffering compounds in milk, especially soluble calcium phosphate, citrate, and bicarbonate (Fox & McSweeney, 1998a), though proteins also contribute to buffering capacity. The stability of proteins in milk can be associated with the molar ratio of calcium and phosphorus (including organic phosphate), which is 0.9 in bovine milk (Walstra et al., 2006). Salts, particularly Na^+, K^+, and Cl^-, are responsible for electrical conductivity of milk. Increase in electrical conductivity may occur due to bacterial fermentation of lactose to lactic acid.

17.2.4 Lactose

Lactose is the primary carbohydrate in bovine milk. It is a disaccharide of galactose and glucose units, linked by a β-1-4 glycosidic bond (4-O-β-D-galactopyranosyl-D-glucopyranose). Lactose is a reducing sugar and exists as α and β anomers in solution. In solution, lactose undergoes mutarotation between α and β forms, involving interchanging of OH and H groups on the reducing group. Mutarotation of lactose depends on temperature and pH. The α and β forms vary in solubility, crystal shape, size, hydration of crystal form, specific rotation, and sweetness. The α-lactose crystallizes as α-monohydrate and α-anhydrous forms, while β-lactose crystallizes into β-anhydride crystal. Some of the physical properties of the two common forms of lactose are given in Table 17.2.

Lactose crystallizes much faster than other sugars like sucrose. It is confusing to talk about supersaturated solutions and which sugar crystallizes first. Crystallization of lactose leads to formation of large crystals, which are associated with sandy texture defect in sweetened condensed milk and ice cream (Fox & McSweeney, 1998b).

17.3 Milk handling

In a typical modern dairy farm, milk is transferred after milking, rapidly cooled to 4 °C in bulk tanks (300–30,000 L), and kept for a maximum of 72 h before transportation to the processing facility. Milk is transported in milk trucks with standard capacity of about 25,000 L/truck and distances that vary from industry to industry. For most dairy industries, the milk hauler/sampler is responsible for an initial quality control (e.g. check for appropriate storage conditions, aroma, and visual appearance of milk) and also collects samples from the bulk tank to be tested at a certified state laboratory and after reception in the plant. A bulk milk hauler/sampler

Table 17.2 Physical and chemical properties of cow milk

Properties	Range
pH (25 °C)	6.5 to 6.7; average 6.6
Acidity	1.3 to 2.0 meq OH^- per 100 mL
	0.14 to 0.16% lactic acid
Redox potential	+0.25 to +0.30 V
Specific gravity	1.023 to 1.040 (20 °C); average 1.032 (20 °C)
Coefficient of viscocity	2.127 mPa.sec
Viscosity	1.5 to 2.0 mPa.sec (20 °C); average value for whole milk 1.45 mPa.sec (27 °C)
Refractive index ($[\alpha]_D^{20}$)	1.3440 to 1.3485
Specific refractive index	0.2075
Electrical conductivity (25 °C)	0.4 to 0.55 A V^{-1} m^{-1}; average 0.5 A V^{-1} m^{-1}
Surface tension	50 to 52 N m^{-1}
Specific heat	3.931 kJ kg^{-1} K^{-1}
Heat capacity[*]	3841 J kg^{-1} K^{-1} at 50 °C
Thermal diffusivity	1.25×10^{-7} m^2 s^{-1}
Thermal conductivity	0.53 W m^{-1} K^{-1} at 20 °C, 0.61 W m^{-1} K^{-1} at 20 °C
Ionic strength	0.067 to 0.080 M; average 0.073 M
Colligative properties	
Freezing point	−0.512 to −0.550 °C; average −0.522 °C
Boiling point	100.15 °C
Osmotic pressure	700 KPa at 20 °C
Specific electrical conductance (25 °C)	0.0040 to 0.0055 ohm^{-1} cm^{-1}; average 0.0050 ohm^{-1} cm^{-1}
Water activity	0.993
Coefficient of cubic expansion	0.08^3 m^3 K^{-1}

[*] Calculated based on the equation: Heat capacity of milk = 2.976 × temperature (°C) + 3692.
Sources: Walstra & Jenness (1984), Sherbon (1999), Singh et al. (1997), Chandan (2006), Walstra et al. (2006).

undergoes a training program and is examined by testing for competency, as prescribed by a regulatory agency, to be licensed. The procedure of milk sampling and care is followed as specified by the *Standard Methods for Examination of Dairy Products* (SMEM; Graham, 2004).

Milk trucks are weighed at arrival and then milk is transferred to large-capacity storage silos. During pumping, a heat exchanger may be used to assure proper temperature of the milk being transferred to the storage silo. The milk, based on its sanitary, microbial, chemical and handling quality, is graded as A (Grade A) or B (manufacturing grade) milk by the US Department of Agriculture (USDA). The standards for Grade A milk are published in the *Grade "A" Pasteurized Milk Ordinance* (PMO; FDA, 2011). The details of the grading, standards, and regulating agencies are provided later on in this chapter.

Milk pricing in the US is a complex system which varies between grades of milk and from industry to industry. The base pricing, especially for Grade A milk, is set administratively through the state and federal milk marketing orders. The federal milk marketing orders are not mandatory, but the dairy producers can request and approve them as authorized by the Agricultural Marketing Agreement Act of 1937. The milk marketing orders utilize "classified pricing" to establish minimum prices for milk and milk components based on their use and "pooling" to set uniform pricing regardless of how milk is used. On the other hand, shelf-stable manufactured dairy product pricing is determined either by the market demand and supply or the dairy price support program. The dairy price support program operates through the USDA's Commodity Credit Corporation (CCC). It maintains the price level and product supply of many manufactured dairy products, including but not restricted to butter, buttermilk, non-fat dry milk, and cheese. A good review of the basic milk pricing concepts has been authored by Jesse and Cropp (2008). Some methods used in base pricing are based on fat and protein content, together with bonuses and/or penalties for milk quality, while more complex pricing systems are adopted by many dairy industries and organizations (Manchester & Blayney, 2001).

17.3.1 Processing of fluid milk

17.3.1.1 Cream separation and standardization

The standard of identity for each dairy product sets different fat and total solids percentage in the final product. Standardization is the process of adjusting the composition, especially the fat content, to the prescribed amount. The process that enables standardization by removing the fat content in milk is cream separation (Figure 17.1). The principle behind cream separation is the lower density of fat compared to the rest of the milk phase. This density difference enables its separation from the rest of the milk by the application of centrifugal force.

The equipment assembly of a cream separator consists of conical disks mounted on top of a bowl (Figure 17.2). The disks are separated by gaps between the plates and have a hole on either side of the cone. Whole milk is pumped from the bottom center of the cone. With centrifugal force produced by the rotation of the disks along with the disk holder, cream moves up to the top and through the holes and skim milk through the sides, to come out of different channels, thus separating them from each other. The cream (approximately 40% fat) can then be used to standardize milk and milk products for fat content. The skim milk stream typically contains less than 0.1% fat.

17.3.1.2 Homogenization

One of the processes that have become standard in the last 50 years for fluid milk and dairy products is homogenization. In the simplest terms, homogenization is subjecting a fluid to extreme mixing energy. This can be done by forcing the fluid through a small hole or gap at high velocity or by very high speed propeller mixing. For fluid milk, homogenization helps control the size of the fat particles this prevents creaming milk fat on storage. Having this control helps prevent the fat particles from agglomerating and floating to the top. Historically, homogenizers were developed early in the 20th century to prevent stealing of valuable fat by whole sellers (Trout 1948). Today most consumers prefer to consume milk products without separation. A few consumers are willing to pay a premium for non homogenized milk, however.

For the most part, homogenizers used in the dairy industry are specialized reciprocating high-pressure pumps with one or two restrictions to flow (valves). Figure 17.3 shows a schematic of a homogenizing valve. The pump propels fluid through the valves at high pressure and volume, causing the fluid to be exposed to very high shear forces and in some cases cavitation.[1] This causes the milk fat globule (MFG) to break apart. Force applied to the valve decreases the size of the gap;

---

[1] Cavitation is the formation and implosion of cavities in a liquid due to rapid changes in pressure. Shock waves due to cavitation are thought to contribute to either the breakdown of the fat globules or to the breaking of clumps of fat globules immediately after the point of maximum shear.

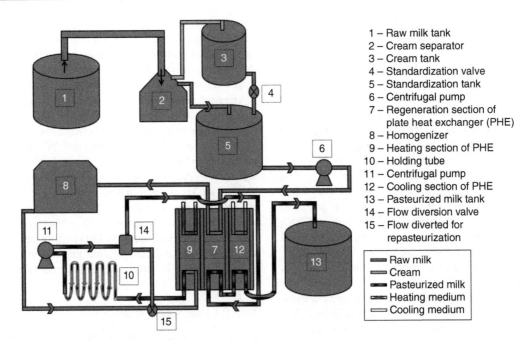

1 – Raw milk tank
2 – Cream separator
3 – Cream tank
4 – Standardization valve
5 – Standardization tank
6 – Centrifugal pump
7 – Regeneration section of
      plate heat exchanger (PHE)
8 – Homogenizer
9 – Heating section of PHE
10 – Holding tube
11 – Centrifugal pump
12 – Cooling section of PHE
13 – Pasteurized milk tank
14 – Flow diversion valve
15 – Flow diverted for
      repasteurization

▭ Raw milk
▭ Cream
▭ Pasteurized milk
▭ Heating medium
▭ Cooling medium

Figure 17.1 Schematic diagram of a typical fluid milk processing unit.

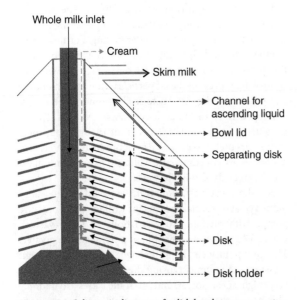

Figure 17.2 Schematic diagram of a disk bowl cream separator.

this in turn increases the pressure and velocity of the suspension of fat globules in the gap. The shear forces thus generated causes the fat globule size to decrease.

Most homogenizers have two valves; the purpose of the first valve is to do the actual size reduction of the globules.

The second valve functions to put a little back pressure on the first, this helps separate the clusters of fat globules that quickly form after the first stage. This also gives a little time for the proteins and MFGMs to form a new fat globule membrane (Walstra, 1975). Sometimes this pressure is reduced or eliminated from the second stage, to encourage these agglomerations. This is done to achieve increased viscosity in creams and other dairy products. The first stage of homogenization usually employs 2000–2500 psi (14–17 MPa) and the second stage 300–1000 psi (4–7 MPa) for milk (see Figure 17.1).

For homogenization to work, the fat must be in the liquid state. For dairy fat, the temperature must be greater than 52 °C for fats to be in liquid state. Higher temperatures result in lower fat viscosity and this in turn results in smaller, more uniform globules. Another benefit of homogenizing milk at elevated temperatures is that inactivation of the enzyme lipase occurs. Lipase is an enzyme that cleaves fatty acids from triglycerides, which can lead to rancid off-flavors in milk. When fat globule size is reduced, the amount of surface that the fat globules have is increased. For milk, the average diameter decreases from 3.3 μm without homogenization to 0.2 μm after homogenization, and the total surface area increases from 80 m^2/L without homogenization to 1250 m^2/L after homogenization. Protein, primarily casein, provides coverage

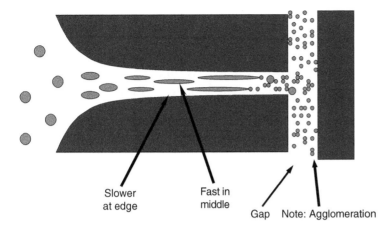

Slower at edge   Fast in middle   Gap   Note: Agglomeration

Figure 17.3 Diagram of a milk homogenization valve.

for the surface generated by homogenization. When milk is homogenized, the tendency for the fat to cluster together and float to the top of container is reduced. Equation 17.1 describes the relationship between fat globule size and the speed of the globule rising to the top.

$$v_s = \frac{-a(\rho_p - \rho_f)d^2}{18\eta_p} \tag{17.1}$$

Equation 17.1. Stokes velocity, the velocity of a particle of density ρ_f and diameter d in a fluid with density ρ_p and viscosity η_p. In the case of creaming due to gravity, a is equal to the force of gravity (9.8 m/s^2).

Factors such as agglomeration of fat and other particles, number of particles, fat globule shape, and the nature of the fat globule surface will also affect the velocity. This formula does describe, in a general way, how the size of the fat globule affects the speed of creaming. Decreasing the fat globule size or increasing the serum viscosity will slow creaming and size will have the stronger effect. Since the fat globules are moving violently during homogenization, individual fat globules may bump into each other and agglomerates or clusters of fat globules are formed in these collisions.

17.3.1.3 Pasteurization

The primary purpose of pasteurization is to inactivate most pathogens (disease-causing bacteria) in milk to make it safer to consume (at least reduction in five log cycles of population of *Mycobacterium paratuberculosis*; Stabel & Lambertz, 2004). Due to food safety concerns, this process is highly regulated all over the world. There are basically three methods used to pasteurize milk: vat

Table 17.3 Times and temperatures for the pasteurization of fluid milk as defined in the CFR (7 CFR 58)

Common name of treatment	Temperature	Time
Low temperature, long time (LTLT)	63 °C (145 °F)	30 minutes
High temperature, short time (HTST)	72 °C (161 °F)	15 seconds
Ultrapasteurization (UP)	138 °C (280 °F)	2 seconds

pasteurization (or low-temperature, long-time (LTLT)); high-temperature, short-time (HTST) pasteurization; and ultra-pasteurization (or ultra high-temperature pasteurization (UHT) if aseptically packaged. The times and temperature requirements are listed in Table 17.3 as defined in the CFR (21 CFR 131).

The regulations further state that every particle of product must reachthese temperatures, at a minimum, and be maintained at or above the temperature for a minimum of these times. The equipment used in pasteurization must be designed and regularly tested to fulfill this mandate as referred to in Appendix H in the PMO (FDA, 2011). The PMO specifies that Grade A milk raw must contain less than 100,000 colony-forming units per milliliter (CFU/mL) of total aerobic bacteria. Pasteurization is designed to kill 5 logs (100,000 CFU/mL) of heat-susceptible bacteria; pathogenic bacteria are heat susceptible. The reason why pasteurized milk is not sterile, and milk needs to be pasteurized, is that thermoduric bacteria endure HTST processing and can later cause spoilage in milk.

The primary method used in the fluid milk industry is HTST. In this process, there are three heat exchange sections: regeneration, heating, and cooling (see Figure 17.1). These sections consist of stacks of corrugated (to enable efficiency by increasing the surface are of heat transfer) stainless steel plates, where the heating medium and product flow in separate alternate plates (Figure 17.4). The plates are formed such that they direct the flow of the product evenly across the surface and maximize the surface area of the plates. The flow in the heat exchangers can be concurrent or countercurrent flow. Concurrent flow is where both the heating medium and the product flow in the same direction. This lowers efficiency heat transfer compared to the countercurrent flow, where the heating medium and the fluid product flow in opposite directions. The regenerator section has pasteurized heated product on one side of the plates and unpasteurized cold product on the other. This section recovers much of the heat from the pasteurized product, increasing the efficiency of the pasteurizer. The only energy that needs to be supplied is that lost to the atmosphere and that due to inefficiencies of the heat transfer. Pressures in these plates are arranged so that if there should be a leak between the plates, the flow would always be from the pasteurized product into the unpasteurized product or the other heat exchange fluid.

The heating section is connected to the holding tube (see Figure 17.1). The heating section brings the temperature up such that at the end of the holding tube, the product is at a temperature higher than the stated pasteurization temperature (i.e. 72 °C) for at least the minimum time (i.e. 15 sec). The heat transfer fluid for this section of the pasteurizer is most usually hot water. To guarantee that every particle of product is heated and held for the prescribed temperature and time, pasteurizers are provided with a holding tube and timing pump. This tube's length is calculated using the flow rate provided by the timing pump to give, at a minimum, the prescribed hold time. The timing pump's speed is adjustable and the flow rate is tested and sealed by regulatory officials. The holding tubes are mounted with an incline so that they will flow continuously back toward the heating section in the case of a shutdown. Pasteurizers are also provided with a set of valves, arranged to divert the product flow back through the pasteurizer when the product flow or temperature settings are not met, called the flow diversion valve (FDV). The goal for all the machinery involved in pasteurization is to guarantee that every particle passing through the system is subjected to the minimum time and temperature set by the regulatory agencies. In the cooling section, the product cooled by the incoming product in the regenerator section is further cooled to the desired exit temperature.

Alternatives to plate heat exchangers can be tube-within-tube heat exchangers, where one fluid passes within a central tube in one direction while the second fluid passes in the other direction in a tube surrounding the central tube. Product can also be heated directly by bringing it into direct contact with steam. The water thus added to the product is removed by vacuum evaporation downstream. Scraped surface heat exchangers are used to

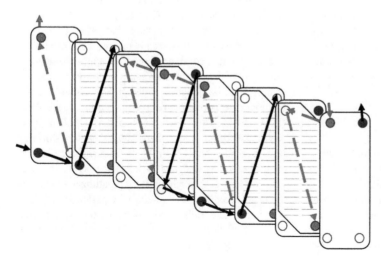

Figure 17.4 Schematic diagram of a plate heat exchanger with end plates and corrugated plates in between. The flow is indicated by the continuous line for fluid product and the dashed line for the heating medium.

heat treat fluids with particulate matter or higher viscosity. They have scrapers that remove the product fouling of the heat exchanger walls.

There are additional methods of pasteurization, including vat, UP, and UHT. Vat pasteurization differs in that the heating fluid passes around the jacket of a specially designed tank. In order to prevent contamination from condensate dripping into the pasteurizing fluid, the temperatures above the fluid in the tank must be above that of the fluid at all times. Since the temperatures are lower for vat pasteurization, the process times are necessarily longer to achieve the same kill of pathogens. The equipment for UP and UHT pasteurization is similar to HTST except for the time and temperatures involved. UP involves higher temperatures (with shorter hold times) than HTST. The primary advantage of UP and UHT is that the product is rendered essentially commercially sterile when combined with aseptic packaging (as with UHT), which allows for shelf stability.

Regulations establish the minimum design and instrumentation for all forms of pasteurization. The sanitary fabrication, construction, and design of dairy equipment have been enforced based on the standards set up by the National Conference on Interstate Milk Shipments (NCIMS) given in the PMO Appendix H (FDA, 2011). Another independent organization that has developed the 3-A standards for the dairy industry is the 3-A Sanitary Stds., Inc. The references to these standards can also be found in the PMO (FDA, 2011).

17.3.1.4 Packaging

At the beginning of the 20th century fluid milk was sold commercially in glass bottles. Glass bottles were replaced by gable topped paperboard containers and by plastic (high-density polyethylene (HDPE)) bottles or jugs and plastic (low-density polyethylene (LDPE)) pouches since the 1960s. A package is defined to protect a food product and preserve its nutritional value. In the dairy industry the package should protect the milk product from mechanical shock, light, and oxygen. The exposure to light and oxygen induces fat oxidation, development of oxidized off-flavors, and loss of vitamins in milk. The different filling systems used in the dairy industry are form fill and seal, aseptic filling, and bottle filling. In the form fill and seal system, the packaging material is made to form a container in the filling machine; the product is filled into this container and then sealed. This method is used for plastic pouches and paper cartons. Aseptic filling is utilized for aseptically packaged milk

products where a sterile product is filled into a sterile container under aseptic conditions.

17.4 Dairy product processing

17.4.1 Fluid milk products

There are a variety of fluid milk products available on the US market. These include the following (along with references for standards of identity in the CFR).

17.4.1.0.1 Whole, reduced-fat (2%), low-fat (1%), and fat-free (skim) milk (21 CFR 131.110)

The percentage of milk fat is adjusted in the milks by cream separation and standardization processes, providing these fluid milk varieties. All of the reduced-fat milks are fortified with vitamins A and D because these vitamins are fat soluble. The standards of identity of whole, reduced-fat, low-fat, and fat-free milks require them to contain >3.25%, >2%, >1%, and <0.1% milk fat by weight. The total solid content of all these milk should be at least 11.5% w/v.

17.4.1.0.2 Lactose-free and lactose-reduced milks

Lactose-free and lactose-reduced milks are produced by the complete or partial hydrolysis of lactose in milk. The hydrolysis of lactose into glucose and galactose is accomplished using B-galactosidase (lactase) enzyme. Lactose intolerance or lactose maldigestion is a condition where people are unable to produce B-galactosidase. Milk treated to be reduced in lactose or free of lactose can be a strategy for living with this condition. It is also possible to recombine milk using materials like whey protein isolate, whole milk protein isolate or vegatable proteins such that the product does not contain lactose. Such recombined milks do not have a standard of identity specified by the FDA.

17.4.1.0.3 Light, light whipping. and heavy cream

Cream is a fat-rich dairy product, obtained from the cream separation from milk and pasteurization. The percentage fat distinguishes the different cream varieties, but cream as a whole should contain more than 18% milk fat. The fat percentage requirements of light, light whipping, and heavy cream as in the CFR are given in

Table 17.4 Standards of identity for some fluid milk products (USGPO, 2013)

Fluid milk product	Standards for composition			
	Fat (by weight)	Total solids (by weight)	Milk solids not fat (by weight)	Section of title 21 CFR
Light cream	>18% and <30%	Not specified	Not specified	131.155
Light whipping cream	>30% and <36%	Not specified	Not specified	131.157
Heavy cream	>36%	Not specified	Not specified	131.150
Half and half	>10.5% and <30%	Not specified	Not specified	131.180
Evaporated milk	>6.5%	>23%	>16.5%	131.130
Concentrated milk	>7.5%	>25.5%	Not specified	131.115
Sweetened condensed milk	>8%	>28%	Not specified	131.120

Table 17.5 Heat treatment times and temperatures for cream for different purposes (7 CFR 58.334)

End use of cream	Heat treatment	Temperature	Time
Cream for butter making	Vat pasteurization	85 °C (165 °F)	30 minutes
	HTST	74 °C (185 °F)	15 seconds
Cream for frozen and plastic creams	Vat pasteurization	88 °C (170 °F)	30 minutes
	HTST	77 °C (190 °F)	15 seconds

Table 17.4. The processing steps in producing cream involve cream separation, homogenization, and pasteurization. The pasteurization temperatures for cream are higher than for milk (Table 17.5), as the higher fat content is known to offer protection to microorganisms (Senhaji, 1977). Appropriate terms have to be added to the product name on the label if the creams are sweetened or flavored.

17.4.1.0.4 Half and half

Half and half is a mixture of milk and cream. It is produced from a series of processes including standardization, homogenization, and pasteurization. The details of the requirements for this product are given in Table 17.4.

17.4.1.0.5 Flavored milk

Flavored milks are milks with or without coloring, nutritive sweeteners, emulsifiers, stabilizers, and fruit or fruit juice. Chocolate milk is a commonplace example of this genera. They are classified under milks in the CFR. These products undergo the same processing steps as for white milk.

17.4.1.0.6 Concentrated, evaporated, and sweetened condensed milk

These dairy products are obtained by the partial removal of water from milk. While sweetened condensed milk (SCM) contains suitable nutritive carbohydrate sweeteners, they are absent in evaporated and concentrated milks. The difference between evaporated and concentrated milk is the sealing of the product in a container and heating before or after packaging to prevent spoilage in evaporated milks. The details of milk fat and milk solids non-fat regulations for concentrated, evaporated, and sweetened condensed milks are given in Table 17.4.

17.4.2 Ice cream processing

Like most products, on the surface ice cream appears to be a simple entity, but on closer examination it is one of the most complex food products that exist. Most foods have a single phase structure that is either fat or water based, or a simple emulsion of fat and water. Ice cream, on the other hand, contains liquid (both water and fat), crystalline (water, fat, and sugar) and gaseous phases (air and water vapor). All of these are important to its production, distribution, and consumption. It contains fat that, in the same product at the same time, exists in both liquid and solid phases. Further, the fat can be crystallized into several forms. Processing of ice cream and frozen desserts is, at its most basic level, the management of these phases. If processing is not carried out in such a way that all the phases are correct, the product will have problems during processing, storage, or transportation or at consumption.

Ice cream is a frozen food made from dairy products, sugars, flavorings, and minor amounts of optional ingredients used to improve texture, enhance shipping and aging characteristics and make manufacturing more efficient. In the US, ice cream is defined in the CFR Title 21, Chapter 1, Subchapter B, Part 135, subpart B, "Requirements for specific standardized frozen desserts." In other countries similar standards exist. It is increasingly important to keep in mind that standardized frozen desserts are a subset of frozen desserts in general. Unstandardized products deviate from standardized products for reasons of expanded creativity, economy, new technology or expedience in manufacture, storage, distribution, marketing, and consumption. A common example of an unstandardized product would be a frozen dessert made in the US using vegetable fat instead of dairy-derived fat.

In the US, people consume about 22 L of ice cream per person per year. In 2013, the global ice cream market is forecast to have a value of $54 billion, an increase of 20.3% since 2008. In 2013, the global ice cream market is forecast to have a volume of 12.7 billion L, an increase of 15.7% since 2008 (Techzone, 2010).

17.4.2.1 Raw materials and their storage

The ingredients used in frozen desserts include, but are not necessarily limited to, air, water, whole milk, skim milk, cream, butter, condensed milk, non-fat dry milk, whey, whey protein powders, caseinates, vegetable-derived protein powders, dairy and vegetable-derived fats, sugars, sugar alcohols, high-potency sweeteners, maltodextrin, eggs, egg yolks, hydrocolloid stabilizers, emulsifiers, flavors, and colors. In addition to these, any number of bulk flavors, inclusions and variegates are added after freezing to produce the variety of frozen dessert products found in the marketplace. All ingredients should be analyzed for microbial contamination and for general quality before they are incorporated into the dessert. Many of these ingredients are raw agricultural products. It is vitally important that these be kept under conditions that discourage the growth of pathogenic or spoilage organisms (cold). Liquid dairy ingredients, being composed of sugars, fats and protein, will separate if not kept adequately agitated.

Certain dairy products, like skim milk and cream, have a tendency to form stable foams. Care should be taken to avoid splashing or whipping these products during storage. Not only does foaming make processing difficult, but in some cases this can cause a severe loss of quality in the final product.

Ingredients that contain fat or other components that are prone to oxidation should be packaged or stored in such a way that they are exposed to the minimum possible oxygen, care should be taken to use these products quickly enough to preclude the oxidation of their components. Raw milk is prone to rancidity, caused by the presence of enzymes (i.e. lipase) in the raw milk. The fat in raw milk is protected from the enzyme by the MFGM. However, this membrane can be damaged by high shear processes such as pumping, violent agitation, splashing, and any other high shear process. Therefore, raw milk should be handled as carefully and gently as possible. Powdered products should be kept under conditions that will maintain them as free flowing as possible. This usually entails keeping the powders as cool and dry as possible and not stacking the bags too high. Frozen ingredients should be kept as cold as possible, with as little temperature fluctuation as possible. When frozen ingredients are thawed, this should be done so that the entire product is maintained at a temperature that discourages microbial growth or it should be done quickly enough that microbial growth is not significant. It should be assumed that microbial growth will happen any time the temperature exceeds $0\,^{\circ}\text{C}$. It is also good practice to assume that all products containing more than 4% water have the ability to support the growth of pathogenic or spoilage organisms or at least maintain their numbers. It is therefore vitally important to pasteurize and/or freeze all ingredients into finished products as soon as possible. Good ingredient rotation will not only result in a superior product but it will minimize the risks involved in processing raw agricultural products into frozen desserts.

17.4.2.2 Preparation for processing

Frozen dessert mixes are made to specifications determined by the product line and regulations (specifications are provided in the CFR; USGPO, 2013). These specifications are called a formula. The formula expresses the components as dry percent and Table 17.6 shows a typical formula for a 10% ice cream mix. For "ice cream," 10% fat is the minimum fat content; ice cream must also weigh at least 4.5 lb/gallon (539.2 grams/L) to comply with US regulations.

The formula is converted daily into a recipe using the specifications and test results of the ingredients that will be used in the product. Table 17.7 shows a recipe developed from this formula. Since the milk solids non-fat and milk fat for dairy ingredients vary considerably depending on the milk supply, the amounts needed in a recipe can vary considerably. If these are not taken into account, variation in the final product will result. Consistency is very important for consumer satisfaction.

In general, specifications for powdered products do not vary as significantly as do many manufactured ingredients, such as sugar or corn syrup. Nonetheless, even if all the ingredients are manufactured to a constant specification, it is good practice to maintain a constant formula and calculate recipes from the formula for each batch. Changes in the form of the ingredients needed for a formula can be based on price and availability. Calculating recipes maintains flexibility in the form of the ingredients used on a day-to-day basis. It is always good practice to analyze and monitor ingredient characteristics in case they change and need to be taken into account in the recipe. A traditional method of calculation

of ice cream mix composition is by the serum point method (description in textbooks including Hyde & Rothwell (1973) and Marshall et al. (2003)). There are many computer programs and spreadsheets that can calculate recipes from a formula. These range from relatively complex systems that can monitor inventories as well as calculate batch sheets to simple single-purpose spreadsheets.

17.4.2.3 Processing steps

The vast majority of frozen desserts are made using the same methods. In order to avoid repetition, the process for ice cream will be described here. The general steps in the process are shown in Figure 17.5.

17.4.2.3.1 Blending

The first step in producing frozen desserts is the production of a pasteurized "mix." In producing a mix, the liquid and dry ingredients must be brought together in such a way that all ingredients are dispersed completely prior to heat treatment. There are several problems that must be overcome in this process. Each ingredient will have different density. The tendency will be for these ingredients to float or sink depending on their density. Ingredients that float to the top or sink to the bottom may not be incorporated into the final mix. This will cause inconsistency in the final product. Some ingredients are so hydroscopic that they have a tendency to clump. In extreme cases, such as with some hydrocolloids and proteins the material will form insoluble clumps that are difficult or impossible to break up. If this happens, these ingredients may not be able to pass through filters or screens and will not be incorporated into the mix, causing quality and consistency issues. The blend of ingredients can have a

Table 17.6 Typical formula for a 10% ice cream mix

	Percent by weight
Milk fat	10.0%
Milk solids non-fat	10.0%
Whey solids	2.0%
Sugar solids	12.5%
36 DE corn syrup solids	5.0%
Emulsifier stabilizer blend	0.5%
Total solids	40.0%

DE, dextrose equivalent.

Table 17.7 Typical recipe for a 100 gallon ice cream mix

Ingredient	Milk fat	Milk solids	Total solids	Pounds for 100 gal batch (928 lb)
Cream	41.50%	5.27%	46.77%	184.16
Milk	3.65%	8.65%	12.30%	443.60
Non-fat dry milk	0.50%	96.50%	97.00%	46.35
Whey powder	0.00%	97.00%	97.00%	19.15
Sugar syrup	0.00%	0.00%	67.50%	172.00
36 DE corn syrup	0.00%	0.00%	80.00%	58.05
Stabilizer blend	0.00%	0.00%	98.00%	4.69

DE, dextrose equivalent.

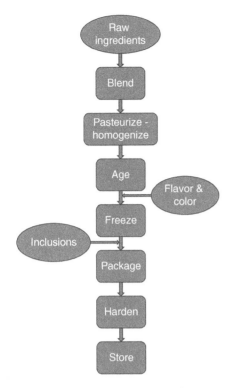

Figure 17.5 Processing steps in ice cream making.

tendency to form stable foams (this is desirable in the final product but undesirable in mixes). The order of addition and the equipment used for blending are designed to minimize or eliminate these problems.

Blending is usually preceded by some method of measuring out the ingredients. This is accomplished, for the liquid ingredients, by weighing or metering the ingredients into a blend tank. This tank should not be violently agitated to prevent the development of rancidity in raw milk. The dry ingredients are most often weighed on scales.

After the dry ingredients are weighed out, they are mixed with the liquid ingredients. As mentioned above, some dry ingredients are so hydrophilic that they form clumps, with very high viscosity or gelled outer layers and essentially dry powder in the interior. In order to keep this from happening, these materials need to be as widely dispersed as possible. There are four basic ways in which this can be done; the most popular and efficient way is using a high shear mixer. A small amount of liquid is brought into the blender, a high shear mixing head is started and the powder is fed, slowly, into the blender. Another method feeds the powder slowly into a stream

of liquid; the powder is fed into the stream through an open funnel. Sometimes a modified pump further mixes the powder and liquid. Perhaps the oldest method and least efficient is to simply add the powder to the top of the tank of liquid ingredients with strong agitation. This tank is often provided with a recirculation pump to aid in distributing the powder while it hydrates. Hard-to-handle products like stabilizer blends containing hydrocolloids can be first dispersed in an ingredient that does not tend to form clumps, like granulated sugar. The mixture can then be added to the blender slowly; the sugar helps keep the particles of the viscous ingredient separated, keeping them from forming clumps. A similar method used is to disperse the powder in liquid sugar in a high shear blender. Since the amount of water in liquid sugar is low, the particles are able to disperse within the syrup before they hydrate. In each case the strategy is to keep the individual particles of dry ingredients far apart from one another in order to wet the entire surface of each particle separately and therefore to disperse all the particles as uniformly as possible in the blend before they fully hydrate.

It is important to be aware that during blending and subsequent storage before pasteurization, many of the ingredients are not dissolved in the liquid but are only dispersed. If given the chance (insufficient agitation, for instance), they will segregate at the top and bottom of the blend tanks. This can prevent important components from being included in the final product. If this problem exists, the final product will be inconsistent. It may also cause the mistaken conclusion that the settled component is underdosed, which in turn may cause loss of money due to adding too much of that ingredient in the future. As blending is the first step in making ice cream, it is very important that it be carried out correctly.

17.4.2.3.2 Pasteurization and homogenization

Similar to fluid milk, the primary purpose of pasteurization is to eliminate the pathogens in the ice cream mix to make the product safe to consume; freezing does not kill pathogens. The times and temperatures for ice cream, which are higher than for fluid milk, are listed in Table 17.8 as defined in the CFR. Temperatures of treatment are higher compared to fluid milk because higher amount of solids in ice cream provide a protective effect to bacteria.

Many frozen dessert mixes are viscous. This can cause the product's flow to be laminar, as opposed to turbulent, for lower viscosity products such as milk. Since the

Table 17.8 Pasteurization – minimum temperature and time for frozen desserts

Temperature	Time
69 °C (155 °F)	30 minutes (vat)
80 °C (175 °F)	25 seconds (HTST)
138 °C (280 °F)	>2 seconds (UHT)

Source: 21 CFR 135 (USGPO, 2013).

residence time for particles near the tube walls is longer for laminar flow than it is for turbulent flow, holding tubes for frozen desserts are designed to hold the product for a longer time (Marshall et al., 2003). Tubular and scraped surface heat exchangers are almost never used in ice cream production. Plate heat exchangers are far more common as are batch pasteurizers. The heating of frozen dessert mixes not only accomplishes the required destruction of pathogens and the reduction in the numbers of spoilage organisms; in products containing fat, the heating melts both the fat and the emulsifiers and allows them to interact. This will become critically important during the aging and freezing of these mixes. Heating aids in the dispersion and hydration of hydrocolloids, allowing for these ingredients to become functional. Proteins can be denatured and they can interact with other proteins, stabilizers and other ingredients with heat. This can help increase viscosity. Exposure of protein and sugars to heat can result in browning reactions that can add flavor to mixes. Heating regimes are often increased above the legal limits for this reason alone.

Homogenization's effects on the fat structure in frozen desserts are the primary reason for its use. The secondary benefit of homogenization includes dispersion of the particles of some stabilizers (primarily microcrystalline cellulose). It can have detrimental effects as well; protein and some hydrocolloids may also be broken down by high shear and temperature. This can cause loss of viscosity, especially when the product is repeatedly homogenized as in reprocessing.

17.4.2.3.3 Aging

Once the product has been pasteurized and cooled, it is placed into a chilled tank for aging. During aging, the following structural changes happen.
• Hydrocolloids and protein become fully hydrated.
• Milk protein desorption from the milk fat globules occurs.
• Fats crystallize.

It is obvious that it would be desirable for production efficiency to be able to freeze the product directly after pasteurization. However, aging improves the quality of the final product. The various arrangements in the emulsion structure of the ice cream mix ensure an acceptable texture for the final product. However, a clear picture of the emulsion rearrangements occurring during aging is still not known. The stabilizers and emulsifiers (hydrocolloids) added to the mix during the blending stage are completely hydrated during the aging step. The cooling to less than 4 °C also causes fat crystallization. The surfactants (emulsifiers) will then adsorb to the partially crystallized fat phase. The proteins forming the MFGM after homogenization are desorbed from the fat surface due to the lowering of the hydrophobic forces at lower temperatures but this is a very minor effect at 10 °C. The main effect is the movement of emulsifiers from the fat on to the surface during fat crystallization thus displacing the protein (Bradford et al., 1991; Danisco 2010; Iversen & Pedersen, 1982; Marshall et al., 2003) and displacement by emulsifiers moving from the fat to the surface of the globule due to fat crystallization. The fat crystals also partially rupture or distort the MFGM, enabling interglobular contact and forming the new stabilized structure (van Boekel & Walstra, 1981).

Since the hydration of protein and hydrocolloids is relatively fast, for products without fat the aging step is less important than it is for products containing fat. In non-fat products, some further hydration of stabilizers and some relaxation of protein can occur; these can make minor improvements in the final product and for processing. In products containing fat, a lack of aging can cause products to be difficult to whip, have poor extrusion properties and to melt improperly. During aging, the fat and emulsifiers (primarily monoglycerides) crystallize. This can be observed in a non-refrigerated tank as an increase in temperature due to the heat given off during fat crystallization. This process can take several hours. Products made without complete crystallization of the fat are prone to churning of the fat. This is evidenced by churned fat (butter) being found in the finished product and inside the freezer and other equipment handling the incompletely aged product. Liquid and partially liquid fat in the high shear environment of pumps, pipes and in the freezer are much more likely to agglomerate together in even larger particles. This uncontrolled agglomeration results in the deposition of fat in the equipment and loss of quality in the finished product.

Monoglyceride emulsifiers are added to frozen dessert mixes to provide controlled agglomeration. These result in the ability of the product to hold more air, extrude more cleanly, maintain shape during melting, avoid collapse and provide better shelf life characteristics. All of these characteristics are related to how the emulsifier behaves during aging. Prior to the crystallization of the fat, the monoglyceride (emulsifier) is primarily dissolved within the fat. As the fat crystallizes, the emulsifier is moved from the interior of the fat particle to its surface by the crystallizing fat (Danisco, 2010). As this is happening, the protein that has been protecting the fat particles from agglomeration is displaced from the surface (emulsifiers have a stronger affinity for the surface of fat than does protein). The presence of emulsifier at the surface of the fat encourages controlled agglomeration of fat during freezing. Mixes made without aging therefore resemble mixes made without emulsification.

One common question is, how long must a product be aged? The actual time necessary depends on many characteristics of the mix and processing environment. The increase in viscosity of the ice cream mix has been taken as the indicator for deciding the time for aging of mix by some researchers (Chang & Hartel, 2002; Dogan & Kayacier, 2007). Cooling rates, mix temperatures, the amount and type of fat in the mix, the type of emulsifiers used and heat transfer properties of the mix all have an effect on the amount of aging necessary. The process of aging is also not linear. Acceptable aging times can be much shorter than optimal aging times. Decisions on how long to age each mix must therefore be a compromise between production efficiency and quality. In the industry there are many rules of thumb and each manufacturer determines the exact minimum aging time experimentally. A good rule of thumb is 4 h.

17.4.2.3.4 Flavoring

In frozen dessert manufacture there are three points at which the flavor of the product is altered. The first point is during formulation and blending. Chocolate ice cream is most often formulated; the cocoa or chocolate is added at blending and pasteurized. Coffee-flavored ice cream is sometimes made this way. Any flavor that does not contain particulates could be made this way but due to the heat treatment, flavors are altered chemically and combine with the other components, which leads to the loss of volatiles. The biggest advantage in this is microbiological; by pasteurizing, any pathogenic bacteria on the flavoring material will be killed. Another advantage is that powdered flavorings like cocoa are

much easier to disperse in the blending process than in the flavor vat. In most frozen dessert operations the mixes are made as white unflavored mix and chocolate mix. These are used as bases for all the flavors made.

The second point where flavor (and color) can be altered is just prior to freezing. This operation is most often done in a flavor vat. A flavor vat is a small tank provided with gentle agitation, placed so that it is convenient to the freezer. These tanks can be provided with refrigeration to maintain temperature while the product awaits freezing. Flavor tanks should be designed to drain completely; this prevents cross-over of one flavor into the next or product loss when rinsing between flavors. Liquid flavorings and colors are most often manually measured into this tank and allowed to mix before sending them to the freezer. Automated metering systems for either dosing flavors into the flavor vat or on line to the freezer have been devised and are used in some operations. This eliminates errors of measurement and, in the case of flavors metered directly to the feed line of the freezer, eliminates a point where product contamination could happen. Flavors added post pasteurization are a contamination risk. It is important that flavorings and colorings be pasteurized or otherwise assured to be free of pathogens. This is especially true of raw agricultural products like strawberry juice.

The third and final place where flavors may be added is just prior to packaging. At this point the flavorings added are most often either large particulate flavors like nuts, fruits and candy or thick syrups like marshmallow or chocolate syrup. The two basic pieces of equipment used at this position are fruit feeders and variegators. Fruit feeders are used to deposit pieces of fruit or nuts into the stream of frozen ice cream. They are designed to be flexible so that they can be switched from one inclusion to another and they are adjustable so that different ingredients can be added at differing rates. These machines distribute the particulates evenly throughout the ice cream flow. Since this equipment is used for many different ingredients, it is especially important that it be kept sanitary and that residue from one flavor is not carried into the next. Since it is used to insert nuts, this equipment must be sanitized to be free of nut particles if it is to be used for non-nut flavors in order to avoid allergen problems.

Variegators are designed to add streams and/or layers of syrups to the ice cream stream. The syrup is pumped from a drum or tote into the variegator. The syrups can be aerated, as with marshmallow, or not, as with chocolate syrup. Some fruit jams can be added into the products using a variegator.

17.4.2.3.5 Freezing

Ice is the characterizing factor for all frozen desserts. There are two basic methods for making frozen desserts: quiescent freezing, where the liquid product is not stirred during freezing, and active freezing, where the product is stirred. Quiescent freezing is generally used to produce molded novelties like ice pops. In this case, the ice pop mix is simply deposited in a mold; the mold is placed in a cryogenic fluid (cold air, cold brine, glycol, liquid nitrogen or other cryogenic heat transfer fluid) and allowed to freeze (with or without stick insertion before complete freezing). The ice crystals start growth at the mold surface and grow towards the center. As ice freezes, it excludes the material dissolved in it. In products without air bubbles, fat droplets or other obstructions, as in ice pops, the ice will push the solutes toward the center and away from the ice crystal surfaces. Since the freezing point of the liquid decreases when the concentration of solutes increases, there will come a point where the ice and the unfrozen liquid reach equilibrium. The number of sites where ice crystals can form is dependent on the temperature at the point where the ice crystals start to form (Drewett & Hartel, 2007; Russell et al., 1999). At lower temperatures, more crystals will form. Also, if there are a large number of points where ice can start to form, like particles of fat or protein, there will be a large number of ice crystals. Ice pops, for instance, have few particles that can initiate ice formation and therefore have a more coarse (icy) texture than ice cream bars. Products with higher fat or protein (particles) will also result in more numerous (smaller) ice crystals (which makes them feel less cold in the mouth). It is the manipulation of the crystals of ice in these products that gives them their textures.

The second and more complicated method for producing frozen desserts is by freezing while stirring (active freezing). The definition of ice cream in the US includes the statement: "Ice cream is a food produced by freezing, while stirring, a pasteurized mix…" (21 CFR 135.110).

There are two basic types of machines that are used to accomplish freezing while stirring: batch and continuous freezers. Batch freezers, as indicated by their name, freeze a measured amount of mix and air to completion, after which the contents of the freezer are emptied. With continuous freezers, mix and air are pumped into the freezer and frozen product is extracted from the freezer in continuous streams.

The ice cream freezer differs little from the original invention by Nancy Johnson in 1843 (US Patent no. 3254). In its essence, an ice cream freezer consists of two parts: a heat exchange surface and some form of scraper to remove frozen product from the surface. Often a beater or dasher is included to aid in mixing the freezing product. Originally and still today, with home freezers the heat exchange surface is cooled using ice and salt. In industrial freezers, the surface is chilled using the direct expansion of freon or ammonia. In the modern continuous freezers the heat exchange surface takes the shape of a cylinder. This cylinder is surrounded by a space where a coolant like ammonia is evaporated. Carbon dioxide or freon may also be used among alternative coolants. Inside this cylinder (called a barrel) is a series of blades that travel around the inside circumference of the barrel. Supporting the blades is a mechanism (called a dasher) that, in addition to supporting the blades, mixes and beats the freezing product to incorporate air. It is the design of the dasher that characterizes the different freezer designs. In continuous freezers, mix and air are metered and pumped into the freezer. In some designs the frozen product is also pumped out of the freezer.

The active freezing process of ice cream and related products is much more complex than for quiescent freezing. During the freezing process the following changes occur.

- The temperature is brought down from approximately $4\,°C$ to $-6\,°C$.
- Approximately 50% of the water in the mix is frozen.
- The concentration of the unfrozen mix rises from about 38% to roughly 55%.
- Air is incorporated into the mix to approximately half the volume and the air bubble size is reduced to an average between 10 and 30 μm (Rohenkohl & Kohlus, 1999).
- Fat globules are destabilized and partially agglomerated into a structure that both supports and somewhat contains the ice crystals and the air bubbles.
- Protein is moved to the surface of the air bubbles, helping to stabilize them (Goff et al., 1999).
- Due to the concentration of the liquid portion of the mix due to freezing and to the number of solid particles imbedded in this liquid, the viscosity of the product increases from a liquid to a semi-solid.

As the mix and air enter the freezer barrel and the mix is cooled, air bubbles are subdivided into smaller and smaller sizes. Mix located near the walls of the freezer begins to freeze and the ice formed at the wall is moved into the center of the volume of the freezer. Initially this melts and gives up the heat of crystallization to the bulk of the mix (Drewett & Hartel, 2007; Hartel, 1996; Russell et al., 1999). At some point the temperature difference between the ice crystals formed at the wall of the freezer and the bulk of the mix is not enough to completely melt the ice crystals being transported into the bulk of the mix. At this point there is both growth of ice crystals in the bulk of the mix and the addition of new ice crystals from the barrel wall. At the same time, air is incorporated and stabilized; this is aided by the shear forces from the dasher, the freeze concentration of the protein, the freeze concentration of the sugars and hydrocolloids, and the increasing concentration of ice and small air bubbles that in turn increases the stiffness of the partly frozen material. The shear forces in the forming product cause fat globules to agglomerate together to form a kind of loose structure with ice crystals, air bubbles and unfrozen mix imbedded within it. This process is similar to the whipping of cream. In frozen desserts, the fat must be concentrated by freezing (for ice cream during freezing the fat concentration in the liquid portion of the product nearly doubles).

17.4.2.3.6 Packaging

Once the product exits the freezer, there is no further formation of new ice crystals. All further ice crystal formation is to existing ice crystals (Drewett & Hartel, 2007). In order to move the frozen product through the lines to packaging, there is a significant pressure placed on the product at the freezer. This can be from the geometry of the dasher or due to a pump (generally called an ice cream pump). This back pressure compresses the air in the air bubbles. The air bubbles will expand somewhat as this pressure is released. Depending on the amount of back pressure, this expansion can be enough to cause air bubble instability in the final product. The amount of air in the final product is expressed as overrun which is calculated using the following formula:

$$\frac{\left(\frac{Weight\ of\ mix}{unit\ volume}\right) - \left(\frac{Weight\ of\ final\ product}{unit\ volume}\right)}{\left(\frac{Weight\ of\ final\ product}{unit\ volume}\right)} * 100$$

Overrun for frozen desserts can range from 0% for quiescently frozen ice pops to around 20% for soft-serve ice cream and super premium ice creams to 100% for hard-pack ice cream to 150% or higher for frozen desserts. The CFR regulations control the overrun in ice creams by mandating not less than 1.6 lb of total solids per gallon (191.7 grams of total solids per liter) and weight of not less than 4.5 lb per gallon (539.2 grams per liter). Batch freezers generally can attain about 50% overrun. As overrun increases, the insulation value of the product increases and the product is perceived as less cold in the mouth. Overrun also affects flavor perception to some extent.

Frozen desserts can be packed in innumerable types and configurations of packages. It is beyond the scope of this chapter to discuss them all. Frozen desserts are subject to all the problems that are found in other similar products. An example of this is when ultraviolet (UV) light from the sun or from display lighting, as with gelato displays or behind clear plastic windows on packaging, affects ice cream quality. The ice cream that is exposed to UV light can develop light oxidized flavors in the areas exposed (Clark et al., 2009).

17.4.2.3.7 Hardening

After packaging, approximately half of the water in the product is not frozen. In this state the product is semi-solid and vulnerable to damage. Hardening is the process of continuing freezing without agitation until the temperature is $-18\,^{\circ}C$ or preferably lower. This process should be done as quickly as possible to avoid the growth of large ice crystals in the product.

As the ice cream exits the freezer there is a distribution of ice crystal sizes (Marshall et al., 2003). The most stable ice crystals are the large ones and the least stable are the smallest ice crystals (Everett, 1988). During hardening ice will form on all the ice crystals available. At any point in time, there is a tendency for the larger ice crystals to grow at the expense of the smaller ones. This process is called Ostwald ripening (Everett, 1988). Further, since water must be supercooled in order to form new crystals (Debenedetti & Stanley, 2003), few ice crystals can form as long as there is other ice present in the liquid portion of the product (at constant, increasing and decreasing temperatures[2]). All the above leads to an increase in ice crystal sizes after the ice cream exits the freezer.

---

[2] This presumes that the temperature is decreasing at a rate within normal processing capabilities. It is possible to form new ice crystals in the presence of ice crystals if the temperature is dropped at a very high rate. This could be done by dropping a small particle of product in liquid helium. The reason for this is, at high rates of cooling water cannot diffuse through the high viscosity mix fast enough to reach the larger ice crystals.

If hardening is done slowly or is delayed, the smaller (more numerous) ice crystals will have disappeared and ice will form on the larger crystals. The result will be product with larger ice crystals, which are more noticeable (cold, coarse) on the tongue. If hardening is done quickly there will be many smaller ice crystals, resulting in a smoother product.

The rate at which the product cools down during hardening is affected by many properties of the product and process conditions. These include the thermal properties of the packaging, the thermal properties of the product, the temperature differential between the product and the cooling medium (most often air), and the flow properties of the cooling medium (Russell et al., 1999). In practice, maintaining the maximum contact between the product and the heat transfer medium and minimizing the distance between them while maximizing the temperature difference will result in the quickest hardening. Some of the methods used to accomplish this can be by direct product contact through cooling plates, by immersion in cooling medium such as liquid nitrogen or most commonly, by directing high-velocity cold air on the product. Where product is palletized before hardening, it is a good procedure to stack the product in such a way that there is maximum contact with high-velocity cold air.

17.4.2.3.8 Storage and shelf life

Once hardened, the product is generally moved into long-term storage. The temperature in storage is generally held near −23 °C. This temperature cycles up and down by several degrees in a daily cycle (called a defrost cycle) that prevents the build-up of frost on the heat transfer surfaces and other surfaces within the storage area. Each time the product temperature increases, the smallest ice crystals in the product will melt and each time the temperature decreases the water released will refreeze on the larger ice crystals. This effect will depend on the magnitude of the change in temperature.[3] With decreasing temperatures, the amount of ice frozen and melted during temperature cycles is decreased. If the temperature rises from −20 °C (−4 °F) to −15 °C (5 °F), about 5% of the water will melt. If the same 5° change starts 5° colder, only 3% of

water will melt. Over time the result of this will be a steady loss of quality.

17.5 US regulations for milk and milk products

The US federal regulations are all the laws and acts passed by the US Congress. These are published in the Federal Register every working day by the executive departments and agencies of the US federal government. All of them are compiled in the US Code of Federal Regulations (CFR) under various titles (1–50), based on the government agency handling them. The CFR is published by the Office of the Federal Register, an agency of the National Archives and Records Administration (NARA). Examples are Title 21 under Food and Drug Administration (FDA), Title 19 under US Customs Services and Title 7 under United States Department of Agriculture (USDA). The Title 21 section 1 to 199 of CFR not only contains standards for different food products but also good manufacturing practices (21 CFR 110), food labeling regulations (21 CFR 101), recall policy (21 CFR 7.40), and nutritional quality guidelines (21 CFR 104) (www.ecfr.gov). These federal laws and regulations ensure the wholesomeness of food, help to inform consumers about the nutritional composition of foods and eliminate economic frauds by dictating the ingredients in foods, tests to be performed on each food, and procedures of analysis.

A comprehensive collection of federal laws, guidelines, and regulations relevant to food and drugs is published by the Food and Drug Law Institute, under the FDA. These include Food and Drug Act of 1906, Federal Food, Drug and Cosmetics Act of 1938, Food Additives Amendment of 1958, Color Additives Amendment of 1960, Nutrition Labeling and Education Act of 1990, Dietary Supplement, Health and Education Act of 1996, and Food Quality and Protection Act of 1996. All these acts have been chronologically referred to in the book published by the Food and Drug Law Institute (Cooper, 2011).

According to the federal regulations, every milk producer, milk distributor, bulk milk hauler/sampler, milk tank truck, milk transportation company, and dairy plant should have a valid permit. The various agencies that work towards establishing and enforcing the regulations and laws for the production, processing, and marketing of milk and milk products in US are (Hui, 1986; Nielsen, 2010):

- United States Food and Drug Administration (FDA)
- United States Department of Agriculture (USDA)

---

[3] An example of this effect can be seen in the common event of transporting a package of ice cream in the car to the home, allowing the temperature of the package of ice cream to nearly melt and then refreezing it in the home freezer. The resulting product is often much less smooth than it would have been had the product been kept cold (as in winter).

- state regulatory agencies
- Environment Protection Agency (EPA)
- US Customs Service
- US Federal Trade Commission (FTC).

17.5.1 FDA

The FDA is a government agency within the Department of Health and Human Services (DHHS). Its function is to regulate the safety of foods, cosmetics, drugs, medical devices, biological products, and radiological products. It enforces laws enacted by the US Congress and regulations developed by it to protect consumers' health, safety, and money. The milk and milk products definitions, standards of identity, quality and fill established by the FDA are published in Title 21 of CFR Section 130–135. These sections of the CFR include descriptions of:

- ingredients the different milk and milk products must contain
- minimum or maximum levels of various ingredients based on their economic value
- list of optional ingredients
- methods of analysis
- nomenclature permitted
- label description.

The FDA is authorized by the Food Drug and Cosmetic Act, Public Health Service Act and Import of Milk Act to regulate the production, procurement, processing, and marketing of milk and milk products. The FDA shares its responsibility with the state regulatory agency, which is the department of health or agriculture to ensure safety, wholesomeness and economic integrity of milk and milk products. The FDA has compiled a model for regulation of sanitation and quality of production and handling of Grade A milk called the *Grade "A" Pasteurized Milk Ordinance* (PMO), which is a consensus of current knowledge and experiences of milk sanitation (FDA, 2011). It is used as the sanitary regulation for milk and milk products for interstate shipment; it is recognized by the public health agencies, the milk industry and many others as the national standard for milk sanitation. According to the PMO, all sampling procedures and laboratory examinations should be in compliance with the current edition of SMEM by the American Public Health Association (Wehr & Frank, 2004) and Official Methods of Analysis by the Association of Official Analytical Chemists (FDA, 2011).

All US states have adopted the PMO as the minimum requirements for Grade A standards. The FDA monitors the state agencies, as all the 50 states are bound by the Memorandum of Understanding with the National Conference on Interstate Milk Shipments. The FDA also trains state inspectors and certifies the producers and dairy plants eligible for shipping milk to other states. The certified producers and dairy plants are included in the Interstate Milk Shippers List, as a part of the Interstate Milk Shippers Program. This program is maintained by a voluntary organization including the representatives of each state, the FDA, and USDA.

The FDA reviews all the food and color additives before manufacturers and distributors can market them. It publishes a list of permitted additives given in a database called "Everything" Added to Food in the United States (EAFUS), published at www.fda.gov/food/ingredientspackaginglabeling/foodadditivesingredients/ucm115326 .htm . It also publishes a list of food contact substances; a food contact substance is defined as "any substance intended for use as a component of materials used in manufacturing, packing, packaging, transporting, or holding food if such use is not intended to have a technical effect in such food" (amended Food, Drug and Cosmetic Act 1998). The FDA also ensures that food labels are truthful and do not mislead consumers, based on the Nutrition Labeling and Education Act 1990. The FDA has a major role in regulating the import of milk and milk products along with the US Customs Services as authorized by the Import of Milk Act 1927.

17.5.2 USDA

The USDA acts as a voluntary grading service for manufactured or processed dairy products, authorized by the Dairy Quality Programs under the Agricultural Marketing Act of 1946 (details about dairy products published in 7 CFR 58). It assists the FDA in regulating safety and quality of milk and milk products by:

- inspecting dairy plants for conformation to "General Specifications for Dairy Plants Approved for USDA Inspection and Grading Services"
- grading, sampling, testing, and certifying products of approved plants
- establishing regulations for manufacturing-grade milk as given in "Milk for Manufacturing Purposes and Its Production and Processing – Recommended Requirements" which is adopted by state agencies to regulate Grade B milk.

Grade B or manufacturing-grade milk does not meet the standards of PMO and can only be used for the manufacture of cheese, butter, and non-fat dairy milk (Womach, 2005).

17.5.3 State agencies

Every state has an authorized state regulatory agency for enacting safety and quality regulations for milk and milk products. The Department of Health or Department of Health Agriculture of the state usually enforces these regulations. The Grade A PMO of the FDA is usually used as the basis for all regulations of Grade A milk. The USDA "Milk for Manufacturing Purposes and Its Production and Processing – Recommended Requirements"is used as the basis for all regulations for manufacturing-grade milk. The state does not necessarily have to use these guidelines for their regulations as they are only voluntary assistance for states and not mandatory. Sometimes state standards may be even more stringent than the federal standards. However, for interstate commerce, it is mandatory to adhere to the standards specified by the PMO for Grade A milk.

17.5.4 Environmental Protection Agency (EPA)

The main function of the EPA with regard to dairy regulations is the establishment of tolerance levels or allowable limits for certain pesticide residues and effluent management. The enforcement of these regulations is carried out by the FDA by collecting and analyzing milk samples for pesticide residues. The EPA also administers the Safe Drinking Water Act of 1974, although it is enforced by the state regulatory agencies. Apart from these, the EPA administers the Federal Water Pollution and Control Act, which sets effluent guidelines and tests various dairy processing plants. The effluent characteristics that are tested include biochemical oxygen demand (BOD), total soluble solids (TSS), and pH. The Title 40 of the CFR Section 405 includes the effluent guidelines and standards for different sections and products manufactured in a dairy industry.

17.5.5 Federal Trade Commission (FTC)

The FTC is a federal agency that develops and administers the regulations on advertising and sales promotion procedures for foods. It is authorized to protect consumers and business persons from anticompetitive behavior and unfair or deceptive business and trade practices by the Federal Trade Commission Act of 1914. The main functions of the FTC are carried out by the Bureau of Consumer Protection. The FTC administers the Fair Packaging and Labeling Act of 1966; however, it is enforced by the FDA. According to this act, the FTC has guidance and

preventive functions and can issue complaints and shut-down orders or sue processing plants for violating the act.

17.5.6 US Customs Service

The US Customs Service ensures that imported products are taxed properly, safe for human consumption and not economically deceptive. It is assisted by the FDA and USDA to carry out its responsibilities and set the various requirements for imported milk and milk products. The regulations under the jurisdiction of the US Customs Service are given in Title 19 of the CFR.

17.6 Sustainability of the dairy industry

A major venture for maintaining sustainability in the dairy industries of the US was put forward by the Innovation Center for US Dairy (affiliated to Dairy Management Inc, a non-profit organization of dairy industry personnel: www.usdairy.com/Sustainability/Pages/Home.aspx). It published the first US dairy sustainability commitment progress report in 2007. Since then, this report had been published in 2010 and 2011. These efforts have been made to achieve environmental, social and economic sustainability in the dairy industry. The sustainability point of view in the dairy industry is providing the consumers with affordable and good-quality products while protecting natural resources and communities. Goals include addressing future issues of resource challenges due to the growing population, environment factors such as water scarcity, greenhouse gas emission, and reduction in available land. Efforts have been made to establish about 11–25% reduction of greenhouse gases associated with the production of 1 gallon of ice cream mix. Also, studies are being initiated to understand the environmental and socio-economic impact of the dairy industry. This broad understanding among the dairy industries should take us a long way in providing economically affordable dairy products while enriching the community and preserving the environment, to create a better tomorrow for future generations.

17.7 Conclusion

The US dairy industry is a pioneer in milk and milk product production and export in the world, with a bright future to look upon. Milk is a highly nutritious addition to the diet, which contains carbohydrates, proteins, fats, minerals, and vitamins. There are a number of processing technologies involved in the production of fluid milk and milk products, including pasteurization, homogenization, freezing and packaging, that have enabled these products to be distributed worldwide for decades. All the milk products and processing technologies are regulated by a number of federal and state agencies to ensure a good-quality product for the consumer. Additionally, a number of programs have been initiated to improve the sustainability of the dairy industry to ensure dairy products for future generations.

References

Barford NM, Krog N, Larsen G, BuchheimW (1991) Effects of emulsifiers on protein-fat interaction in ice cream mix during ageing I: quantitative analyses. *European Journal of Lipid Science and Technology* 93(1): 24–29.

Chandan RC (2006) Milk composition, physical and processing characteristics. In: Chandan RC, White CH, Kilara A, Hui YH (eds) Manufacturing Yogurt and Fermented Milks. Ames, IA: Blackwell Publishing, pp. 17–40.

Chang Y, Hartel RW (2002) Stability of air cells in ice cream during hardening and storage. *Journal of Food Engineering* 55(1): 59–70.

Clark S, Costello M, Bodyfelt FFW, Drake MA (2009) The Sensory Evaluation of Dairy Products. New York: Springer-Verlag.

Cooper RM (2011) Introduction. In: Adams DG, Cooper RM, Hahn MJ, Kahan JS (eds) Food and Drug Law and Regulation, 2nd edn. Washington, DC: Food and Drug Law Institute, pp. 1–22.

Dalgleish DG (2011) On the structural models of bovine casein micelles – review and possible improvements. *Soft Matter* 7(6): 2265–2272.

Danisco (2010). Ageing of Ice Cream. Danisco Cultor Technical Memorandum #2538–1e. Brabrand, Denmark.

de Kruif, CG (1998) Supra-aggregates of casein micelles as a prelude to coagulation. *Journal of Dairy Science* 81(11): 3019–3028.

de Kruif CG, Zhulina EB (1996) κ-casein as a polyelectrolyte brush on the surface of casein micelles. *Colloids and Surfaces A: Physicochemical and Engineering Aspects* 117(1–2): 151–159.

Debenedetti PG, Stanley HE (2003) Supercooled and glassy water. *Physics Today* 56(6): 40–46.

Dogan M, Kayacier A (2007) The effect of ageing at a low temperature on the rheological properties of kahramanmaras-type ice cream mix. *International Journal of Food Properties* 10(1): 19–24.

Drewett EM, Hartel RW (2007) Ice crystallization in a scraped surface freezer. *Journal of Food Engineering* 78(3): 1060–1066.

Everett DH (1988) Basic Principles of Colloid Science. Letchworth: Royal Society of Chemistry.

Evershed RP, Payne S, Sherratt AG et al (2008) Earliest date for milk use in the Near East and Southeastern Europe linked to cattle herding. *Nature* 455(7212): 528–531.

Flynn A, Cashman K (1992) Nutritional aspects of minerals in bovine and human milks. In: Fox PF (ed) Advanced Dairy Chemistry (Vol. 3) – Lactose, Water, Salts and Vitamins, 2nd edn. London: Chapman and Hall, pp. 257–289.

Food and Agriculture Organization (FAO) (2011) Statistical Database of the Food and Agriculture Organization of the United Nations. Rome: FAO. faostat.fao.org/site/339/default.aspx, accessed 25 November 2013.

Food and Drug Administration, Center for Food Safety and Applied Nutrition (2011) Grade "A" Pasteurized Milk Ordinance: 2011 revision. www.fda.gov/downloads/Food/Guidance Regulation/UCM291757.pdf, accessed 25 November 2013.

Fox PF (2003) Milk proteins: general and historical aspects. In: Fox PF, McSweeney PLH (eds) Advanced Dairy Chemistry (Vol. 1): Proteins. New York: Kluwer Academic/Plenum Press, pp. 1–48.

Fox PF, McSweeney PLH (1998a) Dairy Chemistry and Biochemistry. Glasgow: Blackie Academic and Professional.

Fox PF, McSweeney PLH (1998b) Dairy Chemistry and Biochemistry. Glasgow: Blackie Academic and Professional, pp. 21–66.

Goff HD, Verespej E, Smith AK (1999) A study of fat and air structures in ice cream. *International Dairy Journal* 9: 817–829.

Graham T, Maturin L (2004) Sampling dairy and related products. In: Wehr MH, Frank JF (eds) Standard Methods for the Examination of Dairy Products. Washington, DC: American Public Health Association.

Hartel RW (1996) Ice crystallization during the manufacture of ice cream. *Trends in Food Science and Technology* 7(10): 315–321.

Holt C., de Kruif CG, Tuinier R, Timmins PA (2003) Substructure of bovine casein micelles by small-angle X-ray and neutron scattering. *Colloids and Surfaces A: Physicochemical and Engineering Aspects* 213(2–3): 275–284.

Horne DS (1998) Casein interactions: casting light on the black boxes, the structure in dairy products. *International Dairy Journal* 8(3): 171–177.

Hui YH (1986) United States Food Laws, Regulations, and Standards, 2nd edn. Hoboken, NJ: Wiley-Interscience.

Hyde KA, Rothwell J (1973) Ice Cream, 5th edn. Edinburgh: Churchill Livingstone.

International Dairy Federation (IDF) (2012) The world dairy situation. Bulletin of the International Dairy Federation. Cape Town: IDF.

Iversen EK, Pedersen KS (1982) Ageing of ice cream. *Grinsted Technical Paper*, TP 33–le.

Jenkins TC, McGuire MA (2006) Major advances in nutrition: impact on milk composition. *Journal of Dairy Science* **89**(4): 1302–1310.

Jesse E, Cropp B (2008) Basic milk pricing concepts for dairy farmers. *Economic Research* **180**: 200.

Laben RC (1963) Factors esponsible for variation in milk composition. *Journal of Dairy Science* **46**(11): 1293–1301.

Legates JE (1960) Genetic and environmental factors affecting the solids-not-fat composition of milk. *Journal of Dairy Science* **43**(10): 1527–1532.

Loganathan S, Thompson NR (1968) Composition of cows' milk. I: environmental and managerial influences. *Journal of Dairy Science* **51**(12): 1928–1932.

Manchester AC, Blayney DP (2001) Milk pricing in the United States. In: Agricultural Information Bulletin No. 761. Washington, DC: United States Department of Agriculture, Economic Research Service.

Marshall RT, Goff HD, Hartel RW (2003) Ice Cream, 6th edn. New York: Kluwer Academic/Plenum Publishers.

Mulder H, Walstra P (1974) The Milk Fat Globule: Emulsion Science as Applied to Milk Products and Comparable Foods. Farnham Royal, UK: Commonwealth Agricultural Bureau.

Nielsen SS (2010). United States Government regulations and international standards related to food analysis. In: Nielsen SS (ed) Food Analysis, 4th edn. New York: Springer Science, pp. 15–34.

Nierman P (2004) Methods to detect abnormal milk. In: Fitts J (ed) Standard Methods for the Examination of Dairy Products. Washington, DC: American Public Health Association.

Rohenkohl H, Kohlus R (1999) Foaming of ice cream and the time stability of its bubble size distribution. In: Campbell GM, Webb C, Pandiella SS, Niranjan K (eds) Bubbles in Food. St Paul, MN: American Association of Cereal Chemists, pp. 45–53.

Russell AB, Cheney PE, Wantling SD (1999) Influence of freezing conditions on ice crystallisation in ice cream. *Journal of Food Engineering* **39**(2): 179–191.

Salque M, Bogucki PI, Pyzel J et al (2012) Earliest evidence for cheese making in the sixth millennium BC in Northern Europe. *Nature* doi:10.1038/nature11698.

Senhaji AF (1977) The protective effect of fat on the heat resistance of bacteria (II). *International Journal of Food Science and Technology* **12**(3): 217–230.

Sherbon JW (1999) Physical properties of milk. In: Wong NP (ed) Fundamentals of Dairy Chemistry, 3rd edn. Gaithersburg, MD: Aspen Publishers, pp. 414–419.

Singh H, McCarthy OJ, Lucey JA (1997) Physico-chemical properties of milk. In: Fox PF (ed) Advanced Dairy Chemistry, Volume 3: Lactose, Water, Salts and Vitamins, 2nd edn. London: Chapman and Hall, pp. 469–519.

Smith AK, Campbell BE (2007) Microstructure of milk components. In: Tamime AY (ed) Structure of Dairy Products. Oxford: Blackwell Publishing, pp. 59–69.

Stabel J, Lambertz A (2004) Efficacy of pasteurization conditions for the inactivation of mycobacterium avium subsp. paratuberculosis in milk. *Journal of Food Protection* **67**(12): 2719–2726.

Swaisgood HE (2003) Chemistry of the caseins. In: Fox PF, McSweeney PLH (eds) Advanced Dairy Chemistry 1: Proteins. New York: Kluwer Academic/Plenum Press, pp. 139–201.

Sweat VE, Parmelee CE (1978) Measurement of thermal conductivity of dairy products and margarines. *Journal of Food Process Engineering* **2**(3): 187–197.

Techzone (2010) www.techzone360.com/news/2010/06/01/4818616.htm.

Trout GM (1948) The nutritive value of homogenized milk: a review. *Journal of Dairy Science*, **31**(3): 627–655.

USDA Economic Research Services (ERS) (2011) Food Availability Data System. Washington, DC: USDA. www.ers.usda.gov/Data/FoodConsumption/FoodAvailSpreadsheets.htm.

USDA National Agricultural Statistics Service (NASS) (1980) Milk Production 12.12.1980. Washington, DC: USDA-NASS. http://usda01.library.cornell.edu/usda/nass/MilkProd//1980s/1980/MilkProd-12-12-1980.pdf

USDA National Agricultural Statistics Service (NASS) (2011) Milk Production 11.18.2011. Washington, DC: USDA-NASS. http://usda01.library.cornell.edu/usda/nass/MilkProd//2010s/2011/MilkProd-11-18-2011.pdf

USGPO (2013) Code of Federal Regulations (CFR); Title 21 Food and Drugs: Parts 100 to 169. Revised April 1 2013. www.ecfr.gov, accessed 26 November 2013.

van Boekel MAJS, Walstra P (1981) Stability of oil-in-water emulsions with crystals in the disperse phase. *Colloids and Surfaces* **3**(2): 109–118.

Walstra P (1975) Effect of homogenization on fat globule size distribution in milk. *Netherlands Milk and Dairy Journal* **29**(2–3): 279–294.

Walstra P (1990) On the stability of casein micelles. *Journal of Dairy Science* **73**(8): 1965–1979.

Walstra P, Jenness R (1984) Dairy Chemistry and Physics. New York: John Wiley, pp. 186–210.

Walstra P, Wouters JTM, Geurts TJ (2006) Milk – main characteristics/Milk components. In: Dairy Science and Technology. Boca Raton, FL: CRC/Taylor and Francis, pp. 3–108.

Wehr HM, Frank JF (2004) Standard Methods for the Examination of Dairy Products, 17th edn. Washington, DC: American Public Health Association.

Womach J (2005) Agriculture: a glossary of terms, programs, and laws. In: *CRS Report for Congress*: http://crs.ncseonline.org/NLE/CRSreports/05jun/97–905.pdf.

18 Dairy – Fermented Products

R.C. Chandan

Global Technologies, Inc., Minneapolis, Minnesota, USA

18.1 Introduction

Microorganisms employed as starters for production of cultured dairy foods are divided into two types, based on the optimum temperature ranges at which they operate (Hutkins, 2006; Stanley, 2003; Vedamuthu, 2013a). The lactic acid bacteria incubated at temperatures above 35 °C are referred to as thermophilic bacteria and those incubated at 20–30 °C are called mesophilic starters. Yogurt is derived by culturing with thermophilic cultures, which act in symbiosis with each other. In contrast, sour cream or cultured cream is obtained by fermentation with mesophilic lactococci and leuconostocs. Yogurt and sour cream do not involve whey formation and removal from curd. A notable exception is Greek yogurt, which has gained particular attention in recent years (Kilara & Chandan, 2013).

Cheese production involves removal of liquid whey, leading to partial dehydration and concentration of certain milk constituents. Most cheese varieties are produced with mesophilic cultures, but certain varieties use thermophilic bacteria. In addition, other microorganisms, like molds and yeasts, characterize certain cheese varieties. During cheese making, lowering of pH and addition of salt confer a preservative effect, resulting in extension of shelf life as well as safety for consumption. In cheese, the main milk components (proteins, fat, and minerals) are concentrated and protected from rapid deterioration by spoilage microorganisms. For consumers, cheese provides good nutrition, variety, convenience of use, portability, food safety, and novelty of flavors and textures. Modern packaging techniques confer even more extended shelf life, allowing the movement of cheese over long distances from the place of manufacture. There are some

400 varieties of cheese consumed throughout the world. The major varieties have distinctive flavor and texture ascribed primarily to the use of milk of various domesticated animals, discrete microbial cultures, enzymes, and ripening conditions. Their processing procedures influence final chemical composition resulting in distinct fermentation patterns, which in turn develop specific flavors and textures (Chandan & Kapoor, 2011a, 2011b).

At the turn of the last century, developments in melting processes, involving natural cheese of various ages, gave birth to a line of processed cheese products with controlled flavor and texture, and extended shelf life. In addition, various shapes, sizes, configurations, and sliced versions were created to provide varieties with novel applications. The consumer can use these products as ingredients in cooking of several dishes or as a ready-to-eat snack. These products are designed to be consumed as spreads or as slices in sandwiches, and function as a dip or topping on snacks.

Various fermented dairy foods may be consumed in original form or they may be mixed with fruits, grains, and nuts to yield delicious beverages, snacks, desserts, breakfast foods, or a light lunch. A variety of textures and flavors are generated by selection of lactic acid bacteria. A combination of lactic acid bacteria and their strains allows an interesting array of products to suit different occasions of consumption.

This chapter discusses general technical aspects of the manufacture of fermented dairy products with a focus on yogurt, sour cream, Cheddar, and process cheese. Other cultured milks and cheeses are not considered here in any depth. For more extensive treatment of various aspects of fermented dairy products including cheese, the reader is referred to several literature resources

Table 18.1 Total production and per capita sales of yogurt, sour cream and natural cheese in the US in recent years

Year	Yogurt		Sour cream and dips		Cheese (natural)	
	Total production*	Per capita sales**	Total production*	Per capita sales**	Total production*	Per capita sales**
1960	44	0.2	Not available	0.9	1478	Not available
1970	172	0.8	Not available	1.1	2201	11.4
1980	570	2.5	480	1.8	3984	17.5
1990	1055	4.2	625	2.5	6059	24.6
2000	1837	6.5	914	3.2	8258	29.83
2005	3058	10.3	1309	4.4	9189	31.74
2006	3301	11.1	1256	4.2	9525	32.6
2007	3476	11.5	1313	4.4	9777	33.6
2008	3570	11.8	1274	4.2	9913	33.0
2009	3832	12.5	1275	4.2	10,074	32.2
2010	4181	13.5	1270	4.1	10,443	33.1
2011	4272	13.7	1264	4.1	10,597	33.5

* Million pounds.
** Pounds.
Adapted from: IDFA (2012).

(Chandan & Kilara, 2011, 2013; Hill, 2009; Hutkins, 2006; Law & Tamime, 2010; Pannell & Schoenfuss, 2007; Tamime & Robinson, 2007).

18.2 Consumption trends

Trends in the production and consumption of fermented dairy foods (e.g. yogurt, sour cream and dips, and natural cheese) in the US are presented in Table 18.1. Production and consumption of yogurt, sour cream, and cheese have registered significant gains in the time period of 1960–2011. Yogurt production in 2011 was 4272 million pounds with per capita consumption of 13.7 pounds. The growth of yogurt has been especially remarkable during this period. However, compared to Sweden, with per capita consumption of 62.8 pounds, US consumption is modest (Schultz, 2011). Sour cream production of 1264 million pounds with per capita consumption of 4.1 pounds has been fairly flat in recent years. Natural cheese production of 10,597 million pounds and per capita consumption of 33.5 pounds has shown modest growth.

Table 18.2 shows the trends in production and consumption of some cheese varieties in the period 2005–2011. Cheese production and consumption far exceed the production of yogurt and sour cream. The data in the table show that the per capita consumption of natural cheeses has gone up. Process cheese products are derived from natural cheeses and contain water and other

food ingredients. Their consumption is relatively steady compared to natural cheeses. The most popular individual cheese variety in 2011 was Mozzarella, followed by Cheddar cheese. The per capita consumption of all Italian varieties grew.

During 2011, the supermarket sales of natural cheeses was 2271 million pounds (valued at $11,076 million). Cheddar cheese totaled 834 million pounds (valued at $3889 million), followed by Processed American cheese (760 million pounds valued at $2583 million) and Mozzarella cheese (475 million pounds valued at $2190 million) (IDFA, 2012).

18.3 Production of starters for fermented dairy foods

Fermented milk foods with desirable characteristics of flavor, texture, and probiotic profiles can be created by formulating the desired chemical composition of the milk substrate mix, judicious selection of lactic acid bacteria (starter), and fermentation conditions (Chandan, 1982; Chandan & Nauth, 2012; Chandan & Shahani, 1993, 1995). A starter is made up of one or more strains of food-grade microorganisms. Individual microorganisms utilized as a single culture (single or multiple strains), or in combination with other microorganisms, exhibit characteristics impacting the technology of manufacture of fermented milks.

Table 18.2 Per capita consumption (pounds) of various natural and process cheese products in the US in the years 2005–2011

Product	2005	2006	2007	2008	2009	2010	2011
Total natural cheeses	32.5	32.6	33.6	33.0	33.2	33.1	33.5
Cheddar cheese	10.1	10.7	10.8	10.9	11.0	10.3	10.0
Total American cheeses	13.5	3.1	13.3	13.6	14.0	13.3	13.2
Total Italian cheeses	13.2	13.6	13.9	13.8	13.8	14.4	14.8
Total process cheese and products	7.6	7.8	7.5	7.1	7.0	6.7	6.8

Adapted from: IDFA (2012).

Table 18.3 Cultures used for production of yogurt and sour cream

Yogurt	Cultured/sour cream
Required by FDA regulations	*Lactococcus lactis* subsp. *lactis*
Streptococcus thermophilus	*Lactococcus lactis* subsp. *cremoris*
Lactobacillus delbrueckii subsp. *bulgaricus*	*Leuconostoc mesenteroides* subsp. *cremoris*
Optional	*Lactococcus lactis* subsp. *lactis* biovar. *diacetylactis*
Lactobacillus acidophilus	
Lactobacillus casei	
Lactobacillus casei subsp. *rhamnosus*	
Lactobacillus reuteri	
Lactobacillus helveticus	
Lactobacillus gasseri ADH	
Lactobacillus plantarum	
Lactobacillus lactis	
Lactobacillus johnsoni LA1	
Lactobacillus fermentum	
Lactobacillus brevis	
Bifidobacterium longum	
Bifidobacterium breve	
Bifidobacterium bifidum	
Bifidobacterium adolescentis	
Bifidobacterium animalis	
Bifidobacterium infantis	

Modern industrial processes utilize defined lactic acid bacteria as starters for fermented dairy products. For detailed descriptions of starter cultures, the reader is referred to Stanley (2003), Hutkins (2006), and Vedamuthu (2013a, 2013b). Table 18.3 summarizes some of the microorganisms used in the manufacture of yogurt and cultured cream.

The most common lactic acid bacteria employed for fermented dairy foods are: *Lactobacillus delbrueckii* subsp. *bulgaricus*, *Streptococcus thermophilus*, *Lactobacillus acidophilus*, *Lactococcus lactis* subsp. *lactis*, and *Lactococcus lactis* subsp. *cremoris*. They are responsible for the acidic taste arising from lactic acid elaborated by their growth. *Leuconostoc* spp. are used for typical flavor in sour cream. In cheeses, *Lactococcus lactis* subsp. *lactis/cremoris*, *Lactobacillus helveticus*, *Lactobacillus delbrueckii* subsp. *bulgaricus*, and *Streptococcus thermophilus* are employed for acid and distinct flavor development, while *Propionibacterium shermanii* secretes propionate, a natural shelf life extender. Furthermore, it is possible to deliver health-promoting microflora to the consumer of the food. In this regard, yogurt cultures, *Lactobacillus delbrueckii* subsp. *bulgaricus* (LB), *Streptococcus thermophilus* (ST), *Lactobacillus acidophilus*, *Lactobacillus casei*, and *Bifidobacterium* species are notable examples.

The cultures used in the manufacture of natural cheeses are shown in Table 18.4. The cultures are used to provide distinct flavor and textural attributes to a particular cheese variety. The metabolic products of culture growth leave a discrete profile on the sensory properties of cheese. In this regard, in addition to bacteria, edible molds extend color and flavor characteristics, typical of the variety of the cheese (blue, Roquefort, stilton, Camembert, Brie, Gorgonzola).

Adequate production of lactic acid is essential for lowering the pH to a level where critical flavor compounds (acetaldehyde, diacetyl and other compounds) are formed in sufficient quantity. Factors interfering with proper acid development will retard or prevent adequate flavor development. The culture may be incapable of producing adequate amounts of flavor due to a change in fermentation pattern induced by oxygen tension or due to a change in the balance of various bacterial cultures. In certain

Table 18.4 Microbial composition of starters used in the manufacture of major cheese varieties. In some cheeses, two or more starters may be used

Cheese varieties	Microorganism
Cheddar, Colby, Gouda, Edam, Monterey	*Lactococcus lactis* subsp. *lactis* *Lactococcus lactis* subsp. *cremoris*
Cream, Neufchatel, cottage	*Lactococcus lactis* subsp. *lactis* *Lactococcus lactis* subsp. *cremoris* *Lactococcus lactis* subsp. *lactis* biovar. *diacetylactis* *Leuconostoc mesenteroides* subsp. *cremoris*
Swiss, Emmental, Gruyere, Samso, Fontina	*Leuconostoc lactis* subsp. *cremoris* *Lactococcus lactis* subsp. *lactis* biovar. *diacetylactis* *Leuconostoc lactis* subsp. *cremoris* *Lactobacillus delbrueckii* subsp. *bulgaricus* *Lactobacillus delbrueckii* subsp. *lactis* *Lactobacillus casei* subsp. *casei* *Lactobacillus helveticus* *Streptococcus thermophilus* *Propionibacterium freudenreichii* *Propionibacterium shermanii*
Italian cheeses: Mozzarella, Provolone, Romano	*Lactococcus lactis* subsp. *lactis* *Streptococcus thermophilus* *Lactobacillus delbrueckii* subsp. *bulgaricus* *Lactobacillus delbrueckii* subsp. *lactis* *Lactobacillus helveticus*
Brick, Limburger, Muenster, Trappist, Port Salut, St Paulin, Bel Paese, Tilsit	*Streptococcus thermophilus* *Lactobacillus delbrueckii* subsp. *lactis*
Blue-veined cheeses: Roquefort, Bleu, Stilton, Gorgonzola	*Lactococcus lactis* subsp. *lactis* *Lactococcus lactis* subsp. *cremoris* *Penicillium roqueforti*
Camembert, Brie	*Lactococcus lactis* subsp. *lactis* *Lactococcus lactis* subsp. *cremoris* *Streptococcus thermophilus* *Penicillium candidum* *Penicillium caseicolum* *Penicillium camemberti*

From: Chandan and Kapoor 2011a.

instances, acidity may be too high in relation to product standards. This problem may be encountered in yogurt production. High acidity is usually associated with high incubation temperature, long incubation period, or excessive inoculum (Chandan & O'Rell, 2013b).

Any change from the normal fermentation pattern is considered a defect. Insufficient acid development is one of the common defects in lactic cultures. When 1 mL culture inoculated into 10 mL of antibiotic-free, heat-treated milk produces less than 0.7% titratable acidity in 4 h at 35 °C, it is considered a slow starter. Factors contributing to a slow starter may involve milk composition or extraneous inhibitors. Milk from mastitis-infected animals generally does not support the growth of lactic cultures. This effect is ascribed to the infection-induced changes in chemical composition of milk. For example, mastitis milk contains lower concentrations of lactose, higher chloride content and a higher pH than normal milk. Furthermore, a high leukocyte count in mastitis milk inhibits bacterial growth by phagocytic action. Heat treatment restores the culture growth in mastitis milk (Chandan, 1982).

Colostrom and late lactation milk contain non-specific agglutinins, which clump and precipitate sensitive strains of the starter. The agglutinins may possibly retard the rate of acid production by interfering with the transport of lactose and other nutrients. Seasonal variation of the solids-not-fat fraction of milk affects the growth and the balance of strains in culture.

Another cause of slow starters may be attributed to antibiotics in milk. Concentrations as low as 0.005–0.05 international units (IU) of antibiotics per mL of milk, used in mastitis therapy, are high enough to impact partial or full inhibition of the culture. Accordingly, it is imperative that residual antibiotic level in milk be monitored routinely to keep the milk supply suitable for cultured milk manufacture. In addition, prior degradation of milk constituents by microbial contaminants affects the growth of lactic organisms. Careful screening of milk for psychrotrophic organisms is necessary for quality flavor production by lactic cultures.

Many sanitizing chemicals, such as quaternary ammonium compounds, iodine and chlorine compounds, retard acid development by starter cultures. One to five parts per million of these sanitizing compounds are bactericidal to lactic cultures. Proper use of sanitizers includes verification of sanitizer concentration as well as removal of residual sanitizers from the pipes and vessels prior to the use of equipment for processing of fermented dairy products.

Avoiding the use of rancid milk is important not only from the standpoint of culture growth. The free fatty acids generated in rancid milk are inhibitory to culture growth (Chandan, 1982). More significantly, rancid milk would impart an objectionable flavor to the starter and the cultured dairy products derived therefrom.

After continuous use, the starters may change their fermentation activity and consequently produce lower amounts of lactic acid (Chandan, 1982). Bacteriophages (virus-like organisms) may be an important cause of slow acid production by lactic cultures. When the phage has reached a maximum level, all sensitive bacterial cells are infected and lyzed within 30–40 min. When lysis occurs, acid production by the affected culture stops unless some resistant bacteria are present to carry on fermentation. In case of phage attack in cultured milk plants, it is advisable to spray 200–300 ppm chlorine on processing equipment. Additionally, consider fogging the culture rooms with 500–1000 ppm of chlorine. Heat treatment of milk (75 °C for 30 min or 80 °C for 20 sec) is considered adequate to inactivate various phages that attack lactic acid bacteria. Such intensive heat treatment results in too soft curd and is therefore not suitable for production of most cheeses. Since cheese whey is a good source of phages, its reuse should be avoided.

Another step for phage control is to rotate cultures on different production days. Phages are strain specific. Culture rotation results in introducing different strains of the culture from day to day. Cultures containing multiple strains may also be helpful. Phage attack on a given strain would result in loss of acid production by the particular strain, while the other strains are still available to carry on fermentation. Phage activity requires calcium ions. By sequestering calcium with phosphates, it is possible to design and use a phage-resistant medium for propagating cultures in the plant. Accordingly, by using combinations of proper procedures, the probability of trouble with phage can be minimized (Chandan & Nauth, 2012).

Ropiness and gassiness in the bulk starter is another defect resulting from limited proteolytic activity of some starter strains, and is commonly observed with Cheddar cheese cultures. Also, it can be attributed to the presence of proteolytic bacteria in the starter culture. Some spore-formers that survive normal heat treatment of milk may also be involved.

Bitter taste (in the final product) is due to proteolytic activity of some starter strains, and is commonly observed with Cheddar cheese cultures (Chandan, 1982; Chandan & Kapoor, 2013a, 2013b). Proteolysis leads to the formation of bitter peptides, which impart bitterness to cheese.

The factors influencing starter performance have been discussed by Vedamuthu (2013a). The cellular functions (chromosomal and extrachromosal DNA) are encoded in the nuclear materials of the bacterial cell. The genetic materials could be altered by spontaneous or induced mutations, leading to changes in cellular metabolism, which in turn would alter acid generation and flavor production by the starter. The extrachromosomal material consists of plasmids, transposons, and introns. Work on the plasmids of mesophilic lactic acid bacteria has significantly elucidated basic principles relative to acid production, sugar utilization, proteolytic activity, citrate metabolism, bacteriophage resistance, and bacteriocin production (Vedamuthu, 2013a). The plasmids are known to code for several significant functions. Technology for transfer of plasmids has been developed by commercial culture manufacturers to introduce new strains with enhanced bacteriophage resistance, boosted health attributes, ability to accelerate cheese ripening, stability in flavor and texture production, and production of antimicrobial compounds and natural preservatives (Vedamuthu, 2013a).

The starter culture is a crucial component in the production of fermented dairy products of high quality and uniformity. An effective sanitation program, including filtered air and positive pressure in the fermentation area, should significantly control air-borne contamination and contribute to consistently high-quality products.

Commercial cultures may consist of single strains or mixed strains of the same species. Furthermore, they may be mixed strains of a combination of different genus and species of bacteria. They are designed for performance in the production of discrete cultured foods and intended to be used as per instructions issued by the culture manufacturer, which include temperature of incubation, inoculum level, and incubation time. Many plants use frozen direct-to-vat or freeze-dried direct-to-vat cultures for production of cultured dairy foods. However, for cost savings, some large manufacturers prefer to make bulk starters in their own plant from frozen or dry bulk cultures.

The medium for bulk starter production is antibiotic-free, non-fat dry milk reconstituted in water at 10–12% solids level. The medium is heated to 90–95 °C and held for 30–60 min, then cooled to incubation temperature in the vat. For example, the incubation period for yogurt bulk starter ranges from 4 to 6 h and the temperature of 43 °C is maintained by holding hot water in the jacket of the tank. The fermentation must be quiescent (lack of agitation and vibrations) to avoid phase separation in the starter following incubation. The progress of fermentation is monitored by titratable acidity or pH measurements at regular intervals. When the titratable acidity is 0.85–0.90%, (approximate pH 4.4), the fermentation is terminated by turning the agitators on and replacing warm water in the jacket with iced water. Circulating iced water drops the temperature of the starter to 4–5 °C. The starter is now ready to use. Depending on the product to be manufactured, the inoculum size varies from 0.5% to 5%. If necessary, the starter may be stored at 4–5 °C for 3–4 days.

18.4 Biochemical basis of lactic fermentation for flavor and texture generation

The major constituents of milk are altered by interaction with starter organisms. Lactose is first broken down to glucose and galactose, then to lactic acid via the glycolytic pathway by starter bacteria. Galactose is not readily broken down in some strains of lactic acid bacteria. Consequently, it accumulates in fermented products. In Mozzarella cheese, the galactose forms brown coloration on melting. The acid production results in lowering of pH and increase in viscosity until the liquid phase transforms to a solid gel phase in fermented milks. Another function of acid generation is its preservative

effect, which extends shelf life of fermented milks. Furthermore, the starter enzymes are responsible for limited breakdown of proteins, amino acids, and fat, thereby generating key flavor compounds. In this regard, acetaldehyde is regarded as a typical flavor compound in yogurt whereas diacetyl is a distinctive flavor compound in buttermilk and sour cream (Vedamuthu, 2013a, 2013b).

In the production of cheeses, acid formation accelerates the activity of coagulating enzymes (chymosin), leading thereby to the formation of curd. The activity of the starter and other organisms during the aging of cheese leads to breakdown of casein and milk fat, giving particular cheeses their typical flavor and textural characteristics (Fox, 2003b; Singh & Cadwallader, 2008). The reader is referred to extensive literature on cheeses, including a series of articles on types, starter cultures, their chemistry and microbiology, manufacture of various types, and flavor generation (Bottazzi, 2003; Chandan, 2003; Farkye, 2003; Fox, 2003a, 2003b; Fox & McSweeney, 2003; McSweeney, 2003; Olsen, 2003; Stanley, 2003).

18.4.1 General principles of fermented dairy foods manufacture

Fermented milks are generally made from a mix standardized from whole, partially defatted milk, condensed skim milk, cream and/or non-fat dry milk. Yogurt is made from fat-standardized milk fortified with non-fat milk solids. Cultured sour cream is made from 18% fat cream supplemented with some non-fat milk solids for textural improvement. Many cheeses are made from pasteurized milk standardized for fat:solids-not-fat ratio. Standards of identity established by regulatory authorities in each country assure the consumer a defined product. Table 18.5 lists the standards for cultured fluid dairy foods in the US.

All dairy raw materials should be selected for high bacteriological and sensory quality. A general process for fermented dairy foods is illustrated in Figure 18.1.

The following sections summarize the essential parameters prescribed by the US Food and Drug Administration (FDA) regarding composition of yogurt, sour cream, Cheddar cheese, and process cheese.

18.5 Yogurt

Yogurt is a semi-solid fermented product made from a heat-treated standardized milk mix by the activity of

Table 18.5 Essential FDA standards for composition of certain fermented milks in the US

Product	% Milk fat	% Milk solids-non-fat	% Titratable acidity expressed as lactic acid
Yogurt CFR 131.200	Not less than 3.25	Not less than 8.25	Not less than 0.9
Low-fat yogurt CFR 131.203	Not less than 0.5 and not more than 2.0	Not less than 8.25	Not less than 0.9
Non-fat yogurt CFR 131.206	Not more than 0.5	Not less than 8.25	Not less than 0.9
Cultured milk CFR131.112	Not less than 3.25	Not less than 8.25	Not less than 0.5
Cultured low-fat milk* (buttermilk)	Not less than 0.5 and not more than 2.0	Not less than 8.25	Not less than 0.5
Cultured non-fat milk	Not more than 0.5	Not less than 8.25	Not less than 0.5
Cultured (sour) cream CFR 131.160	Not less than 18.0. After the addition of bulky flavors, etc.not less than 14.4	No standard	Not less than 0.5

* Commonly known as cultured buttermilk.
Source: FDA (CFR, 2011).

a characterizing symbiotic blend of *Streptococcus thermophilus* (ST) and *Lactobacillus delbrueckii* subsp. *bulgaricus* (LB) cultures. In addition to mandatory cultures, commercial yogurt contains adjunct cultures, primarily *Lactobacillus acidophilus*, *Lactobacillus casei*, and *Bifidobacterium* spp.

Yogurt is produced from the milk of various animals (cow, water-buffalo, goat, sheep, yak, etc.) in various parts of the world. Cow's milk is the predominant starting material in industrial manufacturing operations in the US. In order to achieve a custard-like semi-solid consistency, the cow's milk is fortified with dried or condensed milk.

Vitamin addition at a level of 2000 IU of vitamin A and 400 IU of vitamin D per quart (846 mL) is allowed. Permissible dairy ingredients are cream, milk, partially skimmed milk, skim milk, alone or in combination. Other optional ingredients include concentrated skim milk, non-fat dry milk, buttermilk, whey, lactose, lactalbumins, lactoglobulins, or whey modified by partial or complete removal of lactose and/or minerals. These ingredients are used to increase the non-fat solids content of yogurt, provided that the ratio of protein to total non-fat solids of the food and the protein efficiency ratio of all protein present shall not be decreased as a result of adding such ingredients. In addition, sweeteners such as sucrose, invert sugar, brown sugar, refiner's syrup, molasses (other than blackstrap), high fructose corn syrup, fructose, fructose syrup, maltose, maltose syrup, dried maltose syrup, malt extract, dried malt extract, malt syrup, dried malt syrup, honey, maple sugar, except table syrup may be used. The regulations allow flavoring ingredients, color additives, and stabilizers.

18.5.1 Raw materials for yogurt

18.5.1.1 Dairy ingredients

In the US, yogurt is a Grade A product (CFR, 2011). Grade A implies that the milk used must come from FDA-supervised Grade A dairy farms and Grade A manufacturing plants as per regulations enunciated in the *Pasteurized Milk Ordinance* (US Department of Health and Human Services, 2011a). To make yogurt mix, milk is supplemented with non-fat dry milk or condensed skim milk to increase the solids-not-fat (SNF). The FDA specification calls for a minimum of 8.25% non-fat milk solids (see Table 18.5). However, the industry uses up to 12% SNF in the yogurt mix to generate a thick, custard-like consistency. The milk fat levels are standardized to 3.25% for full-fat yogurt; low-fat yogurt is manufactured from mix containing 0.5–2% milk fat; non-fat yogurt mix has a milk fat level not exceeding 0.5% (CFR, 2011).

18.5.1.2 Culture

Most of the yogurt is fermented with ST and LB. In addition, optional bacteria are incorporated along with the

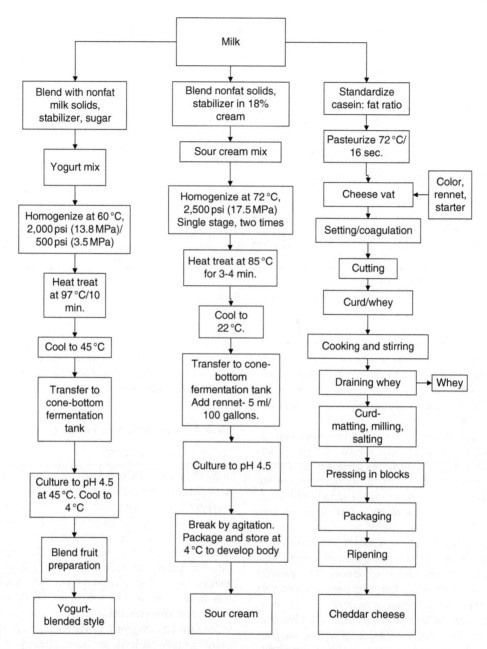

Figure 18.1 General processing outlines for manufacture of yogurt, sour cream, and Cheddar cheese.

starter. *Lactobacillus acidophilus* is commonly used as an additional culture in commercial yogurt. Other cultures commonly added belong to various *Lactobacillus* and *Bifidobacterium* species because of their probiotic benefits. Probiotics are food-grade cultures which upon ingestion have been shown to impart several health benefits (for instance, immune enhancements, gastrointestinal improvements).

Commercial production of yogurt relies heavily on fermentation ability and characteristics imparted by the starter. The starter performance factors are rapid acid development; typical yogurt flavor, body, and texture;

exopolysaccharide-secreting strains to enhance viscosity of yogurt; scale-up possibilities in various production conditions, including compatibility with a variety and levels of ingredients used; fermentation times and temperatures; survival of culture viability during shelf life; exhibit probiotic properties and survival in the human gastrointestinal tract for certain health attributes; and minimum acid production during distribution and storage at 4–10 °C until yogurt is consumed.

Fermentation constitutes the most important step in yogurt manufacture. To optimize parameters for yogurt production, an understanding of factors involved in the growth of yogurt bacteria is important to manage uniformity of product quality and cost-effectiveness of manufacturing operation.

Collaborative growth of ST and LB is a phenomenon unique to yogurt. Yogurt starter organisms display obligate symbiotic relationship during their growth in the milk medium. Although they can grow independently, they utilize each other's metabolites to effect remarkable efficiency in acid production. In general, LB has significantly more cell-bound proteolytic enzyme activity, producing stimulatory peptides and amino acids for ST. The relatively high amino-peptidase and cell-free and cell-bound dipeptidase activity of ST is complementary to the strong proteinase and low peptidase activity of LB. Urease activity of ST produces CO_2, which stimulates LB growth. Concomitant with CO_2 production, urease liberates ammonia, which acts as a weak buffer, to prevent extreme pH conditions that are inhibitory to bacteria. Consequently, milk cultured by ST alone exhibits low titratable acidity (or high pH) of coagulated mass. Formic acid formed by ST, as well as by heat treatment of milk, accelerates LB growth. The rate of acid production by yogurt starter containing both ST and LB is considerably higher than by either of the two organisms grown separately.

Yogurt organisms are micro-aerophilic in nature, which means foaming must be prevented in the production of yogurt. Heat treatment of milk drives out oxygen. It also wipes out competitive flora. Furthermore, heat produces sulfhydryl compounds, which tend to generate reducing conditions in the medium. Lactic cultures prefer such micro-aerophilic conditions for their growth. Accordingly, rate of acid production is considerably higher in heat-treated milk than in raw or pasteurized milk. Although the heat treatment far exceeds time-temperature requirements for proper pasteurization, a legal pasteurization time-temperature regime is still mandated by the FDA in yogurt production.

18.5.1.3 Sweeteners

Yogurt mixes designed for manufacture of yogurt may contain appreciable quantities of sucrose. The sweeteners exert osmotic pressure in the system, leading to progressive inhibition and decline in the rate of acid production by the culture. Acid-producing ability of yogurt culture in mixes containing 8.0% sucrose is acceptable practice in the industry. Sucrose is the major sweetener used in yogurt production. High-intensity sweeteners (e.g. aspartame, sucralose, acesulfame K, steviosides, monk fruit, etc.) are used to produce "light" yogurt containing about 60% calories compared to normal sugar-sweetened yogurt. Low levels of crystalline fructose may be used in conjunction with aspartame and other high-intensity sweeteners to round up and improve overall flavor of "light" yogurt. Generally, commercial yogurts have an average of 4.06% lactose, 1.85% galactose, and 0.05% glucose.

18.5.1.4 Stabilizers

The primary purpose of using a stabilizer in yogurt is to produce smoothness in body and texture, impart gel structure, and reduce wheying off (or syneresis). The stabilizer increases shelf life and provides a reasonable degree of uniformity of the product. Stabilizers function through their ability to form gel structures in water, thereby leaving less free water for syneresis. A good yogurt stabilizer should not impart any flavor, should be effective at low pH values, and should be easily dispersed in the normal working temperatures in a food plant. The stabilizers generally used in yogurt are modified starch, gelatin, whey protein concentrates, and pectin. The stabilizer system used in yogurt mix preparations is generally a combination of starch and gelatin. Whey protein concentrates at 0.5–1% level are also widely used for their water-binding attributes.

18.5.1.5 Fruit preparations for flavoring yogurt

The fruit preparations for blending in yogurt are specially designed to meet the marketing requirements for different types of yogurt. They are generally present at levels of 10–15% by weight in the final product. A majority of the fruit preparations contain natural flavors to boost the fruit aroma and flavor. Flavors and certified colors are usually added to the fruit-for-yogurt preparations for improved eye appeal and better flavor profile. The fruit base should exhibit true color and flavor of the fruit when blended with yogurt, and be easily dispersible in yogurt without causing texture defects, phase separation, or syneresis. The pH of the fruit base should be compatible with yogurt pH (approximately 4.4). The fruit should have zero yeast and mold population in order to prevent spoilage and to extend shelf life. For extensive discussion of fruit for yogurt, the reader is referred to O'Rell and Chandan (2013a).

18.5.2 Processing

Detailed discussion relative to yogurt manufacture is available in the literature (Chandan & O'Rell, 2013a-c; Tamime & Robinson, 2007). The general steps are summarized in the following pages.

18.5.2.1 Mix preparation

A yogurt plant requires a special design to minimize contamination of the products with phage and spoilage organisms. The plant is generally equipped with a receiving room to receive, meter or weigh, and store milk and other raw materials. In addition, facilities include a process and production control laboratory, a dry storage area, a refrigerated storage area, a mix processing room, a fermentation room, and a packaging room.

The mix processing room contains equipment for standardizing and separating milk, pasteurizing and heating, and homogenizing, along with the necessary pipelines, fittings, pumps, valves, and controls. To prevent entry of stray microorganisms, including bacteriophages, the fermentation room (housing fermentation tanks) is isolated from the rest of the plant. Filtered air under positive pressure is supplied to the room to generate clean room conditions. A control laboratory is generally set aside, where culture handling, process control, product composition, and shelf life tests are carried out to insure adherence to regulatory and company standards. Also, a quality control program is established by laboratory personnel. A utility room is required for maintenance and engineering services needed by the plant. The refrigerated storage area is used for holding fruit, finished products, and other heat-labile materials. A dry storage area at ambient temperature is primarily utilized for temperature-stable raw materials and packaging supplies.

Standardization of milk for fat and milk solids-not-fat content results in fat reduction and an increase of approximately 30–35% for other solid components (lactose, protein, mineral, and vitamin content). Nutrient density of yogurt mix is increased in relation to that of milk. Specific gravity changes from 1.03 to 1.04 g/mL at 20 °C. Addition of stabilizers and sweeteners further impacts physical properties. All the additives are blended prior to heat treatment.

18.5.2.2 Heat treatment

Pasteurization equipment consists of a vat, plate heat exchanger or high-temperature, short-time (HTST) pasteurization system. The mix is subjected to much more severe heat treatment than conventional pasteurization temperature-time combinations. Heat treatment at 85 °C for 30 min or 95–99 °C for 7–10 min is an important step in manufacture. The heat treatment:

- produces a relatively sterile medium for the exclusive growth of the starter
- removes air from the medium to produce a more conducive medium for micro-aerophilic lactic cultures to grow
- causes thermal breakdown of milk constituents, especially proteins, releasing peptones and sulfhydryl groups, which provide nutrition and anaerobic conditions for yogurt culture
- produces physical changes in the proteins which have a profound effect on the viscosity of yogurt.

The heat treatment denatures and coagulates whey proteins (β-lactoglobulin in particular) which improves water-binding properties. Whey protein denaturation, of the order of 95%, enhances water absorption capacity, thereby creating a smooth custard-like consistency, high viscosity and stability from whey separation in yogurt. The intense heat treatment during yogurt processing destroys all the pathogenic flora and most vegetative cells of all microorganisms contained therein. In addition, milk enzymes inherently present are inactivated. From a microbiological standpoint, destruction of competitive organisms produces conditions conducive to the growth of desirable yogurt bacteria.

18.5.2.3 Homogenization.

The homogenizer is a high-pressure pump that forces the mix through extremely small orifices. The process is usually conducted by applying pressure in two stages. The first stage pressure, of the order of approximately 14 MPa (2000 psi), reduces the average milk fat globule diameter size from approximately 4 μm (range 0.1–16 μm) to less than 1 μm. The second stage uses 3.5 MPa (500 psi) and is designed to break the clusters of fat globules apart with the objective of inhibiting creaming in milk. Homogenization aids in texture development (partially due to blending of stabilizers) and alleviates the surface creaming and wheying-off problems. Since homogenization treatment reduces the fat globules to an average of less than 1 μm in diameter, no distinct creamy layer (crust) is observed on the surface of yogurt produced from homogenized mix. The homogenizer and heat treatment systems are located in tandem. The homogenized and heat treated mix is brought to 43 °C by pumping through an appropriate heat exchanger. It is then collected in fermentation tanks.

18.5.2.4 Fermentation.

Fermentation tanks for the production of cultured dairy products are generally designed with a cone bottom to facilitate draining of relatively viscous fluids after incubation. The starter is generally inoculated at 5% level. For temperature maintenance at approximately 43 °C (110 °F) during the incubation period, the fermentation vat is usually insulated and covered with an outer surface made of stainless steel. The vat is equipped with a heavy-duty, multispeed agitation system, a manhole containing a sight glass, and appropriate spray balls for clean-in-place (CIP) cleaning. In the CIP process, hot cleaning fluids are circulated through the equipment. The agitator is often of swept surface type for optimum agitation of relatively viscous cultured dairy products. For efficient cooling after culturing, plate or triple-tube heat exchangers are used.

18.5.2.5 Contribution of the culture to yogurt texture and flavor

Proper fermentation with yogurt culture leads to the formation of typical flavor compounds. Lactic acid, acetaldehyde, acetoin, diacetyl, and other carbonyl compounds constitute key flavor compounds of yogurt. The production of flavor by yogurt cultures is a function of time as well as the sugar content of yogurt mix. Acetaldehyde production in yogurt takes place predominantly in the first 1–2 h of incubation. Eventually, an acetaldehyde level of 2–41 ppm develops (Chandan & O'Rell, 2013b). The acetaldehyde level declines in later stages of incubation. Diacetyl varies from 0. to 0.9 ppm, and acetoin varies from 2.2 to 5.7 ppm (Routray & Mishra, 2011). Lactic acid content varies from 0.9% to 1.2%. Only a part of lactose is metabolized and approximately 70% remains intact in fermented yogurt mix. Acetic acid ranges from 50 to 200 ppm (Chandan & O'Rell, 2013b). These key flavor compounds are produced by yogurt bacteria. Among others, diacetyl and acetoin are also metabolic products of carbohydrate metabolism in ST.

The milk coagulum during yogurt production results from the drop in pH due to the activity of the yogurt culture. The ST is responsible for lowering the pH of a yogurt mix to 5.0–5.5 and the LB is primarily responsible for further lowering of the pH to 3.8–4.4 (and production of acetaldehyde).

Several mucopolysaccharide-producing strains of yogurt culture are utilized in the industry because they render yogurt more viscous. The texture of yogurt tends to be coarse or grainy if it is allowed to overferment prior to stirring or if it is disturbed at pH values higher than 4.6 (O'Rell & Chandan, 2013b). Incomplete blending of mix ingredients is an additional cause of a coarse smooth texture. Homogenization treatment and high fat content tend to favor smooth texture. Gassiness in yogurt may be attributed to contamination of the starters with spore-forming *Bacillus* species, coliform, or yeast, producing excessive CO_2 and hydrogen. In comparison with plate heat exchangers, cooling with tube-type heat exchangers causes less damage to yogurt structure. Further, loss of viscosity may be minimized by well-designed booster pumps, metering units, and valves involved in yogurt packaging. The pH of yogurt during refrigerated storage continues to drop by approximately 0.2 pH units. Higher temperature of storage accelerates the drop in pH.

18.5.2.6 Changes in milk constituents

Among the carbohydrate constituents, the lactose content of yogurt mix is generally around 6%. During fermentation, lactose is the primary carbon source, resulting in approximately 30% reduction. However, a significant level of lactose (3.31–4.74%) is maintained in fresh yogurt (Chandan & Nauth, 2012). One mole of lactose (a disaccharide) gives rise to 1 mole of galactose and 1 mole of glucose. Through the glycolytic pathway, glucose is further converted to 2 moles of lactic acid and energy for bacterial growth. Lactic acid production results in coagulation of milk beginning at pH below 5.0 and completing at 4.6. Texture, body, and acid taste of yogurt owe their origin to lactic acid produced during fermentation. Although lactose is in large excess in the fermentation medium, lactic acid build-up beyond 1.5% acts progressively as an inhibitor for further growth of yogurt bacteria. Normally, fermentation is terminated by temperature drop to 4 °C. To achieve this, the yogurt mass is typically pumped through a heat exchanger. To make the texture smooth, a texturizing cone is commonly inserted in the pipe leading to the heat exchanger. At 4 °C, the culture is alive but its activity is drastically limited to allow fairly controlled flavor in marketing channels.

Hydrolysis of milk proteins is easily measured by liberation of $-NH_2$ groups during fermentation. Free amino groups double in yogurt after 24 h. Proteolysis continues during shelf life of yogurt, doubling free amino groups again in 21 days storage at 7 °C. The major amino acids liberated are proline and glycine. The essential amino acids liberated increase 3.8–3.9-fold during storage of yogurt, indicating that various proteolytic enzymes and peptidases remain active throughout the shelf life of yogurt. The proteolytic activity of the two yogurt bacteria is moderate but quite significant in relation to symbiotic growth of the culture and production of flavor compounds (Chandan & Shahani, 1993, 1995).

Both ST and LB are documented in the literature to elaborate different oligosaccharides in yogurt mix medium. As much as 0.2% (by weight) of mucopolysaccharides have been observed after 10 days of storage. In stirred yogurt, drinking yogurt and reduced-fat yogurt, the potential contribution of exopolysaccharides to impart smooth texture, higher viscosity, lower synerisis, and better mechanical handling is noted. Excessive shear during pumping destroys much of the textural advantage because viscosity generated by the mucopolysaccharides is susceptible to shear. Most of the polysaccharides elaborated in yogurt contain glucose and galactose, along with minor quantities of fructose, mannose, arabinose, rhamnose, xylose, arabinose, or N-acetylgalactosamine, individually or in combination. Molecular weight is of the order of 0.5–1 million daltons. An intrinsic viscosity range of 1.5–4.7 dL g^{-1} has been reported for exopolysaccharides of ST and LB. The polysaccharides form a network of filaments, visible using a scanning electron microscope. The bacterial cells are covered by part of the polysaccharide and the filaments bind the cells and milk proteins. Upon shear treatment, the filaments rupture off from the cells, but maintain links with casein micelles. The so-called ropy strains of ST and LB are commercially available. They are especially appropriate for stirred yogurt production.

Other interesting metabolites are produced by yogurt culture. Among them are bacteriocins and several antimicrobial compounds. Benzoic acid (15–30 ppm) in yogurt has been detected and associated with metabolic activity of the culture (Chandan et al., 1977). These metabolites tend to exert a preservative effect by controlling the growth of contaminating spoilage and pathogenic organisms gaining entry post fermentation. As a result, the product attains extension of shelf life and a reasonable degree of safety from food-borne illness. As a consequence of fermentation, yogurt organisms multiply to a count of 10^8 to 10^{10} colony-forming units (CFU) g^{-1}. Yogurt bacteria occupy some 1% of the volume or mass of yogurt (Chandan, 1982). These cells contain cell walls, enzymes, nucleic acids, cellular proteins, lipids, and carbohydrates. Lactase or β-galactosidase has been shown to contribute a major health-related property to yogurt. Clinical studies have concluded that yogurt containing live and active cultures can be consumed by millions of lactose-deficient individuals without developing gastrointestinal distress or diarrhea (Chandan, 2002, 2011, 2013; Chandan & Kilara, 2008).

Yogurt is an excellent dietary source of calcium, phosphorus, magnesium, and zinc for human nutrition. Research has shown that bioavailability of the minerals from yogurt is essentially equal to that from milk (Chandan & Nauth, 2012; Chandan & O'Rell, 2013b). Since yogurt is a low pH product compared to milk, most of the calcium and magnesium occurs in ionic form, which may have positive implications for the bioavailability of the minerals.

During and after fermentation, yogurt bacteria affect the vitamin B content of yogurt. The processing parameters and subsequent storage conditions influence the vitamin content at the time of consumption. Incubation temperature and fermentation time exert a significant balance between vitamin synthesis and utilization of the culture. In general, there is a decrease of vitamin B$_{12}$, biotin, and pantothenic acid and an increase of folic acid during yogurt production. Nevertheless, yogurt is still an excellent source of vitamins inherent to milk.

18.5.3 Manufacturing procedures for different yogurt varieties

18.5.3.1 Plain yogurt

Plain yogurt is the basic style, and forms an integral component of fruit-flavored yogurt. The steps involved in the manufacturing of various types of yogurt are shown in Table 18.6.

Plain yogurt normally contains no added sugar or flavors, in order to offer the consumer natural yogurt flavor for consumption as such or as an option of flavoring with other food materials of the consumer's choice. In addition, it may be used for cooking or for salad preparation with fresh fruits or grated vegetables. In most recipes, plain yogurt is a good substitute for sour cream, providing lower calories and fat alternative. The fat content of yogurt may be standardized to the levels preferred by the market. Labeled as "yogurt," full-fat yogurt is mandated to have a minimum of 3.25% fat. Also, the size of package may be geared to market demand. Polystyrene and polypropylene cups and lids are the chief packaging materials used in the industry.

Table 18.6 Main steps for manufacturing plain, Greek and fruit-flavored refrigerated yogurt

Plain yogurt	Greek style yogurt	Blended style yogurt	Fruit-on-the-bottom style yogurt
1. Standardize mix for fat (0.3–3.25%) and non-fat solids (10–12%) using skim milk, cream, non-fat dry milk and condensed skim milk.	1. Standardize mix for fat (0.3%) and non-fat solids (10–12%) using skim milk, non-fat dry milk and condensed skim milk.	1. Standardize mix for fat (0.3%) and non-fat solids (10–12%) using skim milk, milk, non-fat dry milk and condensed skim milk. Incorporate sugar and stabilizer.	1. Standardize mix for fat (0.3%) and non-fat solids (10–12%) using skim milk, milk, non-fat dry milk and condensed skim milk. Incorporate sugar and stabilizer.
2. Heat-treat the mix at 97 °C for 10 min.	2. Heat-treat the mix at 97 °C for 10 min.	2. Heat-treat the mix at 97 °C for 10 min.	2. Heat-treat the mix at 97 °C for 10 min.
3. Homogenize at 1500 psi, 60 °C. Cool to 45 °C.	3. Homogenize at 1500 psi, 60 °C. Cool to 45 °C.	3. Homogenize at 1500 psi, 60 °C. Cool to 45 °C.	3. Homogenize at 1500 psi, 60 °C. Cool to 45 °C.
4. Transfer to fermentation tank. Inoculate with yogurt starter/culture. Blend.	4. Transfer to fermentation tank. Inoculate with yogurt starter/culture. Blend.	4. Transfer to fermentation tank. Inoculate with yogurt starter/culture. Blend.	4. Transfer to fermentation tank. Inoculate with yogurt starter/culture. Blend.
5. Package in cups. Fill cases and palletize.	5. Incubate at 43 °C until pH reaches 4.5.	5. Incubate at 43 °C until pH reaches 4.5.	5. Using filler, dispense fruit preparation into cups. Pump inoculated yogurt mix to the filler and layer inoculated mix on the top of fruit. Close the lids. Fill cases and palletize.
6. Incubate at 43 °C until pH reaches 4.5.	6. Break curd. Centrifuge/strain enough whey to concentrate solids (~2 times). Cool to 4 °C.	6. Agitate to break curd while cooling to 4 °C.	6. Transfer to fermentation chamber. Incubate at 43 °C until pH reaches 4.5.
7. Blast-cool the cups to 4 °C.	7. For plain Greek yogurt, if required, blend cream to standardize fat content.	7. Add fruit in-line, blend and pump to filler.	7. Blast-cool to 4 °C.
8. Store and distribute at 4 °C.	8. For fruit-on-the bottom Greek yogurt, dispense fruit preparation into cups. Blend cream in plain Greek yogurt to standardize fat and layer Greek yogurt on the top of the fruit.	8. Fill cups, apply lids. Fill cases and palletize.	8. Store and distribute at 4 °C.
	9. For blended style Greek yogurt, blend cream and fruit preparation in non-fat plain Greek yogurt.	9. Blast-cool and store at 4 °C.	
	10. Package into cups, fill cases and palletize.	10. Distribute refrigerated (4 °C).	
	11. Blast-cool to 4 °C. Store and distribute under refrigerated conditions.		

18.5.3.2 Fruit on-the-bottom (sundae) style yogurt

In this type, two-stage fillers are used. Typically, 59 mL (2 oz) of fruit preserves or special fruit preparations are layered at the bottom, followed by 177 mL (6 oz) of inoculated (cultured) yogurt mix on the top. The top layer may consist of yogurt mix containing stabilizers, sweeteners, and the flavor and color indicative of the fruit on the bottom. After placing lids on the cups, incubation and setting of the yogurt take place in the cups (typically at 43 °C). When a desirable pH of 4.4–4.5 is attained, the cups are placed in refrigerated rooms with high-velocity forced air for rapid cooling. At the time of consumption, the fruit and yogurt layers are mixed by the consumer.

18.5.3.3 Blended/Swiss style/stirred style or blended yogurt

Swiss-style yogurt is produced by blending the fruit preparation thoroughly in fully fermented yogurt base obtained after bulk culturing in fermentation tanks.

Stabilizers, especially gelatin, are commonly used in this form of yogurt, unless milk solids-not-fat levels are relatively high (12–14%). In this style, cups are filled with an in-line blended mixture of fermented yogurt base and fruit. Pumping of yogurt-fruit blend can result in considerable loss of viscosity. Upon refrigerated storage for 48 h, the clot is reformed and viscosity is recovered, leading to a fine body and texture. Fruit incorporation at 10–15% is carried out by the use of a fruit feeder. Prior to packaging, the texture of fermented yogurt base can be made smoother by pumping it through a valve or a cone made of stainless steel mesh. The incubation times and temperatures are co-ordinated with the plant schedules. Incubation temperatures lower than 40°C in general tend to impart a slimy or sticky appearance to yogurt. Overstabilized yogurt possesses a solid-like consistency and lacks a refreshing character. Spoon-able yogurt should not have the consistency of a drink; it should melt in the mouth without chewing.

18.5.3.4 Greek style yogurt

This product is characterized by high viscosity and heavy body. The distinct body is obtained by straining/centrifuging some of the whey from plain yogurt (see Table 18.6) to obtain 20–25% solids. In some cases, it is formulated to achieve high solids (using milk protein concentrate, non-fat milk solids, whey protein concentrate, etc.), followed by culturing (Kilara & Chandan, 2013). It is generally a non-fat product, but the fat content can be standardized by blending pasteurized milk or cream to the strained yogurt mass to obtain low-fat or full-fat variations. Another variation consists of adding fruit to market fruit-on-the bottom or blended style Greek yogurt. Commercial Greek yogurt generally contains twice as much protein as regular yogurt.

18.5.3.5 Light yogurt

Light yogurt is generally of the blended/stirred type, which is made without any added sugar. High-intensity sweeteners (commonly aspartame or sucralose) replace the sugar in the formulation. The synthetic sweetener may be added through the fruit preparation. More commonly, a pasteurized solution of the high-intensity sweetener is blended, directly in-line into the fermented base. Light yogurt is generally a blended type yogurt which uses a special fruit preparation containing 30–60% fruit. The fruit preparation (10–12°Brix) is designed for use at 10–18% level (O'Rell & Chandan,

2013a). The major consumer driver for light yogurt consists of significantly lower sugar content and calories.

18.5.3.6 Yogurt beverages

Also called drinking yogurt, yogurt smoothie and yogurt drink, this product is made in a similar procedure as used for stirred style or blended yogurt. It is a low-viscosity yogurt made from low solids mixes. However, fruit preparations generally consist of juices and purees. The stabilizers used are non-thickening types, such as high methoxy pectin, modified starch, gelatin, and certain gums used to impart smooth body and to control whey separation during the product shelf life. A recent trend is to include fructo-oligosaccharide prebiotics, like inulin, and to fortify with a significant daily requirement of most vitamins and minerals. In general, various probiotic strains are included in the product. All the beverages contain live and active cultures to qualify them as a functional food (Schoenfuss & Chandan, 2011).

A variation of drinking yogurt consists of blending yogurt with water and fruit juice, and subjecting the blend to extra shear (homogenization) to reduce viscosity.

Kefir is another fermented milk beverage. Kefir culture containing lactobacilli, leuconostocs, acetic acid bacteria, *Streptococcus thermophilus* and yeasts is used as the starter for this drink, giving it a distinctive flavor. Most kefir products are sweetened with sugar and flavored with fruit juices. In the US, kefir contains no yeast and consequently no alcohol or gas.

18.5.3.7 Aerated yogurt (mousse/whips)

This category of yogurt resembles mousse, in that the product acquires a novel foam-like texture. The aeration process is similar to the ice cream process, but the degree of overrun (extent of air content) is kept low, around 20–50%. Foam formation is facilitated by use of appropriate emulsifiers and the stability of foam is achieved by using gelatin in the formulation. After bulk fermentation and cooling, aerated yogurt is produced with appropriate equipment (Oakes, Tanis, or Mondo), injecting a controlled volume of an inert gas (nitrogen) to create foam in the product. Nitrogen assists in controlling oxidative deterioration. The amount of overrun is related to texture and mouthfeel attributes of the product. It is desirable to insure constant overrun from day to day in order to market a consistent product. The volume of yogurt in the cup is also related to the degree of overrun. Accordingly, overrun control ensures the correct weight of the product in the cup (Schoenfuss & Chandan, 2011).

18.5.3.8 Frozen yogurt

Both soft-serve and hard-frozen yogurts have gained popularity in recent years. The popularity of frozen yogurt has been propelled by its low-fat and non-fat options. Frozen yogurt is a low-acid product, resembling low-fat ice cream in flavor and texture, with only 0.3% titratable acidity. There are no federal legal requirements for frozen yogurt but a few states do have legal definitions.. A common procedure involves blending skim milk fermented with yogurt culture with low-fat ice cream mix to yield frozen yogurt mix. Another procedure uses controlled fermentation of low-fat ice cream to pH ~5.9–6.0. In some instances, the blend is pasteurized to insure destruction of pathogens as well as yogurt bacteria in the resulting low-acid food. To provide live and active yogurt culture in the finished product, frozen culture concentrate must be blended with the pasteurized product. The blend is flavored and frozen in an ice cream freezer or soft ice cream machine similar to frozen dessert manufacture.

18.5.4 Packaging and storage

Most plants attempt to synchronize the packaging lines with the termination of the incubation period. Generally, textural defects in yogurt products are caused by excessive shear during pumping or agitation. Therefore, special pumps are preferred over centrifugal pumps for moving the product after culturing or ripening.

For incorporation of fruit, it is advantageous to use a fruit feeder system. Various packaging machines of suitable speeds (up to 400 cups per minute) are available to package various kinds and sizes of yogurt products.

Yogurt is generally packaged in plastic containers varying in size from 4 oz to 32 oz (113–904 g). The machines use volumetric piston filling. The product is sold by weight and the machines delivering volumetric measure are standardized accordingly. The pumping step of fermented and flavored yogurt base exerts some shear on the body of yogurt.

Cups of various shapes characterize certain brands. Some plants use preformed cups. The cup may be formed in the yogurt plant by injection molding, a process in which beads of plastic are injected into a mold at high temperature and pressure. In this type of packaging, a die-cut foil lid is heat sealed on to the cups. Foil lids are cut into circles and procured by the plants from a supplier along with preformed cups. A plastic overcap may be used. In some cases, partially formed cups are procured and assembled at the plant. Some other plants use roll stock, which is used in form-fill-seal systems of packaging. In this case, cups are fabricated in the plant by a process called thermoforming which involves ramming a plug into a sheet of heated plastic. Multipacks of yogurt are produced by this process.

Following the formation of cups, they are filled with the appropriate volume of yogurt and heat sealed with foil lids. They are then placed in cases and transferred to a refrigerated room for cooling and distribution. In breakfast yogurt, a mixture of granola, nuts, chocolate bits, dry fruit, and cereal is packaged in a small cup and sealed with a foil. Subsequently, the cereal cup is inverted and sealed on the top of the yogurt cup. This package is designed to keep the ingredients isolated from the yogurt until the time of consumption. This system helps to maintain crispness in cereals and nuts, which otherwise would become soggy or interact adversely if mixed with yogurt at the plant level.

Some interesting innovations in yogurt packaging include spoon-in-the-lid and squeezable tubes (Panell & Schoenfuss, 2007). The former adds convenience in eating yogurt, while the squeezable tubes add versatility of use for children. They may be included in lunch boxes for consumption in school. In addition, yogurt-in-tubes is freeze-thaw stable, which adds another dimension of convenience and versatility. The tubes/cups are packaged in cases and stored at 4 °C.

18.5.4.1 Quality control

Quality control programs include controlling product composition, viscosity, color, flavor, body and texture, as well as fermentation process. Daily tests for chemical composition of the product, pH, and sensory quality constitute the core of quality assurance. Product standards of fats, solids, viscosity, pH (or titratable acidity), and organoleptic characteristics should be strictly adhered to (O'Rell & Chandan, 2013c).

The flavor defects can be described as too acid, too weak fruit flavor, unnatural, etc. The sweetness level may be excessive or weak, or may exhibit corn syrup flavor. The ingredients used may impart undesirable flavors such as stale, metallic, old ingredients, oxidized, rancid, or unclean. Lack of control in processing procedures may cause overcooked, caramelized, or excessively sour flavor notes in the product. Proper control of processing parameters and ingredient quality ensure optimum flavor.

Wheying-off or appearance of a water layer on the surface is undesirable and can be controlled by judicious selection of effective stabilizers and by following proper processing conditions. Graininess in texture is caused by premature agitation and malfunctioning texture-smoothening systems. Color defects may be caused by the lack of intensity or authenticity of hue and shade. Proper blending of fruit purees and yogurt mix is necessary for uniformity of color (Tribby, 2009).

To deliver yogurt with the most desirable attributes of flavor and texture to the consumer, it is imperative to enforce strict sanitation programs along with good manufacturing practices. The control of contamination is managed by aggressive sanitation procedures related to equipment, ingredients, and plant environment. The chemical solutions used in CIP cleaning should be checked for their strength and proper temperature. Hypochlorites and iodophors are effective sanitizing compounds for fungal control on contact surfaces and in combating environmental contamination. Hypochlorites at high concentrations are corrosive. Iodophors are preferred for their non-corrosive property as they are effective at relatively low concentrations.

For refrigerated yogurt and yogurt beverages, shelf life expectations generally approximate 6–8 weeks from the date of manufacture, provided temperature during distribution and retail marketing channels does not exceed 7 °C. Lactic acid and some other metabolites produced by the fermentation process protect yogurt from most gram-negative psychrotrophic organisms. In general, most product quality issues in a yogurt plant are related to proliferation of yeasts and molds, which are highly tolerant to low pH and can grow under refrigeration temperatures. Yeast growth during shelf life of the product requires more attention than mold growth. If yeast contamination is not controlled, its growth manifests within 2 weeks of manufacture. To insure maximum shelf life, several manufacturers use potassium sorbate to control yeast and mold growth in the product. Yeast and mold contamination may also arise from starter, packaging materials, fruit preparations, and packaging equipment. Organoleptic examination of yogurt starter may be helpful in eliminating fungal contamination. If warranted, direct microscopic view of the starter may reveal the presence of budding yeast cells or mold mycelium filaments. Plating of the starter on acidified potato dextrose agar would confirm the results. Avoiding contaminated starter for yogurt production is necessary.

Coliform counts are also performed as a measure of sanitation in the plant, which has a profound effect on the shelf life and safety aspects of yogurt. Efficiency of equipment and environmental sanitation can be verified by enumeration techniques involving exposure of poured plates to atmosphere in the plant or making a smear of the contact surfaces of the equipment, followed by plating. Filters on the air circulation system should be changed frequently. Walls and floors should be cleaned and sanitized frequently and regularly.

The packaging materials should be stored in dust-free and humidity-free conditions. The filling room should be fogged with chlorine or iodine regularly. Quality control checks on fruit preparations and flavorings should be performed (spot checking) to ascertain sterility and to eliminate yeast and mold entry via fruit preparation. Refrigerated storage of the fruit flavorings is recommended.

In hard-pack frozen yogurt, a coarse and icy texture may be caused by formation of ice crystals due to fluctuations in storage temperatures. Sandiness may be due to lactose crystals resulting from too high levels of milk solids. A soggy or gummy defect is caused by too high a milk solids-not-fat level or too high a sugar content. A weak body results from too high overrun and insufficient total solids. It is evident that good microbiological quality of all ingredients is necessary for fine organoleptic quality and shelf life of the product.

18.5.5 Live and active status of yogurt culture

Yogurt products enjoy an image of being health-promoting foods. The type of cultures and their viability, as well as active status, are important attributes from a consumer standpoint (Chandan, 2002). Quality control tests are necessary to insure "live and active" status of the culture. As per National Yogurt Association standards (National Yogurt Association, 2008), yogurt must pass an activity test. The cultures must be active at the end of the shelf life stated on the package. Samples of yogurt stored at a temperature between 1 °C and 10 °C for the duration of the stated shelf life are subjected to the activity test. This test is carried out by pasteurizing 12% (w/v) solids non-fat dry milk at 92 °C for 7 min, cooling to 43.3 °C, adding 3% inoculum of the material under test, and fermenting at 43.3 °C for 4 h. The total yogurt organisms are enumerated in the test material both before and after fermentation by a standard procedure (IDF, 2003). The activity test is met if there is an increase of 1 log or more of yogurt culture cells during fermentation.

Generally, at the time of manufacture, live and active refrigerated yogurt should contain not less than 100 million (10^8) CFU/g. Assuming that storage temperature of yogurt through distribution channels and the grocery store is 4–7 °C, a loss of 1 log cycle in culture viability is expected during the period between manufacture and consumption. Therefore, at the time of consumption, yogurt should deliver at least 10 million (10^7) CFU of live yogurt organisms per gram of product. In case yogurt undergoes temperature abuse, it is desirable to manufacture yogurt with even higher counts of viable culture to assure that at consumption stage, the product contains at least 10 million (10^7) CFU/g.

Yogurt products with live and active cultures are associated with several health attributes. The literature describes several strain-specific benefits corroborated by credible clinical studies (Sanders & Marco, 2010). The benefits include stimulation of the immune system, improvement of digestive regularity, reduction of lactose intolerance symptoms, enhancement of calcium absorption, reduction of serum cholesterol and risk of cardiovascular disease, enhancement of resistance to colonization by pathogenic organisms, and restoring the normal balance of gastrointestinal microflora (Chandan, 2011).

18.5.6 Nutrient profile of yogurt

The typical composition and nutrient profile of yogurt is shown in Table 18.7. In general, yogurt contains more protein, calcium, and other nutrients than milk, reflecting the extra solids-not-fat content. Therefore, its nutritional profile is similar to that of milk fortified with non-fat solids. A distinctive dimension of nutrition and health is furnished by the culturing process used for making yogurt. Therefore, active bacterial mass and products of

Table 18.7 Typical nutrient profile of yogurt

Nutrient/100 g	Plain yogurt		Fruit-flavored yogurt		Fruit-flavored yogurt with low-calorie sweetener
	Non-fat[1]	Low fat[2]	Non-fat[3]	Low fat[4]	Low fat[5]
Moisture	85.23	85.07	75.40	75.30	74.10
Calories, kcal	56	63	95	99	105
Protein, g	5.73	5.25	4.40	3.98	4.86
Total fat, g	0.18	1.55	0.20	1.15	1.41
Saturated fatty acids, g	0.116	1.00	0.119	0.742	0.909
Monounsaturated fatty acids, g	0.049	0.426	0.050	0.316	0.387
Polyunsaturated fatty acids, g	0.005	0.044	0.016	0.033	0.040
Cholesterol, mg	2	6	2	5	6
Carbohydrates, g	7.68	7.04	19.00	18.64	2.90
Total dietary fiber, g	0	0	0	0	0
Ash, g	1.18	1.09	1.00	0.93	1.03
Phosphorus, mg	157	144	119	109	133
Magnesium, mg	19	17	15	13	16
Calcium, mg	199	183	152	138	152
Iron, mg	0.09	0.08	0.07	0.06	0.07
Potassium, mg	255	234	194	177	194
Sodium, mg	77	70	58	53	58
Vitamin A, IU	7	51	12	40	443
Thiamin, mg	0.048	0.044	0.040	0.034	0.041
Riboflavin, g	0.234	0.214	0.180	0.162	0.180
Niacin, mg	0.124	0.114	0.100	0.086	0.105
Ascorbic acid, mg	0.9	0.8	0.7	0.6	0.7

From: USDA (2011b).
[1] NDB No. 01118; [2] NDB No. 01117; [3] NDB No. 43261; [4] NDB No. 01120; [5] NDB No. 01203.

the lactic fermentation further distinguish yogurt from milk. As stated earlier, live and active cultures in yogurt products contribute interesting health benefits to the consumers of yogurt that are not reflected in the nutrition label.

18.6 Cultured (or sour) cream

Cultured cream (or sour cream) is manufactured by ripening pasteurized cream of 18% fat content with lactic and aroma-producing bacteria. The FDA has defined sour cream in section 131.160 (CFR, 2011): "Sour cream results from the souring, by lactic acid producing bacteria, of pasteurized cream." It contains not less than 18% milk fat. The fat content may be lower in sour cream products containing bulky ingredients, but it cannot dip below 14.4%. Sour cream must have a titratable acidity of at least 0.5%, expressed as lactic acid. Optional ingredients include stabilizers to improve texture and to avoid wheying-off. The stabilizers help to maintain body and enhance shelf life. In addition, sodium citrate up to 0.1% may be added as a flavor precursor. Salt, rennet, and flavoring ingredients including nutritive sweeteners and fruit products are permitted. Acidified sour cream (CFR Section 131.162) is obtained by direct addition of food-grade acids to pasteurized cream containing 18% milk fat. The process involves no fermentation. All other FDA requirements are identical to those for cultured sour cream.

The fermenting bacteria generate lactic acid to create an acid gel and butter-like (diacetyl) aromatic flavor. The starter for cultured cream is composed of acid producer (*Lactococcus lactis* subsp. *lactis*) and/or *Lactococcus lactis* subsp *cremoris*. The aroma component is provided by *Leuconostoc mesentoides* subsp. *cremoris* and/or *Lactococcus lactis* subsp. *lactis* biovar *diacetylactis*. The lactococci use lactose to produce up to 0.8% lactic acid by a homofermentative pathway, while the aroma producers are heterofermentative. They generate d-lactic acid, ethanol, acetic acid, and CO_2 from lactose. The typical flavor of sour cream is ascribed to diacetyl, which is produced from citrate. Leuconostocs break down acetaldehyde to ethanol. Acetaldehyde gives yogurt its typical flavor (green apple flavor), whereas this flavor is undesirable in sour cream. Some strains of lactococci produce viscosity-generating exopolysaccharides. It is common to utilize starters containing 60% acid producers, 25% acid and viscosity generators, and 15% flavor producer organisms (Goddik, 2012).

Cultured cream is used as an adjunct in baked potatoes and other hot dishes. The texture of sour cream must be maintained (lack of syneresis, no viscosity loss) on hot dishes. It forms a topping on vegetables, salads, fish, meats, and fruits. It can be utilized as a filling in cakes or in soups in place of buttermilk or sweet cream. It is an integral part of many Mexican dishes. It can be dehydrated by spray drying and used as an ingredient wherever its flavor is needed.

18.6.1 Processing

18.6.1.1 Raw materials

The basic raw material of cultured sour cream is cream of high organoleptic and microbiological standards. Cream used in the manufacture of cultured cream should be fresh, with a relatively low bacterial count. During cream separation from milk, the bacteria tend to concentrate in the lighter phase, cream, thereby enhancing its vulnerability to spoilage. Raw cream should not be subjected to turbulent mixing, which leads to activation of lipase, disruption of milk fat globule membrane, and subsequent rancid flavor. Non-fat milk solids are needed to standardize sour cream mix. Condensed skim milk or non-fat dry milk is generally used to increase the solids-not-fat in sour cream mix to 9.0–9.5%. In addition, most sour cream processors use stabilizers to increase viscosity and stability and to prevent whey separation. Stabilizers for sour cream containing 18% milk fat consist of a combination of gelatin, modified starch, Grade A whey products, guar gum, locust bean gum, carrageenan, sodium phosphate, and calcium sulfate at a level of 1.5–1.8%. This stabilizer combination is increased to 1.75–2.0% in low-fat sour cream. The non-fat version uses a blend of modified starch, propylene glycol monoester and additional water-binding agents like microcrystalline cellulose, cellulose gum, and gum acacia. This blend is used at a level of 6.2–6.6% (Goddik, 2012). Sodium citrate is added to act as a precursor of flavoring compound generation during fermentation, thereby enhancing the flavor of sour cream. It aids in the production of minor levels of CO_2 (Born, 2013). This gas needs to be controlled by judicious selection of starter to avoid lid popping or swelling of packages. The coagulating enzyme chymosin (single-strength rennet) at the rate of 5 mL/100 gallons of mix is carefully added for texture improvement. The starter procured from culture suppliers is especially designed for sour cream production.

18.6.1.2 Mix preparation

Cream is blended with non-fat dry milk (or condensed skim milk) to standardize composition to 18.5% milk fat and 27.5% total solids at 4 °C. Next, the stabilizer is incorporated and care is taken to insure hydration of the gums without incorporation of air and excessive foam formation.

18.6.1.3 Homogenization

The mix is preheated to 71 °C for homogenization treatment. The mix is typically homogenized by two single-stage passes at 2500 psi (17.2 MPa) pressure. This step helps to produce firm-bodied sour cream.

18.6.1.4 Heat treatment

Next, the mix is heat treated at 73.9–79.4 °C for 30 min (vat pasteurization) or at 82.2–85 °C for 3–4 min (HTST process) (Born, 2013). The intense heat treatment denatures whey proteins, leading to enhanced water absorption as well as higher viscosity and a smooth thick body after culturing. The pasteurized-homogenized mix is cooled to 21.1–23.9 °C and transferred to a cone-bottom processing vat.

18.6.1.5 Ripening

The pasteurized mix is inoculated with direct set or bulk sour cream starter, similar to yogurt processing. Rennet is also incorporated at 5 mL/100 gallons mix. To avoid grainy texture, rennet is diluted with water (1:40) and gently stirred into the vat. During culturing, the vat contents are not disturbed to avoid consistency problems. The incubation is continued until pH drops to 4.5 prior to breaking. For institutional sale, some manufacturers process cultured cream in individual-size packs. The packages are filled with inoculated mix and incubated to develop desirable acidity and subsequently cooled. In some cases, the packages are filled soon after ripening, followed by cooling. A heavy-bodied product is formed on cooling in the package.

18.6.1.6 Breaking

The set bulk sour cream is broken with slow-sweep agitation (Born, 2013). With the agitator off, the cream is pumped using a positive drive pump through a screen, back pressure valve or similar smoothening device, to a packaging machine. Occasional agitation may be needed to insure that whey pockets are blended back into the main body of the sour cream.

18.6.1.7 Packaging and cooling

Sour cream containing live organisms is packaged into appropriate containers at incubation temperature (21–24 °C). The packages are then transferred to coolers at 4 °C for cooling overnight. This process allows the clot to reform and maximum body firmness is achieved. Factors affecting viscosity of cultured cream are: acidity, mechanical agitation, heat treatment, solids-not-fat content, rennet addition, and homogenization. The lactic cultures *Lactococcus lactis* subsp. *diacetylactis* and *Leuconostoc cremoris* not only provide flavor, but also enhance the smoothness, and to some extent the viscosity, of cultured cream.

18.6.1.8 Hot pack process

In this case, ripened sour cream is blended with a commercial stabilizer to confer heat stability under acidic conditions. The blend is subjected to pasteurization at 74 °C for 30 min or at 85 °C for 1 min to destroy the culture, enzymes, and contaminants in the finished product, followed by packaging while still hot (Chandan, 1982). Accordingly, hot pack processing ensures long shelf life. Packaging in a plastic or metal container with a hermetically sealed lid further ensures prevention of recontamination by microorganisms as well as protection from oxidative deterioration of milk fat in the finished product. The flavor of the hot pack process is not identical to fresh sour cream. It may be postulated that some of the volatile flavor compounds are lost as a result of heat treatment.

18.6.1.9 Sour cream with lower fat content.

Consumer demand for low-fat dairy foods has resulted in several lower fat sour cream products in the marketplace. The standard of identity of sour cream was changed to accommodate products such as sour half and half with 10.5% fat, reduced-fat sour cream (13.5% fat), and non-fat or fat-free sour cream (<0.5 fat/serving of 1 tbsp). These products use extra levels of non-fat solids, whey proteins and stabilizer blends to achieve body and texture comparable to regular sour cream. With extra solids-not-fat in the mix, care should be taken to select a culture with low CO_2 production. Otherwise, the packages would bulge with gas pressure.

18.6.1.10 Filled sour cream

This product may be defined as cultured cream in which part or all of the milk fat has been replaced by vegetable oils or fats. It appears to have advantages over conventional cultured cream in terms of price, caloric value or saturated fatty acid content. Non-fat milk suspension or skim milk is mixed with an appropriate emulsifier and appropriate vegetable oil. Homogenization of the mix and subsequent fermentation result in an acceptable product used mainly in chip dip production. In this regard, a suitable product may be manufactured using a process identical to that for cultured cream with the exception of the starting material.

18.6.1.11 Imitation sour cream

This product is made with vegetable oil and sodium caseinate replacing all the milk fat in its formulation. An imitation sour cream may be prepared by emulsifying a suitable fat in suspensions of casein compounds or soybean protein products. A suitable emulsifier, stabilizer, flavor, and color may be incorporated in the starting mix. It may be fermented or directly acidified.

18.6.1.12 Sour cream dip

Party dips are based on sour cream or filled sour cream or imitation sour cream. The dips are made by blending appropriate seasoning bases. By packaging under refrigerated conditions, the product has a shelf life of 2–3 weeks under refrigerated storage. However, for a shelf life of 3–4 months, a hot pack process is used (Chandan, 1982). To build extra body and stability, 1–2% non-fat dry milk and 0.8–1.0% stabilizer are incorporated at 80 °C in formulating sour cream. The mixture is pasteurized by holding for 10 min, and homogenized at 17.2 kPa to resuspend and produce a smooth product. The seasonings are blended at this stage while the mix is still at 80 °C, followed by hot packing in sealed containers. Upon cooling and storage at 5 °C, a partial vacuum inside the container assists in the prevention of oxidative deterioration to yield an extended shelf life of 3–4 months.

18.6.1.13 Quality control

In general, quality control considerations applicable to refrigerated yogurt are relevant to cultured cream. Table 18.8 summarizes typical quality problems and steps needed to rectify them.

18.7 Cheeses

18.7.1 Types of cheeses

There are more than 400 varieties of cheese recognized as distinctive varieties in the world. Different cheese varieties use different techniques, specific starters and varying acidity development regimes, brine or dry salting and distinct ripening conditions to develop their textures and flavors (Bottazzi, 2003; Chandan & Kapoor, 2011a, 2011b; Hill, 2009). Some cheeses are made without cultures and coagulating enzymes. In such cheese manufacture, curd formation is carried out by adding edible acid directly to hot milk (Chandan, 2003, 2007).

Natural cheese is made directly from milk. Some cheeses, like ricotta, are made from whey or milk-whey blends. Most cheeses are obtained after curding or coagulation of milk, cream, partly skimmed milk, buttermilk or a mixture of these raw materials. The coagulum obtained is stirred and heated, followed by draining of the whey and collecting and pressing the curd. Cheese may be ripened or sold as fresh product. Ripening is also referred to as curing or maturing. It is carried out by storing cheese in cold rooms at 10–18 °C for at least 2–3 months, during which time the development of acid continues and other chemical and bacteriological changes take place to bring about the desired flavor, aroma, body, and texture in the cheese. Based on its body or texture, cheese may be classified as very hard (Romano, Parmesan), hard (Cheddar, Swiss, etc.), medium (Brick and Limburger), or soft (Camembert) (Fox, 2003a; Olson, 2003). Fresh cheeses do not undergo ripening. Hard cheeses are prepared by different traditional methods using lactic acid cultures to develop characteristic flavors. Rennet/chymosin is typically used in the process as the coagulating enzyme. The hard cheeses normally contain salt.

Figure 18.2 shows the classification based on whether the cheese is ripened or not. Figure 18.3 contains classification of cheeses based on moisture content, firmness, and ripening microorganisms.

Dairy ingredients in cheese manufacture are milk, non-fat milk, or cream, used alone or in combination. Milk (CFR 131.110) is defined by the Code of Federal Regulations (CFR, 2011) as the lacteal secretion, practically free from colostrum, obtained by the complete milking of one or more healthy cows, which may be clarified. Fat content may be adjusted by separating part of the fat. Concentrated milk (CFR 131.115), reconstituted milk, and dry whole milk (CFR 131.147) may also be used. Non-fat milk means skim milk, concentrated skim milk, reconstituted

Table 18.8 Typical flavor, body, and texture quality problems in sour cream

Defect	Probable cause	Remedy
Flavor defects		
Insufficient flavor	Low citrate level in milk	Add 0.02–0.05% (up to 0.1%) sodium citrate prior to the mix. Insure flavor producing bacteria in the starter
	Poor acid development	Use correct incubation temperature. Insure pH at breaking is 4.5
	Flavor masked by too much stabilizer	Use correct level and type of stabilizer system
Green/yogurt flavor	Acetaldehyde accumulation	Use proper culture for sour cream
	High incubation temperature and/or excessive incubation time	Follow standard procedure for incubation temperature and time
Oxidized (cardboard) flavor	Copper and iron contamination	Avoid exposure to copper utensils
	Exposure to fluorescent light or sunlight	Protect product from direct sunlight/UV light exposure
Yeast-like/cheesy	Contaminating yeast growth	Sanitation check. Avoid return milk
Rancid flavor	Cream/raw milk contains heat-stable lipases	Do not mix pasteurized and raw dairy ingredients prior to homogenization
	Poor handling of raw milk/cream during pumping	Use fresh cream and follow standard processing procedures
Bitter flavor	High psychrotrophic bacterial counts in milk/cream	Use cream of high microbiological quality
	Excessive proteolytic activity of starter culture	Use right starter designed for sour cream
Unnatural/chemical flavor	Absorbed flavors from environment	Avoid storage/exposure of sour cream to solvents, cleaning supplies, fruit, and vegetables
	Contamination with residual cleaning compounds/sanitizers	Insure absence of residue of cleaning compounds and sanitizers in the pipes and tanks
Body and texture defects		
Weak body	Heat treatment of the mix is insufficient or excessive	Heat treatment should be set at 85°C/3–4 min or 74°C/30 min
	Milk solids-not-fat too low	Fortify with 2–3% non-fat dry milk
	Stabilizer level low	Use appropriate stabilizers-thickeners Insure full hydration Use rennet in sour cream mix Check homogenizer efficiency
	Agitation too severe during and/or after fermentation	Insure quiescent fermentation and control shear after fermentation
Grainy texture	Acidity too high	Exercise rigid acidity control
	Non-fat dry milk or excessive rennet not dispersed properly	Dispersion equipment check Dilute rennet prior to dispersing
	Excessive heat treatment of mix	Insure standard heat treatment
	Screen/back pressure valve left out of line between tank and filler	Use in-line screen to smoothen the texture
	Stabilizer (pectin, alginate, carrageenan) reaction with protein	Insure proper ratio of stabilizer ingredients for sour cream production
	Fat churning	Avoid excessive agitation of mix during setting
Chalky/powdery texture	Too much non-fat dry milk	Check on the quality and quantity of dry milk
Free whey pockets	Temperature/physical shock for packages	Avoid temperature/physical abuse in the plant
	Low solids mix, improper heat treatment of mix, wrong stabilizer or improper blending	Check formulation and processing sheets to insure standard procedures are followed Increase fat content or stabilizer
	Breaking above pH 4.5	Check breaking pH
Gummy body	Too much stabilizer	Check formulation sheet for stabilizer level
	The starter produces excessive polysaccharides	Change starter.

Adapted from Chandan (1982), Chandan and Shahani (1995), Born (2013).

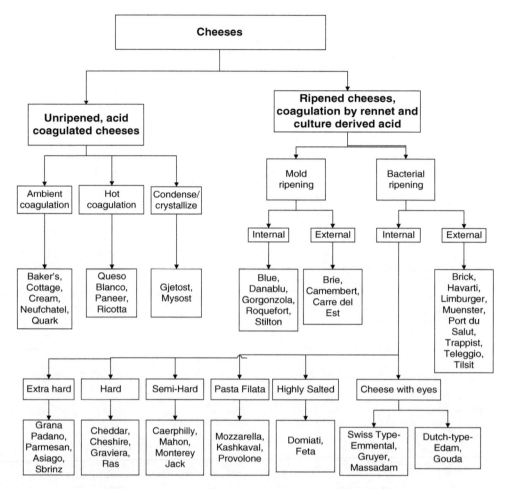

Figure 18.2 A classification of cheeses based on whether they are ripened or not and type of ripening. Source: Chandan and Kapoor (2011a).

skim milk, and non-fat dry milk. Cream means cream, reconstituted cream, dry cream, and plastic cream. Water, in a sufficient quantity to reconstitute concentrated and dry forms, may be added. Cheese can be made from cream, whole milk, reduced-fat, low-fat or non-fat milk or from mixtures thereof. Some cheeses are made from whey, whey cream, or whey-milk mixtures. Furthermore, milk of sheep, goats, water-buffaloes, and other milk producing animals yields distinct color, flavor, and texture profiles (Aneja et al., 2002).

Most cheeses are made from pasteurized milk. FDA regulations require that every particle of milk shall have been heated in properly designed and operated equipment to 63 °C or 72 °C and held continuously at or above that temperature for 30 min or 15 sec,

respectively (CFR 133.3d). Pasteurized dairy ingredients must conform to the phosphatase test for legal pasteurization.

Hydrogen peroxide treatment may be used in lieu of heat treatment (cold pasteurization), followed by a sufficient quantity of catalase preparation to eliminate the hydrogen peroxide. The weight of the hydrogen peroxide shall not exceed 0.05% of the weight of the dairy ingredients and the weight of the catalase shall not exceed 20 parts per million of the weight of dairy ingredients treated (CFR 133.113). If the dairy ingredients are not pasteurized, the cheese is cured at a temperature of not less than 1.7 °C (35 °F) for at least 60 days.

As with all cultured products discussed in this chapter, starters for cheese making are composed of harmless

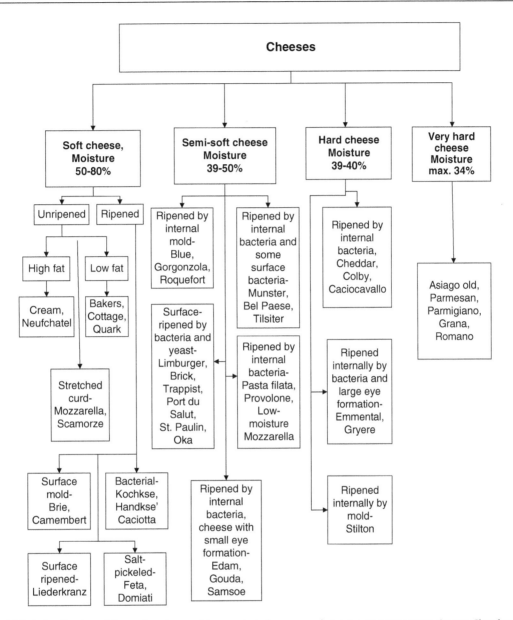

Figure 18.3 A classification of cheeses based on moisture content, firmness, and ripening microorganisms. Source: Chandan and Kapoor (2011a).

cultures. They are used for acid and flavor production during cheese making and curing. Coagulants or clotting enzymes include calf rennet/chymosin and/or other clotting enzymes of animal, plant, or microbial origin.

Calcium chloride, in an amount not more than 0.02% (calculated as anhydrous calcium chloride) of the weight of the dairy ingredients, can be used as a coagulation aid. Ripening aids of enzymes of animal, plant, or microbial origin aid in curing or flavor development. The level of such enzymes cannot exceed 0.1% of weight of the milk used (CFR 133.102). Cheese colors may be used to give characteristic color to certain cheeses and to give light cream color to cheeses made from winter milk, which normally lacks creamy color. Certain lipases and protease preparations are added to cow's milk to accelerate ripening and to simulate the traditional flavor of feta, Romano, and Parmesan cheese.

Slices or cuts of most cheeses in consumer-sized packages may contain an optional mold-inhibiting ingredient consisting of sorbic acid, potassium sorbate, sodium sorbate, or any combination of two or more of these, in an amount not to exceed 0.3% by weight, calculated as sorbic acid (CFR 133.173).

18.7.2 Cheddar cheese

Cheddar cheese, named after a town of the same name in England, is a variety of firm and hard cheese made from standardized milk, which is then cured for as long as 6 months. If no color is used, Cheddar has a cream-like appearance; goat's milk Cheddar is white. The US Code of Federal Regulations requires Cheddar cheese to contain a minimum milk fat content of 50% by weight of the solids, and the maximum moisture content is specified at 39% by weight (CFR, 2011). The production of Cheddar cheese is described in brief below. The reader is referred to Clark and Agarwal (2007) for a comprehensive review of Cheddar cheese and related cheeses. Low-sodium Cheddar cheese is Cheddar cheese that contains not more than 96 mg sodium/pound of finished product. Reduced-fat Cheddar cheese contains 19.2–22.9% fat and 49% moisture.

18.7.2.1 Essentials of Cheddar cheese production

18.7.2.1.1 Standardization of milk

Milk is standardized to a fat:casein ratio of 1.47. Various automated milk standardization instruments are used. For instance, milk analyzers based on infrared technology improve production efficiency as well as the quality and consistency of cheese. Fat content of raw milk can be reduced to a desirable level by the use of a separator. Another way is to add skim milk.

18.7.2.1.2 Heat treatment of milk

Certain cheeses are made from raw milk. Public health concerns dictate that such cheeses must be held at 1.67 °C or higher for at least 60 days before they can be consumed (CFR 133.113). For fresh (unripened) cheeses, the FDA regulations require heat treatment of 71.1 °C for 15 sec or 63 °C for 30 min for proper pasteurization of milk for cheese making (CFR 133.3). Some plants use

"thermization" treatment of 63–65 °C for a few seconds to prevent the spoilage of milk stored over weekends, but the milk requires full pasteurization prior to cheese making. The pasteurized milk is tempered to 31 °C and transferred to a cheese vat.

18.7.2.1.3 Additives

Certain additives are permitted. Calcium chloride, at 0.02% level in milk, accelerates coagulation and improves cheese yield. Cheese color (at the rate of 70 mL/1000 kg of milk) ensures uniformity of cheese color throughout the year. For white Cheddar, no color is added. In this regard, norbixin (anatto seed extract) and carotenoids are used. In certain cheeses, titanium dioxide and chlorophyll-based colorants are permitted in cow's milk to simulate the white appearance of milk of goat, sheep, and water-buffalo.

18.7.2.1.4 Starters

In Cheddar cheese, a starter is used to introduce desirable *Lactococcus* cultures for acidification and to accelerate coagulation of milk. The starter may be prepared in the plant or purchased for direct inoculation of milk. As mentioned earlier, bulk starter containing *Lactococcus lactis* subsp. *lactis/cremoris* is added at the rate of 0.5–2%.

18.7.2.1.5 Coagulation

Coagulation is a key step in cheese making. When the culture growth is indicated by a pH drop of 0.05 pH unit (increase of 0.01% titratable acidity (TA)), rennet/chymosin is added to the vat at the rate of 190 mL/1000 kg of milk after diluting with 10–40 volumes of water.

18.7.2.1.6 Cutting

When the curd is firm enough (usually in 25 min), it is cut using two sets of knives, horizontal and vertical, made of arranged wires separated by 9.5 cm (3/8"). The horizontal and vertical knives are used in succession for three-dimensional cuting of the curd, resulting in cubes of curd suspended in whey liquid. After 5 min, the curd is agitated very gently for 10 min.

18.7.2.1.7 Cooking

The vat contents are cooked at the rate of 1 °C/5 min to reach 39 °C in 30 min. If the whey has higher pH at cutting, cooking is extended to 60 min. Cooking is continued until the pH of whey drops to 6.2–6.3; it may take 75 min to achieve this acidity. The curd should shrink to about the half of the original size observed before cooking. It should be firm with no soft or mushy center. The curd should not stick together when pressed manually.

18.7.2.1.8 Whey drainage

At this point, the Cheddaring process starts by removing 50–70% of the whey. The curd is continuously stirred and the remaining whey continues to drain to dry out the curd.

18.7.2.1.9 Cheddaring

The curd is piled 4–5″ (10.2–12.7 cm) deep along the sides of the vat for matting/congealing. Whey drains slowly through a trench, which is cut in the middle. The front edges of matted curd are cut and layered on the adjacent slabs, which are then cut further into 2–3″ thick strips and piled again. The temperature of the curd is maintained at 30–35 °C. The slabs are turned repeatedly at 5–15-min intervals during a 2-h period.

18.7.2.1.10 Milling

When the pH of whey drops to 5.3–5.4, the slabs are milled and whey is continuously removed. At this stage, the curd slab should peel, reminiscent of chicken breast. Milling involves running the slabs through a milling machine and cutting them into small finger-size curds. After milling, the curds are stirred vigorously to avoid matting. At this stage, the curds should have round and smooth edges. The curd temperature should be maintained at 27–32 °C during milling.

18.7.2.1.11 Salting

After 15–30 min, the curd is sprinkled with salt. To impart the true taste of sodium chloride, the salt used in cheese making should not be iodized. The amount of salt varies from 2 to 3 kg/1000 kg of initial milk. More moist curd needs more salt to drop the moisture within the proper range for Cheddar cheese. Salting results in more whey drainage. The target salt content of cheese is 1.7% (w/w). The salted whey drippings are removed.

18.7.2.1.12 Hooping

Next, a disposable plastic or cheese cloth liner is inserted into a 20 lb hoop and 22 lb of curd is weighed into the cheese hoops, which are usually made of stainless steel. The lid is placed on the top prior to pressing.

18.7.2.1.13 Pressing

The hoops are lined up in the press and pressed for 12–18 h at 10–20 psi (75 kPa). Pressing time is much shorter in automated systems. After pressing, the blocks are removed from the hoops, vacuum packed in films and shrunk skin-tight by dipping in hot water, and transferred to the ripening room.

18.7.2.1.14 Ripening

The purpose of ripening is to generate cheese flavor attributed largely to lipid and protein breakdown caused by bacterial enzymes. Ripening temperature varies from 5–8 °C for slow ripening or 10–16 °C for fast ripening. Slow ripening leads to better flavor control. The ripening period may vary from 3 to 9 months or even longer to achieve a sharper flavor. Cheddar cheese standards are a minimum of 50% fat in dry matter and a maximum of 39% moisture (CFR 133.113). However, for long-hold Cheddar, the moisture should not exceed 36%. For short-hold Cheddar, the moisture content ranges from 37% to 39%. Cheddar cheese yield is generally around 9.5–10.0 kg/100 kg of milk.

18.7.3 Pasteurized process cheese

At the turn of the 20th century, developments in melting processes, involving natural cheese of various ages, gave birth to a line of process cheese products with controlled flavor, texture, and functionality, and extended shelf life. Pasteurized process cheese is the food prepared by comminuting and mixing, with the aid of heat, one or more cheeses of the same or two or more varieties (except cream cheese, Neufchatel cheese, cottage cheese, creamed cottage cheese, cook cheese, hard grating cheese, semi-soft part-skim cheese, part-skim spice cheese, and skim milk cheese) with an emulsifying agent into a plastic homogeneous mass (CFR 133.169). Heating is at not less than 65.5 °C and for not less than 30 sec. Moisture content shall not exceed 1% more than constituent natural cheese, but not exceed 43% (i.e. Colby: less than 40%, Swiss: less than 42%, Limburger: less than 51%). Fat in dry matter is similar to natural cheese, not less than 47% in general (e.g. Swiss: not less than 43%; Gruyere: not less than 45%). Pasteurized process cheese is prepared by melting Cheddar and other types of cheeses, usually the hard-pressed varieties, and emulsifying them with salts, especially citrates and phosphates, and often water. The mixture is heated to kill bacteria present and has, after packing in foil, a long keeping quality.

Pasteurized process cheese food is similar to pasteurized process cheese, except it must contain moisture not exceeding 44% and fat content not less than 23% (CFR 173.173). It may also contain optional dairy ingredients: cream, milk, skim milk, buttermilk, cheese whey solids, anhydrous milk fat, and skim milk cheese for manufacturing. The pH is adjusted to not below 5.0 with vinegar, lactic acid, citric acid, phosphoric acid, or acetic acid. It cannot contain more than 3% emulsifying agents and 0.2% sorbic acid.

Pasteurized process cheese spread is similar to pasteurized process cheese food, but is spreadable at 21 °C. It has moisture content of 44–60% and fat content not less than 20% (CFR 133.179). It may contain optional dairy ingredients, emulsifying agents, and gums (less than 0.5%). Acids may be added to achieve a pH not below 4.0. Sweetening agents may be used (sugar, dextrose, corn sugars). Sorbic acid (less than 0.2%) may be used as a preservative.

Cold pack cheese (club cheese) involves blending without heating various cheeses. Only cheese from pasteurized milk shall be used. Its moisture content is the same as that of individual cheeses and the fat content in dry matter is not less than 47% in most cheeses except Swiss (not less than 43%) and Gruyere (not less than 45%) (CFR 133.123). Cold pack cheese may contain acids to standardize pH to not below 4.5. Sorbic acid (less than 0.3%) can be used as a preservative.

Cold pack cheese food is prepared by comminuting and mixing (without heating) cheeses and other ingredients like cream, milk, skim, buttermilk, whey solids, and anhydrous milk fat. Acids may be added to standardize pH to not less than 4.5 (CFR 124). Sweetening agents (sugar, corn solids) may also be used. Sorbic acid (0.3%) may be used as a preservative. Guar gum or xanthan gum may be used (0.5%). Moisture content cannot exceed 44% and fat content is not less than 23%.

18.7.4 Mechanization in the cheese industry

Automation and mechanization in cheese production have resulted in economies of scale and establishment of mega plants handling several million kilograms of milk per day. The standardization of cheese milk for fat:casein ratio is done by automated systems. Cheese making is done in automated vats in which computer-controlled filling, culture addition, renneting, cutting, stirring, cooking, and emptying operations form an integral part. These vats are enclosed double-jacketed vats with mechanical cutting and stirring devices. The cutting and stirring blades rotate on a shaft vertically or horizontally by a variable speed drive. Computer-controlled sequences in cheese process and in-line pH control further help in saving labor. Many vats are equipped with an automated gel-strength analyzer to standardize the end point of the cutting phase. Curd and whey separation is done on a draining conveyer. In a cheddaring tower, vacuum and air pressure are used for compaction of the curd. The curd is milled and salted in a salting/mellowing conveyer. The salted curd is pressed and formed into 40 lb (or 640 lb) blocks in a block former, followed by mechanized packaging and conveying to the ripening room. Barrel cheese is of the order of 500 lbs per barrel.

18.7.5 Quality control

The US FDA has established compositional standards for natural cheeses (Table 18.9). It is imperative to ascertain that the products produced in a cheese plant adhere to the standards. In this regard, the processing parameters should be standardized for each cheese and standard procedures need to be followed by plant personnel.

The appearance, color, body, and texture as well as flavor are critical consumer attributes of high-quality cheese. Common defects in appearance and color of Cheddar cheese are shown in Table 18.10. Commonly observed defects and remedial measures are also included. The appearance of mold is avoided by use of a mold inhibitor or netamycin dip and effective packaging techniques. In this regard, vacuum packaging in transparent plastic cheese films is very helpful to protect cheese from contact with air.

For judging true flavor, body, and textural characteristics of cheese, it is necessary to temper the cheese sample to 10–15.6 °C. A cheese trier is a double-edged curved knife commonly employed for sampling large blocks of cheese. The trier is penetrated into the cheese block, rotated one-half turn, and pulled out to obtain a long cylindrical piece of cheese known as a plug. This piece is examined for aroma, body, texture, and flavor of the cheese sample. Cheddar cheese should ideally give a waxy and smooth-sided plug which should bend fairly well without snapping but should break slowly upon bending. Flavor, body, and texture are a function of the fermentation pattern. The breakdown of protein and fat leads to chemically simpler and more volatile compounds. Control of bacteriological and biochemical transformations is therefore necessary to insure consistent good quality

Table 18.9 US FDA compositional standards for natural cheeses

Cheese	% Moisture, maximum	% Fat in dry matter	Cheese	% Moisture, maximum	% Fat in dry matter
Asiago, fresh	45	50	Monterey Jack	44	50
Asiago, medium	35	45	Mozzarella, Scamorza	52–60	45
Asiago, old	32	42	Mozzarella, low moisture	45–52	45
Blue	46	50	Mozzarella, part-skim	52–60	30–45
Brick	44	50	Mozzarella, part-skim, low moisture	45–52	30–45
Caciocavalle Siciliano	40	42	Muenster	46	50
Camembert, soft ripened	–	50	Neufchatel	65	20–33[*]
Cheddar	39	50	Nuworld	46	50
Colby	40	50	Parmesan, Reggiano	32	32
Cook, dry curd	80	–	Provolone, Pasta Filata	45	45
Cottage, dry curd	80	0.5[*]	Ricotta	80	11
Cottage, low fat	80	0.5–2[*]	Ricotta, skim milk	80	6–10
Cottage, creamed	80	4[*]	Romano	34	38
Cream	55	33[*]	Roquefort	45	50
Edam	45	40	Samsoe	41	45
Gammelost	52	–	Sap sago	38	–
Gorgonzola	42	50	Semi-soft	39–50	50
Gouda	45	46	Semi-soft, part skim	50	45–50
Granular, stirred curd	39	50	Skim milk cheese for manufacturing	50	–
Gruyere	39	45	Spiced	–	50
Hard cheeses	39	50	Spiced, part-skim	–	20–50
Hard, grating	34	32	Swiss, Emmental	41	43
Limburger	50	50			

[*] % fat in cheese *per se*.
Source: FDA (CFR, 2011), pp. 1–107.

of cheese. The reader is referred to Partridge (2009) for sensory evaluation of Cheddar cheese as a parameter of quality.

Table 18.11 shows common defects in the body and texture of cheese and suggested steps to avoid them. Table 18.12 enumerates flavor defects in Cheddar cheese and suggestions for avoiding them.

In addition, quality factors include functionality and behavior when used in cooking. The melting character and stability to heat (no fat separation) are also desirable attributes, particularly for process cheese varieties.

18.8 Sustainability efforts in whey processing

The dairy industry is making many efforts to mitigate negative environmental impacts of fermented dairy food processing, but deep discussion is beyond the scope of this chapter. However, to that end, whey utilization must not be overlooked.

Whey is the liquid substance obtained by separating the coagulum from milk, cream or skim milk in cheese making (Huffman & Ferreira, 2011). Whey also originates from casein manufacture. Sour cream and yogurt manufacture do not yield whey, except in the case of Greek yogurt where large volumes of whey are generated as a result of the straining/centrifuging step. Whey contains about half of the solids of whole milk. Its composition depends largely on the variety of cheese being made. For every kilogram of cheese produced, approximately 9 kg of whey is generated. The whey solids are valuable additions to the functional properties of various foods, as well as a source of valuable nutrients. Functional and nutritional products have been developed from whey. Therefore, for economic viability, sustainability and

Table 18.10 Problems related to color and appearance of Cheddar cheese

Defect	Probable cause	Remedial measure
Acid-cut, bleached/faded, dull looking portions or entire surface of cheese	Excessive acid development in the whey or at packing stage	Watch acid development carefully
	Non-uniform moisture distribution in the cheese	Take precautions to insure consistent and uniform moisture retention in curd
Mottled appearance, irregularly shaped light and dark areas on cheese surface	Combining curds of different colors, batches or moisture contents	Avoid mixing starter after color addition
	Uneven acid development in curd	Strain the starter before adding to vat
	Growth of yeast and bacteria accompanied by typical fruity flavor and pasty body. H_2O_2 production by microorganisms	Try to cut curd into uniform size particles
		Handle curd carefully to avoid drying during matting and cheddaring steps
Seamy, showing light colored lines around curd pieces. In extreme form cracked cheese	Exudation of fat in curd pieces due to excessive forking	Press curd at 85–90 °F
	Warmer temperature	Allow all the salt to dissolve completely
	Lack of dissolution of salt	Avoid too much forking of curd
		Wash greasy curd at 90 °F and drip dry the curd immediately
White specks or granules or smeared with white powder material	Generally tyrosine if present in aged cheese. Derived from crystallization of calcium lactate if present in young cheese	Ripen at consistent temperature; prevent temperature fluctuations
		Vacuum package
Moldy appearance	Mold growth on the surface	Insure tight seals on cheese packages
		Avoid oxygen in the package by vacuum/flushing with CO_2/N_2 gas

From: Chandan and Kapoor (2011b), Clark and Agarwal (2007).

Table 18.11 Body and texture defects in Cheddar cheese

Defect	Probable cause	Remedial measure
Corky, dry and hard	Lack of acid development	Follow standard production procedures
	Low fat in cheese	
	Overcooked cheese curd	
	Low moisture retention in cheese curd	
	Excessive salt levels	
Crumbly, mealy and short body	Excessive acid production and low moisture retention in cheese	Avoid ripening at high temperatures
		Control acid development and moisture level in cheese
Rubbery or curdy	Lack of curing conditions	Optimize ripening time and temperature
Pasty, sticky or wet	Excessive acid development	Control acid development in relation to time and temperature parameters
	Excessive moisture content	
Weak/soft	Too high fat content	Standardize fat in cheese milk
	Moisture content too high	Cook curd to desirable firmness (higher temperature, longer period)
		Avoid piling curd slabs too high or too soon while cheddaring curd

From: Chandan and Kapoor (2011b).

Table 18.12 Flavor problems in Cheddar cheese

Flavor defect observed	Probable cause	Remedial measure
Bitter taste	Excessive moisture	Use less starter
	Low salt level	Ripen for shorter time and/or lower temperature
	Proteolytic strains of starters and/or contaminating microflora	Check salt level and salting technique
	Excessive acidity	Check starter for purity/suitability
	Poor milk quality and/or sanitary conditions	Improve cleaning and sanitizing practices
	Excessive rennet/chymosin	Control acid and rate of acid production
	Hydrophobic peptides	
High acid/sour	Excessive acidity	Use less starter
	Excessive moisture	Shorten ripening period
	Too much starter level	Follow standard procedure for cutting, cooking and salting steps
	High acid milk used	
	Improper expulsion of whey from curd	
	Low salt level	
Fruity/fermented/ yeasty	Low acidity	Improve sanitation
	Excessive moisture	Check and improve water quality
	Dirty equipment	Follow standard procedure for cheese making and equipment cleaning
	Poor-quality milk	Check salting procedure
Rancid/soapy	Milk lipases activated by improper milk production and handling practices	Check cheese milk for rancid flavor
	Microbial lipases from contaminating microflora	Avoid excessive agitation, foaming and temperature fluctuations of raw milk
	Late lactation milk or accidental homogenization of milk	Avoid microbial contamination of milk and cheese by improving sanitation
Flat/weak/ low flavor	Lack of acid production	Check starter activity
	Low-fat milk used for cheese making	Increase starter level
	Excessively high cooking temperature	Check curing temperature
	Too low ripening temperature/too short ripening period	Extend curing period
		Follow standard procedure for fat standardization in milk and cheese making
Musty/moldy	Extensive mold growth on cheese surface	Seal cheese blocks/barrels to eliminate oxygen entry
Miscellaneous off-flavors: barny, feed, malty, onion, weed	Usually associated with milk production (feed and physiological condition of cows)	Avoid milk with these flavors
		Vacuum pasteurize milk prior to cheese making to volatilize off most of these flavors associated with raw milk

From: Chandan and Kapoor (2011b), Clark and Agarwal (2007).

environmental reasons, whey processing is an integral part of cheese operations. In 2006, the US cheese industry produced 90.5 billion pounds of liquid whey (Ling, 2008) as a by-product.

In the past, whey was not efficiently utilized for human and animal nutrition products but dumped into rivers or sprayed over fields, resulting in wastage of the food resource and causing gross environmental pollution. Now, new technologies have enabled whey conversion into food ingredients of notable commercial significance. Dry sweet whey is produced by drying of defatted fresh whey obtained from Cheddar and Swiss cheese manufacture. It contains all the constituents except water in the same relative proportion as in liquid whey. Dry acid whey is similar to dry sweet whey but is produced by drying of fresh whey obtained from cottage and ricotta cheese manufacture.

The techniques of concentration or reverse osmosis, followed by evaporation and drying, recover all the solids, while the other systems (ultrafiltration, microfiltration, etc.) are fractionating techniques. Concentration reduces the amount of water, thereby lowering shipping costs through reduced bulk and improved keeping quality and providing a product more suitable for direct use in foods. The cost of removing a pound of water in an efficient evaporator may be about one-tenth the cost of removing it in a spray dryer. This consideration has encouraged the development of more uses of whey and whey fractions in concentrated form.

Drying gives maximum concentration, extends storage stability and provides a product amenable to food incorporation. Using an appropriate dryer, dairy processors convert sweet whey into a stable, non-hygroscopic and non-caking product. In this process, high solids whey concentrate is spray dried to a free moisture content of 12–14%, causing lactose to take on a molecule of water and become crystallized. This causes whey solids to convert from a sticky, syrup-like material into a damp powder with good flow characteristics. For drying acid cottage cheese whey, a commercial dryer combines spray drying with through-flow continuous bed drying. The concentrate is spray dried in the hot air chamber to 12–15% moisture. The particles fall to a continuous, porous, stainless steel belt where lactose undergoes rapid crystallization. Crystallization of lactose before final drying is mandatory for drying acid whey. A belt conveys the product to another chamber where the whey is further dried by dehumidified air that moves through the porous bed.

Fractionated whey products are produced by membrane technology (including ultrafiltration, reverse osmosis (RO), electrodialysis, etc.) and ion exchange techniques. The pressure-activated processes separate components on the basis of molecular size and shape. RO is the process in which virtually all species except water are rejected by the membrane. The osmotic pressure of the feed stream in such a system often will be quite high. Consequently, to achieve adequate water flux rates through the membrane, such systems often use hydrostatic operating pressures of 5883.6 kg/cm^2 (600 psi) or greater. Ultrafiltration (UF) is the process in which the membrane is permeable to relatively low molecular weight solutes and solvent (permeate), but is impermeable to higher molecular weight materials (Huffman & Ferreira, 2011). The permeability and selectivity characteristics of these membranes can be controlled during the fabrication process so that they will retain only molecules above a certain molecular weight. Thus, UF is essentially a fractionating process, while RO is effectively a concentrating process. One advantage of UF over other processes is that by varying the amounts of permeate removed, a wide variety of protein concentrates, ranging up to 60% protein, can be obtained. Higher levels can be obtained by simultaneously adding fresh water and concentrating by UF, a process called diafiltration. Permeate (the material that passes through the filter) is used for manufacture of

milk sugar, lactose, by condensing and crystallization. Lactose crystals are harvested and dried in a tumble dryer.

In the ion exchange process, whey is passed through two containers filled with special synthetic resins, which have the ability to exchange ions. In the first container, the special synthetic resins exchange hydrogen ions for cations in the whey. Here, the positive ions of the salt are captured and acid is formed by the release of hydrogen ions. The whey is then passed over the anion exchanger where hydroxyl ions are exchanged for negative ions of the salt, and water is formed. When the mobile ions of the resins are completely replaced by other ions, the resin must be regenerated for further use. This is done by passing an acid (hydrochloric) solution through the cationic exchanger, and a basic solution (NaOH) through the anionic exchanger.

Electrodialysis, a combination of electrolysis and dialysis, is the separation of electrolytes, under the influence of an electric potential through semi-permeable membranes. The driving force is an electric field between the anode (positively charged) and the cathode (negatively charged). Between the anode and the cathode, a number of ion-selective membranes are placed which are permeable only to anions or cations. Every other membrane has a positive charge repelling positive ions and allowing negative ions to pass, and in between there is a negatively charged membrane doing just the opposite. In principle, whey is pumped through every second space between two membranes, and a solution of NaCl (cleaning solution) is pumped through the compartments between the whey streams. The ions move from the whey stream into the cleaning solution where they are retained, because they cannot move any further. The cleaning solution contains minerals, acid, some lactose and small nitrogenous molecules. The membranes are cleaned chemically. Protein molecules remain in the fluid while the minerals are removed. The process results in a protein concentrate.

Lactose is crystallized from condensed whey or from permeate (50–60%, solids) obtained by ultrafiltration fractionation of milk or whey. The supersaturated solution is cooled under specific conditions to crystallize lactose. Lactose crystals are harvested and washed to remove the mother liquor and dried. Crude lactose obtained this way contains about 98% lactose. Food-grade and more refined USP (United States Pharmacopeia) grades are produced from crude lactose by protein precipitation, decolorization with activated carbon and subsequent demineralization. Lactose is further refined by recrystallization or by spray drying.

Whey protein concentrates are products derived from whey by removal of minerals and lactose. The process of protein concentration utilizes ultrafiltration, electrodialysis, and ion exchange technologies. On a dry basis, the protein concentrate contains a minimum of 34% or 50% protein. Whey protein isolate contains at least 92% protein. The manufacture and uses of whey products are discussed in detail by Huffman and Ferreira (2011) and later in this book.

References

Aneja RP, Mathur BN, Chandan RC, Banerjee AK (2002) *Technology of Indian Milk Products*. New Delhi, India: Dairy India Yearbook, pp. 158–182.

Born B (2013) Cultured/sour cream. In: Chandan RC, Kilara A (eds) *Manufacturing Yogurt and Fermented Milks*. Chichester, West Sussex: John Wiley, pp. 381–391.

Bottazzi V (2003) *Manufacture of extra-hard cheeses* In: Cavellaro B, Trugo LC, Finglas PM (eds) *Encyclopedia of Food Sciences and Nutrition, Vol. 2*. New York: Academic Press, pp. 1067–1073.

Chandan RC (1982) Other fermented dairy products. In: Reed G (ed) *Prescott and Dunn's Industrial Microbiology*, 4th edn. Westport, CT: AVI Publishing, pp. 113–184.

Chandan RC (2002) Symposium: Benefits of Live Fermented Milks: present diversity of products. *Proceedings of International Dairy Congress*. Paris, France.

Chandan RC (2003) Soft and special varieties. In: Cavellaro B, Trugo LC, Finglas PM (eds) *Encyclopedia of Food Sciences and Nutrition, Vol. 2*. New York: Academic Press, pp. 1093–1098.

Chandan RC (2007) Cheese varieties made by direct acidification of hot milk. In: Hui YH, Chandan RC, Clark S et al (eds) *Handbook of Food Products Manufacturing, Vol. 1: Principles, Bakery, Beverages, Cereals, Cheese, Confectionary, Fats, Fruits, and Functional Foods*. New York: John Wiley and Interscience Publishers, pp. 635–650.

Chandan RC (2011) Nutritive and health attributes of dairy ingredients. In: Chandan RC, Kilara A (eds) *Dairy Ingredients for Food Processing*. Ames, IA: Wiley-Blackwell Publishing, pp. 387–419.

Chandan RC, Kapoor R (2011a) Principles of cheese technology. In: Chandan RC, Kilara A (eds) *Dairy Ingredients for Food Processing*. Ames, IA: Wiley-Blackwell Publishing, pp. 225–265.

Chandan RC, Kapoor R (2011b) Manufacturing outlines and applications of selected cheese varieties. In: Chandan RC, Kilara A (eds) *Dairy Ingredients for Food Processing*. Ames, IA: Wiley-Blackwell Publishing, pp. 267–316.

Chandan RC, Kilara A (2008) Role of milk and dairy foods in nutrition and health. In: Chandan RC, Kilara A, Shah NP (eds) *Dairy Processing and Quality Assurance*. Ames, IA: Wiley-Blackwell Publishing, pp. 411–428.

Chandan RC, Kilara A (eds) (2011) *Dairy Ingredients for Food Processing*. Ames, IA: Wiley-Blackwell Publishing.

Chandan RC, Kilara A (eds) (2013) *Manufacturing Yogurt and Fermented Milks*. Chichester, West Sussex: John Wiley.

Chandan RC, Nauth KR (2012) Yogurt. In: Hui YH, Chandan RC, Clark S et al (eds) *Handbook of Animal Based Fermented Food and Beverage Technology, Vol. 2: Health, Meat, Milk, Poultry, Seafood and Vegetables*, 2nd edn. Boca Raton, FL: CRC Press, pp. 213–233.

Chandan RC, O'Rell KR (2013a) Ingredients for yogurt manufacture. In: Chandan RC, Kilara A (eds) *Manufacturing Yogurt and Fermented Milks*. Chichester, West Sussex: John Wiley, pp. 217–238.

Chandan RC, O'Rell KR (2013b) Principles of yogurt processing. In: Chandan RC, Kilara A (eds) *Manufacturing Yogurt and Fermented Milks*. Chichester, West Sussex: John Wiley, pp. 239–262.

Chandan RC, Shahani KM (1993) Yogurt. In: Hui YH (ed) *Dairy Science and Technology Handbook, Vol. 2*. New York: VCH Publishers, pp. 1–56.

Chandan RC, Shahani KM (1995) Other fermented dairy products. In: Reed G, Nagodawithana TW (eds) *Biotechnology, Vol. 9*, 2nd edn. Weinheim, Germany: VCH Publishing, pp. 386–418.

Chandan RC, Gordon JF, Morrison A (1977) Natural benzoate content of dairy products. Milchwissenschaft 32(9):534–537.

Clark S, Agarwal S (2007) Cheddar and related hard cheeses. In: Hui YH, Chandan RC, Clark S et al (eds) *Handbook of Food Products Manufacturing, Vol. 1: Health, Meat, Milk, Poultry, Seafood and Vegetables*. New York: John Wiley and Interscience Publishers, pp. 567–594.

Code of Federal Regulations (CFR) (2011) *Title 21, Part 131. Milk and Cream. Part 133. Cheese and Related Cheese Products*. US Department of Health and Human Services, Food and Drug Administration. Washington, DC: GMP Publications, pp. 1–107.

Farkye NY (2003) Chemistry and microbiology of maturation. In: Cavellaro B, Trugo LC, Finglas PM (eds) *Encyclopedia of Food Sciences and Nutrition, Vol. 2*. New York: Academic Press, pp. 1062–1066.

Fox PF (2003a) Cheese: overview. In: Roginski H, Fuquay JW, Fox PF (eds) *Encyclopedia of Dairy Sciences*. London: Academic Press, pp. 252–261.

Fox PF (2003b) Biochemistry of cheese ripening. In: Roginski H, Fuquay JW, Fox PF (eds) *Encyclopedia of Dairy Sciences*. London: Academic Press, pp. 320–326.

Fox PF, McSweeney PLH (2003) Manufacture of hard and semi-hard varieties of cheese. In: Cavellaro B, Trugo LC, Finglas PM (eds) *Encyclopedia of Food Sciences and Nutrition. Vol. 2.*. New York: Academic Press, pp. 1073–1086.

Goddik L (2012) Sour cream and crème fraiche. In: Hui YH, Evranuz EO, Chandan RC et al (eds) *Handbook of Food and Beverage Fermentation Technology, Vol. 2: Health, Meat,*

Milk, Poultry, Seafood and Vegetables, 2nd edn. Boca Raton, FL: CRC Press, pp. 235–246.

Hill AR (2009) Cheese. www.uoguelph.ca/foodscience/cheese-making-technology, accessed 26 November 2013.

Huffman LM, Ferreira LDB (2011) Whey-based ingredients. In: Chandan RC, Kilara A (eds) *Dairy Ingredients for Food Processing*. Ames, IA: Wiley-Blackwell Publishers, pp. 179–198.

Hutkins RW (2006) *Microbiology and Technology of Fermented Foods*. Ames, IA: Blackwell Publishing, pp. 67–106.

International Dairy Federation (IDF) (2003) *Yogurt: Enumeration of Characteristic Organisms. Colony Count Technique at 37°C*. Standard No. 117A. Brussels, Belgium: International Dairy Federation.

International Dairy Foods Association (IDFA) (2012) *Dairy Facts*. Washington, DC: International Dairy Foods Association.

Kilara A, Chandan RC (2013) Greek-style yogurt and related products. In: Chandan RC, Kilara A (eds) *Manufacturing Yogurt and Fermented Milks*. Chichester, West Sussex: John Wiley, pp. 297–318.

Law BA, Tamime, AY (eds) (2010) *Technology of Cheese Making*, 2nd edn. Ames, IA: Wiley-Blackwell.

Ling KC (2008) *Whey to Ethanol: A Biofuel Role For Dairy Cooperatives*. USDA Rural Development Research Report 214. Washington, DC: US Department of Agriculture.

McSweeney PL (2003) Cheeses with "eyes". In: Cavellaro B, Trugo LC, Finglas PM (eds) *Encyclopedia of Food Sciences and Nutrition, Vol. 2*. New York: Academic Press, pp. 1087–1092.

National Yogurt Association (2008) *Live and Active Cultures Seal Program: Procedures and Guidelines*. McLean, VA: National Yogurt Association.

Olson NF (2003) Types of cheese. In: Cavellaro B, Trugo LC, Finglas PM (eds) *Encyclopedia of Food Sciences and Nutrition, Vol. 2*. New York: Academic Press, pp. 1046–1051.

O'Rell KR, Chandan RC (2013a) Yogurt: fruit preparations and flavoring materials. In: Chandan RC, Kilara A (eds) *Manufacturing Yogurt and Fermented Milks*. Chichester, West Sussex: John Wiley, pp. 195–216.

O'Rell KR, Chandan RC (2013b) Manufacture of various types of yogurt. In: Chandan RC, Kilara A (eds) *Manufacturing Yogurt and Fermented Milks*. Chichester, West Sussex: John Wiley, pp. 263–296.

O'Rell KR, Chandan RC (2013c) Yogurt plant: quality assurance. In: Chandan RC, Kilara A (eds) *Manufacturing Yogurt and Fermented Milks*. Chichester, West Sussex: John Wiley, pp. 331–352.

Panell L, Schoenfuss TC (2007) Yogurt. In: Hui YH (ed) Chandan RC, Clark S, Cross N et al (eds) *Handbook of Food Products Manufacturing, Vol. 2: Health, Meat, Milk, Poultry, Seafood and Vegetables*. New York: John Wiley and Interscience Publishers, pp. 647–676.

Partridge J (2009) Cheddar and cheddar-type cheese. In: Clark S, Costello M, Drake MA, Bodyfelt FW (eds) *The Sensory Evaluation of Dairy Products*. New York: Springer, pp. 225–270.

Routray W, Mishra HN (2011) Scientific and technical aspects of yogurt aroma and taste: a review. Comprehensive Reviews in Food Science and Food Safety **10**:(4) 208–220.

Sanders ME, Marco ML (2010) Food formats for effective probiotics. In: Doyle MP, Klaenhammer TR (eds) *Annual Review of Food Science and Technology*. Palo Alto, CA: Annual Reviews, pp. 65–85.

Schoenfuss TC, Chandan RC (2011) Dairy ingredients in dairy food processing. In: Chandan RC, Kilara A (eds) *Dairy Ingredients for Food Processing*. Ames, IA: Wiley-Blackwell, pp. 421–472.

Schultz M (2011) *Dairy Products Profile*. Ames, IA: AgMRC, Iowa State University. www.agmrc.org/commodities__products/livestock/dairy/dairy_products_profile.cfm, accessed 25 November 2013.

Singh T, Cadwallader KR (2008) Cheese. In: Chandan RC, Kilara A, Shah NP (eds) *Dairy Processing and Quality Assurance*. Ames, IA: Wiley-Blackwell, pp. 273–307.

Stanley G (2003) Starter cultures employed in cheese-making. In: Cavellaro B, Trugo LC, Finglas PM (eds) *Encyclopedia of Food Sciences and Nutrition, Vol. 2*. New York: Academic Press, pp. 1051–1056.

Tamime AY, Robinson RK (2007) *Yogurt Science and Technology*, 3rd edn. Boca Raton, FL: CRC Press.

Tribby D (2009) Yogurt. In: Clark S, Costello M, Drake MA, Bodyfelt FW (eds) *The Sensory Evaluation of Dairy Products*. New York: Springer, pp. 191–223.

US Department of Health and Human Services, Public Health Services, Food and Drug Administration (2011) *Grade "A" Pasteurized Milk Ordinance, 2011 Revision*. www.fda.gov/downloads/Food/GuidanceRegulation/UCM291757.pdf, accessed 26 November 2013.

US Department of Agriculture, Agricultural Research Service (2011) National Nutrient Database for Standard Reference, Release 26. NDB # 01004–01050. http://ndb.nal.usda.gov/ndb/search/list, accessed 26 November 2013.

Vedamuthu ER (2013a) Starter cultures for yogurt and fermented milks. In: Chandan RC, Kilara A (eds) *Manufacturing Yogurt and Fermented Milks*. Chichester, West Sussex: John Wiley, pp. 115–147.

Vedamuthu ER (2013b) Other fermented and culture-containing milks. In: Chandan RC, Kilara A (eds) *Manufacturing Yogurt and Fermented Milks*. Chichester, West Sussex: John Wiley, pp. 393–410.

19 Eggs and Egg Products Processing

Jianping Wu

Department of Agricultural, Food and Nutritional Science, University of Alberta, Edmonton, Alberta, Canada

19.1 Introduction

Eggs play an important role in the human diet and nutrition as an affordable nutrient-rich food commodity that contains highly digestible proteins, lipids, minerals, and vitamins (Fisinin et al., 2008). Over the past 35 years, global egg production has grown 203.2%, due to a rapidly increasing demand for proteins in the developing world (Windhorst, 2007). In 1970, the US was the greatest egg producer in the world, contributing over 20% of the global egg production (Windhorst, 2007). However, the US egg production has changed little since then, contributing 9% in 2005. During the same period, China has increased egg production over 20 times, contributing 41.1% of worldwide egg production in 2005. The major egg-producing countries are China, US, India, Japan, and Mexico. The majority of egg consumption in China is in the form of table eggs, as only about 1% of eggs are broken and further processed into egg products. In the US, however, about 30% of eggs are processed into various egg products (i.e. liquid, frozen, and dried egg products) for uses in the food service industry or as ingredients in various food applications. Examples of other egg products include hard-cooked chopped eggs, precooked scrambled eggs or omelets, quiches, precooked egg patties, scrambled egg mixes, and crepes (Froning, 2008).

In addition to food uses, eggs also contain a range of bioactive components that can be used for improving human health and other non-food applications. Extraction and fractionation of bioactive egg components such as lysozyme, avidin, ovotransferrin, ovomucin, antibody (IgY), phospholipids and sialic acid, and development of bioactive peptides from egg proteins represent great opportunities in novel applications of egg components in the future. The objectives of this chapter are to provide an overview of egg formation, structure and chemistry, and then to introduce shell egg and broken egg processing. Perspectives on value-added egg processing for novel applications are also presented. For earlier references on egg science and processing, the reader should refer to Burley and Vadehra (1989), Stadelman and Cotterill (1995), Yamamoto et al. (1997), and Bell and Weaver (2002).

19.2 Shell egg formation

The most common breeds used for laying eggs in North America are White Leghorn (lays white-shell eggs) and Rhode Island Red (lays brown-shell eggs). Pullets (a pullet is a young hen) are first housed in a pullet barn for 19 weeks and then are transferred to another barn where they start to lay eggs for at least 12 months. While there are two sets of reproductive organs, only the left one survives and reaches maturity to produce eggs. The reproductive tract of a laying hen is composed of two parts: the ovary and oviduct (Figure 19.1). The ovary is a cluster of various sizes of developing follicles, each containing an ovum (or a yolk). The follicle contains a highly developed system of blood vessels that carry nourishment such as proteins and lipids to the developing yolk. There are about 13,000–14,000 ova in a mature ovary but most of them gradually degenerate, and only about 2000 accumulate white yolk to grow to about 6 mm in diameter and reach maturity to produce an egg. Each ovum (singular of ova) starts out as a single cell surrounded by a vitelline membrane. As the ovum develops, yolk is added for a period of 7–12 days prior to ovulation (Burley & Vadehra, 1989). Ovulation is the release of the mature ovum from the ovary into the second part of the female

Food Processing: Principles and Applications, Second Edition. Edited by Stephanie Clark, Stephanie Jung, and Buddhi Lamsal.
© 2014 John Wiley & Sons, Ltd. Published 2014 by John Wiley & Sons, Ltd.

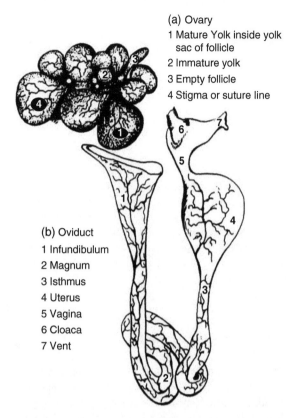

(a) Ovary
1 Mature Yolk inside yolk sac of follicle
2 Immature yolk
3 Empty follicle
4 Stigma or suture line

(b) Oviduct
1 Infundibulum
2 Magnum
3 Isthmus
4 Uterus
5 Vagina
6 Cloaca
7 Vent

Figure 19.1 The reproductive tract of a laying hen is composed of two parts: ovary (a) and oviduct (b). Courtesy of USDA (2000).

reproductive system, the oviduct. The matured follicle is ovulated at intervals of about 24–27 h (Bell, 2002; Burley & Vadehra, 1989).

The oviduct is an organ wherein the yolk passes and the other portions of egg are secreted. In a fully developed laying hen, the oviduct is 40–80 cm long, with an average weight of 40 g, consisting of five parts: infundibulum, magnum, isthmus, uterus, and vagina (see Figure 19.1). With a funnel-shaped structure, the infundibulum is the top portion of the oviduct and opens its ampulla towards the ovary to receive the ovum after it is released from the ovary; the ovum stays approximately 15–30 min here and can become fertilized if sperm are present. The magnum is the longest portion (34 cm long) of the oviduct. The follicle is held there for about 174 min and egg albumen is secreted there to cover the egg yolk. The isthmus is about 11 cm long and is the place for the formation of inner and outer shell membranes in about 75 min. Although the uterus (or the shell gland) is only 10 cm long, the egg is held for about 16–17 h for

the completion of calcification, followed by the deposition of shell pigments and eggshell cuticle on the immobilized egg during the last 1.5 h before oviposition (Nys et al., 2004). The vagina is the last portion of the oviduct and is about 9 cm long. Its function is to carry the egg from the uterus to the cloaca. It takes about 5 min for the egg to pass through this portion (Bell, 2002; Burley & Vadehra, 1989; Stadelma, 1995).

19.3 Structure of eggs

Eggs are composed of three main parts: eggshell, including the shell membranes between the albumen and the inner shell surface, egg white and yolk, representing 9.5%, 63%, and 27.5%, respectively, of the whole egg (Cotterill & Geiger, 1977; Li-Chan et al., 1995). The yolk is located in the center of the egg, surrounded by an albumen layer, which in turn is covered by eggshell membranes and finally a hard eggshell (Hincke et al., 2012; USDA, 2000) (Figure 19.2). The eggshell provides protection against physical damage, microorganisms, and small predators (Hincke et al., 2012).

An eggshell is composed of a thin film layer of cuticle, a calcium carbonate layer, and two shell membranes. There are 7000–17,000 funnel-shaped pore canals distributed unevenly on the shell surface for metabolic gases and water exchange (Dennis et al., 1996). The cuticle is the most external layer of eggs, about 10 μm thick, and it covers the pore canals. It protects the egg from moisture loss and invasion of microorganisms to a certain extent (Board & Hall, 1973; Whittow, 2000). This layer, as well as the outer portion of the calcified shell, contains the eggshell pigments, which serve as camouflage, temperature control, and possibly as factors in parental recognition (Sparks, 1994). The cuticle is a non-calcified organic layer that is deposited on the mineral surface, consisting mainly of proteins, small amount of carbohydrates, and lipids (Hincke et al., 2012). The cuticle layer can be easily removed by soaking eggs in either weak acid solutions or metal chelator-containing solutions or by washing with water (Belyavin & Boorman, 1980). Therefore, washing of eggs often leads to bacterial invasion of the egg.

The calcium carbonate layer is composed of 95% inorganic substance, mainly calcium carbonate, 3.3% protein, and 1.6% moisture (Parson, 1982). It consists of an outer vertical crystal layer, a central palisade layer, and an inner mammillary knob layer (see Figure 19.2). The vertical layer consists of short, thin crystals running in a

(a)

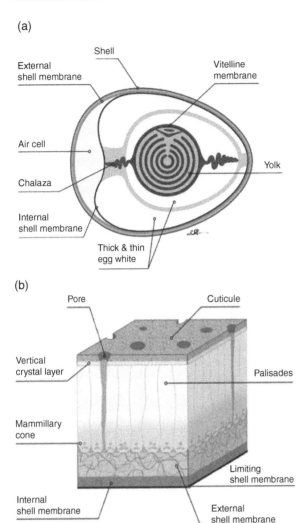

(b)

Figure 19.2 Structures of hen's eggs (a) and eggshell (b) (Hincke et al., 2012). Courtesy of Dr M. Hincke, University of Ottawa.

vertical direction of the shell. The palisade layer, often called the spongy layer, is the thickest layer of eggshell. This layer is made up of groups of columns that are perpendicular to the eggshell surface and extend outward from the mammillary cones, where its crystalline structure is constructed with collagen to form a spongy matrix (Hincke et al., 2012). The mammillary knob layer is composed of a regular array of cones or knobs that are interconnected to the fibers of the outer shell membrane to harden the shell (Parson, 1982). Within the mammillary cone layer, microcrystals of calcite are arranged with a spherulitic texture, which facilitates the propagation of cracks during piping as well as the mobilization of calcium to nourish the embryo by dissolution of highly reactive calcite microcrystals (Nye et al., 2004).

The eggshell membrane is composed of inner and outer membranes with a structure like entangled threads or randomly knitted nets. This structure is important in obstructing invading microorganisms by catching them in the meshwork. The eggshell is critically important in determining egg quality and forms the first line of defense against pathogens.

The albumen or egg white is composed of four distinct layers: an outer thin white next to the shell membrane, an outer thick white layer, an inner thin white, and a chalaziferous or inner thick layer, representing about 23.3%, 57.3%, 16.8%, and 2.7%, respectively, of the egg white (Burley & Vadehra, 1995). The proportions of each layer depend on breed, environmental conditions, size of the egg, and rate of production (Li-Chan et al., 1995). The thick white layer is sandwiched between the outer and inner thin albumen. The viscosity of the thick albumen is much higher than that of thin albumen, mainly due to its high content of ovomucin. The chalaziferous layer is a fibrous layer that directly covers the entire egg yolk. In the long axis of the egg, this layer is twisted at both sides of the yolk membranes, forming a thick rope-like structure named the chalazae cord. This cord is twisted clockwise at the sharpened end of the egg and counterclockwise at the opposite end (see Figure 19.2). The chalazae cord stretches into the thick albumen layer at both sides to suspend the yolk in the center of the egg (Okubo et al., 1997).

The content of yolk, consisting mainly of yellow yolk and less than 2% white yolk, is encircled by a vitelline membrane. The vitelline membrane is composed of an inner layer, a continuous membrane, and an outer layer, with thicknesses of 1.0–3.5, 0.05–0.1, and 3–8.5 μm, respectively. Both the inner and outer layers are three-dimensional meshwork structures consisting of fibers with diameters of 200–600 and 15 nm, respectively, while the continuous membrane is a piled, sheet-like structure consisting of granules with an estimated diameter of 7 nm (Okubo et al., 1997). Yellow yolk consists of deep yellow yolk and light yellow yolk, appearing alternatively and circularly (Romanoff & Romanoff, 1949). The deep yellow yolk is formed in the daytime while the light yellow yolk is formed at night when the protein concentration in the bloodstream is lower than in the daytime.

19.4 Chemical composition of eggs

Avian eggs are an excellent source of nutrients, particularly high-quality proteins, lipids, minerals, and vitamins (Herron & Fernandez, 2004; Kovacs-Nolan et al., 2005a; Li-Chan & Kim, 2008). An egg is composed of approximately 75% water, 12% each proteins and lipids, ~1% carbohydrates and minerals (Burley & Vadehra, 1989; Li-Chan et al., 1995). Most of the proteins are present in the egg white and the egg yolk, amounting to 50% and 44%, respectively; eggshell contains the rest of the proteins. Lipid presents exclusively in egg yolk while albumen has a very low (0.03%, w/w) lipid content. Though not generally part of the human diet, the eggshell forms a rich source of inorganic salts, mainly calcium carbonate and traces of magnesium carbonate and tricalcium phosphate (Li-Chan et al., 1995; Mine, 2002). The presence of novel matrix proteins has been recently characterized (Hincke et al., 2012; Li-Chan & Kim, 2008). The nutritional profile of egg in egg yolk can be modified through diet, leading to "designer eggs" such as "omega-3 eggs" and "vitamin-enriched eggs" with additional health attributes (Fraeye et al., 2012; Surai & Sparks, 2001). Designer eggs continue to grow to meet the demands of health-conscious consumers.

19.4.1 Chemical composition of egg albumen

The major constituents of albumen are water (92%) and protein (~10%), followed by carbohydrates (0.4–0.9%) that exist in free form, usually as glucose, or forming complexes with proteins such as glycoproteins that contain mannose and galactose units. The albumen is also composed of lipid (0.03%) and ash (0.5–0.6%) (Sugino et al., 1997).

Egg albumen can also be considered as a protein system that contains ovomucin fiber in an aqueous solution of globular proteins (Li-Chan & Kim, 2008). Ovoalbumin, a phosphoglycoprotein with a molecular weight of 45 KDa and an isoelectric point (pI) of 4.5, is the most abundant protein in egg white (54–58%) (Huntington & Stein, 2001; Li-Chan et al., 1995; López-Expósito et al., 2008). It is the only protein in albumen that contains free sulfhydryl groups and is a major source of amino acids for the embryo. Ovotransferrin, accounting for about 12% of egg white proteins, is a glycoprotein with a molecular weight of 75 KDa and a pI of 6. It is a disulfide-rich single-chain glycoprotein and belongs to the transferrin family (Li-Chan et al., 1995; Williams et al., 1982). As a member of the transferrin family, it is able to bind iron, and is known for antimicrobial, antifungal and antiviral activities (Wu & Acero, 2012). Ovomucoid, representing 11% of total albumen proteins, is a thermostable glycoprotein and the dominant egg allergen. It belongs to the Kazal family of protease inhibitors, with a molecular weight of 28 KDa and a pI of 4.1 (Kato et al., 1987). Ovomucin is a sulfated glycoprotein that contributes to the gel-like structure of the thick white layer, forming flexible fibers. It is composed of two subunits: α-ovomucin, with a molecular weight of 254 KDa, and β-ovomucin, with a molecular weight of 400–610 KDa (Robinson & Monsey, 1971). Ovomucin represents 2–4% of total egg albumen and its pI is about 4.5–5.0. Lysozyme is an enzyme with a molecular weight of 14.3 KDa and a pI of 5.5; it accounts for 3.5% of total egg white and possesses bacteriostatic, bacteriolytic and bacteriocidal activity (Cunningham et al., 1991). Ovoglobulins consist of two proteins, G2 and G3, with molecular weights between 30 and 45 KDa and a pI of 4.0; these proteins are known for their excellent foaming and beating properties. Ovoinhibitor possesses a molecular weight of 49 KDa and a pI of 1.5; it is capable of inhibiting trypsin and chymotrypsin as well as fungal and bacterial proteases (Huopalahti et al., 2007; Li-Chan & Kim, 2008; Li-Chan et al., 1995).

Other components, including avidin, cystatin, ovoinhibitor, ovostatin, ovoglycoprotein, ovoflavoprotein, and G2 and G3 globulin, are found in the egg white and contain minor levels of carbohydrates, minerals, and lipids (Li-Chan et al., 1995; Mine, 2002).

19.4.2 Chemical composition of egg yolk

Egg yolk represents 36% of the total egg weight. It is composed mainly of 51% water, 16% proteins, 32.6% lipids, 1.7% minerals, and 0.6% carbohydrates. In dry matter, egg yolk is composed of 68% low-density lipoprotein (LDL), 16% high-density liporprotein (HDL), 10% globular proteins (livetins), 4% phosphoprotein (phosvitin), and 2% minor proteins (Huopalahti et al., 2007); lipids represent about 65% and the lipid to protein ratio is about 2 to 1. The egg yolk protein consists of apovitellenin, phosvitin, α- and β-lipovitellin apoproteins, α-livetin (serum albumin), β-livetin (α2-glycoprotein), γ-livetin (γ-globulin), and traces of biotin-binding protein (Li-Chan et al., 1995; Mine, 2002). Lipids are found in the form of lipoproteins and usually are composed of 62% triglycerides, 33% phospholipids and less than 5% cholesterol (Anton, 2007). Phospholipids in egg yolk are very rich in phosphatidylcholine (PC) (76%), which has been recognized as an important nutrient for brain development, liver function, and cancer prevention (Zeisel, 1992). Other phospholipids found are phosphatidylethanolamine (PE) (22%), phosphatidylinositol (PI), phosphatidylserine (PS), sphingomyelin (SM), cardiolipins (CL), lysoPC, and lysoPE, which are present at very low amounts (Sugino et al., 1997). The yellowish color of egg yolk is due to the presence of carotenoids, representing about 1% of the lipids, mainly carotene and xanthophylls (lutein, cryptoxanthin, and zeaxanthin) (Anton, 2007). There has been increased awareness of the role of xanthophylls in human health, in particular the roles of lutein and zeaxanthin in the prevention of age-related macular degeneration (Lesson & Caston, 2004).

The presence of cholesterol in egg yolk leads to controversial public perceptions about egg, which is cholesterol enriched and possibly contributes to the risk of coronary heart diseases (Jones, 2009; Lee & Griffin, 2006; Spence et al., 2012). However, many current studies do not support this hypothesis since the cholesterol metabolism is complicated in the human system and diet is not the sole factor that decides the level of cholesterol in blood. In the largest epidemiological study conducted to date on the relationship between egg consumption and coronary heart disease, consumption of up to one egg per day did not have a substantial overall impact on the risk of coronary heart disease and stroke (Hu et al., 1999, 2001; Qureshi et al., 2006). Although there is a potential to reduce egg yolk cholesterol by genetic selection, the extent of reduction of egg cholesterol by genetic selection is fairly small (9–10 mg), not significant if an average daily intake of cholesterol is about 250 mg (Elkin, 2006; Hargis, 1988; Naber, 1990). If the reduced cholesterol level cannot meet the embryo requirement, the laying hen will simply cease egg production. The discovery of new antioxidants and antihypertensive peptides in eggs might provide new evidence on egg and health (Majumder et al., 2013; Nimalaratne et al., 2011).

19.5 Shell egg processing

Although recognized as the most nutritious food commodity on earth, eggs are well-known source of *Salmonella* that may be present on both the outer surface and in the contents of the egg (Arvanitoyannis et al., 2009; Braden, 2006). The outbreak of more than 1900 illnesses in 11 states in 2010, which led to a voluntary market recall of over 500 million shell eggs nationwide, suggests that *S. enteritidis* is still an important cause of human illness

in the United States (CDC, 2010). *Salmonella enteritidis* (SE) was identified as the cause of infection in 62.5% cases and *S. typhimurium* in 12.9% while other serotypes are responsible for <2% of human infections (EFSA, 2007). Eggs can be contaminated by penetration through the eggshell from the colonized gut or from contaminated faeces during or after oviposition (horizontal transmission), or by direct contamination of the yolk, albumen, eggshell membranes or eggshells before oviposition, originating from the infection of reproductive organs with SE, although it is not yet clear which route is most important for SE to contaminate the egg contents (Gantois et al., 2009).

Salmonella enteritidis can survive in the albumen, but it is effectively inhibited from growing for an extended period of time due to an increased pH from 7.2 to over 9.0 during the initial days after laying, and the presence of a viscous ovomucin also hampers the movement of invading bacteria from egg albumen to egg yolk, as well as several antimicrobial proteins such as ovotransferrin (binding of iron and making iron unavailable for bacteria), avidin (binding of biotin), and lysozyme (disruption of bacterial membranes) (Humphrey, 1994). However, storage of eggs at ambient temperature (above 7.2 °C) for extended periods of time could promote growth and multiplication of SE; therefore, the Food and Drug Administration (FDA) published a final rule in the Federal Register (65 FR 76092), which states that a proposed maximum ambient temperature of 7.2 °C (45 °F), during storage and transportation, would extend the effectiveness of the egg's natural defense against SE and slow the growth rate of this food-borne pathogen (FDA, 2009). However, EU Commission Regulation 589/2008 specifies that "egg should be stored and transported at a constant temperature, and should in general not be refrigerated before sale to the final consumers" (EU, 2008), due mainly to the perception that eggs kept in cold storage are no longer regarded as "fresh." Although this is under debate, all eggs in the US have to be washed, cleaned, sanitized, oiled, dried, refrigerated, stored and transported, before marketing as table eggs or further processing (USDA, 2000, 2006, 2011).

The United States Department of Health and Human Services (HHS) and Department of Agriculture (USDA) and its related agencies have historically led the federal government's efforts to ensure the safety of shell eggs. The HHS, through its authority under the Federal Food, Drug and Cosmetic Act (FFDCA) and the Public Health Service Act (PHSA), has provided oversight of shell egg safety at egg-laying barns (USDA, 2011).

19.5.1 Egg washing

It is assumed that a chicken egg is at its highest quality at the time of laying (Stadelman, 1995a). Therefore, it is critical to properly handle eggs to maintain their highest quality.

Since it is not possible to produce entirely clean eggs, in the US eggs are washed and cleaned to remove stains, dirt, and other surface contaminants to reduce bacterial contamination and prevent the penetration of bacteria into the egg contents, as well as to enhance the appearance to the consumer. In off-line production sites, eggs are transported from farms to a central facility for processing. These center facilities have shell-egg storage rooms to hold eggs for a few days between 10 °C and 16 °C, or below 7.2 °C if over 1 week. The storage room should be adjacent to the empty case storage and transfer rooms for easy transferring of eggs to the washers. In the 1990s, farms with millions of birds made in-line production system more economical and efficient. In the in-line system, eggs are processed the same day that they are laid (Stadelman, 1995b; Zeidler, 2002a). In the Code of Federal Regulations (CFR) Title 7, section 56.76, minimum facility and operating requirements for shell egg grading and packing plants, as well as shell egg cleaning operations, are specified.

During washing, eggs should not be allowed to soak in water to avoid the possibility of bacterial penetration of the shells. Modern egg washers are designed to spray the eggs with water containing a sanitizer along with a detergent (USDA, 2000). Washing water temperature should be maintained at 32.2 °C (90 °F) or higher, and needs to be at least 11 °C (20 °F) warmer than the internal temperature of eggs to prevent the wash water from being drawn into the eggs. As shown in Figure 19.3a, egg loading is separated from egg washing and incoming eggs are conveyed to the washer machine, which is equipped with a series of spray jets and brushes (Figure 19.3b). Rotating of eggs during washing using oscillating brushes provides the scrubbing action. The washing operation must be continuous and fresh water should also maintain continuous overflow. For safety reasons, the wash water has to be changed approximately every 4 h or more often if needed, in order to maintain sanitary conditions, and is mandatory at the end of each shift (Galiş et al., 2013). It is mandatory to use only potable water with an iron content of less than 2 ppm for washing as water high in iron can support bacterial spoilage. The detergents chosen must be listed as approved for use on eggs in the current *Lists of Proprietary and New Food Compounds* (USDA FSIS, Miscellaneous Publication Number 1419); alkaline detergents at amounts sufficient to maintain a pH of 11 are most effective. Water

pH should range between 10 and 11 since *Salmonella* sensitivity to heat increases as pH increases above 10 (Zeidler, 2002a).

Following washing, eggs are then rinsed with water at a temperature at least equal to or warmer than the wash water to remove any adhering dirt and chemicals; the rinsing waster contains a sanitizer, usually chlorine-based compounds such as sodium hypochlorite (not less than 50 ppm or more than 200 ppm of available chlorine or its equivalent) (USDA, 2000).

The main disadvantage of egg washing is the removal of the cuticle layer, the first defense against bacterial contamination (Board & Halls, 1973). To prevent the potential of microbial penetration into the eggs, shell eggs must be sufficiently dried directly after rinsing. Eggs are air dried using electric fans and transported in the conveyors for the oiling process, where they are sprayed with mineral oil, which should be tasteless and colorless. An oiling process is introduced to seal the shell pores to prevent weight loss by evaporation and the escape of carbon dioxide, which slows down the increase in pH and air cell size, preserving egg quality (Stadelman, 1995b; Zeidler, 2002a). Plant personnel segregate broken and dirty eggs after the washing cycle and prior to the candling operation (USDA, 2000). Egg washing, sanitizing, and oiling should be conducted according to the procedures outlined in the current Regulations Governing the Voluntary Grading of Shell Eggs (7 CFR Part 56). All equipment and processing rooms should be thoroughly cleaned at the end of each processing day and should remain reasonably clean throughout the processing shift.

In-shell pasteurized eggs are gaining popularity due to recent incidences of SE contamination. The technology was developed in the late 1980s and commercialized by National Pasteurized Eggs, Inc. (http://www.safeeggs.com/). The key to the technology is to pasteurize eggs without cooking them. Eggs are locked in place in transfer baskets, dipped in two warm water baths with temperatures of 130 °F and 140 °F (54.4 °C and 60 °C) for about 5 h and in a 45 °F (7.2 °C) cold-water bath for about an hour. At the end of the pasteurization process, eggs are dried, coated with food-grade wax to prevent moisture and microbial penetration, and moved to the candling booth for grading and packaging (Zeidler, 2002a). Other methods, such as electrolyzed water, ozone, irradiation, microwave technology, ultraviolet light technology, pulsed light technology, gas plasma technology, ultrasounds, etc. are under development to decontaminate shell eggs (Galiş et al., 2013). The FDA approved irradiation of shell eggs with doses up to 3 kGy (FSIS, 2000).

(a)

(b)

(c)

Figure 19.3 Overview of egg washing machine. Incoming eggs are conveyed to the egg washer (a), eggs are sprayed with washing water and rotated using oscillating brushes to remove dirt without damaging eggshell (b), and egg candling (c). Images (a) and (b), courtesy of Dr Vincent Guyonnet of Burnbrae Farm Ltd., Canada. Image (c), courtesy of Sanovo Technology Group.

19.5.2 Candling and grading

After oiling, eggs are conveyed to the candling area where defective eggs are removed (Vaclavik & Christian, 2008). Candling is a technique that allows the shell and inside of eggs to be viewed without breaking the shell; candling was once used to check incoming eggs for freshness by viewing their internal contents by candlelight, where egg contents could be seen when held up to a candle while being rapidly rotated. Today, commercial eggs may be scanned *en masse*, with bright lights under trays of eggs (Mountney & Parkhurst, 1995; Vaclavik & Christian, 2008) (see Figure 19.3c). During candling, a wand-type pointer electronically marks eggs that are dirty or cracked, or that contain blood spots. New equipment is capable of automatically removing these eggs with minimal oversight. Modern egg washers are able to perform automatic loading, washing, drying, oiling, candling, weighing, and packaging (see Figure 19.3).

After candling, eggs are weighed and then segregated by weight classes and conveyed to the applicable size packing line. In the US there are six weight classes for eggs, differing by 3.0 oz per dozen intervals. Peewee is the smallest category, weighing 15 oz and less per dozen, while jumbo represents the largest category, weighing 30 oz or more per dozen. Each category has a minimum weight for individual cartons as well as for individual eggs in the carton. All peewees, small eggs and many medium eggs are directed to breaking operations, leaving mainly medium, large, extra large, and jumbo categories on the market.

Grading generally involves the sorting of products according to quality, size, weight, and other factors that determine the relative value of the product. Egg grading is the grouping of eggs into lots having similar quality and weight characteristics (USDA, 2000). US standards for quality of individual eggs have been developed on the basis of such interior quality factors as condition of the white and yolk, size of the air cell, and exterior quality factors of cleanliness and soundness of the shell as shown in Table 19.1 (USDA, 2000). Eggs that achieve the quality standards are conveyed to the weighing scales and packed according to their weight in automatic packaging units and then stored in cool rooms before shipping.

19.5.3 Packaging and storage

Packaging and packing are normally carried out in conjunction with the grading operation. After grading, eggs are conveyed to the packaging machine, where they are separated by weight and classified into different categories,

Table 19.1 Summary of US standards for quality of shell eggs

Quality	Remarks
AA	Unbroken shell Air cell less than 1/8″ in depth and clear Firm whites and yolk with no apparent defect
A	Unbroken shell Air cell 3/16″ in depth Egg white clear and reasonably firm and yolk free from apparent defects
B	Slightly abnormal shell Air cell should not exceed 3/8″ in depth Clear egg white but slightly weak Yolk slightly flattened

Courtesy of USDA (2000), non-copyright material.

transported to the automatic packer and packed in cartons (6, 12 or 18 eggs) or filler flats (20 or 30 eggs), and then placed in cardboard cases or wire baskets and moved to the cold storage room for cooling prior to shipment. Packages should be marked with the date of packing, sell-by date codes and plant identification, although some companies choose to ink-jet print the individual eggs as well. Cartons are usually made from recyclable corrugated fiberboard material, which is the most common packing material. Other materials include rigid plastic cases, polystyrene foam cartons, and molded clear plastics; the design and material used are important factors in reducing egg damage during storage and distribution. It is mandatory in the US to keep shell eggs at or below 45 °F (7.2 °C) after processing (during storage, transportation, and marketing) to effectively control SE inside the egg (minimizing the risk of microbial multiplication and reducing the heat resistance of SE) (USDA, 2000); therefore rapid cooling techniques or extended cold storage times are required to meet these refrigeration regulations.

19.6 Further processing of eggs and egg products

Of the 76.2 billion eggs consumed in 2009, 30% were in the form of egg products (eggs removed from their shells) (USDA, 2011). Liquid, frozen, and dried egg products are widely used by the food service industry such as scrambled or made into omelets, or as ingredients in other foods, such as prepared mayonnaise, ice cream, salad dressings, frozen desserts, cream puff, cakes, confections, etc. (FSIS, 2013; Yang & Baldwin, 1995). The term "egg products" refers to eggs that are removed from their shells for processing at facilities called "breaker plants." The processing of egg products includes breaking eggs, filtering, mixing, stabilizing, blending, pasteurizing, cooling, freezing or drying, and packaging. This is done at USDA-inspected plants. The USDA's Food Safety and Inspection Service (FSIS), through its authority under the Egg Products Inspection Act (EPIA), provides oversight of shell eggs after they have left the barn to either be placed in cartons for consumers or sent to an egg product processor (FSIS, 2011).

Shell eggs must be washed and completely dried prior to breaking. Eggs with broken shell membrane are not permitted for human consumption. The egg breaking room should be separate from the egg washing room. The development of highly automatic, computer-controlled egg breaking and separating machines is a key achievement in the production of egg products. The basic egg breaking unit is composed of a cracker and a yolk-albumen separator. The crack head cracks the shells at the center and pulls the two shells apart. The yolk-albumen separator has two cups positioned one above the other. After cracking, the egg content falls into the top cup to retain only yolk while the white slides to the bottom cup (Figure 19.4). Many of these units are attached to a round or oval rotating table which provides great processing capacity, as many as 188,000 eggs per hour in modern egg breaking machines. Breaking machines are cleaned and sanitized after 4 h of operation, and at the end of the final shift of operation. Holding tanks and pipes are made of stainless steel and are cooled by ice water running between the inner and outer walls, keeping the liquid eggs as cold as possible. Strict sanitation and temperature control are the key factors in maintaining low microbial load. After breaking, the edible liquid products are filtered, ingredients (salt, sugar, etc.) are added, blended, standardized, and pasteurized prior to packaging in a separate room or further frozen or dried.

Figure 19.4 Eggs are first cracked and egg contents fall into a yolk-albumen separator where the top cup retains only yolk while whites slide to the bottom cup. Courtesy of Dr Vincent Guyonnet of Burnbrae Farm Ltd., Canada.

19.7 Liquid egg products

Liquid egg white usually has a total solids content of ~12%. The solids content can be affected slightly (usually within 1%) by loss of moisture and egg stocks (smaller eggs have higher solids level while larger eggs from older birds have lower solids level). The pH of liquid whites ranges from 7.6 to 9.3, depending on the amount of carbon dioxide lost from the egg. Liquid egg whites obtained from stored eggs usually have higher pH; as the eggs are stored, the pH will increase due to carbon dioxide loss through the shell during storage. The amount of carbon dioxide loss is the only factor that can affect the pH of liquid egg whites, which depends on the egg temperature, the amount of carbon dioxide in the environment and the degree of egg sealing or oiling (Cotterill & McBee, 1995).

The solids content of pure egg yolk from fresh eggs is about 51.9 ± 0.1% and the pH of pure egg yolk is about 6.0. During storage, water migrating from the white is the major factor affecting the initial egg yolk solids. The total solids range from 44% to 48% from breaking facilities, depending on the amount of egg white adhering to the yolk and moisture migration. The presence of egg whites in egg yolk will also increase the pH. Usually the solids level is standardized to 43–44% solids level by addition of egg white (Cunningham, 1972).

Liquid whole egg is a blend of egg white and egg yolk. The solids level of a blend of egg whites and egg yolk in natural proportions is about 23–25%. Under USDA regulations, the solids level of blended liquid whole egg is standardized to 24.2% (FDA, 2012a); this may require the addition of more yolk than is present in natural proportions in the liquid from shell eggs. The pH of liquid whole egg can vary from 7.0 to 7.6, usually 7.2 (Cotterill & McBee, 1995).

19.8 Pasteurization

After breaking, the liquid eggs should be pasteurized as soon as possible to reduce the possibility of food-borne pathogen contamination or proliferation. The bacterial quality of liquid products depends on the quality of the eggs being broken, the sanitation conditions in the plant, and handling practices. Under the EPIA, egg products must be pasteurized to eliminate *Salmonella* (FSIS, 2013); pasteurization of egg products in the US has been mandatory since 1966 (Cunningham, 1995). The pasteurization process involves a combination of time and temperature in order to reduce the number of viable pathogens, especially *Salmonella*. The USDA now requires that liquid whole egg be heated to at least 60 °C (140 °F) and held for no less than 3.5 min, or at least (134 °F) and held for no less than 3.5 min, or at least 55.6 °C (132 °F) but held for 6.5 min for egg white. Liquid egg products, especially egg whites, are very susceptible to heat treatment, resulting in protein damage and impaired functionality. Adding additives, including carbohydrates such as sucrose, glucose or fructose, or salt at 10% levels can protect susceptible egg proteins from damage and therefore allow higher temperatures, of approximately 3–6 °C before heat damage occurs (Cunningham, 1995). The compositional differences of liquid egg products such as solids level, fat content, and pH, affect the heat resistance of salmonellae and account for the wide range of pasteurization conditions recommended. A higher temperature (60 °C or higher) is needed for pasteurization of egg yolk, as salmonellae are more heat resistant in yolk than in whole egg or egg white. The increased heat resistance in egg yolk is due to the lower pH and higher solid content of egg yolk.

Facilities for pasteurization of egg products must be adequate and of approved construction so that all products are processed as approved. Pasteurization equipment for liquid egg product must include a holding tube, an automatic flow diversion valve, thermal controls, and recording devices to determine compliance with pasteurization standards as set forth by the USDA. The temperature of the heated liquid egg product must

Figure 19.5 Liquid egg products are pasteurized using tubular-type high-temperature-short-time (HTST) systems. Courtesy of Dr Vincent Guyonnet of Burnbrae Farm Ltd., Canada.

be continuously and automatically recorded during the process (FDA, 2012b). The most widely adopted systems for pasteurization in liquid egg products are plate-type or tubular-type high-temperature, short-time (HTST) systems (Figure 19.5). HTST is a continuous method that can process large volumes in a short time; it is composed of a steel plate heat exchanger that heats and cools the egg liquid in a short time. Table 19.2 lists current USDA pasteurization requirements for various egg products. The temperatures and holding times listed in Table 19.2 are minima; the product may be heated to higher temperatures and held for longer periods of time. Pasteurization procedures must insure complete pasteurization, and holding, packaging, facilities, and operations shall be such as to prevent contamination of the product.

After pasteurization, liquid products are cooled and packaged, often shipped in bulk tankers to other plants for various applications, or frozen. Over one-fourth of the total liquid production is frozen (Stadelman & Cotterill, 1995). Freezing causes only minor changes in raw egg white while the formation of gelation upon freezing and storing raw egg yolk below −6 °C is a well-known phenomenon. The extent of gelation in whole egg is less

Table 19.2 Pasteurization requirements

Liquid egg product	Minimum temperature requirements (°F)	Minimum holding time requirements (minutes)
Albumen (without use of chemicals)	134	3.5
	132	6.2
Whole egg	140	3.5
Whole egg blends (less than 2% added non-egg ingredients)	142	3.5
	140	6.2
Fortified whole egg and blends (24–38% egg solids, 2–12% added non-egg ingredients)	144	3.5
	142	6.2
Salt whole egg (with 2% or more salt added)	146	3.5
	144	6.2
Sugar whole egg (2–12% sugar added)	142	3.5
	140	6.2
Plain yolk	142	3.5
	140	6.2
Sugar yolk (2% or more sugar added)	146	3.5
	144	6.2
Salt yolk (2–12% salt added)	146	3.5
	144	6.2

Courtesy of USDA (2000), non-copyright material.

drastic than in yolk. The loss of fluidity makes the gelled product hard to use and hard to mix with ingredients and produces an undesirable appearance. Heating thawed yolk at 45–55 °C for 1 h partially reverses this gelation (Palmer et al., 1970). It is thought that LDL is the primary egg yolk component altered by freezing (Wakamatu et al., 1983). Gelation is easily controlled by adding 10% salt or sugar to egg yolk.

19.8.1 Alternative methods of pasteurization

Other methods of pasteurization may be approved when they give equivalent effects to those specified above.

Ultra-heat treatment (UHT) requires the use of high temperature for a short time. In the case of liquid eggs, this type of pasteurization requires more attention, as egg proteins are more susceptible to temperature denaturation; usually the liquid eggs are heated at temperatures of 70 °C for 1.5 min and then packed in aseptic packages, which increase the shelf life for up to 24 weeks, although the final product still needs to be refrigerated (Cunningham, 1995; Zeidler, 2002b).

Another alternative method is the lactic acid-aluminum sulfate method, which allows pasteurization at temperatures similar to whole egg (62 °C for 3.5–4 min). The maximum stability of most egg white proteins occurs at near neutral pH, with the exception of ovotransferrin (Nakamura & Omori, 1979). To overcome this, aluminum salts such as aluminum sulfate solution are added prior to treatment to stabilize the protein, by forming complexes that are more heat stable than the protein itself. The addition of aluminum salt should be made slowly, with rapid stirring of the whites, to avoid protein coagulation by local high concentrations of acid and aluminum. The pH of the egg white before pasteurization should be 6.6–7.0. This method produces low foaming products so a foaming or whipping aid may be incorporated into the stabilizing solution (Cunningham, 1995).

Another method used is the heat plus hydrogen peroxide method that was developed by Armour and Company. This process combines heat treatment and hydroxide peroxide. Hydrogen peroxide is used as a bactericidal agent, and the egg white can be pasteurized at its normal pH (9.0) at lower temperatures. Egg white is first heated to around 52–53 °C and held for 1.5 min to inactivate natural catalase (thereby eliminating the problem of excessive foam formation); hydrogen peroxide is then added to a final concentration of 0.075–0.1%, and allowed to react for 2 min at the elevated temperature. The egg whites are cooled to 7 °C and catalase is added to remove residual hydroxide peroxide. The advantage of this method is that egg whites can be pasteurized at relatively lower temperatures and therefore the foaming properties of egg white are not affected (Cunningham, 1995). This method is widely used for pasteurization of egg whites.

The heat plus vacuum process is another method used for pasteurization. The use of a vacuum to remove air from egg whites achieves the same microbiological results at lower temperature. The system uses a typical HTST plate pasteurizer equipped with a vacuum chamber. Egg whites are first vacuumed for 17–20 min to eliminate air, and then heated to 57 °C for 3.5 min (Cunningham, 1995).

Non-thermal preservation processes, such as irradiation, high-pressure processing (HPP) and pulsed electric field (PEF) processing, are claimed to result in better quality retention, energy efficiency, and/or longer shelf life, compared to traditional heat processing (Cullen et al., 2012). Proctor et al. (1953) used irradiation for the first time in liquid whole egg and reported that the use of a high voltage of cathode rays could reduce viable *Salmonella*. Irradiation produces free radicals, which increase the oxidation of polyunsaturated fatty acids and cholesterol, change color, and destroy carotenoids in dehydrated egg products (Du & Ahn, 2000). Irradiation has been used in the treatment of liquid egg and can be used in frozen egg products due to its penetration power. Consumer acceptance of irradiated foods primarily dictates the future use of this process in the pasteurization of egg products (Min et al., 2005; Sheldon, 2005).

The PEF technique uses short, high-voltage pulses to break the cell membranes of vegetative microorganisms in liquid media by expanding existing pores (electroporation) or creating new ones (Cullen et al., 2012). The membranes of PEF-treated cells become permeable to small molecules; permeation causes swelling and eventual rupture of the cell membrane. During processing, the product is subjected to the application of short, high-intensity electric fields (in the order of 20–80 kV) by passing through a chamber that consists of two electrodes that deliver pulses for a few milliseconds. Fresh liquid whole egg can have a shelf life of 28 day at 4–6 °C when it has been treated with 10 pulses (2 microseconds per pulse), but changes in viscosity and color were observed (Qin et al., 1995). One of the main advantages of using PEF is that the temperature is not higher than 55 °C (Sheldon, 2005).

High-pressure processing (HPP) treatment generally maintains better taste, appearance, and texture as well as nutrition of the food product (Cruz-Romero et al., 2004). The effect of HHP on eggs was first studied by Bridgman (1914) who observed egg albumen coagulation while pressurized at 600 MPa. Other changes in egg components have been reported, such as improvement of the foaming capacity of egg white due to exposure of SH groups that favors foaming stability (van der Plancken et al., 2007; Yang et al., 2010). HPP reduced *Salmonella enteritidis* when applied to liquid whole egg at 350 MPa and 50 °C at 2-min pulses for four cycles, which indicates that it can be used as a pasteurization method (Bari et al., 2008). Its potential in processing egg products has been extensively reviewed (Juliano et al., 2012). Coagulation of egg albumen can be avoided by adding

7–10% NaCl or sucrose even at pressures of 800 MPa (Iametti et al., 1999).

19.9 Desugarization

Demand for dried egg products for military use was high at the outbreak of World War II. During the 1930s, the US was unable to produce dried egg albumen with satisfactory storage stability and functionality. It was known that the Chinese process involved fermentation of the albumen prior to drying (Sebring, 1995) and that removal of glucose from egg liquids is essential to prevent the Maillard reaction between glucose and protein in producing undesirable color and the glucose-cephalin reaction (a reaction between a cephalin amino group and aldehydes of glucose), responsible for off-flavor development (Sebring, 1995).

The traditional Chinese process uses spontaneous microbial fermentation but is not acceptable due to the health hazard resulting from the growth of pathogenic bacteria. Therefore, controlled bacterial fermentation is preferred to eliminate pathogens and also to reduce the time required for fermentation. A number of strains, including acid-producing organisms such as *Lactobacillus* species, *Streptococcus diacetilactis*, *Klebsiella pneumoniae*, etc., are used for removal of sugar. This method is widely accepted for desugarization of egg whites. The albumen is first pasteurized and then acidified to pH 7.0–7.5 with food-grade citric or lactic acid. The liquid is then inoculated with the proper culture and held at 30–33 °C to prepare inocula. The culture is then frozen for future use as inocula in the fermentation of large batches of egg white. The level of inoculum is usually 10–15% of the total batch weight (Sebring, 1995). Yeast fermentation is also used to desugar whole egg, egg white, and egg yolk. The pH of the liquid egg is adjusted to the range 6.0–7.0, by addition of dilute hydrochloric acid, and controlled fermentation is maintained by adding food-grade baker's yeast (*Saccharomyces cerevisiae*) (FDA, 2012a). Fermentation of whole egg with 0.2–0.4% by weight of moist baker's yeast at 22–23 °C depleted the sugar within 2–4 h (Sebring, 1995).

The glucose oxidase-catalase enzyme system is used exclusively for desugarization of whole egg, other yolk-containing egg products, and egg white. The reaction can be carried out at an elevated temperature of 30–33 °C or at a low temperature of 10 °C. The lower temperature requires a longer reaction time. The optimum pH for glucose oxidase is 6.0; therefore adjustment of pH is not necessary in yolk while pH adjustment with citric or lactic acid may be required for egg white or whole egg. The dose of enzyme is determined by the rate of reaction desired, temperature of egg, activity of enzyme, and the amount of glucose to be removed. Considerable caution is required in adding hydrogen peroxide to the egg because of foaming from evolved oxygen (Stadelman & Cotterill, 1995).

19.10 Dehydration

Dried egg products are used for the production of bakery goods, mayonnaise, salad dressing, pasta, etc. Use of dried eggs facilitates storage and transportation, as they are easy to handle and to formulate. The drying process is mainly performed by spray drying, pan drying, or belt drying. Concentration of liquid egg products before drying is a means of improving thermal efficiency, increasing the capacity of a dryer, and changing product characteristics such as lighter bulk density; liquid eggs can be concentrated using vacuum-type evaporation or ultrafiltration. Spray drying is the most common method used. In this method the liquid egg is atomized by high-pressure nozzles (500–6000 psi) into a stream of hot air for instant removal of water. Flowability can be achieved by adding a free-flowing agent such as sodium silicoaluminate or silicon dioxide, at levels of 2% to 1%, respectively (Bergquist, 1995). The finished dried product must contain not less than 95% total egg solids by weight (FDA, 2012a). The air used for drying is filtered to remove undesirable dust and other contaminants and then is heated to an inlet temperature of 121–232 °C; the hot air is then moved to the spray-drying chamber by an inlet fan. The powder separates from the drying air in the drying chamber and also in a separating device; the air is then removed from the system by an exhaust fan. The dried product removed from the dryer is sometimes cooled and is usually sifted before packaging.

Pan driers are still used for making egg white. Drying on pans to a moisture level of 12–16% will produce a flake-type product with dimensions of 1.5–12.5 mm; material finer than this is sometimes called granular egg white. Pan-dried egg white can also be milled to a fine powder (Bergquist, 1995; Zeidler, 2002b). Belt drying is used in China for making dried whole egg and yolk. The liquid egg is spread as a thin film on a continuous aluminum belt moving through a hot air stream. Another form of belt drying, called "fluff" or "foam" drying, is when the product is whipped into a stable foam and spread in a thin layer on a continuous moving belt passing through heated air. Foam spray drying, a method used for dairy products, has also been used for drying egg products. Egg products produced by this method have a lower bulk density and different particle characteristics than regular spray-dried products (Bergquist, 1995; Zeidler, 2002b).

Heat treatment of dried whites is an approved method for pasteurization (FDA, 2012d). The storage of dried egg white at elevated temperatures has been shown to be an effective means of pasteurization (Cunningham, 1995; Zeidler, 2002b). Albumen is not damaged by storage at 54 °C for as long as 60 days. The moisture content of dried egg white is important and should be around 6.5–8.0% (Froning, 2008). The product should be held in the heat treatment room in a closed container and must be spaced to assure adequate heat penetration and air circulation (FDA, 2012d). The minimum requirements for heat treatment of spray- or pan-dried albumen are as follows:

- spray-dried albumen must be heated throughout to a temperature not less than 130 °F and held continuously at such temperature for not less than 7 days and until it is salmonella negative
- pan-dried albumen must be heated throughout to a temperature of not less than 125 °F and held continuously at such temperature for not less than 5 days and until it is salmonella negative. Mine (1995) reported that this process improves whipping properties of egg white.

19.11 Egg further processing (value-added processing)

The presence of many bioactive components in eggs opens new windows for value-added processing of eggs (Hatta et al., 2008; Huopalahti et al., 2007; Kovacs-Nolan et al., 2005a; Mine, 2007; Seko et al., 1997; Zeisel, 1992). Lysozyme is produced at commercial scale using cation exchange chromatography for various applications. Lysozyme is effective against gram-positive bacteria such as *Bacillus stearothermophilus*, *Clostridium tyrobutyricum*, and *Clostridium thermosaccharolyticum* (Losso et al., 2000). It is estimated that over 100 tonnes of lysozyme are used each year for these purposes (Scott et al., 1987). Lysozyme is a generally recognized as safe (GRAS) protein that has been approved for use in cheese making to prevent the growth of *Clostridium tyrobutyricum*, which causes off-odors and late "blowing" (unwanted fermentation) in some cheeses (Cunningham et al., 1991; Proctor & Cunningham, 1988), in wine and beer production to control lactic acid bacteria such as *Lactobacillus* spp., which is a cause of spoilage (Daeschel et al., 1999; Gerbaux et al., 1997), in meat products such as sausage, salami, pork, beef, or turkey (Bolder, 1997; Hughey et al., 1989), and in oral health care products such as toothpaste, mouthwash, and chewing gum (Sava, 1996; Tenuovo, 2002).

Avidin can be easily recovered as a co-product in lysozyme separation (Durance & Nakai, 1988). Avidin is the best known for its high biotin-binding ability and shows a bacteriostatic effect on bacteria that require biotin (Charter & Lagarde, 2004; Korpela et al., 1984). Ovotransferrin acts as an antimicrobial agent toward bacteria species such as *Pseudomonas* spp., *Escherichia coli*, and *Streptococcus mutans* (Valenti et al., 1983). This activity is largely bacteriostatic, being reversed by the addition of ferric ions; if the protein is saturated with iron, the bacteriostatic effect it has on gram-negative bacteria is overcome (Florence & Rehault, 2007).

Egg yolk antibodies may find wide applications, as previously reviewed (Kovaks-Nolan et al., 2004; Sunwoo et al., 2006). Immunoglobulin Y (IgY), with a molecular weight of 21–25 KDa, is a major antibody in birds. Antibodies combat the infectivity and toxicity of pathogenic antigens by recognizing and binding with the antigen in a specific manner; therefore they are able to enhance the immune system. IgY is transferred from the hen to the embryo through a receptor localized on the surface of the yolk membrane; egg yolk contains about 100 mg IgY/yolk (Hodek & Stiborova, 2003). IgY can be extracted by diluting egg yolk 10 times with water, followed by centrifugation, then sodium sulfate is added to the supernatant and the resulting precipitate is purified by ultrafiltration, alcohol precipitation or salt precipitation. Further precipitation can be obtained by gel filtration or anion exchange chromatography (Ko & Ahn, 2007). IgY can be used in the following immunotherapeutic applications: inhibition and/or prevention of *E. coli*, *Salmonella* spp., *Streptoccocus mutans*, *Helicobacter pylori*, and Crohn's disease, among others (Kovacs-Nolan et al., 2005b). Wen et al. (2012) studied the preparation of IgY against specific influenza B virus with positive results in trials with mice. In addition, IgY can be used as a food ingredient in functional food products such as yogurt or as a bactericidal agent, in sport drinks, and in pharmaceutical products like acne medication (Li-Chan et al., 1995).

Egg proteins have shown great promise as a rich source of various bioactive peptides that may improve human health and prevent diseases. Peptides with antimicrobial, antihypertensive, antioxidative, and immunomodulatory activity have been reported (During et al., 1999; Ibrahim et al., 2000; Lee et al., 2006; Majumder & Wu, 2010; Shen et al., 2010). The identification of these peptides represents a great commercialization opportunity since they have potential as health-promoting food ingredients in different areas, such as prevention and treatment of microbial infections of the gastrointestinal tract, control of microbiological quality of foods and feeds, and extension of shelf life of foods (Korhonen & Rokka, 2011).

19.12 Sustainability

In the 1980s, the major challenge of the egg industry was probably the public perception of the presence of cholesterol in egg yolk and its controversially associated risk of heart diseases. After four decades, the challenges facing the industry are more complicated; debate on the issue of cholesterol continues, and the increasing cost of feed in competition with the use of renewable agriculture products to replace increasingly expensive fossil-based oils and chemicals, concerns about animal welfare using the conventional cage system, globalization, etc. will affect the sustainability of the egg industry. The EU banned the conventional cage system in January 2012, mainly due to animal welfare concerns, while this is the most common commercial housing system for egg-laying hens in the US and other parts of the world. Adoption of a non-cage system will have substantial effects on the cost of production, as well as increased risk for bird-to-bird transmission and internal egg contamination of *Salmonella* and *Campylobacter* (Mench et al., 2011). Sustainability of the egg industry will depend on the community, the industry, the government, international organizations, researchers, and consumers working collectively to address issues of animal welfare, cost of production, environment, food safety, animal health, and human health.

19.13 Conclusion

As the primary animal protein in many parts of the world, egg products will continue to be an important part of our daily diets; new technologies and new methods of egg processing such as non-thermal processing will see applications in the egg industry to improve nutrition, safety, shelf-life and taste of egg products, or to create new egg products. New applications of egg products as functional foods and nutraceuticals and other non-food uses are expected to grow. The egg industry is, however, at a time of rapid change; the industry needs to understand how the change in laying hens practice (removal of conventional cages, etc.) might affect egg safety and egg processing.

Acknowledgments

JW is supported by the Natural Science and Engineering Research Council of Canada (NSERC). The author would like to thank Alexandra Acero-Lopez and Marina Offengenden for their support in preparation of the chapter.

References

Anton M (2007) Composition and structure of hen egg yolk. In: Huopalahti R, López-Fandiño R, Anton M, Schade R (eds) *Bioactive Egg Compounds*. Berlin: Springer, pp. 1–6.

Arvanitoyannis IS, Varzakas TH, Papadopoulos K, Egg D (2009) Eggs. In: Arvanitoyannis IS (ed) *HACCP and ISO 22000: Application to Foods of Animal Origin*. Ames, IA: Wiley-Blackwell, pp. 309–359.

Baker RC, Bruce C (1995) Development of value-added products. In: Stadelman WJ, Cotterill OJ (eds) *Egg Science and Technology*, 4th edn. New York: Food Products Press, pp. 499–521.

Bari ML, Ukuko DO, Mori M, Kawamoto S, Yamamoto K (2008) Effect of hydrostatic pressure pulsing on the inactivation of Salmonella enteritidis in liquid whole egg. *Foodborne Pathogens Disease* 5: 175–182.

Bell D (2002) Formation of the egg. In: Bell D, Weaver WD (eds) *Commercial Chicken Meat and Egg Production*, 5th edn. New York: Springer, pp. 64–69.

Bell D, Weaver WD (eds) (2002) *Commercial Chicken Meat and Egg Production*, 5th edn. New York: Springer.

Belyavin CG, Boorman KN (1980) The influence of the cuticle on egg-shell strength. *British Poultry Science* 21: 295–298.

Benhamou A, Caubet JC, Eigenmann P, Marcos C, Reche M, Urisu,A (2010) State of the art and new horizons in the diagnosis and management of egg allergy. *Allergy* 65: 283–289.

Bergquist DH (1995) Egg dehydration. In: Stadelman WJ, Cotterill OJ (eds) *Egg Science and Technology*, 4th edn. New York: Food Products Press, pp. 335–371.

Board RG, Hall NA (1973) The cuticle: a barrier to liquid and particle penetration of the shell of the hen's egg. *British Poultry Science* 14: 69–97.

Bolder NM (1997) Decontamination of meat and poultry carcasses. *Trends in Food Science and Technology* 8: 221–227.

Braden CR (2006) *Salmonella enterica* serotype Enteritidis and eggs: a national epidemic in the United States. *Clinical Infectious Diseases* 43: 512–517.

Bridgman PW (1914) The coagulation of albumen by pressure. *Journal of Biological Chemistry* 19: 511–512.

Burley RW, Vadehra DV (eds) (1989) *The Avian Egg: Chemistry and Biology*. New York: Wiley.

Centers for Diseases Control and Prevention (CDC) (2010) *Investigation Update: Multistate Outbreak of Human Salmonella Enteritidis Infections Associated with Shell Eggs (December 2, 2010, final update)*. www.cdc.gov/salmonella/enteritidis/, accessed 22 November 2013.

Charter EA, Lagarde G (2004) Natural antimicrobial systems/lysozyme and other proteins in eggs. In: Robinson RK (ed) *Encyclopedia of Food Microbiology*. San Diego, CA: Academic Press, pp. 1582–1587.

Cotterill OJ, Geiger GS (1977) Egg product yield trends from shell eggs. *Poultry Sciences* 56: 1027–1031.

Cotterill OJ, McBee L (1995) Egg breaking. In: In: Stadelman WJ, Cotterill OJ (eds) *Egg Science and Technology*, 4th edn. New York: Food Products Press, pp. 231–260.

Cruz-Romero M, Smiddy M, Hill C, Kerry JP, Kelly AL (2004) Effects of high pressure treatment on physicochemical characteristics of fresh oysters (Crassostrea gigas). *Innovative Food Science and Emerging Technologies* 5: 161–169.

Cullen PJ, Tiwari BJ, Valdramidis V (eds) (2012) *Novel Thermal and Non-Thermal Technologies for Fluid Foods*. San Diego, CA: Academic Press.

Cunningham FE (1995) Egg product pasteurization. In: Stadelman WJ, Cotterill OJ (eds) *Egg Science and Technology*, 4th edn. New York: Food Products Press, pp. 289–316.

Cunningham FE, Proctor VA, Goetsch SJ (1991) Egg-white lysozyme as a food preservative: an overview. *World's Poultry Science Journal* 47: 141–163.

Daeschel MA, Bruslind L, Clawson J (1999) Application of the enzyme lysozyme in brewing. *Master Brewers Association of the America Technical Quarterly* 36: 219–222.

Du M, Ahn DU (2000) Effect of antioxidants and packaging on lipid and cholesterol oxidation and color changes of irradiated egg yolk powder. *Journal of Food Science* 65: 625–629.

Durance TD, Nakai S (1988) Purification of avidin by cation exchange, gel filtration, metal chelate interaction and hydrophobic interaction chromatography. *Canadian Institute of Food Science and Technology Journal* 21: 279–286.

During K, Porsch P, Mahn A, Brinkmann O, Gieffers W (1999) The non-enzymatic microbicidal activity of lysozymes. *FEBS Letters* 449: 93–100.

Elkin RG (2006) Reducing shell egg cholesterol content. I. Overview, genetic approaches, and nutritional strategies. *World's Poultry Science Journal* 62: 665–687.

European Commission (EC) (2008) *Commission Regulation (EC) No 589/2008 of 23 June 2008 laying down detailed rules for implementing Council Regulation (EC) No 1237/2007 as regards marketing standards for eggs. Official Journal of the European Union* L163/6:6.

European Food Safety Authority (EFSA) (2007) The community summary report on trends and sources of zoonoses, zoonotic agents, antimicrobial resistance and foodborne outbreaks in the European Union in 2006. *EFSA J* 130: 34–117.

Fisinin VI, Papazyan TT, Surai PF (2008) Producing specialist poultry products to meet human nutrition requirements: selenium enriched eggs. *World's Poultry Science Journal* 64: 85–98.

Florence B, Rehault S (2007) Compounds with antimicrobial activity. In: Huopalahti R, López-Fandiño R, Anton M, Schade R (eds) *Bioactive Egg Compounds*. Berlin: Springer, pp. 191–198.

Food and Drug Administration (FDA) (2009) *Code of Federal Regulations, Title 21, Parts 16 and 118. Federal Register Final Rule: Guidance for Industry. Prevention of Salmonella Enteritidis in Shell Eggs During Production, Storage and Transportation.* www.fda.gov/downloads/Food/GuidanceComplianceRegulatoryInformation/GuidanceDocuments/FoodSafety/UCM285137.pdf, accessed 22 November 2013.

Food and Drug Administration (FDA) (2012a) *CFR – Code of Federal Regulations Title 21 Part 160: Eggs and Egg Products.* www.accessdata.fda.gov/scripts/cdrh/cfdocs/cfcfr/CFRSearch.cfm?CFRPart=160&showFR=1, accessed 22 November 2013.

Food and Drug Administration (FDA) (2012b) *CFR – Code of Federal Regulations Title 9 Part 590.570: Pasteurization of Liquid Eggs.* www.gpo.gov/fdsys/pkg/CFR–2012–title9–vol2/pdf/CFR–2012–title9–vol2–sec590–570.pdf, accessed 22 November 2013.

Food and Drug Administration (FDA) (2012c) *CFR – Code of Federal Regulations Title 9 Part 590.530- Liquid Egg Cooling.* www.gpo.gov/fdsys/pkg/CFR–2012–title9–vol2/pdf/CFR–2012–title9–vol2–sec590–530.pdf, accessed 22 November 2013.

Food and Drug Administration (FDA) (2012d) *CFR – Code of Federal Regulations Title 9 Part 590.575- Heat Treatment of Dried Whites.* www.gpo.gov/fdsys/pkg/CFR–2010–title9–vol2/pdf/CFR–2010–title9–vol2–sec590–575.pdf, accessed 22 November 2013.

Food Safety and Inspection Service (FSIS) (2000) Irradiation in the production, processing and handling of food – shell eggs, fresh; safe use of ionizing radiation for salmonella reduction. *Federal Registry* 65: 45280–45282.

Food Safety and Inspection Service (FSIS) (2011) *Egg Products and Food Safety.*

Food Safety and Inspection Service (FSIS) (2013) *Egg Products Inspection Act.*

Fraeye I, Bruneel C, Lemahieu C, Buyse J, Muylaert K, Foubert I (2012) Dietary enrichment of eggs with omega-3 fatty acids: a review. *Food Research International* 48: 961–969.

Froning GN (2008) Egg products industry and future perspectives. In: Mine Y (ed) *Egg Bioscience and Biotechnology*. New Jersey: John Wiley, pp. 307–325.

Galiş AM, Marcq C, Marlier D et al. (2013) Control of *Salmonella* contamination of shell eggs – preharvest and postharvest methods: a review. *Comprehensive Reviews in Food Science and Food Safety* 12: 155–182.

Gantois I, Ducatelle R, Pasmans F et al. (2009) Mechanisms of egg contamination by *Salmonella* enteritidis. *FEMS Microbiology Reviews* 33: 718–738.

Gerbaux V, Villa A, Monamy C, Bertrand A (1997) Use of lysozyme to inhibit malolactic fermentation and to stabilize wine after malolactic fermentation. *American Journal of Enology and Viticulture* 48: 49–54.

Hargis PS (1988) Modifying egg yolk cholesterol in the domestic fowl: a review. *World's Poultry Science Journal* 44: 17–29.

Hatta H, Kapoor M, Juneja LR (2008) Bioactive components in egg yolk. In: Mine Y (ed) *Egg Bioscience and Biotechnology*. New Jersey: John Wiley, pp. 185–206.

Herron KL, Fernandez ML (2004) Are the current dietary guidelines regarding egg consumption appropriate? *Journal of Nutrition* 134: 187–190.

Hincke MT, Nys Y, Gautron J, Mann K, Rodriguez-Navarro AB, McKee MD (2012) The eggshell: structure, composition and mineralization. *Frontiers in Bioscience* 17: 1266–1280.

Hodek P, Stiborova M (2003) Chicken antibodies – superior alternative for conventional immunoglobulins. *Proceedings of the Indian National Science Academy* B69: 461–468.

Hu FB, Stampfer MJ, Rimm EB et al. (1999) A prospective study of egg consumption and risk of cardiovascular disease in men and women. *Journal of the American Medical Association* 281: 1387–1394.

Hu FB, Manson JE, Willett, WC (2001) Types of dietary fat and risk of coronary heart disease: a critical review. *Journal of the Americal College of Nutrition* 20: 5–19.

Hughey VL, Wilger PA, Johnson EA (1989) Antibacterial activity of hen white lysozyme against Listeria monocytogens Scott A in foods. *Applied and Environmental Microbiology* 55: 631–638.

Humphrey TJ (1994) Contamination of egg shell and contents with Salmonella enteritidis: a review. *International Journal of Food Microbiology* 21: 31–40.

Huntington JA, Stein PE (2001) Structure and properties of ovalbumin. *Journal of Chromatography B: Biomedical Sciences and Applications* 756: 189–198.

Huopalahti R, López-Fandiño R, Anton M, Schade R (eds) (2007) *Bioactive Egg Compounds*. Berlin: Springer.

Iametti S, Donnizzelli E, Pittia P, Rovere P, Squarcina N, Bonomi F (1999) Characterization of high-pressure-treated egg albumen. *Journal of Agricultural and Food Chemistry* 47: 3611–3616.

Ibrahim HR (2000) Ovotransferrin. In: Naidu AS (ed) *Natural Food Antimicrobial Systems*. Boca Raton, FL: CRC Press, pp. 211–226.

Jones PJH (2009) Dietary cholesterol and the risk of cardiovascular disease in patients: a review of the Harvard Egg Study and other data. *International Journal of Clinical Practice Supplement* 63: 1–8.

Juliano P, Bilbao-Sáinz C, Koutchma T et al. (2012) Shelf-stable egg-based products processed by high pressure thermal sterilization. *Food Engineering Review* 4: 55–67.

Kato I, Schrode J, Kohr WJ, Laskowski Jr M (1987) Chicken ovomucoid: determination of its amino acid sequence, determination of the trypsin reactive site, and preparation of all three of its domains. *Biochemistry* 26: 193–201.

Ko KY, Ahn DU (2007) Preparation of immunoglobulin Y from egg yolk using ammonium sulfate precipitation and ion exchange chromatography. *Journal of Poultry Science* 86: 400–407.

Korhonen HJ, Rokka S (2011) Properties and applications of antimicrobial proteins and peptides from milk and egg. In: Hettiarachchy A, Sato NS, Marshall K, Kannan MR (eds) *Bioactive Food Proteins and Peptides: Applications in Human Health*. Boca Raton, FL: CRC Press, pp. 49–80.

Korpela J, Salonen EM, Kuusela P, Sarvas M, Vaheri A (1984) Binding of avidin to bacteria and to the outer membrane porin of Escherichia coli. *FEMS Microbiology Letters* 22: 3–10.

Kovacs-Nolan J, Phillips M, Mine Y (2005a) Advances in the value of eggs and egg components for human health. *Journal of Agricultural and Food Chemistry* 53: 8421–8431.

Kovacs-Nolan J, Mine Y, Hatta H (2005b) Avian immunoglobulin Y and its application in human health and disease. In: Mine Y, Shahidi F (eds) *Nutraceutical Proteins and Peptides in Health and Disease*. Boca Raton, FL: CRC Press, pp. 161–179.

Lee A, Griffin B (2006) Dietary cholesterol, eggs and coronary heart disease risk in perspective. *Nutrition Bulletin* 31: 21–27.

Lee NY, Cheng JT, Enomoto T, Nakano Y (2006) One peptide derived from hen ovotransferrin as pro-drug to inhibit angiotensin converting enzyme. *Journal of Food and Drug Analysis* 14: 31–35.

Lesson S, Caston LJ (2004) Enrichment of eggs with lutein. *Poultry Science* 83: 1709–1712.

Li-Chan E, Kim HO (2008) Structure and chemical composition of eggs. In: Mine Y (ed) *Egg Bioscience and Biotechnology*. New Jersey: John Wiley, pp. 1–65.

Li-Chan E, Powrie WD, Nakai S (1995) The chemistry of eggs and egg products. In: Stadelman WJ, Cotterill OJ (eds) *Egg Science and Technology*, 4th edn. New York: Food Products Press, pp. 109–160.

López-Expósito I, Chicon R, Belloque J et al (2008) Changes in the ovalbumin proteolysis profile by high pressure and its effect on IgG and IgE binding. *Journal of Agricultural and Food Chemistry* 56: 11809–11816.

Losso JN, Nakai S, Charter EA (2000) Lysozyme. In: Naidu AS (ed) *Natural Food Antimicrobial Systems*. New York: CRC Press, pp. 185–210.

Majumder K, Wu J (2010) A new approach for identification of novel antihypertensive peptides from egg proteins by QSAR and bioinformatics. *Food Research International* 43: 1371–1378.

Majumder K, Panahi S, Kaufman S, Wu J (2013) Fried egg digest decreases blood pressure in spontaneous hypertensive rats. *Journal of Functional Foods* 5: 187–194.

Mench JA, Summer DA, Rose-Molina JT (2011) Sustainability of egg production in the United States – the policy and market context. *Poultry Science* 90: 229–240.

Min BR, Nam KC, Lee EJ, Ko GY, Trampel DW, Ahn DU (2005) Effect of irradiating shell eggs on quality attributes and functional properties of yolk and white. *Poultry Science* 84: 1791–1796.

Mine Y (1995) Recent advances in the understanding of egg white protein functionality. *Trends in Food Science and Technology* 6: 225–232.

Mine Y (2002) Recent advances in egg protein functionality in the food system. *World's Poultry Science Journal* 58: 31–39.

Mine Y (2007) Egg proteins and peptides in human health – chemistry, bioactivity and production. *Current Pharmaceutical Design* 13: 875–884.

Mountney GJ, Parkhurst CR (eds) (1995) *Poultry Product Technology*, 3rd edn. New York: Food Product Press, pp. 353–383.

Naber EC (1990) Cholesterol content of eggs: can and should it be changed? *Feedstuffs* 62: 1, 47, 50–2.

Nakamura R, Omori I (1979) Protective effect of some anions on the heat-denaturation of ovotransferrin. *Agriculture and Biological Chemistry* 43: 2393–2394.

Nimalaratne C, Lopes-Lutz D, Schieber A, Wu J (2011) Free aromatic amino acids in egg yolk show antioxidant properties. *Food Chemistry* 129: 155–161.

Nys Y, Gautron J, Garcia-Ruiz JM, Hincke MT (2004) Avian eggshell mineralization: biochemical and functional characterization of matrix proteins. *Comptes Rendus Palevol* 3: 549–562.

Okubo T, Akachi S, Hatta H (1997) Structure of hen eggs and physiology of egg laying. In: Yamamoto T, Juneja LR, Hatta H, Kim M (eds) *Hen Eggs: Their Basic and Applied Science*. Boca Raton, FL: CRC Press, pp. 1–12.

Palmer HH, Ijichi K, Roff H (1970) Partial thermal reversal of gelation in thawed egg yolk products. *Journal of Food Science* 35: 403–406.

Parson A (1982) Structure of the eggshell. *Poultry Science* 61: 2013–2021.

Proctor BE, Joslin RP, Nickerson JTR, Lockhart EE (1953) Elimination of Salmonella in whole egg powder by cathode ray irradiation of egg magma prior to drying. *Food Technology* 7: 291–296.

Proctor VA, Cunningham FE (1988) The chemistry of lysozyme and its use as a food preservative and a pharmaceutical. *Critical Reviews in Food Science and Nutrition* 26: 359–395.

Qin BL, Pothakamury UR, Vega H, Martin O, Barbosa-Canovas GV, Swanson BG (1995) Food pasteurization using high-intensity pulsed electric fields. *Food Technology* 49: 55–60.

Qureshi AI, Suri FK, Ahmed S et al. (2006) Regular egg consumption does not increase the risk of stroke and cardiovascular diseases. *Medical Science Monitor* 13: CR1–8.

Robinson DS, Monsey JB (1971) Studies on the composition of egg-white ovomucin. *Biochemical Journal* 121: 537–547.

Romanoff AL, Romanoff A (1949) *The Avian Egg*. New York: John Wiley.

Sava G (1996) Pharmacological aspects and therapeutic applications of lysozymes. *Experientia Supplementum* 75: 433–449.

Scott D, Hammer FE, Szalkucki TJ (1987) Bioconversions: enzyme technology. In: Knorr D (ed) *Food Biotechnology*. New York: Marcel Dekker, pp. 413–442.

Sebring M (1995) Desugarization of egg products. In: Stadelman WJ, Cotterill OJ (eds) *Egg Science and Technology*, 4th edn. New York: Food Products Press, pp. 323–331.

Seko A, Koketsu M, Nishizono M et al. (1997) Occurrence of sialylglycopeptide and free sialylglycans in hen's egg yolk. *Biochimica et Biophysica Acta* 1335: 23–32.

Sheldon BW (2005) Techniques for reducing pathogens in eggs. In: Mead GC (ed) *Food Safety Control in the Poultry Industry*. Cambridge: CRC Press, pp. 274–310.

Shen S, Chahal B, Majumder K, You SJ, Wu J (2010) Identification of novel antioxidative peptides derived from a thermolytic hydrolysate of ovotransferrin by LC-MS/MS. *Journal of Agricultural and Food Chemistry* 58: 7664–7672.

Sparks NHC (1994) Shell accessory materials: structure and function. In: Board RG, Fuller R (eds) *Microbiology of the Avian Egg*. London: Chapman and Hall, pp. 25–42.

Spence JD, Jenkins DJA, Davignon J (2012) Egg yolk consumption and carotid plaque. *Atherosclerosis* 224:469–473.

Stadelman WJ (1995a) Egg production practices. In: Stadelman WJ, Cotterill OJ (eds) *Egg Science and Technology*, 4th edn. New York: Food Products Press, pp. 9–37.

Stadelman WJ (1995b) Quality identification of shell eggs. In: Stadelman WJ, Cotterill OJ (eds) *Egg Science and Technology*, 4th edn. New York: Food Products Press, pp. 39–47.

Stadelman WJ, Cotterill OJ (eds) (1995) *Egg Science and Technology*, 4th edn. New York: Food Products Press.

Sugino H, Nitoda T, Juneja LR (1997) General chemical composition of hen eggs. In: Yamamoto T, Juneja LR, Hatta H, Kim M (eds) *Hen Eggs: Their Basic and Applied Science*. Boca Raton, FL: CRC Press, pp. 13–23.

Sunwoo HH, Sadeghi G, Karami H (2006) Natural antimicrobial egg yolk antibody. In: Sim JS, Sunwoo HH (eds) *The Amazing Egg: Nature's Perfect Functional Food for Health*. Edmonton: University of Alberta Publishing, pp. 295–325.

Surai PF, Sparks NHC (2001) Designer eggs: from improvement of egg composition to functional food. *Trends in Food Science and Technology* 12: 7–16.

Tenuovo J (2002) Clinical applications of antimicrobial host proteins lactoperoxidase, lysozyme and lactoferrin in xerostomia: efficacy and safety. *Oral Diseases* 8: 23–29.

Teuber SS, Beyer K, Comstock S, Wallowitz M (2006) The big eight foods: clinical and epidemiological overview. In: Maleki SJ, Burks W, Helm RM (eds) *Food Allergy*. Washington, DC: AMS Press, pp. 49–79.

US Department of Agriculture (USDA) (2000) *USDA Egg Grading Manual*. www.ams.usda.gov/AMSv1.0/getfile?dDocName=STELDEV3004502, accessed 22 November 2013.

US Department of Agriculture (USDA) (2006) *Regulations Governing the Inspection of Eggs and Egg Products (7 CFR Part 57)*. www.ams.usda.gov/AMSv1.0/getfile?dDocName=STELDEV3004691, accessed 27 November 2013.

Vaclavik VA, Christian EW (eds) (2008) *Essentials of Food Science*, 3rd edn. New York: Springer, pp. 205–235.

Valenti P, Antonini G, von Hunolstein C, Visca P, Orsi N, Antonini E (1983) Studies of the antimicrobial activity of ovotransferrin. *International Journal of Tissue Reactions* 5: 97–105.

van der Plancken I, van Loey A, Hendrickx ME (2007) Foaming properties of egg white proteins affected by heat or high pressure treatment. *Journal of Food Engineering* 78: 1410–1426.

Wakamatu T, Sato Y, Saito Y (1983) On sodium chloride action in the gelation process of low density lipoprotein (LDL) from hen egg yolk. *Journal of Food Science* 48: 507–516.

Wen J, Zhao S, He D, Yang Y, Li Y, Zhu S (2012) Preparation and characterization of egg yolk immunoglobulin Y specific to influenza B virus. *Antivirial Research* 93: 154–159.

Whittow GC (ed) (2000) *Sturkie's Avian Physiology*, 5th edn. San Diego, CA: Academic Press.

Williams J, Elleman TC, Kingston IB, Wilkins AG, Kuhn KA (1982) The primary structure of hen ovotransferrin. *European Journal of Biochemistry* 122: 297–303.

Windhorst HW (2007) Changes in the structure of global egg production. *World's Poultry Science Journal* 23: 24–25.

Wu J, Acero A (2012) Ovotransferrin: structure, function and preparation. *Food Research International* 46: 480–487.

Yamamoto T, Juneja LR, Hatta H, Kim M (eds) (1997) *Hen Eggs: Their Basic and Applied Science*. Boca Raton, FL: CRC Press.

Yang RX, Li WZ, Zhu CQ, Zhang Q (2010) Effects of ultra-high hydrostatic pressure on foaming and physical-chemistry properties of egg white. *Journal of Biomedical Science and Engineering* 2: 617–620.

Yang S, Baldwin R (1995) Functional properties of eggs in foods. In: Stadelman WJ, Cotterill OJ (eds) *Egg Science and Technology*, 4th edn. New York: Food Products Press, pp. 405–452.

Zeidler G (2002a) Shell eggs and their nutritional value. In: Bell D, Weaver WD (eds) *Commercial Chicken Meat and Egg Production*, 5th edn. New York: Springer, pp. 1109–1128.

Zeidler G (2002b) Further-processing eggs and egg products. In: Bell D, Weaver WD (eds) *Commercial Chicken Meat and Egg Production*, 5th edn. New York: Springer, pp. 1163–1197.

Zeisel SH (1992) Choline – an important nutrient in brain development, liver function and carcinogenesis. *Journal of the American College of Nutrition* 11: 473–481.

20 Fats and Oils – Plant Based

Amy S. Rasor and Susan E. Duncan

Department of Food Science and Technology, Virginia Tech, Blacksburg, Virginia, USA

20.1 Introduction

Fats and oils are water-insoluble compounds consisting of mainly triacylglycerols: three fatty acids esterified to a glycerol molecule. Products are generally called "fats" when they are solid at room temperature and "oils" when they are liquid at room temperature, although the terms are often used interchangeably. Edible fats and oils contribute to the flavor, texture, aroma, and mouthfeel of foods, while providing nutritive value. Their origin may be animal, plant, or marine.

Plant-based fats and oils are obtained from oilseeds and oil-bearing trees. While used mainly for human consumption, they also find use in animal feed, biodiesel, and industrial applications. They are integral components in a wide range of products such as margarines, shortenings, dressings, confectionery products, baked goods, snack foods, infant formulas, and non-dairy creamers.

20.2 Sources, composition, and uses of plant-based fats and oils

Oils may be obtained from hundreds of plant sources. While some are produced on a large scale and internationally traded as commodities, most are considered minor oils and find specialized use. Recent global production data for the most widely produced commodity plant oils are summarized in Table 20.1. The four predominant oils are palm, soybean, rapeseed/canola, and sunflower seed, in decreasing order of production. Within the last decade, palm has overtaken soybean as the top commodity plant oil. Total production of commodity plant oils has steadily increased over time, and further increases in production are projected (FAS, 2011).

The sources and typical fatty acid composition of selected oils are provided in Table 20.2. In general, three fatty acids dominate: palmitic (C16:0), oleic (C18:1), and linoleic (C18:2n-6). Exceptions are noted in the discussion of individual oils, below.

20.2.1 Major oils

Palm oil is produced from the fruit of the oil palm which is grown in tropical climates, mainly in Malaysia and Indonesia. It is the most widely produced plant oil and tends to be less costly than other commodity oils (Gunstone, 2008b). The crude oil is rich in antioxidants, especially carotenoids. Although most of the carotenoid content is destroyed during the refining process, a red palm oil variety is produced by reducing the distillation temperature and used as a natural colorant (O'Brien, 2004c), cooking oil, or dietary supplement. Palm oil has a high oxidative stability due to its high level of natural antioxidants and its high saturated fatty acid content. It is often fractionated to extend its range of use. Palm oil and its fractions are used in spreads, shortenings, and frying applications.

Soybean oil is produced at the highest level of the oilseeds. A high-protein meal is produced as a valuable co-product. The oil has a relatively low oxidative stability due to its high content of linolenic acid (C18:3) and is often subjected to hydrogenation after refining to improve its oxidative stability. Soybean oil is used in a wide range of products, such as margarine, shortening, salad dressing, mayonnaise, and snack foods. Soybean oil is the most commonly consumed oil in the US.

Traditional rapeseed oil is naturally high in erucic acid (C22:1). Due to health concerns, a low-erucic, low-glucosinolate variety was developed in Canada, known today as canola oil or LEAR (low erucic acid rapeseed).

Food Processing: Principles and Applications, Second Edition. Edited by Stephanie Clark, Stephanie Jung, and Buddhi Lamsal.
© 2014 John Wiley & Sons, Ltd. Published 2014 by John Wiley & Sons, Ltd.

Table 20.1 World production of commodity plant oils, 2006–7 to 2010–11, in million metric tonnes

Oil	2006–7	2007–8	2008–9	2009–10[*]	2010–11[†]
Coconut	3.22	3.53	3.53	3.62	3.68
Cottonseed	5.13	5.22	4.78	4.65	4.98
Olive	2.83	2.78	2.78	3.05	3.01
Palm	37.33	41.08	43.99	45.86	47.97
Palm Kernel	4.44	4.88	5.17	5.50	5.65
Peanut	4.53	4.91	5.00	4.67	4.98
Rapeseed	17.13	18.43	20.49	22.35	22.65
Soybean	36.45	37.72	35.74	38.76	42.13
Sunflower	10.70	10.03	11.99	11.63	11.33
Total	121.75	128.57	133.48	140.08	146.37

[*] Preliminary data.
[†] Forecast data.
Source: FAS (2011).

In the US, canola oil received Generally Recognized As Safe (GRAS) status in 1985, although its use is not permitted in infant formulas. Canola is viewed as a "healthy" oil, as it contains the lowest level of saturated fatty acids of the commodity oils (Gunstone, 2008b), it is a good source of vitamin E due to its high α-tocopherol content, and it has a favorable ratio of omega-6 to omega-3 fatty acids. Canola oil is unique in that it contains a relatively high level of brassicasterol, a phytosterol not found at a significant level in other oilseeds (Gunstone, 2006c). Canola oil is used for cooking and is often a component in salad dressing, mayonnaise, margarine, and shortening.

Traditional sunflower oil contains a high level of linoleic acid (C18:2n-6). Various modification techniques have been used to increase oxidative stability (see section 20.9 for further discussion). High-oleic and mid-oleic varieties were introduced on a commercial scale in the 1980s and 1990s, respectively (Gupta, 2002). The mid-oleic variety (NuSun®; National Sunflower Association, Mandan, ND) is lower in saturated fat content than traditional sunflower oil and has become the most highly produced variety in the US. With a high level of α-tocopherol, sunflower oil is a rich source of vitamin E. Crude sunflower oil contains natural waxes, which may cloud the appearance of the oil during storage, and is typically subjected to a dewaxing procedure.

20.2.2 Lauric oils

Palm kernel oil and coconut oil are obtained from the oil palm and the coconut palm, respectively. Whereas palm oil is derived from the fruit of the oil palm, palm kernel oil is extracted from the kernel. Palm kernel and coconut oils are classified as lauric oils due to their high levels of lauric acid (C12:0) (see Table 20.2). This is unique, as no other commodity plant oil contains more than 1% lauric acid (Pantzaris & Basiron, 2002). Unlike other commodity oils, the lauric oils (particularly coconut oil) contain mainly saturated fatty acids, making them highly resistant to oxidation. Common food applications of the lauric oils include spreads, candies, and non-dairy creamers.

Lauric oils possess melting behavior similar to that of cocoa butter (Pantzaris & Basiron, 2002) – a sharp curve near body temperature – making them suitable for use as cocoa butter alternatives. There are three main types of cocoa butter alternatives, which impart similar snap and gloss to confectionery products and are utilized mainly for economic reasons.

• *Cocoa butter equivalents* (CBE) are typically made from non-hydrogenated fats with similar triacylglycerol composition to that of cocoa butter (e.g. fractionated palm oil). Like chocolate, tempering of CBE is required to obtain the desired crystallization behavior. CBE are often incorporated in products with cocoa butter.
• *Cocoa butter replacers* (CBR) are typically made from hydrogenated, fractionated oils (e.g. cottonseed and soybean). No tempering is required. CBR may be incorporated in products containing cocoa butter, but in a lower proportion versus CBE.
• *Cocoa butter substitutes* (CBS) are mainly produced from hydrogenated palm kernel fractions and hydrogenated coconut oil. Like CBR, no tempering is required. CBS are not used in conjunction with cocoa butter due to their differences in triacylglycerol composition.

Table 20.2 Sources and typical fatty acid composition of selected oils

Oil	Source	Fatty acid composition (% total fatty acids)													
		<C12:0	C12:0	C14:0	C16:0	C16:1	C18:0	C18:1	C18:2n-6	C18:3	C20:0	C20:1	C22:0	C22:1	C24:0
Coconut	*Cocos nucifera* L.	9.6–18.7	45.1–53.2	16.8–21.0	7.5–0.12	ND	2.0–4.0	5.0–10.0	1.0–2.5	ND–0.2	ND–0.2	ND–0.2	ND	ND	ND
Cottonseed	*Gossypium* sp.	ND	ND–0.2	0.6–1.0	21.4–26.4	ND–1.2	2.1–3.3	14.7–21.7	46.7–58.2	ND–0.4	0.2–0.5	ND–0.1	ND–0.6	ND–0.3	ND–0.1
Corn	*Zea mays* L.	ND	ND–0.3	ND–0.3	8.6–16.5	ND–0.5	ND–3.3	20.0–42.2	34.0–65.6	ND–2.0	0.3–1.0	0.2–0.6	ND–0.5	ND–0.3	ND–0.5
Olive	*Olea europaea* L.	ND	ND	0.0–0.05	7.5–20.0	0.3–3.5	0.5–5.0	55.0–83.0	3.5–21.0	ND	0.0–0.6	0.0–0.4	0.0–0.2	ND	0.0–0.2
Palm	*Elais guineensis*	ND	ND–0.5	0.5–2.0	39.3–47.5	ND–0.6	3.5–6.0	36.0–44.0	9.0–12.0	ND–0.5	ND–1.0	ND–0.4	ND–0.2	ND	ND
Palm kernel	*Elais guineensis*	5.0–12.0	45.0–55.0	14.0–18.0	6.5–10.0	ND–0.2	1.0–3.0	12.0–19.0	1.0–3.5	ND–0.2	ND–0.2	ND–0.2	ND–0.2	ND	ND
Rapeseed	*Brassica* sp.	ND	ND	ND–0.2	1.5–6.0	ND–3.0	0.5–3.1	8.0–60.0	11.0–23.0	5.0–13.0	ND–3.0	3.0–15.0	ND–2.0	>2.0–60.0	ND–2.0
Rapeseed, low erucic acid (canola)	*Brassica* sp.	ND	ND	ND–0.2	2.5–7.0	ND–0.6	0.8–3.0	51.0–70.0	15.0–30.0	5.0–14.0	0.2–1.2	0.1–4.3	ND–0.6	ND–2.0	ND–0.3
Sesame	*Sesamum indicum* L.	ND	ND	ND–0.1	7.9–12.0	ND–0.2	4.5–6.7	34.4–45.5	36.9–47.9	0.2–1.0	0.3–0.7	ND–0.3	ND–1.1	ND	ND–0.3
Soybean	*Glycine max* L.	ND	ND–0.1	ND–0.2	8.0–13.5	ND–0.2	2.0–5.4	17–30	48.0–59.0	4.5–11.0	0.1–0.6	ND–0.5	ND–0.7	ND–0.3	ND–0.5
Sunflower	*Helianthus annuus* L.	ND	ND–0.1	ND–0.2	5.0–7.6	ND – 0.3	2.7–6.5	14.0–39.4	48.3–74.0	ND–0.3	0.1–0.5	ND–0.3	0.3–1.5	ND–0.3	ND–0.5
Sunflower, mid-oleic acid (NuSun™)	*Helianthus annuus* L.	ND	ND	ND–1	4.0–5.5	ND–0.05	2.1–5.0	43.1–71.8	18.7–45.3	ND–0.5	0.2–0.4	0.2–0.3	0.6–1.1	ND	0.3–0.4

ND, non-detectable (≤0.05%).
Source: Codex Alimentarius (FAO, 2009a, 2009b).

20.2.3 Minor and specialty oils

Olive oil is produced predominantly in the Mediterranean region. It is highly unsaturated; in particular, it is an excellent source of oleic acid (C18:1) (see Table 20.2). In addition, it contains a high level of polyphenols and natural antioxidants, which contribute to the oil's flavor and stability. Olive oil is available in several grades, based on factors such as method of production, quality, and flavor. While mainly used for cooking, it is also a component of products such as spreads and dressings.

Flaxseed (linseed) oil is unique in its high level (50–60%) of α-linolenic acid (C18:3n-3), an omega-3 fatty acid. Due to its high level of unsaturation, it is easily oxidized; great care must be taken to protect the oil against oxidation (e.g. packaging in amber glass). It also contains lignans, a class of phytoestrogens with antioxidant properties. Interest in flaxseed oil has grown in recent years, and it is typically marketed as a dietary supplement or health food based on its nutritional content. A low-linolenic (~2%) variety has been developed (Linola™, Commonwealth Scientific and Industrial Research Organization, Australia); it is suitable for use as an edible oil (Kochhar, 2002) and is similar in fatty acid profile to sunflower oil.

Specialty oils (e.g. sesame, walnut) are often minimally processed. Many specialty oils are considered "gourmet" products and are purchased at a premium price specifically for their flavor or nutritional content. For example, sesame oil has a distinctive flavor and is widely used in Asian cuisine. Others are used specifically as dietary supplements, and some find use in cosmetics and other industries. Detailed information on dozens of minor and specialty oils has been compiled by Gunstone (2006a).

20.2.4 Alternative sources

Marine algae contain the omega-3 fatty acids eicosapentaenoic acid (C20:5n-3; EPA) and docosahexaenoic acid (C22:6n-3; DHA) found in fish oil (reviewed by Behrens & Kyle, 1996). Cultivation of single-cell oils from microorganisms is now done on a commercial basis via large-scale fermentation. The microalgae *Crypthecodinium cohnii* and *Schizochytrium* sp. yield oils with high levels of DHA, and are often used as dietary supplements, particularly in infant formulas (Ratledge, 2004). Single-cell oils from the microalgae *Phaeodactylum* sp. and *Monodus* sp. continue to be investigated as sources of EPA (Ward & Singh, 2005). Production of single-cell oils is promoted as a sustainable practice.

20.3 Properties of plant-based fats and oils

The properties of fats and oils are highly dependent on their fatty acid and triacylglycerol make-up. In addition to the length of the carbon chain and degree of saturation in a particular fatty acid, the position of fatty acids on the glycerol molecule affects the physiochemical properties and functional characteristics of the fat or oil. While fatty acid composition varies widely between oil sources, some variability exists within species based on factors such as climate, soil quality, growing season, and plant maturity.

20.3.1 Physiochemical properties

Physical properties of fats and oils include melting point, boiling point, smoke point, density, solid fat index, viscosity, refractive index, and color. Melting point is a function of many variables: in particular, melting point increases with increasing fatty acid chain length, complexity of triacylglycerol components, degree of saturation, and content of *trans* fatty acids. Since they are composed of mixtures of triacylglycerols, fats possess a melting range rather than a true melting point; thus, the term "melting point" refers to the end of the melting range.

Solid fat index is a measure of the solid content of a fat at various temperatures, which is an important predictor of melting and crystallization behavior. The solid content of a fat influences its plastic range, the temperature range over which the fat is moldable. Plastic fats possess properties of a both solids and liquids, as the liquid portion is trapped in the solid crystalline network (Hidalgo & Zamora, 2006). Various processing procedures are applied to fats to extend their plastic range, leading to a wider range of use in food applications.

Chemical properties include iodine value, saponification value, peroxide value, and acid value. Iodine value is a measure of the average degree of unsaturation of a fat and is a predictor of its oxidative stability. Saponification value measures the average chain length of fatty acids in a fat. Peroxide value and acid value measure peroxides and free fatty acids present in fat/oil, respectively. Standard methods for determination of physicochemical properties of fats and oils are available (Firestone, 2009; Horwitz & Lattimer, 2005).

20.3.2 Crystallization behavior

Fats are polymorphic: that is, they possess the ability to crystallize in multiple forms. The consistency of the fat is highly dependent upon the form of crystallization.

The α form is the least stable and tends to impart a waxy texture; β' crystals are fine and lead to a smooth product; β crystals, the most stable, are large and have a grainy texture. When modifying fats and oils for use in specific applications, processors often manipulate reaction variables in order to obtain the desired crystallization behavior. Fats in the α form are appropriate for confectionery products; β crystals are desired for baking applications; β' crystals are well suited for margarines and shortenings. Under certain conditions, α crystals may rapidly melt and recrystallize in the more stable β' form; similarly, β' crystals may undergo transformation to the β form. Care must be taken to avoid factors that promote these irreversible changes, particularly avoidance of high temperatures during transport and storage.

Polymorphism is of particular interest in the manufacture of chocolate. Cocoa butter, a key ingredient in chocolate, crystallizes in six different polymorphic forms, designated as I–VI (Wille & Lutton, 1966). Crystallization in the β(V) form is preferred for chocolate products due to the desired gloss and snap that are characteristic of this polymorph (Beckett, 2000). Chocolate must be carefully tempered to achieve crystallization in this form.

20.4 Nutritional areas of interest

Fats and oils are a fundamental component of the human diet. They have more caloric value than proteins and carbohydrates, making them a good energy source. They contain essential fatty acids, which are not synthesized by the human body and must be obtained from food sources. They serve to carry the fat-soluble vitamins A, D, E, and K as well as an array of phytonutrients.

20.4.1 Saturated and *trans* fatty acids

Diets high in total fat content, particularly saturated fat content, are thought to contribute to obesity (Bray & Popkin, 1998). Excess consumption of saturated and *trans* fatty acids has been linked to coronary heart disease (Hu et al., 2001), the leading cause of death in the US. In particular, high intake of *trans* fatty acids is associated with an increase in low-density lipoprotein (LDL) levels and a corresponding decrease in high-density lipoprotein (HDL) levels. *Trans* fatty acids originate from three sources. They are naturally present in lipids obtained from ruminants as a result of biohydrogenation in the rumen. Additionally, they are formed as a result of heating (specifically deodorization) and the industrial process of partial hydrogenation, in baked goods and fried foods. *Trans* fatty acids are chemically induced through hydrogenation to produce margarines and shortenings.

In response to concern over the role of *trans* fats in heart disease, the Food and Drug Administration (FDA) issued a final rule in 2003 that *trans* fat content must be declared on the nutrition facts label (FDA, 2003). Compliance with this rule became mandatory in January 2006, prompting many manufacturers to reformulate products to reduce the levels of *trans* fatty acids. The US Department of Agriculture (USDA) recommends minimal intake of solid and *trans* fats, and replacement of saturated fatty acids with mono- and polyunsaturated fats (USDA, 2010).

20.4.2 Omega-3 and omega-6 fatty acids

The essential fatty acids linoleic acid (C18:2n-6) and α-linolenic acid (C18:3n-3) are precursors of long-chain omega-6 and omega-3 polyunsaturated fatty acids, respectively. Seed oils (e.g. walnut, soybean, rapeseed/canola, and flaxseed) are rich sources of α-linolenic acid, which is linked to decreased risk of cardiovascular disease (Zhao et al., 2004). Linoleic acid is found in all plant-based oils, including soybean and rapeseed/canola. However, linoleic acid is less abundant now than in past years, as processing and seed breeding practices have been employed to reduce the level of this fatty acid in order to obtain oils with greater oxidative stability. Some processing techniques affect the level of essential fatty acids in vegetable oils; in particular, hydrogenation decreases the content of essential fatty acids by converting them to saturated and *trans* fatty acids.

The omega-3 long-chain polyunsaturated fatty acids DHA and EPA are found mainly in (sea) animal sources. They play a role in the secondary prevention of cardiovascular disease, brain development, and reduction of inflammation (Hu & Willett, 2002; Ruxton et al., 2004). A standard recommended daily intake of omega-3 fatty acids has not been established, but it is generally recognized that the US intake is too low (Kris-Etherton et al., 2000). The American Heart Association recommends consumption of two servings of fatty fish per week (AHA, 2011). Often, dietary restrictions (or simply preferences) preclude consumption of fish. Plant-based sources for supplementation of omega-3 fatty acids include flaxseed oil and algal oil. Additionally, products containing plant-based oils are increasingly being fortified with long-chain omega-3 fatty acids; margarines, spreads, and salad oils are good vehicles for their delivery (Muggli, 2006).

Many researchers believe that the ratio of omega-6 to omega-3 fatty acids consumed is much too high, particularly in Western diets (Lands, 1986; Simopoulos, 2008). Soybean and rapeseed/canola oils naturally have a favorable ratio of linoleic acid to α-linolenic acid. However, the balance is upset by processing techniques such as partial hydrogenation. Food manufacturers increasingly consider the omega-6/omega-3 ratio when reformulating products (Hernandez, 2005).

20.5 Degradation of plant-based fats and oils

Crude oils are highly susceptible to degradation via lipolysis and oxidation. Lipolysis, the hydrolysis of free fatty acids from the glycerol molecule, decreases the stability of the oil (Frega et al., 1999; Miyashita & Takagi, 1986). Lipolysis may be catalyzed by enzymes, particularly from microbial sources, or water and heat. Oxidation occurs in the presence of atmospheric oxygen and yields low molecular weight compounds responsible for development of off-flavors and odors. Care must be taken to protect oils from these reactions during refining, modification, and storage.

Plant-based fats and oils are subject to both autoxidation and photo-oxidation. Both mechanisms are selective, acting at points of unsaturation. Thus, susceptibility of fats and oils to oxidation increases as degree of saturation decreases. Oxidation may be promoted by environmental factors (e.g. temperature, light) or components naturally present in the crude oil, such as free fatty acids and trace metals. Both pathways are summarized below; a detailed review is available (Choe & Min, 2006).

20.5.1 Autoxidation

The autoxidation mechanism is a three-step process (Figure 20.1). The initiation step involves formation of an alkyl radical. In the propagation step, alkyl radicals react with atmospheric oxygen to form peroxyl radicals, which in turn react with additional fatty acids to form hydroperoxides and new alkyl radicals. The process continues until the termination step ensues; all free radicals have combined to form new compounds. Oxidation proceeds slowly during the induction period but proceeds rapidly once the supply of antioxidants has been exhausted. Efforts are focused on extending the induction period, which corresponds to the shelf life of the product (Gunstone & Martini, 2010).

Initiation

$RH \rightarrow R\cdot + H\cdot$

Propagation

$R\cdot + O_2 \rightarrow ROO\cdot$

$ROO\cdot + RH \rightarrow ROOH + R\cdot$

Termination

$R\cdot + R\cdot \rightarrow RR$

$R\cdot + ROO\cdot \rightarrow ROOR$

$ROO\cdot + ROO\cdot \rightarrow ROOR + O_2$

Figure 20.1 Mechanism of lipid autoxidation.

Primary oxidation products are hydroperoxides and conjugated dienes. These decompose to form secondary oxidation products, low molecular weight compounds including aldehydes, ketones, and alcohols. Secondary oxidation products reduce the sensory quality of the oil by causing off-odors and off-flavors that are collectively known as oxidative rancidity. These flavors and aromas are detectable at low levels. Repeated consumption of oxidized lipids is reported to be detrimental to human health (Kanner, 2007).

Natural antioxidants (e.g. tocopherols, tocotrienols) are present at varying levels in fats and oils and act as free radical scavengers. They are partially removed as a result of the harsh reaction conditions that occur during processing, but are often added to the finished oils. Synthetic antioxidants that are approved food additives in the US include butylated hydroxytoluene, butylated hydroxyanisole, tertiary butylhydroquinone, and propyl gallate; when two are used in combination, a synergistic relationship ensues. The use of synthetic antioxidants in US food products is regulated by the FDA and is permissible at a combined level of no more than 200 ppm of the total fat content. Citric acid, added after the refining process to chelate trace metals, acts as an antioxidant synergist.

20.5.2 Photo-oxidation

Photo-oxidation occurs when light energy in the form of a photon is transferred to a sensitizer (e.g. chlorophyll), which serves to activate oxygen. The resulting singlet oxygen is highly reactive, readily forming peroxides and free radicals from points of unsaturation in fatty acids. Photo-oxidation occurs at a much faster rate than autoxidation. The photo-oxidation pathway is terminated by singlet oxygen quenchers (e.g. β-carotene) rather than free radical scavengers.

20.6 General handling considerations

Degradation of edible fats and oils cannot be stopped, but can be slowed by taking certain precautions during processing and storage. Where appropriate, care should be taken to avoid:

- **contamination**: equipment surfaces should be cleaned and sanitized regularly to eliminate spoilage micro-organisms and other adulterants, and to avoid build-up of oxidation products
- **oxygen**: some processing steps require an inert atmosphere and may be conducted under vacuum or nitrogen blanket
- **trace metals**: trace metals (e.g. copper) act as pro-oxidants, and their levels may be minimized by the use of stainless steel equipment during processing
- **light, heat, and water**: because light and heat promote reactions that lead to oxidative rancidity (or, in the case of water, hydrolytic rancidity), oils should be stored in a cool, dry, dark location.

20.7 Recovery of oils from their source materials

20.7.1 Preparation of oilseeds for extraction

Figure 20.2 illustrates the general process for extraction of oil from oilseeds. Seeds are pretreated prior to extraction in a series of steps designed to maximize the oil obtained during the extraction process. After harvesting, seeds are cleaned to remove foreign materials, including sticks, leaves, dirt, and stones. Magnets are used to remove traces of metal. Various grades of screens are utilized, with or without aspiration, in succession from coarse to fine in order to remove materials from the oilseeds. When screens are used in conjunction with aspiration, the process is called "scalping" (Kemper, 2005). A further delinting step is necessary for cottonseeds (Anderson, 2005; O'Brien, 2004a). Material that will be stored before processing is dried to approximately 5–10% moisture content, depending on the oilseed variety. Cool, dry

conditions are critical for extended storage, as excessive moisture and heat may lead to degradation (Hernandez, 2005; Wakelyn & Wan, 2006).

A dehulling or decortication step is done for many oilseeds in order to separate the meat from the hull and/or seed coat, thereby increasing the effectiveness of future processing operations. This step is eliminated for smaller seeds (e.g. safflower) because their size precludes effective separation (O'Brien, 2004a). In some cases, seeds are cracked prior to dehulling to facilitate removal of the hull. Next, seeds are subjected to one or more size reduction processes (e.g. cracking, crushing, grinding) to increase surface area and promote the release of oil from the cells.

Seeds generally undergo a conditioning step after size reduction. Seeds are heated to 90–110 °C or 110–150 °C, depending on whether they will be pre-pressed prior to solvent extraction or full pressed, respectively (Kemper, 2005). Addition of steam may be necessary to achieve the optimal moisture level for oil extraction, which varies by seed type (Fellows, 2009). Conditioning serves to denature proteins, deactivate enzymes, inhibit microbial activity, and increase oil yield by facilitating later processing steps. In the case of cottonseed, conditioning destroys gossypol, a mildly toxic pigment.

A flaking step is done for oilseeds. The seeds pass through cylindrical rollers, which rotate in opposite directions, and are flattened to approximately 0.3–0.4 mm thickness. Conditioning of the seeds prior to flaking makes them pliable and facilitates rolling. It is important to achieve flakes of uniform thickness in order to achieve maximum oil yield during solvent extraction (Kemper, 2005). Flakes (especially soybean and cottonseed) may be sent to an expander, a type of extruder that heats and shapes the flakes using steam injection under high pressure. The product is released from the expander in the form of a dense, porous pellet (also known as a collet) that expands upon contacting the air, rupturing the oil cells. Use of an expander leads to a significant increase in extraction yield compared to extraction directly from flakes (Hernandez, 2005). Dry extrusion, which utilizes friction to heat the product, is often used in small-scale facilities to enhance oil yield for seeds that are to be full pressed (Kemper, 2005).

20.7.2 Preparation of tree fruits for extraction

Due to their high moisture content, tree fruits are subject to enzymatic hydrolysis (Gunstone, 2004) and oil must be extracted within hours (e.g. palm fruits) or days (e.g. olives) of processing. Special care must be taken when

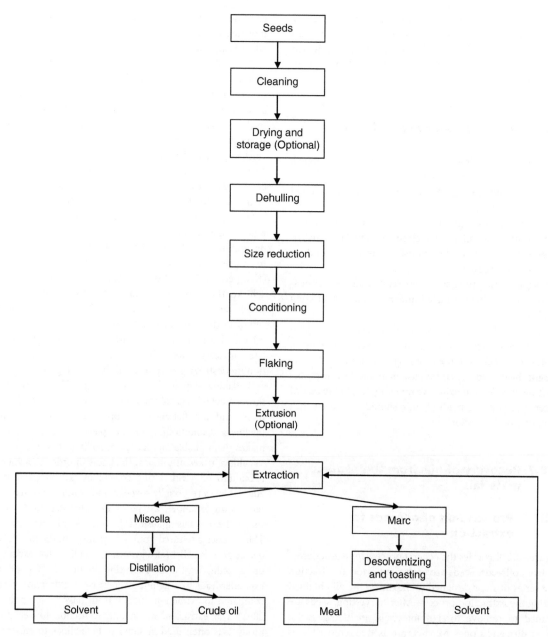

Figure 20.2 Flow diagram for general oilseed extraction process.

handling palm fruits so as not to damage the kernels, which are further processed for their oil. Palm fruits are sterilized immediately after harvesting by application of steam at approximately 120 °C under pressure. This process inactivates lipases and stops bacterial activity in the fruit, while preconditioning the kernels (Fils, 2000). After

sterilization, the fruits are mechanically stripped from their bunches, then heated under agitation in a digester, which aids in separation of the fruit from the kernel. Olives are washed, crushed to form a paste, and mixed to allow some coalescence of the droplets. Coconuts are shelled, cracked, and dried prior to pressing.

20.7.3 Extraction

The goal of extraction is to recover the maximum amount of crude oil of highest purity, while producing a high-value co-product. The co-product consists of solids (usually protein-rich) and residual oil and may be ground into a meal or flour for use in such applications as animal feeds and industrial products. In some cases, the co-product is used as an energy source for refining operations.

There are three main types of extraction processes: mechanical pressing, pre-press solvent extraction, and direct solvent extraction. Oil-bearing fruits are pressed to extract oils. For oilseeds, the extraction procedure is selected based on variables such as oil content of the starting material, size of the extraction facility, and desired specifications for the resulting meal. Throughout the extraction process, it is critical to control moisture and temperature in order to avoid damage to the oil while achieving maximum yield.

20.7.3.1 Pressing

Pressing entails the use of mechanical force to squeeze oil from the solids. The process yields crude oil and a solid "cake" containing approximately 5–10% oil. Pressing may be either a batch or continuous process. Since pressing generates lower yields and sample throughput than solvent extraction, it is not widely used for oilseeds. However, it is still routinely used in small processing facilities and by processors of specialty and/or "natural" oils whose products are often sold at a premium price in health food markets.

Hydraulic batch presses process one group of fruit or seeds at a time, but some models have the capability to process many batches per day. For some processors, hydraulic pressing is an economical alternative to solvent extraction, as capital and operating costs are lower. Olive oil may be separated from its paste by hydraulic pressing or by a combination of centrifugation and filtration (O'Brien, 2004a). Virgin olive oil results from pressing without chemical treatment; it has a lengthy shelf life and is highly stable during frying (Boskou, 2002).

Continuous screw presses ("expellers") are motorized pieces of equipment that consist of a barrel containing a water-jacketed horizontal screw. The pressure resulting from the rotation of the screw removes the oil from the seed. Heat generated by friction increases oil yield by decreasing its viscosity (Fellows, 2009). The water jacket serves to cool the screw so that the optimal extraction temperature may be maintained. Expeller pressing is suitable for processing both fatty fruits (e.g. palm,

avocado) and seeds with high oil content that have been preflaked (e.g. cottonseed, soybean). When pressing palm fruits in this manner, care should be taken to minimize crushing the kernels, which leads to admixture of oils from the fruits and kernels (Fils, 2000).

Cold pressing refers to mechanical pressing that is done without application of heat; the oil temperature typically does not exceed 27 °C. Cold-pressed oils retain more nutrients and flavor compounds than their hot-pressed counterparts. Oils that are commonly cold pressed include olive, flaxseed, and avocado. Cold-pressed oils may simply be filtered without further refining; these products tend to be relatively expensive and appeal to consumers who value "natural" oils that are produced without the use of heat and chemicals.

20.7.3.2 Pre-press solvent extraction

Pre-press solvent extraction may be chosen when the oil content of the product is greater than 30%. The pressure of the expeller is reduced in order to leave a portion of oil in the cake (usually 15–18%) (Johnson, 2008; O'Brien, 2004a), which is then removed via solvent extraction. As a result, less solvent is needed and a smaller extractor may be used, decreasing capital costs. For some seeds with high oil content, use of an expander makes it possible to bypass the pre-press step (Johnson, 2000).

20.7.3.3 Direct solvent extraction

Direct solvent extraction is the most commonly used method of oil recovery. Additionally, it is the most efficient recovery method, removing all but approximately 1% of oil from the starting material. There are two main methods of solvent extraction: immersion of the starting material in solvent or percolation of the solvent through the material (Johnson, 2000; Wang, 2002).

Solvent extraction may be done as a batch, semi-continuous, or continuous process. In batch processes, a single extraction tank processes one batch at a time. Semi-continuous processes utilize multiple batch extractors connected in series. Continuous processes involve constant movement of flakes through an extractor. There are a number of extractor designs, which vary based on the depth of the bed of solids and the direction of the flow (e.g. horizontal, vertical, rotary). Many designs utilize countercurrent flow to minimize the amount of solvent needed; flakes and solvent move through the extractor in opposite directions such that flakes entering the extractor contact the oldest solvent and the freshest solvent is used when extraction is nearly complete. Continuous processes are most widely used, although batch processes are commonly used in production of specialty oils.

The solvent of choice is typically commercial hexane, consisting of a mixture of *n*-hexane and its isomers. Due to its flammability and toxicity, the use of *n*-hexane is a concern; its emissions from US solvent extraction facilities are regulated by the EPA Clean Air Act. Alternative extraction methods have been explored, including various solvents (e.g. isohexane, acetone, alcohols), supercritical carbon dioxide, and aqueous enzymatic processes. The latter methods offer environmental advantages but are currently cost-prohibitive on a commercial scale. In the future, enzyme-assisted extraction may become a viable option, but at present hexane is generally the most economical selection.

Solvent extraction yields two products. The miscella, a mixture of solvent and crude oil, is separated by a series of distillation steps. The marc, consisting of the solids and residual oil, is sent to a desolventizer-toaster which evaporates the solvent by heating; the solids are dried at high temperature and cooled to prepare the meal for its end use (Kemper, 2005). The solvent recovered from both the miscella and the marc is reused for further extraction.

20.8 Refining

Crude oil contains roughly 95% triacylglycerols, accompanied by free fatty acids, mono- and diacylglycerols, phospholipids, sterols, tocols, pigments, and other minor components such as waxes, trace metals, and pesticides. Refining of crude oil is a multistage process consisting of several operations designed to maximize the yield of oil that is free from impurities while keeping the valued components (e.g. sterols, tocols) intact. Refining is the most critical processing step, as poorly refined oils lead to problems in later processing and application steps (O'Brien, 2002; Wakelyn & Wan, 2006), increasing the cost of the end-product. In the US, the National Oil Processors Association (www.nopa.org) publishes an annual yearbook listing trading specifications for refined oils, including physical properties and maximum allowable content of various impurities.

20.8.1 Physical versus chemical refining

There are two types of refining: physical and chemical (Figure 20.3). Physical and chemical refining differ in the manner in which the free fatty acids are removed from the crude oil. Physical refining uses steam distillation to remove the free fatty acids, while chemical refining uses a caustic solution. Physical refining consists of degumming, bleaching, and deodorization. Chemical (alkali) refining is more widely used and consists of an optional degumming step, followed by caustic neutralization, bleaching, and deodorization. Oils that have been through these processes are known as RBD (refined, bleached, and deodorized) oils.

Physical refining is most appropriate for oils that are low in phospholipids and high in free fatty acids. Palm, palm kernel, and coconut oils are typically refined in this manner (O'Brien, 2004a). Physical refining is less costly and involves fewer steps than chemical refining. The pretreatment prior to deodorization must be thorough or a lower-quality oil will result.

Chemical refining yields a high-quality product and is well suited for a wide variety of oils. Cottonseed oil must undergo chemical refining to remove gossypol. The chief disadvantage of chemical refining is the high cost associated with disposal of the waste water from the neutralization step, which is governed by environmental regulations. Additionally, there is greater loss of neutral oil with chemical refining than with physical refining (O'Brien, 2004a).

The general procedures for the refining process are described below. Components removed during each refining step are summarized in Table 20.3.

20.8.2 Degumming

Some crude oils (e.g. soybean, canola, cottonseed) contain up to 3–4% phosphatides. Because they are emulsifiers, phosphatides interfere with refining steps, particularly bleaching (Taylor, 2005), which reduces yield. Degumming is the optional process by which phosphatides are removed from crude oil. Although phosphatides may be removed in future steps (Anderson, 2005), degumming as a first step increases the oil yield and leads to a higher-quality end-product (Farr, 2000). In addition to phosphatides, degumming removes some solid components (e.g. waxes) and trace metals, enhancing the oil's stability. Degumming may be a batch or continuous process.

There are two primary approaches to degumming: water degumming and acid degumming. Water degumming is usually performed as a batch process in the US (O'Brien, 2004a) and involves addition of water to heated oil, followed by agitation. The hydratable phosphatides are converted into gums that are insoluble in the oil and may be removed by centrifugation or filtration. The "gum" contains phospholipids, which are recovered for their by-product value as lecithin, an emulsifier widely used in confectionery products, baked goods, and

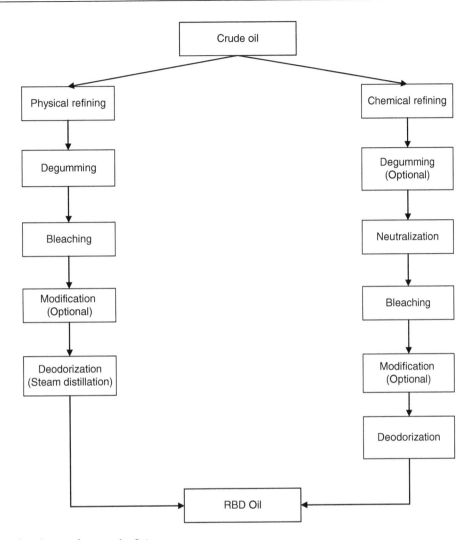

Figure 20.3 Flow diagram for general refining process.

Table 20.3 Components removed during the refining process

Refining step	Targeted component(s) for removal	Other components removed
Degumming	Phosphatides	Filterable solids (e.g. waxes), trace metals
Neutralization	Free fatty acids	Phosphatides, soaps, trace metals, pigments
Bleaching	Pigments	Phosphatides, soaps, trace metals, waxes
Deodorization	Free fatty acids (physical refining)	Sterols, tocopherols, tocotrienols, carotenoids, pigments
	Secondary oxidation products (chemical refining)	

margarines. The quality of lecithin varies depending on the method of degumming utilized.

Treatment with acid is necessary to remove the non-hydratable phosphatides, which are associated with low-quality oils (Anderson, 2005). Acid degumming involves addition of dilute acid (most often phosphoric or citric) to warm oil, followed by rapid heating to break the emulsion. The resultant soapstock may then be removed by centrifugation.

Many modifications to the acid degumming process are available. Dry degumming, effective for oils with low phosphatide content, combines acid degumming and bleaching into one step; the gums are removed by filtration during bleaching. Enzymatic degumming uses a combination of phospholipases and acid to maximize removal of gums. In the super-degumming procedure, the acid/oil mixture is cooled to ambient temperature and agitated with water (Johnson, 2008). Acid refining involves the addition of sodium hydroxide after acid degumming to partially neutralize the acid; as a result, soaps are not formed and some remaining phosphatides are removed as sodium salts. Acid refining is often used as a conditioning step for neutralization.

Alternative methods of degumming include membrane filtration and silica adsorption. Degumming procedures have been reviewed by Cmolik and Pokorny (2000).

20.8.3 Neutralization

The degummed or acid-conditioned oil is mixed with a caustic solution, most commonly sodium hydroxide, at slightly elevated temperature. Neutralization may be conducted as a high-temperature, short-mix or low-temperature, long-mix process; mixing of the oil and caustic is done for seconds or minutes, respectively. Free fatty acids in the oil combine with the caustic in a saponification reaction to form a water-soluble soap, which contains sodium salts of fatty acids. After separation of the oil and soapstock via centrifugation, the oil is washed with hot water in order to remove the remaining traces of soaps. Residual phospholipids, pigments, trace metals, and other minor impurities may also be removed in the process. The oil is dried by low-pressure heating, prior to bleaching.

Modifications to the traditional neutralization procedure are available. For example, in miscella refining, neutralization is conducted immediately after solvent extraction, prior to separation of the oil and solvent. Silica refining eliminates water washing in favor of silica adsorption to remove soaps.

Waste water from the neutralization process may contain soapstock, and is regulated in the US by the EPA Clean Water Act. Treatment and disposal of the soapstock are costly, but degumming prior to neutralization decreases disposal costs (Farr, 2000).

20.8.4 Bleaching

While triacylglycerols are colorless, crude fats and oils contain pigments that impart visible color to the oil. Pigments naturally present in crude oils include carotenoids and chlorophyll which yield yellow/red and green colors, respectively. Color is viewed as an indicator of edible oil quality (Wan & Pakarinen, 1995). Consumers generally desire colorless or lightly colored oils (Taylor, 2005), and bleaching meets this need.

Bleaching is the removal of color compounds resulting from heating the oil with bleaching earth. Some residual soaps, trace metals, and waxes are eliminated in the process. For oils that have not been degummed, the bleaching step serves to remove the phospholipids, but it is difficult to achieve complete removal. Bleaching is the costliest step of the refining process due to the cost of the bleaching earth and its disposal, plus the loss of oil associated with the procedure (Gunstone, 2004).

Bleaching earth is made from mineral-based clays (e.g. calcium bentonite) and may be used in its natural form for oils with low levels of pigments or activated with sulfuric or hydrochloric acid to increase its adsorptive ability (Taylor, 2005). Acid-activated bleaching earth is typically necessary for removal of chlorophyll (Taylor, 2005) and further serves to break down peroxides into products that are subsequently removed in the deodorization process (Wang, 2002). However, the acidity may lead to isomerization of some natural *cis* bonds in unsaturated fatty acids to their *trans* configuration (Gunstone, 2008a).

Activated carbon and amorphous silica hydrogels are often used in conjunction with bleaching earth due to their affinity for polycyclic compounds and soaps, respectively (Taylor, 2005). Because they are much more costly than bleaching earth, they are typically used in small amounts. Silica gel may be added to the oil prior to addition of bleaching clay in order to first remove soaps and phospholipids, making the bleaching clay more effective for removal of color compounds (Anderson, 2005). The proportions of bleaching clay and other adsorbents are selected based on the properties of the oil to be bleached.

Bleaching may be done in an atmospheric batch process or as a batch, semi-continuous, or continuous vacuum process. Processing under vacuum is preferred, due to the reduction of oxygen that promotes degradation. Continuous processes are most widely used; some utilize countercurrent flow for more efficient use of the bleaching earth, reducing processing costs.

The bleaching procedure involves addition of the bleaching earth to heated oil, followed by agitation. A holding period is employed to ensure complete bleaching, followed by filtration to remove the spent bleaching earth. There is some loss of oil in the spent earth, which may be treated with steam to ensure more complete removal of oil (Wakelyn & Wan, 2006). The effectiveness of the bleaching treatment is affected by variables such as temperature, time, and type of bleaching earth (Zschau, 2000). The standard for measurement of complete bleaching is a peroxide level of zero, rather than a particular color (Hernandez, 2005).

Bleaching earth is considered a hazardous material, as the spent earth is prone to self-combustion upon exposure to atmospheric oxygen, so disposal must be undertaken in a proper manner. The disposal of spent bleaching clay is regulated by the EPA Resource Conservation and Recovery Act.

20.8.5 Deodorization

Deodorization is the low-pressure heat treatment of oil with steam in order to remove impurities responsible for off-flavors and odors (e.g. free fatty acids, secondary oxidation products). Oxidative stability of the oil is significantly increased as a result (Pokorny, 2006).

The goal of deodorization is to produce oil with neutral flavor and odor. Deodorization is normally performed as the last step prior to packaging or transport (Wakelyn & Wan, 2006). In the case of RBD oils, the deodorization step occurs after bleaching. For oils subjected to modification procedures, deodorization follows modification.

Oil is de-aerated, heated to an intermediate temperature, then heated under low pressure to the specified deodorization temperature (generally 200–250 °C). Stripping steam is injected into the oil, removing water-soluble compounds in the course of the distillation process. In addition to the impurities mentioned above, sterols, tocopherols, and tocotrienols are partially removed. These high-value compounds may be obtained from the distillate, and recovered by condensation for addition to foods or use as nutraceuticals.

Deodorization may be a done as a batch, semi-continuous, or continuous process. Semi-continuous and continuous set-ups are most widely used, while batch processes are still used in small or older facilities. Thin-film deodorization is a continuous process that yields excellent results. It involves exposure of a high surface area of oil, reducing the temperature and time needed for removal of impurities.

The harsh conditions associated with deodorization lead to additional reactions. Degradation of carotenoids causes a heat-bleaching effect. Formation of *trans* isomers of linolenic and linoleic acid occurs during deodorization and is dependent on reaction time and temperature (Hénon et al., 1999; O'Keefe et al., 1993). *Trans* isomers formed during deodorization are geometrical isomers, whereas those formed during partial hydrogenation are positional isomers (Duchateau et al., 1996) and those formed by rumen metabolism are both geometrical and positional isomers (Pariza et al., 2001). Deodorization may also cause hydrolysis or polymerization. Therefore, it is critical to weigh the benefits and costs when selecting deodorization conditions. The variables to control include temperature, time, pressure, and amount of steam. The appropriate levels may vary based on oil type, deodorizer design, and refining process (Kellens & de Greyt, 2000). Oils that are physically refined require harsher conditions than those that are chemically refined.

20.9 Modification of plant-based fats and oils

Modification of fats and oils is widely employed to obtain products with properties best suited to their end use. Reasons for modification include increasing oxidative stability, alteration of fatty acid composition, and improvement of textural characteristics. Gunstone (1998) classified modification techniques in two categories: technological methods include blending, fractionation, hydrogenation, and interesterification, while biological methods include plant breeding and genetic engineering.

20.9.1 Blending

Blending is the simplest method of modification and involves mixing oils and fats to yield products with properties desired for specific food applications (e.g. shortenings and spreads). Blending may be used to improve the nutritional or physical properties of the product (Gunstone, 2006b). Typically, a solid base-stock is heated and blended into a liquid oil to yield a product with a wide plastic range. Agitation is required in order to ensure uniform heat (O'Brien, 2002). A wide variety of products may be used as base-stocks, including those that have been hydrogenated, interesterified, or fractionated. Blending has been utilized to produce plastic fats free of *trans* fatty acids (Jeyarani & Reddy, 2003).

20.9.2 Fractionation

Fractionation is the process used to separate oils and fats into two fractions of different triacylglycerol composition. Fractionation is purely a physical process; no chemical changes are involved. The resultant products have a wider range of use than the starting material. The low-melting fraction is known as "olein", while the high-melting fraction is called "stearin." Oleins have high oxidative stability and are well suited for use as salad oils or in frying applications. Stearins are widely used in the manufacture of margarines, shortenings, and confectionery products, often as cocoa butter substitutes, which are less costly than cocoa butter but possess similar melting characteristics. Fractions may undergo further modification, including blending and interesterification; additionally, individual fractions may be subject to refractionation ("double fractionation").

There are three types of fractionation: dry fractionation, solvent fractionation, and aqueous detergent fractionation. Dry fractionation is the most common and least expensive process. While the procedures for solvent fractionation and aqueous detergent fractionation are more costly, the added cost is warranted for products sold at a premium price, such as cocoa butter substitutes.

20.9.2.1 Dry fractionation

Dry fractionation involves the crystallization of solids from liquid oil, followed by separation of the components via filtration or centrifugation. Crystallization occurs in three stages: cooling to supersaturation, nucleation, and crystal growth. Cooling is done gradually so as to achieve stable crystals that may be easily filtered. Because crystallization is an exothermic reaction, agitation is necessary to distribute the heat, although care must be taken to avoid disturbing the crystals (Gunstone, 2004).

Winterization is the most common form of dry fractionation and is typically used for removal of triacylglycerols with high melting points (i.e. saturated triacylglycerols), yielding oil that remains clear under refrigerated conditions. Products of winterization are known as salad oils and are widely used in preparation of dressings and mayonnaise. Cottonseed, partially hydrogenated soybean, and canola oils are commonly winterized.

The traditional winterization procedure involves placement of the oil into deep tanks in a refrigerated room (approximately 5 °C), holding for 2–3 days to allow proper crystallization, then filtration of the saturated crystals (Wakelyn & Wan, 2006). In the modern procedure, oil is cooled to refrigeration temperature by passing through heat exchangers, then held in tanks with mild agitation during the crystal growth period, which may last several days. Removal of the crystals is most often accomplished with vacuum filtration (Johnson, 2008).

Dewaxing may be viewed as a form of winterization. Some oils (e.g. corn, sunflower) have high levels of waxes, which impart a hazy appearance after settling. Dewaxing is done to improve the clarity of the oil. The oil is cooled slowly and a filter aid is added. After crystallization and settling of the waxes, they are removed using filtration.

20.9.2.2 Solvent fractionation

Solvent fractionation utilizes an organic solvent (e.g. acetone, hexane) and results in better separation of the fractions and shorter reaction times. There is some variability in the products based on the choice of solvent. Solvent fractionation is most commonly done for palm oil, but is also applied to coconut oil, palm kernel oil, partially hydrogenated soybean oil, and others. Palm oil is typically subjected to double fractionation. Palm oleins, removed in the first stage, are used as frying oils and exhibit excellent oxidative stability. Palm stearins and mid-fractions, separated in the second stage, are widely used in margarines and confectionery products.

20.9.2.3 Aqueous detergent fractionation

In aqueous detergent fractionation, a wetting agent (e.g. sodium lauryl sulfate) and an electrolyte (e.g. magnesium sulfate) are used to precipitate crystals in the aqueous phase (Kellens & Hendrix, 2000). The crystals are then removed by centrifugation and the fractions are heated and washed to remove the wetting agent and electrolyte. A drying step completes the process. Excellent separation of the fractions may be achieved, but the process is not widely used due to the high cost and potential for contamination of the oil with detergent (Kellens et al., 2007).

20.9.3 Hydrogenation

Hydrogenation results from reaction of a liquid oil with hydrogen gas in the presence of a catalyst at elevated temperature. Hydrogen atoms are added to the double bonds of unsaturated fatty acids, yielding a product with increased saturated fatty acid content. There are two predominant reasons for hydrogenation: hydrogenation improves the oxidative stability of the product via decreasing the content of polyunsaturated fatty acids, extending shelf life; and the conversion of liquid oils to plastic fats improves their functionality for use in specific applications.

In general, hydrogenation is conducted within the temperature range of 160–200 °C. A solid catalyst (most often nickel) is added to the oil, followed by introduction of hydrogen gas with agitation. The gaseous hydrogen is dissolved in the liquid oil and diffuses to the surface of the catalyst (O'Brien, 2004a), where the reaction occurs. Because hydrogenation is an exothermic process, even distribution of the heat is required, often in the form of circulating cool water through the reactor coils (Anderson, 2005).

The reaction is typically stopped when the melting point of the product reaches the desired value (Pokorny, 2006). Because melting temperature increases with degree of saturation and level of *trans* fatty acids, the product is higher in solid content than the starting material. A side reaction is the isomerization of *cis* double bonds to their *trans* configuration; this particularly affects oleic acid (C18:1). Because melting temperature increases with degree of saturation and level of *trans* fatty acids, products of hydrogenation are higher in solid content than their starting materials and are therefore more appropriate for use in products such as shortenings and spreads.

Variables affecting the results include composition of the starting oil, amounts of hydrogen and catalyst, selectivity of the catalyst, reaction temperature, time, pressure, and level of agitation. In particular, an abundance of hydrogen, which results from increasing agitation and pressure, causes hydrogenation to be predominant and isomerization to be minor (Gunstone, 2004).

Hydrogenation is most often done as a batch process, although some semi-continuous processes are used (Gunstone, 2004). It is necessary to start with well-refined oil, as traces of gums, soaps, and metals may hinder the activity of the catalyst (Johnson, 2008). Three degrees of hydrogenation may be achieved: brush hydrogenation, partial hydrogenation, and full hydrogenation.

20.9.3.1 Brush hydrogenation

In brush hydrogenation, a minor amount of hydrogenation is done in order to improve the stability of liquid oil without conversion to its solid form (Hastert, 2000). This highly selective form of hydrogenation involves manipulation of reaction conditions to target the most unsaturated fatty acids (Wright & Marangoni, 2006). Brush hydrogenation

is typically applied to oils containing high levels of linolenic acid, especially soybean and canola (Gunstone, 2008a). Isomerization from the *cis* to *trans* form may be decreased compared to the standard hydrogenation process, particularly with a lower reaction temperature and a selective catalyst (e.g. platinum) (Beers, 2007).

20.9.3.2 Partial hydrogenation

Partial hydrogenation is commonly used for soybean oil and other oils with significant levels of linoleic acid (C18:2n-6) (Gunstone, 2008a). Some partially hydrogenated oils are suitable for frying applications. As the degree of hydrogenation increases, oils become more plastic, making them suitable for use in spreads, shortenings, and confectionery applications.

20.9.3.3 Full hydrogenation

Full hydrogenation results in the virtual absence of unsaturated and *trans* fatty acids. Fully hydrogenated oils are typically solid at ambient temperature and are used as base-stocks for blending and interesterification.

20.9.3.4 Post bleaching

A post-bleaching step is typically conducted following hydrogenation. The purposes of post bleaching are to remove traces of catalyst remaining in the oil, colors that may have developed during hydrogenation, and primary and secondary oxidation products (O'Brien, 2002). Phosphoric or citric acid is added prior to the bleaching earth in order to chelate the metal catalyst. The bleaching procedure proceeds as previously described.

20.9.4 Interesterification

Interesterification is one of the least common methods of modifying fats and oils and is more commonly used in conjunction with other methods. It involves the rearrangement of fatty acid esters on the glycerol molecule; this redistribution may occur both within and between triacylglycerols. While the composition of triacylglycerols is altered, the individual fatty acid content is unchanged. The process yields products with different physical characteristics (e.g. melting point range, crystallization behavior, plastic behavior) than the starting product(s). Interesterification is more costly than hydrogenation, but a higher degree of specificity can be achieved, and the products are free of *trans* fatty acids (Wright & Marangoni, 2006).

Interesterification may be utilized for single oils or, most commonly, blends of fats and oils, including those that have been previously modified. Usually, liquid oils are interesterified with solid fats or fractions to yield plastic fats suitable for use in applications such as shortenings and margarines. The use of starting materials with wide melting ranges yields products with an extended plastic range (Hidalgo & Zamora, 2006). Common base-stocks include partially or fully hydrogenated oils, lauric oils, and stearin fractions. Products of interesterification may exhibit less oxidative stability than their sources, possibly due to changes in the levels of pro- and antioxidants resulting from the procedure (Gunstone, 2008a).

Interesterification may be done as a batch, semi-continuous, or continuous process. The two principal types of interesterification differ based on the catalyst utilized; both chemical and enzymatic catalysts may be used.

20.9.4.1 Chemical interesterification

Chemical interesterification may be either random or directed and requires a chemical catalyst (most often a metal alkoxide), which must be soluble in the oil. The steps in random chemical interesterification are pretreatment of the oil, reaction with the catalyst, and deactivation of the catalyst. The reaction should be performed at ambient or slightly elevated temperature (Wang, 2002); temperatures above 50 °C may cause polymerization and degradation (Richards, 2006). The reaction proceeds until random equilibrium is reached; fatty acids are randomly distributed on the glycerol molecules.

Directed interesterification is used to selectively increase the level of specific triacylglycerol component(s). While the reaction proceeds, the temperature is decreased below the melting point of the desired component(s), causing crystallization of triacylglycerols with the highest melting points. The equilibrium shifts, leading to increased production of saturated triacylglycerols (Wright & Marangoni, 2006). The resulting product has an extended plastic range (Richards, 2006). Directed interesterification, most often done as a continuous process (O'Brien, 2004a), requires a longer reaction time than random chemical interesterification.

20.9.4.2 Enzymatic interesterification

Enzymatic interesterification is catalyzed by lipases, which may be obtained from bacteria, yeasts, or fungi (Richards, 2006). Lipases may either be random or specific to particular fatty acids or positions on the glycerol backbone (Wright & Marangoni, 2006). Hence, products of the reaction vary based on the lipase utilized. Drawbacks to enzymatic interesterification include higher cost, longer reaction time, and the need for more careful monitoring (Lampert, 2000). However, there is less waste, and purification of the products is simpler. Batch processes are typically used for enzymatic interesterification.

20.9.4.3 Prebleaching and post bleaching

A prebleaching step (drying under vacuum, followed by neutralization) prior to interesterification is necessary for oils with high peroxide values (Lampert, 2000). Post bleaching is done following interesterification in order to remove both chemical and enzymatic catalysts.

20.9.5 Biological methods of oil modification

There are four methods of modifying fatty acid profiles of oils: traditional plant breeding, *in vitro* mutagenesis, natural mutation, and genetic engineering. Traditional plant breeding methods have been used to produce canola oil and high-oleic safflower oil (McKeon, 2005) as well as a mid-oleic variety of peanut oil (Gorbet & Shokes, 2002). *In vitro* mutagenesis is typically used to remove an unwanted characteristic (McKeon, 2005) and has been used in conjunction with plant breeding to yield soybean oils with altered fatty acid profiles, high-oleic sunflower oils, and Linola™ oil. Natural mutation, followed by plant breeding, has led to high-oleic cottonseed, peanut, and rapeseed lines. While traditional breeding methods may transfer up to thousands of genes at a time, advances in biotechnology have led to the ability to isolate specific gene(s) for transfer (O'Brien, 2004c).

Genetic engineering of oilseeds was initially done to improve resistance to pests and herbicides. In recent years, efforts have focused on altering the fatty acid profile of oilseeds, both for nutritional purposes and to obtain desired physical properties in the resultant oils, such as increased oxidative stability. However, there is a high cost associated with commercialization of genetically modified plants, and commercial success has historically been difficult to achieve.

A growing area of interest is the development of transgenic plants that contain long-chain omega-3 and omega-6 fatty acids, especially the omega-3 fatty acids traditionally found in marine oils (Murphy, 2006). For example, Qi et al. (2004) successfully incorporated eicosapentaenoic acid (C20:5n-3) and arachidonic acid (C20:4n-6) into *Arabidopsis thaliana*. This research area has been thoroughly reviewed by Venegas-Caleron et al. (2010).

20.10 Packaging and postprocessing handling

Oil and fat products intended for consumer use are packaged immediately after deodorization. According to O'Brien and Baird (2000), "the ideal container provides product protection, cost effectiveness, end use application, and appealing appearance."

Packaging materials include paper, aluminum, and various plastics. The materials must be kept free from microbial contamination prior to filling. Liquid oils are typically packaged in clear polyethylene terephthalate (PET) containers ranging in size from 16 to 64 fluid ounces or in larger polyvinyl chloride (PVC) opaque jugs (O'Brien, 2000). Clear containers promote oxidation but are generally preferred by consumers. In the case of plastic containers, the potential exists for transfer of flavor and aroma compounds between the oil and container (O'Brien & Baird, 2000). "Scalping" refers to absorption of these compounds by the container, while "leaching" describes transfer of flavors and aromas from the packaging to the oil. Glass or metal packaging is typically used for specialty oils.

Packaging for other products is often chosen based on the shelf life necessary for the product. For example, shortening cans consist of a composite of multiple layers of paper and plastic, with the aim of extending the shelf life (Caudill, 2005). Packages may be filled by level, volume, or net weight (Caudill, 2005). After filling, nitrogen blanketing or sparging may be used to extend the product's shelf life prior to opening of the container. Temperature control is especially important during transport and storage to maintain the quality of the product.

20.11 Margarine processing

Margarine is a solid water-in-oil emulsion that was invented in France in 1869 as a substitute for butter. Margarine is produced by blending base-stock(s) with water, milk solids, emulsifiers, and other ingredients to yield a product with similar characteristics to butter.

To be classified as margarine in the US, a product must contain at least 80% fat by weight (CFR, 2010b).

The principal varieties of margarine are:
- *solid (stick) margarine*, usually available in 1 lb packages consisting of four sticks
- *soft (tub) margarine*, which has a lower melting point than solid margarine due to a higher content of polyunsaturated fatty acids
- *whipped (tub) margarine*, injected with nitrogen to yield a product with increased volume
- *liquid margarine*, available in squeezable plastic bottles. Products with reduced fat levels are widely available and are classified as "spreads." While spreads often use the same fat blends as margarines, their water content is higher, and thus they require the use of thickeners as well as additional emulsifiers, preservatives, and coloring agents (O'Brien, 2004b). Some spreads currently marketed are fortified with nutrients such as long-chain omega-3 fatty acids, as well as esters of plant sterols and stanols (saturated derivatives of sterols).

20.11.1 Ingredients and their functionality

Standards for the production of margarine are listed in the Code of Federal Regulations (CFR 21.166.110), including ingredients and their allowable levels (CFR, 2010b).

20.11.1.1 Mandatory ingredients

Margarine must contain one or more fats or oils of plant, animal, or marine origin. Plant-based oils commonly used include soybean, cottonseed, canola, and palm oils. These oils have often been subjected to modification procedures (e.g. fractionation, hydrogenation) in order to obtain the desired functional characteristics, especially spreadability and melting behavior. Two or more base-stocks may be blended to provide properties from both. Interesterification may be utilized to produce margarines with zero *trans* fatty acids. The consistency of the final product is highly dependent upon the fats and oils selected.

One or more aqueous phase ingredients are required, which must be pasteurized and may be cultured. Water, milk, milk products, and edible proteins (e.g. whey, soy) are permitted. Most manufacturers use spray-dried whey, which promotes browning during pan-frying, similar to butter (Chrysan, 2005).

Vitamin A is required at a level of at least 15,000 IU/lb. β-Carotene is usually added until the desired color is reached, followed by addition of colorless esters of vitamin A until the required level is achieved (O'Brien, 2004a).

20.11.1.2 Optional ingredients

Natural and artificial flavorings are permitted, some of which simulate the flavor of butter (e.g. diacetyl). Salt and sweeteners contribute to the flavor profile of the product, and salt also possesses antimicrobial properties. Acidulants or alkalizers may be added to adjust the pH of the product, which affects flavor perception. Color additives include synthetic β-carotene as well as natural extracts that are rich sources of carotenoids (e.g. annatto, red palm oil).

Emulsifiers act at the interface of the oil and aqueous phases, stabilizing the emulsion. A combination of lecithin, mono- and diacylglycerols is typically used (Chrysan, 2005; O'Brien, 2004b). Preservatives include sorbic acid, used as an antimicrobial agent, and ethylene-diamine tetra-acetate (EDTA), which chelates trace metals. Antioxidants are permitted, but the level of natural antioxidants in the fats is typically sufficient. Margarine may be fortified with vitamin D; if this is chosen, the level of fortification must be at least 1500 IU/lb.

20.11.2 Structure

Margarine consists of a continuous crystalline network; the liquid oil and solid fat crystals contribute structural support, while the aqueous phase is dispersed throughout the system (Heertje, 1993). The fat crystals provide stability for the aqueous phase. The aqueous droplets generally range in size from 1 to 20 μm (Bumbalough, 2000). The resultant product possesses plastic properties.

20.11.3 Production

The general process for margarine production is shown in Figure 20.4. The equipment used in all production steps should be stainless steel or plated nickel, which resists corrosion (O'Brien, 2004a). Since consumers desire products that are spreadable upon removal from the refrigerator, maintain their solid state at room temperature, and rapidly melt in the mouth, solid fat index values in these temperature ranges are monitored during production.

20.11.3.1 Preparation of the emulsion

Either a batch or continuous process may be used. In general, the selected base-stocks are warmed to liquid state, then blended with the fat-soluble ingredients with agitation. The aqueous phase ingredients are weighed, pasteurized, and mixed in a separate tank. Finally, the oil and aqueous phase ingredients are combined with agitation and mild heat, yielding a semi-solid emulsion.

20.11.3.2 Solidification

The emulsion is fed into a scraped surface heat exchanger (e.g. Votator®, Waukesha Cherry-Burrel, Delavan, WI), which is a jacketed chilling cylinder utilizing blades to continuously scrape the walls, promoting adequate heat transfer. For soft (tub) and whipped products, nitrogen is injected at low or high levels, respectively, to reduce the product's density and improve its appearance. The emulsion is rapidly chilled to refrigerated temperature in less than 30 sec (O'Brien, 2004a). This super-cooling produces small nuclei for crystallization.

Super-cooling promotes the formation of α crystals, which rapidly transform to the β′ form. The β′ form is desired for margarine, as the three-dimensional network of fine crystals traps a high volume of liquid oil, resulting in a product with plastic properties. The β′ crystals may further transform to the β form if the fat blend has that crystallization tendency. Large β crystals are undesirable, as they produce a coarse texture. This transformation is irreversible in the completed product; however, melting and recrystallization may be done during production (O'Brie, 2004a).

20.11.3.3 Plasticization and tempering

The mixture is pumped from the scraped surface heat exchanger to a closed vessel. For soft (tub) and whipped margarines, a "working unit" is utilized. The working unit consists of a revolving shaft with finger-like projections that interact with corresponding projections on the wall of the tank, providing consistent mixing during the crystallization process. An aging tube is used without agitation for stick products, as hardness is desired. Crystallization is allowed to proceed slowly until equilibrium is reached. These processes serve as methods of tempering the final product, therefore extending its plastic range. Liquid margarines are tempered with a period of holding and agitation, so as to increase their stability (O'Brien, 2004a).

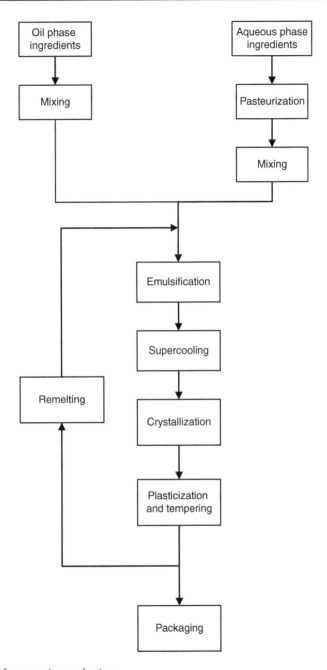

Figure 20.4 Flow diagram for margarine production.

20.11.4 Packaging

Two methods are used to package stick margarine. The sticks may first be formed, then wrapped in paper or foil wrappers. Alternatively, the wrappers may be placed into a mold, for injection of the semi-solid product into the wrapper (Bumbalough, 2000; O'Brien, 2004a). For tub, whipped, and liquid margarines, the products are filled directly into their respective packages. Extensive labeling requirements for margarine are provided in CFR 21.166.40 (CFR, 2010c). In particular, the word "oleomargarine" or "margarine" must appear in prominent type on the package. Specific nutrient content claims and health claims are regulated by the FDA.

20.12 Mayonnaise processing

Mayonnaise is a creamy, semi-solid food product used as a dressing and sandwich spread. Mayonnaise is an oil-in-water emulsion and is unique in that the dispersed phase (oil) is present at approximately seven times more than the aqueous phase. Because the emulsion is thermodynamically unstable (Ghosh & Rousseau, 2010), the process for producing a stable emulsion is delicate. Specific commercial processing techniques are patented.

Recent trends in mayonnaise production include flavored products, fortification with omega-3 fatty acids and plant sterols and stanols, and the use of nontraditional oils (e.g. olive).

20.12.1 Ingredients and their functionality

Requirements for the production of mayonnaise are listed in the Code of Federal Regulations (CFR 21.169.140); in particular, ingredients and their allowable levels are provided (CFR, 2010c).

20.12.1.1 Mandatory ingredients

Mayonnaise must contain one or more vegetable oils, one or more acidifying ingredients, and one or more egg yolk-containing ingredients (CFR, 2010c). While vegetable oils must make up at least 65% by weight of the product, most commercial mayonnaise contains approximately 80% oil. Salad oils are utilized, with clarity at refrigerated temperatures, since crystallization causes breakdown of the emulsion. Oils commonly used in the production of mayonnaise include soybean, canola, and cottonseed. Soybean and canola oils may be subjected to brush hydrogenation to increase their oxidative stability.

Permissible acidifying ingredients are vinegar, lemon juice, and lime juice. Vinegar may be diluted with water so that the final acetic acid content is at least 2.5% by weight. Vinegar contributes to the flavor of the product and functions as a preservative against spoilage. Lemon and/or lime juice may be diluted with water so that the final citric acid content is at least 2.5% by weight.

Egg yolk-containing ingredients include egg yolks or whole eggs in liquid, frozen, or dried form; liquid or frozen egg whites may also be added. Egg yolk contains phospholipids, which serve as emulsifiers. The amount of egg yolk necessary to stabilize the emulsion varies based on the amount and make-up of the oil and aqueous phases, as well as the size of the oil droplets (Hill & Krishnamurthy, 2005). Egg yolk-containing ingredients affect the color, flavor, and texture of the product.

20.12.1.2 Optional ingredients

Salt and sweeteners contribute to the flavor of mayonnaise, act as preservatives against microbial spoilage, and contribute to the stability of the emulsion. Spices and natural flavorings typically include mustard flour, paprika, and white pepper; saffron, turmeric, and others that simulate the color of egg yolk are excluded. Monosodium glutamate is permissible but rarely used due to health concerns.

Citric acid and/or malic acid may be added in conjunction with vinegar at a level of no more than 25% of the weight of the acetic acid in the vinegar. Citric acid serves to chelate trace metals, which act as pro-oxidants, and malic acid contributes to the flavor of the product and increases shelf life (Moustafa, 1995). Sequestrants, including calcium disodium EDTA, bind calcium ions and are used to preserve color and flavor. Crystallization inhibitors, (e.g. oxystearin, lecithin) may be added to the oil.

20.12.2 Structure

The aqueous phase contains the acidifying ingredients, flavorings, and seasonings, while the egg yolk solids are present at the interface between the dispersed (oil) and aqueous phases. The oil droplets are closely packed (Heertje, 1993) in an arrangement resembling honeycomb, with droplet size ranging from 2 to 25 μm (Langton et al., 1999). The stability of the emulsion and texture of the final product are influenced by many factors, including the size of the oil droplets, temperature and quality of ingredients, and the equipment and techniques used for preparation (Hill & Krishnamurthy, 2005; Lluch et al., 2003).

20.12.3 Preparation

A batch, semi-continuous, or continuous process may be used. In batch and semi-continuous processes, the seasonings and egg yolks are beaten in a Dixie mixer with a portion of the vinegar. The oil is slowly added, beating constantly, followed by addition of the remainder of the vinegar (Hill & Krishnamurthy, 2005). The coarse emulsion is then passed through a Charlotte colloid mill, which reduces the size of the oil droplets, yielding a smooth, creamy texture (Hill & Krishnamurthy, 2005; O'Brien, 2004b). Continuous operations are usually two-stage processes. First, the premixed ingredients are pumped to an in-line mixer, where a coarse emulsion is formed. Next, the mixture is fed into a colloid mill for further emulsification. Finally, the product is pumped to a filler for packaging.

20.12.4 Packaging and handling

Although mayonnaise has an extended shelf life, the emulsion is subject to breakdown. Care must be taken to avoid excessive heat, freezing, or agitation, as these may break the emulsion, leading to separation of the oil and water phases. Further, since mayonnaise has a high oil content, it is subject to oxidation. The risk of oxidation may be minimized by utilizing antioxidants and high-quality oils in production, adding carbon dioxide or nitrogen to the mixture during preparation of the emulsion, and minimizing headspace in the container while filling.

In recent years, glass jars have been abandoned in favor of plastic packaging. Mayonnaise is typically packaged in PET containers in the form of the traditional jar shape, a rectangular shape, or a squeezable bottle.

Products must be labeled "mayonnaise" and the label must include a list of all ingredients contained in the product (CFR, 2010c).

20.13 Sustainability

Sustainability has been pursued in many areas of the production of plant-based fats and oils. Energy expenditures in extraction and refining operations have been reduced by heat recovery techniques as well as the use of solid waste from starting materials to fuel downstream operations. Improvements in design and engineering, specifically solvent recovery systems and countercurrent flow techniques, have achieved reduction in hexane emissions. Physical refining is becoming increasingly popular due to the reduction in waste water versus chemical refining (Gunstone, 2008b). Efforts are directed toward the promotion of sustainable agricultural practices for oil-bearing crops, particularly the oil palm. Finally, single-cell oils and transgenic crops continue to be explored as sustainable alternative sources of long-chain omega-3 fatty acids.

References

American Heart Association (AHA) (2011) *Fish 101.* www.heart.org/HEARTORG/GettingHealthy/NutritionCenter/Fish-101_UCM_305986_Article.jsp, accessed 26 November 2013.

Anderson D (2005) A primer on oils processing technology. In: Shahidi F (ed) *Bailey's Industrial Oil and Fat Products. Volume 5: Edible Oil and Fat Products: Processing Technologies*, 6th edn. Hoboken, NJ: John Wiley, pp. 1–56.

Beckett ST (2000) *The Science of Chocolate.* Cambridge: RSC Publishing.

Beers AEW (2007) Low trans hydrogenation of edible oils. *Lipid Technology* **19**(3):56–58.

Behrens PW, Kyle DJ (1996) Microalgae as a source of fatty acids. *Journal of Food Lipids* **3**(4): 259–72.

Boskou D (2002) Olive oil. In: Gunstone FD (ed) *Vegetable Oils in Food Technology: Composition, Properties, and Uses.* Oxford: Blackwell, pp. 244–277.

Bray GA, Popkin BM (1998) Dietary fat intake does affect obesity! *American Journal of Clinical Nutrition* **68**(6): 1157–1173.

Bumbalough J (2000) Margarine types and preparation technology. In: O'Brien RD, Farr WE, Wan PJ (eds) *Introduction to Fats and Oils Technology*, 2nd edn. Champaign, IL: American Oil Chemists Society Press, pp. 452–462.

Caudill V (2005) Packaging. In: Shahidi F (ed) *Bailey's Industrial Oil and Fat Products. Volume 5: Edible Oil and Fat Products: Processing Technologies*, 6th edn. Hoboken, NJ: John Wiley, pp. 231–266.

CFR (2010a) Labeling of margarine. In: *Code of Federal Regulations, 4–1–10 ed. 21 CFR 166.40.* Washington, DC: Federal Register, pp. 584–585.

CFR (2010b) Margarine. In: *Code of Federal Regulations, 4–1–10 ed. 21 CFR 166.40.* Washington, DC: Federal Register, pp. 585–587.

CFR (2010c) Mayonnaise. In: *Code of Federal Regulations, 4–1–10 ed. 21 CFR 166.40.* Washington, DC: Federal Register, pp. 592–594.

Choe E, Min DB (2006) Mechanisms and factors for edible oil oxidation. *Comprehensive Review of Food Science and Food Safety* **5**(4): 169–186.

Chrysan MM (2005) Margarines and spreads. In: Shahidi F (ed) *Bailey's Industrial Oil and Fat Products. Volume 5: Edible Oil and Fat Products: Processing Technologies*, 6th edn. Hoboken, NJ: John Wiley, pp. 33–82.

Cmolik J, Pokorny J (2000) Physical refining of edible oils. *European Journal of Lipid Science and Technology* **102**(7): 472–486.

Duchateau GSMJE, van Oosten HJ, Vasconcellos MA (1996) Analysis of *cis*- and *trans*-fatty acid isomers in hydrogenated and refined vegetable oils by capillary gas–liquid chromatography. *Journal of the American Oil Chemists Society* **73**(3): 275–282.

Farr WE (2000) Refining of fats and oils. In: O'Brien RD, Farr WE, Wan PJ (eds) *Introduction to Fats and Oils Technology*, 2nd edn. Champaign, IL: American Oil Chemists Society Press, pp. 136–57.

Fellows PJ (2009) Separation and concentration of food components. In: *Food Processing Technology: Principles and Practice*, 3rd edn. Boca Raton, FL: CRC Press, pp. 188– 228.

Fils JM (2000) The production of oils. In: Hamm W, Hamilton RJ (eds) *Edible Oil Processing*. Sheffield,: Sheffield Academic Press, pp. 47–78.

Firestone D (ed) (2009) *Official Methods and Recommended Practices of the AOCS*, 6th edn. Champaign, IL: American Oil Chemists' Society Press.

Food and Agriculture Organization (FAO) (2009a) *Codex standard for named vegetable oils*. CODEX STAN 210–1999. Rome: FAO, pp. 1–13.

Food and Agriculture Organization (FAO) (2009b) *Codex standard for olive oils and olive pomace oils*. CODEX STAN 33–1981. Rome: FAO, pp. 1–8.

Food and Drug Administration (FDA) (2003) *Food Labeling: Trans Fatty Acids in Nutrition Labeling, Nutrient Content Claims, and Health Claims. Final Rule*. Washington, DC: Federal Register **68**(133): 41433–506.

Foreign Agricultural Service (FAS) (2011) *Oilseeds: World Markets and Trade*. Circular Series, 07–11. Washington, DC: USDA Foreign Agricultural Service.

Frega N, Mozzon M, Lercker G (1999) Effects of free fatty acids on oxidative stability of vegetable oil. *Journal of the American Oil Chemists Society* **76**(3): 325–329.

Ghosh S, Rousseau D (2010) Emulsion breakdown in foods and beverages. In: Skibsted LH, Risbo J, Andersen ML (eds) *Chemical Deterioration and Physical Instability of Food and Beverages*. Cambridge: Woodhead Publishing, pp. 260–295.

Gorbet DW, Shokes FM (2002) Registration of 'C-99R' peanut. *Crop Science* **42**(6): 2207.

Gunstone FD (1998) Movements toward tailor–made fats. Prog Lipid Res **37**(5):277–305. doi: 10.1016/S0163–7827(98) 00012–5.

Gunstone FD (2004) Extraction, refining and processing. In: *The Chemistry f Oils and Fats: Sources, Composition, Properties and Uses*. Oxford: Blackwell, pp. 36–49.

Gunstone FD (2006a) Minor specialty oils. In: Shahidi F (ed) *Nutraceutical and Specialty Lipids and Their Co-Products*. Boca Raton, FL: CRC Press, pp. 91–126.

Gunstone FD (2006b) Modified oils. In: Shahidi F (ed) *Nutraceutical and Specialty Lipids and Their Co-Products*. Boca Raton, FL: CRC Press, pp. 313–328.

Gunstone FD (2006c) Vegetable sources of lipids. In: *Modifying Lipids for Use in Food*. Cambridge: Woodhead, pp. 11–27.

Gunstone FD (2008a) Extraction, refining, and modification processes. In: *Oils and Fats in the Food Industry*. Chichester: Wiley-Blackwell, pp. 26–36.

Gunstone FD (2008b) The major sources of oils and fats. In: *Oils and Fats in the Food Industry*. Chichester: Wiley-Blackwell, pp. 11–25.

Gunstone FD, Martini S (2010) Chemical and physical deterioration of bulk oils and shortenings, spreads and frying oils. In: Skibsted LH, Risbo J, Andersen ML (eds) *Chemical Deterioration and Physical Instability of Food and Beverages*. Cambridge: Woodhead Publishing, pp. 413–438.

Gupta MK (2002) Sunflower oil. In: *Vegetable Oils in Food Technology: Composition, Properties, and Uses*. Oxford: Blackwell, pp. 128–156.

Hastert RC (2000) Hydrogenation. In: O'Brien RD, Farr WE, Wan PJ (eds) *Introduction to Fats and Oils Technology*, 2nd edn. Champaign, IL: American Oil Chemists Society Press, pp. 179–193.

Heertje I (1993) Structure and function of food products: a review. *Food Structure* **12**(3): 343–364.

Hénon G, Kemény Z, Recseg K, Zwobada F, Kovari K (1999) Deodorization of vegetable oils. Part I: Modelling the geometrical isomerization of polyunsaturated fatty acids. *Journal of the American Oil Chemists Society* **76**(1): 73–81.

Hernandez E (2005) Production, processing, and refining of oils. In: Akoh CC, Lai OM (eds) *Healthful Lipids*. Urbana, IL: American Oil Chemists Society Press, pp. 48–64.

Hidalgo FJ, Zamora R (2006) Fats: physical properties. In: Hui YH (ed) *Handbook of Food Science, Technology, and Engineering. Volume 1: Food Science: Properties and Products*. Boca Raton, FL: CRC Press, pp. 9-1–9-27.

Hill SE, Krishnamurthy RG (2005) Cooking oils, salad oils, and dressings. In: Shahidi F (ed) *Bailey's Industrial Oil and Fat Products. Volume 5: Edible Oil and Fat Products: Processing Technologies*, 6th edn. Hoboken, NJ: John Wiley, pp. 175–206.

Horwitz W, Lattimer GW (eds) (2005) *Official Methods of Analysis of AOAC International*, 18th edn. Arlington, VA: AOAC International.

Hu FB, Willett WC (2002) Optimal diets for prevention of coronary heart disease. *Journal of the American Medical Association* **288**(20): 2569–2578.

Hu FB, Manson JE, Willett WC (2001) Types of dietary fat and risk of coronary heart disease: a critical review. *Journal of the American College of Nutrition* **20**(1): 5–19.

Jeyarani T, Reddy SY (2003) Preparation of plastic fats with zero trans FA from palm oil. *Journal of the American Oil Chemists Society* **80**(11): 1107–1113.

Johnson LA (2000) Recovery of fats and oils from plant and animal sources. In: O'Brien RD, Farr WE, Wan PJ (eds) *Introduction to Fats and Oils Technology*, 2nd edn. Champaign, IL: American Oil Chemists Society Press, pp. 108–135.

Johnson LA (2008) Recovery, refining, converting, and stabilizing edible fats and oils. In: Akoh CC, Min DB (eds) *Food Lipids: Chemistry, Nutrition, and Biotechnology*, 3rd edn. Boca Raton, FL: CRC Press, pp. 205–244.

Kanner J (2007) Dietary advanced lipid oxidation endproducts are risk factors to human health. *Molecular Nutrition and Food Research* **51**(9): 1094–1101.

Kellens M, de Greyt W (2000) Deodorization. In: O'Brien RD, Farr WE, Wan PJ (eds) *Introduction to Fats and Oils Technology*, 2nd edn. Champaign, IL: American Oil Chemists Society Press, pp. 235–268.

Kellens M, Hendrix M (2000) Fractionation. In: O'Brien RD, Farr WE, Wan PJ (eds) *Introduction to Fats and Oils Technology*, 2nd edn. Champaign, IL: American Oil Chemists Society Press, pp. 194–207.

Kellens M, Gibon V, Hendrix M, de Greyt W (2007) Palm oil fractionation. *European Journal of Lipid Science and Technology* **109**(4): 336–349.

Kemper TG (2005) Oil extraction. In: Shahidi F (ed) *Bailey's Industrial Oil and Fat Products. Volume 5: Edible Oil and Fat Products: Processing Technologies*, 6th edn. Hoboken, NJ: John Wiley, pp. 57–98.

Kochhar SP (2002) Sesame, rice-bran and flaxseed oils. In: Gunstone FD (ed) *Vegetable Oils in Food Technology: Composition, Properties, and Uses*. Oxford: Blackwell, pp. 297–326.

Kris-Etherton, PM, Taylor DS, Yu-Poth S et al. (2000) Polyunsaturated fatty acids in the food chain in the United States. *American Journal of Clinical Nutrition* **71**(S): 179S–188S.

Lampert D (2000) Processes and products of interesterification. In: O'Brien RD, Farr WE, Wan PJ (eds) *Introduction to Fats and Oils Technology*, 2nd edn. Champaign, IL: American Oil Chemists Society Press, pp. 208–234.

Lands WEM (1986) *Fish and Human Health*. Orlando, FL: Academic Press, pp. 1–170.

Langton ME, Jordansson E, Altskar A, Sorenson C, Hermansson A (1999) Microstructure and image analysis of mayonnaises. *Food Hydrocolloids* **13**(2):113–125.

Lluch MA, Hernando I, Perez-Munuera I (2003) Lipids in food structures. In: Sikorski ZE, Kolakowska A (eds) *Chemical and Functional Properties of Food Lipids*. Boca Raton, FL: CRC Press, pp. 9–28.

McKeon TA (2005) Transgenic oils. In: Shahidi F (ed) *Bailey's Industrial Oil and Fat Products. Volume 5: Edible Oil and Fat Products: Processing Technologies*, 6th edn. Hoboken, NJ: John Wiley, pp. 155–174.

Miyashita K, Takagi T (1986) Study on the oxidative rate and prooxidant activity of free fatty acids. *Journal of the American Oil Chemists Society* **63**(10): 1380–1384.

Moustafa A (1995) Salad oil, mayonnaise, and salad dressings. In: Erickson DR (ed) *Practical Handbook of Soybean Processing and Utilization*. Champaign, IL: American Oil Chemists Society Press, pp. 314–338.

Muggli R (2006) Fortified foods: a way to correct low intakes of EPA and DHA. In: Akoh CC (ed) *Handbook of Functional Lipids*. Boca Raton, FL: CRC Press, pp. 389–402.

Murphy DJ (2006) Plant breeding to change lipid composition for use in food. In: Gunstone FD (ed) *Modifying Lipids for Use in Food*. Cambridge: Woodhead Publishing, pp. 273–305.

O'Brien RD (2000) Liquid oils technology. In: O'Brien RD, Farr WE, Wan PJ (eds) *Introduction to Fats and Oils Technology*, 2nd edn. Champaign, IL: American Oil Chemists Society Press, pp. 463–85.

O'Brien RD (2002) Cottonseed oil. In: Gunstone FD (ed) *Vegetable Oils in Food Technology: Composition, Properties, and Uses*. Oxford: Blackwell, pp. 203–30.

O'Brien RD (2004a) Fats and oils processing. In: O'Brien RD (ed) *Fats and Oils: Formulating and Processing for Applications*, 2nd edn. Boca Raton, FL: CRC Press, pp. 57–174.

O'Brien RD (2004) Margarine. In: O'Brien RD (ed) *Fats and Oils: Formulating and Processing for Applications*, 2nd edn. Boca Raton, FL: CRC Press, pp. 383–400.

O'Brien RD (2004b) Raw materials. In: O'Brien RD (ed) *Fats and Oils: Formulating and Processing for Applications*, 2nd edn. Boca Raton, FL: CRC Press, pp. 1–56.

O'Brien RD, Baird L (2000) Packaging and bulk handling of edible fats and oils. In: O'Brien RD, Farr WE, Wan PJ (eds) *Introduction to Fats and Oils Technology*, 2nd edn. Champaign, IL: American Oil Chemists Society Press, pp. 269–284.

O'Keefe SF, Wiley VA, Wright D (1993) Effect of temperature on linolenic acid loss and 18:3 Δ9-*cis*, Δ12-*cis*, Δ15-*trans* formation in soybean oil. *Journal of the American Oil Chemists Society* **70**(9): 915–917.

Pantzaris TP, Basiron Y (2002) The lauric (coconut and palmkernel) oils. In: Gunstone FD (ed) *Vegetable Oils in Food Technology: Composition, Properties, and Uses*. Oxford: Blackwell, pp. 157–202.

Pariza MW, Park Y, Cook ME (2001) The biologically active isomers of conjugated linoleic acid. *Progress in Lipid Research* **40**(4): 283–298.

Pokorny J (2006) Fats, and oils: science and applications. In: Hui YH (ed) *Handbook of Food Science, Technology, and Engineering. Volume 1: Food Science: Properties and Products*. Boca Raton, FL: CRC Press, pp. 34-1–34-21.

Qi B, Fraser T, Mugford S et al. (2004) Production of very long chain polyunsaturated omega-3 and omega-6 fatty acids in plants. *Nature Biotechnology* **22**(6): 739–745.

Ratledge C (2004) Fatty acid biosynthesis in microorganisms being used for single cell oil production. *Biochimie* **86**(11): 807–815.

Richards MP (2006) Lipid chemistry and biochemistry. In: Hui YH (ed) *Handbook of Food Science, Technology, and Engineering. Volume 1: Food Science: Properties and Products*. Boca Raton, FL: CRC Press, pp. 8-1–8-21.

Ruxton CHS, Reed SC, Simpson MJA, Millington KJ (2004) The health benefits of omega-3 polyunsaturated fatty acids: a review of the evidence. *Journal of Human Nutrition and Diet* **17**(5): 449–459.

Simopoulos AP (2008) The importance of the omega-6/omega-3 fatty acid ratio in cardiovascular disease and other chronic diseases. *Experimental Biology and Medicine* **233**(6): 674–688.

Taylor DR (2005) Bleaching. In: Shahidi F (ed) *Bailey's Industrial Oil and Fat Products. Volume 5: Edible Oil and Fat Products: Processing Technologies*, 6th edn. Hoboken, NJ: John Wiley, pp. 285–340.

USDA (2010) *Dietary Guidelines for Americans 2010*. www.health.gov/dietaryguidelines/dga2010/DietaryGuidelines2010.pdf, accessed 26 November 2013.

Venegas-Caleron M, Sayanova O, Napier JA (2010) An alternative to fish oils: metabolic engineering of oil-seed crops to produce omega-3 long chain polyunsaturated fatty acids. *Progress in Lipid Research* **49**(2): 108–119.

Wakelyn PJ, Wan PJ (2006) Solvent extraction to obtain edible oil products. In: Akoh CC (ed) *Handbook of Functional Lipids*. Boca Raton, FL: CRC Press, pp. 89–131.

Wan PJ, Pakarinen DR (1995) Comparison of visual and automated colorimeter for refined and bleached cottonseed oils. *Journal of the American Oil Chemists Society* **72**(4): 455–458.

Wang T (2002) Soybean oil. In: Gunstone FD (ed) *Vegetable Oils in Food Technology: Composition, Properties, and Uses*. Oxford: Blackwell, pp. 18–58.

Ward OP, Singh A (2005) Omega-3/6 fatty acids: alternative sources of production. *Process Biochemistry* **40**(12): 3627–3652.

Wille RL, Lutton ES (1966) Polymorphism of cocoa butter. *Journal of the American Oil Chemists Society* **43**(8): 491–496.

Wright AJ, Marangoni AG (2006) Physical properties of fats and oils. In: Akoh CC (ed) *Handbook of Functional Lipids*. Boca Raton, FL: CRC Press, pp. 135–162.

Zhao G, Etherton TD, Martin KR, West SG, Gillies PJ, Kris-Etherton PM (2004) Dietary α-linolenic acid reduces inflammatory and lipid cardiovascular risk factors in hypercholesterolemic men and women. *Journal of Nutrition* **134**(11): 2991–2997.

Zschau W (2000) Bleaching. In: O'Brien RD, Farr WE, Wan PJ (eds) *Introduction to Fats and Oils Technology*, 2nd edn. Champaign, IL: American Oil Chemists Society Press, pp. 158–178.

21 Fats and Oils – Animal Based

Stephen L. Woodgate[1] and Johan T. van der Veen[2]

[1]Beacon Research, Clipston, Leicestershire, UK
[2]Ten Kate Holding, Musselkanaal, The Netherlands

21.1 Introduction

Lipids are of vital importance. Without lipids there is no life, we cannot think, and we experience no sense. Lipids are vital components in cell membranes and for the construction of vitamins and glycolipids used as intracellular messengers. The main categories of lipids are the triglycerides or fats produced from plant, marine, and animal sources. Fats are major taste, flavor, and texture components in daily foodstuffs and provide a significant part of our energy needs. Animal fats provide important amounts of fats used in the world today, both for foods and for many non-food applications.

Animal fats are produced by a process termed rendering, which has been developed over the last 200 years or so from very basic cooking and melting processes to refined industrial processes that are essentially following the same process steps. Rendering is used to convert land animal slaughter by-products into marketable products, including edible and inedible animal fats and proteins for food, agricultural, and industrial use. Animal fat is a versatile, sustainable, and natural basis for many products. Wordwide, 172 million tonnes of vegetable and animal oils and fats are produced annually, from which approximately 25 million tonnes (14%) are estimated to be of animal origin (REA, 2013). These animal fats are mainly categorized as tallow, lard, fish oil, and butter.

This chapter will focus on fats derived from land animals slaughtered for the production of meat for human consumption. It will follow the sourcing of raw materials, the processing standards applied, and the necessary legislation. European systems will be used to illustrate the main features of animal-based fat production and to enable the differences between specific types of process systems (such as food, inedible, and technical) to be emphasized. Aspects of safety, quality, applications, nutrition, and sustainability will also be considered. Woodgate and van der Veen (2004) published a review of the fat processing sector, which comprised a comprehensive synopsis of regulatory controls currently in place and those foreseen for the future as well as providing statistical data.

21.2 Raw materials

In Europe, about 16 million tonnes by-products are processed annually by fat processors and renderers, with Germany, France, UK, and Spain as the main processing countries. About 10 million tonnes (63%) are fit for the food chain and are classified according to EU regulations as food or low-risk animal by-products. Details of the categorization of food-grade and animal by-products will be covered in later sections. However, certain raw materials such as specified risk animal by-products, which are only suitable for technical applications or energy generation by combustion, amount to 6 million tonnes. From these slaughter by-products, about 2.8 million tonnes of animal fat and 4 million tonnes of animal proteins are produced annually in Europe (EFPRA, 2011). Rendering plants operate either in conjuction with slaughterhouses, the so-called integrated rendering plants, or as independent plants specialized in the off-site collection and processing of meat by-products. Sanitation and utilization of slaughter by-products are not easy, due to the biological instability, microbial activity, high water content, and enzymatic activities in the raw material arising from the slaughter. In this respect, rapid processing and cooling are essential to preserve the high quality to make the maximum utilization possible.

Food Processing: Principles and Applications, Second Edition. Edited by Stephanie Clark, Stephanie Jung, and Buddhi Lamsal.
© 2014 John Wiley & Sons, Ltd. Published 2014 by John Wiley & Sons, Ltd.

The utilization of meat by-products for food is often dependent on traditions, culture, and religion. Regulatory restrictions are also important, because some countries limit the use of certain by-products for food safety or quality reasons. The utilization of slaughter by-products can run through several pathways or industries depending on the type of raw material. For example, hides and skin are generally valorized by the leather or gelatin industry. In Europe, there is also a specialized industry processing blood for the production of blood plasma, hemoglobin , and blood meal. Bones, making up about 15% of the carcass weight, are utilized for gelatin or in the rendering industry for the production of animal fats and processed animal protein. Furthermore, other animal raw materials such as feathers, hair and other species' by-products are processed on separate or dedicated production lines, often by species. In the same way, fat processing to produce fats for human foods is operated as a dedicated and individual sector.

21.3 Land animals

Farmed terrestrial animals are used mainly for the production of food for human consumption. Two types of production are common. First, there are animal production systems for the continued harvesting of products from live animals, such as dairy products and eggs. The second is meat production, which involves the slaughter of animals to yield meat and edible by-products and fats, together with associated inedible animal by-products not intended for direct human consumption.

A flow chart describing the steps in the harvesting of raw fat from animals (using a pig as an example) for further processing into edible fat is shown in Figure 21.1.

There are some significant differences among the raw materials harvested from various species or groups of animals, which result in the fat products having different characteristics. In the majority of cases the fats

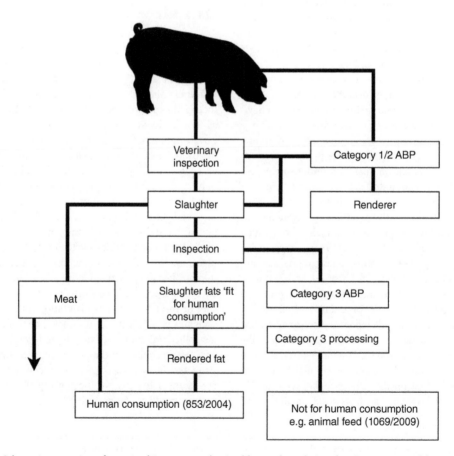

Figure 21.1 Schematic separation of an animal into meat and animal by-products (using the pig as an example).

Table 21.1 Composition of porcine and bovine animals (Jayathilakan et al., 2011)

	Pigs		Cattle	
	%	kg	%	kg
Market live weight		100		600
Whole carcass	77.5	77.5	63.0	378.0
Blood	3.0	3.0	18.0	4.0
Fatty tissue	3.0	3.0	4.0	24.0
Hide or skin	6.0	6.0	6.0	36.0
Organs	7.0	7.0	16.0	96.0
Head	5.9	5.9		
Viscera	10.0	10.0	16.0	96.0
Feet	2.0	2.0	2.0	12.0
Tail	0.1	0.1	0.1	6.0
Brain	0.1	0.1	0.1	6.0

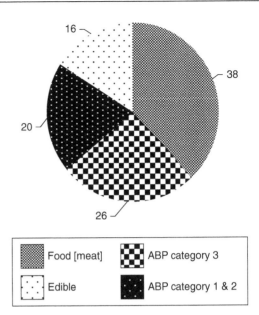

Food [meat]	ABP category 3
Edible	ABP category 1 & 2

Figure 21.2 Cattle.

from the different species are processed separately and the resulting products are named in relation to the source material.

The majority of by-products in the meat industry are produced during slaughtering. Slaughterhouse by-products consist of the portion of an animal that cannot be sold as meat or used as meat products. Slaughter by-products, which are not sold with the meat, are collected and processed by fat processors or renderers. Up to 50% of the live weight of some animals are considered to be by-products. This raw material consists of water, fat, protein, and minerals. Fat processors or renderers sanitize, separate and purify these components for utilization and creation of added value for the animal chain. Efficient utilization of the by-products is very important for the profitability and sustainability of the meat industry. It has been estimated that 7.5–11.4% of the gross income from the slaughter industry comes from pig and bovine by-products (Jayathilakan et al., 2011). Therefore, improved utilization of meat by-products (Table 21.1) can result in a higher sustainable economy for the entire animal production chain.

21.3.1 Bovine

The average slaughter weight of a bovine animal is about 550 kg. In the EU, almost 40% of the animal is utilized as meat or meat products. About 62% of the animal is specified as slaughter by-product (Figure 21.2). Edible

slaughter by-products fit for human consumption amount to approximately 16% and animal by-products suitable for the feed chain make up a quarter of the total weight; 20% of the weight is defined as specified risk material (SRM). Fats derived from SRM cannot be used in any food or animal feed applications and are used in alternative applications such as renewable fuels or converted to biodiesel for use in road transport.

21.3.2 Porcine

The average slaughter weight of a pig is about 110 kg. In the EU, 62% of the pig is utilized as meat or meat products whereas 38% is determined as slaughter by-product (Figure 21.3). Edible slaughter by-products fit for human consumption amount to 15%, and animal by-products fit for the feed chain make up about 20% of the total weight. A small part (4%) is condemned material due to veterinarian rejection of animals or meat by-products just before or during slaughter and is processed separately and used in the oleochemical industry or for energy generation.

21.3.3 Avian

A variety of poultry is slaughtered in Europe, from small birds like chickens and turkeys to big birds like ostrich. Most are chickens and the average slaughter weight of a

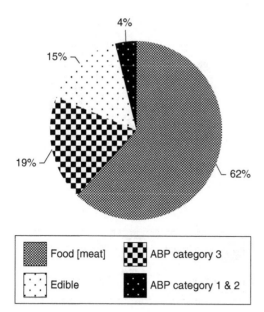

Figure 21.3 Pig.

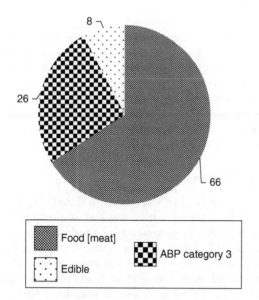

Figure 21.4 Chicken.

chicken is about 2 kg. In the EU, 66% of the chicken is utilized as meat or meat products; 33% is determined as slaughter by-product. Edible slaughter by-products fit for human consumption, for example skins, amount to about 8% and animal by-products fit for the feed chain make about 26% of the total weight (Figure 21.4).

21.4 Processing methods

The process of extraction of animal fats from raw material is termed melting or rendering. "Rendering" is an old word, which can mean different things to different people. In its simplest form, rendering means "to render open" (or split) – by heat processing – raw material into a solid (proteins) and a liquid (fat, liquid at elevated temperatures). While in theory, this covers all aspects of animal by-product processing, in the practical world rendering has, in many cases, become synonymous with the processing of inedible animal by-products. However, rendering can also be used to describe the processing of edible grade by-products, and in these circumstances edible rendering should be clearly stated, although many still prefer to use the alternative term, "fat processing."

There are two main systems of rendering, described as either "wet" or "dry" systems, with the latter being further divided into natural fat and added fat systems. However, this is still rather an oversimplification and in reality many types of processes are in existence throughout the world, and many have been altered and adapted in accordance with technical advances and legislative changes over the years.

In general, most rendering processes refer to the processing of fat-rich or bone-rich raw materials. A simplified generic process description for the processing of high-fat raw material is shown in Figure 21.5.

For wet melting (Figure 21.6), the heat applied is only enough to melt the fat, and both the "protein" and "fat" still contain water after decanting. The water is evaporated or separated in subsequent steps, with the final products being protein meal and rendered animal fat. Wet melting is preferably applied for the processing of edible fat-rich material, like cutting fat, back fat, and leaf fat. The main disadvantage of wet melting is that water is added by means of steam

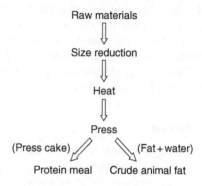

Figure 21.5 Generic process flowchart.

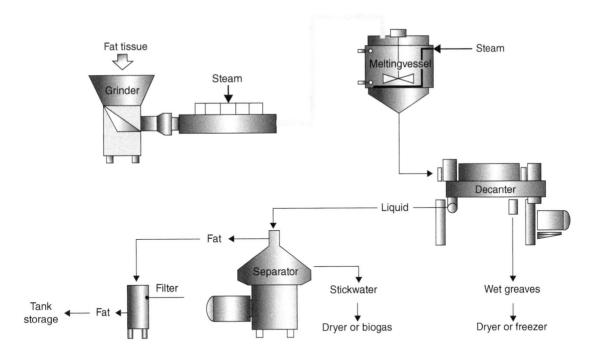

Figure 21.6 Wet rendering process.

Box 21.1 Different processing options for rendering and fat processing

System type: batch or continuous
Process fat level: natural fat or added fat or defatted
Process condition: atmospheric or vacuum or pressure

---

injection and much water has to be removed by physical separation and drying. On the other hand, wet melting applies less heat energy (up to 95 °C), which results in a higher quality fat and protein compared to dry rendering. In dry rendering, the raw material is boiled in its own fat or added fat until most of the water is evaporated. After evaporation of the water, physical separation takes place like sieving, decanting and pressing, to separate and purify the components, protein and fat. Dry rendering can be characterized as a frying process, usually with temperature between 115 °C and 135 °C. More detailed diagrams for the two main processing systems are shown in Figures 21.6 and 21.7.

There are many hybrid processes of the wet and dry methods used in the industry and these can be most simply described in terms of system type, fat level, and process condition. The possible options and/or combinations are shown in Box 21.1.

It is difficult to generalize about the main systems in use in the EU. However, it is realistic to say that continuous systems are now most prevalent, while defatted, natural fat, and added fat systems are equally commonly used. In terms of process, atmospheric and pressure systems are the most common, with stand-alone vacuum systems now being obsolete.

Fat processing or rendering is a serious business activity used to create added value in the animal chain. The European Association of Fat Processors was merged with the European Association of Renderers into a new organization, the European Fat Processors and Renderers Association (EFPRA), in 2001 to inform the public about the industrial application of animal slaughter by-products. Renderers processing dead stock and SRM have an important sanitation duty. Furthermore, the rendering industry is processing raw material sourced from approved animals on separate locations. Fat processors and renderers aim to be transparent in the animal chain and to act as responsible entrepreneurs to create added value. The business of fat processors is historically determined by the processing of edible animal slaughter by-products exclusively from approved animals for human consumption. The fat processing activity is divided into the production of animal fats and proteins

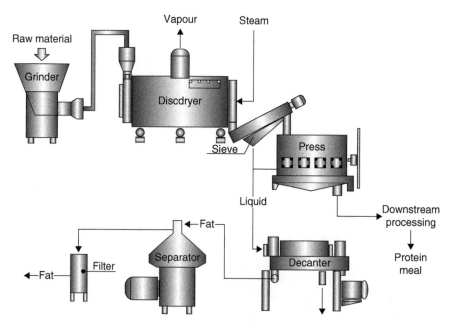

Figure 21.7 Dry rendering process.

fit for human consumption in accordance with EU food regulations and the processing of animal by-products not intended for human consumption.

Fat processing is historically associated with species-specific processing and the production of high-grade animal fats for specific markets like the bakery industry, calf milk replacers, and pet food. Within the framework of species-specific processing, the production of bovine fat, lard, and poultry fat can be distinguished. Animal fats are also supplied to the oleochemical industry for the production of detergents, cosmetics and technical products, and to the biodiesel industry.

The animal proteins produced are wet frozen or dried and used for production of gelatin or foodstuffs (e.g. meat products or as an ingredient for pet food). Fat processing is characterised by its fresh fat-rich raw material, like fat tissues, and relative mild processing conditions in order to preserve the product properties, which are essential for the applications. The freshness, quality, and source of the raw materials are important aspects of fat processing.

Fat tissues are collected daily from slaughterhouses and the fat tissues (internal) from animals slaughtered today are melted on the same evening. The principle of quality control is focused on the animal chain and based on the food safety system, Hazard Analysis Critical Control Point (HACCP). HACCP is explained in more details in the Codex Alimentarius, *General Principles of Food*

Hygiene (1969). The raw materials for fat processors are subject to the same quality and food safety inspections and monitoring program on environmental contaminants, growth hormones and veterinary drugs as is compulsory for meat. Fat processors are able to pay for slaughter by-products, therefore creating an added value for slaughterhouses.

The difference between renderers and fat processors was partly minimized by the introduction of a new European animal by-product regulation in 2002 which gave the animal by-product industry an important impulse to make strategic choices for processing for the food/pet food chain or for alternative uses. However, harmonization of European legislation remains a subject for attention. Some member states are still going further than the European rules, which has lead to serious disruption of the European market and complication of regulations. The clearest example of differences in opinion is the European acceptance of the use of animal fats in feed documented by the European Food Safety Authority (EFSA), where Germany and France have a partial prohibition on the use of certain animal fats in feed, even when it is fit for human consumption. However, the imposition of stricter national legislation by one or more member states in an attempt to add further safeguards for food safety is totally ineffective to protect consumers, because there is free trade of meat and meat by-products in the EU.

In Europe, the European Commission and authorities in each member state should listen to the opinion of the EFSA and harmonize the European regulations accordingly between the member states.

21.5 EU legislation

21.5.1 Edible animal fat, fit for human consumption

In Europe, legislation regulating the food sector includes the regulation on hygiene for foodstuffs (852/2004), together with the regulation with specific hygiene rules for food of animal origin (853/2004). The specific hygiene rules for edible fat processing are described in 853/2004, annex 2, section 7, Rendered Animal Fats and Greaves. Movement documentation, combined with HACCP systems at the various processing locations, ensures that a fully traceable system is in place from farm to fork. Typical edible animal fats produced by fat processors are poultry fat, lard, and premier jus (bovine fat). The maximal legal limits for animal fat produced in the EU for use in foodstuffs in the EU are clearly specified for each species and are described in Table 21.2.

21.5.2 Animal by-product regulation, fit for animal feed

The animal by-product regulation (ABPR 1069/2009), together with the implementation regulation 142/2011,

regulates the rendering industry in Europe for derived products fit for animal feed. In addition, codes of practice have been adopted by various sectors in the feed and feed ingredient industry, as retail pressures have added to the legislative requirements.

Category 3 material is derived exclusively from approved animals, slaughtered fit for human consumption, but is considered as an animal by-product not intended for human consumption.. The animal by-products of category 3 material have to be processed in conformance with rigorous production methods and sanitizing processing conditions. The final derived products, animal fats and proteins, are applied in the animal feed chain, but also used in the oleochemical industry and for the production of biodiesel. The rules permit downgrading of materials to a lower category, but upgrading is strictly prohibited. Furthermore, the processing of category 3 material has to be carried out in production locations separate from the other categories to avoid the possibility of cross-contamination.

21.5.3 Animal by-product regulation, technical applications

The ABPR (1069/2009), together with the implementation regulation (142/2011), also regulates the rendering industry for raw materials and derived products for technical applications. Animal by-products only for technical application are defined in the regulation as category 1 and category 2 material. Category 1 material consists mainly of specified risk material but also includes animal

Table 21.2 EU legal specification for edible fats

	Ruminants			Porcine animals			Other animal fat	
	Edible tallow		Tallow for refining	Edible fat		Lard and other fat for refining	Edible	For refining
	Premier jus*	Other		Lard†	Other			
FFA (m/m% oleic acid) maximum	0.75	1.25	3.0	0.75	1.25	2.0	1.25	3.0
Peroxide Maximum	4 meq/kg	4 meq/kg	6 meq/kg	4 meq/kg	4 meq/kg	6 meq/kg	4 meq/kg	10 meq/kg
Total insoluble impurities	Maximum 0.15%			Maximum 0.5%				
Odor, taste, color	Normal							

*Rendered animal fat obtained by low-temperature rendering of fresh fat from the heart, caul, kidneys and mesentery of bovine animals, and fat from cutting rooms.
†Rendered animal fat obtained from the adipose tissues of porcine animals.

by-products contaminated with environmental contaminants. Catgeory 2 material comprises condemned slaughterhouse by-product and farm dead stock. Category 1 and 2 material have to be processed in accordance with the prescribed production methods and processing conditions. With a few exceptions, category 1 material has to be processed according to a specific processing method, termed method 1. The specific conditions for method 1 comprise: cooking raw material <50 mm, under pressure of 3 bar at a temperature of 133 °C for a minimum of 20 min. As a result, category 1 material is only permitted to be used for energy generation. Category 2 material may also be used for technical applications.

21.5.4 Transmissible spongiform encephalopathy regulations

The transmissible spongiform encephalopathy regulation (TSE Regulation 999/2001) affects several aspects of the food chain in the EU. In particular, it proscribes the organs that are considered to be SRM in bovines, ovines, and caprines. SRM comprises parts or organs of bovine, ovine, and caprine animals that are considered to be a high risk to animal and human health. The main tissues defined as SRM are brain, vertebral column, thymus, and parts of intestines.These are regarded as category 1 animal by-products by the corresponding animal by-products regulations and must be removed from the food chain and used by only approved methods not considered to be a risk to human and animal health or the environment.

The same TSE regulations also lay down conditions under which certain animal proteins (derived from category 3 materials only) may be used in the food chain now and in the future.

21.5.5 Renewable energy

The Renewable Energy Directive (RED) (2009/28), is an important regulation for Europe as it attempts to convert European society to a more sustainable future. There are implications here for all fats, and animal fats in particular, that may be used as fuels or ingredients for fuels. Animal fats have excellent sustainability credentials (see section 10) and as such are suitable raw materials for fatty acid methyl ester (FAME)-based biodiesel or can be used as a substitute for fossil fuel in steam raising boilers or thermal oxidizers.

21.6 Safety

21.6.1 Food safety slaughter by-products

Safety and quality assurance are chain orientated and start, therefore, before the actual production process. Raw materials are sourced from approved animals which originate from EU-registered and veterinary-controlled suppliers, including controlled farming. HACCP food safety principles apply to all aspects of the valid regulations, including the slaughterhouse, on to the processing plant and finally to the destination of the products. Movement documentation, combined with HACCP systems at the various processing locations, ensures that a fully traceable system is in place from farm to fork. In addition, codes of practice have been adopted by various sectors in the industry. Due to food scare incidents, food recalls and retail pressure, the feed industry has introduced extensive compulsory rules for the feed chain, including feed ingredient suppliers. These codes are based on HACCP and quality assurance to guarantee the safety of products for the food chain. Part of the quality assurance system is a monitoring program for undesired substances in animal fats . Animal fats are frequently examined by both government and processing companies for levels of pesticides, heavy metals, antibiotics, dioxin, dioxin-like polychlorinated biphenyls (PCBs), and polycyclic aromatic hydrocarbons (PAHs).

The rendering industry is governed by strict and comprehensive regulations. Food safety, based on the HACCP principles, is an important part of the regulations, especially concerning time, temperature, and particle size profiles. Sanitation, hygiene, and separation of activities are prominant aspects in the detailed implementation regulations. The veterinary authorities of the member states check and inspect the meat chain continuously for application of European rules. Additionally, the European Commission audits member states with Food and Veterinary Office (FVO) check on compliance with implemented legislation. Within the European Union, the EFSA evaluates and detemines the safety of processes and products. The EFSA emerged out of the former food safety authority, the Scientific Steering Committee (SSC). It is composed of a multidisciplinary team of independent experts from universities and research institutes. The EFSA determines the risks and the EC manages the risks.

In general, for the processing of slaughter by-products, sanitation conditions to guarantee safe use of derived products are prescribed clearly in the European legislation. For the safe use of melted animal fats, e.g. lard, for food applications for import into the EU and export to third countries, the minimal process conditions are prescibed in European Directive (92/118/EC), as 70 °C for 30 min, 90 °C for 15 min or a minimum of 80 ˚C in a continuous system.

21.6.2 Animal proteins and bovine spongiform encephalopathy

The incidence of bovine spongiform encephalopathy (BSE) in Europe has drastically reduced due to measures taken by the European Commission to ban the use of animal protein meals for, first, ruminants and second, all farmed animals at the height of the BSE crisis in cattle.

After the discovery of BSE in the early 1990s in the UK, the European Commission dictated different measures to control and eradicate the disease. SRM was defined, e.g. bovine brain and spinal cord, and removed from of the food chain. Taylor et al. (1995, 1997) investigated empirically the inactivation condition of prions from BSE and scrapie. These experiments were based on industry-related processes and the results for animal proteins were translated into the well-known inactivation conditions of 133°/20'/3 bar and transcribed into the EU regulations current at the time. In 1998, Schreuder confirmed the efficacy of pressure cooking in inactivating BSE in his study related to rendering procedures (Schreuder et al., 1998).

As a precautionary measure, all animal proteins were prohibited from the feed chain for farmed animals in 2000. Additionally, 100% BSE monitoring was introduced in the EU on all slaughtered bovine animals above 30 months of age at the same time. The aim was to prevent any infected material entering the food or feed chain. With the introduction of animal by-product regulation 1774/2002 in 2002, the processing of animals that died from a disease and SRM (category 1) was strictly enforced by requiring completely separate processing from the processing of animal by-products from approved animals fit for human consumption (category 3).

The measures were very succesful and led to a drastic reduction of BSE cases in Europe. BSE disease is now contained and under control. Twenty years after the first BSE case, the time has come for a gradual relaxation of some unnecessary precutionary measures, which are an economic burden for the meat industry. The industry and the European Commission are preparing to lift the feed ban for non-ruminant animal protein for fish feed in 2013 and for pig and poultry feed in 2014. The EFSA (2011) has positively evaluated the safe use of porcine and avian protein for animal feed in the EU.

Currently, in the EU, animal proteins are still only used for limited appliations with a low added value. The feed industry had to replace animal proteins in feed formulations with vegetable proteins, considered to be less sustainable with a higher land use. Animal protein is a sustainable and safe by-product of the rendering industry with a high added value for the feed industry and the reapproval of animal protein in the EU, following a step-wise approach, will put the EU more in line with other world practices with regard to the use of processed animal proteins.

21.6.3 Animal fats: tallow and bovine spongiform encephalopathy

In general, tallow is accepted as safe regarding BSE transmission, when it is sourced from healthy animals and treated with an appropriate purification process. The numbers of empirical publications concerning the significance of tallow in the transmission of BSE are limited (Appel et al., 2001b). Experimental challenge infecting cattle via dietary tallow has never been attempted. Taylor observed in his empirical study on brain injection of mice that industrial tallow, even using the lowest time-temperature combination and a source that was highly infective, was free from detectable BSE infectivity (Taylor et al., 1995). It is still striking that the inactivation results for animal proteins from the Taylor study were readily translated into legislation but that the results regarding tallow from the same study are still contested, especially since the national food safety authorities of Germany and France have questioned the possible significance of tallow in TSE transmission. Inquiries on the variation in incidence of BSE have indicated that BSE incidence is not consistent with the use of tallow in feed. The Japanese/Dutch BSE inquiries have cleared tallow as a possible route for transmission. The research team concluded that the calf milk replacers hypothesis was not confirmed by facts (Dutch Ministry of Agriculture,, 2003). The chance that BSE contamination of bovine animals via this route had taken place was considered on the basis of the research results to be practically out of the question. In 20% of the Dutch BSE cases, the calves were only fed with cow's milk.

A study on tallow investigated its use in oleochemical processes on the heat stability of prions in water, lipid, and lipid-water mixtures (Appel et al., 2001a). There were some questions about the efficiency of prion inactivation in a hydrophobic environment with very low water activity. This resulted in a second research project

Table 21.3 EU dioxin limit values for animal fats (EC Regulation 1259/2011)

Food product	Sum of dioxins and furans (WHO-PCDD/F-TEQ)	Sum of dioxins, and dioxin-like polychlorinated biphenyls s(WHO-PCDD/F-PCB-TEQ)
Vegetable oil and fats	0.75 pg/g fat	1.25 pg/g fat
Animal fats from:	2.5 pg/g fat	4.0 pg/g fat
• bovine animals and sheep	1.75 pg/g fat	3.0 pg/g fat
• poultry	1.0 pg/g fat	1.25 pg/g fat
• pigs		
Mixed animal fats	1.5 pg/g fat	2.50 pg/g fat
Marine oil	1.75 pg/g fat	6.0 pg/g fat

(Appel et al., 2001b), together with APAG (European Oleochemicals and Allied Products Group), on the efficiency of inactivation under complete oleochemical processing conditions, which demonstrated effective inactivation of prions in spiked tallow.

In conclusion, the EU Scientific Steering Committee (SSC) has in several scientific opinions established that tallow in feed is safe to use under certain conditions (maximum 0.15% insoluble impurities, SRM removal, food/feed grade). To silence the discussion on the use of tallow, the Commission asked the EFSA to carry out a quantitative risk assessment of the residual BSE risk in bovine-derived products covering tallow, gelatin, and dicalcium phosphate. This opinion was published in 2005 and indicated again that animal fats are safe to use in feed (EFSA, 2005). Furthermore, both the SSC and the EFSA have reported that there is no evidence of natural occurrence of TSE in non-ruminant farmed animals producing food, such as pigs and poultry.

21.6.4 Animal fat and dioxin

In the last decade, Europe suffered from several incidents of dioxin contamination of feed and foods, due to fraudulent and careless acts in the feed ingredient industry. Dioxins and dioxin-like compounds are a diverse range of chemically related compounds that are persistent environmental pollutants. Dioxins are found throughout the world in the environment generated by the burning of waste, forest fires, and volcano eruptions. Certain types of dioxins are highly toxic. In 1999, the industry was faced with a criminal act in Beglium, where transformator oil with high dioxin levels was mixed with used cooking oil and used in feed formulations. A second incident occurred in Ireland, where bread meal for feed application was contaminated with high dioxin levels, due to a drying process with unsuitable

combustion conditions. The most recent dioxin incident occurred in 2011 in Germany. Vegetable fatty acids, a by-product from the biodiesel industry, with excessive dioxin values were fraudulently mixed with fats destined for the production of feed. These incidents have led to stricter legislation and control of the feed chain. Monitoring systems for dioxins are further enhanced by a new Commission regulation that came into force in 2012.

Fats in general are never free of dioxin and dioxin-like PCBs, because dioxins are present in the environment and consumed by animals. For this reason, maximum background limits of dioxin and dioxin-like PCBs are determined for animal and vegetable oils and fats (Table 21.3). Even for fats sourced from different species, separate maximum dioxin limits are fixed, with porcine the lowest and bovine the highest limit. The actual dioxin levels are collected from the meat industry and the difference between species is breeding inside in a farm (pigs) or outside on pasture land (bovine animals).

Unfortunately, all fats generally have the ability to solubilize dioxins and dioxin PCBs very well. In the case of a dioxin incident in the feed industry, as a consequence, the dioxin level in animal fats will increase. From time to time animal fats are criticized as a risk for dioxin. However, we should remember that dioxin in animal fat is the result of an incidence and not the cause of the problem. In this respect, monitoring dioxins in animal fat is an indicator that the feed industry is executing its responsibility.

21.7 Characteristics and quality

21.7.1 Characteristics of animal fats

During the last 15 years, the processing of slaughter by-products has changed considerably. Only the parts of animals fit for human consumption are used for the production of animal fats and proteins. The production is separated per animal species, so that more distinguished products like tallow, lard, and poultry fat are on the market.

Chemically, animal fats are triglycerides in which glycerol is esterified with three fatty acids. The chemical differences between the different animal fats are due to the fatty acid composition. The applicability of fats and oils is for the most part detemined by the fatty acid composition. Animal fats have typically a higher content of long-chain fatty acids and/or saturated fatty acids. Saturated fatty acids have a high melting point and are oxidatively and thermally stable. Thermal stability is specifically important for frying and baking applications to prevent rancid, off-flavors and toxic polymerization products caused by prolonging oxidation processes. The oxidation speed from saturated to unsaturated fatty acids increases quickly from C18:0 to C18:1 (10 times faster) and C18:0 to C18:2 (100 times faster). Unsaturated fats, which are more present in vegetable oils, are much more sensitve to oxidation and polymerization, which makes vegetable oils less suitable for frying, cooking, and baking.

Animal fats are either solid or liquid at ambient temperature. In fact, when the product is liquid, it should be classified as an oil, e.g. poultry fat or lard oil. Palm oil and animal fats have comparable properties and are therefore often used as substitutes. However, palm oil is, just like lard and tallow, solid at 20 °C, so it should be classified as fat and not as an oil. The melting point is dependent on chain length and degree of saturation of the fatty acids. For most animal fats there is not a clear melting point but a melting range. This melting range is important for certain applications where texture and consistency are significant, such as the use of lard as a leavening agent in pastry products.

The fatty acid profiles from different types of animal fat are compared with some vegetable and fish oils in Table 21.4. The fatty acid profiles are important for certain applications and typically for the different species. The chain length of the fatty acid is described by the first number. The amount of unsaturated bonds is given as the number after the colon. In general, compared to vegetable fats, animal fats are more solid and have more saturated fatty acids (SFA). Vegetable oils are more rich in mono- and polyunsaturated fatty acids (MUFA and PUFA). Fish oils in general have a high content of palmitic acids (C16:0), but are rather unsaturated due to long-chain PUFA. However, fish oils vary considerably in fatty acid composition per species. In addition, butter

fat is comparable to tallow, except that butter has a higher content of medium-chain fatty acids (C8–C14). For tallow, the rumen produces a higher content of stearic acid (C18:0) by natural hydrogenation of polyunsaturated fattty acids. This also results in a low level of natural *trans* fatty acids in tallow. In contrast to ruminants, the body fat from pigs is mainly produced directly from feedstuffs. Lard is rich in MUFA oleic acid C18:1 and has also a significant amount of the PUFA linoleic acid (C18:2).

21.7.2 Animal fat quality for human food and animal feed

For animal fats, a number of characteristics are of importance in determining the quality of food and feed. Odor is not easy to measure, but is significant for acceptance for food and for feed, as calves and pets are very demanding about odor. A low percentage of free fatty acids (FFA), in combination with a low peroxide value (POV), indicates a fresh product. FFAs reduce digestibility. Fatty acids are generated by microbial activity and therefore the FFA value is an indicator of the freshness of the material. The peroxide value is a measure of the oxidation of fatty acids. The POV gives an indication of the degree of rancidity of the product. Hydroperoxides formed by oxidation are further turned into aldehydes and ketones, which are responsible for the odor and flavor of rancid fats. Oxidation can further lead to toxic polymerization by-products. Adding antioxidants can slow down the oxidation process in fats. Furthermore, moisture and insoluble impurities (II) should be as low as possible in animal fats. Moisture is undesirable because it facilitates the formation of FFA. Typically, freshly melted edible animal fats derived by fat processors have the following commercial qualities: FFA maximum 0.50%, POV maximum 4 meq/kg, moisture maximum 0.20% and II maximum 0.02%. Customers in the food industry can additionally ask for a taste specification, typical odor, nuclear magnetic resonance (NMR) specifications, iodine value, color, slip melting point, and the absence of crystals. In this respect, NMR profiles are used to express the solid fat content (SFC) for certain temperatures. Furthermore, iodine values are specified to distinguish between hard or soft quality lard.

The commercial quality specifications of animal fats, e.g. required for calf milk replacers and pet food, are traditionally even stricter than the legal food specifications. For instance, II have not been accepted for a long time for the production of calf milk replacers, as in practice they

Table 21.4 Fatty acid profiles for animal and vegetable fats

Fatty acid		Coconut oil %	Palm oil %	Rapeseed oil %	Soybean oil %	Sunflower oil %	Olive oil %	Butter fat %	Beef tallow %	Pork lard %	Chicken fat %	Lard oil %	Fish oil %
C-4:0	Butyric acid							3.5					
C-6:0	Capronic acid	0.5						2					
C-8:0	Caprylic acid	7						1.5					
C-10:0	Capric acid	6						3					
C-12:0	Lauric acid	46						3.5			1		
C-14:0	Myristic acid	18.5	1			1		11	3.5	1.5	1	1.5	6
C-14:1	Myristoleic acid							1	0.5				0.5
C-15:0	Pentadecanoic acid							1	1				0.5
C-15:1	Pentadecenoic acid												0.5
C-16:0	Palmitic acid	9.5	43	4.5	10.5	6	11.5	28	27	25.5	23	20.5	13.5
C-16:1	Palmitoleic acid			0.5			1	2.5	2.5	2.5	5	3.5	7.5
C-16:2	Hexadecadienic acid												0.5
C-16:3	Hexadecatrienic acid												0.5
C-16:4	Hexadecatetraenic acid												1
C-17:0	Margaric acid							1	2.5				0.5
C-17:1	Heptadecenoic acid							0.5	0.5	0.5		0.5	
C-18:0	Stearic acid	3	5	1.5	4	4	2.5	10	22	17.0	5.5	6	2.5
C-18:1	Oleic acid	7.5	38.5	58	22	20.5	75	25	36.5	40	40	51.5	14
C-18:2	Linoleic acid	2	11	20	54.5	68	9	3	2.5	12	21	13.5	1.5
C-18:3	Linolenic acid			9.0	7.5		1	0.5	0.5	1.0	2.5	1	1
C-18:4	Octadecatetraenoic acid												3
C-19:0	Nonadecanoic acid												0.5
C-20:0	Arachic acid		0.5	0.5	0.5		0.5	0.5					0.5
C-20:1	Gadoleic acid			2				0.5	0.5	1	0.5	1	11.5
C-20:2	Eicosadienoic acid									0.5		0.5	
C-20:5	Eicosapentaenoic acid												8.5
C-22:0	Behenic acid			0.5	0.5								
C-22:1	Erucic acid			2.5									14
C-22:5	Docosapentaenoic acid												1
C-22:6	Docosahexaenoic acid												8.5
Saturated fatty acids		90.5	49.5	7	15.5	11	14.5	65	56	44.5	30.5	28	24
Monounsaturated fatty acids		7.5	38.5	63	22	20.5	76	29.5	40.5	43.5	45.5	56.5	48
Polyunsaturated fatty acids		2	11	29	62	68	9.5	3.5	3	13.5	23.5	15	25.5

may block the nozzles of spray dryers. Subsequently, animal fats fit for human consumption are also utilized for feed applications, such as artificial calf milk, pet food, and increasingly for compound feeds. Typical animal fats derived by the rendering industry destined for feed or technical applications have a wider spread in quality and range from FFA 2–15%, POV maximum 10 Meq/kg, moisture maximum 0.20%, and II maximum 0.15%. Normally, quality specifications are agreed in a contract between supplier and customer.

21.8 Applications

Worldwide, annually 172 million tonnes of natural fats and oils of animal and vegetable origin are produced; 134 million tonnes of this production are consumed by humans (78%) while 7 million tonnes are consumed by animals. Technical applications in the oleochemical industry are estimated at 14 million tonnes, and the biodiesel industry 17 million tonnes. In recent years biodiesel utilization of natural fats and oils has grown steadily.

Fats and oils sourced from animals worldwide are estimated at 25 million tonnes. In Europe, nearly 3 million tonnes of animal fats are produced annually. The main

outlets for animal fats are feed (26%), oleo and soaps (22%), energy (19%), biodiesel (15%), and pet food (13%) (EFPRA, 2011). Efficient utilization of slaughter by-products has a direct impact on the economy and the environment. EU member states have regulatory limitations that restrict use of certain animal by-products for reasons of food safety or quality.

In the past in Europe there were strong fluctuating prices for animal fats due to political issues, market developements, and animal diseases. The animal fat market price peaked (up to 50%) several times, in 1998 due to the threat of an import ban of animal fats from the US, in 2001 induced by measures regarding BSE, and in 2002 caused by swine fewer. In 2004, the UK was faced with a lard crisis. Lard is traditionally used as an ingredient in mince pies and Chistmas puddings. In the run-up to Chistmas there was a national shortage of lard and the supermarkets shelves were empty, caused by a higher demand for cheap cuts of pork in Eastern Europe. The price of lard peaked for a short period, far above vegetable fat prices. Normally the animal fat price follows the palm oil price on a slightly lower level. The recent price development keeps pace with the overall strong increasing prices for agricultural products (Figure 21.8). For the future, volatile prices remaining on a high level for the long term are expected for animal fats.

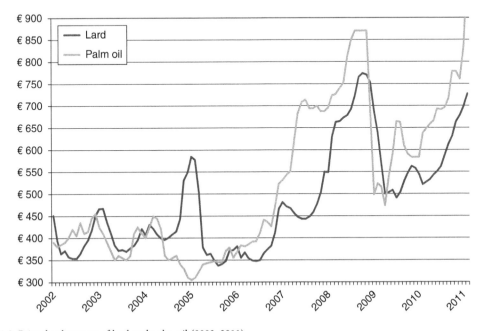

Figure 21.8 Price development of lard and palm oil (2002–2011).

21.8.1 Human food

Edible meat by-products contain many essential nutrients (Jayathilakan et al., 2011). Animal fats play an important role in a balanced diet and in the manufacture of food products, contributing to texture and palatability. They are a valuable source of concentrated energy and essential fatty acids needed for growth and development. In fact, lard has been suggested as an excellent alternative to cow's milk fat in infant formula due to the fact that its fatty acid profile is close to that of breast milk, and lard is easily absorbed and digested. The flavor-enhancing properties of bovine fat are the reason for its application as a frying agent, for example frying fish and chips in Belgium and the UK. Lard has been used for hundreds of years as a major fat for cooking. Traditionally, it is used in bread or pastry making to assist the leavening process and to soften the crumb. The soft consistency and crystalline character make lard the most suitable shortening for pastry. At the usual lower mixing temperatures of pastry, lard retains its plastic properties, while other fats become too hard. Lard is used in the bakery industry for its color, flakiness, flavor, and tenderness. Lard and bovine fat are typically used for food in southern Europe and Asia.

In some food applications, the natural flavor of fresh melted animal fat is required to realize the typical quality and characteristics of the final food product. In other food applications, the typical odor requires removal beforehand by a refining or deodorizing procedure. Refining is a technology widely used to purify vegetable oils. In the refining process undesirable substances, like color, odor, POV, FFA, etc., are removed by degumming, neutralization, bleaching, and deodorization. Deodorization can also be executed to remove only the flavor components.

21.8.2 Animal feed

Natural animal fats are essential in numerous pet food products, milk replacers, and compound feeds. Animal fat is an indispensable ingredient in a balanced diet for domestic animals. It not only improves the appearance and taste of pet food, but also promotes the utilization of vitamins. Adequate addition of animal fat produces a healthy and glossy fur and improves general performance, health, and well-being. Animal fats provide concentrated energy as well as fatty acids esential for growth. High-quality animal fats are important in diets for pets and young farm animals, such as calves that are fed with milk replacers. The total feed composition affects feed intake, gut health, and digesibility of all nutrients.

The pig is what it eats. About 80% of the body fat of pigs is produced directly from feedstuffs. The fatty acid composition of the fat used for feeding monogastric animals has a direct influence on the composition of the fat stored by the animal. Fats are especially stored under the skin and around specific organs, like kidneys. But muscular tissue also contains fat, producing more tender and tasty meat. Soft fats, like rapeseed oil, should be avoided in pig feed. Rapeseed oil instead of pig fat in feed results in lower yields of bacon and cooked ham and alters the quality of a number of meat products (Claudi-Magnussen, 2007; Claudi-Magnussen & Jacobsen, 2005). Unsaturated fat sources result in higher unsaturated and soft fat in the carcass and more oxidation of pork chops. These changes in fatty acid profile after a dietary fat change are rapid and occur within 2 weeks (Wiseman & Agunbiade, 1998). The alternative use of vegetable oils with lower melting point results in the so-called "weak" meat, which is not acceptable to consumers.

The fatty acid composition and the melting point of fat for feed applications are important features for the production of high-quality feed pellets and a good consistency of meat. For animal fats, good nutritional aspects (e.g. linoleic acid), digestibility and high energy density play an important role in determining feed composition. One gram of fat yields 9.4 kcal of heat when combusted (gross energy) (Doppenberg & van der Aar, 2010). This is 2.5 times higher than the caloric value of 1 g of sugar. The digestibility of fats depends on the chain length of fatty acids, the degree of saturation, the position of the fatty acid on the glycerol molecule and the presence of an emulsifier. Fats and oils with a higher proportion of unsaturated acids will have better digestibility and fats with a lower ratio (U:S ratio) will have reduced digestibility (Table 21.5).

Pig feeds contain a significant proportion of grain or oil seed by-products, which tend to be rather unsaturated. Added fat as a single source is not preferable. In general,

Table 21.5 Digestibility of fatty acids (Doppenberg & van der Aar, 2010)

Fatty acid		Digestibility (%)
Stearic acid	C18:0	33–58
Palmitic acid	C16:0	78–85
Palmitinic acid	C16:1	78–85
Oleic acid	C18:1	82–87
Linoleic acid	C18:2	87–90
Myristic acid	C14:0	88–90

Table 21.6 Digestibility of animal fats versus palm oil in pigs (Doppenberg & van der Aar, 2010)

	Lard	Tallow	Palm oil
Fat digestion (%)	92.4	91.4	90.7

2% animal fat (tallow/lard) can be used in feed as a single fat source without falling below the minimum U:S ratio of 2.25, which is required for pigs and layers (Doppenberg & van der Aar, 2010). However, a 2% addition of palm oil is not possible, because it requires adding an unsaturated oil to meet the minimum digestibility ratio. In vegetable fats and oils, SFA is predominantly found on the sn-1 and sn-3 position of the glycerol. For animal fats the preferred position of SFA is in the middle on sn-2. Pancreatic lipases in the digestive system hydrolyze fatty acids on the outside positions sn-1 and sn-3, leaving a monoglyceride with fatty acid on sn-2. Fatty acids on sn-2 are more easily absorbed as monoglycerides than solely as fatty acids. Approximately 80% of the monoglycerides are absorbed intact, which is especially beneficial for animal fats with SFA on sn-2 (Ratnayake & Galli, 2009). Feed experiments with pigs confirmed that the digestibility of lard is significantly higher than the digestibility of palm oil (Table 21.6) (Doppenburg & van der Aar, 2010).

A higher proportion of FFA results in a reduction of fat digestibility. This is caused by a deficiency of monoglycerides, which improve emulsification of fatty acids and absorption in the small intestines. Accordingly, emulsifiers like lecithins are used in pig feed, poultry feed, and milk replacers to improve fat digestion. Improved digestibility of proteins is also suggested as a result of fat addition, due to the high energy concentration in the feed, reduced feed intake, and losses by feed consumption. Feed grinding and palleting also increase the digestibility of nutrients due to a reduction of particle size. Fat is added to create a firm pellet, which decreases dust and therefore prevents waste formation of meals. A firm pellet is important to improve feed intake, but also leads to higher production capacities (up to 15%) and a higher feed performance. In compound feeds up to 4% of fat is added. Feeding animal fats have a positive influence on meat quality and taste. Feed producers prefer animal fats on account of the positive effect on the crystallization characteristics for the production of calf milk replacers and compound feeds. Better economics are also relevant in the case of feeding animal fats, i.e. lower feed costs, higher feed performance, and higher return on animal by-products.

The veal industry has indicated that the use of animal fat in milk replacers results in a 1–3% higher yield or 2–5 kg extra per animal. Furthermore, deskinning of calves with the use of animal fat in feed is much easier without loss of meat. The main purpose of using animal fats in feed, however, is increasing the energy content of the feed economically.

21.8.3 Technical applications

Animal fats are more and more utilized as bio-based raw materials in the oleochemical industry. Animal fats are excellent raw materials for technical applications where thermal and oxidative stability are important, such as surfactants, lubricants, and biodiesel. The world market share of natural ingredients will grow, driven by consumers' "green" awareness, and will probably increase from the current 10% growth to 30% in 2030.

Traditionally, animal fats are used for the production of soaps, shampoos, and cosmetic products. Nowadays oleochemical products made out of animal fats and oils are numerous and diverse. Applications are found in personal care products (25%), lubricants, plastics, cleaning agents, coatings, glues, softeners, emulsifiers, additives, rubber, paper, paint, etc. A recent development is the production of renewable biofunctional building blocks created from animal fatty acids for the production of plastics or biopolymers. Such components can produce materials with properties like water repellence, hydrolytic stability, and flexibility. Oleic acids, which are especially abundant in animal fats, are very suitable as a raw material for conversion to biopolymers.

Animal fats may also be physically modified by fractionation to create an oleine fraction, e.g. lard oil. Lard oil is applied as a biolubricant or as a rolling oil in the metal industry, due to its thermal stable properties. Chemically modified lard oil, for example by sulfonation, is applied as fat liquor to soften and lubricate leather.

Recently, animal fats and oils have also been used to produce biodiesel or biofuel, as they are a sustainable source for the production of biodiesel (Renewable Energy Directive 2009/28). Biodiesel may be used as a transport biofuel to replace normal diesel from fossil sources or as an addition to fossil-based fuels up to 5%. The use of biofuels is commonly now obligated by the authorities of different member states in Europe. The high melting properties of animal fats can lead to a higher cold filter plugging point (CFPP), which is solved by mixing of raw materials or by using alternative biodiesel processes. In 2010, animal fats used in biodiesel production in Europe amounted to approximately 350,000 tonnes (EFPRA, 2011).

21.9 Health aspects

Lipids are of vital importance to life. The main category of lipids is the triglycerides or fats. Triglycerides are constructed from one glyceride molecule and three fatty acids. Every triglyceride has its own combination of fatty acids determining the functional properties of the fat or oil. One fatty acid is not the same as another so in general, we can differentiate between saturated (SFA), monounsaturated (MUFA), and polyunsaturated fatty acids (PUFA). Fats are major components in daily foodstuffs and provide for a significant part of our energy needs. Body fats can easily be used to provide energy for metabolic processes or increasing body temperature. Apart from that, fats are a sustainable energy source that play an essential role in the human body. Fats have many biological functions, such as storage of fat-soluble vitamins (A, D, E, and K), supply of essential fatty acids, construction components of cell walls, and protection of organs. Fatty acids are used for the synthesis of polar lipids such as phospholipids and glycolipids for the production of lipid bilayers of cell membranes and intracellular messengers. Additionally, fatty acids regulate enzyme activity and the expression of genes related to lipid metabolism.

In spite of all these positive biological functions, animal fats have a negative reputation concerning supposed obesity and increasing cholesterol. Are animal fats bad or is it bad science? If you take Charles Darwin as a starting point to determine the appropriateness of animal fat for the human species, the conclusion is probably that the genes of the best adapted individuals or species have survived a long history of animal foods being part of the menu. Nevertheless, recommendations related to the intake of saturated fat and cholesterol have been introduced by health councils, independent scientific advisory bodies providing governments with advice in the field of public health. These restrictions have lead to an avoidance of animal fats, in the belief that animal fats are full of saturated fatty acids and cholesterol, which is not the case. For example, lard contains 60% unsaturated fatty acids. To condemn animal fats because of their cholesterol and SFA content is simply not justified, because vegetable fats also contain SFA and generalization is scientifically not acceptable. One would hope that these recommendations are backed up with solid and clear evidence but unfortunately, we have to deal with fragile science with

many publications indicating that animal fats are not harmful to human health (Colombani, 2006; Hoenselaar, 2012; Parodi, 2011; Ravnskov, 2008; Siri-Tarino et al., 2010a, 2010b; Stanley, 2010; Taubes, 2001). In several meta-analyses, which give a summary of studies on one topic, scientists have come to the conclusion that dietary cholesterol has a negligible impact on blood cholesterol of healthy humans (Colombani, 2006).

The relation between dietary fats and cardiovascular disease (CVD) has been extensively investigated (Siri-Tarino et al., 2010a). The lipid SFA hypothesis was contested in critical reviews which has led to extensive scientific disagreement (Colombani, 2006; Hoenselaar, 2012; Parodi, 2011; Ravnskov, 2008; Stanley, 2010; Taubes, 2001). Two meta-analysis studies have concluded that SFAs increase the level of "bad" LDL-cholesterol, with concerns for developing coronary heart disease (CHD). However, SFAs also increase the level of "healthy" HDL-cholesterol, which reduces the risk profile. Research has concluded that the ratio of LDL:HDL-cholesterol, related to the different fatty acids, is the strongest indicator for the development of CHD and is much more powerful than either the total cholesterol or LDL-cholesterol level. But even this cannot be disconnected from the other ingredients in dietary food. In the last 10 years many researchers have studied the effects of SFA on human health. The majority looked at the relationship of SFA intake and the incidence of cardiovascular disease, stroke or CHD and found no connection. Recent prominent scientific examples are the follow-up to the Nurses' Health study (Oh et al., 2005), the Japanse Study (Kirihara et al., 2010) and the meta-analysis of Siri-Tarino et al. (2010a).

In everyday life it is presumed that SFAs will be replaced by carbohydrates. Replacing SFAs by carbohydrates does not lower the CHD risk (Micha & Mozaffarian, 2010; Siri-Tarino et al., 2010b). Moreover, there are differences between the different types of SFA and the effect on the total cholesterol/HDL ratio (Jutt et al., 2002; Mensink et al., 2003). In general, unsaturated fatty acids reduce the "bad" LDL-cholesterol and are therefore thought of as "good"; *trans* fatty acids, present in hydrogenated vegetable fats, which increase the level of LDL but also reduce the "healthy" HDL level, are indicated as being "bad." In a meta-analysis of 60 trials, it was estimated that *trans* fatty acids have a factor of 7.3 more adverse effect on CHD compared to saturated fats (Mensink et al., 2003). In initial research, more than 40 years ago, on the influence of SFA intake on blood lipids, unfortunately *trans* fatty acids were not measured or taken into account (Reiser, 1973). Instead of natural

SFAs, many researchers used hydrogenated vegetable oils and the negative effect of *trans* fats was erroneously assigned to saturated fats.

Fats have more calories per weight unit than carbohydrates. However, there is no evidence that fat per ingested calorie is more fattening than carbohydrates. The increase in average body weight in many countries since the introduction of low-fat diets seems to point towards the opposite. It is suggested that changes in food, for example high sugar consumption connected to the quick transformation of sugar into body fat reserves, move much faster than alterations in genetic material as the possible cause of prosperity diseases. The traditional SFA paradigm has driven government bodies and non-profit organizations to urge the public to avoid eating saturated fats. Scientific progress is strongly indicating that the time has come to accept that animal fats are natural components in the human diet. No food is either good or bad as such, only the dose makes it harmful. An optimal and balanced ratio between the different food ingredients related to age and exercise is the key to improving health and body weight.

21.10 Sustainability

If an animal is slaughtered for meat, we should try to use all parts of the animal. This began centuries ago, when native Americans used the entire buffalo and lived in harmony with nature. Not only in the meat industry but in every line of business, the development of waste should be avoided. In the ideal situation, waste does not exist. If residues or by-products can be turned into useful products, then we can create a very sustainable operation. In fact, this is exactly the case for rendering animal by-products in the meat industry. The main product of slaughtering is meat. The by-products are considered as offal or slaughter waste. The fat processing and rendering industry is specialized in converting offal or slaughter by-products into products with added value. All meat by-products, bones, skins, intestines, slaughter fats, etc., are converted into derived products that can be used in food, feed, oleochemical products or biodiesel. Pollution is prevented and by-products are recycled into useful products with a higher value.

The utlitization of animal fats and proteins sourced from by-products from slaughter or meat cutting is considered very sustainable (Elferink, 2009), and the European Commission has recognized rendering activity as highly sustainable for the use of animal fats as a raw material for the production of biodiesel. In the RED,

typical greenhouse gas emission savings for biofuels and bioliquids are compared to fossil fuels. Animal fats are listed as one of the most sustainable raw materials for biodiesel with a potential carbon dioxide reduction of 88% compared to fossil fuel. Vegetable fats, such as palm oil and rapeseed, have produced 36% and 45% CO_2 reductions, respectively. In 2017, the EC plans to draw a line for sustainable raw materials at a minimum of 50% CO_2 reduction compared to fossil fuel. As a consequence, materials with a lower than 50% CO_2 reduction will not be classified as sustainable.

The carbon footprint is also an important indicator for the degree of sustainability. The carbon footprint of a product is the sum of all greenhouse gas emissions in the production chain of the product. There are no official carbon footprint calculation rules available. Taking into account the upstream emmisons, allocated on economic values, the carbon footprint of animal fat is calculated to be within the range of 0.32–0.74 kg CO_2 eq per kg (Pensioen & Blonk, 2010). The footprint of animal fat is much lower than that of vegetable oil. Blonk calculated a carbon footprint for palm oil of 3.2–3.5 kg CO_2/kg and soybean oil 1.8–2.4 kg CO_2/kg, excluding the additional carbon footprint for vegetable oils for land use changes and land conversion. For vegetable oils, the CO_2 impact for additional land use changes and land conversion should also be taken into acoount. According to Blonk, these additional emmisons are considerable for vegetable oils (Pensioen & Blonk, 2010). For animal by-products, the additional footprint for land use changes and land conversion is considered very low due to the low economic value and animal by-products being a residue. However, the EU calculation models for land use changes and land conversion are still under discussion.

21.11 Conclusion

Without lipids there is no life; we cannot think and we cannot sense. We exist, thanks to lipids. Lipids play an essential role in the human body as building blocks for cell membranes, phospholipids, and sphingolipids, which are responsible for signal transmission and cell recognition. Animal fats are natural available lipids and are an essential part of our daily food. At the moment, there is no consensus among scientists about the health consequences of SFAs, which has led to extensive discussions and scientific disagreement. Recent scientific publications are explaining that there is a discrepancy between the scientific literature and the dietary advice for the intake of SFAs

(Hoenselaar, 2011). Scientific progress is indicating that the time has come to rehabilitate animal fats as a natural ingredient in the human diet. In a balanced ratio related to age and exercise, moderate animal fat intake can be a healthy part of our diet. Furthermore, animal fat is a sustainable and natural basis for many products.

Rendering is used to convert slaughter by-products into marketable products, like animal fats and proteins, with a higher added value. Animal fats provide an important proportion of the fats used in the world today, both for foods and for many non-food applications. Natural animal fats are essential ingredients for pet food and animal feed. They provide concentrated energy as well as fatty acids essential for growth. The utlization of animal fats and proteins sourced from by-products from slaughter or meat cutting is considered very sustainable with a high reduction of carbon dioxide. Pollution is prevented and by-products are recycled into useful products with a higher value. Efficient utilization of slaughter by-products is very important for the profitability and sustainability of the meat industry.

References

Appel TR, Wolff M, Von Rheinbaben FM, Heinzel M, Riesner D (2001a) Heat stability of prion rods and recombinant prion protein in water, lipid and water-lipid mixtures. *Journal of General Virology* **82**: 465–473.

Appel TR, Riesner D, Von Rheinbaben FM (2001b) Safety of oleochemical products derived from beef tallow or bone fat regarding prions, *Eurpean Journal of Lipid Science and Technology* **103**: 713–721.

Claudi-Magnussen C (2007) *Avoid rapeseed in pig feed.* Slagteriernes Forskningsinstitut, Danish Meat Association.

Claudi-Magnussen C, Jacobsen T (2005) Rapeseed oil lowers the quality. *Orientering* **8**. Roskilde, Denmark: Danish Meat Research Institute.

Codex Alimentarius (1969) *General Principles of Food Hygiene.* www.codexalimentarius.org, accessed 27 November 2013.

Colombani PC (2006) Animal fats: evil in the diet or natural food for humans? *Lipid Technology* **18**(7): 151–154.

Doppenberg J, van der Aar PJ (2010) *Facts About Fats: A Review of the Feeding Value of Fats and Oils in Feed for Swine And Poultry.* Lelystad, Netherlands: Schothorst Feed Research.

Dutch Ministry of Agriculture (2003) *Japanese BSE investigation report (unofficial English translation of the official Japanese report).* The Hague, Holland: Dutch Ministry of Agriculture.

EC Regulation 1259/2011 Dioxin regulation. http://eur-lex.europa.eu/LexUriServ/LexUriServ.do?uri=OJ:L:2011:320:0018:0023:EN:PDF, accessed 27 November 2013.

EFPRA (2011) www.efpra.eu, accessed 27 November 2013.

EFSA (2005) Assessment of the human and animal BSE risk posed by tallow with respect to the residual BSE risk. Opinion of the scientific panel on biological hazards of the European Food Safety Authority. *EFSA Journal* **221**: 1–45.

EFSA (2011) Opinion on the revision of the quantitative risk assessment (QRA) of the BSE risk posed by processed animal proteins (PAPs). *EFSA Journal* **9**(1): 1947.

Elferink EV (2009) Meat, Milk and Eggs: analysis of the animal food environmental relations. PhD study. Groningen, Netherlands: University of Groningen.

Hoenselaar (2012) Review: saturated fat and cardiovascular disease: the discrepancy between the scientific literature and dietary advice. *Nutrition* **28**: 118–123.

Jayathilakan K, Khudsia Sultana, Radhakrishna K, Bawa AS (2011) Utilization of by-products and waste materials from meat, poultry and fish processing industries: a review. *Journal of Food Science and Technology* **49**: 278–293.

Jutt JT, Baer DJ, Clevidence BA, Kris-Etherton P, Muesing RA, Iwane M (2002) Dietary cis and trans monounsaturated and saturated FA and plasmalipids and lipoproteins in men. *Lipids* **37**(2): 123–131.

Kirihara Y, Hiroyasu I, Yatsuya NT et al. (2010) Dietary intake of saturated fatty acids and mortality from cardiovascular disease in Japanese: the Japan Collaborative Cohort Study for Evaluation of Cancer Risk Study. *American Journal of Clinical Nutrition* **92**(4): 759–765.

Mensink RP, Zock PL, Kester AD, Katan MB (2003) Effects of dietary fatty acids and carbohydrates on the ratio of serum total to HDL cholesterol and on serum lipids and apolipoproteins: a meta analysis of 60 controlles trials. *American Journal of Clinical Nutrition* **77**: 1146–1155.

Micha R, Mozaffarian D (2010) Saturated fat and cardiometabolic risk factors, coronary heart disease, stroke, and diabetes: a fresh look at the evidence. *Lipids* **45**(10): 893–905.

Oh K, Hu FB, Manson JE, Stamfer MJ, Willet WC (2005) Dietary fat intake and risk of coronary heart disease in women: 20 years of follow-up of the Nurses Health Study. *American Journal of Epidemiology* **161**(7): 672–679.

Parodi PW (2011) *Fats and Fatty Acids in Human Nutrition: Report of an Expert Consultation.* FAO Food and Nutrition Paper 91. Rome: Food and Agriculture Organization.

Pensioen T, Blonk H (2010) *Carbon Footprints of Animal Fats and Vegetable Oils. C4.0.* Gouda, The Netherlands: Blonk Milieu Advies BV.

Ratnayake WM, Galli C (2009) Fat and fatty acid terminology, methods of analysis and fat digestion and metabolism: a background review paper. *Annals of Nutrition and Metabolism* **55**: 8–43.

Ravnskov U (2008) The fallacies of the lipid hypothesis. *Scandinavian Cardiovascular Journal* **42**: 236–239.

REA (2013) www.rea.co.uk/rea/en/markets/oilsandfats/world production, accessed 27 November 2013.

Regulation (EC) 852/2004 Regulation on the hygiene of foodstuffs.http://eur-lex.europa.eu/LexUriServ/LexUriServ.do?uri=OJ:L:2004:139:0001:0054:en:PDF, accessed 27 November 2013.

Regulation (EC) 853/2004 Regulation on the specific hygiene rules for food of animal origin. http://eur-lex.europa.eu/LexUriServ/LexUriServ.do?uri=OJ:L:2004:139:0055:0205:EN:PDF, accessed 27 November 2013.

Regulation (EC) 999/2001 (TSE regulation). http://eur-lex.europa.eu/LexUriServ/LexUriServ.do?uri=CONSLEG:2001R0999:20080426:EN:PDF, accessed 27 November 2013.

Regulation (EC) 1069/2009 (ABP regulation). http://eur-lex.europa.eu/LexUriServ/LexUriServ.do?uri=OJ:L:2009:300:0001:0033:EN:PDF, accessed 27 November 2013.

Regulation (EC) 142/2011 (ABP regulation-implementing regulation). http://eur-lex.europa.eu/LexUriServ/LexUriServ.do?uri=OJ:L:2011:054:0001:0254:EN:PDF, accessed 27 November 2013.

Reiser R (1973) Saturated fat in the diet and serum cholesterol concentration: a critical examination of literature. *American Journal of Clinical Nutrition* **26**: 525–555.

Renewable Energy Directive (RED) 2009/28. http://eur-lex.europa.eu/LexUriServ/LexUriServ.do?uri=OJ:L:2009:140:0016:0062:en:PDF, EC Regulation 1259/2011 Dioxin regulation. Available at: http://eur-lex.europa.eu/LexUriServ/LexUriServ.do?uri=OJ:L:2011:320:0018:0023:EN:PDF

Schreuder BEC, Geertsma RE, van Keulen LJ et al. (1998) Studies on the efficacy of hyperbaric rendering procedures in inactivating bovine spongiform encephalopathy (BSE) and scrapie agents. *Veterinary Record* **142**: 474–480.

Siri-Tarino PW, Sun Q, Hu FB, Krauss RM (2010a) Meta-analysis of prospective cohort studies evaluating the association of saturated fat with cardiovascular disease. *American Journal of Clinical Nutrition* **91**(3): 535–546.

Siri-Tarino PW, Sun Q, Hu FB, Krauss RM (2010b) Saturated fat, carbohydrates, and cardiovascular disease. *American Journal of Clinical Nutrition* **91**(3): 502–509.

Stanley J (2010) How good is the evidence for the lipid hypothesis? *Lipid Technology* **22**(2): 39–41.

Taubes G (2001) Nutrition: the soft science of dietary fat. *Science* **291**(5513): 2536–2545.

Taylor DM, Woodgate SL, Atkinson AJ (1995) Inactivation of the BSE agent by rendering procedures. *Veterinary Record* **137**(S): 605–610.

Taylor DM, Woodgate SL, Fleetwood AJ, Cawthorne RJG (1997) The effect of rendering procedures on scrapie agent. *Veterinary Record* **141**(S): 643–649.

Wiseman J, Agunbiade JA (1998) The influence of changes in dietary fats and oils on fatty acid profiles of carcass fat in finishing pigs. *Livestock Production Science* **54**: 217–227.

Woodgate S, van der Veen JT (2004) The use of fat processing and rendering in the European Union animal production industry. *Biotechnology, Agronomy, Society and Environment* **8**(4): 283–294.

22 Aquatic Food Products

Mahmoudreza Ovissipour,[1] Barbara Rasco,[1] and Gleyn Bledsoe[2]

[1]*School of Food Science, Washington State University, Pullman, Washington, USA*
[2]*College of Agricultural and Life Sciences, University of Idaho, Moscow, Idaho, USA*

22.1 Introduction

Due to the unique features of aquatic foods, this chapter will not emphasize processing techniques that are covered elsewhere in this text for other food products. The emphasis in this chapter is processes unique to seafood, including cured and smoked seafood, harvest and production incorporating the following factors.

• **Product quality**: aquatic foods are often sold at a premium price and high-quality products are required. Aquatic animals should be healthy and with few or no cosmetic defects. Animals must not be treated with any prohibited additives or drugs or have been exposed to high levels of environmental contaminants. Delivery of live animals to retail markets provides the highest quality for the consumer and highest market price.

• **Resource management and fisheries sustainability**: aquatic foods must be harvested or cultivated under conditions that limit environmental impact. Processing operations should be conducted to maximize recovery of food materials and useful by-products. Overharvesting impacts their long-term sustainability, and international recognition and enforcement of total allowable catch provisions for high seas and inland fisheries have met with some success. Aquaculture in support of restoration efforts and as a component of community-wide strategies to address overharvesting has been successful for some fisheries.

• **Animal welfare**: best practices in animal agriculture and fisheries are to ensure that animals are raised, harvested, impounded, transported, and slaughtered in humane ways, limiting discomfort.

22.2 Aquatic plants and animals as food

From prehistoric times aquatic animals were among the first animal protein sources harvested by humans for food. Ancient civilizations were established near water sources because of the readily available food resources. Commercial fishing was part of Egyptian culture and the design of small fishing boats from that period has changed little until today. Records of Asian and European cultures document fish usage back many thousands of years, with fish cultivation and fish ranching in ponds dating back to more than 2000 years ago in China. Ancient fish ponds are present throughout Polynesia, across Asia into Europe, and in the New World. These practices provided fresh fish, mollusks, and crustacea throughout the year, allowed for storage of migratory fish for longer periods, and provided fish when weather conditions made fishing dangerous. Impoundment provided a method to grow fish to an attractive size and provided popular fish when they were out of season. The native peoples of coastal North America tracked migratory patterns of salmon, smelt, herring, seals, and whales that provided much of their food. These seasonal fisheries remain important to this day.

Aquatic food species are incredibly diverse. More than 350 species of Mollusca (e.g. clams, oysters, snails, octopi), Arthropoda (e.g. lobsters, crabs, shrimp, and crayfish), reptiles (e.g. turtle, alligators), Amphibia (frogs), Gastropoda (whelks), holothurians (sea cucumbers), echinoderms (sea urchin), finfish, aquatic plants, and marine mammals are used as food. These myriad tissues may be refrigerated, frozen, dehydrated, thermally

Food Processing: Principles and Applications, Second Edition. Edited by Stephanie Clark, Stephanie Jung, and Buddhi Lamsal.
© 2014 John Wiley & Sons, Ltd. Published 2014 by John Wiley & Sons, Ltd.

processed, high-pressure processed, pickled, salted, and fermented or consumed live or raw. A wide range of tissues besides muscle are consumed, including internal organs, skin, and blood. For many molluscan species such as scallops, muscle and roe are often eaten together. Animals including oysters, clams, limpets, and many small fish are consumed whole without evisceration. Aquatic animals provide special ingredients with unique flavor or textural properties such as shark fin and cartilage, fish gelatin, and fish maw. Nutritional supplements or drug components are recovered from aquatic foods including fish liver oil (vitamin D), fish oil (containing unique bioactive components), bone, and shell (calcium). Polysaccharides, oils, and colorants are recovered from marine plants. Polysaccharides are used as food ingredients, lipids as a source of fatty acids, colorants as food additives. Examples of the variety of aquatic products available and sold are shown in Table 22.1.

22.3 Cultivation, harvesting, and live handling – reducing stress and maintaining quality

Aquaculture is the husbandry of aquatic animals and plants and has been practiced since the earliest records of human history. It is a rapidly growing aspect of the food industry in many parts of the world. In Norway, Bangladesh, Canada, Chile, China, Ecuador, Thailand, Vietnam, and Japan, the marine aquaculture industries contribute substantially to a positive balance of trade and comprise a major part of their food production sector. World aquaculture is dominated by the Asia-Pacific region, which accounts for about 89% of production in terms of quantity and 77% in terms of value (FAO, 2008). More than half of global aquaculture production in 2006 was freshwater finfish (27.8 mmt), mollusks (14.1 mmt), and crustaceans (4.5 mmt) (FAO, 2008).

In the US, freshwater culture of fish (primarily catfish, tilapia and more recently sturgeon, paddlefish, perches, basses, mollusks, and crustaceans) represents a growing agricultural industry (DeVoe & Mount, 1989). The US aquaculture industry has failed to capture a significant share of the potential domestic or global market because of regulatory restrictions placed upon the industry. Hatcheries in support of ocean ranching of salmonid species and molluscan shellfish primarily clams and oysters are successful marine aquaculture activities.

Although farming of aquatic animals and plants is a practice equally as old if not older than farming on land,

as a modern industry, aquaculture provides a relatively minor source of food compared to agriculture and traditional capture fisheries. Total fish production in the world exceeds 150 mmt; half is cultivated for a value of > $100 billion at farm gate. Aquaculture continues to be the fastest growing animal food-producing sector and fortunately outpaces population growth, with the per capita supply from aquaculture increasing from 0.7 kg in 1970 to 7.8 kg in 2006, growing at an average annual rate of 7%. In China, 90% of the aquatic foods consumed are from aquaculture, with aquaculture currently providing about one-third of the food fish supply for the rest of the world.

Many of our important capture fisheries are seasonal, and aquaculture provides a year-round supply of fish products. For example, aquaculture provides a more stable supply of salmon, and through the use of hatcheries, provides for a higher sustainable yield for wild-caught salmon. Aquaculture provides job opportunities in rural communities, especially in developing countries with suitable water basins. Many of these jobs are culturally appropriate for women who otherwise would not have employment. Water sources which are not suitable for human consumption can often be used for production of aquatic organisms that are then suitable for food or feed. Further, cultivation of aquatic organisms provides a method to clean up the water column and improve overall water quality, by employing filter-feeders such as mussels to remove harmful bacteria from the water. Aquaculture is one of the best methods to preserve threatened and endangered aquatic animals and can provide a basis for maintaining genetic diversity through captive broodstock programs. For example, the beluga sturgeon (*Huso huso*) is one of the best known sturgeon fishes in the Caspian Sea and is at risk of extinction (Ovissipour & Rasco, 2011); the Fisheries Organization of Iran has led efforts to develop aquaculture systems to restore the fishery.

22.3.1 Finfish culture in ponds

According to the Food and Agriculture Organization (FAO) (2008), 57% of total aquaculture involves finfish cultivated in marine or fresh water. Water source availability and maintaining water quality, species biology, feeding regimes, and economics of production, harvesting and marketing must be considered. Cultivation methods can be extensive, semi-intensive, intensive or highly intensive, depending upon the stocking density of the fish.

Commercial practice in much of the world is focused on intensively farming fish and this involves consideration of the numerous abiotic and biotic factors necessary

Table 22.1 A summary of the variety of aquatic foods

Tissue or body part	Animal or plant source	Examples of foods
Whole animal; sometimes includes shells of small clams live or fresh	Mollusk	Clams, oysters, octopus, limpet
	Crustacea	Small crabs, shrimp, crawfish
	Gastropods	Abalone, whelks
	Holoturians	Sea cucumber
	Echinoderm	Sea urchin
	Thalassinideans	Mud shrimp
	Fish (Family: Cyprinodontidae, Engraulidae, Clupeidae and others)	Sardine, anchovy, smelt, kilka and a variety of other small fish
Plants	Marine plants	Sea vegetables, e.g. nori
		Food additives (agar, carrageenan)
Liver/hepatopancreas	Finfish	Cod, pollock, salmon, sturgeon
	Marine mammals	Seal, walrus, whale
	Crustacea	Lobster, crab
Tongue	Finfish	Cod, halibut, larger pelagic fishes
	Marine mammals	Seal, whale
Heart/kidney	Finfish	Salmon, tuna, sturgeon
	Marine mammals	Seal, whale
Stomach/throat	Finfish	Cod and other finfish
Brain/head	Finfish	Salmon, cod, halibut, tuna, sturgeon, smaller whole fish
	Marine mammals	Whale, seals
Spinal column, notochord	Finfish	Sturgeon, some billfish and other large fish
Eyes	Finfish	Salmon, tuna, carp; a variety of smaller and larger fishes
Gills	Finfish	Salmon, other fish served whole or in stews or soups
Gonadal tissue, roe, eggs	Finfish	Sturgeon, herring, salmon, mullet, lumpfish, capelin, catfish, carp, cod, shad and others
	Echinoderm	Sea urchin
	Mollusk	Scallop, abalone, oysters, clams
	Crustacea	Crab, lobster, shrimp
	Reptiles	Turtle
Male reproductive organs and tissues including milt	Finfish	Sturgeon, sharks
	Marine mammals	Seal, whale
Skin and scales	Finfish	Trout, salmon, rockfish, other fishes consumed whole or as skin-on fillets or steaks
Connective tissue and fin	Finfish	Shark (cartilage), shark fin, gelatins from various sources
Adipose tissue or depot fat	Finfish	Sturgeon, cod, tuna, salmon, other larger cultivated fishes, pelagic and anadramous fishes
	Crustacea	Shrimp, lobster, crab
	Marine mammals	Seal, whale
Oil, rendered	Finfish	Herring, salmon, shark, cod, tuna, menhaden, blue whiting, anchovy
	Crustacea	Shrimps, krill
	Marine mammals	Whale, seals, sea lions, walrus
Body fluids	Mollusks	Squid, octopus (ink)
	Finfish	Blood
Fin	Finfish	Sharks, rays

(Continued)

Table 22.1 (*Continued*)

Tissue or body part	Animal or plant source	Examples of foods
Shell	Mollusks	Oyster (as a nutritional supplement)
Exoskeleton	Crustacea	Shrimps, krill, smaller lobsters
Bone	Finfish	Smelt, herring, anchovy, kilka, other smaller fishes consumed whole, bones from larger fish are present in some canned products (salmon, mackerel)
Pyloric caeca	Finfish	Sturgeon, cod, carps, tuna, other whole finfish used in soups or stews or as a source of enzymes

to make the rearing environment suitable with all of these factors influencing both the behavior and culture performance of the fish to a greater or lesser extent.

Pond culture of fish is quantitatively the most important form of fish production for freshwater species. Ponds are often extensive production systems, but vary in degree of stocking densities and feed input. The water sources for ponds may be precipitation and direct surface run-off, surface water from rivers or canals, ground or well water. Most often, several sources of water are used in combination. Ponds are open systems and this poses a number of disadvantages. For example, ponds are dynamic ecosystems influenced by factors that are uncontrollable and unpredictable, e.g. weather, bird predation, animal predation including other fish, introduction of undesirable biota that could introduce disease, increased oxygen demand in the pond, and competition for feed. As such, it is difficult to exert rigorous control over the rearing environment to which the fish are exposed.

Despite these issues, polyculture of warm-water species, such as carps, tilapia, catfish, mullet, milk fish and shrimp, is conducted in open ponds. Providing food for fish in ponds can be a very low-technology endeavor, such as adding animal manures to the pond to increase algal production which serves as a feed for herbivorous fish. This system works well for low-density cultivation. Intensive operations require the use of supplemental feed and often supplemental oxygenation and water treatment to maintain pond water quality.

Fish quality is affected by culture conditions and diet. Fish can develop a benthic or muddy flavor when raised in earthen ponds from consumption of natural algae and cyanobacteria in the aquatic environment and actinomycetes and fungi in pond sediments. Chrysophytes, cryptophytes, and dinoflagellates produce several fishy-smelling fatty acid derivatives, whereas aliphatic hydrocarbon metabolites produced by green algae are associated with grassy odors (Tucker, 2000). The terpenoids geosmin and methylisoborneol and sulfides are important metabolites produced by cyanobacteria associated with off-flavors in farmed fish which sensitive tasters can detect at the low part per billion level. Fish are commonly depurated, or removed for 1–2 weeks to a source of clean water, so that undesirable flavors will be removed.

The flesh of farmed fish and shellfish reared in open systems may harbor parasites and bacteria that can cause diseases in fish or humans. In addition, pesticides and other industrial pollutants that enter the ponds in run-off or precipitation, or have accumulated in pond sediments over a period of many years, can enter the flesh through absorption by skin, gills or through ingestion, and pose potential safety problems. Many organic pollutants are very persistent in the environment and can be present at relatively high levels in carnivorous or omnivorous fishes. Petroleum off-flavors may develop in fish that have been exposed to polluted waters and other industrial activities that may lead to discharge into pond water. This is why water and sediment testing is critical for siting aquaculture operations and these factors are monitored routinely under aquaculture best practice programs.

22.3.2 Pen and cage culture systems

Cages and pens are common in both marine and freshwater aquaculture. Both are located in water over which the culturist would have little control. Water exchange occurs through the net or mesh-screen walls of the cages or pens. Local currents or tidal flow influence the rates of water exchange and removal of metabolic wastes and fecal matter. Unfortunately, cages can be damaged by storms or strong tides resulting in fish escape. Sites may also be vulnerable to deterioration in water quality resulting from coastal water pollution. Fish in cages require supplemental feed, although most obtain some natural food. Fish in cages are susceptible to predation from other fishes, reptiles, amphibians, mammals and birds, although for submerged cages predation may be easier to control than it is in open pond systems.

One advantage of cages over ponds is that they can be placed in any open body of water – fjords, bays, lakes, ponds, reservoirs and quarry pits, rivers, and streams. An additional advantage of cages over ponds is the ease with which most farm operations can be carried out. Cages are relatively easy to establish and the cage culture of fish is often a low-input form of farming with a high economic return. Monitoring and harvesting of the fish are relatively easy. Demand feeders are common with fish obtaining some food from the natural environment.

22.3.3 Tanks and raceways

Land-based systems composed of tanks or raceways are common for the cultivation of coldwater fishes such as rainbow trout and other salmonids, sturgeons and many higher value fish (Figure 22.1). With a raceway, it is possible to control temperature, pH, dissolved oxygen level, feeding rate, and light regime to a greater extent than in a pond or cage. Tanks and raceways may be components of either flow-through (open) or recirculating (closed) culture systems. Flow-through systems require large quantities of water that are discharged after passing through the rearing units, whereas in a recirculating

system, more than 90% of the water may be recycled. A unique feature of tank and raceway systems is the potential to cultivate fish in conjunction with a horticultural crops (aquaponics) with the fish providing a source of nitrogen for the plants.

Rearing systems incorporating tanks and raceways have a number of advantages over ponds and cages including simplified observation of the fish, better control of feeding and harvest, the ability to deliver disease treatments more precisely, and the possibility of collecting and treating wastes. These systems tend to be more secure than ponds and cages, with the risk of damage to fish from predation or loss from escape being reduced compared to open ponds. Disadvantages include the increased risk of fish loss due to mechanical failure of pumping or oxygenation systems, and higher energy consumption. Flow-through systems require a large volume of high-quality water for best fish muscle quality.

In both flow-through and recirculation systems, water pretreatment may include settlement of suspended solids, filtering to remove particles of various sizes, treatment with ozone or ultraviolet (UV) light to kill microorganisms, heating or cooling, and aeration or degassing. The effluent leaving the rearing units will also usually be treated prior to discharge. Effluent treatment will often incorporate removal of solids, such as waste feed and fish feces, treatments to remove excess dissolved nutrients and metabolic wastes and to destroy potential pathogens. Recirculating systems have very sophisticated water treatment systems, since a very large proportion of the water is reconditioned and reused. Recirculating systems usually

Figure 22.1 A large raceway based cultivation system for rainbow trout (*Oncorhynchus mykiss*) in southern Idaho. Photo credit: Gleyn Bledsoe.

have filtration systems to remove solid wastes, convert ammonia to nitrite and nitrate using biofilters, remove carbon dioxide, oxygenate the water, regulate pH, and include some form of disinfection and temperature control. A primary role of the biofilters is to convert nitrite to nitrate. Temperature control with biofilters is important; 20–26 °C tends to be most suitable and for cultivation of warm-water fishes this is generally not a problem, but for cultivation of cold-water fishes, where water temperatures are often less than 15 °C, maintaining the effective operation of biofilters can be problematic. Off-flavor in recirculating systems is a common and persistent problem. Fish raised in recirculating systems may require depuration, often for a number of weeks. Feeding rates are usually reduced and water changes increased so that off-flavors do not develop within the purging system.

In highly intensive systems such as many recirculation systems, stocking density of the fish is high and this can cause stress. Fin deterioration, weight loss, disease, decreased growth, and decreased muscle quality can occur. Higher quality rations, particularly diets high in vitamin C, can mediate stress and improve product quality. Maintaining high-quality water, sufficient space and sufficient food of high nutritional quality is important in high stocking density systems. In addition, the grower should limit disturbance and physical handling. It is also important to keep the fish isolated from potential sources of infection.

22.3.4 Disease prevention and treatment

Biosecurity is important for disease control and prophylaxis. Since a major source of disease-causing organisms for fish is the water flowing into the rearing units, semi-closed or recirculating systems offer significant advantages over open systems such as ponds and cages. Water is treated by filtration and/or disinfection to remove or inactivate a wide range of bacteria, viruses, and protozoa. Chlorination, ozonation or UV radiation are common. Application of compounds such as malachite green, formalin, gentian violet, and antibiotics (chloramphenicol, fluoroquinolones, etc.) for disease control in aquaculture operations is effective, but the use of these compounds is now either greatly restricted or banned. Malachite green or its reduced form leucomalachite green, a highly effective fungicide (e.g. against *Saprolegnia* spp.), has been used as a general hatchery disinfectant and for treating external parasitic infections on fish eggs, fish and shellfish but is no longer permitted for food fish in many developed countries; its detection has led to increased residue monitoring. Both forms can persist in edible fish tissues for a number of days (Table 22.2) (Mitrowska & Posyniak, 2004, 2005) and in the aquatic environment for extended periods of time.

Chloramphenicol is a broad-spectrum antibiotic that is not approved for use in food-producing animals in the US, being reserved for humans for treatment of certain serious, potentially life-threatening infections. Antimicrobial doses correlate with a low incidence (1:30,000 to 1:50,000) of a serious and potentially fatal blood disorder, idiosyncratic aplastic anemia, in humans. However, the use of prohibited antimicrobial substances occurs, with measurable levels of the antimicrobials listed in Table 22.3 having been found in cultivated products between 2001 and 2007, the latest data available at the time of publication.

Table 22.2 Persistence of malachite green in the muscle tissue of juvenile and adult fishes following exposure in a treatment bath

Species	MG concentration in bath (mg/L)	Bath duration (h)	Water temperature (°C)	Bath pH	MG sampling time following bath treatment (days)	MG concentration in muscle (µg/kg)
Eel (4.1 g)	0.10	24	25	6.9	62, 80, 100	2, bdt, bdt
Eel (0.3 g)	0.15	24	25	6.9	330	bdt
Channel catfish (600 g)	0.80	1	21	7.1	14, 42	12, bdt
Channel catfish (580 g)	0.80	1	22	7.2	14	6
Rainbow trout (1350 g)	1.00	1	21	7.0	1	73
Rainbow trout (1350 g)	1.00	1	12	7.8	5	15
Rainbow trout (0.1 g)	0.20	72	9.7	–	40, 140	1, bdt
Rainbow trout (0.1 g)	0.20	144	13	–	300	bdt

bdt, below the detection threshold of the analytical method.
Data from Mitrowska and Posyniak (2005).

Table 22.3 Prohibited antimicrobial substances detected in aquacultured species

Species	Residue
Crab (Family: Portunidae)	Chloramphenicol
Catfish (*Ictalurus* spp.), Basa (*Pangasius* spp.), Tilapia (*Oreochromis* spp.)	Malachite green, fluoroquinolones (includes ciprofloxacin and enrofloxacin), gentian violet
Walking catfish (*Clarias* spp.)	Malachite green
	Fluoroquinolones (includes ciprofloxacin and enrofloxacin)
Shrimp (Family: Penaeidae)	Chloramphenicol, malachite green, fluoroquinolones, nitrofurans, gentian violet
Dace (*Leuciscus* spp.)	Malachite green, gentian violet
Eel (Suborder: Anguilloidei)	Malachite green, gentian violet

Detected in cultivated products (2001–2007): www.fda.gov.

Facilities on rearing farms may be a source of infection, and to prevent this, facilities are disinfected, often with formalin but more commonly with commercial sanitizing agents approved for use in food and animal rearing facilities. Fish transferred from outside the operation are commonly quarantined for some period of time and are often bathed with brine to remove external parasites.

22.4 Animal welfare issues in fisheries

Increased public concern about the treatment of animals influences product choices. The culture environment determines the health, well-being, and performance of farmed fish. Welfare of farmed animals, including fish, encompasses everything from general husbandry and disease prevention, rearing environment, feeds and feeding methods, to transport and slaughter methods (Frewer et al., 2005). Interactions between environmental factors have an impact on the physiology and behavior of fish and biotic and abiotic parameters of a culture environment must be established to ensure high standards of welfare, productivity, and quality within a population of farmed fish.

Societal concerns over causing animals unnecessary pain and suffering focus attention on the most humane methods for handling live animals during cultivation,

capture, and handling. The first question is if aquatic animals experience pain. It is evident that fish do react to stress but it is not yet clearly understood how indicative these responses are of pain (Lagler et al., 1977). Blood cortisol, blood glucose, and ammonia levels can all measurably fluctuate in fish exposed to stressors. Fish also produce endogenous opioids in response to stressful stimuli, similar to the morphine-like chemicals produced in higher vertebrates. However, whether pain accompanies these measurable responses to stress is unclear. Fish and mammals share neurotransmitters that in mammals transmit sensations perceived as pain in the neocortex of the mammalian brain. However, fish lack a neocortex. For example, at the dentist one may receive a local anesthetic to deaden pain; the patient can feel sensations as the teeth are drilled or pulled, but they are not painful. Local anesthetic blocks the perception of pain in the neocortex. Furthermore, in mammals, nervous impulses triggering sensations of pain are transmitted through the spinal column in the fibers of the spinothalamic tract, a structure that is entirely lacking in fish. Fish also lack nociceptors, which are specialized pain receptors found in higher vertebrates (Stoskopf, 1992).

Sudden, intense or prolonged changes in holding conditions can impact fish health, reduce the productivity and product quality of the stock, and thereby the profitability of the farming enterprise. Fish are among the animals most susceptible to stress and stress-induced disease, which economics alone dictates commercial harvesting and aquaculture would take measures to reduce, including provision of sufficient oxygen, a proper temperature range during transport including avoiding extreme temperature fluctuations, and removal of risks that could result in injury from crowding or mechanical injury from nets or transport containers.

22.5 Harvesting methods and effect on quality

Capture techniques include the use of nets (e.g. seine, trawl, and gillnet), hooks and lines, traps, and even harpoons for larger fish. These harvest methods cause stress and lead to reduced muscle glycogen from the fish struggling before they are brought on board. Best practices dictate that fish be removed from the water, and killed and bled as soon as possible after capture and that the fish not be permitted to remain in nets or on lines for long periods of time. Decreasing temperature and providing moderate amounts of carbon dioxide as an anesthesia prior to slaughter can reduce the stress to fish and improve muscle quality. Stunning larger fish with a blow to the head is effective, although some animal welfare advocates, out of ignorance, particularly in Europe, consider this to be cruel and are in favor of electroshock, which instead is a more stressful method even under the best of conditions for freshwater fish (Holiman & Reynolds, 2002), resulting in hemorrhaging, muscle bruising and gaping, and is clearly highly stressful for large marine fish such as tuna (Soto et al., 2006). In aquaculture, harvest techniques are generally less stressful than for wild harvest of the same species since capture occurs under conditions that are usually easier to control, although some stress always occurs.

Changes in muscle quality can be greatly affected by postharvest handling. During rigor mortis, the ultimate pH of the fish tissue is higher than that of meat, generally pH 6.4–6.6, so little glycogen remains in tissues following slaughter. Higher pH is one of the major reasons why fish muscle is more susceptible to microbial spoilage than land animals. In addition, fish are cold-adapted animals or poikilotherms, so the enzymes in their muscle and viscera may be active at refrigeration temperatures, as are the cold-adapted microbes on gills and surface of fish, which can more easily grow at lower temperatures, causing spoilage (Bledsoe & Rasco, 2006).

Another important factor affecting muscle quality following harvest results from animal migratory and reproductive behaviors. For example, during the spawning season, salmon migrate from the sea into rivers and finally their natal stream to spawn. During this period, they avoid eating, and they must mobilize their energy reserves (adipose fat, muscle fat, and muscle protein) for migration and for production of roe or milt. Muscles deteriorate and become soft, depending upon the species, gender, and run conditions for the fish. Fish are of highest quality in the ocean prior to capture or at the mouth of the river prior to upstream migration than at the spawning grounds, which may be more than 1000 miles upstream.

Regardless of the condition of the animal prior to harvest, all aquatic animals should be refrigerated or cooled, commonly on ice or in ice slurries, as soon as possible. Finfish should be bled by removing or cutting the gill rakers or by cutting a major artery without damaging the heart since this will allow the blood to be pumped out of the fish. This is a common practice with harvesting cultivated salmon which are first anesthetized with CO_2, the gill rakers are then cut with care not to cut the heart, and the fish are bled out in an ice slurry. When many valuable species such as bluefin and big eye tuna are

harvested, the fishers immediately bleed the fish, often by cutting the caudal artery and using a small pump to remove the blood from the flesh. Removing blood results in a more uniform color of the meat, reduces growth of spoilage flora, improves flavor and extends shelf life from reduced lipid oxidation. Live marine mollusks and crustacea should be kept in cold seawater or salt water ice, or protected from direct contact on the surface with fresh water ice, since exposure to fresh water ice will kill them.

22.6 Reducing stress in live handling

Various techniques are used to reduce stress for aquatic animals during live shipment and subsequent storage with temperature control, immobilization, internal baffles in transport tanks, and maintenance of proper levels of oxygen being the most critical factors. With proper handling, it is possible to keep fish in good condition for a number of weeks in a retail setting.

For invertebrates such as molluscan shellfish, crustaceans, and other invertebrates, keeping the animals moist and reducing respiration rates by wrapping them in seaweed or paper or liners, and also decreasing the temperature slightly, are recommended. In some cases, these aquatic organisms are placed into a tank with circulating water. Fish and crustaceans are commonly transported in temperature-controlled tanks with air or oxygen circulation. In larger shipping units, baffles are placed in tanks to keep water levels more constant and to reduce wave motion, since fish can experience difficulties with balance and equilibrium and become "seasick" during transit. For handling and transporting live fish, several important factors should be taken into consideration.

22.6.1 Controlling bacterial growth

In the majority of cases, the root of any problem associated with live shipping of aquatic animals involves the environment into which they have been placed. Bacterial growth can flare up during shipment causing illness, death, and significant quality problems. Incorrect temperature control can cause both pathogenic and spoilage bacteria to multiply rapidly during shipment in addition to parasites such as *Lernaea* spp., which are common in warm-water fish and can cause wounds in fish skin, providing susceptible sites for bacterial contamination. Bacterial growth also causes a reduction in water quality, which stresses the fish, making them more susceptible to disease. Bacterial growth is a problem during shipping and handling of fresh fish and is the major cause of quality loss. Certain pathogens such as *Listeria monocytogenes* and *Clostridium botulinum* can grow at refrigerated temperatures. Viruses remain viable in contaminated molluscan shellfish during live shipment and refrigerated storage.

22.6.2 Water quality

Attention to water quality is an essential consideration for handling and transport of live aquatic animals. One of the most important strategies in aquaculture settings is to take fish off feed for a short period so that the digestive tract is empty prior to transport.

22.6.3 Oxygen

Providing sufficient oxygen in the water during shipping by providing air or compressed air or oxygen is usually required for shipping fish any distance. Using oxygen- or air-permeable packaging has become more popular for shipping fish in water-filled bags and is commonly used for retail sale. Air or oxygen is pumped in and dispersed through filters or airstones to widely disperse the gas throughout the holding tank.

22.6.4 Ammonia

Fish excrete ammonia and other nitrogenous wastes into the water during shipping, handling, and on-site storage. Stress results from gill damage and changes in respiration in response to changing pH. When acute ammonia build-up occurs in the holding water, hemorrhagic damage to gill filaments is observed. Obviously, a fish suffering this type of damage will not be capable of normal respiration and will not survive for very long. Zeolite- and carbon-based filters are among the most effective ways to remove ammonia from water during shipping. Incorporating a system for adding new fresh water is also effective.

Purging fish prior to shipment reduces the amount of ammonia produced from residual feed or feces being released into the water during shipment. However, ammonia excretion across the gill membranes continues.

22.6.5 pH

The pH of the holding water decreases (becomes more acidic) during shipment. In some ways this is beneficial because it converts ammonia to a non-toxic form. But at pH of 5.5 or less, fish skin will become increasingly irritated and may even burn slightly, and eye and gill injury can occur. The fish will produce more mucus to protect itself, and market acceptability will be decreased.

22.6.6 Salinity

Salt level is important for shipping marine fish, and generally water directly from the marine environment or artificial seawater is used. Salt (0.1–0.3%) is commonly added to shipping tanks for fresh marine fish. This is necessary to reduce osmotic stress to the fish.

22.6.7 Temperature

Temperature during shipment is critical. Temperature abuse is a common problem with shipments that are removed from a cargo plane and allowed to sit on the tarmac for an extended period of time. Another problem is with the inadvertent freezing of shipments of live or fresh aquatic products that are marked "keep refrigerated." Either scenario will kill live animals or rapidly decrease the quality of fresh product.

When possible, live fish are shipped at a reduced temperature to lower their metabolic rate, reduce the need for oxygen, and reduce ammonia excretion. However, temperature acclimation is important prior to shipment. Even for carp which can tolerate a wide temperature range, being able to live under ice during the winter and survive in hot climates in shallow ponds in water temperatures >32 °C, a slow acclimation period is necessary. Live fish destined for high-value markets in Asia are often shipped in specialized air transport containers after first being placed in individual plastic trays in sea water. The temperature of that water is gradually reduced until the fish become somewhat dormant, and are then kept moist with seawater spray with the container maintained at the appropriate, often proprietary temperature.

22.6.8 Live holding and shipment

There are different methods for transferring live aquatic organisms depending upon the species, distance to be transported, harvesting system, and the animal's size (Figure 22.2). The most important issues are sorting and separating different species, and within a species sorting into different size categories. For lobsters, one of the best methods is to immobilize the claws with stretchable bands, and then separate the animals into individual trays, stacking the trays under cold and moist conditions. Lobsters can be layered with wet newspaper or as an

Figure 22.2 Transferring and transporting live sturgeon at a farm site with brailer, crane, and transport trucks. Photo credit: Barbara Rasco.

alternative seaweed, although the use of seaweed is becoming less common, inside a plastic liner in styrofoam or cardboard boxes. Chill-packs are added to keep the animals cool but should not be in direct contact with them. In this manner, lobsters can be transported by air from eastern Canada or the north eastern US to Japan or other points in east Asia in about 48 h, with very high survival rates. Similar methods are used to ship Dungeness, king, and blue swimming crabs. Very valuable giant prawns may be shipped in individual tubes to ensure that these animals are in perfect cosmetic condition for sale in east Asian markets. The antennae are protected from breakage by carefully removing them and packing them separately in sealed tubes, reattaching the antennae with glue when the animals are brought to the market. Using these strategies, it is possible to ship very high-value fish, mollusks, and crustaceans so that they reach the market with a value five times higher compared to the fresh or frozen. Another alternative is to employ a small-scale recirculation system within a water-filled container to keep animals alive for long-distance shipping for high-value finfish; care must be taken to control increased ammonia levels during transportation.

Transporting live mollusks with special care dates in Europe to Roman times (Bertram, 1865) when oysters were transported live from coastal waters and then transported inland using snow for refrigeration, and then held in seawater aquaria until needed. The great scallop (*Pecten maximus*) from northern Europe, ranging in size from 10.5 to 14 cm, are packed in layers in Styrofoam boxes (5–12 kg of scallops per box) with moist paper or wood fibers under and on top of them. Placement must be done carefully to prevent damage and boxes must not be overloaded so that shells are not cracked under pressure. Some exporters add packaged water to keep the temperature low. This relatively simple method will deliver live scallops in good condition without significant product loss. A "dry" shipping method may also be used in which the temperature is kept low to decrease metabolism, leading to decreased oxygen demand.

Keeping scallops and other mollusks in a moist condition helps their gills exchange oxygen and limits dehydration. High oxygen-permeable bags facilitate oxygen transfer and reduce stress and potentially bacterial levels. Shipping mollusks requires management of moisture level (dry shipment) or water quality, oxygen, temperature, ammonia production, and carbon dioxide.

22.7 Fishing methods

Trawling with nets is the fishing method that provides the largest harvest volumes (Figure 22.3) of wild harvested finfish and shrimp. Cod, pollock, whiting, anchovies, menhaden, and krill captured on the high seas, and shrimp and flatfish harvested in gulf regions are commonly harvested in this manner.

Larger and higher value fish are harvested by trolling using a hook and line, including swordfish, halibut, tuna, large cod, and salmon. Pacific cod and large tuna are

Figure 22.3 Alaskan pollock coming on board a large Alaskan factory trawler for fillet, block, surimi, and roe production. Photo credit: Chris Bledsoe.

harvested by long lines that may fly as many as 27,000 hooks per set. The harvested fish are commonly headed, gutted, and frozen shipboard within 2.5 h at −30 °C in a pre-rigor state, which is preferred for the Japanese market. Troll-caught fish are harvested by hooks and lines trolled behind a boat at speeds appropriate to the targeted species and tend to have a higher quality than those caught by net since troll-caught fish are not subject to scale loss and net marking, and are not crushed or crowded as the net is brought on board.

Bleeding the fish immediately after landing is important and improves muscle color and quality, storage stability, and flavor. This is an extremely common method and one which was taught to one of the authors by his commercial fishermen father and grandfather over 65 years ago and is still required for all troll-caught salmon. Maintaining the highest degree of sanitation to prevent microbial contamination and immediate cooling are keys to keeping fish quality high. This is particularly important for fish with a very active metabolism such as albacore and other tunas. Tuna are fast-moving fish and mobilize substantial energy and oxygen to maintain their muscles. Due to this, their muscle temperature can be 12 °C or more higher than the ambient water temperature. After harvesting, they immediately need to be chilled, otherwise burnt flesh (the *yake niku* phenomenon) can occur.

Ice and refrigerated sea water (RSW) are commonly used with finfish, which have been killed. RSW is made by mechanically chilling sea water. Benefits of RSW include greater speed of cooling, reduced pressure upon the fish leading to less textural damage, lower holding temperature, greater economy in handling the fish due to time and labor saved, and longer effective storage life of the fish. It can also freeze the eggs, damage the fish, and contains many contaminants. The real advantage of RSW compared with freshwater ice appears to be that the brine temperature can be maintained at about −1 °C, which is just above the freezing point of fish. Although it is now generally regarded that the storage life of whole fish is longer in RSW than in ice, shelf life is limited by the uptake of water and salt, particularly with lean fish species, and by an enhancement of oxidative rancidity often observed in RSW stored fish. A problem common with certain fatty fish such as anchovy, herring, kilka (*Clupenella* spp.), etc. held in RSW is the eventual growth of spoilage bacteria in the brine if the brine is not changed frequently enough, producing foul odors which can be imparted to the fish. High activity of digestive enzyme and their release from the viscera into the brine and fat oxidation can be problems with such fish.

Slush ice is often appropriate for small boats that lack an onboard mechanical refrigeration system. In this case, slush ice with a ratio of 3:1:1 (fish, ice, and sea water) is used. With this system, it is important that ice and sea water be mixed together just prior to loading with fish and that efficient circulation be maintained. Also, the temperature of the slush ice must not drop to below the point at which the fish muscle will partially freeze which will cause gaping when the fish are filleted. The mixture should also be dense enough to support the weight of the individual fish so that they do not pile up on the bottom of the tank. In general, the use of RSW or slush ice is not recommended for long fishing trips unless the eventual market is a filleted product since both methods tend to bleach skin and gills, and cloud the eyes which will reduce acceptability on the dressed-fish market.

For cultured fish, the stress after harvesting is usually less and resulting quality can be higher. To further improve the quality, cultured fish are fasted for one or more days (depending on water temperature and fish breed) before harvesting. This reduces the metabolic activity of digestive enzymes. During the rigor mortis in whole fish, if digestive enzymes are active, these enzymes can migrate to fish flesh adjacent to the abdominal cavity, causing deterioration of the fish meat. Fasting fish before harvest is an important factor in live fish shipment. Because the digestive tract is empty, this decreases the amount of ammonia produced by the fish in the shipping tank, maintaining higher water quality during shipment, reducing fish stress and improving survival. Fasting is particularly important if fish are to be shipped at a high density in tanks.

Beside stress and harvest method, seasonal variation plays an important role in fish quality after harvest. In wild fish, when captured from wild sources during the breeding season, there are several changes which occur. For Alaskan pollock (*Theragra chalcogramma*), the fish harvested during the breeding season for their roe (processed form is called *mentaiko* in Japanese) have poorer quality flesh compared to fish harvested outside the breeding season. The same phenomenon may be observed for salmon. During the spawning season, male fish have higher flesh quality than female fish, because the female has mobilized muscle fat for egg production. The flesh of the salmon harvested for roe is often of little use as food but can be used for a number of other purposes, including the production of fishmeal and protein hydrolyzates.

Another important factor that can affect fish quality after harvest is the diet for cultured fish. Cultured fish are usually fed several different diets during their lifetime: a larval feed which is often a live diet of phytoplankton and zooplankton (artemia (brine shrimp), rotifers, etc.), then a grow-out diet the composition of which will vary depending upon the desired growth rate of the fish, and then a finishing diet which may contain carotenoid pigments to impart a desirable red color for those species where that is an expectation, and lipids for a desired flavor profile, for example vegetable oil or chicken fat, or a desired nutritional profile (polyunsaturated fatty acids, DHA, EPA).

22.7.1 Preparation of fish for storage

In some instances, fish are frozen or otherwise preserved "in the round"; that is, the fish are not cleaned or eviscerated. This may be a cultural preference, as in the case of herring and other smaller fish for particular Asian and Russian markets, or it may be that the fish will be slacked (partially thawed) at a later time and further processed. These might include herring (*Clupea* spp.), rock sole (*Lepidopsetta bilineata*), Alaskan pollock, and Pacific whiting (*Merluccius productus*). The herring are typically frozen post harvest as rapidly as practical near the harvest areas and are later slacked out at roe processing facilities using several exchanges of salt water. The roe is removed from the females and further processed and packed in brine. The resultant roe skeins are called *kosunoko* in Japanese. The carcasses are then normally reduced to fishmeal and herring oil. The freezing of the herring is used not only to preserve the fish, but also as part of the processing protocol, serving a function of "conditioning" the fish.

Rock sole are also harvested primarily for their roe and the frozen fish are later headed and sliced diagonally to be used for soups and other preparations, again primarily in Japan where the "rock sole with roe" is a favorite seafood dish.

Alaskan pollock and Pacific whiting are frozen as soon after harvest as practical, most commonly at sea in the harvest areas. The frozen fish are then shipped to countries where labor is less expensive. There the fish are slacked out and then filleted, repacked and frozen in blocks or fillet forms, and often re-exported to markets as "twice-frozen" fish. The same process is used for other species as well. The end quality of the fish is certainly not as good as it would have been had it been immediately processed following harvest in the harvest area, but is often acceptable for applications such as breaded "fishwiches" and similar processed foods.

Several species of tuna (*Thunnus* spp.) are also frozen in the round and later further processed, commonly in canned form.

To insure better quality, particularly for more valuable species, the fish should be dressed immediately after harvest. Dressing typically includes eviscerating and removal of the gills, although the gills may be retained for specific markets such as the Asian fresh market where the buyers want the gills retained as an indicator of freshness. The head and fins may also be removed or retained dependent upon the targeted market. Fresh fish typically have bright red gills whereas the gills tend to become pale and eventually brown as decomposition progresses. The clearness of eyes is also used as an indicator, as they become clouded and shrunken as time from harvest increases. A third common indicator is the turgor or flaccidity of the fish muscle. All these indicators are affected or accelerated by temperature abuse.

It is preferable that evisceration and processing be accomplished in a cool environment, preferably at less than 15 °C, with 7–10 °C being even better. The facility should be sanitary to maintain the quality and safety of the fish and care taken to maintain the cosmetic appearance of the fish to the greatest extent possible. Care must be taken not to rupture the gut contents and contaminate the edible portion with digestive juices which contain hydrolytic enzymes that could cause the fish to deteriorate upon spoilage, or with pathogenic or spoilage microorganisms. Again, bleeding the fish at harvest is commonly done to insure that the quality of the fish is maintained and spoilage is minimized.

Filleting fish is done by hand or mechanical methods. The method selected depends upon several factors including the size of the operation, available skilled workers, labor costs and the marginal value of small yield differences. Commercial skinning and filleting machines can process over 180 fish per minute (Baader, 2012) and these units are common on at-sea processing vessels as well as many shore plants. Skinning machines can be set to remove the dark muscle along the lateral line (deep skin) or to leave this intact. The dark muscle is considered to be undesirable in fish nuggets and in some minced fish items because it promotes off-flavor development from oxidation of fat due in part to presence of higher amounts of heme iron in the dark muscle. Many fish with relatively thin fillets and white-colored flesh are visually examined for the presence of intramuscular parasites by candling whereby the fish are passed over a lighted table or belt and visually inspected (Figure 22.4). Parasites and other defects that can be seen are removed by hand trimming. There are also some automated systems that detect parasites and other defects and remove these using blades or water jets.

Similarly, manual or machine removal of the exoskeleton from shrimp is commonly employed. Often the digestive tract or "veins" of shrimp are removed through an incision in the exoskeleton and the shrimp sold with (deveined) or without (deveined shell off) the exoskeleton. Shrimp are also sold head off or head on, tail off or tail on. For other crustacea, such as the larger crabs and lobsters, the exoskeleton commonly remains through to the point of retail sale. It may be partially removed as is the case with red king crab (*Paralithodes camtschaticus*) and other large crabs collectively referred to as king crabs (*Paralithodes* and *Lithodes* spp.), bairdi (*Chionoecetes bairdi*), snow crab (*C. oppilio*), and Dungeness crab (*Cancer magister*) which are commonly sold as sections or clusters (one half of the legs still attached to one another) or as individual legs. For sales of crabmeat, however, the crab is first cooked whole or in sections, and the meat is removed ("shaken") or picked from the exoskeleton commonly by hand. The crab meat is then packed into containers, often cans, plastic tubs or bags, or retort pouches, and then pasteurized and chilled or simply frozen.

Shellfish are commonly sold shell on but the shell can be removed (shucked) by hand or with the adductor muscle released through the use of high hydrostatic pressure processing (HHP). B-thiaminase in the adductor of mollusks is barosensitive and inactivated by moderate

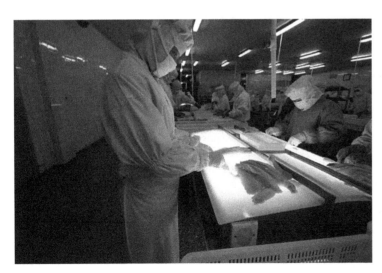

Figure 22.4 Filleting (*background*) and candling (*foreground*) of fish fillets. Photo credit: Gleyn Bledsoe.

pressure and this will cause a clean release of the muscle, mantle, and digestive tract from the shell. Many shellfish, such as oysters and clams, may be first subjected to steam to make shucking easier, but this is not always required.

22.8 Refrigerated products

Harvested fish and seafood materials undergo a series of handling operations from catch sites until the product is delivered to the end user. Fish are highly perishable starting from the point of harvest. Several factors predispose fresh fish to rapid quality degradation once they are harvested, and these include removal from the marine or aquatic environment, high moisture content of fish, activities of microorganisms inside the gut and intestines, highly active enzymes in muscles and digestive tract, and physical damage resulting from the use of improper harvesting tools and procedures, and rough handling practices.

Losses (wastage) occur when otherwise suitable fish are removed from the food supply chain, and this may occur due to spoilage, discard, or physical damage when the product is considered completely unacceptable for use as food and or even as an industrial raw material. Quality loss in fresh fish occurs when the product's value to the end user is compromised and downgraded due to a reduction in the attributes that are important to the end user. Quality loss often results in a lower unit market price for fish products and a concomitant reduction in profits to the producer. There are several common grading methods that may quantifiably describe degrees of quality or loss of quality in some aquatic products such as surimi gel strength and conductivity-based systems such as the "Torry Meter" that was developed in Scotland at the former Torry Research Station. Given the high volume of globally traded fresh fish produced, quality losses are a major contributor to the total economic loss in the fresh fish and seafood industry. It should be noted that warm-water fish are less susceptible to temperature-driven decomposition immediately after harvest than are cold-water species, but still must be chilled and maintained at chilled or frozen temperature to maintain quality and safety.

Rapid removal of heat after harvest and maintenance of the cool or cold chain is usually the most effective strategy for fish and seafood preservation and quality maintenance. Most fish are chilled with ice, ice/water slurries, refrigerated seawater, and/or mechanical refrigeration as soon after harvest as practical. Refrigerated temperatures are within the range of 0 °C to <10 °C. While fish flesh typically begins freezing at −2 °C, fresh fish are still commonly held at 0 °C or higher due to the difficulty in maintaining tight temperature control aboard vessels and many processing facilities as well as the adverse effects of partial freezing of some parts of the fish, particularly roe and gills.

Chilling reduces the rate of chemical and biochemical changes as well as microbial growth. The temperature at which most commercially important fish and seafood materials live is relatively low, thus chilling does not have as great an effect as the chilling of meat from warm-blooded animals. Rigor mortis usually begins within 1–2 h after harvest, depending on the species and ambient temperature. In large fish, rigor mortis begins more slowly than in smaller fish. In addition at higher temperatures, rigor happens sooner. Fish usually pass through rigor after a few hours, and should pass this stage before fillets are frozen, to avoid toughening and shrinkage, and to reduce drip loss during the thawing. After rigor, fish should be handled carefully to avoid gaping (separation of muscle structure). Gaping is an important quality consideration in finfish harvested from cold water. At higher temperatures (more than 17 °C), gaping is commonly observed (Bledsoe & Rasco, 2006). Gaping can also result from low intramuscular pH or from physical damage to muscle tissue during slaughter or handling operations.

Muscle fibers contract at physiological temperatures. At lower temperatures the amount of contraction decreases and around 10–20 °C, it is at its lowest. At temperatures lower than 10 °C, contraction increases again which causes a quality defect known as cold shortening and toughening of the muscle which is sometimes observed in large fishes, but less so than in the muscles of terrestrial animals. This phenomenon occurs at lower temperatures because the sarcoplasmic reticulum and the mitochondria cannot store sufficient calcium ions to prevent contraction. It also happens in fish which are frozen in a pre-rigor stage. Because adenosine triphosphate (ATP) is not depleted in the muscle cells when tissue is frozen pre-rigor, the muscle fibers contract rapidly during thawing, releasing large amounts of tissue fluids. There are several methods for short-term storing of fish: crushed ice, slush ice, and champagne ice (ice slurry through which air is bubbled to cause circulation of the cooling to and from the fish surfaces thereby improving heat removal), although high-quality mechanical refrigeration is the recommended method. In some methods such as slush ice and champagne ice, additives such as carbon dioxide, air, salt, organic acids, sugar, antimicrobials, and

Table 22.4 Approximate shelf life of different raw fish (in days)

Product	Temperature			
	0 °C	5 °C	10 °C	16 °C
Cod, fresh	14–16	7	4	1
Herring, fresh	10	4	–	–
Salmon, fresh	12	5	–	1
Plaice	18	–	8	–
Halibut, fresh	14	–	–	1
Salted herring	364	–	–	90–120
Finnan haddie	28	–	–	2
Cod, dried salted	364	–	–	120–180

Adapted from Rasco and Bledsoe (2006).

antioxidant agents could be applied to increase shelf life. Table 22.4 provides the shelf life of different fishes at various temperatures.

22.9 Freezing and frozen products

At best, the freezing process only preserves the initial quality, which will never improve beyond the quality of the original raw material. Recommendations on freezing conditions are provided in Table 22.5. In reality, although rapid freezing and cold storage at temperatures below −25 °C have shown superior quality retention, the US consumer still continues to perceive that frozen products are inferior to fresh products in terms of quality although frozen at sea product is of consistently high quality if handled properly. For frozen groundfish such as cod, the selection of packaging materials is important for quality retention and moisture- and oxygen-impermeable packaging materials are essential.

22.9.1 Freezing methods

Several methods of freezing are employed in freezing and holding of aquatic food products. Common amongst these are blast or forced air, direct contact/plate, and immersion/cryogenic. There are several variants of each system, a few of which will be described here. For freezing, the fish may be in direct contact with the refrigerant such as nitrogen, carbon dioxide, or freon substitutes or in surface-contact freezers, through contact with refrigerated surfaces such as plates, belts, drums, or shelves. For some

seafood products that are marketed in a uniform form, such as blocks of fish fillet, fish mince and fish pastes (e.g. surimi), and some whole fish, contact plate freezers are employed. Blast freezers are perhaps the most common as they are easiest to construct and maintain. In their simplest form, they consist of an insulated cabinet or room, an evaporator circulating the refrigerant through the freezer, and a fan or fans to circulate air that makes contact with the fish and removes the heat from the same. More sophisticated are spiral freezers in which the product is transported through the cold air on an open link conveyor belt.

Contact and plate freezing are used in many applications such as on-board factory processors, but are also found in many shore facilities and secondary processors because they are compact, efficient, and have lower operating costs. In a contact freezer, product is placed on a plate or between two metal plates that contain circulating refrigerant or may be otherwise cooled. These plates press against the fish or package containing it and remove heat from the product. The contact surfaces may be stationary, moving, or adjustable to make contact with at least two surfaces of the product.

The product may be submerged in a refrigerating medium such as brine to be frozen or in the vapor of the refrigerant, as in the case with cryogenic freezing, a form of immersion freezing in liquid nitrogen or carbon dioxide that is a popular method for freezing high-value seafood such as shrimp and molluscan shellfish. In this method the freezing rate is rapid, and for some products this can cause cracks or splits in the food so careful monitoring is required. Carbon dioxide is sometimes preferred to liquid nitrogen since it often has a lower cost and freezing with it results in less mechanical damage. After the product is cryogenically frozen, it is packaged in plastic and allowed to equilibrate to the frozen storage temperature before it is transferred to a storage freezer.

A novel method of freezing, pressure-shift freezing, is under investigation. This technique produces smaller and more uniform ice crystals than conventional freezing. By increasing the pressure to 200 MPa, the freezing point of water is depressed, allowing the product to be cooled to about −20 °C without the water freezing. When the pressure is released, the water rapidly freezes with uniform nucleation with small ice crystals throughout the product.

The key element to any freezing system is the refrigerant or medium that removes and transports the heat away from the product. Common commercial refrigerants include the natural refrigerants ammonia (R717), carbon

Table 22.5 Recommended freezing methods for different forms of fish muscle

Product	Type of freezer	Temperature (°C)	Air velocity (m/sec)
Fish, whole	Air-blast/batch	−30 to −40	17
	Air-blast/continuous, plate, brine	−40	
Fish, fillets	Tunnel	−30	–
	Plate	−40 to −50	
Fish, bulk or fillets	Cryogenic – nitrogen	−196	–
	Cryogenic – carbon dioxide	−78.5	

Data from Rasco and Bledsoe (2006).

dioxide (R744), and non-halogenated hydrocarbons as well as several fluorocarbons (FCs) and hydrofluorocarbons (HFCs). Previous to the 1990s, the most common refrigerants used in the fishing industry were chlorofluorocarbons (CFCs) and hydrochlorofluorocarbons (HCFCs), particularly R12 and R22, often referred to as freons (a trademarked brand of the Dupont Chemical Company). Similarly, the R number combination is a system to identify individual refrigerants that was also developed by Dupont.

Concern regarding depletion of the earth's ozone layer by CFCs and HCFCs in the 1980s led to the cessation of manufacturing the very efficient R12 in 1995, as well as several related refrigerants, in the US and R22 is to be phased out by 2020. Most other nations followed suit, although R12 is still manufactured in some developing countries. Several replacement blends and newly formulated FCs and HFCs are now being used with R404A, a HFC "nearly azeotropic" blend, designed as a replacement for R22 and R502 CFC (ASHRAE, 2010).

All fish products should be frozen as quickly as possible. Research has shown that rapid freezing reduces the size of ice crystals in the flesh, which results in a better product texture. It is also recommended that the product should only be frozen to an internal temperature equivalent to the intended cold storage temperature. Fish products frozen to temperatures below the intended cold storage temperature will tend to experience an increase in the relative size of ice crystals when the temperature equilibrates to the higher frozen storage temperature. This can result in textural changes in the finished product along with drip loss, texture, color, and flavor changes. As for other types of frozen food, fluctuation in cold storage temperature will result in ice crystal growth and lead to quality loss. Low temperature fluctuation is more

important than the actual storage temperature, within a range. Thus holding a product at −18 °C where the refrigeration system can maintain that temperature ± 1 °C will result in a much better product than holding it at −20 °C in a unit that cannot maintain that temperature ± 4 °C (Rasco & Bledsoe, 2006).

Dehydration and weight losses are the major problems in frozen aquatic food products. "Freezer burn" is a defect that develops on the surface of exposed frozen fish, particularly on unwrapped or unglazed steaks and fillets after a few weeks of frozen storage. Freezer burn is evident by patches of light-colored tissue, developed due to vapor pressure gradients within the product and between the product and external environments during temperature fluctuations. By using a tight water vapor-impermeable film over the fish, one can reduce freezer burn. Freezer burn results in textural toughening, discoloration, and oxidative rancidity. Products packaged simply within sealed bags (air packs) are not recommended unless the frozen fish has been first glazed (covered with a protective covering of ice) that is preferentially lost during freezer cycling so that the fish does not become desiccated. Glazing is a common practice for salmon, shrimp, and scallops.

Cold storage holding temperatures should be consistent and as low as possible. In general, temperatures less than −18 °C are recommended. Retention of high quality has been shown for non-fatty species such as cod and haddock stored at −18 °C from 3 to 5 months but fatty species such as mackerel will only retain high quality for 2–3 months at this temperature. In comparison, storage at −30 °C will retain high quality for 8–10 months for lean species and 6 months for fatty species (Table 22.6). Japanese cold storage facilities often maintain even lower temperatures (nearing −40 °C) and may hold salmon for up to 2 years. Japanese at-sea harvest/processors of

Table 22.6 Shelf life of frozen aquatic food products (in months)

Product	Temperature		
	−12°C/ 10°F	−18C°/ 0°F	−24°C/ −12°F
Fatty fish, glazed	3	5	>9
Lean fish, whole	4	9	>12
Lean fish, fillets	–	6	9
Crustacea, in shell	4	6	>12
Shrimp, cooked, peeled	2	5	>9
Clams, oysters, mussels	4	6	>9

Data from Rasco and Bledsoe (2006).

high-value sashimi-grade tuna hold the fish at temperatures of −80 °C. The resultant frozen tuna have been known to sell for extremely high prices. A 752 lb Pacific bluefin tuna was sold at a Japanese auction on 6 January 2011 for $396,000 (Erickson-Davis, 2011).

The seemingly ever-increasing cost of energy is affecting refrigerated warehouses, as they look for ways to reduce total energy consumption. Refrigeration requires energy and the colder the temperature, the greater the requirement for energy to hold that temperature. Thus the operators of refrigerated storage are attempting to raise temperatures or take other means to reduce their energy costs. This will have an effect on cost of storage and may also affect the quality of stored fishery products. Therefore, it is important for individuals using such storage to insure good refrigeration procedures, know the actual storage temperatures that their product is held under, and be able to monitor the actual temperature and fluctuations of their product while in storage. The use of independent electronic temperature recorders is highly recommended.

22.9.2 Blast or forced air freezing

This remains the most common method of initially freezing fish in most parts of the world, particularly in shore-based facilities. In the simplest form, fish are placed on racks or suspended from frames and cold air, normally −30 °C or colder, is circulated around the fish with a velocity of 10–15 m/sec or greater, for a period of time until it becomes frozen sufficiently to commercially reduce water, enzymatic and other chemical contributors

to quality degradation. An alternative to this system is to enclose the system in an insulated cabinet or tunnel. The fish are loaded on racks or "trucks" in the latter case and the trucks travel through the "tunnel freezer," which uses countercurrent flow of cold air to freeze the product. The countercurrent flow maximizes the driving force, the difference between the product and coolant temperature, for removal of heat. Conveyor belts are often used in lieu of trucks, particularly in the case of secondarily processed individually quick-frozen (IQF) products. Yet another variant uses a spiral, linked conveyor to transfer the product through the freezing cabinet. Such spiral freezers are relatively expensive compared to other blast freezing methods, but are significantly more energy efficient and occupy a much smaller "footprint," making them preferable where space is an issue. They are most commonly found in shore plants although they have been installed aboard factory trawlers (at-sea processors) as well.

A major drawback of blast freezers is that they tend to desiccate (remove water from) the product and the resultant water loss can contribute to degradation of quality as well as economic loss through reduction in weight. One method to reduce this loss that has been applied by at least one manufacturer of spiral freezers is to use impingement freezing as the product enters the freezing cabinet. This is accomplished by employing multiple high-velocity refrigerated air impingement jets to quick-freeze food products such as hamburger patties and fish or chicken filets. Ultra high molecular weight polyethylene blocks are provided with a multiplicity of internal jet nozzles. These blocks are located in air ducts located above and below a conveyor belt so that the topsides and bottomsides of food products are blasted with the high-velocity jets of refrigerated air so as to break up the boundary layer around the products to effect a much higher rate of heat transfer. A dual conveyor line employs direct-drive centrifugal fans to deliver high-pressure refrigerated air to the air ducts. The air ducts are spaced apart so that return air can freely return to be re-refrigerated after impinging on the food products. Blast frozen fish and fillets are commonly glazed after freezing.

22.9.3 Fluidized bed freezing

This is an adaptation of blast freezing and has several advantages compared to blast freezing systems, including more efficient heat transfer, faster freezing rates, and lower product dehydration. Although initially designed for IQF vegetables such as peas and shelled corn, small fisheries products such as shrimp can also be frozen with this system. The product is carried through the equipment on a mesh belt through which chilled air is pumped upward from below the belt at a rate sufficient to partially lift or suspend the product off the belt. An air velocity of at least 2 m/sec is necessary to fluidize the product and an air temperature of −35 °C is commonly employed. The bed depth depends upon the ease of fluidization and this in turn depends on size, shape, and uniformity of the product being frozen. Some fluidized-bed freezers employ a two-stage freezing technique, wherein the first stage consists of ordinary air-blast freezing to set the surface of the product and the second stage consists of fluidized-bed freezing.

22.9.4 Contact/plate freezing

In contact freezing, fish products are kept in contact with metal surfaces such as a stainless steel drum, solid belts or plates, which have been cooled with refrigerant at −30 to −40 °C or colder. The most commonly used contact freezers use plates, predominantly aluminum, through which refrigerant is circulated and are collectively referred to as plate freezers. The plates may be fixed, as is the case with "Dole" freezers, or may be movable using hydraulic or other means so as to bring the plates into contact with the product at both the top and bottom, referred to as "contact" plate freezers. The equipment consists of a stack of horizontal cold plates with intervening spaces to accommodate single layers of the packaged product, facilitating quick heat transfer. The plates may also be vertically aligned (vertical plate freezer). This system is not as common as horizontal plate freezers but may be advantageous in applications using smaller fish or other products. With this method, freezing time is relatively short due to the direct contact on two surfaces, most commonly both the upper and lower surfaces of the product, with the refrigerated plates. There are commonly multiple plates in a single freezer. These freezers are used for fish, shrimp, and surimi blocks or panned headed and gutted fish, although they are also used for whole fish such as salmon or halibut. Contact freezers may also be used as an initial freezing step for wet, sticky products such as raw-peeled shrimp, scallops, mussels, and delicate fish fillets, prior to these being transferred into a spiral freezer for the final freezing step.

The advantages of this method are high product quality and low space requirements, making it an appropriate method for shipboard use. Both Dole and contact plate freezers may be enclosed in insulated cabinets to enhance freezing rates and reduce total required energy consumption. The product to be frozen may also be placed in metal frames that will provide for heat removal from the sides of the product "block" liners.

22.9.5 Liquid immersion freezing

Immersion methods solve the problem of handling and freezing wet, soft and sticky products such as raw shrimp, mussels, scallops, fish fingers, etc., which tend to form clumps or adhere to freezing equipment. Refrigerants used for liquid immersion freezing should be non-toxic, inexpensive, and stable, and should have low viscosity, low vapor pressure, low freezing point, and reasonably high thermal conductivity. They should not adversely affect sensory characteristics, particularly taste, of the product. One of the most efficient methods previously was to immerse the material in purified chlorofluoro-carbon, commonly RF12 (food-grade R12 refrigerant. However, after an international movement to phase out CFCs due to perceived environmental problems with ozone depletion, aqueous solutions of propylene glycol, glycerol, sodium chloride, calcium chloride, and mixtures of sugars and salts are now used, but with limited success. Ethyl alcohol is even used for the freezing of "green" (uncooked) crab for the Japanese market but obviously is a very dangerous medium to employ. Immersion freezing uses calcium chloride brine to freeze whole fish such as salmon. The fish do not make direct contact with the fluid in this case, but are vertically suspended in the fluid in special, heat-conducting plastic sleeves or bags.

22.9.6 Cryogenic freezing

Liquid nitrogen or carbon dioxide is used as the freezing medium in this system, which is also commonly referred to as an immersion system although when the system is working properly, the product actually is immersed in the vapor of the cryogenic rather in the liquid itself. Liquid nitrogen at -196 °C is sprayed into the freezer and may additionally be circulated by small fans. The cryogenic agent partially evaporates immediately upon leaving the spray nozzles and upon contact with the food. Heat removal from the product occurs by evaporative cooling of the cryogens. The rate of freezing for cryogenic methods is much more rapid than with conventional air-blast freezing or plate freezing, although only moderately greater than that obtained with fluidized-bed or liquid immersion freezing. Lower time requirement, high throughputs (less floor space), flexibility (adaptable to different products), low dehydration loss (<1%), absence of contact with oxygen and associated oxidative changes result in improved quality in terms of flavor and appearance, and minimal drip. However, care must be taken that product does not freeze so quickly as to burst or fracture. The advantages of the system are low capital and maintenance costs, but these are more than offset by the disadvantage of the system's relative high operating costs, primarily the consumable refrigerants.

22.9.7 Glazing

Glazing is an effective and economical means of protecting frozen fish and shellfish from desiccation during storage and transport. Glazing forms a protective aqueous-based coating over the frozen product, which retards moisture loss and prevents oxidative rancidity. Glazing is done by spraying cold water over frozen product or by dipping frozen product in cold (as close to freezing as possible) water for a short period of time. Under these conditions, the glaze solution freezes almost instantly to the surface of the product. For whole fish, the extent of glaze is 5–10% of its weight, or sometimes more, providing a coating of ice of 0.5–2 mm thickness. For shrimp and smaller items, the level of glaze can be substantial, 20% or more by weight; because of this addition of water, seafood should be sold on the basis of net weight (absent glaze) and not upon delivered weight of the frozen food. The addition of excess glaze is a common form of economic fraud.

Sometimes, cryoprotectants are included in the glaze, such as sugar, commonly fructose or corn syrup solids, salt, disodium acid phosphate, sodium carbonate and calcium lactate, alginate solution, antioxidants such as ascorbic and citric acids, glutamic acid, and monosodium glutamate. The addition of any of these compounds should not be used to cover signs of decomposition and these should be listed as components on the product label because they become a constituent of the food. However, for products imported into the US this is often not done.

22.9.8 Water retention

It is unethical, and in some markets such as the USA it is illegal, to sell fish at a moisture level higher than the fish naturally contains since the product will be considered either adulterated or misbranded, or both, depending upon the jurisdiction. Treatment of fish, shrimp, and some molluscan shellfish with polyphosphates prior to freezing helps improve the water-holding capacity of myofibrillar proteins and reduce oxidative rancidity by chelating metal ions. However, overuse is a form of economic fraud since it provides an opportunity for retention of excess water. Overuse of phosphate in fish, particularly blocks, is observed by soap and foam formation when the block is thawed out, an undesirable metallic, bitter or salty flavor, excess drip, and muscle toughness. Excessive shrinkage during cooking is also a sign of excessive phosphate use. Because of the problems with the sensory characteristics of polyphosphate-treated products and abuse, other technologies have been developed for moisture retention in fish. Injection of acid or base solubilized and hydrolyzed myofibrillar proteins into fish of the same species is a new method for water retention in fish and one that is difficult if not impossible to detect analytically. Overuse of these protein-derived additives is common and is observed as excessive shrinkage of fish during cooking. Similarly, addition of polysaccharides for moisture retention and calcium to increase firmness is permissible if the use is to maintain product quality, not to give the fish characteristics that are better than it naturally possesses.

22.9.9 Quality changes during freezing

Frozen fish under proper storage conditions will not spoil due to bacterial growth since microorganisms cannot grow at commercial freezer temperatures. However, poikilothermic enzymes, either endogenous or from bacterial sources, remain active. Degradative processes from proteases, lipases and phospholipases can occur, with oxidative rancidity being the most critical. The freezing process, prolonged frozen storage, and thawing conditions have an adverse influence on muscle quality. Denaturation of structural proteins during freezing and frozen storage can be explained by three theories: an increase in solute concentration (salt, protein); cell dehydration; and auto-oxidative changes that alter the balance of protein–protein and protein–water interactions.

To assess freezing-induced quality changes, it is necessary to understand fish composition, freezing rate, temperature, and storage duration. Fish myofibrillar proteins are more sensitive to denaturation during frozen storage than those of terrestrial animals. The denatured proteins are incapable of absorbing the exudate drip water, resulting in a tough texture for the frozen fish. However, cephalopod (squid or octopus) myofibrillar proteins may be even less resistant to freeze-induced denaturation (Moral et al., 2002).

Another important issue is textural changes in fish muscle during frozen storage. As a consequence of frozen storage and associated denaturation of proteins, the texture of fish gradually changes from the usually soft, springy, moist succulence of fresh or recently frozen fish to unacceptably firm, hard, fibrous, woody, spongy, or dry. The problem is more acute for the white flaky fillets of fish such as cod, haddock, and hake. Whole fish when thawed show fewer textural changes than fillets, basically as a result of the presence of the backbone, which serves as a structural support for the flesh. Changes in texture are correlated with decreases in the water-holding capacity of the myofibrils, affecting sensory and functional attributes of the muscle. Changes in firmness, juiciness, and fibrous nature are significant textural changes in fish during frozen storage.

Generally, fresh fish has a fresh seaweed odor, which can be retained even after freezing and frozen storage. During frozen storage, off-odors and off-flavors gradually develop, which become noticeable as the storage period advances. Characteristic off-odors and flavors of frozen lean fish and shellfish are variously described as acid, bitter, turnipy, cardboard or musty. Those of fatty fish are typically rancid, oxidized, painty, or linseed oil-like. Lipid oxidation and generation of a variety of compounds are the major reasons for flavor changes during frozen storage, particularly of fatty fish such as herring and mackerel. In the case of low-fat white fish such as cod, fewer oxidative changes are observed during frozen storage. Oxidation reactions in frozen fish depend on the accessibility to oxygen, storage temperature, and the nature of the tissue (Davidek et al., 1990).

The pigments of fish and shellfish tend to fade and become duller or change in hue during frozen storage and this change in appearance generally reduces sensory acceptability. For example, prolonged frozen storage of rainbow trout fillets resulted in increased lightness, redness, and yellowness, and decreased hue values. The carotenoid pigments of salmonids are sensitive to light, heat, and oxygen, but are stable after 6 months of storage at $-20\,°C$ (No & Storebakken, 1991). If freezing is carried out after smoke treatment, fewer changes in quality occur compared with fillets subjected to smoking after to freezing. Tuna develop discoloration during frozen storage, reportedly due to oxidation of myoglobin to metmyoglobin.

22.10 Surimi and surimi analog products

Surimi is a type of fish protein gel composed of myofibrillar proteins recovered by washing minced fish to remove soluble materials such as sarcoplasmic proteins and other impurities. It is the primary ingredient for the manufacture of other food products. Initially surimi provided a method to convert small, lower valued fish into an intermediate product with relatively longer shelf life by adding preservatives to it, most commonly sugar. Surimi from some species, such as Pacific whiting, may also require the addition of inhibiting agents to counter enzymatic activity that may be present in the fish tissue as a result of the *Kudoa* sp. parasites. The most common inhibitory additives are derived from whey or egg albumen. Bovine blood plasma was previously used for exports to Japan and Korea, but this has been discontinued due to concerns about bovine spongiform encephalitis (BSE).

Large-scale commercial production of surimi only became possible because of the development of freezing technology in the 1950s, coupled with large-volume, low relative value, high seas fisheries such as that of Alaskan pollock (*Theragra chalcogramma*) and various whiting and hake fisheries (*Merluccius* spp.) beginning in the 1980s. Surimi can be made out of just about any fish, including some mollusks such as squid, but the limiting factor is normally the cost of producing the surimi, which can often exceed the market value of the finished product (Sankar & Ramachandran, 2002).

Surimi is made on board processing vessels or in shore plants located relatively close to the harvest areas from which the fish mince is recovered and washed. It is made by removing sarcoplasmic proteins, adding cryopreservatives (sucrose, sorbitol, polyphosphates), extruding into "sleeves' or blocks, frozen in plate freezers and stored for later conversion into a variety of products that may include boiled, grilled or fried fish patties, cakes, balls, sausages, etc. Three of the most common analog products are:

- **kamaboko**: washed fish flesh with flavorants, gelling agents (starch) and colorants (usually pink, not uniformly but as a coating on the outside layer) added, which is shaped into a roll and steamed. It is most commonly served sliced in udon, a type of noodle soup
- **chikuwa**: broiled kamaboko. Flavored fish paste is shaped on a skewer into an open cylinder and broiled
- **satum-age**: fried flavored fish paste.

For western markets, crab meat analog, commonly referred to as "imitation crab," is the most common item, which is also labeled as kanibo, "imitation crab," "krab" or another fanciful name. The popularity of kanibo increases when there are economic hardships or when the price of crab meat, particularly king crab (red king crab: *Paralithodes camtschatica*; golden or brown king crab: *Lithodes aequispina*) meat, is very high. The imitation crab is made by sheeting or extruding surimi combined with flavorings, starches and other ingredients, including salt, and cutting it into thin strips mimicking crustacean muscle. This is then rolled into logs and set, usually with a colored coating on the outside surface. The extruded fibers are also cut, or cut and mixed with a paste to make imitation crab with differing textural properties. The salt is added at a level of approximately 3% and causes the surimi-based mixture to set with the addition of heat. Imitation scallop, lobster, and shrimp are produced in a similar manner. Experiments using surimi as the basis for sausages, vegetable or mushroom analogs, other meat analog products, gel desserts and foams have met with limited commercial success. Surimi quality is normally graded according to method of production and resultant rheological characteristics such as gel strength (Rasco & Bledsoe, 2006).

22.11 Curing, brining, smoking, and dehydration

22.11.1 Dehydration

Drying is an ancient method to preserve fish. Currently about 25% of the world's fish harvest is dried and all types of aquatic foods are dried, including many finfish (mostly low-oil fish), shrimp, mollusks including clams and scallops, octopi and cuttlefish, coelenterates (jellyfish), and marine plants (Figure 22.5). Dried fish, principally cod, was the primary source of protein for ocean voyages and during naval exploration from Europe during the 16th to 19th centuries. Dried fish is light and easy to transport and does not deteriorate at room temperature. Dried cod is the basis for cuisines in the Caribbean and Atlantic Islands, parts of Central and South America, and throughout North Africa. Dried fish are a popular snack throughout Asia, and flaked dehydrated fish (*katsuobushi* or dried bonito flakes) are important condiments in Japanese cooking.

Solar drying remains a popular technique for many products. In this case, the fish may be eviscerated, filleted or cut into portions with or without the skin, then placed on racks in the sun to dry. Larger fish such as tuna are cut into smaller portions and flaked following drying. Scallop adductor muscle is usually dried without digestive tract or gonadal tissue. However, clams are commonly dried whole. The same is true for octopus and cuttlefish. Jellyfish may be first cured in alum, dried, and then used in soups and salads.

Much fish drying is still done in the open air or in drying shacks oriented to take advantage of prevailing winds. Unsalted dried cod (stockfish) from the Norwegian Lofoten Islands is famous for its high quality, as is dried salmon from Alaska and the Columbia River.

Aquatic foods are also dehydrated using atmospheric, vacuum, and freeze drying. Freeze drying is common for small fish or shrimp used for soup mixes and snacks.

Figure 22.5 A variety of dried fish, abalone, shellfish, sea cucumber, and small fish in Hong Kong. Photo credit: Gleyn Bledsoe. For color details, please see color plate section.

Most marine plants are dried in commercial driers using a variety of techniques. Nori for sushi and miso soup is dried in thin sheets. Kelp and other algae from which polysaccharides are extracted are commonly dried prior to further processing.

Fish may be brined (as described below), and then dried. Bacalau is salted dried cod commonly prepared by solar drying, which has been freshened out to remove most of the salt. It is usually the base in a stew, soup, casserole or croquette or simply cooked in a white sauce. In contrast, the unsalted but dried stockfish is used to make lutefisk, a traditional Scandinavian Christmas dish that is prepared by rehydrating the stockfish with several exchanges of water and then treating it with lye. Both of these foods are considered by most people to be an acquired taste.

22.11.2 Cured and salted products

Curing involves two steps: adding salt and curing; when the salt reaches equilibrium inside the tissue, tissue fluids are expressed and a surface pellicle may be formed. Salting may be accomplished by submerging the product in a salt solution (brining) or by adding salt directly to the fish (kedging) and layering it in an appropriate container. The fish should be held at refrigerated temperature regardless of which system is used while it completes the curing process. Following curing, many products are often treated with smoke, with or without added heat. Smoking adds flavor and if heat is added, this modifies the texture of the product and pasteurizes the food. Some further dehydration occurs during the smoking process.

Curing in all its forms is the oldest of the techniques available to preserve fish, with evidence of curing in archeological finds dating back over 20,000 years in Spain, and from ancient Egypt, China, and Rome. Descriptions of fish curing and dehydration date from the 9th to the 11th centuries in Scandinavia. Simple techniques such as air drying, the use of salt and smoking (and combinations of these processes) still have applications today.

22.11.3 Brining

The addition of salt is the initial step in the production of cured, smoked, salted, fermented, and dehydrated products. The salt removes water from the tissues by osmosis, with salt diffusing into the tissues and reducing water activity. Curing follows brining or kedging and raises the soluble solids content of the fish muscle, roe or other tissue to a level at which spoilage organisms are less likely

to propagate. This is accomplished primarily by encouraging tissue fluids to be expressed from the salted food. Drying and smoking may follow.

Numerous factors influence the rate at which fish absorbs salt.

- **Fat content**: the higher the fat content of the fish muscle, the lower the rate of salt diffusion into muscle tissue. Consequently, a more concentrated brine or a longer curing time is needed for brining or kedging fatty fish.
- **Skin**: salt does not diffuse through fish skin, so a longer brining time is required for skin-on fillets. For this reason, the fish are salted on the flesh side only in kedging and are layered alternating skin and flesh side up so that the fillets are always flesh to flesh and skin to skin.
- **Fish size and shape**: generally, smaller fish and pieces of smaller physical dimensions will reach the desired salt concentration more quickly than larger fish and larger pieces of fish because of the higher surface area for salt diffusion.
- **Previously frozen**: tissue damage occurs during freezing, so salt will be taken up more quickly in previously frozen fish compared to fresh fish of the same species (and dimensions). Salt diffusion is temperature dependent, so it is critical that all fish in the same batch be at the same temperature prior to salting. It is easier to brine fish uniformly if they are completely thawed out.
- **Agitation**: stirring the brine assists with mass transfer of salt into fish flesh.
- **Brine strength**: brining time decreases with increasing brine strength. The authors recommend that a saturated salt brine be prepared and allowed to equilibrate prior to use; this can take several hours or can be accelerated using a brine generator. The latter is a non-corrosive tank or other container filled with salt and through which water is diffused or circulated so as to rapidly reach saturation. The brine solution may be diluted with water if a lower strength is desired. Brining strengths typically range from 60% to 100% salt saturation. As salt uptake is directly related to concentration, time, temperature and specific characteristics of the product being brined, many technicians use saturated brine at a standard temperature, and product of specific dimensions to minimize the variables that need to be considered in developing uniform results. Often, a more dilute brine is used for cold-smoked fish (see description of process below), lightly flavored hot-smoked fish, and roes.
- **Addition of other components to the brine**: adding sugar or herbs including liquid smoke to the brine will impede salt uptake and increase brining time.
- **Temperature**: salt uptake is faster at higher temperatures but for food safety reasons, the authors recommend that brining be conducted at as low a temperature as

possible, not higher than 65 °F (18 °C). Sometimes morphological properties of the tissue such as the closed surface pores on fish eggs require use of a higher brine temperature at least initially during the brining process.

- **Submersion**: full submersion of fish into brine will help brining uniformity.
- **Density**: fish should be suspended in brine and not touching or overlapping each other otherwise salt uptake will not be uniform.
- **Thickness and tissue structure**: for kedging whole fillets with dry salt. Initially no salt is applied to the tail (caudal region) of a split fish or large fillet. After a period of time, the fish are removed, restacked, and resalted. If this is not done, the tail portion will be too salty.

22.11.4 Curing

Cured products are ready to eat without further cooking. One of the most common cured products is gravlax, which is made by adding sugar, spices, salt and herbs, and sometimes an alcoholic beverage such as vodka or Aquavit, to salmon fillets and allowing the fish to cure for several days under refrigeration. *Teijin* is a popular cured food in Japan, made by dry salting or brining, or injecting brine into, salmon for a short period. Lox is prepared in a similar manner but after curing, the fish fillets are smoked for a short period at <32 °C. These foods are not usually cooked before consumption so food safety in production and proper temperature control during distribution are important to maintain food safety. Other fish, such as black cod, may be cured and lightly smoked, then cooked.

After salting, fish are drained by placing the fish on inclined or perforated surfaces and allowing a surface film or pellicle to form. This step reduces water activity, and for roe products, like black caviar, may be sufficient to reduce the water activity to <0.92. The draining step provides a desirable texture and color.

Heavily cured cod and other large fish are prepared by first splitting the headed fish from head to tail along the backbone, and removing the backbone. By splitting the fish, salt penetration is more easily controlled. Salting is done, kedging stacking the fish in layers, again flesh to flesh and skin side to skin side, alternating each layer with a layer of salt. The pile should be restacked and resalted periodically to assure a consistent cure. After this, the fish are in the "green-cure" state, at which point the water content has been reduced to two-thirds of its original amount and the salt has penetrated throughout the fish until the internal fluids are saturated. These green-cured fish are

dried to a final water content of 25–38%. The fish are hung to dry, either in the sun or a breeze, or more likely by artificial means such as warm air circulating within an indoor drying chamber. The final salt content may be as much as one-third of the weight of the finished product.

Fatty fish such as herring, mackerel, or anchovies are well suited for brining or curing. First, the viscera are removed, and then the remaining whole fish are packed into barrels or casks, alternating a layer of fish with a layer of salt. The salt removes water from the fish by osmotic pressure, forming a brine. Removing water from the fish causes the fish to shrink considerably. Consequently, after about 10 days of curing, the barrel must be repacked by draining off the excess brine and then adding additional cured fish to make up for the shrinkage. Additional brine or even wine may have to be added to displace air introduced as a result of disturbing the barrel. The barrel is tightly capped and sealed. At this point, the fish have been stabilized and are suitable for storage.

Fish sauce (*nuoc mam* in Vietnamese and *nam pla* in Thai), a popular condiment in south east Asia, is prepared in a similar manner. Small fish, often whole, such as anchovies, are layered with salt in a perforated crock. A combination of curing and fermentation is conducted at ambient temperatures, often >25 °C. Proteolysis occurs due to the activity of endogenous and microbial enzymes and the fish liquefy. This liquid may or may not be clarified with a resulting sauce of 20–40% salt.

Fish roe is salted and cured. Fish roe and products made from fish roe such as *ikura*, *sugiko*, sturgeon caviar, *tobiko*, etc. are sensitive to heat treatments and denature at 70 °C or higher. In addition, some marine plants such as brown kelp (*Laminaria japonica*) and *wakame* (*Undaria pinnatifida* and *U. peterseniana*) are salted and dehydrated as an ingredient in miso soup and salads. Other specialized products such as whole herring skeins (*kazunoko*) and herring roe on kelp (*kazunoko kombu*) may be cured with several exchanges of brine and stored in saturated brine until used.

After brining, fish are hung or laid on racks for drying, smoking, and pasteurization (unless the foods are ready to eat). If these operations cannot be conducted within 2 h after removal from the brine, fish should be stored in a refrigerator at 4 °C or below. Fish are smoked with as much flesh as possible exposed to the smoke source. For example, a split fish should be hung or presented in such a way that the split halves are exposed. Depending upon the size and cut, fish are hung from hooks or placed on racks in the smokehouse. For small, headed and split fish, rods may be used to thread fish through the tail, head, gills, or mouth. There should be sufficient space between the fish to ensure circulation of the smoke, removal of water, and uniform dehydration. Racks are used for pieces of large fish such as chunks, steaks, blocks, sides, or fillets that do not have skin or skeletal structure to support hanging. Racks should be made of large mesh screens or other materials that allow even exposure to smoke and air circulation.

Forced drying (raising the temperature and drawing a current of air through the smokehouse) reduces the overall time required for the smoking process. Drying time depends on such factors as air circulation, temperature, and the relative humidity of the air. It usually takes several hours for dehydration, and in some cases a number of days. Fish should be dried in a cool place with circulating air from a blower or fan. Fish flesh, like that of other animals, is primarily composed of protein. When surface proteins are dried too fast a hard skin forms on the surface (case hardening), impeding moisture transfer.

22.11.5 Smoked fish

Smoked fish is a very popular and high-value food in Europe, North and South America, with new markets developing in Asia. During the preparation of smoked fish, the fish is first brined or kedged, cured, and then smoked. Often both a natural smoke extract or artificial smoke flavor and treatment with smoke from a wood fire are used. The artificial smoke is typically added to the brine as may be brown sugar, spices, and other additives that affect the color, texture, appearance, and taste of the end-product. The type of wood used is critical and is usually fruitwood (apple, cherry), mesquite, or alder. Resinous woods, particularly resinous softwoods, are avoided. Smoking remains an art, and the exact techniques used for brining, curing and smoking fish are closely guarded secrets.

Fish is either cold smoked or hot smoked. For cold-smoked fish, the fish is smoked at a temperature of 90 °F (32 °C) or less for several minutes to hours or days. Traditionally smoked dried salmon produced in native American communities in the Pacific Northwest and Alaska are a cold-smoked dehydrated product (e.g. a form of salmon jerky) that is made by treating thin strips of fish with salt, or salt and sugar, and then suspending these strips on racks over wood fires in drying sheds until the fish strips are somewhat pliable but no longer moist.

The style of cold-smoked salmon from Europe, a lox style, is prepared by exposing cured salmon to wood smoke between roughly 50–90 °F, depending upon the processor, for a short period of time, no longer than a couple of hours, until the proper flavor is obtained. This is done over wood fires inside kilns, some of which are quite large, or in mechanical smokehouses. Cold-smoked herring, mackerel, cod, halibut and other finfish, oysters, and scallops are prepared in a similar manner.

Hot-smoked fish is prepared by smoking and heating fish after it has been salted and cured. Smoking and heating can be conducted simultaneously or sequentially. The objective of heating is to pasteurize the fish so as to kill vegetative pathogens, principally *Listeria monocytogenes*. Hot-smoked fish can be quite moist or heated under dehydrating conditions, producing a hot-smoked jerky. A popular form of jerky is salmon candy, a pliable hot-smoked salmon strip cured with both sugar and salt.

Any hot- or cold-smoked product must be stored under refrigeration. The only exceptions are jerkies or dried fish where the water activity is low enough so that pathogen growth is not likely (<0.85).

Smoked fish and oysters may also be canned. In this case, smoked fish or oysters are placed into a retort pouch or can and subjected to a commercial sterilization process.

22.12 Additives and edible coatings

22.12.1 Edible coatings

There has been interest in the development of edible coatings from polysaccharides, proteins, and lipids to extend the shelf life of foods. Such coatings can retain the quality of fresh, frozen, and processed muscle foods, including fish, by retarding moisture loss, reducing lipid oxidation and discoloration, enhancing product appearance in retail packages, and also as carriers of food additives such as antimicrobials, such as nisin, and antioxidant agents (Gennadios et al., 1997). Edible films made from fish myofibrillar proteins or gelatin have been developed and are used in glazes for fish fillets or minced foods.

22.12.2 Additives

Seafood myofibrillar proteins readily denature even at refrigeration temperature and may lose up to 80% of their water-binding capacity within 5 days, while similar changes to beef muscle take in excess of 45 days and require higher temperatures (>20 °C). Cryoprotectants are compounds added to protect the protein functionality during frozen storage and are critical for surimi manufacture. Cryoprotectants improve the quality and extend the shelf life of frozen foods. Cryoprotectants include sucrose, sorbitol, sodium lactate, sugars, and hydrocolloids (alginate, carrageenan, carboxymethyl cellulose, guar gum, and xanthan gum). Fish protein hydrolyzate is a cryoprotectant. These are prepared by either acid or base hydrolysis of myofibrillar proteins recovered from fish frames or mince and are injected into fish muscle for moisture retention.

Additives are helpful but the following must be taken into consideration.

• No substance should be used either to disguise any damage or quality inferiority or to make the product look better or of a higher value than it naturally possesses.

• The use of additives is acceptable, as long as they are used for suitable and approved applications and at limits at or less than those established by the regulation of the countries in which the product is to be sold, and no more than is needed to perform the necessary function under the specific conditions for such use.

• The additive must safe and of food-grade quality.

22.12.2.1 Polyphosphates

Polyphosphates are widely used additives that improve the eating quality of many foods, particularly meat and fish products. A wide variety of phosphates are used in foods such as simple phosphates, pyrophosphates containing two phosphate units, tripolyphosphates containing three units, or polyphosphates containing more than three phosphate units (Aitken, 2001). The phosphate salts usually used in surimi are sodium tripolyphosphate (STPP), sodium pyrophosphate (SPP), sodium hexametaphosphate (SHMP), tetrasodium pyrophosphate (TSPP), tetrapotassium pyrophosphate (TPP), and trisodium phosphate (TSP).

The main value of polyphosphates lies in improving the water retention in fish muscle protein. Polyphosphate treatment of fish before freezing or chilling often reduces the amount of thaw drip under conventional and modified atmosphere packaging (MAP) (Alvarez et al., 1996), inhibits the growth of bacteria in fish stored in ice (Kim et al., 1995), retards the oxidation of unsaturated fatty acid in seafood products by chelation of heavy metals and cryoprotection through chelation of prooxidative metal ions (Dziezak, 1990), inhibits fluid losses during shipment and prior to sale, reduces the viscosity of the paste, allowing better machinability, and improves texture and tenderness (Goncalves & Ribeiro, 2008). Increased water retention improvement by phosphates is achieved through muscle fiber expansion (swelling) caused by electrostatic repulsion, which allows more

water to be immobilized in the myofibril lattices (Offer & Trinick, 1983). The effectiveness of phosphates on functional properties of meat products depends on the type of phosphate, the amount used, and the specific food products.

Phosphates are generally applied by immersion, spray, injection or tumbling into phosphate solution at different concentrations. Dry addition is also used in minced meat systems. Among the current systems, the most efficient way to apply phosphates is through vacuum tumbling as long as protein extraction is controlled.

22.12.3 Carbon monoxide

Carbon monoxide (CO) is a colorless, odorless, and tasteless gas produced by incompletely combusted cellulose-based materials. CO enhances the pink or red color of seafood through the formation of a stable cherry red carboxymyoglobin (Huang et al., 2006) in tuna, tilapia, basa, and shrimp. It has been found that treatment of tuna fillets using CO reduces total plate count bacteria significantly (Huang et al., 2006).

Carbon monoxide binds to heme proteins including hemoglobin, myoglobin, and neuroglobin, found in brain and neural tissue (Brunori & Vallone, 2007). In the US, modified atmospheres with low levels of CO, up to 0.4%, are used commercially as an antioxidant in muscle foods, as it has the ability to inhibit metmyoglobin formation. Packaging of meat with filtered smoke containing 30–40% CO is permitted for pretreatment of fish (Kristinsson et al., 2006a, 2006b) but its use in Taiwan and Japan is restricted (Huang et al., 2006). CO may promote metmyoglobin reduction, even when oxygen is present (Lanier et al., 1978). Lipid oxidation and browning effects are reduced and the shelf life prolonged. CO bound to heme inhibits the heme-catalyzed reaction between lipids and oxygen leading to rancidity (Warriss, 2000). CO possesses some antimicrobial activity (Huang et al., 2006).

22.13 Roes and caviar

Aquatic animals are consumed through the world in various forms. One of the most important non-muscular tissues from aquatic animals is roe. Skeins of eggs or individual "singled out" eggs are salt-cured and preserved, sometimes dried (Bledsoe et al., 2003).

22.13.1 Different types of caviar

There are many aquatic animals that can be used as a source of caviar during the breeding season. A list of some of the aquatic animals from which roe is recovered is presented in Table 22.7.

The most widely recognized and valued caviar is made from sturgeon harvested from the Caspian Sea. Only sturgeon caviar can be labeled in the US as "caviar." Caviar from other fish or aquatic animal species must be identified with a qualifying term that includes the common name of the fish, i.e. "salmon" caviar. More than 20 species of sturgeon are harvested through the world for caviar production. The most famous caviars are produced from Russian and Iranian great sturgeon (Beluga; *Huso huso*), Osetra (*Acipenser guldenstadtii*), and Sevruga (*A. stellatus*). In the US, caviar is produced from cultured white sturgeon (*A. transmontanus*). Due to the price and reduction of the availability of Caspian Sea caviar, greater attention is being paid to caviar products from other species of fish.

22.13.2 Fish roe processing

Recovery of roe from fish can vary a great deal and is dependent upon the species, method of reproduction, stage of maturity, feeding regimes, various environmental conditions, and desired product form. For female pink salmon, the yield of eggs is approximately 15% and for gravid sturgeon it can be as high as 25%.

The first process is separating the roe skeins. This step is performed by hand or mechanical screening or by an enzymatic process to remove the connective tissues (Bledsoe et al., 2003). Following singling out or "screening," the eggs are washed in a chilled, dilute (2–3%) salt solution. The eggs are then salted; for black caviar and other caviars made from fish with small sized eggs, this is typically accomplished by the direct addition of flake or finely powdered salt. For salmon eggs (*ikura*), a brine is generally used. Following this, the eggs are drained, then packaged, often in plastic buckets or tubs or even glass containers with lug type closures. Metal with slip-on closures are used for black caviar; plastic, wooden, and bamboo boxes are also used. Salmon or sturgeon roe of lower quality may be pressed, mixed with butter, prepared into spreads or pastes, or added to sausages and sauces. An attempt has been made to market the adipose fat from cultivated sturgeon as an ingredient in shampoos and body creams.

Whole skein products are also made from kutum (*Rutilus frisii* Kutum), a Caspian Sea fish. When this is

Table 22.7 A partial list of aquatic species with edible roe

Fish	Scientific names
Sturgeons	Beluga (*Huso huso*)
	Osetra (*Acipenser guldenstadtii*)
	Sevruga (*A. stellatus*)
	Kaluga sturgeons (*H. dauricus, A. dauricus*)
	Ship (*A. nudiventris*)
	Siberian (*A. baerii*)
	White sturgeon (*A. transmontanus*)
Paddlefish and related species	Paddlefish (*Polydon spathula*)
	Shovel-nose catfish (*Hemisorubim platyrhynchos*)
"Whitefish" roes	Atlantic whitefish (*Coregonus huntsman*)
	Lake whitefish (*Coregonus clupeaformis*)
	Mountain or Rocky Mountain whitefish (*Prosopium williamsoni*) Other fish with small light-colored roe:
	Lavaret (*Coregonus lavartus*)
	Roach (*Rutilus rutilus*)
	Caspian roach (kutum) (*Rutilus frisii kutum*)
	Perch (*Perca fluviatilis*)
	Pacific herring (*Clupea pallsii*)
	Atlantic or Baltic herring (*Clupea harengus*)
	Smelt (*Spirinchus lancerolatus*)
	Burbot (*Lota lota*)
Herring (kazunoko and kazunoko kombo)	Pacific herring (*Clupea pallasii*)
	Baltic herring (*Clupea harengus*)
Pollock	*Theragra chalcogramma* (mentaiko)
Flying fish (tobiko)	*Cheilopogon furcatus*
Capelin	*Mallotus villosus*
Smelt	*Spirinchus lancerolatus*
Gadoids	Cod – *Gadus morhua* and other *Gadus* sp.
	Whiting or hake – *Merluccius* sp.
	Lumpfish (*Cyclopterus lumpus*)
Tuna	*Thunnus* spp.
Salmonids (ikura, sujiko)	Chum salmon (*Oncorhynchus keta*)
	Pink salmon (*O. gorbuscha*)
	Coho salmon (*O. kisutch*)
	Sockeye salmon (*O. nerka*)
	Chinook (*O. tshawytscha*)
	Atlantic salmon (*Salmo salar*)
	Arctic char (*Salvelinus alpinus*)
Catfish	Channel catfish (*Ictalurus punctatus*)
	Basa (*Pangasius gigas*)
	Clarias (*Clarias batrachus*)

Table 22.7 (*Continued*)

Fish	Scientific names
Kutum	Kutum (*Rutilus frisii kutum*)
Gars	*Lepisosteus spatula*
	Lepisosreus osseus
	Lepisosteus productus
Mullet	Mullet (*Mugil cephalus*)
Pikes	*Esox ircius*
Shad	*Alosa sapidissima*
Barbels	*Barbus barbus*
Killifishes	*Fundulus diaphanous*
	Fundulus hetcrocitus
Croelefishes	*Paranthias furcifer*
Cabezon	*Scorpaenichthys marmoratas*
Blenny	*Stichaeus grigorfewi*
Puffers	*Fugu rubripes rubripes*
	Fugu niphobles
	Lngoceplinlus sceleratus
	Arorhron hispidus
	Sphneroides maculates
Mollusks	Queen scallop (*Aequipecten opercularis*)
	Sea scallop (*Placopecten magellanicus*) and others
Crustacea	Prawn (*Penaeus monodon*) and others
	Shrimp (*Pandalus borealis, Crangon crangon, Indicus* sp.) and others
	Swimming crabs – blue crab (*Callinectes sapidus*), three spot crab (*Portunus sanguinolentus*) and others
	Dungeness crab (*Cancer magister*)
	Lobsters (*Homerus americanus*) and others
	Red crawfish (*Procambarus clarkii*) and others
Echinoderms	Green sea urchin (*Strongylocentrotus pucheriius*)
	Red sea urchin (*S. franciscanus*)
	Purple urchin (*S. intermedius*)
	Pseudocentrotus depressus
	Heliocidaris crassispina
Holothurians	Sea cucumber (*Stichopu* spp.) and others

brined in a concentrated brine (18–25% salt), the egg color will change from yellow to brown, and this is called *eshpel*. *Sujiko* is made by salting skeins of salmon eggs with the inclusion of sodium nitrite.

Fish roes are heat labile and conventional pasteurization causes deterioration of product texture. Certain

countries permit the addition of borax for preservation and to impart sweetness but this is prohibited in the US (Bledsoe et al., 2003; Shin et al., 2010).

Addition of polysaccharides is used for some shelf-stable products like lower quality lumpfish caviar to reduce water activity. Packaged processed roes are refrigerated, superchilled, and sometimes frozen.

Whole skeins of roe of larger fish are sometimes smoked. Breaded and fried whole skeins of cod or shad roe skeins are popular foods in northern Europe. Shad, introduced into the rivers of the north eastern and western US, are sold fresh in the spring and fried. Shad roe is thermally processed and sold as a canned food; it is not usually brined. Atlantic hake (*Merluccius hubbsi*) roes are boiled in water and seasoned, which are a traditional food in Uruguay during autumn.

Mentaiko is processed Alaskan or walleye pollock roe (*Theragra chalcogramma*) presented as whole, matched pairs of skeins, often with the oviduct intact, and cured with or without additives besides salt. There are over 20 grades of *mentaiko*, depending on cosmetic appearance, with defects such as hemorrhages or bruising, crushed roe skeins, large veins or unattractive veining, fracture of the oviduct connection between the two skeins, paired skeins of non-uniform size, and skeins with different color or no longer connected together, reducing the overall value of the product. Pollock roe is also used in the production of salad dressing and is now being introduced to European markets in caviar form (Bledsoe, 2012).

Rock sole (*Lepidopsetta bilineata*) with roe is an example of a finfish that is sold intact or diagonally sliced with its roe. Similarly, whole crabs (several genera) with roe, shrimp, crawfish, and lobster with roe (coral), and scallops with roe are very valuable foods, particularly if these are sold live. Freshness and cosmetic appearance are critical, with colorful roes in contrast with fresh white muscle. She crab soup has long been a favorite in Europe and around the world.

Crustacea and mollusks in roe are also an important part of the live fish trade. Sea urchin roe can be served raw, steamed, baked or sautéed. *Uni* is brined sea urchin gonadal tissue treated with alum for sushi. A uniformly bright orange product is the most desirable. The salted sea urchin gonadal tissue may also be fermented to produce a paste (*neri uni*). Sea cucumber (Stichopu spp.) are normally eviscerated, cooked, cleaned and dried.

Dried roe products are considered by many to be an acquired taste but are still economically important and popular in many cuisines. For example, mullet (*Mugil cephalus*) roe is salted and dried (*karasumi*) for consumption in Japanese and other Asian cuisines. The dried mullet roe has a yellowish red color and a rubbery chewy mouthfeel due to the large quantity of wax esters. *Botargo* is a popular Mediterranean roe product that is commonly made out of grey mullet (*Mugil* spp.) roe but also tuna and swordfish prepared in a similar manner. It is also called "Sardinian's caviar" and is prepared from skeins of eggs that are cured in sea salt for several weeks until they are firm textured, often as hard as boards. After curing, the skeins are typically enclosed in colored or natural beeswax, but not always. *Botargo* is sliced or grated and served often with raw garlic or in pasta dishes.

22.14 Other non-muscle tissues used as food

The differences in muscle structure between aquatic and terrestrial animals are due to the need for swimming and buoyancy. Fish are supported by a mass of water, so the muscle fibers require less structural support than those in land animals. Because of this, fish muscle tends to have less connective tissue than muscles from terrestrial animals, resulting in a more tender texture (Kristinsson & Rasco, 2000a). Fish muscle lacks the linear, tubular structure found in most terrestrial animals, but instead is formed of interlocking "w"-shaped sheets called myomeres or myotomes. The sheets are separated from one another by connective tissue. About 70–80% of fish muscle is composed of structural proteins which are soluble in cold neutral salt solutions of fairly high ionic strength. The remaining 20–30% contains sarcoplasmic proteins that are soluble in water and dilute buffers, with a final part of the structural protein, 2–3%, being insoluble connective tissue proteins (Spinelli & Dassow, 1982). Myofibrillar proteins are the primary food proteins of fish, comprising 66–77% of the total protein in fish meat. In addition, fish muscle tends to be predominantly composed of either "white" or "red" meat. The white meat is generally more abundant and contains fewer lipids than the dark meat (Kristinsson & Rasco, 2000a). White meat is also associated with slow-moving fish, particularly bottom or demersal species, while red muscle is red due to the presence of a high hemoglobin content, but is also affected by diet, particularly of salmonids, that includes carotenoids. The red muscle is used for continuous swimming and is found in many predator species including salmonids, tuna, and several other mid-water or upper water species. The white muscled fish tend to move much less and are only capable of high speeds for relatively short periods.

In addition to muscle tissue, there are other proteins, oil and microelements in large quantities in fish tissues, which can be used nutritionally by humans. The yields of by-product fractions from various fish are provided in Table 22.8.

One of the non-muscle tissues is fish blood. Although cattle and swine blood and their fractions have been well used, there is relatively little information about fish blood and its fractions as food ingredients. Fish blood or its fractions contain high-quality protein and heme iron and may be useful in nutrition with functional properties making them suitable for food or feed applications.

Fish gills are consumed fresh or following dehydration. To clean them, fresh gills are soaked for a day in water. They are sun dried on racks, which takes 3–4 days, giving a yield of 15%. Half gills from big fishes are cut and made into a set of two pieces. Usually, gills are recovered from large fish such as ghol (*Protonibea diacanthus*), sailfish (*Istiophorus platypterus*), skates (e.g. *Raja raja*) and edible rays, and India is a major exporter to several east Asian countries. As previously mentioned, Russians can fish gills, fins, and blood to make a soup stock.

Fish fin is eaten as a fried food (*karrage*) (Nagai & Suzuki, 2000), and a dried or fermented food (*hongtak*) (Mizuta et al., 2003) in some Asian countries. *Hongtak* is a fermented product made in Korea from the pectoral fin of a common skate (*Raja kenojei*). Shark fin is in high demand as an ingredient in popular Asian dishes, in particular soup, which has led to overfishing and the deplorable practice of shark "finning" whereby the shark are captured, but only their fins are harvested and the remainder of the fish is released without utilization. This has resulted in a ban on its harvest or sale in many locales, recently in Oregon, and likely soon in the rest of the US and in many other countries.

Shark liver oil contains a large amount of squalene, vitamin A, and long-chain omega-3 polyunsaturated fatty acids. Shark skin is also sun dried as a food, and used for leather production; shark cartilage and shark cartilage chondroitin have food and drug uses.

22.14.1 Fish cartilage and skin

Sharks and rays are elasmobranch fish, characterized as having an internal skeleton containing only cartilage and no ossified bone. The fins of these fish have unique

Table 22.8 By-product yields for selected fish

Fish	The amount of by-product (% animal weight)							References
	Head	Viscera	Trim	Skin	Roe	Milt	Backbone	
Sardine	30							Souissi et al. (2007)
Cod	20.2	5.6	8.2	4.2	0.7	1.3	9.7	Falch et al. (2006a)
Saithe	15.3	7.2	8.8	4.8	0.3	0.2	9.9	Falch et al. (2006a)
Haddock	18.9	6.2	9.3	4.5	0.7	0.1	10.6	Falch et al. (2006a)
Tusk	17.9	9.9	21.2	6.4	2.0		8.4	Falch et al. (2006a)
Ling cod	18.6	3.3	?	?	1.7	?		Falch et al. (2006b)
Atlantic salmon	10.0	14.0		5.0			10.0	Sandnes et al. (2003)

thickening and textural properties and are used as ingredients in soups and traditional dishes, primarily in east Asia, an example of which is salmon skin *temaki*. The use of fish skin for the production of gelatin is practical and could add value to processing wastes (Kristinsson & Rasco, 2000a, 2000b). Fish cartilage, fins and even gills are recovered as food ingredients, supplements, and as a source of gelatin as well as a variety of products derived from hydrolyzates (described below). Sturgeon have a high concentration of cartilage and a notochord which provides a source of chondroitin sulfate. The skins of shark, rays, sturgeons, salmon, and hagfish (*Eptatretus* spp. and *Myxine* spp.) are tanned into leather for high-end accessories, wallets and handbags, and specialty footwear. Hagfish, also commonly called "slime eels," are a rather unpleasant fish by many standards and products manufactured from their skins are marketed as "eel skin." A distant relative, the lamprey (*Geotria australis, Petromyzon marinus, Entosphenus tridentatus*), is a tradititional source of food for many native Americans and a favorite in several European countries.

22.14.2 Chitin and chitosan

Chitin, poly-(164)-N-acetyl-d-glucosamine, is a cellulose-like biopolymer found in a wide range of products in nature (Shahidi et al., 1999), including the exoskeleton of many marine creatures, and is the second most plentiful natural polymer on earth after cellulose (Ornum, 1992). Chitin is an aminopolysaccharide that is a major bioresource, with an estimated annual potential availability of 100 billion tonnes. Marine organisms, which include lobster, crab, cuttlefish, shrimp, and prawn, are richer in chitin compared to terrestrial organisms such as insects and fungi from which chitin can also be obtained.

Chitosan is a linear polysaccharide consisting of (1,4)-linked 2-amino-deoxy-β-d-glucan, that is an alkaline or enzymatically deacetylated chitin derivative. The functional properties of chitosans differ with crustacean species and how it is recovered. Lower molecular weight fractions of chitin, chitosan, and their oligomers have different applications, such as wound-healing agents, hypocholesterolemic agents, antitumor, and antiulcer agents, and as a component of coatings included in food preservation possessing antimicrobial properties (Shahidi et al., 1999).

22.14.3 Gelatin

Gelatin is hydrolyzed collagen and is an important industrial biopolymer because of its utility, particularly as a food ingredient and in the pharmaceutical, photographic and cosmetics industries. Functional applications in food formulations include water holding, thickening, colloid stabilization, crystallization control, film formation, whipping, emulsification and melt-in-the-mouth perception. The global demand for gelatin has been increasing over the years. Although pork and bovine gelatin may be preferred to fish gelatin because of superior functional properties (Choi & Regenstein, 2000), concerns over BSE in bovine animals and increased market demand for Halal and Kosher foods provide opportunities for aquatic foods to fill this need. Gelatin is recovered by a series of extractions conducted on fish frames or skin with the recovered gelatin dehydrated.

22.14.4 Bone

Bone comprises a significant portion of seafood processing wastes. The backbone is one of the major by-product fractions, yielding about 15% of the fish weight, which contains 60–70% minerals, mainly calcium phosphate and hydroxyapatite, and about 30% proteins that may be used for human consumption (Gildberg et al., 2002). Bone is a composite of organic (primarily collagen) and inorganic (bioapatite, $Ca_{10}(PO_4)_3(OH)_2$) components, as well as lipids and water.

22.14.5 Shell and exoskeleton

Aquatic invertebrates have a shell or exoskeleton. These animals comprise about 60% of total landings and over 80% of the value of the total world fishery (FAO, 2008). These shells are high in calcium, and oyster shell in particular is recovered, cleaned and powdered, and used as a nutritional supplement. Exoskeletons are the primary source of chitin.

22.14.6 Isinglass

Isinglass is recovered from the dried air bladder of fish and has been used for several hundred years to clarify wine and beer. It is primarily comprised of a collagen which is soluble in weak organic acid. Isinglass is available as a powder, paste, or a highly viscous liquid and is added directly to beverages to cause aggregation of the yeast cells and other insoluble components, which then sediment to the bottom of the container or can then be removed by filtration.

22.15 Fish meal and protein hydrolyzates, and fish oil

22.15.1 Fish meal and protein hydrolyzates

World fish production has remained relatively constant and presently stands at 143 mmt (FAO, 2008). While wild harvests have declined in several areas of the world, other regions, such as the Alaska North Pacific region, have addressed these issues and several species, such as Pacific salmon and Alaskan pollock, are now considered to be healthy and sustainable. The decline in some wild harvest volumes is in contrast to aquaculture harvest, which continues to grow throughout the world. Cultivated fish are mainly for human consumption whereas fish from wild harvests are for both human consumption and fish meal production. Roughly 75% of the total harvested fish (edible) is used directly as human food and 25% for non-food uses such as poultry and other terrestrial animal feeds, including pet food, as well as for fish feed. High-quality fish meal is made by heating fish byproducts and unusable fish, decanting oil, and recovering the solids by centrifugation or fractionation. Bones and scales are removed. The solids are dried using various technologies based upon the value of material and availability. Fish meal may be pressed to remove most of the water and then vacuum or otherwise dried to produce the commercial product. Some higher value hydrolyzates may be spray dried to preserve protein quality. Freeze drying is also used for specialty diet ingredients. There is a worldwide shortage of fish meal currently and the cost of it is increasing. The most common sources remain menhaden, anchovy, blue whiting, and sardine, with significant quantities of high-quality white fish meal coming as a by-product from the Alaskan pollock fishery.

Fish protein hydrolyzates (FPH) production is often an excellent method to recover the protein and oil to produce value-added products (Kristinsson & Rasco, 2000a), organic fertilizers and soil conditioners with technologies available for waste management in remote areas through controlled acid or enzymic hydrolysis. Controlled enzymatic hydrolysis of a broad range of protein ingredients of good quality can be produced from undesirable raw materials (Kristinsson & Rasco, 2000a-c; Ovissipour et al., 2009a-d, 2010a,b). Fish hydrolyzates using a number of enzymes have been described that have different catalytic activities, different reaction pH, and temperature stabilities. Various mineral and organic acids are also employed for hydrolysis, but materials made in this fashion are not generally appropriate for food or feed ingredient use.

The choice of enzymes depends not only on the end-product but also on cost. Mixing fish viscera into the slurry is the least expensive method. The fish enzymes are active at very low temperatures, which minimizes microbiological and quality problems, but will decrease recovery because of the low relative activity. Addition of commercially available endopeptidases and exopeptidases is common, and reaction conditions can be controlled to yield FPHs with very different properties. Important production parameters are isolation, precipitation, drying or dehydration, concentration and modification (enzymatic, alkaline, acid hydrolysis, chemical), and environmental parameters including temperature, pH, and ionic strength.

Fish protein hydrolyzates can have many food applications, i.e. water-binding and oil-holding capacity, viscosity, foaming properties, emulsifying properties, gelation, and solubility. Functional properties depend on different factors such as protein source, and production and environmental parameters. High solubility over a wide range of pH is important for many food applications as it influences other functional properties, such as emulsifying and foaming. FPH tend to have high water-holding capacities and are thus useful for certain food formulations. However, if dehydrated, FPH are highly hygroscopic, and need to be packaged to exclude oxygen and moisture and held at relatively low temperatures. These materials have reasonable emulsifying and foam-forming properties. FPH can facilitate emulsion formation and improve stability. Peptides may adsorb to the surface of freshly formed oil droplets during homogenization and form a protective membrane that prevents droplets from coalescing, but they are often not as good emulsifiers as dairy, egg and soy proteins (Sathivel et al., 2004, 2005). Weak foams are commonly observed when food proteins are hydrolyzed. However, the advantage of using hydrolyzed proteins as foaming agents is their insensitivity to change in pH.

The sensory properties of FPH depend in part on the fish used to make them. Bitterness is a concern due to the presence of medium-sized peptides with hydrophobic amino acid residues. Various FPHs and peptides exhibit *in vitro* bioactivities such as antioxidative, immunomodulatory, antihypertensive, anticancer, and antithrombotic activities (Kim & Mendis, 2006).

22.15.2 Fish oil

Polyunsaturated fatty acids (ω-3 PUFAs) play an important role in neurological and cardiac health. Fish oil is the major source of these fatty acids, and fish lipids are recovered from various fisheries as a food ingredient or supplement. The quantity and fatty acid profile in marine organisms are dependent upon genus, species, environmental conditions, season (in pelagic fish: 12–15% in winter, 3–5% in summer), sexual maturity, spawning cycle, diet, and age. Marine lipids are composed of neutral lipids comprising triacylglycerols, phospholipids, sterols, wax esters, and some unusual lipids such as glyceryl esters, glycolipids, sulfolipids, and hydrocarbons.

Marine fish are commonly classified by the fat content of their fillets as lean (under 3% fat), medium (3–7% fat),

and high fat (over 7% fat). In lean fish, lipids are deposited in the liver, muscle, and gonads, while fatty fish fat is mostly subcutaneous. Lean fish include sole and demersal flatfish, catfishes, tilapia, and carp, halibut, cod, hakes, and pollock. The flesh of high-fat fish (e.g. herring, sardine, anchovy, salmon) is usually pigmented and has a prominent lateral line with accompanying high-fat dark muscle. In an individual fish, lipid content increases from tail to head, with higher level of fat deposition adjacent to the abdominal cavity.

Oil is recovered following cooking of the fish tissue to release fat followed by decantation, centrifugation or mechanical pressing. Solvent extraction and supercritical fluid extraction are also used. The latter can increase oil recovery to as much as 95% from high-fat fish without oxidation of PUFAs. Extraction using proteases to hydrolyze muscle tissue can also enhance release, and the resultant recovery, of oil. Following extraction, oil is refined to remove non-triglyceride components, color, and off-flavor components. Of note, the high cost of petroleum fuels for fishing vessels has made it more economical to burn the fish oil in the vessels' diesel engines (at levels of 5–10%) than to tanker it to fish oil markets.

22.16 Sustainability

With a dramatically increasing world population and a world catch of fish of more than 100 mmt per year, there is obviously an increased need to use our sea resources with more intelligence and foresight (Ovissipour & Ghomi, 2009) to provide a sustainable source of food and food protein. Aquatic food products will become increasingly important as a food source; their intrinsic diversity in terms of species, product forms, and harvest or cultivation conditions makes this a product category more resilient to environmental impacts than terrestrial agriculture. Aquaculture provides the only untapped source for a world-wide increase in the production of animal protein, whether as a food in and of itself or as a feed ingredient for other food animals. Marine and most freshwater capture fisheries are at or exceed sustainable harvest levels, the major exception to this being Pacific salmon and Alaskan pollock resources which have been well managed in recent history.

Food production of all kinds is restricted by the availability of water and climate change appears to be exacerbating the deterioration of important farming areas across the globe. Global terrestrial animal production is effectively "maxed out" due to water limitations either directly, from the inability to produce sufficient feed, or from demand from other water users who are restricting water usage by agricultural activities. Aquaculture, unlike other farming practices, uses water but does not "use it up"; further, aquaculture can be coupled with production of food plants in greenhouses (aquaponics) providing nutrients to the plants. Cultivation of fish or shellfish in irrigation canals, ponds or open water can remove suspended organic matter and potentially detrimental materials from the water column, thus improving overall water quality. Aquaculture provides the underpinnings for sustainable fisheries into the future: as a method of production for animal protein, to enhance native stocks through hatcheries, maintain biodiversity through cataloging and maintaining broodstocks and germplasm, by stock enhancement, and restoration activities to support aquatic food animals in their native habitat.

22.17 Summary

Foods from the aquatic environment are diverse, with hundreds of different species consumed, many of which have only entered international trade in the last two decades. Fish and aquatic animals and plants provide a source for organic fertilizers and potentially for nature-derived fuels not yet tapped. Every possible technology used in the food science field is employed in the processing of aquatic foods, and an understanding of aquatic foods and their production methods, particularly aquaculture, will become more essential for all food technologists in the future.

References

Aitken A (2001) *Polyphosphates in Fish Processing*. Torry, Aberdeen: Ministry of Agriculture, Fisheries and Food – Torry Research Station. Retrieved August 16, 2009: www.fao.org/wairdocs/tan/x5909e/x5909e00.htm#Contents, accessed 27 November 2013.

Alvarez JA, Pozo R, Pastoriza L (1996) Effect of a cyroprotectant agent (sodium tripolyphosphate) on hake slices preserved in modified atmosphere packing. *Food Science Technology International* **2**: 177–181.

ASHRAE (2010) *The 2010 ASHRAE Handbook*. Atlanta, GA: American Society of Heating, Refrigerating and Air Conditioning Engineers.

Baader (2012) *Baader Food Processing Machinery*. www.baader.com, accessed 27 November 2013.

Bertram JG (1865) *The Harvest of the Sea*. London: self-published.

Bledsoe CD (2012) Aquatic Foods International, personal discussion.

Bledsoe G, Rasco B (2006) Caviar and fish roe. In: Hui YH (ed) *Handbook of Food Science, Technology and Engineering*. New York: CRC Press, pp. 161–172.

Bledsoe GE, Bledsoe CD, Rasco B (2003) Caviars and fish roe products. *Critical Reviews in Food Science and Nutrition* **43**(3): 317–356.

Brunori M, Vallone B (2007) Neuroglobin, seven years after. *Cellular Molecular Life Sciences* **64**: 1259–1268.

Choi SS, Regenstein JM (2000) Physicochemical and sensory characteristics of fish gelatin. *Journal of Food and Chemical Toxicology* **65**: 194–199.

Davidek I, Velisek J, Pokorny J (1990) Chemical changes during food processing. In: *Developments in Food Science*. Amsterdam: Elsevier, p. 107.

DeVoe MR, Mount AS (1989) An analysis of ten state aquaculture leasing systems: issues and strategies. *Journal of Shellfish Research* **8**(1): 233–239.

Dziezak JD (1990) Phosphates improve many foods. *Food Technology* **44**(4): 80–91.

Erickson-Davis M (2011) Bluefin tuna gets record price ($396,000) at Japanese auction. http://news.mongabay.com/2011/0106-morgan_tuna_record.html, accessed 27 November 2013.

Falch E, Rustad T, Aursand M (2006a) By-products from gadiform species as raw material for production of marine lipids as ingredients in food or feed. *Process Biochemistry* **41**(3): 666–674.

Falch E, Rustad T, Jonsdottir R et al. (2006b) Geographical and seasonal differences in lipid composition and relative weight of by-products from gadiform species. *Journal of Food Composition and Analysis* **19**(6): 727–736.

Food and Agriculture Organization (FAO) (2008) *Year Book of Fishery Statistics*. Rome: Food and Agricultural Organization, p. 98.

Frewer LJ, Kole A, van de Kroon SM, Lauwere C (2005) Consumer attitudes towards the development of animal-friendly husbandry systems. *Journal of Agricultural and Environmental Ethics* **18**: 345–367.

Gennadios A, Hana MA, Kurth LB (1997) Application of edible coatings on meats, poultry and sea foods: a review. *Lebensmittel-Wissenschaft-Und Technologie* **30**: 337–350.

Gildberg A, Arnesen JA, Carlehog M (2002) Utilisation of cod backbone by biochemical fractionation. *Process Biochemistry* **38**: 475–480.

Goncalves AA, Ribeiro JLD (2008) Do phosphates improve the seafood quality? Reality and legislation. *Pan-American Journal of Aquatic Sciences* **3**(3): 237–247.

Holiman FM, Rynolds JB (2002) Electroshock induced injury in juvenile white sturgeon. *North American Journal of Fisheries Management* **22**(2): 494–499.

Huang YR, Shiau CY, Hung YC, Hwang DF (2006) Change of hygienic quality and freshness in tuna treated with electrolyzed water and carbon monoxide gas during refrigerated and frozen storage. *Journal of Food Science* **71**: M127–M133.

Kim CR, Hearnsberger JO, Vickery AR, White CH, Marshall DL (1995) Extending shelf life of refrigerated catfish fillets using sodium acetate and monopotassium phosphate. *Journal of Food Protection* **58**: 644–647.

Kim SK, Mendis E (2006) Bioactive compounds from marine processing byproducts – a review. *Food Research International* **39**: 383–393.

Kristinsson HG, Rasco BA (2000a) Fish protein hydrolysates: production, biochemical, and functional properties. *Critical Reviews in Food Science and Nutrition* **40**: 43–81.

Kristinsson HG, Rasco BA (2000b) Biochemical and functional properties of Atlantic salmon (*Salmo salar*) muscle proteins hydrolyzed with various alkaline proteases. *Journal of Agricultural Food Chemistry* **48**(3): 657–666.

Kristinsson HG, Rasco BA (2000c) Hydrolysis of salmon muscle proteins by an endogenous enzyme mixture extracted from Atlantic salmon (*Salmo salar*) pyloric caeca. *Journal of Food Biochemistry* **24**: 177–187.

Kristinsson HG, Balaban M, Otwell WS (2006a) The influence of carbon monoxide and filtered smoke on fish muscle color. In: Otwell WS, Kristinsson HG, Balaban M (eds) *Modified Atmospheric Processing and Packaging of Fish: Filtered Smokes, Carbon Monoxide and Reduced Oxygen Packaging.* Ames, IA: Blackwell, pp. 29–52.

Kristinsson HG, Balaban M, Otwell WS (2006b) Microbial and quality consequences of aquatic foods treated with carbon monoxide or filtered wood smoke. In: Otwell WS, Kristinsson HG, Balaban M (eds) *Modified Atmospheric Processing and Packaging of Fish: Filtered Smokes, Carbon Monoxide and Reduced Oxygen Packaging.* Ames, IA: Blackwell, pp. 65–86.

Lagler KF, Bardach JE, Miller RE, Passino DRM (1977) *Ichthyology*, 2nd edn. New York: John Wiley, p. 26.

Lanier TC, Carpenter JA, Toledo RT, Reagan JO (1978) Metmyoglobin reduction in beef systems as affected by aerobic, anaerobic and carbon monoxide-containing environments. *Journal of Food Science* **43**: 1788–1792.

Mitrowska K, Posyniak A (2004) Determination of malachite green and its metabolite, leucomalachite green, in fish muscle by liquid chromatography. *Bulletin of the Veterinary Institute in Pulawy* **48**: 173–176.

Mitrowska K, Posyniak A (2005) Malachite green: pharmacological and toxicological aspects and residue control (in Polish). *Medycyna Weterynaryjna* **61**: 742–745.

Mizuta S, Hwang JH, Yoshinaka R (2003) Molecular species of collagen in pectoral fin cartilage of skate (*Raja kenojei*). *Food Chemistry* **80**: 1–7.

Moral A, Morales J, Ruíz-Capillas C, Montero P (2002) Muscle protein solubility of some cephalopods (pota and octopus) during frozen storage. *Journal of the Science of Food and Agriculture* **82**(6): 663–668.

Nagai T, Suzuki N (2000) Isolation of collagen from fish waste material – skin, bone and fins. *Food Chemistry* **68**: 277–281.

No HK, Storebakken T (1991) Color stability of rainbow trout fillets during frozen storage. *Journal of Food Science* **56**: 969–972.

Offer G, Trinick J (1983) A unifying hypothesis for the mechanism of change in the water holding capacity of meat. *Journal of the Science of Food and Agriculture* **34**: 1018–1023.

Ornum JV (1992) Shrimp waste – must it be wasted? *Infofish International* **6**: 48–52.

Ovissipour M, Ghomi MR (2009) *Biotechnology in Seafood Production.* Tehran, Iran: Islamic Azad University, pp. 1–198.

Ovissipour M, Rasco B (2011) Fatty acid and amino acid profiles of domestic and wild beluga (*Huso huso*) roe and impact on fertilization ratio. *Aquaculture Research and Development* **2**(3): 113.

Ovissipour M, Abedian A, Motamedzadegan A, Rasco B, Safari R, Shahiri H (2009a) The effect of enzymatic hydrolysis time and temperature on the properties of protein/hydrolysates from Persian sturgeon (*Acipenser persicus*) viscera. *Journal of Food Chemistry* **115**: 238–242.

Ovissipour M, Safari R, Motamedzadegan A, Shabanpour B (2009b) Chemical and biochemical hydrolysis of Persian sturgeon (*Acipenser persicus*) visceral protein. *Food and Bioprocess Technology* **5**(2): 460–465.

Ovissipour M, Taghiof M, Motamedzadegan A, Rasco B, Esmaeili Mulla A (2009c) Optimization of enzymatic hydrolysis of visceral waste proteins of beluga sturgeons (I) using Alcalase. *International Aquatic Research* **1**: 31–38.

Ovissipour M, Safari R, Motamedzadegan A et al. (2009d) Use of hydrolysates from Yellowfin tuna (*Thunnus albacares*) fisheries by-products as a nitrogen source for bacteria growth media. *International Aquatic Research* **1**: 73–77.

Ovissipour M, Abedian A, Motamedzadegan A, Nazari RM (2010a) Optimization of enzymatic hydrolysis of visceral waste proteins of Yellowfin tuna (*Thunnus albacares*). *Food and Bioprocess Technology* **5**(2): 696–705.

Ovissipour M, Benjakul S, Safari R, Motamedzadegan A (2010b) Fish protein hydrolysates production from Yellowfin tuna (*Thunnus albacares*) head using Alcalase and Protamex. *International Aquatic Research* **2**: 87–95.

Rasco BA, Bledsoe GE (2006) Surimi and surimi analogs. In: Hui YH (ed) *Handbook of Food Science and Technology.* New York: Marcel Dekker, p. 160.

Sandnes K, Pedersen K, Hagen H (2003) *Kontinuerlig enzymprosessering av ferske marine biprodukter (Continuous enzymatic processing of fresh marine bi products). Report no. 4503/108.* Trondheim, Norway: Stiftelsen RUBIN, p. 29.

Sankar V, Ramachandran A (2002) Rheological characteristics of suwari and kamaboko gels made of surimi from Indian major carps. *Journal of the Science of Food and Agriculture* **82**(9): 1021–1027.

Sathivel S, Bechtel PJ, Babbitt J, Prinyawiwatkul W, Negulescu II, Reppond KD (2004) Properties of protein powders from arrowtooth flounder (*Atheresthes stomias*) and herring (*Clupea harengus*) byproduct. *Journal of Agricultural and Food Chemistry* **52**: 5040–5046.

Sathivel S, Bechtel PJ, Babbitt J, Prinyawiwatkul W, Patterson M (2005) Functional, nutritional, and rheological properties of protein powders from arrowtooth flounder and their application in mayonnaise. *Journal of Food Science* **70**: 57–63.

Shahidi F, Kamil J, Arachchi V, Jeon J (1999) Food applications of chitin and chitosans. *Trends in Food Science and Technology* **10**: 37–51.

Shin JH, Oliveira AC, Rasco B (2010) Quality attributes and microbial storage stability of caviar from cultivated white sturgeon (*Acipenser transmontanus*). *Journal of Food Science* **75**: 43–48.

Soto F, Villarejo JA, Mateo A, Roca-Dorda J, de la Gandara F, Garcia A (2006) Preliminary experiences in the development of blue-fin tuna *Thunnus thynnus* (L., 1758) electroslaughtering techniques in rearing cages. *Aquaculture Engineering* **34**: 83–91.

Souissi N, Bougatef A, Triki-Ellouz Y, Moncef N (2007) Biochemical and functional properties of Sardinella (*Sardinella aurita*) by-product hydrolysates. *Food Technology and Biotechnology* **45**(2): 187–194.

Spinelli J, Dassow JA (1982) Fish proteins: their modification and potential uses in the food industry. In: Martin RE, Flick GJ, Hebard CE, Ward DR (eds) *Chemistry and Biochemistry of Marine Food Products*. Westport, CT: AVI Publishing Company, pp. 13–25.

Stoskopf MK (1992) *Fish Medicine*. Philadelphia: WBSaunders, pp. 82–83.

Tucker CS (2000) Off-flavor problems in aquaculture. *Reviews in Fisheries Science* **8**: 45–88.

Warriss PD (2000) The chemical composition and structure of meat. In: Warriss PD (ed) *Meat Science: An Introductory Text*. Bristol: CABI Publishing, pp. 37–67.

23 Meats – Beef and Pork Based

Robert Maddock

Department of Animal Sciences, North Dakota State University, Fargo, North Dakota, USA

23.1 Introduction

Beef and pork are two commonly consumed meats that can be processed in many different ways. The primary reason for processing beef and pork is to improve consumer acceptability and convenience of preparation, and to extend shelf life. To achieve this, processing usually accomplishes the following.

1. Remove bones, if applicable.
2. Make connective tissue less objectionable, by removal of extremely tough or inedible pieces.
3. Ensure the fat to lean ratio is appropriate.
4. Maintain the nutritional value of the beef and pork.

Most people associate "meat" with skeletal muscle, but there are also sources of meat that are primarily internal organs. Skeletal muscle makes up most of the meat that is processed and consumed from cattle and pigs but there are several uses and processes for variety meats. Variety meats include the liver, heart, tongue, kidneys, tripe (edible part of the bovine stomach), oxtail, brains, intestines, and blood. Variety meats can be minimally processed before being sold to consumers or used in the formulation of value-added or processed meats.

Beef is meat from bovine (*Bos*) species, especially domestic cattle (*Bos taurus* and *B. indicus*). Beef is a common meat in many cultures, especially the Americas and Europe. It is also an important meat source for cultures from the Middle East, Australia, Argentina, Europe, Africa, parts of east Asia, and south east Asia. Beef is the third most widely eaten meat in the world, accounting for about 25% of meat production worldwide (USDA, 2011a). The US, Brazil, and China are the world's three largest consumers of beef in total; however, on an annual per capita basis, Argentines eate the most beef, at 64.6 kg per person in 2010; people in the US ate 40.2 kg, while those in the EU ate 16.9 kg (USDA, 2011a).

Beef comes from all classes of cattle, but can vary greatly in quality and processing characteristics depending upon the class of cattle and management of the cattle prior to slaughter and processing. The classes of cattle include steers, which are young castrated males; heifers, which are young females that have not had a calf or been pregnant; cows, which are mature females that have produced offspring; bulls, which are mature uncastrated males. There are other classifications of cattle such as bullocks (young bulls) and heiferettes (young cows that have had one or two calves). Worldwide, there are other terms used to define classes of cattle, such as oxen, which can refer to any castrated male beef, but often is used to indicate a more mature steer, and stag, which refers to an uncastrated male.

Pork is meat from pigs (*Sus scrofa* or *Sus scrofa domesticus*), and in many parts of the world is referred to as pigmeat. Pork is often referred to as the most commonly consumed meat in the world despite many cultures that prohibit its consumption. Pork is primarily sourced from young castrated males, called barrows, and young females that have not produced offspring, called gilts. Cull sows, which are mature females that have produced offspring, and boars, which are intact males, are also sources of pork for processing.

23.2 Beef and pork characteristics and quality

23.2.1 Composition of beef and pork

Beef and pork are composed of the same basic components as any other meat, namely protein, water, fat, and a small amount of ash (inorganic matter such as minerals) and very little carbohydrate. The composition of beef and

Food Processing: Principles and Applications, Second Edition. Edited by Stephanie Clark, Stephanie Jung, and Buddhi Lamsal.
© 2014 John Wiley & Sons, Ltd. Published 2014 by John Wiley & Sons, Ltd.

Table 23.1 Composition of beef and pork per g/100 g (USDA, 2011b)

Product	Water	Protein	Fat	Ash	Calories (kJ)
Beef (lean)	75.0	22.3	1.8	1.2	116
Beef carcass	54.7	16.5	28.0	0.8	323
Pork (lean)	75.1	22.8	1.2	1.0	112
Pork carcass	41.1	11.2	47.0	0.6	472
Beef fat (subcutaneous)	4.0	1.5	94.0	0.1	854
Pork fat (back fat)	7.7	2.9	88.7	0.7	812

pork can vary greatly, with protein generally being the least variable and fat and water content being inversely related (Kauffman, 2011). For example, 85% lean beef trimming has a composition of 18% protein, 15% fat, and 64% water, with the remaining 3% being ash and carbohydrate, whereas 75% lean beef trimming has a composition of 16% protein, 25% fat, 56% water, and 3% ash/carbohydrate. Lean pork has a composition that is approximately 75% water, 20% protein, and 2–5% fat, with the remaining amounts being ash and carbohydrate. Approximate values for these meat components are given in Table 23.1.

23.2.2 The relationship between beef and pork color, pH, and water-holding capacity

Consumers equate "bright" colors of beef and pork with freshness and quality. Therefore, meat processors attempt to provide consumers with bright-colored beef and pork while still maintaining an acceptable shelf life.

Meat color is determined primarily by the state of the muscle pigment myoglobin (Mancini & Hunt, 2005). Myoglobin is a protein that binds and stores oxygen in living muscle tissue, and does the same thing in meat. When oxygen is present, myoglobin is in the state called oxymyoglobin, and has a bright red color. When there is no oxygen present, myoglobin has a dark color, which is often characterized as being "purple," which is the color of vacuum-packaged meat. Other molecules can bind to myoglobin to form various colors; for example, carbon monoxide binding results in a bright red color, whereas nitric oxides from meat curing process will cause the familiar pink cured meat color. When myoglobin has been exposed to heat or light, it can become oxidized and form metmyoglobin, which has a brown color.

Metmyoglobin can be reduced under certain rare conditions to myoglobin, and regain a bright red color. Other factors that affect meat color include light reflectance, which is influenced by water-holding capacity of the meat. Water-holding capacity is subsequently affected greatly by pH. Beef and pork with a pH of greater than 5.8 are often referred to as being dark, firm, and dry (DFD). Beef and pork that is DFD is generally considered to have lessened consumer acceptance (Wulf et al., 2002). Beef and pork with a pH of less than 5.3 are often pale in color, and referred to as being pale, soft, and exudative (PSE) (Warner et al., 1997).

23.2.3 Beef and pork tenderness

Tenderness is the most important palatability factor influencing enjoyment of meat. For beef and pork, tenderness is defined as the amount of force it takes to bite into and chew a piece of meat. Tenderness of meat is often measured using mechanical means that shear through or compress a meat sample. Two common instruments used to measure tenderness are Warner–Bratzler shear force and slice shear force. Both instruments measure the tenderness of cooked meat samples. The Warner–Bratzler shear force involves removing 1.27 cm cores parallel to the muscle fibers and shearing them across the fibers using a dull, V-shaped blade attached to a scale that measures the force required to cut through the sample. Slice shear force uses a flat blade to slice through a single 1 cm thick, 5 cm long sample (Shackelford et al., 1999). Protocols for shear force determination were developed by a committee of the American Meat Science Association and are found in the bulletin *Research Guidelines for Cookery, Sensory Evaluation and Instrumental Tenderness Measurements of Fresh Meat* (AMSA, 1995).

Tenderness of meat is influenced by four primary factors, briefly described here.
- **Protein effects**. Proteins that allow living muscle to do work are called contractile proteins and are primarily made up of actin and myosin. After the animal is slaughtered, these contractile proteins become disrupted over time, mostly due to the action of muscle enzymes called calpains (Goll et al., 2008).
- **Sarcomere length**. The distance between Z-lines in a myofibril is the sarcomere length. In living muscles, contraction results when Z-lines are moved closer together; that is, contracted muscles have a shorter sarcomere length. After slaughter, the sarcomere length becomes fixed at the time of rigor mortis. Shorter sarcomeres result in tougher meat (Wheeler et al., 2000).

• **Connective tissue**. The amount and type of connective tissue is very important in beef and pork tenderness. In general, meat from older animals has greater amounts of connective tissue than that from younger animals (Koohmaraie et al., 2002).

• **Composition of the meat**. The amount of fat, water, and protein in meat will affect tenderness. While this is not always a major factor, in general greater amounts of intramuscular fat (marbling) are associated with greater tenderness (Savell & Cross, 1988).

23.3 General categories of beef and pork processing

Processing of beef and pork is the utilization of the carcass in a way that makes the products more accessible for consumption or further processing. Following slaughter and dressing, beef and pork carcasses are chilled and initial processing involves cutting of carcasses into smaller pieces. These first, large cuts are referred to as primals and subprimals in the US and as joints in other parts of the world. Further processing of beef and pork involves continuing to break down primals and subprimals into cuts suitable for retail sale and consumption, grinding, making of sausages, cured meats, marinated meats, or cooked meats, and many other less common processing practices.

There are many ways to classify the various processing methods of beef and pork but the two major considerations are whole muscle versus ground, and fresh versus cooked. This gives essentially four major categories of beef and pork products: whole muscle fresh, whole muscle cooked, ground fresh, and ground cooked. In addition, there are several subcategories related to the addition of non-meat ingredients that include functional ingredients, such as salt or nitrites, or flavorings, such as spices and seasoning. There are other ingredient categories such as binders and extenders that will not be covered in depth here but exist to improve product quality or reduce costs. Alternatively, beef and pork products could be classified as whole muscle and ground, with subcategories of cooked, cured, sausage, and many others. Table 23.2 provides a general overview of the different sorts of beef and pork processing commonly undertaken.

23.3.1 Fresh whole-muscle beef and pork processing

Fresh beef and pork processing results in products that are sold to consumers while still in the raw or uncooked state. There are two broad categories of fresh beef and pork:

Table 23.2 An overview of common processing techniques for beef and pork

Fresh or cooked meat	Ground or whole muscle	Other ingredients	Examples
Fresh beef and pork processing	Whole muscle	No added ingredients	Steaks, chops, roasts
		Added ingredients	Seasoned or marinated steaks, roasts or chops
	Ground	No added ingredients	Ground beef, ground pork
		Added ingredients	Fresh sausage, breakfast sausage
	Restructured		Restructured beef roasts
	Tenderized		Needle tenderized steak and chops, cube steaks, cutlets
Cooked beef and pork processing	Whole muscle	No added ingredients	Cooked beef or pork roast
		Added ingredients (uncured)	Marinated and cooked roast beef or pork, jerky
		Added ingredients (cured)	Ham, bacon, dried beef, jerky
	Ground	No added ingredients	Cooked ground beef or pork patties, precooked meatballs
		Added ingredients (uncured)	Pizza toppings, some sausages
		Added ingredients (cured)	Most sausages, hot dogs
	Restructured meat		Restructured cooked roasts
Fermented, dry and semi-dry	Ground		Pepperoni, summer sausage
Dried	Whole muscle	Added ingredients (uncured)	Some jerky
		Added ingredients (cured)	Country-cured ham, prosciutto

whole-muscle cuts, and ground or minced. Each of these categories can be further broken into those products that have no added ingredients and those with ingredients added. For example, a whole beef roast could be sold at retail, or could have a solution or flavorings added to alter the eating quality. Ground beef and pork could have ingredients added to make them into fresh sausages. In addition, both whole-muscle cuts and minced meat are used as ingredients in further processing.

The "rules" of meat cutting were likely developed by our caveman ancestors and have not changed much since.
- Separate thick from thin, for example the rib and loin from the plate and flank.
- Separate tough from tender, for example the round from the sirloin.
- Separate more valuable from less valuable. Value of cut is often tied to tenderness, but not always. For example, the flank steak has much higher value than the rest of the flank.
- Always cut retail across the grain of the meat. This is not always done for every cut, but is pretty closely followed. The initial processing of beef and pork involves the fabrication of cuts from the whole carcase or carcass sides (Figure 23.1); this is referred to as "breaking" carcasses. The next step in most beef processing, and some pork is called "boning" as many of the bones are removed to allow for easier further processing. The resulting large pieces are referred to as primals, which can be further processed into subprimals. In large meat processing operations, it is subprimals that are often sold to other businesses for further processing. The subprimals of beef are found in Table 23.3, along with typical percentage yields. Beef subprimals are often categorized into end cuts such as beef chucks and beef rounds, middle meats, such as the beef rib and loin, and thin or rough cuts such as the beef plate, flank, brisket, and shanks. Table 23.4 lists the primal cuts of pork, which broadly fall into two categories: the "four lean cuts" and the fatty and bony cuts. The four lean cuts contain much of the muscle in the carcass, while the other cuts have greater fat and bone content.

Before reaching consumers, whole muscles from beef and pork are typically broken or further processed into retail cuts. Processing typically includes trimming of excess fat, and fabrication into retail cuts such as steak, chops, and roasts. Retail meat cutting often occurs at the point of sale, such as market and groceries, or at the wholesale level, where processors, often referred to as purveyors, cut retail products for the hotel, restaurant, and institutional markets. Some retail portioning also occurs at the large processor level, where steaks, chops,

and roasts are cut and packaged for consumers. This processing method is referred to as case-ready packaging and requires that retail cuts be shipped and stored in packages that limit oxygen exposure to ensure adequate shelf life of the product.

23.3.1.1 Institutional meat purchase specifications

Institutional meat purchase specifications (IMPS) are a series of meat product specifications maintained by the United States Department of Agriculture's Agricultural Marketing Service (NAMP, 2007). These specifications are developed as a voluntary consensus that describes the anatomical location including muscles, bone, and fat present in various wholesale cuts of pork and beef. The IMPS also provide specification for various fresh and cooked meat cuts products. Large-volume purchasers such as federal, state and local government agencies, schools, restaurants, hotels, and other food service users reference the IMPS for procuring meat products.

23.3.1.2 Value cuts

Producer groups in the US, especially the National Cattlemen's Beef Association and the National Pork Board, have investigated alternative cut fabrication to increase the variety of retail cuts available to consumers. These cuts are often referred to as "value cuts" and are generally steak or chops that are fabricated from muscles and cuts that were previously used as roasts or other lower value meats such as shanks or shoulders (von Seggren et al., 2005). The most common of these value cuts is the flat iron steak, which comes from the beef clod, a large cut from the beef shoulder that was typically marketed as a boneless shoulder roast. The flat iron, specifically, is the infraspinatus muscle, which is very tender but has a large band of connective tissue that prevents easy use by consumers. Research in the area showed an economical method to remove the connective tissue, resulting in a retail cut that is now more consumer friendly.

Ultimately, the decision to change cut fabrication for beef and pork carcasses has to be driven by greater profit potential. This may come from the ability to market various muscles at a greater price or to meet the needs of valued customers. In general, the "low hanging fruit" of value cuts have been found and are being produced. For example, removal of the teres major (shoulder tender) from the beef shoulder clod is now common, and cutting of the flat iron is also routine for many processors. These muscles are fairly easy to remove from the associated subprimal,

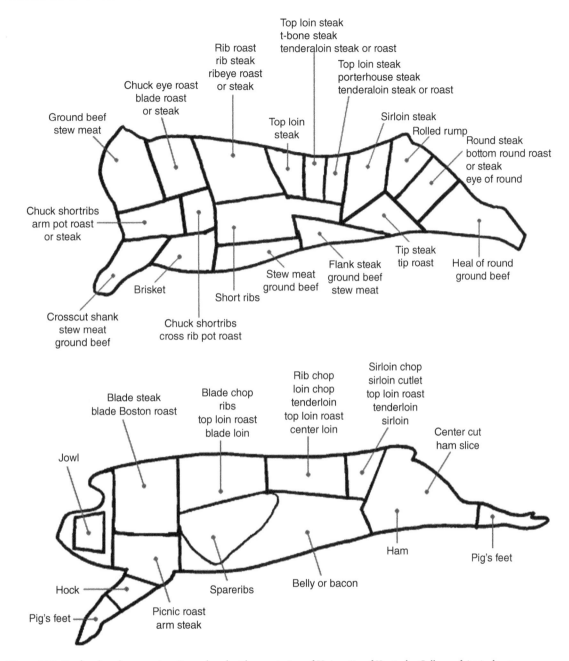

Figure 23.1 Beef and pork processing. Reproduced with permission of University of Kentucky College of Agriculture, Food and Environment, Agripedia.

especially the shoulder tender, which does not require much modification of the processing line or training of personnel, nor does it devalue the rest of the cut. The next step in anatomical deboning is removal of whole muscles without concern for traditional breaking into primals. Research shows that tenderness variation within muscle should provide guidance as to the separation point of cuts, especially in beef (Searls et al., 2005).

Table 23.3 Beef wholesale and subprimal cuts and their percentage of hot carcass weight

Wholesale cut	Subprimal cut (boxed beef)	Percentage of hot carcass weight
Round	Top (Inside) round, closely trimmed (IMPS 168)	5.3
Round	Gooseneck round, closely trimmed (IMPS 170)	6.8
Round	Knuckle (round tip), peeled (IMPS 167A)	2.7
Loin	Top butt (top sirloin), closely trimmed (IMPS 184)	3.0
Loin	Ball tip (bottom sirloin) (IMPS 185B)	0.5
Loin	Tri-tip (bottom sirloin) (IMPS 185C)	0.6
Loin	Flap (bottom sirloin) (IMPS 185A)	0.8
Loin	Tenderloin, peeled (IMPS 189A)	1.5
Loin	Strip loin, 0×1, closely trimmed (IMPS 180)	2.8
Rib	Ribeye, 2×2, boneless (IMPS 112A)	3.4
Rib	Back rib (IMPS 124) and short rib (IMPS 123B)	2.0
Chuck	Shoulder clod, closely trimmed (IMPS 114A)	4.7
Chuck	Chuck roll (IMPS 116A)	6.7
Flank	Flank steak (IMPS 193)	0.5
Plate	Outside skirt (IMPS 121C) and inside skirt (IMPS 121D)	1.3
Brisket	Brisket, deckle-off (IMPS 120)	2.3
Various	Special trimmings	2.4
Various	80% Trimmings	8.7
Various	50% Trimmings	5.7
	Total fat	19.8
	Total bone	13.6
	Total shrink	4.9

Table 23.4 Pork primal cuts

Wholesale cut	Retail cuts or application of each cut
Four lean cuts of pork	
Ham	Fresh ham, bone-in ham, ham steak, boneless ham
Loin	Boneless loin, bone-in loin, blade chops, rib chops, loin chops, sirloin chops
Boston butt	Boneless boston, shoulder steaks
Picnic shoulder	Shoulder meat, cushion meat
Other pork cuts	
Belly	Side pork, fresh pork belly, bacon
Hocks	Pork hocks
Feet	Pig feet
Jowl	Jowl bacon
Fat back	Lard, pork fat
Ribs	Spareribs, backribs

23.3.2 Beef and pork bind values

Bind refers to the ability of meat pieces to attach to each other, resulting in textures and products that meet consumer expectations. For example, consumers expect cooked sausages to have a cohesive texture, and chunked and formed boneless hams to have an appearance that is similar to whole-muscle hams. There are three essential parts to achieving bind:
- intrinsic binding proteins must be brought to the meat surface to form a protein-protein interaction
- the meat must be pliable so it can be formed
- the proteins must be coagulated (bound) by cooking or heating (Pearson & Gillett, 1996).

Meat proteins, especially the contractile proteins actin and myosin, are brought to the surface of the meat ingredients through salt addition and mechanical action. The proteins that cause bind are often referred to as salt-soluble proteins and are made up of the contractile muscle fiber proteins, myosin and actin. Salt is required, as these proteins are not soluble in water; the salt changes the ionic strength of the water and allows proteins in the meat to come to the surface. The most common salt used to achieve bind is simple table salt, sodium chloride (NaCl), but potassium chloride can be used in the production of low-sodium products. This is referred to as protein extraction. Massaging or tumbling of raw materials increases the amount of protein that comes to the surface. The surface of the meat pieces becoming "shiny" after the addition of salt and mechanical action (tumbling or massaging) are the muscle proteins that allow the meat to bind. In addition, tumbling or massaging also improves the pliability of the raw meat, allowing it to be formed. Finally, when exposed to cooking temperature, the proteins will bind together, ideally forming a bond that is almost undetectable in formed meat products.

There are alternatives to using muscle proteins to bind meat, such as alginates, transglutaminase enzyme (Kuraishi et al., 1997), or other products called cold binders. These bind meat pieces together by forming a layer between the pieces, causing them to bind, or by linking proteins on the meat surface. The most common examples of whole-muscle formed products are formed hams, turkey breast, and beef roasts.

23.3.2.1 Fresh whole muscle product with added ingredients

Whole-muscle cuts can also have added ingredients to increase options available to consumers. The simplest ingredients added to beef and pork whole-muscle cuts are seasoning and spices that are applied to the surface of the meat and then packaged. These products allow processors to add flavors to products without greatly increasing processing complexity. For example, lemon pepper pork tenderloin is produced by simply adding spices to the pork. Another strategy besides adding ingredients to whole-muscle cuts is the practice of "enhancing" or injecting solutions into fresh meats. A typical enhancement solution, utilized to improve palatability traits, includes water, salt, phosphates, and possibly an antioxidant (Hoffman, 2006). In addition, it is possible to add tenderizing enzymes, such as those derived from tropical plants such as papain from papaya or bromelin from pineapples, to tenderize tougher cuts of meat (Ashie et al., 2002).

23.3.2.1.1 Preseasoned or flavored beef and pork

The simple addition of spices or flavoring to whole-muscle cuts is another common industry practice and is referred to as "preseasoning." One example of preseasoned and marinated product is pork tenderloins that have been marinated and have flavors such as pepper and citrus added, which are common in many markets in the US. Flavorings or marinades can be added as part of an injection or added via tumbling. Flavorings can also be simply added to a container and absorbed by meat over time. Seasonings added to whole muscle include spices,

herbs, and other flavorings, and the marinades can be simple salt and water solutions. However, seasonings often have an acid component, such as dilute acetic acid (vinegar) or citric acids added to improve tenderness.

23.3.2.2 Fresh whole muscle tenderized

A common method used to improve tenderness of almost any beef or pork cut is mechanical tenderization (Loucks et al., 1984). Needles penetrating the cut disrupt muscle fibers and connective tissue. The primary issue with mechanical tenderization is a possible reduction in the safety of steaks and roasts, which may not be cooked to temperatures high enough to kill bacteria introduced by the needles during tenderization. Therefore, mechanical tenderization is most appropriate for those cuts that will be fully cooked. It is also the best option for cuts with large amounts of connective tissue, for example round steaks. Mechanical tenderization also occurs at the retail and restaurant level to help ensure the tenderness of product being served. Enzymatic tenderization can also be used to improve the tenderness of tough cuts of meat. In general, enzymes are derived from tropical plants or from bacterial or fungal sources (Wang et al., 1958).

23.3.3 Fresh ground

There are two categories of minced or ground meats: those that do not have added ingredients, such as ground beef or pork, and those that have ingredients added to typically form a sausage product, such as fresh pork or beef sausage. A large percentage of beef, up to 55% of beef consumed in the US, and a smaller percentage of pork are minced or ground to form ground beef or pork.

Grinding is a processing technique that utilizes tougher or less valuable meat sources and grinds or minces the meat to improve tenderness and ease of use by consumers. While the most common source of meat for grinding is "trimmings," which are small pieces of meat produced during primal and subprimal cut fabrication, other sources of meat such as beef plates (meat and fat over the ribs), beef flanks or navels, shanks, and pork shoulders are also often used when producing ground meat. In addition, the "premium" ground meats such as ground chuck, sirloin, or round are produced at a higher cost. Moreover, ground meats are often produced from cull animals such as cows, bulls, sows, and boars. These cull animals tend to be older and have less desirable palatability characteristics, such as being tougher, so grinding is an easy method to improve consumer acceptance of these meat sources.

The production of ground beef has been drastically changed in recent years, mostly in response to safety concerns, and the need to test and hold product. However, even if food safety is the most important consideration of ground beef production, efficiency and yield are essential. Safe production of ground beef has created the need for many processors to rethink process design to ensure safety while still being profitable. Reports of food-borne illness are published by the US government, but it is difficult to determine the source of many outbreaks. Pathogens associated with meats, and especially ground meats, are a major concern (CDC, 2011).

For larger processors, the standard design for ground beef processing is: receiving of trimmings, either from an in-plant source or from the outside; a quality control check; initial grinding; fat testing; final grinding; bone collection; forming and/or packaging. For some processors, the addition of boneless, lean beef trimmings may also occur. At various points the beef is put through metal detectors, and other quality assurance evaluations, such as temperature checks, visual evaluation of finished product, and microbial testing for common pathogens such as *E. coli* and *Salmonella*, take place.

23.3.3.1 Fresh ground beef and pork with added ingredients

The addition of ingredients to ground meat is a common processing method that results in products typically referred to as sausages, but also includes other products such as meatballs, pizza toppings, and others. Salt, specifically sodium chloride, is the essential ingredient added to turn ground beef and pork into sausage. Salt is added at levels of 0.5–2.5% of the weight of the meat, with the product being mixed to give the proper flavor and consistency. In addition, spices and other flavoring are added to provide for an almost unlimited variety of products. Some common fresh sausages are bratwurst, pork sausage, beef sausage, breakfast sausages and many others that vary by culture and custom.

Fresh sausage is a very common type of processed meat product and is the easiest to make. Fresh sausages must be kept refrigerated and are cooked by the consumer before consumption. Sometimes, fresh sausages are cooked at the processing plant before sale; these sausages are called "fresh cooked."

Regulations for fresh sausage manufactured in the US (USDA, 2005) include the following.
• Fresh sausages must not contain any cure (nitrate or nitrite).

• Fresh sausage must not contain any phosphates.
• Fresh pork sausage may be 50% total fat.
• Fresh pork sausage may have the chemical antioxidants butylated hydroxyanisole (BHA) and butylated hydroxytoluene (BHT) or propyl gallate added at a level of 0.01% of the weight of the fat in the formulation.
• Fresh beef sausage may be 30% fat.
• Water or ice may be added to fresh sausage at the level of 3%.
• Whole hog sausage is a fresh sausage made from the meat of an entire hog in the proportions normally found in a hog carcass; that is, the entire pig carcass, after the removal of skin, bones and other inedible tissue, is ground and made into sausage.
• Italian sausage is limited to 35% fat, and must contain fennel and/or anise.

The process of making fresh sausage is relatively simple and involves: grinding of raw meat; addition of salt, spices, and other ingredients; mixing to extract proteins and disperse the non-meat ingredients; and stuffing of the raw meat into packaging or casings. The final product is then either sold directly to consumers at small markets or packaged for distribution.

The use of casings is also important in many meat products. Casings are the material that encloses the filling of a sausage or other formed meat product. Casings are divided into two categories: natural and artificial. Artificial casings, such as collagen, cellulose, plastic, and lately extruded casings, are manufactured via various methods. Natural casings are made from the internal organs of livestock, such as both the large and small intestines, bladder, and occasionally stomach of pigs, cattle, and sheep. Collagen casings are made primarily from beef hides. The collagen in the hides is removed chemically and the resulting product extruded to form a tube. Collagen casings have the advantage of a large variety of sizes that are consistent in size, strength, and other attributes. Casing sizes can vary depending upon product, with sizes from 12 to 35 mm widely available. Cellulose and plastic casings are generally inedible and meant to be removed from the product before consumption. Cellulose casings are derived from plant material and plastic casings from petroleum sources. Some common products packaged in cellulose casings include large-diameter sausages such as salami or summer sausage. Plastic casings are often used to form meat during the cooking process and then removed before being offered to consumers. Products made with plastic casings include skinless franks, and many deli meats such as turkey rolls or boneless ham. Natural casings are the submucosa portion of the internal organs, primarily intestines that have been stripped of the outer fat and mucosa. Natural casings are washed and packaged in salt for storage and are often used in products such as bratwurst or smoked sausages.

23.3.4 Cooked and precooked, uncured, whole-muscle cuts

Precooking is a common method of adding convenience and value to whole-muscle meat products. In many cases, the shelf life and safety of meat products can be improved through precooking, as the cooking process will destroy most microbes that can cause illness and spoilage. It is possible to precook almost any meat product and achieve customer satisfaction with product quality if done properly. Fully cooked meat products are a growing segment of meat processing and offer potential for increasing profit by utilizing low-valued muscles to make higher valued items. Classically, deli-style roast beef is the product made by precooking beef; however, there is great potential to precook other types of beef and pork.

Precooking whole-muscle products can involve several technical hurdles, the most common being to avoid introduction of off-flavors or other product defects (Boles & Shand, 2001). The most common off-flavor associated with precooked meats in general, and beef in particular, is warmed-over flavor (WOF), often described as the "cardboard-like" flavor that occurs when meat is cooked and reheated (Gene & Pearson, 1979). It is possible to minimize off-flavors in precooked beef products through knowledge about factors that cause off-flavors in beef. Preventing WOF and other defects requires several quality control procedures noted here.

• **Degree of doneness (final cooking temperature)**. Most off-flavors in precooked beef items are the result of the oxidation of fat, causing rancidity (Gene & Pearson, 1979). The addition of heat during cooking is the primary source of oxidation which leads to rancid flavors. Lesser doneness results in less fat oxidation and the potential for fewer off-flavors (Wadhwani et al., 2010). Depending upon the product, meat should be cooked to the lowest degree of doneness that is feasible.

• **Ingredients**. Many ingredients are strong antioxidants and prevent the formation of WOF in cooked beef and pork. Most notably, nitrite (cure) is a very effective antioxidant, which is why hams and sausages generally do not have problems with oxidation flavors. However, nitrite is not an option for precooked fresh meat items, as the nitrite is not an effective antioxidant until after heat treatments have been applied. In addition, regulations prevent the addition of nitrites and nitrates to fresh meat, as this will change meat color. Some common ingredients are also good antioxidants, including phosphates, organic acids (lactic acid, citric acid, lemon juices), and several other products commercially available specifically to prevent oxidation. WOF development can also be decreased by the addition of natural antioxidants such as vitamin E or chemical antioxidants such as BHA and BHT. It is important to note that BHA and BHT are limited in the US to only fresh pork sausage, and are not permitted in cooked meat products.

• **Raw product handling**. It is not possible to start with abused or low-quality raw product and end with a high-quality finished product. Temperature abuse allows for the rapid growth of bacteria, which are able to cause oxidation, protein degradation, and other spoilage issues. On the other hand, product that has freezer burn, which is a condition caused by sublimation of water from the meat surface, is also highly oxidized and will result in many off-flavors. The simple practice of maintaining proper cold storage of meat and preventing temperature abuse can improve the flavor of cooked beef products.

23.3.4.1 Cured whole muscle cuts

Cured meat is a large category that includes ham, bacon, cured beef, and any other meat product that is whole muscle and has nitrite or nitrates in the ingredients. Meats are cured using the following common ingredients: water, salt, sugar, and nitrite. The curing ingredients can be added to the meat either in a dry rub form or as brine. A "brine" is when the dry ingredients are mixed with a specific amount of water to make a solution that can either cover or be injected into the meat. In addition, it is common for cured meats to have phosphates and erythorbates. Sometimes spices or flavorings, such as black pepper or honey, are added either to the surface of the meat or to the brine.

There are several ways to cure meat.

• **Dry curing**. This is the oldest method of meat curing. Dry curing is obtained by rubbing the surface of the meat (usually ham or bacon) with salt, sugar, and cure (usually nitrates). This process takes a long time (3 months to a year) and results in a great amount of shrinkage, up to 45%, and loss of product to drying out the surface. Examples of dry-cured products are country-cured hams and bacons, prosciutto hams, and country-cured bacons. Dry curing has become a very specialized industry and is limited in scope in the US. This type of product has a very intense flavor and is highly valued by chefs.

• **Cover curing**. Cover curing involves mixing of a brine containing, salt, sugar, and nitrites or nitrates in water, placing the meat in a large vat and covering the meat with the cure. This type of curing also takes a long time (up to 6 weeks for large cuts such as pork hams, but a few days for thin cuts such as bacon). Cover curing improves the rate of ingredient penetration into the meat and results in much less loss of product weight due to shrinkage and spoilage.

• **Injection curing**. Injection curing involves injecting brine, usually including a curing agent, such as sodium nitrite, salt, and usually a sweetener directly into the meat. This greatly reduces the amount of time it takes to cure meat, as the curing process happens within a few hours, and is the most common type of curing in the US. Injection curing is further broken into single-stitch or multiple-stitch injection. Single-stitch injection uses a small hand-held injector, whereas a multiple-stitch injector uses a large injection machine with many needles.

• **Artery injection**. Some plants will inject hams by placing a needle into the femoral artery and use the existing blood vessels to distribute the cure. This is an uncommon process, as artery injection requires specialized equipment and knowledge.

• **Combination curing**. This is a combination of injection and dry curing or cover curing. Typically, the meat is injected and then rubbed with a dry cure to enhance flavor and alter the appearance of the product. By injecting the meat, time to finish is decreased from weeks to a few days for thicker cuts such as hams.

23.3.5 Cooked ground beef and pork including cooked sausages

Cooked and/or emulsified sausages are the most widely available consumer sausage products (hot dogs). Cooked sausages are fully cooked prior to sale, but still must be kept refrigerated to prevent spoilage. Cooked sausages are sometimes reheated, for example, smoked sausages, but are also consumed without further cooking (bologna).

The process involved in producing cooked ground beef and pork starts in a similar manner to uncooked sausages. Meat trimmings are ground to the appropriate size, mixed with salt, spices, and other ingredients, and stuffed into casings or forms for cooking. A key difference is that nitrite, usually in the form of sodium nitrite, is commonly added to the product formulation. The addition of nitrite provides many benefits, including stabilizing color, preventing oxidation and formation of off-flavors, and preventing the growth of some pathogenic bacteria, especially *Clostridium* species.

Regulations and common facts for cooked beef and pork sausage products (USDA, 2005) include the following.

• Cooked sausages almost always contain cure (nitrite), which is added at 156 ppm nitrite (0.000156% of fresh weight of the product).

• Maximum fat in hot dogs is 30%, and the maximum water plus fat is 40%.

• Hot dogs and bologna are essentially the same, but stuffed into different-sized casings.

• Cooked sausages will often have phosphates added to prevent excessive cooking loss.

• Fully cooked sausages must be cooked to a minimum of 148 °F (82.2 °C), with most cooked to around 155 °F (86.1 °C).

• Water or ice is added to most cooked sausages, as moisture is lost during the cooking process. Regulations require that either the additional weight added by water is lost during the cooking process, or that water is included on the label. Ice may be added to emulsion type sausages to keep temperatures low during emulsification, especially when using a piece of equipment called a bowl chopper. During the emulsification process, mechanical energy of the emulsifier increases meat temperature, which may cause the fat in the meat to liquefy. If the fat liquefies, the meat will not form an emulsion.

23.3.6 Fermented sausages

Fermented sausages are the oldest type of sausages produced. Fermented sausages require that the pH of the product be lowered from normal meat pH of approximately 5.6 to less than 5.0, and potentially to 4.6, usually through bacterial action (USDA, 2005). There are several commercial starter cultures of bacteria available for meat fermentation, usually strains of *Lactobacillus* and/ or *Staphylococcus*. The lower pH results in a distinct flavor and increased storage life. Fermented sausages fall into two broad categories: dry and semi-dry. True dried sausages are usually made from pork and rely on pH decline and dryness to ensure safety, as they are usually never cooked. The process of drying sausages can take from 4 to 6 weeks at the shortest, to over a year for some products. Semi-dry sausages are often cooked after the pH declines to less than 5.0 and then dried, usually in specialized drying rooms that have controlled temperature and humidity to prevent the growth of yeasts on the product surface, before sale or consumption.

Some additional facts about fermented sausages.

• Fermented sausages fall into two broad categories: semi-dry, which includes summer sausages (cervelat, thuringer, etc.) and snack sticks; dry sausages, which include pepperoni and salamis.

• Fermented sausages are generally shelf stable and may or may not be cooked before consumption.

• Dried sausages are generally never cooked; they rely on fermentation and drying to prevent microbial growth.

• If pork is used in dried sausages it must be "certified" to ensure the destruction of trichina worms. Pork is "certified" if it has been subjected to certain processes that kill trichina, such as freezing. The United States Department of Agriculture has regulations for certified pork (CFR, 1990).

• Sausages are fermented by the addition of starter cultures or using the natural microbes in the meat. Using starter cultures provides a much more consistent product than natural microbes. Old-time sausage makers would use a few pounds of an existing batch to spread microbes into a new batch. This was called "backslopping" and is potentially dangerous, as it can spread pathogens as well as other microbes.

• When making fermented sausages, simple sugars (dextrose) are added to provide an energy source for the microbes.

23.3.7 Dried whole muscle product

Jerky is a dried meat product that usually can be stored at room temperature. Jerky can be either whole muscle or ground and formed, with the process being substantially different depending upon the product. All jerky products typically have a moisture to protein ratio of 0.75:1 or less and can be made from any species, for example pork jerky or lamb jerky. Products may be cured or uncured, air or oven dried, and may be smoked or unsmoked. Some important considerations about jerky are as follows.

• All jerky products should have a moisture to protein ratio of 0.75:1 or less (USDA, 2005). This ensures that bacterial growth is very limited and it allows the product to be stored at room temperature. Jerky products may be cured or uncured, air or oven dried, and may be smoked or unsmoked,.

• Jerky is produced from a single piece of meat.

• Jerky, chunked and formed is produced from chunks which are molded and formed and cut into strips. This allows for the addition of flavorings and for more precise portioning. Jerky ground and formed or chopped and formed is produced where the meat block may be finely ground before being molded and formed and cut into strips.

• Jerky may be dried at any stage of the process.

The process of making whole-muscle jerky is fairly simple. Large pieces of meat, usually beef round or shoulder, or pork hams, or large cuts of other species, are sliced into thin strips and mixed with a marinade. The marinade typically contains salt at levels of 10–20% and may contain a sweetener, such as sucrose, dextrose, or corn syrup solid, usually at a level of approximately 50% of the salt and may have other flavors or seasonings added such as pepper or teriyaki. In addition, the marinade may contain cure (nitrite) to prevent bacterial growth and to serve as an antioxidant. The meat may simply be allowed to soak in the marinade for several hours, or the meat and marinade may be tumbled to decrease production time. After the meat has absorbed the marinade, it is laid out and smoked or dried, usually in an oven or smokehouse or possibly in specialized drying equipment for drying foods, such as a dehydrator. After drying, the jerky can be portioned and packaged.

23.4 Equipment needed in beef and pork processing

Equipment used in beef and pork processing is varied and includes everything from small utensils such as knives and containers, to large pieces capable of processing several thousand pounds of meat per hour. Beef and pork processing equipment should have several characteristics related to ability to clean and sanitize, durability, efficiency, and ease of operator use. In general, meat processing equipment is made of stainless steel or other, smooth, easy-to-clean construction. Most types of equipment are available in various sizes. For example, small processors may use a meat grinder with the capacity to hold 20 kg of fresh product, whereas large processors may have equipment that will hold 2500 kg. However, the primary principles for each type of equipment are similar, not matter what the size.

23.4.1 Grinder

The meat grinder is used extensively in many beef and pork processing techniques; almost all sausage products start with grinding, and fresh ground beef and pork make up a large portion of meat products. Grinders have several components, which include the hopper, auger (worm), blade, and plate.

Some important grinder information is summarized below.

• The grinder auger must be correctly inserted and tightened. If the auger is loose, meat will not be pushed into the plate and blade correctly, and will be turned to mush.
• As the grinder works, the plate and blades wear together. If the plate and blades are not matched, they will wear differently and not grind as effectively. Make sure to always keep plates and blades together, especially during cleaning.
• Make sure there is meat or water in the grinder before starting. The plate and blade will heat up almost immediately if there is no liquid or meat for lubrication. If the blade gets hot, it will lose "temper," become soft, and be useless.

23.4.2 Emulsifier/bowl chopper

Emulsifiers and bowl choppers achieve the same end-product, which is very finely chopped meat, called an emulsion. Emulsified products can hold a large amount of water and fat. Emulsifiers work by passing meat and the additional ingredients, usually salt, a sweetener, a curing agent such as sodium nitrite, and a cure accelerator such as sodium erythorbate, through a very rapidly spinning blade, resulting in meat pieces that are very small.

23.4.3 Mixer

There are various mixers used in beef and pork processing, most commonly to add ingredients to ground or minced meats. Grinders also can be used to combine ground beef or pork with varying fat levels to achieve a desired content in the final product.

23.4.4 Stuffer

Stuffers force meat pieces into packaging or casings. Stuffers can be piston type, in which a plunger forces meat through a small opening into the stuffing horn, or vacuum stuffers, which pull meat into a screw or paddle type drive assembly using vacuum and then push the meat through the stuffing horn.

23.4.5 Tumbler

Tumblers are used to apply spices and other flavors to the surface of fresh meat, and to extract proteins to the meat surface to aid in binding of formed products such as boneless hams. Meat surfaces in tumblers are subjected to friction, which causes contractile proteins to be extracted, facilitating bind. Most tumblers are essentially containers with baffles on the internal surface that rotate; however, some tumblers resemble large drums without baffles. Meat in the tumbler will move up the side and fall to the bottom repeatedly. Tumblers commonly use a vacuum pump to remove air from the chamber before starting, which reduces the amount of time required to tumble the product.

23.4.6 Massagers

Massagers work in a similar manner to tumblers but rather than relying on the meat falling to extract proteins, paddles within the massager provide friction to the meat surface.

23.4.7 Injector

Injectors are used to insert solutions into the interior of meat products. Injectors are commonly used to make hams, bacons, turkey, and other products such as deep marinated beef. They can have a single needle or multiple needles, and for larger capacities manufacturers may use a multistitch injector, which has several rows of needles and can process several thousand pounds per hour.

23.4.8 Slicer

The slicer uses a very sharp blade to slice various meats such as deli ham and roast beef. Slicers used in large processing plants are often automated and linked to portioning and packaging machines. Deli slicers are used by smaller or local processors to slice beef and pork for sale or sometimes to slice raw meats for jerky processing.

23.4.9 Smokehouse

The smokehouse is an oven that has a smoke generator attached. Smokehouses can range in size from those used by small or home processors that can hold 20–50 kg of meat to large commercial smokehouses that can process several thousand kilograms. Most commercial smokehouses can control humidity, air temperature, air movement, and smoke application. Modern smokehouses are computer controlled and can cook beef and pork to exacting specifications. Smoke generation is natural in many smokehouses, with small chips of wood being burned at low temperatures, the smoke being forced into the oven via fans. Wood chips are generally from hardwood sources, with hickory being the most common. It is possible to utilize wood from other sources, such as fruit trees, but this source is expensive and used for specialty products. Another option is the use of liquid smoke, which is made from the burning of hardwoods, with the volatile smoke components captured in water. Liquid smoke is available from ingredient suppliers and is not made by most processors. It is applied to the surface of meat in the smokehouse via nozzles supplied by a holding tank. Liquid smoke does not penetrate the meat surface as natural smoke will, resulting in smoke flavor only on the surface of the beef and pork.

23.5 Beef and pork processing and HACCP

Hazard Analysis Critical Control Point (HACCP) systems are required in the US for wholesale or commercial meat processors, with the exception being some retail stores that are allowed to operate under exemption from HACCP (CFR, 1990). Beef and pork processors follow HACCP systems developed by the National Advisory Committee for Microbiological Criteria for Food. Beef and pork processors follow HACCP with seven principles (Tompkin, 1990) along with additional information such as a flowchart that details steps in processing and a detailed description of the product. Additional regulations for meat processors include the development and implementation of sanitation standard operating procedures (SSOPs), which are detailed procedures for the cleaning and sanitation of processing plant facilities, equipment, and other product contact surfaces. In addition, the USDA has standards for processing plant facilities that include building materials, water testing, employee practices, and food handling. Several of these standards are addressed as Good Manufacturing Practices (GMPs). At various times, the USDA has issued additional rules regarding specific production practices to improve food safety. For example, control of *Listeria* species on cooked, ready-to-eat meat products was detailed with expectation for processors to routinely test for *Listeria* in the processing plant environment and make changes to sanitation or processing procedures if a positive *Listeria* test was found.

23.6 Sustainability

Sustainability in meat processing is commonly defined as reducing the use of resources during production and processing. The issue of sustainability in beef and pork processing usually involves the modern production of livestock, or utilizing the best science and production practices available. Beef and pork are available from many different production systems, including those that may be resource intensive and produce beef and pork in an economically efficient manner and extensive systems which require longer growing times but utilize fewer resources. For example, increasing the use of forages and rangelands for growing cattle rather than using high-grain diets can decrease the use of resources while potentially increasing economic efficiency of production. Choosing beef and pork from either system is a matter of personal preference.

Sustainability at the processor level usually involves reducing the use of resources such as water, electricity, and packaging materials, and utilizing the entire carcass in an efficient manner. Improvements in sustainability are generally driven by increased economic returns for producers and processors rather than government intervention. The other issue driving changes in sustainability is the demands of downstream customers. Some end users of beef and pork are increasing pressure on producers and processors to utilize fewer resources during production.

Beef and pork processing embrace the issue of sustainability by utilizing the entire carcass. During the slaughter and butchering process all of the carcass components are utilized in an efficient manner. The offal, which includes the internal organs, is even used as a food source in many ways, while trimmed fat and bone are utilized in a manner to maximize the use of each component.

References

American Meat Science Association (AMSA) (1995) *Research Guidelines For Cookery, Sensory Evaluation And Instrumental Tenderness Measurements Of Fresh Meat.* Savoy, IL: American Meat Science Association.

Ashie I, Sorensen TL, Nielsen PM (2002) Effects of papain and a microbial enzyme on meat proteins and beef tenderness. *Journal of Food Science* 67: 2138–2142.

Boles J, Shand P (2001) Meat cut and injection level affects the tenderness and cook yield of processed roast beef. *Meat Science* 59: 259–265.

Centers for Disease Control (CDC) (2011) *Foodborne Diseases Active Surveillance Network (FoodNet). FoodNet Surveillance Report for 2009 (Final Report).* Atlanta, GA: US Department of Health and Human Services.

Code of Federal Regulations (CFR) (1990) *Animals and Animal Products.* Washington, DC: Office of the Federal Register, Government Printing Office, 9: 212–220.

Gene J, Pearson A (1979) Role of phospholipids and triglycerides in warmed-over flavor development in meat model systems. *Journal of Food Science* 44: 1285–1290.

Goll D, Neti G, Mares SW, Thompson VF (2008) Myofibrillar protein turnover: the proteasome and the calpains. *Journal of Animal Science* 86: E19–E35.

Hoffman L (2006) Sensory and physical characteristics of enhanced vs. non-enhanced meat from mature cows. *Meat Science* 72: 195–202.

Kauffman RG (2011) Meat composition. In: Hui YH (ed) *Handbook of Meat and Meat Processing*, 2nd edn. New York: CRC Press, pp. 45–61.

Koohmaraie M, Kent MP, Shackelford SD, Veiseth E, Wheeler TL (2002) Meat tenderness and muscle growth: is there any relationship? *Meat Science* 62: 345–352.

Kuraishi C, Sakamota J, Katsutoshi Y, Susa Y, Kuhara C, Soeda T (1997) Production of restructured meat using microbial transglutaminase without salt or cooking. *Journal of Food Protection* 62: 488–490.

Loucks L, Ray FE, Berry BW, Leighton EA, Gray DG (1984) Effect of mechanical tenderization and cooking treatments upon product attributes of pre-rigor and post-rigor beef roasts. *Journal of Animal Science* 58: 626–630.

Mancini R, Hunt M (2005) Current research in meat color. *Meat Science* 71: 235–257.

North American Meat Processors Association (NAMP) (2007) The Meat Buyer's Guide. Reston, VA: North American Meat Processors Association.

Pearson A, Gillett T (1996) *Processed Meats*. New York: Chapman and Hall.

Savell J, Cross H (1988) The role of fat in the palatability of beef, pork, and lamb. In: *Designing Foods: Animal Product Options in the Diet*. Washington, DC: National Academy Press, pp. 345–350.

Searls GA, Maddock RJ, Wulf DM (2005) Intramuscular tenderness variation within four muscles of the beef chuck. *Journal of Animal Science* 83: 2835–2842.

Shackelford S, Wheeler TL, Koohmaraie M (1999) Evaluation of slice shear force as an objective method of assessing beef longissimus tenderness. *Journal of Animal Science* 77: 2693–2699.

Tompkin R (1990) The use of HACCP in the production of meat and poultry products. *Journal of Food Protection* 53: 795–803.

US Department of Agriculture (USDA) (2005) *Food Standards and Labeling Policy Book*. http://www.fsis.usda.gov/OPPDE/larc/Policies/Labeling_Policy_Book_082005.pdf, accessed 27 November 2013.

US Department of Agriculture (USDA) (2011a) *Livestock and Poultry: World Markets and Trade*. www.fas.usda.gov/psdonline/circulars/livestock_poultry.pdf, accessed 27 November 2013.

US Department of Agriculture (USDA) (2011b) *USDA National Nutrient Database for Standard Reference, Release 24*. www.ars.usda.gov/ba/bhnrc/ndl, accessed 27 November 2013.

von Seggren D, Calkins CR, Johnson DD, Brickler JE, Gwartney BL (2005) Muscle profiling: characterizing the muscles of the beef chuck and round. *Meat Science* 71: 39–51.

Wadhwani R, Murdia LK, Cornforth DP (2010) Effect of muscle type and cooking temperature on liver-like off-flavour of five beef chuck muscles. *International Journal of Food Science and Technology* 45: 1277–1283.

Wang H, Birkner ML, Parsons J, Ginger B (1958) Studies on enzymatic tenderization of meat III. Histological and panel analyses of enzyme preparation from three distinct sources. *Journal of Food Science* 23: 423–438.

Warner R, Kauffman RG, Greaser ML (1997) Muscle protein changes post mortem in relation to pork quality traits. *Meat Science* 45: 339–352.

Wheeler T, Shackelford SD, Koohmaraie M (2000) Variation in proteolysis, sarcomere length, collagen content, and tenderness among major pork muscles. *Journal of Animal Science* 78: 958–965.

Wulf DR, Emnett S, Leheska JM, Moeller SJ (2002) Relationships among glycolytic potential, dark cutting (dark, firm, and dry) beef, and cooked beef palatability. *Journal of Animal Science* 80: 1895–1903.

24 Poultry Processing and Products

Douglas P. Smith
Prestage Department of Poultry Science, North Carolina State University, Raleigh, North Carolina, USA

24.1 Poultry processing

Primary and further processing of poultry is a multibillion dollar industry in the US. The majority of poultry raised and commercially processed in the US is composed of chickens, turkeys, and ducks, with more than 9 billion birds processed in 2010 (Table 24.1). Total retail product weight exceeded 58 billion pounds, with an estimated retail value of $100 billion.

The general process of turning live birds into meat is similar for all three species, but details of the process can be distinctly different for specific steps. Differences are due to bird size and physiology differences, requiring unique equipment and species-specific methods. Poultry meat is further processed into many forms as a result of customer and consumer demands that developed in the 1980s (Figure 24.1). The industry was able to respond to this demand since the profit margins were higher for further processed items than for raw whole carcasses or parts. Also, equipment had been developed to automate further processing and packaging on a large scale, and otherwise defective whole carcasses not marketable as raw Grade A could be further processed and sold as value-added. Since the vast majority of poultry meat is represented by broiler chickens, the following sections will describe broiler processing. Two further sections will address differences specific to turkey and duck processing.

The basic steps of broiler chicken processing are shown in Box 24.1. The processing plant is divided into a slaughter area, a separate evisceration area, a chilling area, and, depending on the products produced, further processing areas. In addition to the processing areas, plants contain employee areas such as locker rooms, breakrooms, and restrooms; office areas for management and a separate space for USDA employees; accessory areas such as coolers, shipping and receiving docks, warehouse storage, maintenance and repair, and box making; and waste (offal) and waste water processing. The commonality among plants is that live birds are processed or further processed into food, but any individual plant is uniquely different from any other plant.

24.1.1 Preprocessing

Prior to processing, several steps are taken to prepare the live birds. First, feed is removed from broilers approximately 8–12 h before slaughter. Feed withdrawal is important to reduce gastrointestinal contents within the bird, which reduces the chance of ingesta or fecal contamination during processing. One negative aspect of feed withdrawal is that birds lose weight, which reduces payments to the contracted growers. Microbiological problems occur from overly long withdrawal as contamination increases due to gut fragility during evisceration. Also, pH of the crop increases, which encourages the growth of *Salmonella*, adding to potential pathogen contamination. Per animal welfare guidelines, birds are not fasted more than 24 h.

The poultry house is cleared of equipment prior to arrival of the catching crew, then birds are loaded into metal multilevel coops. These coops are loaded onto trucks and birds are transported to the processing plant. Upon arrival, the birds are weighed as a group and then are held in a holding shed to provide some minimal protection from the environment prior to unloading.

Food Processing: Principles and Applications, Second Edition. Edited by Stephanie Clark, Stephanie Jung, and Buddhi Lamsal.
© 2014 John Wiley & Sons, Ltd. Published 2014 by John Wiley & Sons, Ltd.

Table 24.1 Live weight, number of animals, and pounds inspected of chickens, turkeys, and ducks slaughtered under USDA inspection in 1990, 2000, and 2010 (data from USDA National Agricultural Statistics Service)

Species	Year of production	Live weight (individual animal average, in pounds)	Number of animals inspected (in millions)	Pounds inspected (in billions)
Chicken	1990	4.4	6017.6	26.4
	2000	5.0	8424.2	42.2
	2010	5.7	8789.1	50.1
Turkey	1990	21.3	271.2	5.8
	2000	25.6	268.0	6.9
	2010	29.1	242.6	7.1
Duck	1990	6.6	20.8	0.137
	2000	6.6	24.5	0.162
	2010	6.8	23.6	0.162

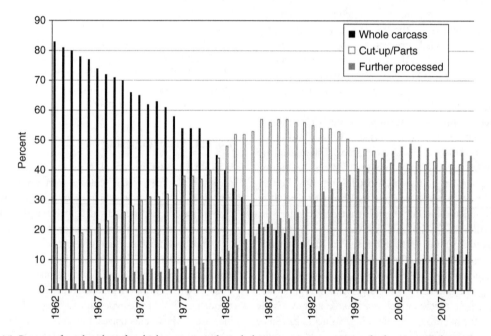

Figure 24.1 Percent of market share for chickens expressed as whole carcass, cut-up parts, or further processed meat by year from 1962 to 2010 (data from the National Chicken Council, Washington, DC).

24.1.2 Primary processing

24.1.2.1 Slaughter

The coops are unloaded from the truck and a machine tilts and dumps the entire coop onto a conveyor that transports the birds into the plant to the live hang area. Employees pick up each bird and hang them upside down by the feet on a stainless shackle. Shackles are spaced 6 inches apart and are attached to rollers on a continuous track; depending on size of the plant and speed of the line,

there may be several thousand shackles. Birds remain on these shackles for the duration of steps in the slaughter area, approximately 6–7 min total. The live hang room is the beginning of the slaughter area.

On the shackle line, the birds first pass through an electrical stunning device. A non-conductive fiberglass trough of water (fresh or salted) with a metal mesh grate immersed in the water hangs below the shackles. As the bird is pulled through, the head touches the water and immersed grate. An electrical control box has one

---

Box 24.1 Major steps of broiler processing and further processing

1. Preprocessing	2. Slaughter	3. Evisceration
Feed withdrawal	Live hanging	Oil gland removal
House preparation	Stunning	Venter
Loading	Killing	Opener
Holding	Bleeding	Eviscerator
Unloading	Scalding	Inspection – traditional
	Picking	Viscera removal/giblet
	Head removal	harvest
	Hock cutting	Crop removal
		Lung removal
		Carcass washing IOBW
		(Inspection – HIMP)
		Chilling

4. Secondary processing (no ingredients or heat treatment)	5. Further processing (ingredients or heat applied or both)	6. Packaging and shipping
Cut-up	Marination	Packing
Deboning	Coating	Labeling
Portioning	Cooking	Palletizing
Grinding	Glazing	Transport
Mechanical separation	Freezing	

HIMP, HACCP-based Inspection Models Project; IOBW, inside-outside bird washer.

---

electrode connected to the metal grate in the water, and the other electrode is connected to a metal bar that touches the shackle above the bird's feet. The body of the bird connects the electrical circuit and the bird is stunned by the current (approximately 10–20 mA) passing through the head and body. Stunning as conducted in the US is reversible, and should the bird be removed from the line immediately after stunning, it will regain full consciousness and mobility in approximately 5 min. In other countries higher voltages and amperages are used to irreversibly stun birds. EU processors electrocute poultry in stunners as an alternative method for stunning and killing (although birds are still bled immediately after electrocution). As a result of higher amperage (150 mA), poultry in the EU is subject to more carcass damage, especially broken wings and breast hemorrhages.

Immediately after exiting the stunner trough, an automated kill machine cuts the throat of the bird. Metal guide bars hold the head and neck against a spinning circular blade; some machines have two blades in an effort to cut both sides of the neck. In the US, birds slaughtered for human consumption must be killed by exsanguination (blood removal). Other forms of slaughter are approved,

including decapitation and pithing, which also provides blood loss. Using an automated killer or neck cutter requires a manual back-up method. A person will manually cut the neck of any birds that may have been missed by the machine.

After the neck is cut, exsanguination takes approximately 2 min. Only about 50% of the blood is lost during exsanguination. The blood loss is necessary to allow the bird to fully expire and ensure there is no blood pressure in skin or muscles so no bruising or discoloration occurs. For birds that still have any blood pressure, the skin will turn pink or red during scalding, a condition termed "cadaver," and these carcasses will be condemned as not suitable for human consumption. Blood, because of its high nutrient content, is collected in a trough or pan and is not allowed to enter the waste water stream. It is usually sold to a renderer for processing into blood meal.

After exsanguination, the carcasses are scalded. This requires that the carcasses are pulled through troughs or baths of hot water. Some plants utilize multiple baths at various temperatures, and all have agitation to increase the hot water moving through feathers and achieving better skin contact. Feather follicles in the skin relax in hot water and allow better feather removal. Plants utilize many time and temperature schemes during scalding, but are grouped into two primary methods. The first scheme uses lower temperature, typically 125 °F (69.4 °C), for up to 2 min. This temperature preserves the outer layer of skin and is used by processors that either market whole birds based upon skin color (bright yellow bird skin, for example) and for those further processors that do not apply batter and breading to skin-on parts (the outer cuticle that is left on does not favor breading adhesion). This lower temperature has traditionally been termed "soft scald." Higher temperature scalds, up to 145 °F (80.5 °C), for up to 2 min (but usually less), are termed "hard scald." The cuticle is removed, along with most skin color, which allows better breading adhesion for further processors.

Immediately after exiting the scalders, the carcasses pass into picking machines. Banks or rows of picker fingers (which are blunt, hard rubber, ribbed, and approximately 4 inches long and 0.5 inches wide) are mounted on metal plates that spin at high speed and are horizontal to both sides of the carcass. Two banks are sufficient for smaller broilers, although some processors use additional specialty pickers to concentrate on the neck or leg regions. These spinning fingers exert tremendous pressure and force on the carcass and strip off the feathers quickly. A downside to this process is broken wings, and in some cases broken legs. The picking process is typically accomplished in less than 2 min. After the picker, a guide bar with a hook at the end is positioned to pull the heads off the carcasses.

24.1.2.2 Evisceration

At this time the carcasses must be passed into the evisceration area of the plant. US Department of Agriculture (USDA) Food Safety Inspection Service (FSIS) rules traditionally requried a physical separation or partition between slaughter and evisceration. Carcasses must be removed from the slaughter shackle line, pass through the partition, and be rehung on a second continuous shackle line for evisceration. The yellow feet are cut from the carcass as part of removing the bird from the slaughter shackle line. The rehang process may be manual with employees hanging the carcasses onto the evisceration line upside down by the hock after the feet are cut and the carcass drops onto a table or conveyor. Alternatively, some plants use an automated foot cutter and rehang machine; as feet are cut, the machine holds the carcass then slides the hocks onto the evisceration shackles.

The evisceration area organization and equipment may vary between processing plant and company, usually based on type of evisceration system (group of processing machines offered by different vendors) and type of federal inspection system. Based on production, a typical plant will have one slaughter line feeding into two evisceration lines. For example, a 140 bird per minute slaughter line will supply two 70 bird per minute evisceration lines. In one traditional inspection system, two USDA FSIS inspectors are placed on each evisceration line; in another system a 180 bird per minute slaughter line feeds two 90 bird per minute evisceration lines with three USDA FSIS inspectors per line. A third, more recent innovation is a 140 bird per minute slaughter line and one evisceration line (also 140 birds per minute) with two inspectors on the line just prior to chilling. A traditional system employs a series of machines to remove the viscera from the bird but leave it attached to the carcass for inspection. Newer systems remove the viscera and either clamp it above the carcass or drop the viscera onto a tray below the carcass. Both the carcass and viscera must be observed during inspection to determine if the carcass is suitable for human consumption.

Regardless of the equipment system and inspection regime, carcasses entering the evisceration area go through a series of steps to produce a carcass suitable for chilling. Following rehanging, the carcasses have the oil or preen gland removed from the base of the tail. A multistep viscera removal process is accomplished by three pieces of equipment in sequence. First, the venter uses a probe to locate the cloaca, then a circular knife descends to cut the cloaca and attached colon free of the carcass. Then the opener uses a small blade to open the abdominal cavity from the cloacal opening (enlarged by the venter) to the tip of the keel bone. Lastly, the eviscerator drops a triangular paddle into the opened carcass to scoop out the viscera from the abdominal cavity.

In traditional inspection plants, which comprise more than 80% of operations in the US, the USDA FSIS inspectors examine the carcass and viscera for any sign of disease and abnormality. Inspection stations on the shackle line have specific requirements, including adjustable chairs, amount of light, and mirrors behind the carcasses to give inspectors the opportunity to view the back of the carcass. Inspectors view approximately 30 carcasses per minute but this number varies depending on the type of plant and line speed. In newer HACCP-based Inspection Models Project (HIMP) systems, one inspector per evisceration line is placed just prior to the chiller. Signs of seven major disease signs for broiler chickens are observed and recorded: synovitis, sepsis, leukosis, infectious process, air sacculitis, tumors, and skin lesions (by FSIS in traditional plants, by plant personnel in newer inspection systems). A number of other minor diseases and disorders may also affect product disposition, and other diseases specific to other species, such as osteomyelitis for turkeys. The inspectors designate disposition of each affected carcass, from washing to trimming of parts to whole carcass condemnation. Condemnation rates vary among plants and species, but approximately 1% of total pounds of poultry inspected are condemned.

A newer inspection system, termed *Salmonella* Initiative Program (SIP), is a hybrid program that enables plants to conduct experiments and request variations from normal processing procedures. For example, some plants have increased line speeds after providing data that the practice is not detrimental to food safety. Plants have also received exemptions for refrigerated product exceeding 40 °F after processing after showing the product did not contain higher numbers of pathogens nor promote their growth. In return for FSIS exemptions, plants must conduct pathogen testing of carcasses at several areas during processing to show they are in control of the process and provide the data to the FSIS.

After inspection, the viscera are separated into edible (giblets, or heart, liver, and gizzard) and inedible portions (intestines, spleen, gallbladder, etc.). The inedible viscera, heads, feet (if not sold), feathers, and blood are termed offal. Offal is collected and sent to a rendering plant for conversion into feed additives. For giblet processing, the gizzards are split open and the lining peeled away; hearts are trimmed of aorta, and livers have the gallbladder cut or peeled. These giblets are then placed in a small chiller separate from carcasses. After chilling, some may be sent to be packed inside whole carcasses, some are destined for pet food, and some may be packed for raw or frozen supermarket cases.

Carcasses, after inspection, pass into a series of other machines along the shackle line. A lung machine vacuums out lung tissue. A cropper machine pushes a turning probe through the carcass and out the neck, winding the crop tissue around it and removing it from the carcass. A neck breaker dislocates the neck at the base. If the carcass is eventually to be sent for deboning, many plants leave the neck attached; necks may be labeled as white meat if left attached to the frame after deboning, which adds value to mechanically deboned meat.

The final important piece of equipment in evisceration is the inside-outside bird washer (IOBW). As the name describes, pressurized water is sprayed on the exterior and interior of each carcass to remove any extraneous material or fecal contamination. The FSIS has established a policy of zero tolerance for fecal material entering the chilling tank, so this is the last point to remove feces and ingesta. Other equipment in evisceration also employs low-volume water sprays that are typically chlorinated, including the opener, eviscerator, and cropper, but the IOBW is the primary for washing the carcass. The drawback of using IOBWs is water usage as each machine may use 40 gallons of water per minute. During a 480 min work day, with an IOBW on each of two evisceration lines, nearly 20,000 gallons of water would be used.

24.1.2.3 Chilling

Poultry chilling in the US is usually (more than 95% of plants) conducted with immersion chilling. There is typically some separation between the evisceration and chilling areas, but not always a physical barrier as found between slaughter and evisceration. FSIS regulations for chilling require that broilers be chilled to 40 °F (22.2 °C) or less within 4 h of slaughter. Larger birds have up to 8 h for chilling. Carcasses drop off the evisceration line shackles into large vats that may be 8–12 feet wide, 8–10 feet deep, and several hundred feet in total length. Some processors use a smaller tank termed a prechiller prior to the main chilling tank, and some use a small chiller at the end of chilling to apply an antimicrobial rinse.

The two main types of chillers are drag and auger chillers, referring to the method used to move carcasses through the tank. Drag chillers use bars spaced a few inches apart that form a grid that projects into the tank from top to bottom. These grids are several feet apart and are attached to a track above the tank. Motors move this series of grids along the tank to push carcasses along. Auger chillers use a central auger in the tank that spans the entire tank length; motors turn this large screw that pushes the carcasses through the tank. Both chiller designs add fresh water at the exit end so carcasses are exposed to the cleanest water just before exiting the tank. Chillers employ cold water systems termed rechillers that recycle the cold water and usually apply chlorine. Air nozzles at the bottom of the tank agitate the water and keep carcasses moving and assist in heat transfer. Dwell time for an efficient chiller filled with broilers is approximately 1 h. Some plants that debone carcasses use longer chill times of 2–3 h to allow carcasses to pass through rigor mortis so meat can be removed from the carcass upon exiting the chiller. Bacterial cross-contamination occurs with immersion chilling but overall numbers of bacteria are decreased by the washing action of the chillers.

Very few US plants have adopted air chilling as is practiced in the EU. Air-chilled birds remain on shackles after evisceration and are transported into rooms that are refrigerated, requiring a dwell of 2.5 h or longer for smaller carcasses and much longer for turkeys. Air is a less efficient method of heat transfer, resulting in longer chill times, and much longer shackle lines and many cold air compressors and fans are required for carcass heat removal. Cross-contamination of bacteria is less likely with air chilling but still occurs, although in a different pattern than with immersion chilling. Generally air-chilled carcasses have less overall contamination per carcass, but carcasses that are contaminated have higher numbers of pathogens since there is no washing effect from immersion chillers.

24.1.3 Secondary or further processing (raw)

There are many potential product form outcomes for carcasses exiting the chiller. The carcass may be packaged for sale, with or without giblets, but this market form has decreased over the years. Most carcasses are cut into parts, meat is deboned from the carcass, or both. Secondary processing usually refers to further processing of the raw carcass into value-added product forms. Therefore cut-up and deboning are both considered areas of secondary processing. Although whole carcasses were the main product form prior to 1980, customer demands for more convenient food products drove the poultry processing industry toward cut-up parts and boneless, skinless meat. Just as important, consumers were willing to spend more for convenience foods and value-added products. Processors also found an outlet for USDA non-Grade A carcasses. If a part was trimmed due to defects, the rest of the carcass could be cut into parts and sold at a higher price per pound. Automated cut-up equipment and improved packaging systems were becoming available. All of these factors combined have greatly diminished the whole carcass product form, and if not for the recent increase in rotisserie chicken products, whole carcasses would be nearly extinct in the market.

24.1.3.1 Cut-up

The simplest and most original cut-up device was simply a person with a knife. To meet increased demand, more automation was needed, and an early device still in use in some plants is a circular saw. A trained employee with appropriate protective gear can cut a chicken into eight pieces in a few seconds. Even that rate is not fast enough to produce several hundred thousand pounds per shift, so fully automated lines were developed and installed

prior to 1990. Whole carcasses are hung on a continuous modified shackle line system with multiple stations with knives or circular blades; some recent advanced systems are computer controlled. Virtually any cut can be programmed, from a simple cut to separate the front half from the back half of a whole carcass, to a 13-piece cut (two wings, two drums, four thigh pieces, and five breast pieces). Precision machines require consistently sized carcasses, so scales are placed on the shackle line and an unloading mechanism sorts those that are too large or too small. Some plants use a combination of methods, with a machine to cut front halves from whole carcasses, then remove the wings, while the leg quarters are cut on other machines or by hand.

A more recent development is the placement of a hybrid cut-up and deboning line at the chiller exit. Chilling times are increased to ensure carcasses have more time to enter and exit rigor mortis. Cutting meat from the bone or wings from front halves too early (less than 3–4 h) may result in tougher breast meat. Carcasses are immediately placed on a cone line as they exit the chiller, which is described further in the next section. Wings are removed by manual knife cuts during the breast meat deboning process, and drums and thighs may be removed or left on the carcass and deboned.

Wings have become a popular snack item in the past few years. Although a market exists for whole wings, the "buffalo" wing market requires cut-up wings. They may be cut manually, but a wing wheel machine that resembles a ferris wheel cuts whole wings into drummettes, flats, and wing tips as it revolves. A human loader continuously adds a fresh supply of whole wings as the cut wing portions fall into different containers or conveyor belts.

24.1.3.2 Deboning

As in the cut-up department, a person with a knife was sufficient for early deboning practices. Most deboned meat is the breast (fillets and tenders), with some thigh meat also available to the retail market. To keep up with demand, a continuous conveyor track with deboning cones mounted approximately every 24 inches is utilized. A deboning cone is a blunt cone shape approximately 4 inches tall and 3 inches in diameter, composed of either white plastic or stainless steel. The line is 30–50 feet long and customized for other lengths as needed. As the cones move, a line of people with specific assignments occupy one side of the line. One person loads either front halves or whole carcasses, others then cut the wings loose from the shoulders, others strip the whole breast (with skin still attached to the butterfly fillets) from the front half. Subsequent employees manually pull the tenders from the front half, and others separate the butterfly fillets from the skin. If whole carcasses are deboned then the thigh and leg meat is cut free from the carcass, and skin separated from the meat for some products. There are multiple variations of this process, with a typical alternative process where wings are left on the front half to facilitate breast removal, then trimmed off later with the skin.

There are several versions of automated deboning machines for removing breast meat. One is a modified version of a cone line, with stations that mechanically cut wings, cut and remove the fillets, and slide the tenders free. Another system uses a modified shackle system, where stations use robotic knives to cut the wings, fillets, and tenders from the front half frame. Frames, or skeletal remains after deboning, may be sold as offal but are usually collected for mechanically separated meat production.

24.1.3.3 Portioning

Portioning is conducted on deboned meat, usually breast fillets. A person with a knife and a scale or template was the original portioner, but now much is done by machines. One machine relies on a hard plastic three-dimensional template. A fillet is placed in it, and a cover pressures the meat into place while slicing any excess that is outside the mold. Another type of machine has a loading conveyor belt with laser guides for manual placement of fillets or other boneless meat. As the meat enters the machine, it is imaged and a computer program determines the optimum cut to maximize required product. Cuts are made by robotic arms with high-pressure water spray nozzles that cut the meat into precise patterns as determined by the computer.

Meat from either machine is usually weighed to finish the portioning and sorting process. Weight scanning conveyors with multiple drop stations may be used to sort boneless meat by weight, or meat may be weighed manually.

Portioning is important to supply meat at a certain weight, length, width, and thickness. Although it does not seem important other than for marketing, fast food retail operators rely on portion-controlled meats as a food safety method. For example, a breaded fillet of a certain size is placed in a fryer with a set temperature and a timer set for 3.5 min; an overweight or overly thick portion may not reach a safe internal temperature with the programmed settings.

24.1.3.4 Grinding

A growing market segment for poultry meat is ground product. Boneless skinless meat (breast meat, dark meat, or both) is coarsely ground through a plate. Particle size of the grind is determined by plate size and is normally ¼ to 3/16 inch (6.4–9.6 mm). Most of this product is tray packed for the retail market. It has become more popular due to the typically lower fat content than ground beef or pork. The grinding is similar to what is used for poultry sausages, without the additives or casing process. One drawback to this product is the typically higher prevalence of pathogens, including *Salmonella*. The product is not more contaminated than any other raw poultry meat product, but the grinding and batching process tends to spread a few cells over and into a large quantity of ground product. Proper cooking is also essential since bacterial cells are distributed throughout the product thickness. The grinding process also produces more nutrients for bacteria as the muscle sarcoplasm (especially water and protein) is freely available which could shorten shelf life.

24.1.3.5 Mechanical separation

Deboned meat trim or scraps, skeletal frames, necks, and any edible portion of a carcass, with or without bones, may be sent for mechanical separation. Common acronyms include MSC (mechanically separated chicken meat), MST (mechanically separated turkey meat), and MDP (mechanically deboned poultry meat). In the separator machine, meat is first ground, then forced under high pressure (when the meat and bones are reduced to small particles in a paste) through screens that remove bone and most connective tissue. The pressure quickly heats the meat so cooling systems are used to keep it from cooking during the process.

The resulting meat paste is usually bulked packed into 40 or 50 pound boxes or 2000 pound totes. It may be left fresh or frozen after packing. The majority of this meat is destined for further processors to be formulated into products such as hot dogs.

A combination of the heat during the process, the grinding process itself, and the high levels of heme from bone marrow can cause contamination and shelf life issues for this product. Although bacterial spoilage is an issue, the fat and heme present can also result in a relatively quick chemical rancidity. Some buyers request a curing agent be added to the meat paste during production. A powdered form of salt, sodium nitrate and nitrite is added at low levels, which provides a redder color and controls some bacteria, such as *Clostridium* spp., and extends shelf life.

Table 24.2 Representative further processed poultry product retail forms from raw poultry portion(s) and processing step(s)

| Raw poultry portion | Further processing step | | | | Marinated | | | Batter/ breaded (B/B) | Parfried | Fully cooked (F/C) | | | | Glazed | Individually quick frozen (IQF) | Finished product form |
	Cut-up	Deboned and skinned (B/S)	Chopped/ formed	Ground	Needle injection	Vacuum tumbled	Ribbon blender			Fryer	Oven	Water	Canning			
Broiler carcass (whole) chicken	✓															Rotisserie
Broiler wing	✓	✓			✓			✓	✓					✓	✓	Hot wings
Broiler fillet or tender	✓	✓				✓		✓							✓	Chicken tender
Broiler fillet, tender, thigh	✓	✓	✓				✓	✓		✓					✓	Chicken nugget or patty
Broiler fillet, tender chicken patty		✓	✓				✓				✓				✓	Roasted
Broiler thigh, drum chicken		✓				✓					✓			✓	✓	Teriyaki
Broiler fillet, thigh	✓	✓				✓					✓				✓	Chicken fajita strips
Broiler fillet, thigh sausage		✓				✓					✓					Chicken
Broiler split breast	✓				✓										✓	IQF split breast
Breeder chicken fillet, tender chicken		✓				✓							✓			Canned
Turkey carcass (whole)	✓				✓	✓									✓	Frozen turkey
Turkey fillet, thigh	✓	✓		✓												Ground turkey
Turkey fillet, thigh burgers				✓		✓					✓					Turkey
Turkey fillet, tender	✓	✓	✓			✓	✓					✓				Turkey ham
Duckling fillet	✓	✓			✓	✓								✓	✓	Duck breast

24.1.4 **further processing by ingredient addition and/or heat treatment**

Raw whole carcasses, parts, deboned meat, and similar products still retain market share in the US, but the growth of market share is from further processed items. Further processing is in this context the addition of ingredients,and/or heat treatment to poultry meat to create a variety of value-added products. Poultry companies involved with further processing usually have several hundred product labels. Products may be produced and labeled under the company's own name brands, private labeled for retail sale, and produced for the fast food or hotel-restaurant-institute wholesale market. Some processed forms (as shown in Table 24.2) include marinated, chopped and formed, breaded, glazed, parfried, oven roasted, fried, chargrilled, and individually quick frozen (IQF). Major product forms include patties (breaded or roasted), nuggets, tenders, fillets, wings, drums, and thighs, prepared and either parfried (partially cooked in oil for less than a minute) or cooked in many different possible forms. There are many ways to categorize and describe further processed poultry products, but due to the complexity and overlap of processes it is difficult to separate further process products into simple descriptions. Examples of production lines for further processed items are shown in Figure 24.2.

24.1.4.1 Marination

The majority of further processed meat is marinated and is typically one of the first and most extensive of operations in further processing plants. Marination, in the simplest form, is addition of liquid solution to meat for improving functional quality. The functional properties provided include increased water-holding capacity of the meat, with perceived increase of juiciness by the consumer. The added weight of the water is positive for the producer as a yield increase. Other functional properties may include a slight increase in tenderness, and a perceived increase in flavor from the salt and other added ingredients if spices, broth, or juices are added. Basic marinades include water, salt, and sodium phosphates. A typical mixture would be 90% water, 6% salt, and 4% phosphate added at up to 10% of raw product weight. More marinade could be added by reducing the phosphate content for reasons discussed in the following paragraph. More complex marinades may include spices, citrus juices, chicken broth, vinegar, or any flavor additive required for the formulation.

Mixing the marinade solution is somewhat different than other solutions; first, the water should be chilled, then phosphate slowly added while stirring, and the salt and other ingredients added after the phosphates have dissolved. A limiting factor when formulating and adding phosphates is that in the cooked and finished product, the USDA stipulates that phosphate may not exceed 0.5% by weight. The total amount of marinade added to a product must be displayed on the final label. There is no limit to the type of ingredients that can be added to marinades (other than practical limitations such as flavor and ability to dissolve in solution) as long as the ingredient is Generally Recognized as Safe (GRAS).

24.1.4.2 Meat: bone-in and whole muscle

Whole carcasses may be further processed and in fact are an important "new" product form for producers as the resulting form is rotisserie chicken. Parts, such as split breasts, drums, thighs, and wings, may also be marinated. Different product types receive different marinade levels, and also require different types of marination equipment. Whole carcasses or bone-in, skin-on parts are injected with needles in an injector machine. A bed of needles injects marinade under pressure as the meat passes along a conveyor. The needles are on springs so the needles do not penetrate into bones. Injectors can marinade product at a high rate of throughput and a higher rate of pick-up, sometimes exceeding 20% by weight of raw product. A potential downside is that surface bacteria may be carried into the interior of the meat. Typical marinades for injection are no different from marinades described previously. The amount of marinade pick-up must be taken into account to keep final product phosphate content below 0.5% as described above.

Boneless skinless meat, usually fillets, tenders, and thighs, are marinated in a vacuum tumbler. A large vessel roughly resembling a barrel, containing from 50 to 10,000 pounds, is filled with meat and marinade, sealed, and a vacuum is created. Larger vessels are double-walled so that refrigerant can be added to keep the contents cool during marination. The vessel rotates to provide a tumbling action for approximately 20 min. Baffles inside the vessel promote further movement to maximize marinade pickup. Skin-on carcasses and parts cannot be vacuum tumbled without the loss of skin from the tumbling movement.

24.1.4.3 Meat: muscle formulation

Boneless meat, including any that is unfit for other product forms (for example, minor downgrade whole-muscle parts, trimmings, etc.), is reduced to smaller particle size (approximately one-eighth to one-fourth inch, 3.2–6.4

I. **Fully cooked chicken nuggets**

II. **Individually Quick Frozen (IQF) Split breasts**

III. **Parfried chicken tenders**

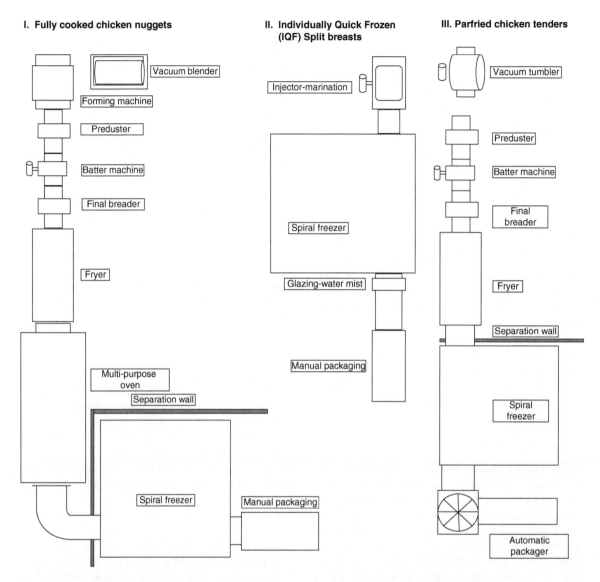

Figure 24.2 Examples of equipment lines in a plant for further processed products (not to scale).

mm) by many methods. Equipment includes meat grinders for larger particle size product, and comminutors or bowl choppers for much smaller particle products. The USDA FSIS allows skin to be incorporated "in natural proportions" to the meat block, or approximately 20% by weight. The skin obviously adds weight at a cheap price, but also has some binding capacity, and adds flavor mainly through its lipid component. A number of other ingredients may be added for functional reasons. Salt acts as a binding promoter by extracting myofibrillar proteins from the muscle and is also a flavor enhancer. Cornstarch increases binding; sugars add color and flavor; gums add binding capability and may enhance texture; and soy proteins add flavor and nutritional value to formulations.

Spices and flavoring agents may be added at this stage depending on product specifications. Salt is a common additive, as are various peppers, vinegars, fruit and vegetable juices and extracts (such as lemon juice). The mixture of small particle size meat and ingredients is referred to as a meat batter. Ingredients, along with the meat block (bulk, reduced particle-size raw meat), are typically mixed in a ribbon blender. The blender allows fine mixing control, and may be operated under vacuum if marinade is added. Some of these blenders also may be double-walled to provide cooling capability or directly incorporate carbon dioxide into the meat block for temperature control.

Processing methods include forming machines. The meat block, in bulk quantities, is added to the machine's hopper, and the meat mixture is forced by pressure into molds, then stamped out onto a conveyor belt for additional processing (breading, glazing, etc.). Molds may be simple round nugget or patty sized and filled with batter, or shaped for specialty markets such as stars or dinosaur shapes for children. Other molds replicate a whole-muscle fillet or tender shapes and are filled with larger meat chunks to form a higher value product. Once breaded, these appear to be whole-muscle fillets. A different product and process is turkey meat chunks in a matrix of meat batter placed in plastic bags, then cooked in the bag to form a turkey ham and oven-roasted turkey breasts.

24.1.4.4 Coatings

Numerous, widely different coatings are applied during further processing at different processing points. The simplest is a water spray for parts or deboned meat that coats the product immediately after it is frozen. The spray forms an ice glaze on the IQF product which reduces sticking once bagged, adds an attractive sheen to the product, and adds weight (although the average weight gained is deducted from the label weight.

Perhaps the most popular coating system for poultry is batter and breading, resulting in one of the best known poultry products in the US, fried chicken. The basic process is a three-step system, with raw or marinated parts placed in a dry predust, followed by a wet batter coating, then a final dry breading. A typical system adds approximately 20–30% by weight to the incoming product. The USDA FSIS allows up to 40% breading addition, but the product must be labeled as a fritter if pick-up exceeds 40%. Above 60% and the product is labeled as other than chicken, such as chicken meat-flavored breading. Higher pick-ups may be achieved by thickening the batter, or adding passes of the product through multiple predust and batter applicators. There is incredible variety in the ingredients used in these coatings; types of grains (although usually corn or wheat based), salt and other spice content, color additives, and particle size all contribute to different outcomes in finished product. Predusts and breadings with smaller particle sizes result in a finer,

more homogenous coating and even coloration, and probably smoother and slightly chewy mouthfeel. A large particle-size breader will produce more uneven coloration and texture, providing a more colorful and crunchier coating.

Equipment to add bread coatings utilizes a continuous conveyor line of stainless steel wire to move product through the system. The preduster is loaded with the dry mix, and it falls or sifts onto the product from above, and the wire conveyor also pulls the product across a bed of predust to coat the bottom. The product then enters the batter machine with a pool of batter coating the product, with some machines also cascading batter from above the product. A dry mix is combined with water at the machine and is typically refrigerated to both thicken the batter for increased pick-up and to extend the shelf life of the recirculated liquid. The product is then conveyed into the final breader machine, usually the same machine as used for the predust. Additional in-line batter and breader machines may be used to increase the pick-up for frittered products. A different type of predust or final breader machine is barrel shaped. It is in-line, mounted at an angle, and turns as product enters the top and exits the bottom. These are usually used for smaller whole-muscle pieces to be frittered, such as tenders, or specialty items such as chicken livers. Another specialty product is tempura batter, a unique, thick application for products such as corndogs or Asian products. The battered product falls immediately into a fryer to set the batter, and may be parfried or fully cooked.

Other coatings are glazes for oven-roasted products. Whole-muscle or formed fillets are placed on an oven line for cooking. Prior to oven entrance, product is dipped into a glaze, a slightly thickened water-based liquid usually with added sugar. This imparts additional weight but mainly has an esthetic effect, adding a slight color and glossy appearance. Other products may have a sauce glaze applied after cooking and before freezing, such as hot wing items.

24.1.4.5 Heating

Because there are many products and preparation methods, there are many ways to heat or cook poultry meat. More meat is cooked by the further processor currently than before as more major fast food customers are requiring fully cooked items due to food safety concerns in their restaurants. For heating and cooking, ovens and fryers are designed to be used for in-line processing to keep product moving through the heating process. Water cookers also have a system to pull items through the system. Steam kettles and rack ovens (similar to the smokehouse cookers) used for some items such as whole spent hens and turkey loaves are batch only. In-line microwave cookers are still used but are rare.

Fryers may be used to parfry or fully cook poultry items. Fryers currently sold are indirect heat models, where a non-flammable liquid is heated external to the fryer body, then pumped to the fryer and circulated through tubes in the bottom of the fryer, which heats the frying oil. Older model fryers used direct heat. Gas jets under the fryer were ignited and these heated the tubes in the fryer bed. However, these were discontinued because they were too dangerous to operate due to fire hazard and resulted in several plant fires. For parfrying, batter/breaded items dwell for 30–60 sec to set the coating adhesion and color. Fully cooked item dwell depends on the size and type of product. Most products are not fully cooked in fryers any longer due to poor yield as fluid escapes, health-conscious consumers requiring lower fat products, and the greatly reduced oil quality from the long dwell and escaping fluids and breading crumbs or fines. Stainless wire belting pulls product along the fryer bed. An overhead belt operating at the same speed as the bottom belt is also used so that items floating in the oil are moved along and out of the fryer. The hot oil is circulated, filtered to remove crumbs and fines, and reused until the quality is degraded. Eventually the accumulation of free fatty acids and oxidation products renders the oil unusable. Fryer oil temperature is closely monitored to keep product cooking at a temperature lower than the flashpoint of the oil and extend the life of the oil.

The typical in-line oven cooking system is between incoming product and a freezer. Product may arrive at the oven belt from a batter/breader line, from a marination line, from a fryer used for less than a minute to set the coating, from a forming machine, or a bulk loader placing product on line. Product enters the oven to reach a particular temperature and then is immediately frozen to increase yield. Some roasted products may be chilled and packaged without the freezing step.

There are several general in-line oven types and many variations among the designs. Simplest is the tunnel oven heated by gas jets, electrical heat, or steam. More recent designs combine heating systems to maximize yield and product quality. Special batter and breaded coating systems can be cooked in an oven without a fryer by combining or alternating dry and moist heat as the product passes through the oven, along with fans for convection heating. Another type is a spiral oven, where the wire conveyor spirals upward. This design promotes a smaller footprint within the plant and also requires less energy for heating product. Another oven utilized for non-breaded items conveys product through on plates that are heated, while heated plates also contact the product from above. The plates sandwich the product above and below, acting as a griddle press.

Traditional cooking technology is still used, such as batch cooking whole birds in steam kettles. The meat is deboned afterward, while broth is captured and sold as another product. Whole birds may also be bagged and passed through water cookers or placed in racks and wheeled into batch ovens. Turkey loaves and formed roasts are also bagged and cooked in the same type of oven.

Poultry meat is also canned. Meat from old birds, usually breeders past their laying capability, is processed and deboned. The canning process is the same as for low-acid foods, with similar time and temperature requirements to eliminate pathogens and spores. The meat from old birds is used because it is tough and contains more connective tissue. The tender meat from younger birds would dissipate into solution inside the can from the salt and harsh canning procedures (extreme heat and pressure over time).

24.1.4.6 Freezing

Very few further processed poultry are sold or marketed as fresh. Rotisserie chickens or roasted parts and some cooked-in-bag products such as turkey loaves are refrigerated rather than frozen. Most other products are frozen, and many options are utilized. Tunnel freezers with conveyors pass product through forced cold air impingement at −35°F (−37.2°C). Older versions still in use force carbon dioxide into the tunnel to create powder (snow) or pellets that are in direct contact with product to produce a quick freeze. Even faster freeze is accomplished with liquid nitrogen with an effective temperature below −150°F (−101.1°C). Sometimes the tunnel freezer is used as the sole method for freezing, but may also be used as a prefreezer to crust freeze cooked product and maximize yield, before another freezer such as a spiral freezer brings the product to a range of −20°F to 0°F. Another method is a plate freezer, similar to the plate oven, except the plates are cold and provide fast heat transfer. These freezers are normally used for smaller, flatter items such as boneless skinless beasts, tenders, or patties. For higher volume throughput, further processors are employing in-line spiral freezers. These freezers utilize fans and large condensers to quickly freeze products. The footprint to throughput ratio is small, and less energy is needed than for tunnel freezers with equal freezing capacity.

24.1.5 packaging and labeling

Processed and further processed items are packed into initial packaging on the line, and then into secondary containers. There are literally hundreds of different types of packages. Basic types include foam trays with plastic overwrap, plastic bags, resealable pouches, shrink-wrap bags, and paper cartons.

Raw or fresh poultry is typically packaged in one of three types of systems/materials: ambient, vacuum, or modified atmosphere packaging (MAP). Ambient atmosphere packages are polystyrene foam trays with stretch film which are common for parts and deboned meat. An oxygen barrier is not used so as to minimize *Shewanella putrafaciens* growth and accompanying noxious odors associated with the product. Vacuum packaging can be used for whole carcasses (bags) or for parts or deboned meat (thermoform pouches, either single-use or resealable). Vacuum packages can extend shelf life to a limited degree by reducing aerobic bacteria growth (although growth of anaerobes and facultative anaerobes is promoted). MAP is used for parts, deboned meat, or ground product. Some shelf life extension is accomplished via low oxygen (20–30% CO_2, balance nitrogen), high oxygen (75–80% O_2, balance CO_2) or high carbon dioxide (60% or higher CO_2) flushes, depending on the product type.

Many types of systems are available for further processed and frozen products. Shrink-wrap bags or pouches, form-fill bags, rigid trays with sealed film overlay, cook-in bags, cook-in sealed trays, resealable bags, pouches, or trays are all currently used for various items. The plastics used are oxygen barriers, and some also incorporate oxygen or odor scavengers to protect products. Packaging for value-added products must balance consumer convenience with product protection.

Packing may be done by hand, but many processors use automated weighing and bagging equipment for smaller size items such as nuggets and tenders. The product falls into a sorter and then into a number of bins that have scales incorporated; each bin drops its contents into bags that are sealed as they exit the machine. Cook-in bags are used for raw poultry destined for water or batch cooking, usually whole birds or loaves. Regardless of pack method, once in the initial package, multiple units are placed into secondary containers, typically a cardboard box. Boxes are stacked and palletized in preparation for storage and shipment. Some products, such as mechanically separated meat sold to other further processors, may be placed in plastic liners inside large cardboard containers that contain 2000 pounds and are preattached to pallets.

Labels are approved according to federal regulations for inspected products. Labels are required for both initial and secondary packaging, depending on the end use of the product. Labels must contain information such as product name, net weight, ingredients, manufacturer name and address, nutritional information, and safe handling instructions, and the plant (P) number assigned by the USDA. Barcodes are added to packages for inventory purposes of the manufacturer and retailer but are not required by the FSIS. However, barcodes and distribution information are extremely helpful when conducting product recalls.

24.1.6 Shipping

Poultry products, whether fresh or frozen, raw or further processed, must be shipped from the plant to commercial locations. Important factors for maximizing product quality and profit include moving the packaged products immediately into coolers or freezers. Second, the products must be rotated quickly and efficiently out of holding and onto transportation. Trucks transport most items due to flexibility of movement and, until recently, relatively cheap fuel costs. Processors with railroad access may send lower profit bulk items with long shelf life, such as frozen leg quarters, via railcar to seaports for export. Third, the truck trailer or railcar must be inspected and found to be in good working condition and clean before product is loaded. Lastly, the carrier must be required to carry an insurance policy in case of accidents or refrigeration equipment failure.

Most non-further processed poultry produced for domestic consumption is shipped "fresh"; poultry may be chilled to 26 °F (−3 °C) before it is considered to be frozen by the USDA. The crust-frozen raw product retains a longer shelf life when shipped long distances. Further processed items, especially battered and breaded items, are shipped frozen. More than 19% of US poultry production was exported throughout the world in 2012, with major markets in Asia, the Russian Federation, and Mexico. Practically all export shipments are frozen non-further processed parts or mechanically separated chicken (MSC). Approximately 12% of US turkey production is exported.

Data from the USDA National Agricultural Statistics Service showed that in 2010 approximately 10% of broiler pounds processed and 59% of turkey pounds processed were frozen. Duck meat is stored for longer periods, so although 43 million pounds were processed in 2010, there were 73 million pounds in frozen storage (excess from the previous year's production).

24.2 Turkey processing

The overall procedures for transforming live animals into meat are similar for turkeys as those previously described for broilers, with notable exceptions. Males (toms) and females (hens) are sexed at the hatchery and raised separately. Larger males (up to approximately 20 kg) are grown longer to maximize breast meat yield that is deboned for further processing. Lighter and younger females (approximately 5–10 kg) are grown more for the whole carcass market, including retail holiday sales. Preharvest steps are generally similar to broilers, except the feed withdrawal program may differ as many turkeys are still on natural light versus artificial lighting, requiring adjustments to the program. Achieving an 8–12 h withdrawal is more difficult as it is determined by the last feeding time, which depends on day length. Also, live turkeys are loaded onto trucks by herding the animals up ramps from the house to the trailer then manually placed in the cage.

Unloading of turkeys would seem to be a simple process, and not associated with stunning, but recent developments are significantly changing these portions of turkey processing operations. The traditional system involves the truck trailer pulling into a loading dock with different heights of dock on both sides of the truck. The slaughter shackle line extends onto these loading docks, and workers pull the live turkeys from the trailer coops and hang them upside down on the shackles. The work is especially difficult for unloading large tom turkeys so the unloaders are highly paid, but working conditions are most unpleasant as the area is basically a covered shed without heat or air conditioning. Injuries are common to both workers and the turkeys, and since the shackle line is very long the turkeys are on the line for several minutes prior to stunning. Because of the labor, animal welfare, and quality issues, some turkey processors have implemented various types of gas stunning technology. Gas, or modified atmosphere, stunning is conducted by applying gas or gas blends (typically 30% carbon dioxide and 60% argon in air, or 90% argon in air, or 40% carbon dioxide, 30% oxygen, 30% nitrogen) to turkeys in a confined space to render them unconscious. The procedure for some systems has the advantage of stunning birds before hanging, alleviating problems described earlier. Meat quality is improved by inflicting less stress on birds and also from fewer broken wings due to flapping after hanging. Negatives to the system include higher costs and proper implementation; if not conducted correctly, gas stunning stresses the birds as they react to low oxygen

levels. Gas stunning may be conducted by placing a shroud over the entire truck trailer and pumping in gas; individual coops may be removed from the trailer and placed in a chamber, and, for smaller hens, they may still be hung on shackles and then the line enters a chamber filled with gas.

Turkey processing after stunning is generally similar to broilers, but with more people used and less automation to eviscerate the turkeys. The lines run much slower with fewer animals processed per day compared to broilers at 140 birds per minute: depending on the inspection system, either 45 birds per minute for carcass weights above 16 pounds or 51 per minute for carcass weights below 16 pounds. A different shackle may be used to hang the neck as well as the feet, allowing easier access to the internal cavity. Chilling times are twice as long (8 h to ≤40 °F (4.4 °C)) due to the considerably larger carcass size. Alternatively, when carcasses are processed for a further processor, a short immersion chilling time is used and carcasses are immediately deboned, then only the meat is chilled. Turkey further processing is also generally similar to broilers, with the main difference in market forms. Major market segments include the traditional whole frozen carcass and formulated turkey loaves and ham products, but not many batter and breaded parts or formed items.

24.3 Duck processing

Ducks or ducklings (less than 8 weeks old) are processed similar to broilers. They are grown similarly and reach market weight of 6 pounds live weight by 35–40 days. For processing, they are loaded onto truck trailers more similar to turkeys as they are walked onto multilevel trailers. They are unloaded by walking them off the truck after arrival at the plant. Live birds are hung on the shackle line by the feet and go through slaughter, evisceration, and chilling equipment similar to broilers. Although an electrical stunning system is used, ducks are much harder to stun since they have little exposed tissue on the head to conduct electricity, and their feathers and thicker, fattier skin also impede electrical flow. Gas stunning has not been successful with ducks due to their diving reflex and ability to not breathe for several minutes. Feather picking is also much more difficult because of the physiology of the skin and feathers. Multiple passes through various specialty pickers are required, and after picking the carcasses pass through a series of hot paraffin and cold water baths. The paraffin sticks to the carcass and small pinfeathers, and when removed by workers most of the pinfeathers and hairs stick to the wax. During evisceration, some plants have more people and less equipment, while other plants are more automated.

Duck further processing has increased significantly during the past 20 years. Previously duckling was sold as a whole frozen carcass for the holiday or specialty market. Now duckling carcasses are deboned and meat is marinated, battered and breaded or glazed, parfried or cooked, just as broiler meat is currently further processed.

Duckling processing yields valuable by-products not usually captured by the broiler or turkey processors. Feathers are collected, washed and dried, sorted and baled at the plant. Although it is a cyclical market, in some years feathers are worth more per pound than duck meat. Tongues are harvested from duck heads for specialty Asian markets, and are always worth more per pound than duck meat. Feet are collected (similar to chicken paws), but are worth more than broiler feet. Livers are collected and may be used for paté or foie gras. Abdominal (leaf) fat is harvested from carcasses, cooked with water in steam kettles to render the fat, then skimmed and cooled. It is packaged as high-value lard sought by gourmet chefs for pastry. Approximate retail prices in 2012 for various duck products were (on a dollar per pound basis): raw whole carcass $3.00–3.50; tongues $17.50; fat $7.25–13.00; raw liver $5.75; and feet $3.25. Another specialty market is the head-on, feet-on (HOFO) market, where carcasses are sold with head and feet intact. To improve carcass appearance, birds are pithed (killed by insertion of a metal rod into the brain through the roof of the mouth) or have a knife inserted into the throat through the mouth to cut the carotid arteries and jugular veins. This occurs after stunning, and does not leave a visible external cut on the neck. Since the heads and feet remain, the birds are not USDA inspected. The non-inspected HOFO carcasses are exported or sold locally; all quality standards are negotiated between the plant and customers.

24.4 Microbiology and food safety

Extensive information, including several textbooks, is available on the subject of poultry microbiology. In general, regardless of the type of poultry processed, shelf life extension and pathogen control are dependent on low numbers of microbes on live birds and proper handling during processing and distribution. Heavily contaminated birds entering the plant or temperature abuse of the processed carcass and products create opportunity for both diminished shelf life and increased pathogen levels.

24.4.1 Spoilage

Fresh poultry rarely exceeds 2 weeks of refrigerated storage before microbial overgrowth degrades the quality to the point of inedibility. Several genera are typically reported to cause poultry spoilage, including *Pseudomonas*, *Acinetobacter*, *Lactobacillus*, and *Shewenella* species. Fresh poultry that is marinated has also shown a propensity to spoil from yeast overgrowth. Frozen poultry, even without microbial spoilage, usually spoils after several months due to chemical spoilage (rancidity) as the monounsaturated and polyunsaturated fats common in poultry meat undergo oxidation.

Microbial spoilage of fresh poultry is accelerated, as previously noted, by higher loads of bacteria remaining on the product from live conditions or introduced to the product during processing. After processing, periods of higher temperature conditions (temperature abuse above 40 °F) promote bacterial growth of psychrotrophic and psychrophilic bacteria, which shortens the time to spoilage (Figure 24.3). Chilling followed by refrigerated conditions select for these bacteria, which is a cost of controlling pathogens that are typically mesophiles.

24.4.2 Food safety

There are several bacteria that cause human illness and are associated with consumption of poultry meat or products. The US Department of Health and Human Services Centers for Disease Control and Prevention (CDC) collect and publish human food-borne illness data. In 2011, the CDC published estimates that the five leading causes of food-borne illness were norovirus (58%), *Salmonella* spp. (11%), *Clostridium perfringens* (10%), *Campylobacter* spp. (9%), and *Staphylococcus aureus* (3%). Norovirus is not normally associated with poultry but the four bacteria have been implicated in consumption of poultry products. Another important bacterium associated with further processed poultry is *Listeria monocytogenes*, which the CDC estimated is the third leading cause of death from food contaminated with bacteria.

Salmonella (non-typhoid) has been closely associated with poultry consumption for many years. The CDC estimated that 17% of salmonellosis in the US is due to poultry consumption. *Salmonella* has more than 2000 serovars, and survives in many different environments outside the intestinal tract since it is a mesophile and facultative anaerobe. It readily colonizes bird intestines (without any deleterious effect on the animal) and the housing environment. Many control methods have been

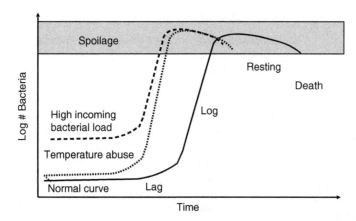

Figure 24.3 Example of a bacterial growth curve (*solid line*) showing increase of numbers of bacteria over time divided into lag, log, resting, and death phases of a typical microbial population present on a raw refrigerated poultry product, from 0 to 20 days. Other lines represent the decrease in shelf life due to shortened lag times from product subjected to temperature abuse (*dotted*) or containing higher initial numbers of bacteria (*dashed*).

developed to control *Salmonella* contamination, but so far it has been impossible to eradicate from poultry.

Clostridium perfringens is a ubiquitous bacterium found in live poultry environments. In high numbers, it causes enteritis in young chickens and poults. As a mesophilic spore former, it is almost impossible to eliminate from poultry. Live poultry can carry vegetative cells and spores, and after cooking the spores may vegetate. In low numbers, *C. perfringens* is unlikely to cause illness, but in mishandled poultry leftovers or temperature-abused cooked product, numbers quickly develop to levels causing human illness.

Campylobacter jejuni and *E. coli* are commonly found in poultry intestines. Outside of intestines, the organism is found in contaminated water and milk, but is not considered hardy in most environments. It requires a micro-aerophilic environment and therefore does not survive long in aerobic conditions. However, it is very motile and can travel to better conditions if sufficient moisture is present. The USDA began requiring poultry plants to monitor and report *Campylobacter* incidence in 2011–2012, depending on plant size. This action was taken since the CDC estimated that 24% of campylobacteriosis in the US is due to poultry consumption. Cold temperature, dry conditions, and acidic environment reportedly help control incidence, but these conditions do not normally exist in US poultry processing plants.

Staphylococcus aureus can be found on live poultry, but is also commonly associated with humans who can be carriers. It is associated with poultry mainly due to postcook contamination of dishes containing poultry such as chicken salad or catered poultry. It survives well at higher salt levels and is a facultative anaerobe, and can be an environmental contaminant. *S. aureus* toxin is heat stable, so contaminated foods can still cause illness after cooking or reheating.

Listeria monocytogenes has been associated with several foods, including processed and fully cooked poultry. Since it does not survive cooking it is introduced via the environment post cook. *Listeria*, a psychrophile, survives very well in cool, moist environments, which characterize poultry processing plant fully cooked packaging rooms. The organism lives in the refrigeration condensers, drain pans under the condensers, and floor drains, and therefore has the opportunity to contaminate product via condensation or during cleaning in breaks.

24.5 Sustainable poultry production and processing

More people are choosing to grow and process their own animals for meat in the US, especially poultry in the south east. North Carolina has become a leading state in small poultry flock growing and processing. In 2012, approximately 100,000 poultry were slaughtered at two small federally inspected processing plants. At least four mobile processing units (MPU) are available in the state for rental, producing an additional unknown number of processed poultry for sale to neighbors, farmer's markets, and local restaurants. The state allows up to 20,000 poultry per farm to be grown and slaughtered, exempt from federal inspection as long as product is not placed in interstate commerce. The NC Department of Agriculture conducts occasional inspections and records review but does not inspect each animal.

Breeds other than commercially available broilers and turkeys are used because they seem better adapted

Table 24.3 Domestic production and exports of broiler meat (×1000 metric tonnes) of the top four producers in world broiler production, 1997–2012 (data from USDA Foreign Agricultural Service)

	Country	1997	2002	2007	2012
Domestic production	USA	14,952	14,467	16,536	16,476
	China	10,400	9558	11,500	13,700
	EU	8177	6850	8111	9480
	Brazil	4562	7355	10,305	12,750
Exports	USA	2565	2177	2618	3211
	China	367	438	358	400
	EU	944	850	623	1080
	Brazil	665	1588	2922	3478

to pasture or mobile coop conditions than the white-feathered commercial hybrids. The pasture breeds are slower growing, may have slightly better disease resistance, and are more mobile for foraging, possibly because of slower growth and lower body weights.

Carcass yields from these birds are generally lower compared to commercial hybrids for breast meat, but have similar yields for other parts (wings, drums, thighs, back). Meat texture has generally been reported to be more tender from commercial hybrid strains than from pasture raised or *Label Rouge*-type broilers. Reports on meat color have been varied with no particular pattern observed between multiple studies. Similarly, subjective sensory panels evaluating meat for color, flavor, texture, juiciness, and other profiles have found mixed results between commercial hybrids and pasture or *Label Rouge* broiler meat. Although anecdotal claims are made that pasture-raised or *Label Rouge* broiler meat is superior to commercially available meat, no clear trend exists in scientific studies.

Resources are available to consumers interested in sustainable poultry rearing and processing (see Further reading). The sustainable poultry market segment is expected to grow due to consumer awareness of these products and willingness of new small farmers to add poultry to their diversified products. Also, major commercial processors are addressing sustainability in various ways. Most are considering or have already implemented measures that reduce or eliminate antibiotics given to birds during rearing, promoting environmentally positive uses of poultry litter, and reducing waste at the plant, particularly packaging waste. Of course, these measures also reduce live bird and processing costs, as well as appealing to sustainability-conscious consumers.

24.6 Conclusion

The US is the world's largest producer of poultry and second largest exporter (see Table 24.3). Other countries such as Brazil and China are rapidly expanding their production capabilities. While China's production has been mostly to support domestic consumption, Brazil already exceeds the US as the world's largest poultry exporter. This was accomplished by exporting more than 27% of production, compared to the US exporting 19% of domestic production.

In the US, per capita consumption of poultry has exceeded both beef and pork consumption, and is gaining on total red meat consumption. The poultry processing industry has continued to grow as a business and currently employs more than 300,000 people. The industry should remain at or near the top of world production in the foreseeable future just by supplying the demands of the expanding US population. Constraints on future US expansion include rising feed costs, labor shortages, and availability of water.

Further reading

General processing

Barbut S (2002) *Poultry Products Processing: An Industry Guide.* Boca Raton, FL: CRC Press.

Mead GC (ed) (2004) *Poultry Meat Processing and Quality.* Boca Raton, FL: CRC Press.

Mountney GC, Parkhurst CR (1995) *Poultry Products Technology*, 3rd edn. New York: Haworth Press.

Owens CM, Alvarado CZ, Sams AR (eds) (2010) *Poultry Meat Processing*, 2nd edn. Boca Raton, FL: CRC Press.

Regulatory

Code of Federal Regulations (CFR) (2013) *Poultry Products Inspection Regulations, Title 9, Volume* **2***(3), Part 381.* www.ecfr.gov/cgi-bin/text-idx?SID=99fb21766d250d438199bccb-d74aca15&tpl=/ecfrbrowse/Title09/9tab_02.tpl, accessed 28 November 2013.

US Department of Agriculture (USDA) Food Safety Inspection Service (FSIS) *Regulations, Directives and Notices.* www.fsis.usda.gov/wps/portal/fsis/topics/regulations, accessed 9 December 2013.

Statistics

US Department of Agriculture (USDA) Foreign Agricultural Service (2013) *Livestock and Poultry: World Markets and Trade.* http://usda.mannlib.cornell.edu/MannUsda/viewDocumentInfo.do?documentID=1488, accessed 28 November 2013.

US Department of Agriculture (USDA) National Agricultural Statistics Service (2013) *Poultry Production and Value.* http://usda.mannlib.cornell.edu/MannUsda/viewDocumentInfo.do?documentID=1130, accessed 28 November 2013.

Spoilage/food safety

Cunningham FE, Cox NA (1987) *The Microbiology of Poultry Meat Products.* Orlando, FL: Academic Press.

Mead GC (ed) (2005) *Food Safety Control in the Poultry Industry.* Boca Raton, FL: CRC Press.

Scallan E, Hoekstra RM, Angulo FJ et al. (2011) Foodborne illness acquired in the United States – major pathogens. *Emerging Infectious Diseases* **17**: 7–15.

Sustainability

Castellini C, Mugnai C, dal Basco A (2002) Effect of organic production system on broiler carcass and meat quality. *Meat Science* **60**: 219–225.

Fanatico AC (2003) *Small-Scale Poultry Processing.* Appropriate Technology Transfer for Rural Areas (ATTRA). www.attra.ncat.org/attra-pub/PDF/poultryprocess.pdf, accessed 28 November 2013.

Fanatico AC, Cavitt LC, Pillai PB, Emmert JL, Owens CM (2005) Evaluation of slower-growing broiler genotypes grown with and without outdoor access: meat quality. *Poultry Science* **84**: 1785–1790.

Fanatico AC, Pillai PB, Cavitt LC, Owens CM, Emmert JL (2005) Evaluation of slower-growing broiler genotypes grown with and without outdoor access: growth performance and carcass yield. *Poultry Science* **84**: 1321–1327.

Farmer LJ, Perry GC, Lewis PD, Nute GR, Piggott JR, Patterson RLS (1997) Responses of two genotypes of chicken to the diets and stocking densities of conventional UK and label rouge production systems: II. Sensory attributes. *Meat Science* **47**: 77–93.

Grashorn MA, Serini C (2006) Quality of chicken meat from conventional and organic production. Proceedings of the 12th European Poultry Conference, Verona, Italy. www.cabi.org/animalscience/Uploads/File/AnimalScience/additionalFiles/WPSAVerona/10237.pdf, accessed 28 November 2013.

Husak RL, Sebranek JG, Bregendahl K (2008) A survey of commercially available broilers marketed as organic, free-range, and conventional broilers for cooked meat yields, meat composition, and relative value. *Poultry Science* **87**: 2367–2376.

Lewis PD, Perry GC, Farmer LJ, Patterson RLS (1997) Responses of two genotypes of chicken to the diets and stocking densities typical of UK and 'label rouge' production systems: I. Performance, behavior, and carcass composition. *Meat Science* **45**: 501–516.

Ponte PI, Rosado CM, Crespo JP et al. (2008) Pasture intake improves the performance and meat sensory attributes of free-range broilers. *Poultry Science* **87**: 71–79.

Sandercock DA, Nute GR, Hocking PM (2009) Quantifying the effects of genetic selection and genetic variation for body size, carcass composition, and meat quality in the domestic fowl (*Gallus domesticus*). *Poultry Science* **88**(5): 923–931.

Smith DP (2012) Pastured broiler processing yields and meat color. *Journal of Applied Poultry Research* **21**: 651–656.

Smith DP, Northcutt JK, Steinberg EL (2012) Meat quality and sensory attributes of a conventional and a Label Rouge-type broiler strain obtained at retail. *Poultry Science* **91**(6): 1489–1495.

Index

Food Processing: Principles and Applications, Second Edition. Edited by Stephanie Clark, Stephanie Jung, and Buddhi Lamsal.
© 2014 John Wiley & Sons, Ltd. Published 2014 by John Wiley & Sons, Ltd.

CPSIA information can be obtained
at www.ICGtesting.com
Printed in the USA
BVHW01*0915260418
514458BV00005B/14/P